Building Institution
Kim Förster

Architecture

Kim Förster (Dr. sc. ETH), researches and teaches at the University of Manchester with a focus on knowledge and cultural production, institutional as well as environmental history of architecture. An architectural historian, he served as Associate Director of Research at the Canadian Centre for Architecture from 2016 to 2018. In 2010, he was a visiting scholar at Columbia University in New York.

BUILDING INSTITUTION

The Institute for
Architecture and
Urban Studies,
New York 1967–1985

Kim Förster

[transcript]

Imprint

Published with the support of the Graham Foundation for Advanced Studies in the Fine Arts and the Swiss National Science Foundation.

Graham Foundation

Schweizerischer Nationalfonds

Bibliographic information published by the Deutsche Nationalbibliothek
The Deutsche Nationalbibliothek lists this publication in the Deutsche Nationalbibliografie; detailed bibliographic data are available on the internet at https://dnb.dnb.de

First published in 2024 by transcript Verlag, Bielefeld

Cover and graphic design: Till Sperrle, Berlin
Cover illustrations: front cover, Cameron Armstrong and Theodore Liebman; back cover, IAUS, *Casabella* 359/360, November/December, 1971, 101; interior cover and dust jacket, Massimo and Lella Vignelli papers, Vignelli Center for Design Studies, Rochester Institute of Technology, Rochester, NY
Copyediting and proofreading: Nicola Heine, Berlin
Image editing: Achim Hatzius, Berlin
Typesetting: Jan Gerbach, Bielefeld
Production management: Die Produktion, Düsseldorf
Print-ISBN 978-3-8376-6518-5
PDF-ISBN 978-3-8394-6518-9
https://doi.org/10.14361/9783839465189
ISSN of series: 2702-8070
eISSN of series: 2702-8089

Printed on permanent acid-free text paper.

Contents

Introduction: Institutional and Cultural History

The Institute for Architecture and Urban Studies (IAUS), realized as a distinctive institution of architecture culture, was—for the eighteen years it operated in New York—already a legend in its own time.[1] Founded in 1967 by architect Peter Eisenman, the Institute received support in terms of premises, personnel, and finances, etc. from major institutions such as the Museum of Modern Art (MoMA) and the Department of Architecture at Cornell University, which eventually conferred legitimacy on it, with strong support from Arthur Drexler and Colin Rowe, respectively. Conceived as an interface between academia and architectural practice, the Institute was officially registered with the School Board of the State University of New York from its inception and served as an alternative educational organization that offered both its Fellows and students from multiple universities the opportunity to acquire practical experience by working on actual projects. And yet it was not easy to establish. In the late 1960s and early 1970s, Eisenman assembled an entire circle of people around him, several of whom, with the support of the Chicago-based Graham Foundation for Advanced Studies in the Fine Arts, were made Fellows. These were (in chronological order): Emilio Ambasz (who was made a Fellow in 1968 but left the Institute soon after to become a curator at MoMA), William Ellis (1968), Kenneth Frampton (1970), Stanford Anderson (1971), Peter Wolf (1972, coequal with Eisenman for years as chairman of the Board of Fellows), Mario Gandelsonas (1972), and lastly Diana

1 In its by-laws, the formal abbreviation for the Institute for Architecture and Urban Studies was "Institute," a term also used by the Fellows. This name, in capitalized form, is therefore used in this book. The acronym IAUS, which existed early on, did not become common until the second half of the 1970s, around its tenth anniversary, mainly as a brand name in public relations.

Agrest (1973).[2] Over the years, the Institute came to work in diverse groupings and with varying emphases—research and design, education, culture, and publishing, adapting to changing circumstances and sociocultural contexts—surviving until 1985.

The Institute's history shows that in its founding years, despite being a rather small organization comprising only a few Fellows and Research Associates as well as a secretary and several administrative assistants, it was extremely successful at weaving itself into existing architecture networks in New York and on the East Coast and using its institutional relationships and especially its Board of Trustees for legal, political, and economic gain.[3] Because the Institute was able to acquire research and design contracts almost immediately—personal relationships helped to secure relatively small contracts from municipal planning offices at first and larger contracts from both state and federal agencies soon thereafter—its budget grew quickly, and it was able to expand. In 1970, a high-paying government contract from the U.S. Department of Housing and Urban Development (HUD) enabled the Institute to move from its small office space on 47th Street to more spacious and prestigious premises: a two-story loft on 40th Street overlooking Bryant Park in Midtown Manhattan, which once housed the publisher of Le Corbusier's *When the Cathedrals Were White* and was closely identified with the Institute from then on. After winning its only construction contract as an architecture firm from the Urban Development Corporation (UDC) of New York State in 1972, before a change in the political landscape put an end to public housing projects, the Institute's vision and values, strategy, and culture underwent a series of significant transformations. In the years that followed, especially in its heyday from the 1974–75 academic year onwards, the Institute, as a 501 (c) (3) non-profit organization, gained in importance, offering programs for undergraduate and high school students as well as internships, organizing thematic, group, and solo exhibitions, and holding lecture series every night during the semester for a period of six

2 With funds from the Graham Foundation, the Institute initially awarded individuals the title of Visiting Fellow for one year to prove themselves. In addition to those appointed as Fellows, Joseph Rykwert and Anthony Vidler were also granted Visiting Fellow status in 1970, but they left the Institute after a short time.

3 The concept of institution is defined in social, cultural, and historical studies; see "Institution," in Raymond Williams, *Keywords. A Vocabulary of Culture and Society* (New York: Oxford University Press, 1976), 139–140; John Searle, "What is an Institution?" *Journal of Institutional Economics* 1, no. 1 (2005), 1–22; and *Stanford Encyclopedia of Philosophy*, "Social Institutions,", last modified April 9, 2019, http://plato.stanford.edu/entries/social-institutions/ (last accessed: May 31, 2023). Institutional critique emerged in the 1960s and 1970s in the visual arts, less so in architecture; cf. Stefan Nowotny and Gerald Raunig, *Instituierende Praxen. Bruchlinien der Institutionskritik* (Vienna: transversal texts, 2016). Even though the Institute succeeded in positioning itself vis-à-vis existing institutions, especially museums and universities, its existence and the nature of its work was not understood as institutional critique. The institutional analysis approach was popularized in France by sociologist Rémi Hess, and then used for school education. I have applied it to architecture in my narrative on the Institute's various educational programs; see George Lapassade, *Gruppen, Organisationen, Institutionen* (Stuttgart: Klett Verlag, [1967] 1972).

years until spring 1980. Now enjoying support from the New York State Councils as well as the National Endowments for the Arts and the Humanities, the Institute served as a home for aspiring architects and scholars from the United States and abroad, particularly Europe, to a somewhat lesser extent from Latin America, and eventually Asia, providing a workplace that facilitated a novel kind of practice in education and culture and, along with research and design, increasingly combined its teaching, lecturing, exhibiting, and, of course, publishing activities.

The Institute's significant contribution to instigating a shift toward cultural production in architecture stands at the center of the historiography presented in this book, which combines an institutional analysis and a cultural critique of architecture.[4] Even though the Institute displayed a stable, successful pattern of institutional and financial growth in the further course of the decade, based on revenue from tuition fees, cultural funding, and philanthropic efforts, and was able to continuously expand its social and institutional networks to encompass liberal art colleges and schools of architecture at Ivy League universities, provide its own education program as an architecture school, and offer individual architects and academics as well as established art and architecture publishers the opportunity to produce, curate, and edit content within the scope of lecture series, exhibitions and publications, its existence was always at risk.[5] It was its published output, most notably the ambitious architecture journal *Oppositions*, which was launched in 1973 and distributed by MIT Press starting in 1976, but also the monthly architecture newspaper *Skyline*, the quarterly art journal *October*, the comprehensive IAUS Exhibition Catalogues and the exclusive Oppositions Books series, that brought the Institute much acclaim, reaching readers across North America, as well as internationally. Having found its way onto the bookshelves of architecture firms and university libraries around the world, *Oppositions* and Oppositions Books are perhaps the Institute's most enduring and robust cultural products. Its theory-heavy, jargon-laden publications portrayed the Institute as an architecture "think tank" that aimed to influence both the profession and

4 Here, I rely particularly on Pierre Bourdieu's seminal work on the field of cultural production and a symbolic economy that valorizes individual producers and their artworks and apply this to architecture in my narrative on the Institute's lecture series, exhibitions, and publications; see Pierre Bourdieu, "The Market of Symbolic Goods," *Poetics* 14, no. 1–2, ([1971] 1983), 13–44; "The Field of Cultural Production, or: the Economic World Reversed," *Poetics* 12, no. 4–5 (1983), 311–356; republished as "The Field of Cultural Production, or: the Economic World Reversed," in Pierre Bourdieu, *The Field of Cultural Production. Essays on Art and Literature* (New York: Columbia University Press, 1994), 29–73. Interestingly, Bourdieu's analysis and the terms he used were echoed in the reflections on autonomous and critical practice by Institute Fellows, see Peter Eisenman, "A Critical Practice: American Architecture in the Last Decade of the Twentieth Century," in *Education of an Architect*, eds. Elizabeth Diller, Diane Lewis, and Kim Shkapich, (New York: Rizzoli International, 1988), 190–193.

5 Bruno Latour's actor-network theory has been discussed in regard to the sociology of art, see Niels Albertsen and Bülent Diken, "Artworks' Networks. Field, System or Mediators?" *Theory, Culture & Society* 21, no. 3 (2004), 35–58.

the discipline—a cultural myth that persists to this day.[6] As the Institute became increasingly institutionalized and professionalized throughout the 1970s, while being able to reinvent itself several times with the rise of neoliberalism and a postmodern cultural logic, it sought not so much to rehearse a modernist approach and attitude, but rather to initiate not only an epistemological, but also a historiographic paradigm shift—or even a postmodern turn.

Over the years, the Institute was continually associated with Peter Eisenman; Philip Johnson, who emerged as the Institute's gray eminence even spoke of the "Eisenman Institute." With Eisenman as its long-time director, the Institute had a charismatic, intellectually ambitious, and also entrepreneurially savvy leader at its helm. Not entirely selfless, he took up ideas about establishing an institute that were floated at the time and made them his own. Instead of starting his own firm, Eisenman launched the Institute as a kind of start-up (a move he ultimately made from necessity, after being denied a permanent teaching position at Princeton) with a workforce made up of students and Fellows, and as a new work environment for himself and others, one that allowed him to focus on his abstract house designs (1967–77) and theoretical texts. As a "project maker" and "auto-entrepreneur," he knew how to initiate large-scale projects and, above all, manage the Institute's affairs. The Institute's later success as a cultural venture in the field of architecture based on its capacity to produce and disseminate new architectural knowledge can be attributed to Eisenman's success as an "impresario" and "publicist" in building and continuously expanding the inner circle of Fellows which, despite the idiosyncratic constellations of the group and its dynamics, grew to include:[7] Leland Taliaferro (1974), Julia Bloomfield (1975), Andrew MacNair (1975), Carla Skodinski (1977), Frederieke Taylor (1977), Anthony Vidler (1977), Suzanne Frank (1978), Stephen Potters (1979), and Myles Weintraub (1979); further additions to the Fellowship in the early 1980s, when the first long-time Fellows began to step back, being Deborah Berke (1980), Silvia Kolbowski (1980), Lawrence Kutnicki (1981) Rosalind Krauss (1981), Joan Ockman (1981), Robert Silman (1981), Joan Copjec (1982), Douglas Crimp (1982), Christian Hubert (1982), Annette Michelson (1982), and Lindsay Stamm Shapiro (1982). Until he himself stepped down as the Institute director in 1982, after which the Institute collapsed like a house of cards, Eisenman single-handedly oversaw its day-to-day operations, only consulting the Board of Trustees when this became unavoidable, and redesigned its

6 Documents from the early 1970s indicate that the Institute thought of itself as a "think tank" at the same time as it was trying to make money by producing theory. Eisenman, with his characteristic subtlety, repeatedly referred to the Institute as a "halfway house" because of the position it took between academia and architectural practice, thus adding another provocative meaning to the Institute with this play on words; in American, "halfway house" colloquially stands for an open psychiatric ward or rehabilitation clinic.

7 For this characterization of Eisenman, see Joan Ockman, "Resurrecting the Avant-Garde: The History and Programme of Oppositions," in *ArchitectuReproduction*, ed. Beatriz Colomina (New York: Princeton Architectural Press, 1988), 180–199, here 183.

bureaucratic structure, strategic orientation, and programmatic content. At the same time, the wider circle of the Institute was continuously expanded in the second half of the 1970s to include international architects such as Rem Koolhaas, Bernard Tschumi, Grahame Shane, Rafael Moneo, Arata Isozaki, Giorgio Ciucci, Massimo Scolari, Aldo Rossi, and others, all of whom stayed for some time as Visiting Fellows before moving on. Not only the Institute's international presence, which it maintained by editing journals and curating the American contribution to the 1976 Biennale di Venezia, but also the commitment of the Visiting Fellows to the Institute contributed to its international recognition. Ultimately, Eisenman, as well as many Fellows and Visiting Fellows, used the Institute to make a name for himself and build an international career.

This book is the first to examine the Institute's eventful, tumultuous, and varied history, which encompasses its formation and organization, the restructuring of its activities, and reciprocal relationships—particularly the shift from construction to cultural production—in terms of its contributions to the new economy of attention and to complex mechanisms of marketing or self-marketing with implications for education, culture, and discourse, and the key role it played in the early careers of its protagonists and their canonization in the present day.[8]

Research Status

While the protagonists of postmodernism, among them also the main figures of the Institute, and their contributions to the built environment and architectural debate, as well as the Institute's specific media, such as exhibitions and periodicals, have previously received attention in historiographical research, as the next generations of architects, theorists, and historians felt the need to inscribe themselves in or dissociate themselves from this legacy, and other museums and universities that were active at the time have also been historicized, the Institute, as a project office, an educational and cultural institution, and as a publishing house, has not yet been systematically studied—in spite of a few but promising attempts and despite the fact that leading and subsidiary Fellows have repeatedly asserted and underlined its importance and enduring relevance.[9] This is all the more surprising given the role it played in institutionalizing a particular strand of architecture

8 The architect and philosopher Georg Franck, drawing on both Bourdieu's and Karl Marx's concept of capital, has examined the economy of attention in relation to deconstructivism, i.e., the architecture culture of the 1980s, but not that of the 1970s, cf. Georg Franck, *Ökonomie der Aufmerksamkeit* (Munich and Vienna: Carl Hanser Verlag, 1998) and "*Ökonomie* der Aufmerksamkeit," in *Perspektiven metropolitaner Kultur*, ed. Ursula Keller (Frankfurt: Suhrkamp, 2000), 101–118; also Georg Franck, *Mentaler Kapitalismus. Eine politische Ökonomie des Geistes* (Munich and Vienna: Carl Hanser Verlag, 2005).

9 Suzanne Frank, "Institute for Architecture and Urban Studies. New York, New York," in *Encyclopedia of 20th Century Architecture*, ed. R. Stephen Sennott (New York: Fitzroy Dearborn, 2006), 677–678; Harry Francis Mallgrave, *Modern Architectural Theory. A Historical Survey, 1673–1968* (Cambridge: Cambridge University Press, 2005) and *An Introduction to Architectural Theory. 1968 to the Present* (Oxford: Wiley-Blackwell, 2011).

culture in North America, and soon across the globe, in terms of institutionalized postmodern beliefs and practices. Peter Eisenman has in particular laid claim to this, repeatedly making bold comparisons between the Institute's influence and that of the Bauhaus in Dessau (in the period from 1925 to 1931) during the Weimar Republic, and claiming that it shares a lineage with contemporary schools of architecture, notably the Cooper Union in New York under John Hejduk (1975–2000), where Eisenman himself taught design after 1968, the Architectural Association (AA) in London under Alvin Boyarsky (1971–90), and the Istituto Universitario di Architettura di Venezia (IUAV), in particular the circle of neo-Marxist architects, historians, and critics around Manfredo Tafuri (1967–80).[10] From an institutional and cultural-historical perspective, however, the Institute eludes such comparisons with other schools of architecture established in their respective national educational system by virtue of its exuberance and changeability, and because of its erratic and ultimately unfulfilled institutional constitution. The historical desideratum, therefore, consists of nothing less than an exploration of the structure and work of the Institute, how it defined and presented itself and became relevant and powerful in the context of the North American educational system, the intellectual and artistic life of New York, and the academic and commercial publishing landscape in North America, especially on the East Coast of the United States, if not in the transatlantic, transpacific, and global cultural spheres.

Thus far, the Institute's history has been told primarily by Eisenman or by people from the Institute's inner circle; surprisingly, these were all women who initially served as Institute staff and were then granted Fellow status based on their merits (less so from the *Oppositions* editorial board, which, in addition to Eisenman, first encompassed Kenneth Frampton and Mario Gandelsonas, and later Anthony Vidler). Joan Ockman began this work in 1988 with a well-informed and incisive essay on the institutional strategies, discourses, and materialities associated with *Oppositions* in the anthology *ArchitectuReproduction* (she herself had worked for *Oppositions*, first as an intern in 1976 and later, in the early 1980s, as an associate editor).[11] Then, in a 1995 *Casabella* article, Ockman wrote about the intellectual confrontation between Eisenman and Tafuri (whose book *The Sphere and the Labyrinth* she supervised as an executive editor at Oppositions Books), while stressing the differences between the IAUS and the IUAV in terms of institutional

10 Peter Eisenman (in conversation with Alvin Boyarsky), "The Institute in Theory and Practice," (January 20, 1976) in *Supercritical: Peter Eisenman & Rem Koolhaas*, Brett Steele (London: AA Publications, 2007), 83–87; on the AA, see Irene Sunwoo, "Pedagogy's Progress: Alvin Boyarsky's International Institute of Design," *Grey Room*, no. 34 (Winter 2009), 28–57, and "From the 'Well-Laid Table' to the 'Market Place:' The Architectural Association Unit System," *Journal of Architectural Education* 65, no. 2 (March 2012), 24–41; on the IUAV, see Andrew Leach, "Choosing History. Manfredo Tafuri, Criticality and the Limits of Architecture," *The Journal of Architecture* 10, no. 3 (2005), 235–244, and "Imagining Critique, or the Problematic Legacy of the Venice School," in *The Missed Encounter of Radical Philosophy with Architecture*, ed. Nadir Lahiji (London: Bloomsbury, 2014), 95–112.

11 Ockman, 1988.

structures, missions, and cultures.[12] This was followed by Frederieke Taylor (director of development at the Institute from 1976 to 1980), who wrote a term paper on the Institute's "Exhibition Program", for the first time historicizing not just solo or group shows, for a course at Columbia University, but instead formulating a fairly comprehensive history.[13] In 2010, Suzanne Frank (the Institute's official librarian from 1976), penned a self-published combined historical account and personal memoir of her time at the Institute, including twenty-seven interviews held over a number of years with other Fellows and friends.[14] Julia Bloomfield (managing editor of *Oppositions* from 1974) produced an insider's account of the Institute's editorial work for a commemorative book on Kurt Forster published in 2010.[15] And finally, in 2012, Diana Agrest (who was head of the design studio in the undergraduate program from 1975, long before becoming an editor of *Oppositions* herself in 1984), released a documentary film with the support of the Graham Foundation. The film was about the Institute's avant-garde, if not political, ambitions, and consisted partly of her own Super 8 footage from her time at the Institute and partly of contemporary interviews with Fellows, contributors, and eyewitnesses (albeit with a strong focus on those individuals who have since made a successful career for themselves in architecture).[16] For a long time, historical knowledge of the Institute has been shaped by first-hand personal accounts and various attempts at documenting an oral history, rather than archival work.

More than any other project, program, or production, it is the Institute's publications—most notably *Oppositions*—that have secured it a firm footing within architecture history. Considerable credit for this can certainly be attributed to K. Michael Hays's *Oppositions Reader* from 1999, which reprinted a selection of essays from the twenty-six issues of the journal of ideas and criticism and thus allowed *Oppositions* to be reread and assigned to students of

12 Joan Ockman, "Venice and New York," *Casabella* 59, no. 619/20, (1995), 56–73.

13 Frederieke Taylor, "Appendix C: Frederieke Taylor on Exhibitions," in *IAUS. An Insider's Memoir (with 27 Other Insider's Accounts)*, Suzanne Frank (New York: self-published, 2010), 315–322.

14 Suzanne Frank, *IAUS. An Insider's Memoir (with 27 Other Insider's Accounts)* (New York: self-published, 2010); see Cesare Birignani, "Feature: Talking Heads. Team Vitruvius," *The Architects' Newspaper* (April 6, 2011), https://www.archpaper.com/2011/04/talking-heads/ (last accessed: May 31, 2023).

15 Julia Bloomfield, "A Tale of Two Institutes: Thoughts on Publication Worlds," in *Art History on the Move: Festschrift für Kurt W. Forster*, eds. Nanni Baltzer, Jacqueline Burckhardt, Marie Stauffer, and Philip Ursprung (Zurich: Diaphanes, 2010), 66–83.

16 *The Making of an Avant-garde* (2013, dir. Diana Agrest).The interviews are with Peter Eisenman, Diana Agrest [interviewing herself], Kenneth Frampton, Mario Gandelsonas, Emilio Ambasz, Anthony Vidler, Richard Meier, Charles Gwathmey, Rem Koolhaas, Frank Gehry, Mark Wigley, Robert Stern, Barbara Jakobson, Deborah Berke, Bernard Tschumi, Joan Ockman, Julia Bloomfield, Peter Wolf, Frederieke Taylor, Stan Allen, Suzanne Stephens, Paul Lewis, Lucia Allais, etc. see Belmont Freeman, "The Moment for Something to Happen," *Places* (January 13, 2014), https://placesjournal.org/article/the-moment-for-something-to-happen/ (last accessed: May 31, 2023).

architecture.[17] Hays's introduction to the anthology, however, ensured that the fundamental distinction between "history" and "theory" propagated by the journal, a distinction also drawn in the conception and development of new master's and doctoral programs, went largely unquestioned. The same can be said of the various reviews and essays that followed the publication of the *Reader* which served to consolidate this position. (Apparently, the social function of criticism, to which *Oppositions*, as its title implies, was committed, did not play a major role in the journal.)[18] After studies of the actual editorial work involved in the making of *Oppositions*, the historicization of which formed the basis for further research, the research focus remained on the journal:[19] in 2008, Louis Martin published an account of the prehistory of journal-making at the Institute and in 2010, Lucia Allais followed with genealogical research and a critique of the production of theory at the Institute, with a focus on *Oppositions*.[20] Despite this increased interest, the editorial activities and labor involved in making the other publications that were conceived and produced at the Institute—for example, *October*, the quarterly art theory journal, *Skyline*, a monthly architecture newspaper with a cultural calendar, the IAUS Exhibition Catalogues, a series comprising original archival material and essays, and Oppositions Books, an ambitious and luxuriously designed book series—have, with few exceptions, received little historiographical attention, despite the fact that these publications played an important and trendsetting role for the Institute and, more broadly, for the discourses of architecture and art history.

When it comes to the institutional significance and other activities of the Institute's Fellows, however, the current state of knowledge remains cursory.

17 K. Michael Hays, ed., *Oppositions Reader* (New York: Princeton Architectural Press, 1998).

18 K. Michael Hays, "The Oppositions of Autonomy and History," in Hays, 1998, IX–XV.; see also Mitchell Schwartzer, "History and Theory in Architectural Periodicals. Assembling Oppositions," *Journal of the Society of Architectural Historians* 3 (September 1999), 342–348; Daniel Sherer, "Architecture in the Labyrinth. Theory and Criticism in the Unites States: 'Oppositions,' 'Assemblage,' 'Any' (1973–1999)," *Zodiac*, no. 20 (1999), 36–63; Ralph Stern, "Oppositions Revisited—The Oppositions Reader," *Kritische Berichte*, no. 3 (1999), 65–72. At the same time, in the early 1980s, there had been some reflection on critical historiography against the background of the reception of Marxist and poststructuralist approaches in *Oppositions* (especially by authors of the IUAV) and at the Institute itself (initiated by the younger generation formed in the *ReVisions* group).

19 Joan Ockman, ed., *Architecture Criticism Ideology* (New York: Princeton Architectural Press, 1985).

20 Louis Martin, "Notes on the Origins of Oppositions," in *Architectural Periodicals in the 1960s and 1970s. Towards a Factual, Intellectual and Material History*, eds. Alexis Sornin, Hélène Jannière, and France Vanlaethem (Montréal: IRHA Institut de recherche en histoire de l'architecture, 2008), 147–169; Lucia Allais, "The Real and the Theoretical, 1968," *Perspecta*, no. 42 (2010): "The Real," 27–41. Here, Allais presented a narrative that addressed, among other things, the unlikelihood of early unrealized projects in order to call Eisenman's motivation and interest into question, and then analyzed the beginnings and conditions of theory production at the Institute,

Major research projects at leading American schools of architecture have yielded insights into the indirect contexts, premises, and overall conditions governing architecture discourse in the 1960s and 1970s. Such studies have dealt explicitly with the architecture journal as an essential medium, one capable of being produced quickly and independently, as well as with the availability of utopian, modernist concepts in the early postmodern era, and with conceptual approaches to the relationship between writing and architecture.[21] But even though the Institute is typically mentioned in this context—and characterized as being one of the dominant actors shaping American architecture history and most notably the emergence of postmodernism—these research projects only managed to overcome the prevailing myth of the Institute as a "think tank" within the field of architecture—a new avant-garde, a new school, or a movement—to a limited extent. In most cases, the Institute's own institutionalism, or the very institutionality to which it was exposed, is not even broached.[22] It is also worth mentioning that in the various historiographies of American urban renewal and housing after 1968, the Institute plays only a minor role, despite the prototype for low-rise housing that was researched and designed there.[23] Meanwhile, in the historiography of architecture education in America, and even worldwide, the Institute's role in transforming postmodern architecture pedagogy has been

21 The 2000s saw innovative research conducted as part of doctoral programs at Princeton (directed by Beatriz Colomina), Columbia University (Reinhold Martin), and the UCLA (Sylvia Lavin), partnering with the CCA in Montréal as an archive and museum. The projects "Clip Stamp Fold," "Utopia's Ghost," and "Take Note" each resulted in exhibitions and/or books—all of which are important resources. The exhibition "Clip Stamp Fold" opened at the Storefront for Art and Architecture in New York (November 14, 2006, to January 31, 2007), before going on display at the CCA (April 12 to September 9, 2007), as part of Documenta 12 in Kassel that same year, and then traveling around the globe; see Beatriz Colomina and Craig Buckley, eds., *Clip Stamp Fold. The Radical Architecture of Little Magazines. 196X to 197X* (Barcelona: Actar, 2010). Public events on publishing were organized at Storefront as part of "Clip Stamp Fold," e.g., former editors of *Oppositions*, *October*, and *Skyline* were invited to three of the panel discussions; see https://vimeo.com/user1360843 (last accessed: May 31, 2023). The exhibition "Utopia's Ghost. Postmodernism Reconsidered" was on only display at the CCA (February 28 to May 25, 2008); see Martin, 2010. The exhibition "Take Note" was also on display at the CCA (February 4 to May 30, 2010); see Sylvia Lavin, "IAUS. Institute for Architecture and Urban Studies" *Log*, no. 13/14 (Fall 2008), 53–66.

22 Yannik Porsché, Ronny Scholz, and Jaspal Naveel Singh, "Introducing Institutionality," in *Institutionality. Postdisciplinary Studies in Discourse*, eds. Porsché, Scholz, and Singh (Cham: Palgrave Macmillan, 2022), 1–28. If institutionality refers to the omnipresence of institutions in modern society, the Institute was accordingly confronted with institutional enactments, characterizations, transformations, and resistances. However, we should not make the mistake of equating the Institute, or even architecture, with institutions such as the church, the monarchy, the caste system, the patriarchy, the nation-state, the judiciary, the prison, or the police.

23 To date, the relationship between architecture, planning, and society in the United States of the 1960s and 1970s has been historicized. On the quasi-welfare state urban development policies of New York under Mayor John Lindsay (1966–73), see Mariana Mogilevich, "Designing the Urban: Space and Politics in Lindsay's New York," PhD diss., Harvard University, 2012; on the U.S. government's biopolitical research under Presidents Lyndon B. Johnson and Richard M. Nixon, see Joy Knoblauch, "Going Soft: Architecture and the Human Sciences in Search of New Institutional Forms (1963–1974)," PhD diss., Princeton University, 2012.

subject to little study, even though teaching was the Institute's central field of activity—and its financial foundation—for many years. When it has been studied at all, the focus has been on the content of the Institute's teaching activities, rather than its didactics, methods, or goals.[24] And similarly, the Institute's cultural production has yet to be thoroughly examined. As probably the least tangible and thus the most ephemeral of the Fellows' contributions and the most difficult to chronicle, the Institute's cultural production contributed significantly to the transformation of New York's architecture and art scenes and their symbolic economies and helped to raise the market value of not only the key figures who exhibited, facilitated, and lectured at the Institute, but also of the Institute itself. The hypothesis propounded by Tafuri in 1976, in an essay that was more a piece of architecture criticism than architecture history, that the Institute was one of those "well-defined cultural spaces" of the New York architecture scene "entrusted with the task of pleasurably entertaining a highly select audience" has not been further analyzed—neither by Tafuri himself, nor in architecture historiography.[25]

Main Argument

The main concern and ultimate goal of this institutional and cultural history of the Institute—if the Institute can be understood at different levels as a group, an organization, or also as an institution, following the tenets of French institutional

24 See Joan Ockman, ed., *Architecture School: Three Centuries of Educating Architects in North America* (Cambridge: MIT Press, 2012). This publication includes a chapter on post-1968 architecture education by Mary McLeod, see Mary McLeod, "The End of Innocence. From Political Activism to Postmodernism," in *Architecture School: Three Centuries of Educating Architects in North America*, ed. Joan Ockman (Cambridge: MIT Press, 2012), 160–201. A more recent academic research project at Princeton University, "Radical Pedagogies: Action-Reaction-Interaction" (directed by Beatriz Colomina), was exhibited at the Venice Architecture Biennale in 2014 and at the 7th Warsaw "Under Construction" Festival in 2015. The Institute is listed in the timeline and on the world map as one of the case studies, but was not further discussed; cf. Beatriz Colomina, with Esther Choi, Ignacio Gonzalez Galán, and Anna-Maria Meister, "Radical Pedagogies in Architectural Education," *Architectural Review* (September 28, 2012), http://www.architectural-review.com/essays/radical-pedagogies-in-architectural-education/8636066.article (last accessed: May 31, 2023); Beatriz Colomina and Evangolos Kotsioris, with Ignacio Gonzalez Galán, and Anna-Maria Meister, "The Radical Pedagogies Project," *Volume* 45 (2015): "Learning, Insert," 2–5; see also Beatriz Colomina, Ignacio G. Galán, Evangelos Kotsioris, and Anna-Maria Meister, eds., *Radical Pedagogies* (Cambridge: MIT Press, 2022).

25 Manfredo Tafuri, "The Ashes of Jefferson," in *The Sphere and the Labyrinth: Avant-Gardes and Architecture from Piranesi to the 1970s*, trans. Pellegrino d'Arcierno and Robert Connolly (Cambridge: MIT Press, 1987), 291–303. Tafuri first published this text in French with an emphasis on cultural production in the original, see Manfredo Tafuri, "Les cendres de Jefferson," *L'Architecture d'Aujourd'hui*, no. 186 (August/September 1976): "New York in White and Gray," 53–72. Nor have researchers yet adopted the broader notion of reading the Institute's activities in terms of Max Horkheimer's and Theodor W. Adorno's arguments concerning the culture industry, i.e., the commercial marketing of culture as entertainment with the triumph of television and advertising in the United States in the post-war period; see Sandro Marpillero, quoted in George Baird, "A Reflection on the End of Assemblage," *Assemblage*, no. 41 (April 2001), 11; see also Theodor W. Adorno and Max Horkheimer, "The Culture Industry: Enlightenment or Mass Deception," in *Dialectic of Enlightenment* (New York: Herder and Herder, [1944] 1972), 120–176.

analysis—is to comprehensively, exemplarily, and yet systematically explore its day-to-day activities, organizational structure, and broader context, drawing on perspectives from architecture studies and the humanities and incorporating critical geography and historiography, institutional and cultural sociology, and literary and cultural studies.[26] A carefully crafted, precisely formulated historiographical study of the Institute qua *institution*, the first of its kind, focuses not only on its most tangible product, its publications, but also considers all the incredibly multifaceted projects, programs, and products, both material and immaterial, that the Institute launched between 1967 and 1985, when it closed its doors forever due to a lack of financial and political support. This approach deviates from classical art history and architecture history approaches, which center on prominent individuals or objects of material culture and argue on the basis of styles, epochs, ideas, and protagonists. Manfredo Tafuri has intimated that a cultural critique of the Institute might be read as a cultural space—this book seeks to answer the question of how. Beyond this, the task of writing a genealogical-archaeological narrative of cultural production using the example of the Institute involves excavating the specific aspects, conditions, elements, and limitations that have shaped the Institute's history. If we apply Pierre Bourdieu's theory of cultural production—which he developed in the 1970s and 1980s in relation to nineteenth-century French bourgeois art and literature—to American, or even globalized architecture and thus update it, then the major break from a society-oriented architectural practice toward a postmodern one driven by the principles of a symbolic economy resulted in the establishment of new architectural knowledge, derived from artistic and literary references, and of a new power structure.[27] This book demonstrates that this architecture culture was interspersed with fragments of theory and positioned in relation to architectural modernism, bringing with it a new vocabulary and metaphors that functioned as a new system of reference for contemporary architectural practice. If the Institute distinguished itself by refashioning cultural production in architecture and by strengthening architecture culture (rather than just architecture) as an autonomous practice while making it economically viable, this means that we must establish a new narrative about the Institute and support that narrative with ample evidence from the institutional archives. This is the only way we can achieve an understanding of the Institute's influence on North American architectural discourse, on architecture education in light of the transformation, economization, and corporatization of higher education in the United States, and on

26 Lapassade, [1967] 1972. The Institute's work and structure changed over the course of its existence and displayed characteristics of all three types. The debate about institutions has taken on new forms with the curatorial turn; see Paul O'Neill, Lucy Steeds, and Mick Wilson, *How Institutions Think: Between Contemporary Art and Curatorial Discourse* (Cambridge: MIT Press, 2017); see also Paul O'Neill, "The Curatorial Turn: From Practice to Discourse," in *The Biennial Reader*, eds. Jelena Filipovic, et al. (Ostfildern: Hatje Cantz, 2010), 240–259.

27 Bourdieu, [1971] 1983; see also Franck, 1998 and 2000.

the architecture culture of the neoliberal age on a global scale.[28] In conjunction with the emergence of a post-Fordist accumulation regime, the Institute, as argued in this book, heralded a change in architecture and architecture culture, in what would in epistemological terms be called a paradigm shift, away from post-war or late modernism to postmodernism, and it did so quite powerfully, by actuating and enforcing an autopoietic and yet commercial system. After the collapse of the great utopias, briefly reanimated once again in the United States under President Lyndon B. Johnson in the mid-1960s with the Great Society, the Institute's pedagogical, cultural, and discursive practice was marked by an economic pragmatism, characterized by a project-based organization, and asserted by particular interests. The capitalizing on culture, in turn, must be viewed in the context of the dramatic developments of the 1970s: the commercialization of the national and global education market, the blossoming of federal cultural policy and cultural patronage, the expansion of the publishing landscape for both academic and popular books, journals, and magazines, and the emergence of an art market for architectural projects, drawings, and models.

The Institute's History

There are a number of parallel, competing myths about the founding of the Institute, including the 1964 Conference of Architects for the Study of the Environment (CASE) and the 1967 MoMA exhibition "The New City: Architecture and Urban Renewal," which heralded the end of Peter Eisenman's working relationship with Princeton University. For the purposes of this historiographical study, it is important to contextualize these myths in both the history of architecture and the history of urban studies as expressions of a paradigm shift in late capitalism. The novelty of an approach to architecture history that is grounded in institutional analysis and critique, however, is that the Institute's founding narratives encompass not only socio-cultural dimensions but also and above all, as the first chapter of *Building Institution* will show, legal, political, and economic ones. In this regard, the name that was chosen, "The Institute for Architecture and Urban Studies," was significant in that it was both a positioning and a provocation, implying that the grouping's skills and aspirations lay in two fields of knowledge and work: first, the Institute was obviously seeking to redefine the role of "Architecture" (with a capital "A") in society, after a decade marked by professional differentiation and interdisciplinary work; and second, the Institute was also seeking to capitalize on the fundability and popularity of the brand new discipline of "Urban Studies" and stake a central claim for architecture practice. For in the early years, the Institute was able to carry out research and planning, and ultimately design large-scale projects, some of which were highly remunerated, on behalf of public authorities. In this

28 In this context, architecture culture is not confined to the traditional understanding of building culture, but instead refers to all activities, objects, phenomena, and structures related to architecture.

context, the Institute's history underscores that, contrary to how it was institutionalized at the time, how it portrayed itself, and how it was perceived by others, it was anything but autonomous and radical. What it did instead was present itself in a communicative context that was self-legitimizing and self-referential, constantly oscillating between tradition and innovation. The Institute's leadership offered its services to various planning offices and organizations at different levels of scale and was met with initial success, benefiting from the fact that urban policy under Mayor John V. Lindsay (a liberal Republican) operated along welfare-state principles, and that Governor Nelson A. Rockefeller was using housing policy to moderate a tense social situation. The Institute collaborated with the revitalized New York City Planning Commission (CPC), the Urban Design Group (UDG)—which like the Institute was founded in 1967—the U.S. Department of Housing and Urban Development (HUD)—which was responsible for large-scale urban renewal based on the Model Cities Program—and the New York State Urban Development Corporation (UDC)—founded in 1968 with Edward J. Logue at the helm and tasked with improving the urban situation on a large scale as part of an effort to prevent further racial unrest. With its Fellows, Visiting Fellows, postgraduate Research Associates, as well as students and interns, the Institute had a cheap labor force that it could use for these projects. Ultimately, the Institute was commissioned to actually build new housing that was originally intended for the inner city and suburbs across New York State, and also became involved in the design, but this was never implemented on a mass scale.

As part of a history of knowledge, of the discipline, its concepts, and its methodology, the journal *Oppositions*, first launched in 1973, provided the Institute's Fellows with the opportunity to establish themselves as critics, historians, and theorists, in the role of editors and authors. The journal signaled the Institute's claim to interpretive authority when it came to renegotiating the role of the architect in the theoretical and historiographical discourse on architecture. As the second, third, and especially fourth chapters of *Building Institution* show, the changes at the Institute meant that these architects no longer wished to be perceived as "mediators" but instead as intellectuals and artists—a demand that was to have far-reaching consequences, both discursively and socio-culturally, for the relationship between architecture and society. The education, as well as the culture provided by the Institute at the time, were instrumental in cultivating the next generation. The events, lecture series, and exhibitions hosted by the Institute soon established it as an arena for clashes between the figures grouped around such labels as the "Whites" and the "Grays," and the symbolic economy that accompanied them.[29] That is to say, the

29 The symbolic economy of New York's architecture scene was legendarily reinstituted in the early 1970s by a polemical debate between two camps, Peter Eisenman's "Whites" and Robert Stern's "Grays," each working from different historical references: the classicist formal language on the one hand and the modernist idiom on the other. See Manfredo Tafuri, "American Graffiti. Five x Five = Twenty-five," trans. Victor Caliandro, *Oppositions* 5 (Summer 1976), 35–72; see also

Institute's premises became a physical site of confrontation in the fierce conflict between formalist and historicist positions at the time. The good-natured feud and undeterred collaboration between Peter Eisenman and his adversary Robert Stern, then president of the Architectural League in New York, enabled the Institute to thrive as a venue where a pluralism of different approaches was possible—from the realist, pragmatist, modernist, neo-rationalist, and populist to the idealist—a place where varying perspectives and stylistic orientations could be debated. However, these debates staged by the Institute largely ignored other topics that had fundamental ramifications for architecture and urban environments: topics with arguably more sociopolitical relevance, such as the conservative turn in American politics and society under the administration of President Richard M. Nixon, the diverse range of global crises that marked the 1970s in particular, the emerging environmental crisis and especially the oil and energy crisis that culminated in 1973, and the New York financial and fiscal crisis of 1974–75, along with the urban crisis that accompanied it. In this regard, the Institute, which took the offensive in setting itself apart from other figures and institutions by proclaiming to be the last stronghold of architectural modernism in North America, if not the world, was ultimately, as argued within the pages of this book, one of the trailblazers of architectural postmodernism in the United States and beyond—conceived in this sense not merely as a discursive phenomenon, but also as a cultural formation with all the receptivities, uncertainties, and ambiguities that this entails.

This, even more than 1968 with its impact on architecture education within universities, is the moment when, at the beginning of the 1974–75 academic year, the Institute, having understood that its special niche of both architecture production and theory production was not financially viable—especially after the government's moratorium on public housing in 1973—decided to turn its perceived weakness into a strength and reinvent itself. Architecture history has largely overlooked the fact that, faced with the major political, economic, and social changes of the mid-1970s, the Institute's leadership decided to increase its focus on education, culture, and publishing in order to disseminate new architectural knowledge. This act of repositioning and restructuring through what was only later theorized as cultural production allowed the Institute, thanks to its tax-exempt status, to develop a more complex business model based on a sophisticated "Educational Program" with multiple offerings, a "Public Program," including an extensive "Evening Program" with a wide range of lecture series every night of the week and a professional "Exhibition Program" of externally curated and in-house produced exhibitions, as well as a diversified "Publication Program." While the Institute, having

Nadia Watson, "The Whites vs. the Grays: Re-Examining the 1970s Avant-Garde," *Fabrications* (July 2005), 55–69 and Reinhold Martin, "Language, c. 1973," in *Utopia's Ghost. Architecture and Postmodernism, Again* (Minneapolis: University of Minnesota Press, 2010), 43–67. Camps were formed at the CASE conferences, with CASE 8 of the New York subdivision being the first to bring together the "Five Architects" who formed the inner circle of the "Whites".

built up strong networks with its "Development Program," media outreach, and public relations, had a graphic identity from the outset, it was the printed materials created for these programs by New York graphic designer Massimo Vignelli that represented its new institutional identity—an approach that was developed at the very moment that "urban branding" was first being implemented in New York with Milton Glaser's "I ♥ NY" campaign and announced that the political economy of the city was about to change decisively.[30] This move, archival records show, was accompanied by the comprehensive branding of all of the Institute's educational and cultural projects, programs, and products, with the promise of generating further income. Whether intentionally or not, this fetishization and reification of culture meant that the Institute also contributed significantly to the constantly increasing, more broadly conceived medialization and commercialization of postmodern architecture, otherwise critiqued by literary and cultural studies, as works of art—and thus to the emergence of a "celebrity culture" in architecture, a development that would later be referred to as "starchitecture."

After navigating phases of near-bankruptcy, the deciding factor in ensuring the Institute's success in the second half of the 1970s, culminating in its tenth anniversary in 1977, and indeed in its continued existence, was that—despite criticism of its withdrawal into an ivory tower, its isolation, and its detachment—it continued to receive significant assistance from liberal arts colleges and universities, endorsement from government foundations within the framework of the American Bicentennial, and support from academic and commercial publishers. A socio-analysis and discourse analysis will focus on the fact that the Institute favored a negative definition of itself, especially vis-à-vis official schools of architecture, longstanding cultural institutions such as MoMA, and newer ones such as P.S.1, and publishing houses. In doing so, the Institute's leadership refused to fully institutionalize it, not least because of its limited resources: at no point did the Institute ever offer an accredited degree program, only once did it ever fully open up to the general public, and it always tended to seek the backing of the publishing industry. From an archaeological-genealogical perspective, the Institute can thus be more aptly described as a pedagogical, curatorial, and editorial practice within the nexus of postmodern architecture, as an educational offer, as a stage event, and as a communication medium—especially if one follows Manfredo Tafuri's line of reasoning about the formation of "well-defined cultural spaces" for the New York architecture scene to celebrate its own existence.[31] By specializing in cultural production aimed at both professional and metropolitan audiences, what the Institute promoted was, in Tafuri's words, the

30 See Miriam Greenberg, "The Battle to Brand New York: 1975–1985," in *Branding New York* (New York: Routledge, 2008), 131–260; see also McLain Clutter, *Imaginary Apparatus. New York City and Its Mediated Representation* (Zurich: Park Books, 2015).

31 See Tafuri, 1987, 293; see also Ullrich Schwarz, "Another look-anOther gaze. Zur Architekturtheorie von Peter Eisenman," in *Aura und Exzess. Zur Überwindung der Metaphysik der Architektur*, ed. Schwarz (Vienna: Passagen Verlag), 1995, 11–34.

formation of "new circuits of production and use."[32] This development in architecture under the changing societal conditions linked to post-Fordism was not entirely dissimilar to what later came to be characterized as "immaterial labor," albeit politically different.[33] The Institute nevertheless provided a basis for conveying, integrating, and testing new ways of working and new role models for architects and academics, always seeking profit-making opportunities while cutting costs. What the Institute amounted to, subsequent to the CASE conference of 1964, and in parallel with the "New York Five" of 1972—other formations initiated by Eisenman, the latter being more of a media event—was a grouping that, thanks to its innovative admixture of social and discursive practices, was able to foster certain positions and anticipate new positionings in the world of art and architecture and in the public sphere.[34] However, this would facilitate the emergence, production, and valorization of the "neo-avant-garde," a phenomenon that drew inspiration in form, but not in content, from the movements and icons of architectural and artistic modernism.[35]

Methods and Methodology

This book offers a novel take on the historiography of architecture culture through the lens of the Institute. It draws on architecture history, literary and cultural studies, and institutional and cultural sociology while providing a solid footing for the paradigm shift of the 1960s, '70s, and '80s in terms of institutionalizing and institutionalized tendencies. *Building Institution* is based on the research I conducted as part of my doctoral dissertation project at the Institute for the History and Theory of Architecture (gta) at ETH Zurich (2007–11):[36] com-

32 Ibid.

33 Cf. Maurizio Lazzarato, "Immaterial Labor," in *Radical Thought in Italy. A Potential Politics*, eds. Paolo Virno and Michael Hardt (Minneapolis: University of Minnesota Press, 1996), 133–147. For Lazzarato, however, immaterial labor focused on subversion, that is, altering social structures rather than consolidating them.

34 After the publication of *Five Architects*, the term "New York Five" was disseminated by the press, see Paul Goldberger, "Architecture's '5' Make Their Ideas Felt," *The New York Times* (November 26, 1973), 33.

35 See Manfredo Tafuri, "L'Architecture dans le Boudoir. The Language of Criticism and the Criticism of Language," trans. Victor Caliandro, *Oppositions* 3 (May 1974), 37–62; see also Esra Akcan, "Manfredo Tafuri's Theory of the Architectural Avant-garde," *The Journal of Architecture* 7 (Summer 2002), 135–167.

36 The situation of the Institute archives is not entirely clear. In the course of my oral history research, I came across various accounts according to which the original archives were handed over to the bailiff and auctioned off in the course of the difficult closure of the Institute in May 1985 as part of foreclosure proceedings, thus becoming the property of one or more of the parties involved. As part of my archival research, I worked primarily at the Canadian Centre for Architecture, where documents relating to the Institute have been located since 1998, first in the Peter Eisenman Collection (fonds 143), and second in the IAUS Collection (fonds 57). In addition, in the early 2000s, the CCA began to create an oral history of the Institute, which already includes some interviews with protagonists and contemporaries that have already been conducted, although the concept for this and the interest behind it were ultimately directed

prehensive archival research undertaken during a several-month residency as a doctoral researcher at the CCA (2009), and in holdings of numerous other institutions on the East Coast of the United States; an extensive body of oral history totaling over one hundred interviews with more than eighty people involved in the Institute, including former Fellows, Visiting Fellows, staff, students, and interns, with board members, collaborators, and other contemporaries, conducted during a year-long stay as a visiting scholar at Columbia University in New York (2009–10) financed by a grant from the Swiss National Science Foundation (SNSF), which offer insights into individual motivations and contributions; and a close reading of the Institute's numerous publications, almost all of which can be accessed in North American archives, museums, and universities, and other relevant literature.[37] The focus of the analysis is on the available text corpus of the Institute, which includes a large number of original (or photocopied) circulated institutional documents (charters, by-laws, agendas, meeting minutes, official correspondence, internal memos, handwritten notes, working papers, concept papers, position papers, proposals, reports, leaflets, brochures, budget plans, financial reports, press releases, press reviews, etc.), and on visual representations (architectural, graphical, photographic) and institutional imaginaries connected to its projects, programs, and products. To avoid reproducing oft-repeated narratives and to de-mythologize the Institute's history with the aim of opening up meaningful insights into the broader institutional, even postmodern

mainly at Eisenman, not the Institute (fonds archives institutionelles / archives orales AO 04 – Louis Martin: Institute for Architecture and Urban Studies, 1998–2003). As of 2016, other documents relating to the Institute are also in the Kenneth Frampton Collection (fonds AP 197) at the CCA, particularly those relating to the Institute's only building project, which Frampton was in charge of. Other collections and records from archives and museums (MoMA, Walker Art Center), universities (primarily Sarah Lawrence College, but also Columbia University, Princeton University, and Yale University, such as the documents relating to a variety of public events in the Robert A.M. Stern Collection and the photographs of events by Dorothy Alexander in the Beinecke Library), research centers (Getty Research Institute and the graphic designs of Massimo Vignelli in the Vignelli Center for Design Studies at Rensselaer Institute of Technology), and other institutions in the United States, government agencies (Internal Revenue Service), foundations (New York Council on the Arts, National Endowment for the Humanities), associations (American Institute of Architects), etc. In addition, there were numerous private archives of individuals (Deborah Berke, David Buege, Peter Eisenman, Suzanne Frank, Deborah Gans, Peter Greenberg, Jessica Helfand, Margot Jacqz, Jonathan Kirschenfeld, Lawrence Kutnicki, Andrew MacNair, Patrick Pinnell, Stephen Potters, Massimo Scolari, Robert Silman, Suzanne Stephens, Mimi Shanley Taft, Frederieke Taylor). Although there is no single archive of the Institute, the archived materials together with documents provided by individuals are comprehensive. Many of the protagonists had not opened their private archives for my research, similar to some institutions (Graham Foundation) and publishers (MIT Press), etc.

37 As associate director of research at CCA from 2016 to 2018, I again had direct access to the archival holdings. In this capacity, I assisted PhD students and postdoctoral fellows working on the holdings in their research and activated Kenneth Frampton's then newly acquired private archive of research by curating an event with Frampton, an oral history interview, and the exhibition "Educating Architects" (May 31 to September 24, 2017) about four of the courses taught by Frampton at Columbia in the 1970s and 1980s, https://www.cca.qc.ca/en/events/49514/educating-architects-four-courses-by-kenneth-frampton (last accessed: May 31, 2023). I also conducted further research in the United States at that time.

culture and its flexible, entirely precarious modes of production, the individual chapters will focus on the Institute's history of cultural production alongside the history of its reception: the Institute conceptualized, realized, and staged as an architectural project of institution-building. A comprehensive institutional analysis in terms of research and design, education, culture, and publishing is necessary because the Institute's history cannot necessarily be broken down into its component parts and easily incorporated into architecture history with a conventional monographic study or master narrative—the sheer number of people involved and their widely divergent values, ideas, and motivations are evidence of this.[38] Additionally, the story of the Institute, seen from an epistemological perspective, encompasses strategies of a very different kind—strategies that would come to influence not only architecture but many other fields as well, ranging from urban politics and culture to architecture education and publishing, to the art market and criticism. This incompatibility, or even incomprehensibility, of the Institute's activities, is already evident when one defines the overarching themes, a process that was accomplished with the immediate history of its reception. As these smaller counter-narratives—both the subject of research in terms of the circulating legends and the state of research—show, the Institute managed to attract the attention, admiration, and regard—part appreciation, part criticism—of European architecture historians and critics with international standing, such as Reyner Banham and Manfredo Tafuri, almost immediately after its founding. Meanwhile, architecture journalists—including Ada Louise Huxtable and Paul Goldberger of *The New York Times*, and later Brian Brace Taylor and Michael Sorkin—who were based in New York and had broader insight into local politics and economics, commented on the Institute's activities with more regularity.[39] Even after the Institute's ultimate demise in 1985, North American architects and academics continued to engage with its legacy—both artistic and intellectual—its cadres, and its intrigues. Opponents of Peter Eisenman and his circle of friends repeatedly made themselves heard in North America's architecture press and in the mid-1990s cast themselves once again in the role of the opposition in *Progressive Architecture*, with the Institute's place in history at stake.[40] And thus, perhaps more than anything else, this is what the Institute bore witness

38 Some quite promising attempts were made, failed, and were finally reduced to a narrative revolving around a few characters that oscillated between a theatrical monologue and a chamber play; see Louis Martin, "The Search for a Theory in Architecture. Anglo-American Debates, 1957–1976," PhD diss., Princeton University, 2002; see Colin Brent Epp, "The Education of Rosalind Krauss, Peter Eisenman and Other Americans: Why the Fantasy of Postmodernism Still Remains," PhD diss., University of British Columbia, 2007.

39 Reyner Banham, "Vitruvius over Manhattan," *New Society* (December 7, 1967), 827–828; Tafuri, 1976.

40 Richard Plunz and Kenneth Kaplan, "On 'Style'," *Precis* (Fall 1984), 33–43; Diane Ghirardo, "Eisenman's Bogus Avantgarde," *Progressive Architecture* (November 1994), 70–73; Peter Eisenman et al., "Eisenman (and Company) Respond," *Progressive Architecture* (February 1995), 88–91; Vincent Pecora, "Eisenman and Friends," *Progressive Architecture* (May 1995), 13–15, 26.

to over the nearly two decades of its existence: the extent to which architecture, knowledge, and power are interwoven, illustrated by the example of New York.[41]

This book's method and methodology is to outline and discuss, on the basis of the various roles played by the Institute throughout its existence from 1967 to 1985, the changing relationships between architecture and culture, knowledge, politics, and economics in their respective contexts, i.e., their local conditions and conventions, and to document and highlight the significance and implications of these changes for a globalizing world. To this end, this institutional analysis will examine the Institute's microhistory against the backdrop of the broader socio-cultural contexts at the time. One key finding is that cultural production at the Institute enabled structures for material and, to an even greater extent, immaterial labor to be tested and established in the architecture-specific marketplace of culture, academia, and art, that not only incorporated but promoted broader developments. This needs to be seen, according to the core argument of this study, in relation to the changes in education and culture that were associated with architecture in general—a dynamic that continues to resonate to this day. Grounding this historical and at the same time critical perspective on cultural production in the social sciences and the humanities in order to inform contemporary architecture studies also allows this work to interrogate the sociocultural phenomena prevalent at the time, i.e., the postmodern order of collective interpretation and knowledge formation. This book, as far as the institution of architecture is concerned, ideally depicts the degree to which the Institute, by virtue of its postmodern plurality, heterogeneity, and diversity, helped redefine, alongside the neoliberal political and economic shifts in the mid-1970s, not only the "economy" of production and reception but also the "politics" of mediatization and interpretation in architecture. The form taken by this historiography, proceeding on the basis of an analysis of the everyday institutional practices at the Institute, their social and contextual contingency, and a critique of the discursive and material culture, while employing collective biography as its primary method, differs from a narrative grounded in biography, from a purely psychoanalytic method that would focus exclusively on Eisenman.[42] In such a narrative, there is a danger that Eisenman would simply be foregrounded as a self-promoter and puppet master, standing atop the stage provided by the Institute.[43] Yet this would miss the chance to offer a more nuanced and complex history of the Institute's design and function, work and significance, as a group, an organization, and even an institution,

41 Michel Foucault commented on postmodernism in one of his rare interviews about architecture, which was actually published in *Skyline*, i.e., from within the Institute; see Michel Foucault, "Space, Knowledge, and Power," *Skyline* (March 1982), 16–20.

42 Wilhelm Heinz Schröder, "Kollektive Biographien in der historischen Sozialforschung: Eine Einführung," in *Lebenslauf und Gesellschaft. Zum Einsatz von kollektiven Biographien in der historischen Sozialforschung*, ed. Schröder (Stuttgart: Klett-Cotta, 1985), 7–17.

43 Alex Tzonis and Liane Lefaivre, "The Narcissist Phase in Architecture," *The Harvard Architecture Review*, no. 1 (Spring 1978), 53–63; republished in German, see "Die narzisstische Phase der Architektur," *Arch+*, no. 42 (December 1978), 51–57.

not just for those who were directly involved, and even a critical assessment and appraisal of its impact and lasting influence on the development of the discipline and the profession.

Book Structure

This book about the life and career of the Institute, how it was enacted, characterized, transformed, and also resisted, is divided into four chapters that cover in detail the four main roles and activities of the Institute as a group of Fellows, as a non-profit organization, and as an architecture institution, with its shifting focus.[44] This four-part structure, which portrays the various developments in sequence and in parallel, allows the Institute to be understood as a "cultural space" rather than an architecture firm, distinct from other educational and cultural institutions in New York and on the East Coast of the United States—museums and universities, but also counter-institutional practices such as alternative art spaces—as well as from academic and commercial publishers. The first chapter examines the Institute's deliberate design and skillful networking as a "project office," in addition to the founding acts, and indeed institutionalization processes, and the research and design work it performed for various agencies (municipal, state, and federal) with student labor. This includes in particular the "Streets Project" for HUD (1970–72) and the "Low-Rise Housing" for the UDC (1972–73), neither of which would have been possible without the Institute's association with MoMA. The second chapter then looks at the Institute's first reinvention of itself as an "architecture school" and its fundamental restructuring. This starts with the 1974–75 academic year, when an "Undergraduate Program in Architecture," taught and supervised by Fellows, was offered for students from liberal arts colleges on the East Coast, followed by a series of alternative educational offerings for different audiences, an "Internship Program," an "Undergraduate Program in Planning," a "High School Program," "Design-and-Study Options," and finally, as a commercial alternative competing with established schools of architecture, the "Advanced Design Workshop." The third chapter focuses on the Institute's emergence as a "cultural space," again in the 1974–75 academic year, and the expansion of existing activities into fully fledged programs with funding available for culture. On the one hand, these included a comprehensive "Evening Program" with a range of lecture series, simply titled "Architecture" and comparable to those offered by larger institutions in the context of lifelong learning, with a focus on architecture history and theory, urban planning, art, and design. This was relaunched in 1977–78 as an "Open Plan" program in the humanities. On the other hand, the Institute also offered an "Exhibition Program" with group and solo exhibitions that made contemporary drawings and models their subject while integrating them into the art market, as well as historiographical exhibitions that emphasized

44 See Porsché, Scholz, and Singh, 2022, 2.

European and American modernist positions—both formats that served the attention economy even more than publications and were marketed and advertised nationwide. The fourth chapter covers the Institute's activities as a "publishing imprint," and the writing, editorial, and publishing work of the intellectually ambitious Fellows—not just *Oppositions*, but the whole portfolio of formats developed in the second half of the 1970s and published in collaboration with MIT Press, some of them later on with Rizzoli International: *October* (1976), *Skyline* (1978), IAUS Exhibitions Catalogues (1979), and Oppositions Books (1982).

As a contribution to architecture history that foregrounds socio-analysis and discourse analysis, *Building Institute* takes a dual approach to the manifold practices: institutional structure and organization, project, program and product conception, and the productions themselves, in addition to research and design, education, culture, and publishing.[45] From an epistemological perspective, this book—conceived as a collective biography of institutional services rather than an intellectual biography of individual positions—is equally concerned with the development of the Institute's strategic direction and organization structure; with the interplay of fields of activity and the day-to-day work of its Fellows, Visiting Fellows, staff, students, and interns; and with the networks established at various stages of the Institute's history. From a historiographical perspective, this work triangulates central developments in North American and globalized architecture culture—above all as they relate to the emergence, restriction, and concretization of particular ideas, concepts, and forms of knowledge that shape the conception, planning, and implementation of major projects, programs, and products—with an eye toward the Institute's contribution as a mediator in the regeneration of the inner city as a residential and recreational space or workplace; toward the Institute's interventions in the technocratic organization and regulation of housing; toward the Institute's particular social function and role in the North American educational system and its humanistic ideal of education; toward the Institute's specific method of finding solutions to problems, always undertaken in interaction with the discipline or profession of architecture as an autonomous, sometimes critical practice *vis-à-vis* the problems of a "public environment;" and toward the development of an intellectual and artistic position at the Institute, an architectonic approach that was both sculptural and iconographic, accompanied by theoretical reflection and a historical justification of its own foundations and conditions.[46]

45 The institutional analysis approach was originally developed in sociology, inspired by and in distinction from the discourse analysis approach; see Michael Foucault, *The Archaeology of Knowledge* (New York: Pantheon Books [1969], 1972).

46 The work of Eisenman, as well as that of the other longtime Fellows, notably Frampton, Gandelsonas, Agrest, and Vidler—whether built or written—is therefore not explicitly the subject of this study but is cited whenever it comes to the interactions between institutional structures and individual careers.

Building Institute depicts the changing context and limitations of the Institute's impact on the institution of architecture, if not society as a whole, conditioned by socio-economic and political transformations of the 1960s, '70s, and '80s, and constrained by notions of class, race, ethnicity, and gender prevalent in North American society at the time. From the initial hopes and promises of conducting research and design that was socially relevant, especially for disadvantaged populations, until budget cuts for public housing under the Nixon administration put an end to this, to the reinvention of the contribution of education and culture as two important areas of the information and service society, both subject to commodification,[47] to its increasing prominence and establishment as a "postmodern salon" for a globalizing architecture culture that was self-referential and yet highly influential on a national and international scale.[48] While each of the four chapters highlights both the internal synergies and the network synergies that were created and exploited, they can also be read as stand-alone (his)stories that offer new insights into moments of change, opportunities, and failures. But only when read as a whole do the four chapters provide a full picture of the Institute as a major player in a shifting architecture culture that has become differentiated, commodified, and globalized by its actions: a picture of its—despite or perhaps because of its constant reinvention and repositioning—constantly asserted impact and significance, not only in North America but also beyond in terms of its transatlantic aesthetics, discourse, education, and culture. This is especially true of the Institute's involvement in postmodernism, which in turn encompasses far more than just the Institute's course offerings, lecture series and exhibitions, and publications. This institutional and cultural history of the Institute as an architectural project—more of an institution than a building—in analytical and critical terms serves as a lens through which we can understand the processes of institutionalization, professionalization, and differentiation inherent in architecture since the 1970s, and how they persist in their updated permutations and the revival of postmodernism to this day.

Ultimately, though, *Building Institution* will have to come to terms not only with the Institute's agency and achievements, the buildings that were erected, the students and interns that were trained, the cultural, social, and economic

47 Eisenman advertised this complexity and the contradictions of the Institute early on; see Eisenman, 2007. In his 2010 monograph *Utopia's Ghosts*, architecture historian Reinhold Martin, drawing on Gilles Deleuze and Félix Guattari's concept of the machine, referred to the Institute only once, stating that Eisenman had created a "discourse machine" with *Oppositions* and a "teaching machine" with the Institute; see Martin, 2010, 66. The Institute's cultural production and publications, however, were equally groundbreaking.

48 For the transcripts of a two-day symposium which Eisenman organized in Charlottesville at the University of Pennsylvania on 12 and 13 November 1982, following his directorship at the Institute, see Jacquelin Robertson, *The Charlottesville Tapes* (New York: Rizzoli International, 1985); republished in German, see Jacquelin Robertson and Stanley Tigerman, *Der postmoderne Salon. Architekten über Architekten* (Basel: Birkhäuser, 1991).

capital that was accumulated, and the titles that were published, but also the discontinuities, transformations, reinventions, and endings, etc. For the gradual demise of the Institute is no less relevant to architecture history, nor is the 1981 effort to transform it into what was to become the Philip Johnson Center for Architecture, a new institution that was to be founded and named after New York architecture's patron and the Institute's main sponsor, whose fascist past was by then already common knowledge. These plans ultimately fell through due to a lack of financial and political support. Ultimately, the Institute was a complex network that helped to cultivate a new academic and architecture elite in the United States. While this development was already viewed quite critically by outsiders and the local architecture press in its day, a reassessment from a feminist, if not intersectional perspective would be necessary today, especially in the wake of the #MeToo-movement, despite the fact that a generation of women rose to prominence there in administrative, editorial, curatorial, and teaching positions.[49] Within the framework of such a historiography of the Institute, not only the founding narratives and major breaks and ruptures will be given new significance, but so too will the institutional power relations between the inner and outer circle of Fellows, Visiting Fellows, staff, students, friends, and enemies; various neglected, inglorious impasses, events that shed light on the Institute's rise and fall, its successes and failures.

49 Apparently, some names of male architects involved in the Institute in 2018 have appeared on the "Shitty Men in Architecture" spreadsheet, see https://archinect.com/forum/thread/150054690/shitty-men-in-architecture-spreadsheet and https://www.archpaper.com/2018/03/shitty-architecture-men-list-address-abuse-in-architecture/ (last accessed: May 31, 2023).

Adopted September 29, 1967

THE INSTITUTE FOR ARCHITECTURE AND URBAN STUDIES. Voted, That

1. A provisional charter, valid for a term of 5 years, is granted incorporating Gibson Danes, Arthur Drexler, Peter Eisenman, John Entenza, and Burnham Kelly and their associates and successors as an educational corporation under the corporate name of The Institute for Architecture and Urban Studies, to be located in the borough of Manhattan, city, county and State of New York.

2. The purposes for which such corporation is to be formed are:

a. To encourage and develop the study of architecture and design and their relation to urban environments by furnishing instruction and research facilities at the graduate and post-graduate level;

b. Through collaboration with public and private agencies, to perform research and planning activities with a view to draw upon any available resources of any university, of the Museum of Modern Art, and any municipality; and

c. To provide continuing education to the public through seminars, lectures, publications and exhibitions.

3. The persons named as incorporators shall constitute the first board of trustees. The board shall have power to adopt bylaws, including th therein provisions fixing the method of election and the term of office of trustees, and shall have power also, by vote of two-thirds of all the members of the board of trustees, to change the number of trustees to be not more than 25 nor less than 5.

4. The corporation hereby created shall be a nonstock corporation organized and operated exclusively for educational purposes, and no part of its earnings or net income shall inure to the benefit of any individual, and no officer, member, or employee of the corporation shall receive or

Fig. 1

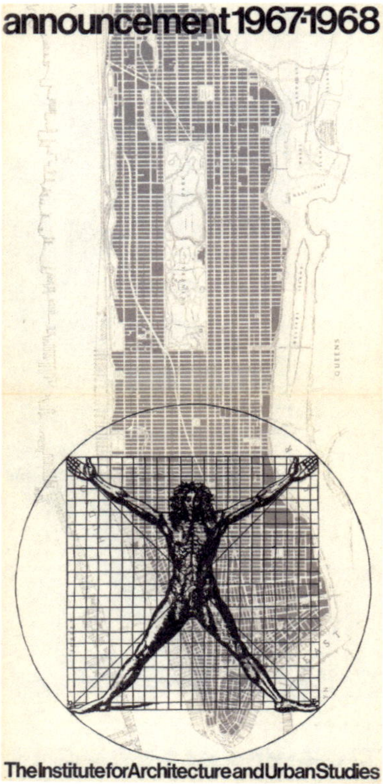
announcement 1967-1968
The Institute for Architecture and Urban Studies
Fig. 2

NEW URBAN SETTLEMENTS analytical phase
Fig. 3

Form approved
Budget Bureau No. 63-R1111

HUD-6800
Rev. 3-67

U. S. DEPARTMENT OF HOUSING AND URBAN DEVELOPMENT

URBAN RENEWAL DEMONSTRATION PROJECT BUDGET

(See instructions on reverse side)

NAME OF PUBLIC BODY: The Institute for Architecture and Urban Studies

PROJECT NUMBER *(If known)*:

BUDGET NUMBER:

Line No.	ACCOUNT CLASSIFICATION NO.	DESCRIPTION	TO BE COMPLETED BY PUBLIC BODY — USE ONLY FOR REVISED BUDGETS: LATEST APPROVED BUDGET DATED ____, 19__ (a)	ACTUAL COSTS TO DATE (b)	OUTSTANDING AND ESTIMATED COMMITMENTS (c)	BUDGET REQUESTED *(Col. (b) + Col. (c))* (d)	TO BE FILLED IN BY HUD — BUDGET APPROVED BY HUD (e)
1	D 1410	ADMINISTRATIVE COSTS *(Sum of Lines 2 through 5)*	$	$	$	$238,535.00	$
2	D 1410.2	Staff Salaries				187,000.00	
3	D 1410.7	Employee Benefit Contributions				6,385.00	
4	D 1410.91	Travel				4,000.00	
5	D 1410.93	Other Administrative Costs				41,150.00	
6	D 1430	Contract Services *(Specify below)*[1]	$	$	$	$ 48,400.00	$
		a. See attached sheet					
		b.					
		c.					
		d.					
		e.					
7	D 1460	Other Project Costs *(Specify below)*[1]	$	$	$	$	$
		a.					
		b.					
8	D 1475	Nonexpendable Equipment				6,300.00	
9	D 1502	Writing and Publishing Project Reports				32,000.00	
10		Subtotal *(Sum of Lines 1 and 6 through 9)*	$	$	$	$325,235.00	$
11		Contingencies				5,000.00	
12		TOTAL PROJECT BUDGET *(Sum of Lines 10 and 11)*	$	$	$	$330,235.00	$

Approval of the Urban Renewal Demonstration Project Budget in the amounts shown in Column (d) is hereby requested.

Oct. 14, 1969

Date *Signature of Authorized Officer*

Title

The Urban Renewal Demonstration Project Budget is hereby approved in the amounts shown in Column (e). The authorized activities shall be completed by ____________, 19____.

Date *Signature*

Title

[1] If additional space is needed, show complete listing on attached sheets.

Fig. 4

MINUTES OF MEETING

5 October 1971

Those Present:

Peter Eisenman

Arthur Drexler

Lily Auchincloss

Armand Bartos

George Dudley

John Entenza

Frank Stanton

THE CHAIRMAN MR. DREXLER OPENED THE MEETING

MR. Drexler moved to accept the Minutes of THE LAST MEETING

The motion was carried.

MR. DREXLER THEN PROPOSED RICHARD MEIER AS A NEW MEMBER OF THE BOARD OF TRUSTEES AND THE MOTION WAS CARRIED UNANIMOUSLY.
~~The Trustees voted to adopt Richard~~ MR. Meier ~~as a new member of~~ WAS THEN ASKED TO PARTICIPATE IN THE MEETING, AND
the Board ~~and~~ welcomed him.

MR. DREXLER THEN ASKED MR. EISENMAN TO PRESENT THE DIRECTOR'S REPORT
~~Projects in Progress~~

Eisenman presented ~~the Annual~~ HIS Report to the Trustees. ~~explaining~~ HE FIRST REVIEWED THE ACTIVITIES OF THE ~~that it contains an introduction to what has been done in the~~ past year: the new Fellows, the new Graham Foundation Visiting Fellows, public ~~activities~~ events last year and proposed for this year, and a discussion of projects. UNDER THE DISCUSSION OF PROJECTS ~~The faculty for this year and a summary of research and design projects are under the discussion of projects.~~

Eisenman explained that the RESEARCH program has now been broken down into ~~four areas:~~ FOUR BASIC TYPES

Urban Components	Streets as Urban Components ~~- moving into final phase~~ Housing Special building types
Urban Settlements	Linear City STUDIES ~~supported by GBDF~~ Development in Israel - forwarded to Prime Minister's Office Program Alternatives to University Based New Towns - submitted to UDC Low Rise High Density Suburban Land Settlements ~~-~~ submitted to Ford Foundation (AS part of ~~whole~~ a comprehensive plan to study suburban land settlements being developed by Kenneth Frampton)

Fig. 5

Fig. 6

Fig. 7

Fig. 8

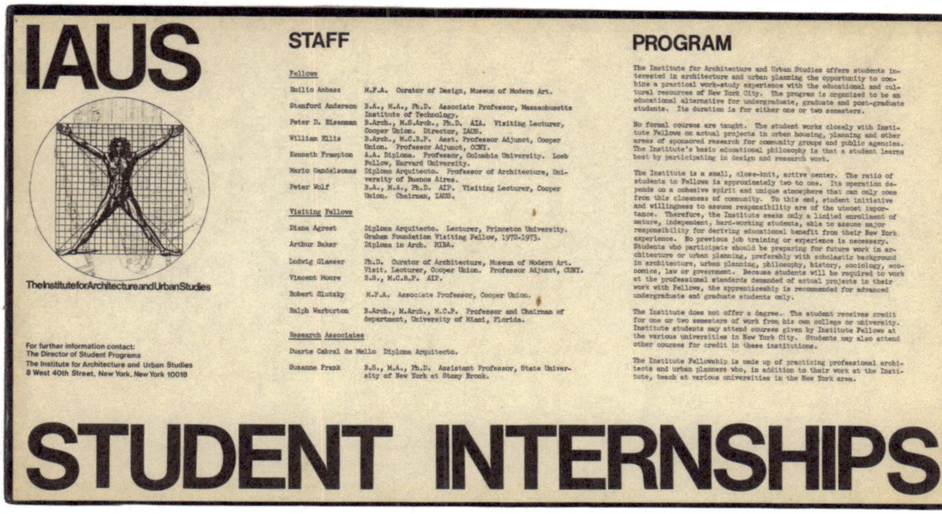
IAUS
TheInstituteforArchitectureandUrbanStudies
For further information contact:
The Director of Student Programs
The Institute for Architecture and Urban Studies
8 West 40th Street, New York, New York 10018
STAFF
Fellows
Emilio Ambasz M.F.A. Curator of Design, Museum of Modern Art.
Stanford Anderson B.A., M.A., Ph.D. Associate Professor, Massachusetts Institute of Technology.
Peter D. Eisenman B.Arch., M.S.Arch., Ph.D. AIA. Visiting Lecturer, Cooper Union. Director, IAUS.
William Ellis B.Arch., M.C.R.P. Asst. Professor Adjunct, Cooper Union. Professor Adjunct, CCNY.
Kenneth Frampton A.A. Diploma. Professor, Columbia University. Loeb Fellow, Harvard University.
Mario Gandelsonas Diploma Arquitecto. Professor of Architecture, University of Buenos Aires.
Peter Wolf B.A., M.A., Ph.D. AIP. Visiting Lecturer, Cooper Union. Chairman, IAUS.
Visiting Fellows
Diana Agrest Diploma Arquitecto. Lecturer, Princeton University. Graham Foundation Visiting Fellow, 1972-1973.
Arthur Baker Diploma in Arch. RIBA.
Ludwig Glaeser Ph.D. Curator of Architecture, Museum of Modern Art. Visit. Lecturer, Cooper Union. Professor Adjunct, CUNY.
Vincent Moore B.S., M.C.R.P. AIP.
Robert Slutzky M.F.A. Associate Professor, Cooper Union.
Ralph Warburton B.Arch., M.Arch., M.C.P. Professor and Chairman of department, University of Miami, Florida.
Research Associates
Duarte Cabral de Mello Diploma Arquitecto.
Suzanne Frank B.S., M.A., Ph.D. Assistant Professor, State University of New York at Stony Brook.
PROGRAM
The Institute for Architecture and Urban Studies offers students interested in architecture and urban planning the opportunity to combine a practical work-study experience with the educational and cultural resources of New York City. The program is organized to be an educational alternative for undergraduate, graduate and post-graduate students. Its duration is for either one or two semesters.
No formal courses are taught. The student works closely with Institute Fellows on actual projects in urban housing, planning and other areas of sponsored research for community groups and public agencies. The Institute's basic educational philosophy is that a student learns best by participating in design and research work.
The Institute is a small, close-knit, active center. The ratio of students to Fellows is approximately two to one. Its operation depends on a cohesive spirit and unique atmosphere that can only come from this closeness of community. To this end, student initiative and willingness to assume responsibility are of the utmost importance. Therefore, the Institute seeks only a limited enrollment of mature, independent, hard-working students, able to assume major responsibility for deriving educational benefit from their New York experience. No previous job training or experience is necessary. Students who participate should be preparing for future work in architecture or urban planning, preferably with scholastic background in architecture, urban planning, philosophy, history, sociology, economics, law or government. Because students will be required to work at the professional standards demanded of actual projects in their work with Fellows, the apprenticeship is recommended for advanced undergraduate and graduate students only.
The Institute does not offer a degree. The student receives credit for one or two semesters of work from his own college or university. Institute students may attend courses given by Institute Fellows at the various universities in New York City. Students may also attend other courses for credit in these institutions.
The Institute Fellowship is made up of practicing professional architects and urban planners who, in addition to their work at the Institute, teach at various universities in the New York area.
STUDENT INTERNSHIPS

Fig. 9

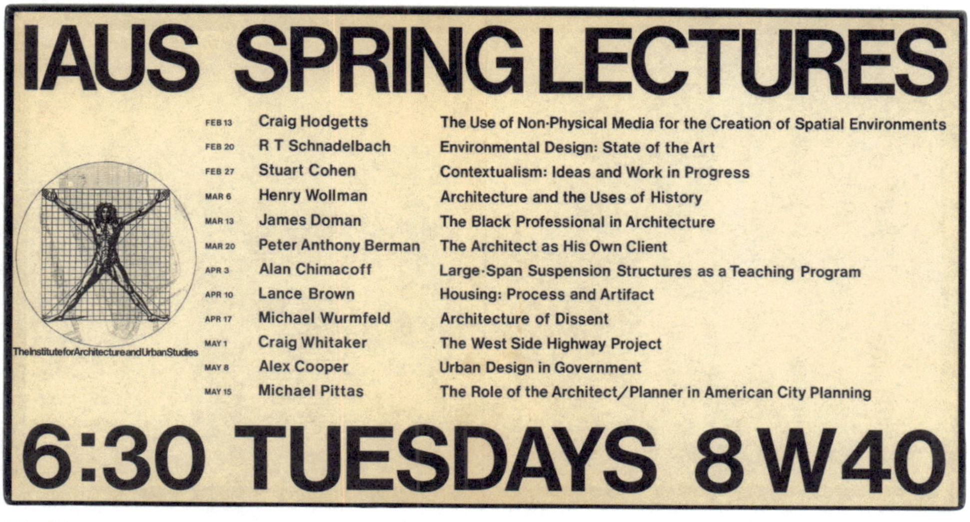
IAUS SPRING LECTURES
FEB 13 Craig Hodgetts The Use of Non-Physical Media for the Creation of Spatial Environments
FEB 20 R T Schnadelbach Environmental Design: State of the Art
FEB 27 Stuart Cohen Contextualism: Ideas and Work in Progress
MAR 6 Henry Wollman Architecture and the Uses of History
MAR 13 James Doman The Black Professional in Architecture
MAR 20 Peter Anthony Berman The Architect as His Own Client
APR 3 Alan Chimacoff Large-Span Suspension Structures as a Teaching Program
APR 10 Lance Brown Housing: Process and Artifact
APR 17 Michael Wurmfeld Architecture of Dissent
MAY 1 Craig Whitaker The West Side Highway Project
MAY 8 Alex Cooper Urban Design in Government
MAY 15 Michael Pittas The Role of the Architect/Planner in American City Planning
TheInstituteforArchitectureandUrbanStudies
6:30 TUESDAYS 8 W 40

Fig. 10

Fig. 11

Fig. 12

Wednesdays at 6.45pm

EVENING LECTURES / HABITAT & URBAN FORM

IAUS The Institute for Architecture & Urban Studies

Eight West Fortieth Street New York

FEBRUARY	9:	BRIAN RICHARDS	TRAVELATION - THE FUTURE OF AN ILLUSION
	16:	WERNER SELIGMANN	LOW RISE HIGH DENSITY DEVELOPMENT IN ITHACA, N.Y.
	23:	ANTHONY VIDLER	COMMUNITY FORM - IDEAL VERSUS REALITY
MARCH	1:	PAUL RUDOLPH	HOUSING AS ARCHITECTURE
	8:	FRED KOETTER/JERRY WELLS	MODULAR HOUSING FOR THE MASS MARKET
	15:	D. AGREST/M. GANDELSONAS	A CRITIQUE OF CURRENT URBAN PLANNING MODELS
	22:	JOHN DE MONCHAUX	ANGLO-AMERICAN PLANNING AND THE MARK III NEW TOWNS
	29:	F. JIMENEZ/A. UNGER	THE PROSPECTS FOR MASS HOUSING IN LATIN AMERICA
APRIL	5:	JACQUELIN ROBERTSON	MID-TOWN MANHATTAN - THE EVOLUTION OF A NEW CODE
	12:	GIANCARLO DE CARLO	LEGITIMIZING ARCHITECTURE
	19:	STANFORD ANDERSON	THE STREET AS AN ARTIFACT IN EUROPE AND AMERICA
	26:	O. MATTHIAS UNGERS	STRATEGIES FOR REGIONAL DEVELOPMENT
MAY	3:	SHADRACH WOODS	CRITERIA FOR MASS HOUSING - NEW YORK AND BERLIN
	10:	THEODORE LIEBMAN	UDC AND THE EVOLUTION OF A HOUSING POLICY
	17:	FRANCOISE CHOAY	URBANISM AND SEMIOLOGY
	24:	COLIN ROWE	COLLAGE CITY

Fig. 13

ORGANIZATIONAL STRUCTURE PHASE 3

STREETS AS COMPONENTS OF THE URBAN ENVIRONMENT
N.Y. D-13

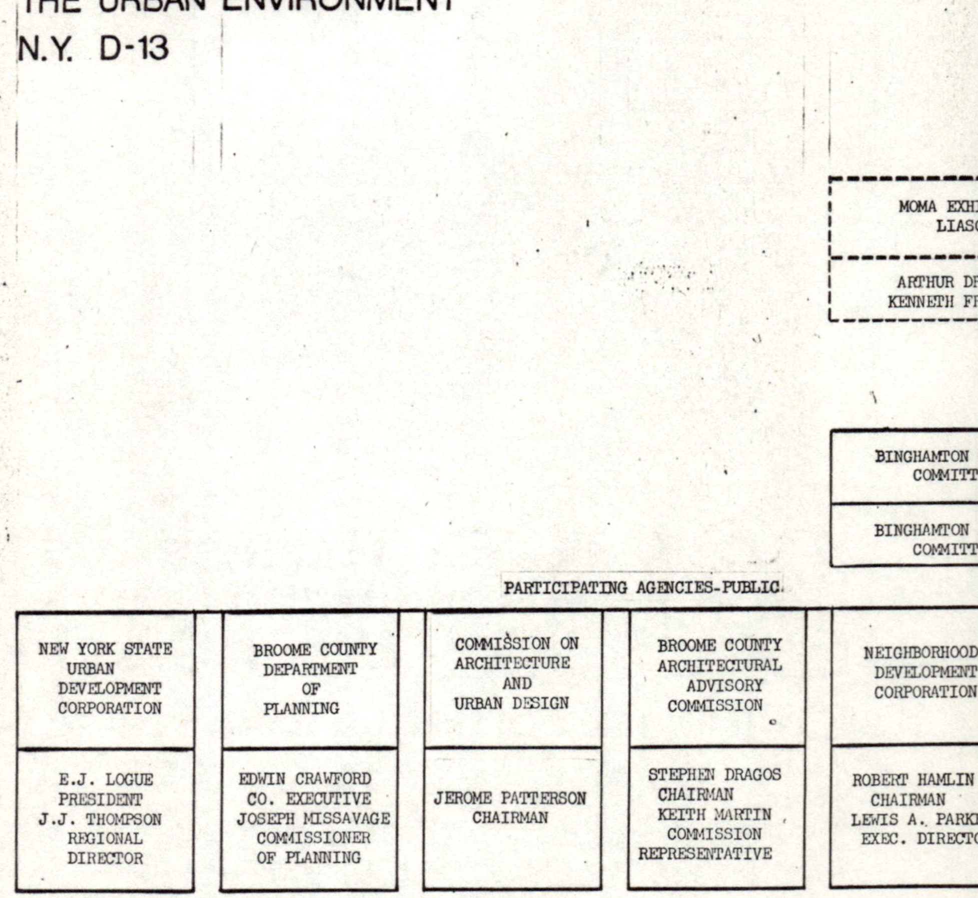

Fig. 14

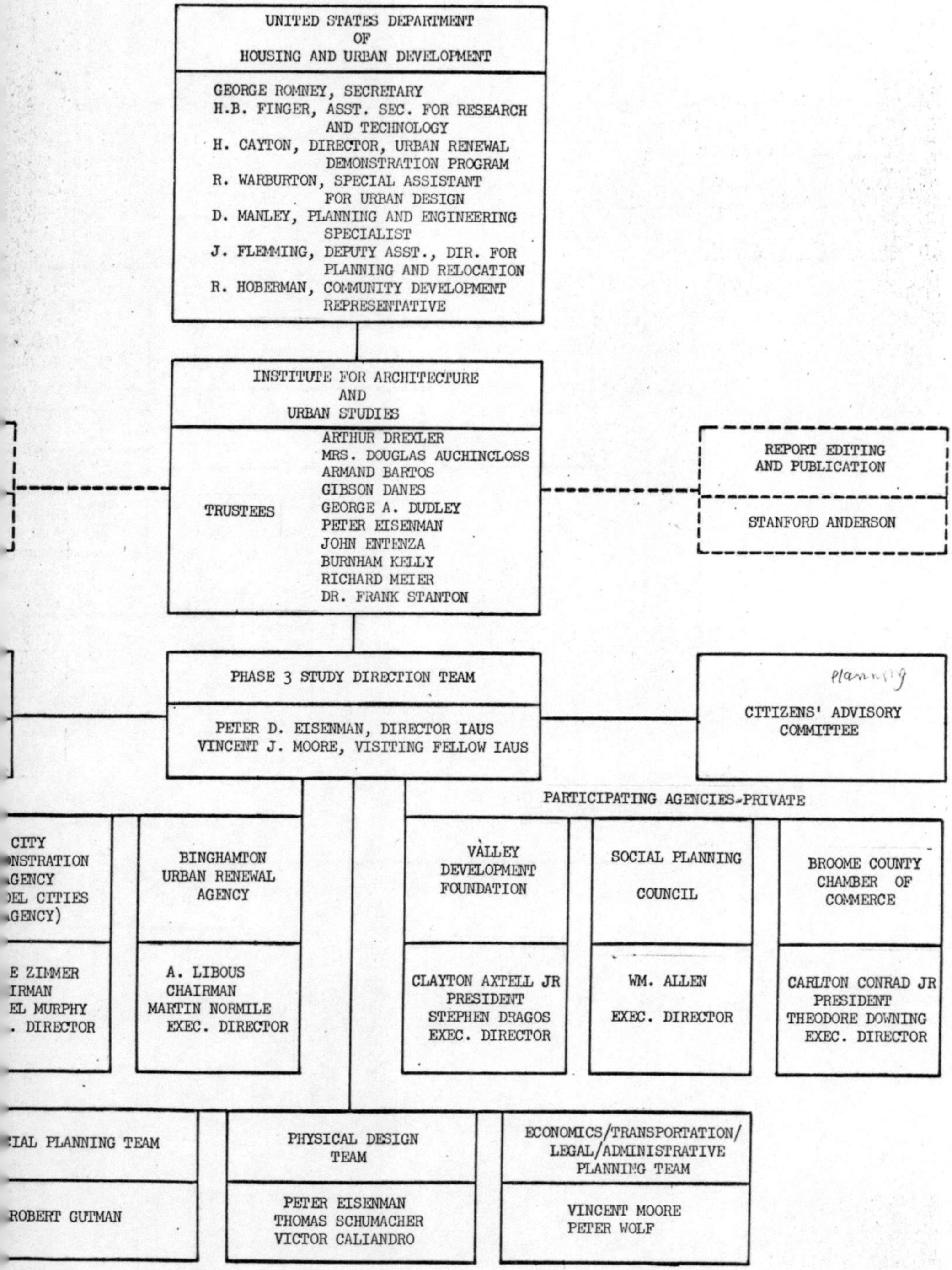
UNITED STATES DEPARTMENT OF HOUSING AND URBAN DEVELOPMENT
GEORGE ROMNEY, SECRETARY
H.B. FINGER, ASST. SEC. FOR RESEARCH AND TECHNOLOGY
H. CAYTON, DIRECTOR, URBAN RENEWAL DEMONSTRATION PROGRAM
R. WARBURTON, SPECIAL ASSISTANT FOR URBAN DESIGN
D. MANLEY, PLANNING AND ENGINEERING SPECIALIST
J. FLEMMING, DEPUTY ASST., DIR. FOR PLANNING AND RELOCATION
R. HOBERMAN, COMMUNITY DEVELOPMENT REPRESENTATIVE
INSTITUTE FOR ARCHITECTURE AND URBAN STUDIES
TRUSTEES
ARTHUR DREXLER
MRS. DOUGLAS AUCHINCLOSS
ARMAND BARTOS
GIBSON DANES
GEORGE A. DUDLEY
PETER EISENMAN
JOHN ENTENZA
BURNHAM KELLY
RICHARD MEIER
DR. FRANK STANTON
REPORT EDITING AND PUBLICATION
STANFORD ANDERSON
PHASE 3 STUDY DIRECTION TEAM
PETER D. EISENMAN, DIRECTOR IAUS
VINCENT J. MOORE, VISITING FELLOW IAUS
Planning
CITIZENS' ADVISORY COMMITTEE
PARTICIPATING AGENCIES-PRIVATE
CITY
NSTRATION
GENCY
DEL CITIES
GENCY)
E ZIMMER
IRMAN
EL MURPHY
. DIRECTOR
BINGHAMTON URBAN RENEWAL AGENCY
A. LIBOUS
CHAIRMAN
MARTIN NORMILE
EXEC. DIRECTOR
VALLEY DEVELOPMENT FOUNDATION
CLAYTON AXTELL JR
PRESIDENT
STEPHEN DRAGOS
EXEC. DIRECTOR
SOCIAL PLANNING COUNCIL
WM. ALLEN
EXEC. DIRECTOR
BROOME COUNTY CHAMBER OF COMMERCE
CARLTON CONRAD JR
PRESIDENT
THEODORE DOWNING
EXEC. DIRECTOR
IAL PLANNING TEAM
ROBERT GUTMAN
PHYSICAL DESIGN TEAM
PETER EISENMAN
THOMAS SCHUMACHER
VICTOR CALIANDRO
ECONOMICS/TRANSPORTATION/LEGAL/ADMINISTRATIVE PLANNING TEAM
VINCENT MOORE
PETER WOLF

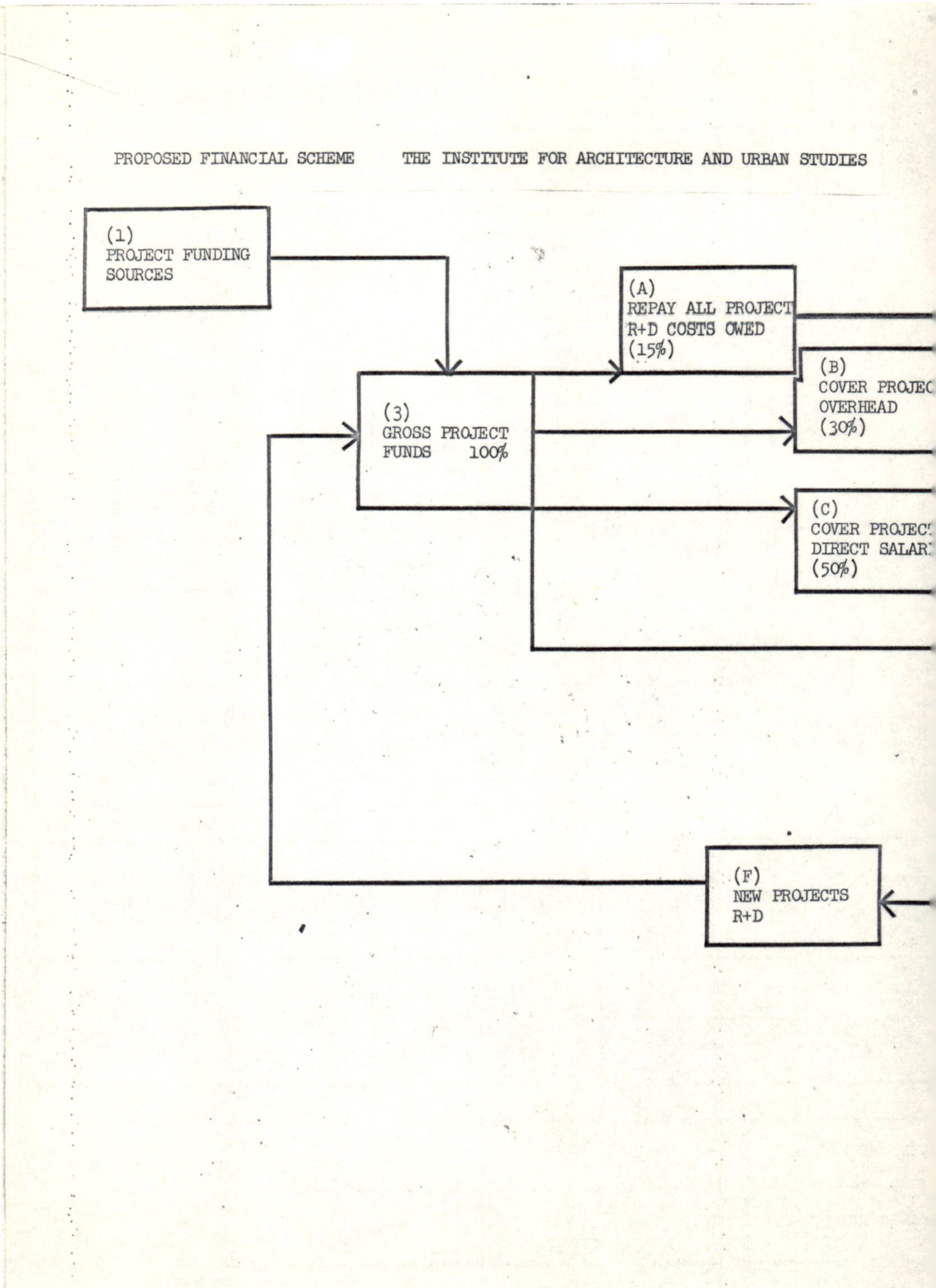
PROPOSED FINANCIAL SCHEME
THE INSTITUTE FOR ARCHITECTURE AND URBAN STUDIES
(1) PROJECT FUNDING SOURCES
(A) REPAY ALL PROJECT R+D COSTS OWED (15%)
(B) COVER PROJEC OVERHEAD (30%)
(3) GROSS PROJECT FUNDS 100%
(C) COVER PROJEC DIRECT SALAR (50%)
(F) NEW PROJECTS R+D

Fig. 15

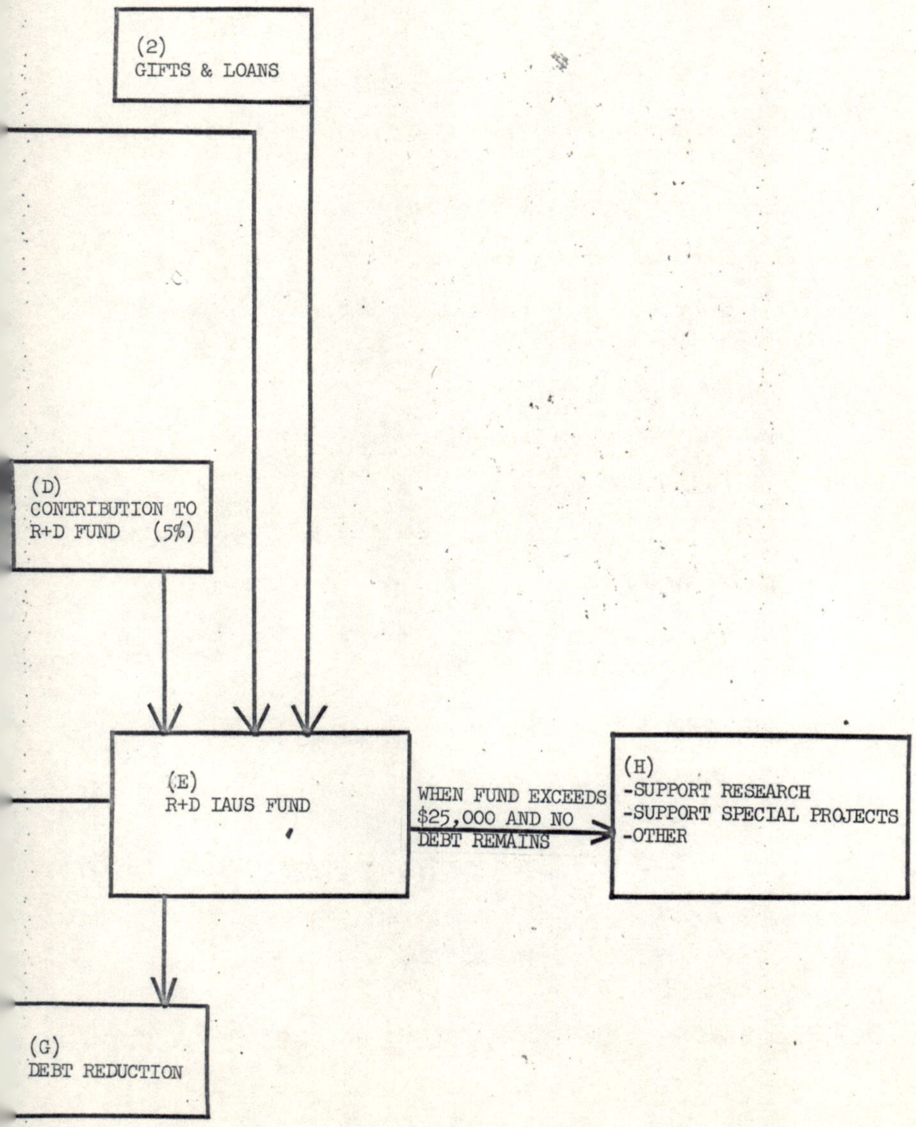
(2)
GIFTS & LOANS
(D)
CONTRIBUTION TO
R+D FUND (5%)
(E)
R+D IAUS FUND
WHEN FUND EXCEEDS
$25,000 AND NO
DEBT REMAINS
(H)
-SUPPORT RESEARCH
-SUPPORT SPECIAL PROJECTS
-OTHER
(G)
DEBT REDUCTION

WHAT IS LRHD HOUSING?

1. Direct access to ground (walk up)
2. Adequate semi-private and public open space for each dwelling
3. Private open space for each dwelling
4. High level of privacy <u>and</u> security
5. Unit mix of 1 - 5 bedrooms, with maximum feasible accommodation of families
6. Car park near dwellings
7. Density of 40-60 dwellings per acre
8. Four floors maximum

TWO BLOCKS MAXIMUM.

[illegible].

BEDROOM MIX FOR 3's AND 4's ONLY A SMATTERING OF
NOT INTERESTED IN TOWERS

BUILDER DEVELOPER MANAGER.

NO CROWDED

10 ACRES. NOT REAL-SITE 500 UNITS

POLICY.

NEW YORK CITY BLOCK.

PARKING

CHILD CARE FACILITIES. 1st FLOOR APARTMENT

Fig. 16

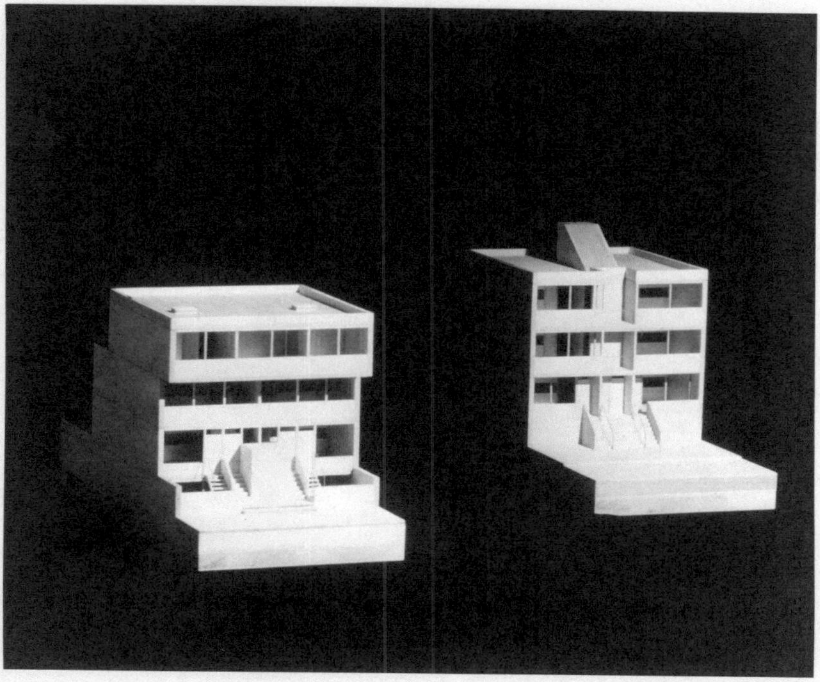

Fig. 17

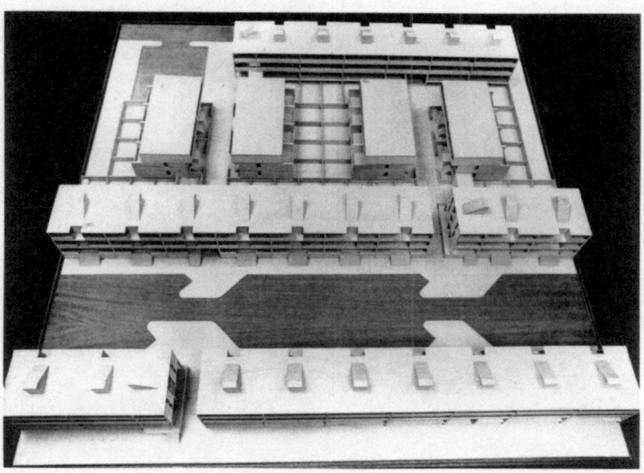

Fig. 18

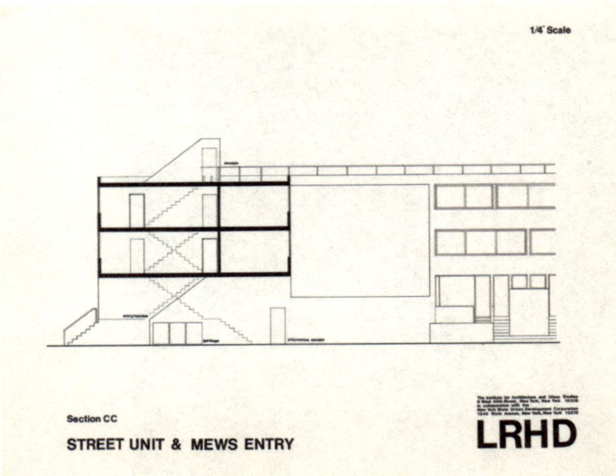

Fig. 19

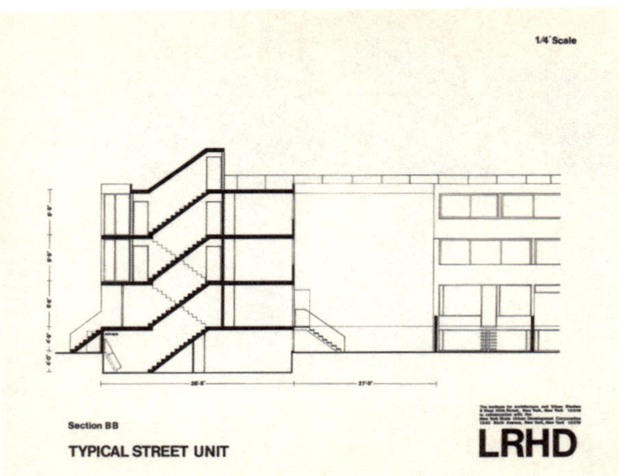

Fig. 20

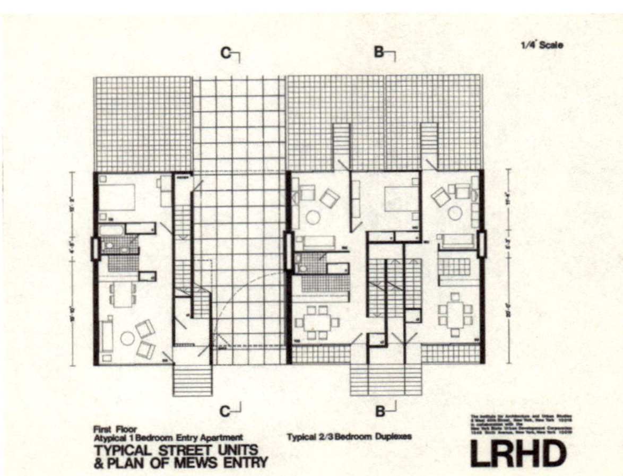

Fig. 21

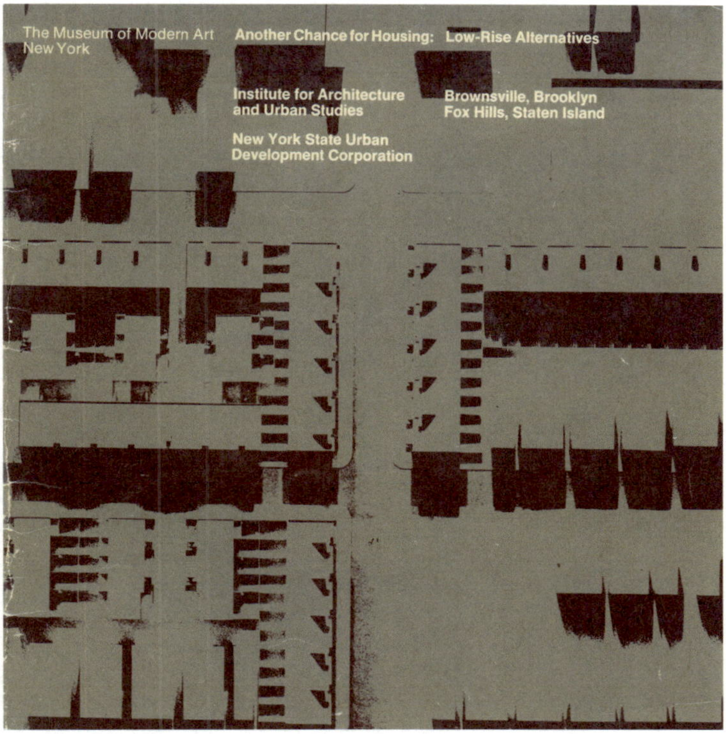
The Museum of Modern Art
New York
Another Chance for Housing: Low-Rise Alternatives
Institute for Architecture
and Urban Studies
New York State Urban
Development Corporation
Brownsville, Brooklyn
Fox Hills, Staten Island
Fig. 22

Fig. 23

28

Typical View of Cluster Element Fox Hills, Staten Island, New York Rendering by Craig Hodgetts

Fig. 24

Typical View from Cul-de-Sac Mews to Street   Marcus Garvey Park Village Urban Renewal, New York City   Rendering by Craig Hodgetts

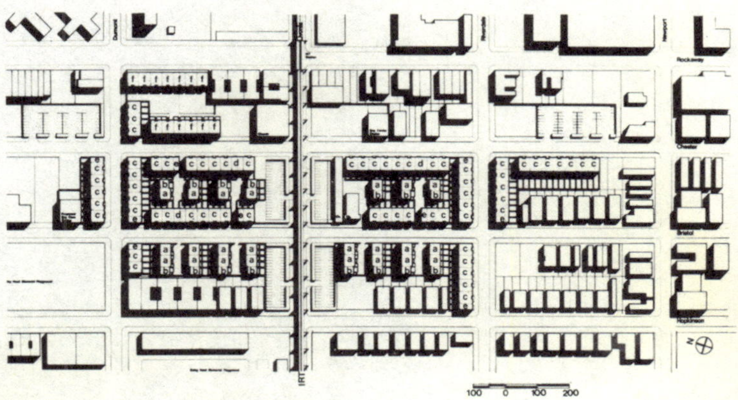

Site Plan, Marcus Garvey Park Village Urban Renewal. Letters A through F show the location of the unit types shown on pages 22 and 23.

Fig. 25

Fig. 26

Fig. 27

1. Project Office

The provisional by-laws of the Institute for Architecture and Urban Studies, adopted on September 29, 1967, approved by the trustees, and valid for five years, marked its official foundation. The by-laws laid out three objectives, which already set out the path for the next seventeen and a half years:[50] first, "to encourage and develop the study of architecture and design and their relation to urban environments by furnishing instruction and research facilities at the graduate and postgraduate level," second, "through collaboration with public and private agencies, to perform research and planning activities with a view to drawing upon any available resources of any university, of the Museum of Modern Art (MoMA), and any municipality," and third, "to provide continuing education to the public through seminars, lectures, publications and exhibitions." This programmatic focus on different areas of activity, combining research and design, education, culture, and publishing, reveals the extent to which the Institute planned to capture architectural discourse, create networks, and exploit synergies, i.e., to redesign architecture in general. The original idea was that, given the prevailing socio-economic and political trends of the time, the Institute could carry out consultancy work for urban planning projects in New York. Right from its founding, the Institute as a service provider was concerned with the production and dissemination of knowledge, but without a concrete vision or mission statement of how practice orientation could engage with the new knowledge system of urban studies. Neither the young architect Peter

50 IAUS, provisional by-laws, September 29, 1967. Source: CCA Montréal, IAUS fonds: A.2-3 / ARCH401124. Emilio Ambasz noted in our oral history interview that it was he who wrote a first draft of the Institute's by-laws while he was still a student of Eisenman at Princeton.

Eisenman, the founding director of the Institute, nor any of his early collaborators had much practical experience. But Eisenman's stance of always having to challenge the discipline allowed him to mediate between the profession and academia. Wisely enough, Eisenman explicitly stated at the time that the Institute did not intend to compete with the existing schools of architecture. Conforming to the cultural climate, the Institute immediately positioned itself and its research fields and education offerings as an alternative institution for research and design, education, and culture—an "anti-institution" as it were—in the East Coast university landscape with its affiliation to MoMA, and in New York public life. As a "newcomer" with a focus on adult education, however, it was in competition with other institutions such as the Architectural League, which had already been offering a public program of events since the 1960s.[51] The *Princeton Report* by Robert Geddes and Bernard Spring, or more precisely "A Study of Education for Environmental Design," published in 1967, established 'lifelong learning' as a new buzzword in architecture as well, so that architecture education also received a broader focus within the framework of a nationwide educational reform.[52] The Institute's unique selling point as a new actor vis-à-vis other groups, organizations, and institutions was that it combined research and design work with cultural production and public relations as an instrument of educational policy early on.

Founding Narratives

Shortly thereafter, on October 15, 1967, the *New York Times* featured the Institute as a newly formed institution that was poised to radically change the profession and discipline of architecture.[53] The article, a single column by Steven L. Roberts running to nearly a full page, represented Eisenman's public debut as Institute director in the country's leading daily newspaper a good two weeks after the official launch and compellingly demonstrated that the Institute's founding act was to inscribe itself in the social reality of the United States. The name "Institute," formulated in reference to institutes within or close to universities, may have been a misleading choice for a novel facility such as this. The seriousness of its creative and academic claims still had to be proven;

51 Robert Stern et al. "Architectural Culture: Discourse," in *New York 1960. Architecture and Urbanism Between the Second World War and the Bicentennial*, eds. Robert Stern, Thomas Mellins, David Fishman (New York: The Monacelli Press, 1995), 1205–1211, here 1209. In retrospect, Stern saw the Institute as competing with existing institutions: "During the early 1970s the Architectural League was challenged in its role as the city's most vital forum for architectural experiment and discourse by a newcomer, the Institute for Architecture and Urban Studies."

52 Robert Geddes and Bernard Spring, *A Study for Environmental Design* (Princeton: Princeton University, 1967).

53 Steven L. Roberts, "School Is Formed for Urban Design," *The New York Times* (October 15, 1967), 52.

nevertheless, the Institute had asserted competence in two separate fields by the act of naming alone. In the context of an institutional and intellectual history, the Institute displays different networks, both in terms of the actual work done there, the social relations, power structures, and micro-economies, as well as the socio-political context on a local, national, and international level. Once the Institute had been chartered by the School Board of the State University of New York, the attention brought by the *New York Times* proved instrumental in enabling it to become active with research and design projects. From the beginning, Eisenman knew how to use media exposure and the public visibility it brought to the Institute (and to himself) to acquire commissions, plan, and oversee the budget, and build the institution. Alongside the *New York Times*—with whose architecture critic Ada Louise Huxtable he soon established ties—the architectural press became one of the Institute's most important allies. One particular characteristic of the founding phase was that the group of people he gathered around him—Fellows, trustees, and staff—was rather loosely organized and still quite manageable. It was the organization itself that facilitated the close interweaving of individual and collective developments. Roberts' article was not only an institutional portrait but also a biographical one of Eisenman, whose picture illustrated it. Eisenman, just thirty-five years old, was still at the beginning of his professional career and seized the opportunity that presented itself. Since returning from England, where he had earned his doctorate with a thesis on *The Formal Basis of Modern Architecture* in 1961, he had spent several years researching and teaching, primarily at Princeton.[54] The dramatic twist was that Eisenman was denied a tenured position in 1967, and so he made a virtue out of necessity. With his move from rural New Jersey to New York, he had finally arrived in the metropolis, the city of dreams, the much-vaunted capital of capital, where he henceforth appeared as an intellectual, posed as an artist, and sold himself as an entrepreneur. This city that was to give birth to a new globalized architectural culture was the ideal breeding ground that, after deindustrialization, was to undergo a comprehensive transformation during the course of culturalization—and the Institute found itself in the midst of it. In the following decade, New York was to undergo a regime change from a welfare state to a new neoliberal politics and economy on its way to becoming a global city—with diverse and profound effects on architecture and the city itself.[55]

54 The publication of Eisenman's doctoral dissertation was a long time coming: it was first published in German in 2005 by gta Verlag of ETH Zurich; the original English edition also found a Swiss publisher in Lars Müller; see Peter Eisenman, *Die formale Grundlegung der modernen Architektur* (Zurich: gta Verlag, 2005), and *The Formal Basis of Modern Architecture* (Baden: Lars Müller Publishers, 2006).

55 Kim Moody, *From Welfare State to Real Estate: Regime Change in New York City from 1974 to the Present* (New York: The New Press, 2007); David Harvey, *The Condition of Postmodernity: An Enquiry into the Origins of Cultural Change* (Cambridge: Blackwell, 1989).

Eisenman succeeded in establishing his Institute because of this transformation and made a name for himself as a driving force in New York and on the East Coast. The Institute as a project was also about promoting and asserting a new generation that wanted to understand architecture and the city as an intellectual and artistic artifact and from a global perspective. However, although the Institute was installed and presented as a counter-architecture alternative to the established schools of architecture and the conventional understanding of architectural practice, in the context of an institutional analysis and critique of cultural production, similar to an archaeology of postmodernism in architecture, including early forms of the debates about "autonomy" and "criticality," the Institute was by no means as radical and independent from the beginning as has been repeatedly claimed.[56] In terms of the balance and oscillation between innovation and tradition, the Institute, as a nationally recognized educational institution from its inception, should rather be considered in relation to established institutions and networks, the older generation of architects and academics, and its sponsors, against the backdrop of the prevailing political and economic situation and the philanthropic culture in the United States in the late 1960s. In this way, a better understanding will emerge of all the discursive and institutional strategies that were successfully employed under Eisenman's direction to legitimize the Institute and establish its position.

One of the founding narratives of the Institute, first circulated with the *New York Times* article, therefore concerns its close connection with MoMA. Not only did Arthur Drexler, director of the MoMA Department of Architecture and Design since 1956, actively support Eisenman in the founding of the Institute throughout 1967, MoMA was even responsible for renting its first premises on 5 East 47th Street, a small office floor with two offices and a meeting room at the back, and a large studio space at the front facing the street, and initially paid the rent. This, despite claims to the contrary, is evidence of a certain degree of dependency.[57] The annual report of the Modern, as the museum was then commonly known, initially described the Institute as "an outcome of the Department's continuing concern with urban problems."[58] Eisenman

56 The Institute and subsequent groups launched by Eisenman, such as the ANY Corporation and its conference series (1991–2000), are often seen today as hotbeds of debates about "autonomy" and "criticality," which Eisenman gave built form to with his own projects; see Robert Somol, *Autonomy and Ideology: Positioning an Avant-garde in America* (New York: Monacelli Press, 1997), and Hays, 1998, IX–XV; see also Eisenman, 1988, and Brett Steele, *Supercritical: Peter Eisenman & Rem Koolhaas* (London: AA Publications, 2007).

57 The fact that the lease was dated August 22, 1967, indicates that its occupation anticipated the Institute's foundation.

58 MoMA, annual report 1967–1969. Source: The MoMA. The Institute was thus an offshoot of MoMA. In contrast, Stern described the Institute as resulting from "The New City" exhibition, "an outgrowth of the Museum of Modern Art's provocative exhibition The New City," cf. Stern et al., 1995, 1209.

received further support for his new project to create an alternative to academia and the classical architecture firm from Colin Rowe, his former mentor from his time at Cambridge, who at the time was a professor of urban design at Cornell University's architecture department.[59] Apparently, Rowe had assured Eisenman that he would bring some of his best students to New York, as well as teach at the Institute himself. Despite their different interests and approaches, Eisenman, Drexler, and Rowe shared similar intentions and understandings of architecture, public relations, and pedagogy: Eisenman, who needed a new job, wanted to use the Institute to establish a permanent footing from which to work as an architect and theorist;[60] Drexler was in the process of making MoMA's exhibition operations more socially relevant and wanted to use the Institute to gain influence over New York planning;[61] and Rowe, who had been teaching in Upstate New York since 1962, wanted to use the Institute to move the second year of his Urban Design Program to the metropolis, where the College of Architecture, Art and Planning was already active with its New York Studio as an early off-campus program.

The same article also provided a second founding narrative that has been reproduced many times since. It places the founding of the Institute in a direct context with the exhibition "The New City: Architecture and Urban Renewal" which was shown at MoMA in the spring of 1967.[62] With this exhibition, the largest of the year, Drexler originally intended to initiate a debate on urban redevelopment in the context of de- and post-industrialization, pursuing macroeconomic and biopolitical goals as official policy. To this end, he eventually invited

59 Rowe had once studied art history under Rudolph Wittkower and at the Warburg Institute in London and had shaped neo-Palladianism in Britain in the postwar period; see Anthony Vidler, "Mannerist Modernism. Colin Rowe," in *Histories of the Immediate Present* (Cambridge: MIT Press, 2008), 61–104; in more biographical accounts on Eisenman, reference is made to Rowe's role as mentor, see Werner Oechslin, "'Out of History'? 'Formal Basis of Modern Architecture'," in Eisenman, 2005, 12–61, here 33ff.; Eisenman himself emphasizes in interviews the great influence Rowe had on him, see Peter Eisenman, Kenneth Frampton, Mario Gandelsonas, Anthony Vidler, "New York—Barcelona—Milan," in Colomina and Buckley, 2010, 58–69, here 60.

60 In early 1967, a personnel decision was pending at Princeton: Eisenman and Michael Graves were both competing for a permanent position; both had comparable qualifications, but only one position was up for grabs. Robert Geddes, the new dean of the architecture department, ultimately chose Graves as a shoo-in.

61 It is unclear to what extent Drexler himself wanted to create a professional alternative at the Institute. Ambasz, a former student of Eisenman at Princeton, who was at the Institute from 1968 and worked as a young curator at MoMA at the end of the 1960s, mentioned in our oral history interview that Drexler's position as director of the Department of Architecture and Design was repeatedly up for debate. At first, Drexler's position was publicly advertised, and then powerful trustees, John Hightower and David Rockefeller, tried to remove him.

62 MoMA, ed., *The New City: Architecture and Urban Renewal* [Exhib. Cat.] (New York: The Museum of Modern Art, 1967); MoMA, "Press Release," no. 10: "The New City: Architecture and Urban Renewal", February 24, 1967, https://www.moma.org/momaorg/shared/pdfs/docs/press_archives/3838/releases/MOMA_1967_Jan-June_0012_10.pdf?2010 (last accessed: May 31, 2023). The MoMA Archives contains a folder (CUR 818) on "The New City"-exhibition.

four teams of architects from prestigious Ivy League universities on the East Coast to present their approaches to urban renewal. The exhibition carried a certain poignancy, as at the same time these urban renewal policies were being debated, race riots were escalating in American cities due to persistent inequalities between Black and white people, e.g., in Watts, Los Angeles, in 1965 and in Detroit in 1967. While racial inequalities across the United States were becoming increasingly apparent, the New York neighborhood of Harlem, of all places, was chosen as the experimental field for MoMA's architectural and planning solutions, i.e., that part of Manhattan that was most heavily populated and historically shaped by the African American community. The featured urban design interventions, each developed specifically for the exhibition by teams from Princeton, Columbia University, Cornell University, and the Massachusetts Institute of Technology (MIT), were intended to address the renewal of the neighborhood in a very fundamental and less contextual way.[63] Yet when Drexler began curating the exhibition in 1965, he had originally been thinking in much more urbanistic terms, focusing more on urban infrastructure to improve the quality of life. The redesign and quality of public space in New York was actually to be addressed in five thematic areas: "Housing," "Parks and Playgrounds," "Schools," "Urban Transportation," and "Highways."[64] Ultimately, however, the exhibition turned out to be much more architectural, which was also attributable to the role of Eisenman, who had been selected by Drexler early on as a contact and cooperation partner. Subsequently, Eisenman had a great influence on the conception of the exhibition and also dominated the selection and composition of the teams.[65] Finally, as shown in a diagram, Eisenman presented some of the members of the Conference of Architects for the Study of the Environment (CASE), i.e., the very group or organization that he had co-initiated in 1965—a precursor to the Institute, in other words, which existed in parallel for some time—with the lofty

63 Drexler long planned to commission a fifth team to conduct a study, composed of historians, sociologists, planners, etc., see Arthur Drexler, letter to Burnham Kelly, February 25, 1966. Source: The MoMA Archives: CUR 818.

64 Arthur Drexler, letter to Rene D'Haroncourt, December 7, 1965. Source: The MoMA Archives: CUR 818. The exhibition was subsequently to be dedicated to the theme of "New Towns." Elizabeth Kessler had already been commissioned in 1965 to report on a seminar on European new towns; see Elizabeth Kessler, "A Report to the MoMA on the European New Town Seminar" January 28, 1966. Source: The MoMA Archives: CUR 818. After a change of title from "New Towns" to "The New City," the plan was to invite sociologists, anthropologists, planners, as well as writers and critics to contribute.

65 The original plan for Drexler and the deans of the schools of architecture was to assemble the teams, but Eisenman emphatically offered to assist in the selection, see Peter Eisenman, letter to Arthur Drexler, January 20, 1966; Robert Geddes, letter to Arthur Drexler, January 20, 1966. Source: The MoMA Archives: CUR 818. The exhibition, with its bird's-eye view, was redolent of previous planning efforts, for urban renewal in Harlem was nothing new, beginning with William Lescaze and his redevelopment plan of 1944 and continuing through Philip Johnson and Robert Stern; see William Richards, *Revolt and Reform in Architecture's Academy. Urban Renewal, Race and the Rise of Design in the Public Interest* (New York and London: Routledge, 2017), 51.

goal of formulating an American response to the *Congrès Internationaux d'Architecture Moderne* (CIAM).[66] The fact that MoMA received considerable support for this exhibition from the New York City Planning Commission (CPC), newly established under Mayor John V. Lindsay, already demonstrated the powerful alliance of architecture, education, culture, and political economy that was to shape the Institute's work in its early days. The city of New York co-sponsored "The New City," and Lindsay, a Republican who espoused liberal ideas, even spoke at the opening; he had made urban development policy a central election issue in 1966 and was convinced that he could change the metropolis for the better, even in socially turbulent and economically lean times.[67]

For their contributions to "The New City" exhibition, Drexler had given each of the four participating university teams specific tasks in different areas of Harlem, Ward Island, and Randall Island to ensure that they delivered fundamentally different problem-solving approaches to urban renewal. What emerged from the individual, textbook solutions was that they primarily pursued formal and morphological approaches, while largely ignoring socio-political and economic issues:[68] Princeton University (led by Peter Eisenman and Michael Graves), for example, experimented with a waterfront megastructure along the Hudson River that was designed to house a convention center, a research laboratory, and an aquarium, in addition to service facilities and hotels; Cornell University (Colin Rowe and Thomas Schumacher) proposed modifying the street grid to create a modernist urban landscape with large-scale housing that was clearly reminiscent of Le Corbusier's "tower in the park," among others; Columbia University (Jaquelin T. Robertson, Richard Weinstein and Giovanni Pasanella), on the other hand, planned to employ a technically novel mega-surface development constructed over the Harlem, Hudson, and New Haven railroad lines for new housing, which was then tied to air rights rather than to land prices; and MIT (Stanford Anderson, Robert Goodman and Henry Millon) worked with new earthfill embankments on the East River to build low-cost, small-scale housing for local residents.

66 In the United States, many former protagonists of the Bauhaus and CIAM had found a new home after WWII and held influential positions at universities or worked successfully as architects; see Kenneth Frampton and Alessandra Latour, "Notes on American Architectural Education from the End of the Nineteenth Century until the 1970's," *Lotus International*, no. 27 (1980): "Architecture in the American University," 5–39. At the time, it was readily overlooked that, as a professional organization, this produced far more than conferences and publications; see Andreas Kalpakci, "Making CIAM: The Organizational Techniques of the Moderns, 1928–1959," PhD diss., ETH Zurich, 2017.

67 Sam Roberts, ed., *America's Mayor: John V. Lindsay and the Reinvention of New York* (New York: Columbia University Press, 2010); Mogilevich, 2012.

68 For the composition of the four teams that participated in "The New City" exhibition, see MoMA: Members of the Princeton University Team / Cornell University Team / Columbia University Team / M.I.T. Team, https://www.moma.org/momaorg/shared/pdfs/docs/press_archives/3844/releases/MOMA_1967_Jan-June_0018.pdf?2010 (last accessed: May 31, 2023).

Overall, “The New City” exhibition, with its avant-garde stances, was clearly in the tradition of modernist urban planning; i.e., with approaches that argued partly morphologically, partly functionally, and differed quite markedly from those propagated, for example, by Robert Venturi’s *Complexity and Contradiction in Architecture*, which was published in 1966 as the first volume of a new MoMA series of books on architectural theory and which, as an early example of postmodern architecture, addresses the urban context formally and aesthetically.[69] “The New City,” on the other hand, in three of its four prototypes of what should have been considered good urbanism, allowed large parts of Harlem to fall victim to large-scale clear-cut redevelopment, without regard for its historically evolved structure, let alone its residents.[70] The only exception was the MIT team’s proposal, which, in addition to its small scale, also relied on on-site cooperation with local planners and experts. For Anderson, Goodman, and Millon had explicitly set themselves the goal of leaving the existing social and architectural structures in place as far as possible and not renovating and modernizing the brownstones that are typical of Harlem until the residents’ relocation housing became available. Moreover, it soon became apparent that the MoMA exhibition, for whatever reason, fundamentally failed to engage with current debates. For example, it lacked a position on the fact that Harlem had long since been discovered as a profitable development area for urban and private investment; on the fundamental critique of urban renewal, which had already been voiced in the early 1960s by urban critic Jane Jacobs in relation to the impending clear-cut redevelopment of Greenwich Village; or on local initiatives to educate and empower the African American community, such as the Storefront movement or the Architects Renewal Committee in Harlem (ARCH).[71]

Measured by the number of visitors, “The New City” was not a major event. However, the exhibition was widely reviewed in the daily and trade press.

69 Robert Venturi, *Complexity and Contradiction in Architecture* (New York: MoMA, 1966).

70 Michael Schwarting, “The Institute of [sic!] Architecture and Urban Design [sic!], New York City—1967: The Museum of Modern Art exhibition: The New City: Architecture and Urban Renewal—1967,” *Arc 2 citta* (July 10, 2012), http://www.arcduecitta.it/2012/07/archduecitta-magazine/ (last accessed: May 31, 2023). The exhibition has been criticized in retrospect for being dedicated to urban renewal at a time when the policy was already considered to be a failure in the United States and was seen as being socially destructive and racist.

71 The Architects’ Renewal Committee in Harlem (ARCH), founded in 1963 as the first community design center (CDC) and an extension of the housing commission of the New York chapter of the American Institute of Architects (AIA), was initially an all-white organization that was transformed in a short period of time and by 1967 consisted entirely of African American members; see Anthony Schuman, “Community Engagement. Architecture’s Evolving Social Vocation,” in *Architecture School: Three Centuries of Educating Architects in North America*, ed. Joan Ockman (Cambridge: MIT Press, 2012), 252–259; see also Jesko Fezer, “Soft Cops und Anwaltsplanung: Planungsbeteiligung oder die Politik der Methode (1962–1973),” in *Wer gestaltet die Gestaltung? Praxis, Theorie und Geschichte des partizipatorischen Designs*, eds. Claudia Mareis, Matthias Held, Gesche Joost (Bielefeld: transcript, 2015), 43–64.

Huxtable, the first professional and female architecture critic writing for the *New York Times*, published a thoroughly positive review on the day of the opening, in which she emphasized above all the good intentions and didactic qualities of the exhibition—based on the insight "that esthetics [sic!] and practical problem-solving are inseparable."[72] As appreciative as she was obliging, Huxtable stated that American schools of architecture now provided a solid education in the methods and problems of urban design and planning. While she certainly saw the exhibition as groundbreaking for cultural life and urban policy, she criticized the lack of a sense of reality in the architectural projects it showed, which were supported neither by construction plans nor by financing plans and, moreover, did not offer any approaches to solving higher-level social or urban issues. A much less sympathetic review appeared under the title "The Museum of Modern Art Discovers Harlem" in the March issue of the *Architectural Forum.*[73] Here, architect C. Richard Hatch echoed Huxtable's criticisms, but what was much more fundamental in his view was the fact that the four projects on display disregarded urban reality and lacked an understanding of what was specific to the location. In his opinion, the exhibition thus missed the opportunity to put pressure on the government to find long overdue answers to real inner-city problems and then put them into practice. Moreover, MoMA did not provide a utopia for a better life, as "the proposals all lack the vision of social space and purpose." In concrete terms, Hatch then called for better housing for the poorer sections of the population, as well as measures against speculation and the displacement of residents.

Finally, the two founding narratives, one organizational, and the other programmatic, attracted attention abroad when the young British architecture historian Reyner Banham published a scathing commentary on the founding of the Institute in the British weekly *New Society* in late 1967.[74] In his column, titled "Vitruvius over Manhattan," which was later criticized in a letter to the editor for its intemperate exaggeration of the Institute's role in local planning discourse, Banham touched on both its close association with MoMA and its explicit ties to

72 Ada Louise Huxtable, "Planning the New City. Modern Museum Exhibits Projects that Link Esthetics and Sociology," *The New York Times* (January 24, 1967), 39 & 45.

73 C. Richard Hatch, "The Museum of Modern Art Discovers Harlem," *The Architectural Forum* (March 1967), 38–47. Hatch was once a founding member and executive director of ARCH who, based on this experience, had been proposed as an outside consultant to "The New City" by Robert Goodman prior to the exhibition, but was ultimately not brought in. In 1967, Hatch published articles in relevant journals at the time about the organization's actions' explicitly opposing government and private housing programs; see C. Richard Hatch, "Renewal in Harlem," *Zodiac*, no. 17 (1967), 196–198; "Planning for Change. Towards Neighborhood Design and Urban Politics in the Public. Schools," *Perspecta*, no. 11 (1967), 43.

74 Banham, 1967. The title of the essay alludes to the Institute's first logo, designed by Eisenman, and thus to the self-image of the new architectural institution, as well as to the American architect's longing for European tradition.

"The New City" exhibition. In doing so, he criticized the Institute, which he had visited shortly after it was founded, in his characteristically polemic style. As an event of high culture, he opined, the exhibition was pure window dressing that obscured the real problems. Aside from the forms, which were perhaps visually appealing from an architectural point of view, he criticized it as "even nigglingly mischievous" from a socioeconomic perspective, especially projects like the megastructure of the Princeton team headed by Eisenman and Graves, "where they replaced manufacturing trades that create the kind of semiskilled jobs Harlem needs, with office and laboratory type installations that simply make more paper jobs for Mister Charlie."[75] From a distinctly European, albeit working-class, perspective, Banham, who had previously made his mark in Britain as a spokesman for the Independent Group and theorist of Brutalism, was amused by the high regard in which formalism was still held in the United States. He also noted that formal and morphological approaches in American schools of architecture were apparently seen as less frivolous and more responsible. In the end, the only positive thing he had to say about the exhibition was that it had, for once, focused the attention of the New York art public on architecture.

The Institute had found its place in precisely this legacy and the active role played by Eisenman in the curation, conception, and coordination of "The New City" exhibition, especially when it came to preparing the lists of participants, continued to be crucial as it enabled him to network and establish valuable contacts not only with New York art and architecture communities, but also with the broader realms of politics and business. This would later benefit him in his role as Institute director—both personally and professionally. Both the production and reception history of the "The New City" exhibition showed that when the Institute was founded, the New York art and architecture community was divided between formalism and activism, Europe, and America, the real and the theoretical, architectural and cultural production, and ultimately between a waning modernism and an incipient postmodernism. This also tallies with the legend that Eisenman and Drexler had already conceived the plan for a completely new institution for architecture in New York during the preparations for the exhibition. Neither archival research nor oral history can satisfactorily verify who ultimately came up with the idea for the Institute. The only thing that is certain is that Eisenman was able to use the exhibition to demonstrate his interest in urban planning and urban policy issues and use the Institute to assert expertise in these matters in the future. The idea of architecture and the city projected in Princeton University's contribution, however, was truly novel in that they viewed Manhattan from an urban economy point of view and started from a largely de-industrializing urban space that was yet to be repurposed and

75 Ibid., individual exhibits from "The New City" exhibition were brought to the Institute in October 1967, turning it, at least briefly, into MoMA's archive and a storage space for the urban planning ideas of modernism.

upgraded. For in addition to the construction of a mega-structure as a new site of consumption rather than production, they also proposed the transformation of former industrial and rail yards and their "gentrification" through new recreational and cultural facilities, stores, and cafes. Overall, however, the proposal also showed how problematic the role of architects in urban renewal could be and how little it was reflected. In Harlem in particular, the four proposals made in the exhibition would have led to massive changes in the building fabric and urban structure, accompanied by the displacement of low-income, primarily African-American residents. The main beneficiaries would have been the target group or clientele of the Ivy League architects, i.e., the white middle class of a new information, knowledge, and service society. Compared with the other three contributions to the "New City" exhibition, the intervention proposed by the Princeton University team was a realistic and pragmatic, if not revanchist, form of urban renewal, formulated in more radical-utopian, technological, and progressive terms.

Architecture *and* Urban Studies

The deciding success factor in establishing the Institute was its name and the associated dual claim to professional competence in the disciplines of "Architecture" *and* "Urban Studies" for its future fields of activity. On the one hand, Urban Studies had been flourishing as an academic discipline at American universities for several years. Viewed as a distinctively American research approach to urban phenomena, it was distinguished by its interdisciplinary nature, as noted in a theme issue of *The American Behavioral Scientist* in 1963.[76] In their editorial to this issue, editors Robert Gutman and David Popenoe, two young sociologists from Rutgers University, pointed out that the emergence of new urban knowledge at the intersection of the disciplines of history, economics, geography, political science, sociology, etc., coincided with the emergence of an almost completely urbanized society. From the perspective of the history of science, the Institute's dual focus on architecture and urban studies may be explained by the fact that Gutman, who began researching the interaction of architecture and sociology in 1965 with a grant from the Russel Sage Foundation and was subsequently invited to Princeton University's school of architecture under the new dean Robert Geddes, served as a discussion partner and possibly advisor to Eisenman, his colleague and friend at Princeton, in the run-up to the Institute's founding in 1967

76 Robert Gutman and David Popenoe, eds., *The American Behavioral Scientist* 6, no. 6 (1963): "Urban Studies". The issue of the then still young interdisciplinary journal in the field of social science outlined the as yet brief history of urban studies, its framework, and its objectives. Individual articles were constitutive for the further development of the fledgling subdiscipline, providing an overview of the research literature, formulating the object of research, defining pedagogical practice, conceptualizing the interdisciplinary agenda, discussing the relationship to urban planning as an urban service, and describing the institutional work completed to date.

and helped him generate his ideas, especially during the founding period.[77] The founding of the Institute—which occurred around the same time as the development of community design centers in the United States and the institutionalization of advocacy planning at American universities—as a new cultural and epistemological space might therefore best be approached from this angle, considering the extent to which a new educational institution was actually created here, as Gutman suggests, one which actively turned to architectural and urban research, developed new concepts and methods, and, through its teaching, produced a new type of architect and planner, trained in both theory and empiricism, whose role was to devise innovative solutions to urban problems.[78]

In addition to the circulating academic, disciplinary, and institutional concepts of a new kind of research, the Institute can also be explained by the political, economic, and social contexts of the design profession. The political changes that took place in the context of the Great Society proclaimed by President Lyndon B. Johnson brought about a new upswing for designers, especially in the second half of the 1960s, and this had far-reaching consequences for the professional understanding of architects, planners, and urban designers, who found new and socio-politically relevant tasks in government-sponsored urban renewal and public housing projects. It was a decade when people still believed in the power of architecture, planning, and urban design to make a difference to social development. Faced with the boom in urban design against the backdrop of a Fordist mode of production, architects and planners developed new approaches to urban politics by applying psychological insights, for example, or implementing and promoting democratic participation. They also laid claim to possessing the tools and visions necessary to influence the future development of the city.[79] In New York, Mayor Lindsay championed an urban planning approach that actively shaped issues of demographic and economic change, the rediscovery of the inner city as a residential area, and changes in the composition of the city's population, thereby also providing a tool to overcome racial,

77 Robert Gutman, "Urban Studies as a Field of Research," *The American Behavioral Scientist* 6, no. 6 (1963): "Urban Studies," 11–16. If Gutman's fairly normative conception of science, which focused on the metropolis as a social system in order to distinguish urban studies from sociology as its parent discipline and to define its object of research, had been followed, architecture and urban studies would have focused on three aspects: "the goals appropriate for metropolitan development," "the nature of social organization and social processes of metropolitan regions," and "the means through which metropolitan policy is implied."

78 On the history and positions of CDC and advocacy planning, see *An Architecture*, no. 19–21: "Community Design."

79 One example of the boom in planning is Philip Johnson's urban planning project for a community of 150,000 for Harlem from 1966, which was published in the catalogue of "The New City" exhibition. Strangely enough, this urban plan envisioned the construction of a fortified new housing development in the middle of an African American neighborhood with a wall of high-rises as a large-scale urban renewal project in the wake of a clear-cut redevelopment, see MoMA, 1967, 17.

ethnic, and class divides. While Lindsay focused on self-governance and equal opportunities for the African-American population, he also worked closely with the real estate industry.[80] For example, at the initiative of the Lindsay administration, a report was commissioned that established strategies for urban planning, urban renewal, and neighborhood preservation. In addition, the New York City Planning Commission (CPC), headed by Donald H. Elliott, a real estate and land-use attorney, was given new life with the primary goal of creating housing for all social classes. And a Mayor's Task Force on Urban Design was established, headed by William S. Paley, CEO of Columbia Broadcasting Corporation, which included four architects (Philip Johnson, I.M. Pei, Jaquelin Robertson, and Robert Stern) and initiated a new urban development policy that, in the face of the decline of urban architecture, focused entirely on urban planning and, in particular, sought to make New York more livable by establishing a group of planning specialists who were to work closely with the CPC but were also given design powers, especially through the implementation of new zoning regulations.[81]

This was the setting in which the Urban Design Group (UDG) was founded, shortly before the Institute was established in September 1967. While the Institute, as an institution associated with MoMA, was more architecturally and culturally oriented from the outset, the UDG was a more practical and politically active group around Jonathan Barnett, Jaquelin Robertson, Richard Weinstein, and Myles Weintraub. The UDG, as the CPC's "corps d'elite," was charged with overseeing, linking, and coordinating all areas of New York development policy (e.g., policy approaches, land use planning, and architectural projects).[82] A neighborhood plan was established for the Twin Parks urban renewal area in the Bronx, designated in 1963, with the goal of preserving the physical and social diversity of the neighborhood and testifying to the fact that New York urban policy was indeed making an effort to address current social issues through architecture and urban design.[83] The Institute was well positioned from the start and, as clearly indicated by its choice of name, sought to enter and engage in this type

80 On the history of urban planning in New York, see Robert Stern et al., "Death by Development," in Stern et al., 1995, 61–134.

81 Stern et al., 1995, 92–93. The UDC worked on different aspects of land use planning, planned unit development, starting from the block and the street as design principles, or the special district plan, with the aim of combining uses in the same block; see UDC, *Planned Unit Development* (New York: City Planning Department, 1968); see also Stern et al., 1995, 390.

82 Jonathan Barnett, *Urban Design as Public Policy. Practical Methods for Improving Cities* (New York: Architectural Record Books, 1974).

83 The founding of UDG and IAUS were viewed as parallel events in professional circles, see "Three Institutes Are Formed to Study Urban Problems," *Architectural Record* (December 1967), 54. The comparison was later taken up by Brian Brace Tayler to highlight the cultural focus of the Institute; see Brian Brace Taylor, "Self Service Skyline," *L'Architecture d'Aujourd'hui*, no. 186 (August/September 1976), 42–46.

of work by combining research and design as a consulting project office offering various services.[84] This is the context in which Eisenman's statement, quoted in the *New York Times* article, that "architects have abrogated their responsibility to deal with social problems," should be seen. With his rhetoric of "radical chic," Eisenman was at least ostensibly critical of society, while at the same time advertising on his own behalf: "The Institute," he said, "will thus try to make the study of architecture more relevant to social ideas and problems."[85] In a socio-politically turbulent climate, the Institute director used all the right language of contemporary political discourse, initially relying on urbanist themes and multidisciplinary approaches to assert the Institute's relevance and exert its influence.[86] But there was no indication that Eisenman might be the right person. Ultimately, however, his statement was based on ambiguities and ambivalences that conveyed a sense of confusion surrounding his perspective and thus aimed at nothing less than disorienting his readers. This diagnosis—that the connection between architecture and society had been neglected—was on the one hand a thoroughly factual analysis of the prevailing trend of modern post-war architecture, but on the other hand, it was also a rather cynical statement. For even then, Eisenman's credo was an architectural and urban formalism, which he had already displayed in his dissertation with his formalist reading of selected buildings by Le Corbusier, Alvar Aalto, Frank Lloyd Wright, and Giuseppe Terragni, and especially in the Jersey Corridor Project (1965)—a twenty mile, linear urban development he had designed with Michael Graves while still at Princeton and published in a special double issue of *Life Magazine* on the fate of the American city.[87] And it was Eisenman's formalism, borrowing from art or linguistics, coupled with his narcissism, that was subsequently to be his personal contribution to the Institute and that would shape his life's work: a truly postmodern style of thought and practice. If Eisenman attracted the attention of the architectural public as a postmodern project maker, the construction of the Institute, which

84 Barnett was to become a cooperating partner with the Institute when it came to issues urban planning and real estate.

85 Roberts, "School Is Formed for Urban Design," 52.

86 In architecture history, the extent to which the Institute under Eisenman's direction initially worked on urban research and design projects has hardly been addressed. In retrospect, Richard Plunz and Kenneth Kaplan criticized its early "chic radicalism," pointing out that the latter criticized the formalism of his professors, but that only a short time later the same reproach could be levelled at him with regard to his substantive contributions to the Institute; see Plunz and Kaplan, 1984, 36f. In their essay on New York architectural culture, Stern et al. debate in particular Eisenman's statements made in the 1967 *New York Times* article, by highlighting that one of the Institute's aims would have been to correct the errors of architectural modernism; see Stern et al., 1995, 1209 (second edition). Lucia Allais is one of the few architecture historians to critically examine the myth of the Institute by debunking the founding narrative of *Oppositions*, see Allais, 2010. In her essay, she reproduced Eisenman's portrait from the *New York Times*, showing that as founding director, he alone determined the programmatic direction.

87 Peter Eisenman and Michael Graves, "Jersey Corridor Project," *Life Magazine* (December 24, 1965): "The U.S. City: Its Greatness Is at Stake."

always portrayed itself as both a non-professional and non-academic institution, initially undertaken in cooperation with MoMA and staffed and legitimized by Cornell University, later by other universities, can itself be seen as an architectural project, but one that was subject to conditions and realities that changed over time as a result of newly emerging configurations, shifting collaborations dictated by what was deemed convenient, and new opportunities.

1.1 Institutionalizing a Network

The Institute's early years from the academic year 1967–68 to 1973–74, i.e., before the creation of the journal *Oppositions*, were characterized by urban and architectural consulting activities and projects commissioned by municipal, state, and federal planning authorities. As a locally active, yet internationally networked group, the Institute worked more or less successfully on various research and design projects with an architectural, at times thoroughly interdisciplinary approach, reaching for whatever public support was available. The good connections to MoMA, especially via Peter Eisenman's personal relationship with Arthur Drexler, proved to be instrumental. Without this close affiliation, and the promise to exhibit, the Institute would not have existed, at least not in this form.[88] The Institute's work in the initial phase consisted of proposing new designs for urban renewal and housing based on ownership or public-private cooperations, with the pedagogical mandate of providing students with work experience on real projects, while at the same time using them as a labor force. In addition to the concrete research and design project work, the start-up period was also characterized by the structuring and hierarchization of the Institute's organization, the increasing institutionalization and differentiation of its work, and further networking with and positioning vis-à-vis other institutions, before the declared goal of actual establishing itself as a group, if only in one case, was realized.

The founding of the Institute in the fall of 1967 as, by its own account, a unique institution in the field of architecture can be read quite differently, depending on whether one focuses on an individual or a collective biographical narrative, i.e., primarily as a biography of Eisenman, including in his role as Institute director, or as a biography of the Institute as a group, which, especially in the early years, was inevitably shaped by Eisenman and those involved in the project from the start, but over time also came to include further Institute

88 MoMA has been criticized for its contribution to the museumization and depoliticization of modern art from Europe in the postwar period, yet the Institute's historical and theoretical treatment of European architectural modernism in the long 1970s can be seen quite similarly; see Serge Guilbaut, *How New York Stole the Idea of Modern Art. Abstract Expressionism, Freedom, and the Cold War* (Chicago: University of Chicago Press, 1983).

Fellows, staff members, friends, and colleagues, building on a complex analysis of power and relationships embedded in the changing social and historical context. Such a contextualization of the construction of careers, whether personal or institutional, allows for multiple readings of the new architectural institution as a connection for the many mechanisms and structures of the local architectural scene, which at the time was transitioning to a more globalized architecture culture, influencing both the developments of the American academic landscape and New York metropolitan society. The history of the Institute that will be told here, combining institutional analysis and critique, examines not only the three levels of organization and program, day-to-day work, and integration within American society but also the self-image and the public image of this particular grouping, as well as the history of its transmission and reception in architecture history. In addition, it will demonstrate the newly emerging opportunities for architects and academics to work meaningfully and successfully within and beyond architecture firms and schools of architecture in New York in the late 1960s, throughout the 1970s, and then in the early 1980s. A psychoanalytical interpretation of the Institute's establishment would underscore the fact that Eisenman founded a new, institution-like workspace for himself and others—and in doing so, was not always acting in a completely self-determined manner.[89] Eisenman's actions as an entrepreneurial subject—which, according to Michel Foucault and a history of individualization and governmentality, can be understood as a descendant of homo economicus—were characterized by strategic thinking.[90] In the course of his subjectification, Eisenman took the idea of an alternative institution in architecture, which was already in the wind at the time, and made it big.[91] Moreover, he repeatedly showed great talent in rallying the right people around him. It is striking that the Institute, which in its early years was still just a small circle of architects and academics, has always made

89 The following generation of architecture scholars in the United States interpreted the Institute's history quite differently. In our expert interview, architecture theorist Mark Wigley suggested a psychoanalytic interpretation of the Institute's founding.

90 Robert Gutman, "Architecture: The Entrepreneurial Profession," *Progressive Architecture* (May 1977), 55–58. The work of architects is to be seen more in the role of entrepreneurs than intellectuals or artists, see Ulrich Bröckling, *Das unternehmerische Selbst: Soziologie einer Subjektivierungsform* (Frankfurt: Suhrkamp, 2007).

91 Emilio Ambasz, previously a student of Eisenman at Princeton, for example, claimed for himself the idea of a new institution outside the academy, and Stanford Anderson, Eisenman's companion and confidant since their paths crossed at Columbia University in the 1960s, where Anderson was a PhD candidate and Eisenman a graduate student, characterizes him as a soufflé maker, i.e., as someone who, if you apply this image of the high art of cooking to architecture, breathes life into ideas—but also knows how to blow things up. In the oral history of the Institute, for which I interviewed protagonists and contemporaries, many spoke about their own contribution to the Institute, but also about Eisenman and their personal relationship with him. It is an established fact that in 1966, before the founding of the Institute, Eisenman received a grant from the Graham Foundation for a project titled "Universitas Project." Under the same title, Ambasz organized an international conference at MoMA in 1972, initially with the support of the Institute and, again, with a Graham Foundation grant.

itself look bigger and been portrayed as larger than it ever was—a peculiarity it shared for instance with the Bauhaus of the Weimar Republic, or with other contemporary schools of architecture such as the Architectural Association in London or the Cooper Union in New York.

Despite provisions to the contrary, however, a socio- and discourse-analytical interpretation indicates that, from the very outset, the Institute had a strong institutional basis and established good networks in its efforts to ascribe a new meaning to architecture as a form of work and organization, discourse, and art. When it was founded in the fall of 1967, the Institute was officially recognized as an educational institution by the Board of Regents of the State University of New York (SUNY), albeit provisionally for five years, and was thus also officially assigned a social function. Legal, political, and economic aspects initially played a role for the quasi-academic Institute as it repeatedly asserted its autonomy and independence. The Institute's status as a hybrid of a professional, educational, and ultimately cultural institution was also legitimized by the composition of the initial five-member Board of Trustees, which, in addition to Eisenman and Drexler, consisted of representatives of established institutions: Gibson Danes, dean of visual arts at SUNY's Purchase College; John Entenza, director of the Graham Foundation for Advanced Studies in the Fine Arts; and Burnham Kelly, dean of the College of Architecture at Cornell University. At the first meeting of the Board of Trustees in early October 1967, leadership positions were assigned, and their duties and responsibilities defined. Here Eisenman was appointed the first director of the new architectural institution, initially for a term of only two years. Officially authorized, he thus assumed institutional responsibility for day-to-day operations and reported to the Board of Trustees at biannual meetings. These, in turn, legitimized the Institute's activities, facilitated funding, and represented external relations. Although he was required to implement the directives of the Board of Trustees, much of the institutional power was concentrated in Eisenman, who, in addition to serving as Institute director, was also elected president of the Institute. Subsequently, Drexler served as chairman and treasurer of the Institute, Danes as secretary, and both Entenza and Kelley as vice presidents. A crucial factor for the Institute's work was that personal, institutional, intellectual, and political interests and business strategies always influenced, conditioned, and overlapped each other.

On October 13, 1967, Eisenman finally received the seal for the Institute as a registered company, thus cementing its foundation. According to its by-laws, however, it was a non-profit company that could not be listed on the stock exchange or make a profit. As start-up capital, Drexler had acquired private donations from among MoMA's trustees in the summer of 1967; the Pinewood Foundation of Armand and Celeste Bartos provided US$30,000 in start-up funding; Mrs. Douglas "Lily" Auchincloss, an early and longtime supporter, provided five original drawings by Le Corbusier as a permanent loan, with the artwork serving as a capital contribution; and an anonymous donor also gave the Institute twenty-six shares of Corning Class stock valued at US$9000. Cash flow from current operations was assured as the Institute drew revenue from tuition and contracts.[92] In the first fiscal year, the Institute's budget of US$50,000 was still quite modest and manageable, with rent and personnel costs accounting for most of it and hardly any material costs. Eisenman paid himself a director's salary of US$15,000 and hired a secretary, Louise Joseph, who was employed from October 1967 to June 1973, making her one of the few permanent employees for a long time. He also received a Graham Foundation grant as an individual for the second year running in 1967–68 for the purpose of analyzing individual buildings down to their very structure along the lines of the Italian architect Giuseppe Terragni, after which he refined his own formal, or as he termed it, "rational" approach.[93] A press release issued by MoMA when the Institute was founded noted that it relied on outside capital and commissions for "research and development projects from municipal, state, and federal agencies," which promoted the new actor's public performance.[94] In early 1968, when the Institute was granted legal status as a non-profit

92 Another founding narrative is that Eisenman also accepted funds from the CIA. In interviews, he repeatedly told the story of how, shortly after the article about the founding appeared in the *New York Times*, he was contacted by a CIA employee and, after auditioning once with Drexler in Washington D.C., accepted a not inconsiderable sum of cash per year. In return, according to his own statement, he compiled and passed on a list of the names of all the people who frequented the Institute at the time; see Peter Eisenman, "The Agency Interview: The Institute for Architecture and Urban Studies," *Perspecta*, no. 45, (2012): "Agency," 59–66. This narrative is revealing in that it not only references the practices of governmental action and the absurdity of intelligence surveillance, but also underscores Eisenman's patriotism, his desire for power, if not his fixation with lists. He boasts of having done everything for the Institute (and for money) and of having exploited his roles as Institute director and host, knowing full well that he was putting his relationships and friendships at risk. Eisenman, who usually presented himself as a politically "middle of the road" character, emphasizes his position of power, as well as his dissociation from faculty and students and later from the supposedly neo-Marxist approaches of his peers, Fellows, friends, and colleagues. To better appreciate the cultural and social significance of this, it should be remembered that in the preceding decade, at least according to the argumentation of historian Francis Stoner Saunders, the CIA had at least indirectly helped found abstract art in the United States as a strategic move in the Cold War, in order to use the cultural power of American artists to impress cadre people in the USSR—an intelligence activity that was apparently repeated in the late 1960s but this time applied to abstract architecture at the Institute. See Francis Stoner Saunders, *The Cultural Cold War: The CIA and the World of Arts and Letters* (New York: The New Press, 2000). In the end, Eisenman and the Institute only served for a few years and never really as an instrument of American intelligence in the fight against communism, possibly because there was not much to report.

93 Eisenman was not really interested in determining Terragni's "time and place" in modern architecture. This had already happened a few years earlier at MoMA as part of the exhibition "The Modern Movement in Italy: Architecture and Design" (August 18 to September 6, 1954), curated by Ada Louise Huxtable, an exhibition of the Department of Circulating Exhibitions that focused on Nervi but did not leave out Terragni as an architect of Fascist Italy. *Casa del Fascio* (1932–36) in Como and *Casa Rustica* (1933–35) were on display, see MoMA, Press Release no. 71, August 18, 1954, https://www.moma.org/momaorg/shared/pdfs/docs/press_archives/1856/releases/MOMA_1954_0077_71.pdf (last accessed: May 31, 2023)

94 MoMA, Press Release, n.d. Source: The MoMA Archives, New York: PI II.B.579.

corporation by the Internal Revenue Service as the national tax authority with explicit tax-exempt status as a 501 (c) (3) organization, this made it financially independent from MoMA, if not autonomous, allowing the Institute's leadership to subsequently apply directly for grants and enjoy special tax status.[95]

Researching/Teaching

In its first two years, the Institute was commissioned by the City of New York to work on several urban research projects under Eisenman and Rowe that were linked to a thoroughly innovative teaching approach. The Institute exploited its collaboration with Cornell University by promising students practical experience in New York, thus playing up its standing as an alternative place of education.[96] The idea behind the not entirely unorthodox pedagogical experiment was that "especially talented graduate students" from the "Urban Design Program" would spend their second year working on urban planning projects at the Institute instead of on fictional assignments at their home university. In the 1967–68 academic year, Rowe brought four students—Stephen Potters and Michael Schwarting, both graduate students, and William Ellis and Jon Stoumen, both associated with the School of Architecture—to the Institute. Cornell's Dean Kelly had had to assure the students that they would receive credit for their involvement with the Institute without knowing whether the Institute would even meet the requirements.[97] The university's commitment was backed by the fact that the architecture faculty could now add an interesting graduate program to its New York Program, which had offered an attractive alternative for undergraduate students for the past five years, at little expense, especially since its investment was limited: the university hired only Rowe as a lecturer, whose salary it had to pay, and waived its tuition fees of US$4,000 per student. These were passed on to the Institute. While Institute director Eisenman was to benefit from Rowe's expertise and experience, pedagogical and conceptual differences between them quickly became apparent, not to mention personal ones. While Rowe wanted to teach his students contextualism, a formal, yet topological and typological approach, using New York as an example, Eisenman set his mind on conveying formalism as well.

The Institute's first commission, valued at US$15,000, was from the CPC to conduct a morphological analysis of a section of the Bronx and to submit

95 In the United States, a 501 (c) (3) organization is the most conventional category for nonprofit organizations and refers to the following organization type: religious, educational, charitable, scientific, literary, testing for public safety, to foster national or international amateur sports competition, or prevention of cruelty to children or animals.

96 CCA's Peter Eisenman fonds contains a folder with original documents on the collaboration between the Institute and Cornell University. Source: CCA Montréal, Peter Eisenman fonds: PDE-105-T - Eisenman Education & Teaching. Cornell 1967–1969.

97 Peter Eisenman, letter to Burnham Kelly, July 17, 1967. Source: The MoMA Archives, New York.

an urban design proposal for it. Here, Eisenman's liaison at CPC was Jaquelin Robertson, who had also been involved in MoMA's "The New City" exhibition for the Columbia team and was one of the founding members of the UDG. The year-long project comprised two phases, with the first examining the urban layout, topography, transportation infrastructure, etc. of Kingsbridge Heights-Jerome Park. In the second phase, Rowe's students were then tasked with preparing a more detailed analysis of a subsection of the study area that ran along Webster Avenue and developing planning proposals. Working under guidance, they produced figure-ground diagrams of the spatial conditions following the approach taught at Cornell at the time, which provided a basis for formal interventions. But while Rowe, who was still pursuing his commitments in Ithaca, New York in parallel with his teaching at the Institute, had to commute to New York two days a week, Eisenman, who was present at the Institute the entire time, had a formative influence on the project. Another factor, which was apparently not seen as an issue at the time, was that the research team was entirely Caucasian, even though the study area was primarily an African-American neighborhood.

The approach taken by the Institute in its teaching, research and design activities seemed unrealistic and artificial from the outside. Early in the academic year, Dean Kelly began to have doubts as to whether the Institute's expertise and equipment would even enable it to take on concrete planning tasks with the Cornell students involved and immediately communicated his doubts in a letter to Eisenman.[98] Later, Kelly even felt compelled to renegotiate what was publicly portrayed as a joint venture. Banham's reporting was also critical of the Institute, particularly the formalism practiced there. At the same time, however, he saw the Institute's firm belief in architecture as being its greatest potential.[99] For despite his distrust of some of the attitudes displayed there, he placed his hopes precisely in Eisenman's conviction that students needed to be taken out of their school context and confronted with real-world issues: "The fundamental virtue of the Institute, however, is that it can tackle [...] substantial problems [...] and must come forward with workable solutions to them," the qualifier "workable" being crucial here. Banham was enthusiastic about the potential he attributed to the Institute, "that it might yet prove to be a workable bridge between what are at present the utterly alien and non-communicating worlds of academic culture and expediency planning." On the positive side, he saw that the Institute had an interesting mix of faculty with very different approaches in the form of Rowe, but also Robert Gutman, who occasionally helped out in the early years and taught architecture sociology there. Moreover, from his point of view, the Institute was already well connected, not least because Eisenman's Cambridge past also gave him connections to the

98 In CCA's Eisenman fonds, there is correspondence between Eisenman and Burnham Kelly from the academic years 1967–68 and 1968–69 which characterizes the cooperation between the Institute and Cornell University. Source: CCA Montréal, Peter Eisenman fonds: PDE-105-T.

99 Banham, 1967.

British architect James Stirling, who was then a visiting professor of design at Yale University in New Haven, Connecticut, and who initially gave regular guest lectures at the Institute. Eisenman's preferences for architectural positions from Great Britain would come into full play during the founding years.

In his article, Banham also commented on the image the Institute was portraying of itself. Aware of the importance of an identity, whether for a company or for an institution, Eisenman had chosen the Vitruvian Man as the Institute's first logo. But instead of Leonardo da Vinci's well-known 1492 version, he chose a much more boastful version, deliberately selected from Cesare Cesariano's 1521 Vitruvian edition, which depicted the ideal image of man with an erect phallus.[100] With this shocking, even pornographic image, in Banham's eyes an affectedly frivolous, pretentious, and excessive inscription in a humanist tradition of architecture, the Institute experienced its first branding which was furthermore reproduced for years to come on all kinds of official promotional materials (brochures, posters, ads in the *New York Times*, even sweatshirts). A drawing of the supposedly well-proportioned *homo ad quadratum*, and *homo ad circulum* was even placed immediately on one side of the revolving door that separated the Institute's conference room from the rest of the office floor, where the students sat; on the other side was a wallpaper of Le Corbusier's Modulor as a modernist interpretation of man as the measure of all things. Banham's tongue-in-cheek interpretation of this reference and the juxtaposition of the two drawings was as a kind of religious profession of faith in the traditional values of architecture; he concluded that a Vitruvian order could not so easily be imposed on New York's urban grid, but that the attempt alone would have been nevertheless worthwhile since it would at least have shaken up the two disciplines of architecture and planning in their constant crisis. It has not been documented how Banham's first international coverage was received at the Institute.

As the research at the Institute was translated into designs over the course of the academic year, it quickly became clear that the two principal architects had quite different ideas not only of urban design but also and especially about didactics. While Rowe's unique approach was to break up the existing city blocks, Eisenman took a far more radical, even destructive approach. He proposed, as he had done with the Princeton team for "The New City" exhibition, to deconstruct the existing street patterns and redesign them using large geometric shapes that would have been visible, for the most part, only from a bird's eye view—an unparalleled provocation for everyone else working on the project. By

100 Banham commented quite cynically on the choice of logo, since for him "the Vitriuvian man, for example, [was] not the fairly familiar version drawn by Leonardo da Vinci, but the mannered and rather campy one from Cesariano's more obscure 1521 edition of Vitruvius." See Banham, 1967, 828; see also Rudolf Wittkower, *Architectural Principles in the Age of Humanism* (London: Warburg Institute / University of London, 1949). By "campy" Banham was most probably referring to Susan Sontag's 1964 essay "Notes on Camp."

the end of 1967–68, those involved could no longer hide the fact that the ambitious teaching and learning goals had not been achieved and that the inaugural research and design project was ultimately nothing more than a purely abstract, not to mention theoretical exercise. In the end, the students received a master's degree in urban design from Cornell University, but it remained unclear what significance their participation in the project may have had for their later careers (Ellis, Potters, and Schwarting eventually stayed at the Institute for some years) and, more importantly, what significance it would have had for the neighborhood. Instead of providing students with hands-on work experience at the Institute by being involved in a project throughout all phases of planning, from conception to realization, the project ended with a summary of the results submitted to the CPC in a report. This report was conceived and laid out as a manuscript for publication, but this did not materialize, and the project participants' essays were not produced.[101] The CPC had originally intended to include the study's findings in the Bronx portion of the *Plan for New York City* published in 1969, i.e., the official planning document for the five boroughs, but in the end, the Institute failed to deliver.

After just one year of the Institute's existence, it became clear that this balancing act between office and school would not be easy when it came to reconciling the expectations of contractual partners and partner universities. In order to inform stakeholders about the organizational structure and the goals of the Institute and to promote its research work and study program, the Institute created a first prospectus, with the new logo, which was sent to architecture schools, foundations, as well as public and private planning authorities.[102] The prospectus reiterated the Institute's claim to have a positive impact on both education and the profession: "The Institute seeks to amplify both the present system of architectural education and the process of physical planning by bridging the gap between the theoretical world of the university and the pragmatic world of the planning agencies." In 1968, then, Institute director Eisenman's main concern was to acquire new commissions, rather than to attract new students. At the same time, more faculty members were to be hired and the Institute was to be networked with other universities. The close ties to Cornell University alone, and to Rowe in particular, were by now seen as problematic in obtaining

101 IAUS, ed., *Kingsbridge Heights* (New York: The Institute for Architecture and Urban Studies, 1968). Source: private archive of Stephen Potters. Schwarting prepared many plans for the report, including the one that summarized all the individual designs; Ellis wrote much of the text. He years later criticized Rowe's contribution to the Kingsbridge Heights study as "extremely abstract" in an essay in *Oppositions*, see William Ellis, "Type and Context in Urbanism: Colin Rowe's Contextualism," *Oppositions* 18 (Fall 1979), 2–27, here 13–14. Schwarting's plan was also reprinted in this essay, see figure 25–29.

102 IAUS, "The Institute for Architecture and Urban Studies, 67–68," Source: CCA Montréal, IAUS fonds: C.3-1 / ARCH153783.

further research contracts in the field of urban renewal. In the of spring 1968 "Director's Report," Eisenman reported that he had already had initial talks with representatives of various universities and, mobilizing his existing network, had established contacts with Columbia University, New York University, Rutgers University, and Cambridge University. In addition, Eisenman indicated that there were up to four research projects on the horizon for 1968, with a broad range of potential clients: a planning and case study on 110th Street in Harlem with the U.S. Department of Housing and Urban Development (HUD) in cooperation with the CPC and MoMA; an urban design project for Baltimore West, Maryland, under the Model Cities Program, as well as a study of public housing for low-income households in Brochester, Maryland, with the Baltimore Multi-Purpose Council in cooperation with Alexander Ewing and Associates; a design study and advocacy planning in Harlem, with the New York Urban League, an African American civil rights organization; and finally, an urban design project in Newburgh, New York, with Hunter College and New York University.

At one point in 1968, the unlikely cooperation with the New York Urban League was perhaps most promising in this regard.[103] Meeting minutes reveal that a central question at the time was how the Institute would fit into the Urban League's image: as a "brain trust" functioning as a "program planning development department," or as an educational institution training "Black students" with "the unique kid" later serving as an expert and facilitator. As part of the so-called Harlem Plan, two specific projects were outlined in late July 1968:[104] the preparation of a model block study for a prototypical future Harlem, and the launch of a new educational mechanism modeled on a Harlem "street academy" that would focus on teaching the fundamentals of "physical design" and relevance to the urban ghetto. The Institute's interpretation of the social situation in Harlem was quite progressive, as not "a race but a class problem," with a distinction being made between the "have and have nots." One of the long-term goals that were outlined was to train "Black architects" to create a "Black architecture." Criticism was raised in these meetings that this would not be enough to solve the situation. Livingston "Leroy" Wingate, the executive director of the New York Urban League, was obviously more interested in integration at this point, in placing African American youth in white educational institutions such as the Institute or Columbia University in order to communicate the problems and needs of Harlem.

103 CCA's IAUS fonds contains a folder with original documents on the cooperation between the Institute and the New York Urban League. Source: CCA Montréal, IAUS fonds: B.2-2.

104 In her 2012 *Perspecta* article on the Institute's founding years, contrasting the Institute's research and theory production, Lucia Allais accuses Eisenman of opportunism, see Allais, 2010. Allais points out that the letter "U" in the IAUS acronym signified that urban studies were a lucrative source of revenue for an architecture institution in the late 1960s. According to her reasoning, Eisenman intended to enter the unlikely cooperation with Wingate's New York Urban League solely to fund the Institute. She calculates that the budget for the project, titled "Harlem Plan," would have increased the Institute's overall budget by 150 percent.

And despite this mutual appreciation and interest in cooperating on several levels, the situation changed over the course of the summer and negotiations broke down, with the last meetings probably taking place in September 1968, possibly because of Wingate's further politicization and eventual radicalization, and possibly because it was to be funded by white funds, but also because "key whites" were again to play a key role alongside the "Black middle class."

The Institute's flirtation with the New York Urban League was not the only avenue sought in the wake of the race riots that followed the assassination of Martin Luther King Jr. and the Housing and Urban Development Act, signed into law on August 1, 1968, to capitalize on political will and help improve the social condition of African Americans. In the late summer of 1968, Eisenman was also in contact with George W. Broadfield, the program development consultant for the National Association for the Advancement of Colored People, the oldest and most influential Black civil rights organization working for political, social, and economic equality. A proposal emerged for a planning and development agency within the NAACP. In general, federally funded low-income housing projects under the Model Cities Program, a core element of President Lyndon B. Johnson's "Great Society" and its "War on Poverty," were very attractive to the Institute. Eisenman also continued to pursue CPC assignments, for example, for a land-use study—the first to use computers for planning, for a project to revitalize waterfront and brownfield sites in New York, and for new town planning efforts outside the city. All of these initiatives in 1968, however, failed to produce results. Although Eisenman was accountable to the trustees, he acted largely alone (Ellis advanced to become his closest confidant at the time). Despite his radical departure, however, he acted above all pragmatically, never missing an opportunity that came his way. By offering research and design services as core competencies of the Institute, he explicitly positioned it as an intermediary between cooperating partners. Ultimately, however, the Institute as a framework for action was always about a grammar of governance (or self-governance), about gaining economic leverage and political power, and about securing power within the Institute.

The further institutionalization of the Institute took place at various levels and for various purposes: to achieve better networking, to create better structures, and to ensure better work. As early as the 1967–68 fiscal year, an Advisory Board was established to advise the Institute's director on matters of research and teaching, publications, premises, and resources. Armand Bartos and Lily Auchincloss, among others, were represented here as MoMA trustees and major donors to the Institute. The Board of Trustees was successively expanded over time: first of all in 1968 with the addition of George Dudley, who worked for the Rockefeller Foundation which supported architectural and urban projects. In principle, individuals who either had sufficient private capital to invest or whose position was expected to generate new sources of funding were admitted to the

Board of Trustees. The Institute's by-laws stipulated a minimum of five and a maximum of twenty-five trustees. Eisenman was ultimately able to continue building and expanding the Institute as a group of people primarily thanks to funds from the Graham Foundation. In a personal letter to John Entenza in the summer of 1968, he requested two projects: first, the establishment of a fund to pay a stipend for architects and academics to be invited to the Institute as Visiting Fellows for a year; second, funding for a book series, comparable to that of MoMA (which ultimately also only made it to one volume) which would initially feature books by Rowe and Eisenman. The Chicago-based Graham Foundation, whose funding profile was a perfect match for the Institute's work, supported Eisenman unconditionally, at least for his first request. The book series, on the other hand, was not funded. But by providing funds to establish a Graham Fellowship at the Institute, the private foundation summarily turned it into a kind of field office in New York. Although recipients had to reapply for the grant each year, the foundation went on to fund the inner circle of the Institute to the tune of US$10,000 per year until 1973. Eisenman, meanwhile, secured the right to personally select the Visiting Fellows—without having to justify himself to anyone.[105]

With these strategic moves, Eisenman laid the foundation for the Institute's growth and later success, the acquisition of longer-term, more complex research projects, an economization of creative and intellectual work, education, and culture, and ultimately the capacity to influence the zeitgeist, thinking, and practice of an entire generation. Whatever others may have thought of Eisenman's changing attitudes and abilities as Institute director, he undoubtedly succeeded in assembling a new group that he initially saw as working in parallel to CASE, but which gradually became its de facto replacement.[106] In the meantime, CASE had split into several regional subdivisions, with the subgroup of members from New York and Princeton meeting several times at the Institute in early 1968; among other things, they read and built on the Athens Charter

105 The list of Graham Foundation Visiting Fellows at the Institute included: Emilio Ambasz, Ludwig Glaeser, Robert Gutman, Robert Slutzky (all 1968–69), Kenneth Frampton, Joseph Rykwert (both 1969–70), Stanford Anderson (1970–71), Mario Gandelsonas (1971–72), and Diana Agrest (1972–73).

106 CASE was founded in 1964 as a network of young architects and academics who had only recently been hired at schools of architecture on the East Coast of the United States; see Stanford Anderson, "CASE and MIT. Engagement," in *A Second Modernism. MIT, Architecture and the 'Techno-Social' Moment*, ed. Arindam Dutta (Cambridge: MIT Press, 2013), 578–651. According to an organizational chart dated April 4, 1965, the central committee was composed of: Stanford Anderson, Peter Eisenman, Kenneth Frampton, Michael Graves, Robert Kliment, Richard Meier, Henry Millon, Giovanni Passanella, Jaquelin Robertson, Colin Rowe, and Thomas Vreeland. In addition to sections on "Politics of Architecture," "Psychology of Architecture," "Creative Process," "Education," and "Mass," CASE envisioned launching its own journal. Eisenman stated in the interview that he had flown in Frampton, who had previously worked as technical editor at *Architectural Design*, from London especially for this purpose. In May 2015, a conference entitled "Revisiting CASE" was held at MIT to mark the 50th anniversary of the network's founding, https://www.youtube.com/watch?v=2VwLZLp6Dsg (last accessed: May 31, 2023).

and the Team X Primer.[107] Enmities and friendships, understandings and misunderstandings within CASE caused the network to split into several camps that together formed—to paraphrase the epistemologist Ludwik Fleck—a new community of thought with competing, but also mutually supportive thought-collectives and thought-styles. If CASE and the Institute were heterogeneous groups, however, not only was a generational change initiated but—if the argumentation of one of Fleck's students, the philosopher of science Thomas S. Kuhn, who had been teaching at Princeton University since 1964, about the structures of revolutions in the sciences can also be applied to "architecture" and "urban studies"—also a profound paradigm shift towards a postmodern discursive formation and cultural configuration.[108]

Over the years, Eisenman benefited enormously at the Institute from his contacts in Europe, at first primarily with people from Great Britain, then Spain and Italy. This network was further expanded in June 1968 when he participated in the Design Conference organized by Banham in Aspen, Colorado, on the theme of "America and Europe."[109] Beginning in 1968, Eisenman initially used Graham Foundation grant money to bring old acquaintances, good friends, and former students to the Institute. Their role was to support his project by teaching and participating in the research and design projects or even bringing their own projects to the Institute. Work and personal relationships thus became intertwined in a very specific way. In the academic year 1968–69, funds were divided among four Graham Foundation Visiting Fellows, an illustrious title that, in the Institute's public relations, alluded to the entrepreneurial culture in the United States and the attention economy associated with cultural philanthropy: Emilio Ambasz, who until recently had studied under Eisenman at Princeton and was now assistant professor there, and who was already associate curator of design at MoMA; Robert Gutman, who taught sociology of architecture at Rutgers and Princeton; Robert Slutzky, a New York painter who was assistant professor of architecture at Cooper Union; and Ludwig Glaeser, an art historian who was curator at the Department of Architecture and Design at MoMA and had recently become director of the Mies van der Rohe Archives there. With the Graham Foundation's support, Eisenman was thus able to draw on a pool of people in unique, powerful positions—he was himself an entrepreneur in this—who took responsibility for others, without immediately granting

107 With the Institute, Eisenman created a new group that allowed him and others to do what they wanted, as Alvin Boyarsky, the director of the AA in London once provocatively put it in a joint conversation in the mid-1970s; see Eisenman, 2007.

108 Ludwik Fleck, *The Genesis and Development of a Scientific Fact* (Chicago: University of Chicago Press [1935] 1979); Thomas S. Kuhn, *The Structure of Scientific Revolutions* (Chicago: University of Chicago Press: 1962).

109 Reyner Banham, ed., *The Aspen Papers: Twenty Years of Design Theory from the International Design Conference in Aspen* (New York: Praeger, 1974).

them a permanent position. Through its organizational structure and mode of operation, the Institute, which was in line with the meritocratic ideal from the outset, marked the transition to new flexible forms of work in architecture. From the mid-1970s, these new forms—under neoliberal auspices—would increasingly come to shape the work of architects and academics, intellectuals, and cultural producers alike.

In the fall semester of 1968, after other collaborations had not materialized to the extent that Eisenman had hoped for, the Institute initially continued its collaboration with Cornell University. Despite the Institute's failure to meet expectations from the first commission, Eisenman again received a US$10,000 contract from the CPC, this time in conjunction with the UDC (liaison: Jonathan Barnett) to prepare a case study on land use and development potential for three Manhattan neighborhoods. In addition, the Institute received a first grant of US$30,000 from the National Endowment for the Arts (NEA) for 1968–69 through the National Council on the Arts. The NEA awarded a matching grant for a research project on the urban street, the purpose of which was to investigate "the visual and functional role of the city street as a basic element of urban design."[110] The Institute had tapped into a new source of revenue early on, namely the art foundation, which otherwise supported art spaces and art projects. The foundation was to become increasingly important to the Institute's programming—at times even to its institution-building. Based on concrete legal, political, and economic requirements, the mandate from the city officials actually aimed "to propose a new physical zoning envelope, to enhance and preserve the quality of the street in Manhattan." However, under the Institute's direction, it quickly became "a series of prototypical design studies on the street with a specific street case study as a demonstration model." Here, for the first time, the Institute's affiliation with MoMA came into full play, both conceptually and in terms of cultural policy; a fact sheet on the Institute's activities stated that the original plan was to display the results in an exhibition titled "Street, Arcades, Gallerias." This never materialized.

Rowe invited four students from Cornell University to the Institute for the 1968–69 academic year: Jack C. Dobson, Stephen Quick, Roswell Sanford Jr., and Terrance Williams. In addition, Ellis, who by then was studying urban and

110 According to NEA's press release, the National Council on the Arts hoped that its decision to fund the "Street Project" would highlight the development potential of urban streets: "The city street is one of the most prevalent but, at the same time, most underdeveloped urban open spaces in our cities. The redeveloped city street could serve as a principal organizing element for structuring activities in local areas of the city as well as linking precarious areas of the city together socially. It is hoped that the study will reveal the potentials of the American city street. It is long overdue and may prove of great value to planners and urban dwellers." National Council on the Arts / National Endowment for the Arts, Press Release, Washington D.C., n.d. Source: The MoMA Archives, New York: PI II.B.579.

regional planning at Rutgers, stayed on for another year because his previous year at the Institute had not been recognized. He was joined by a graduate student of Gutman's who was pursuing a PhD in urban sociology. The Institute's semester assignment was to work in parallel on the two research projects, one real and one theoretical, about the urban street, starting with a single case study, the two avenues on either side of Central Park. For the first time, a kind of curriculum was established: a seminar on "The Street," led by Gutman, was scheduled for the fall semester and a seminar on "Modern Architecture: Some Problems," led by Rowe, was scheduled for the spring semester. The four Visiting Fellows were listed as critics. The institutional network expanded, with Eisenman himself now teaching a design studio at Cooper Union from 1968–69, where he offered a course on "Syntactic Structures in Architecture and Design" that was made mandatory for students in the Institute. The Institute's faculty also included Rowe's teaching assistant at Cornell, Alexander Caragonne, who was to supervise the Institute's students. But the plan to exploit synergies in content, staffing, and workload, and ultimately satisfy both the CPC and the NEA proved difficult.

For the practical experimental arrangement of carrying out commissioned work with students failed during the 1968–69 academic year, not only for institutional but also for conceptual and personal reasons. The Institute bore contractual responsibility towards its clients and donors and had assumed pedagogical obligations towards the university and its students. Consequentially, Eisenman and Rowe, being the two faculty members in charge at the Institute, again interpreted the two research assignments differently from the very beginning, so that two camps emerged. While Rowe analyzed historical examples and developed a concrete proposal for the so-called Speiregen Report for the NEA, Eisenman envisioned a study of formal properties that he also wanted to use as a grant proposal for further research and design projects. Rhetorically deft, both sides strove to contrast the topos of the real with the topos of the theoretical. However, from the outset, the students felt forced to follow the contextual approach as it was taught at their university. The Cornell team was in the end characterized by great integrity and loyalty. Overall, the power struggle between Eisenman and Rowe (and Caragonne) had a negative impact on the Institute's teaching. Although the disagreements were initially negotiated quietly, ultimately the issue was not just one of interpretive authority, but of professional dominance. The divergences and ultimately the rift between the Institute director, faculty, and students made work on the two projects almost impossible and put the Institute to its first severe test. The students, as potentially the weakest link in the chain, were the ones who suffered. After two months of standstill, they felt compelled to stop their work altogether at the end of the year. In other words, they went on strike.

In the spring of 1969, the Cornell students finally rebelled against the Institute's director and demanded more professional vocational training. The rebellion at the Institute, however, was not a countercultural act like the more socio-politically motivated student revolts that took place in 1968 and 1969 at universities such as Harvard, Berkeley, and Stanford, where students spoke out against the racial and social inequalities that were clearly emerging in American society. There were also protests at schools of architecture at the time, e.g., at Columbia University, Yale University, and Cornell University.[111] Yet these events seemingly passed the Institute by without a trace, not least because it was an almost exclusively white (and male) institution. In correspondence and conversations between the Cornell students and Dean Kelly, the latter offered a thoughtful and well-reasoned summary of why the students were so concerned about their futures and career opportunities: what was at stake was nothing less than their degrees, the access code to higher positions in contemporary society. The accusation that they had acquired only useless knowledge at the Institute weighed just as heavily as the criticism of Eisenman's management style and his lack of pedagogical competence. Thus, it was the students who declared the Institute's experimental arrangement which Drexler, Eisenman, and Rowe had devised a failure and likened their situation to "working as draftsmen." In contrast to its official status, they viewed the Institute less as a school than an office.

The Institute students' rebellion had far-reaching consequences. Rowe first voiced fundamental criticism of the structure of architecture education to his employer.[112] In his view, the quality of the graduate program at Cornell had suffered and students were burned out at the Institute. Drexler and Kelly then tried to resolve the conflict in their own way.[113] While Kelly spoke of a "clash of personalities and politics" and, as dean, defended his students and faculty, Drexler called the students to MoMA. The incident was eventually settled in a heavy-handed manner, with Eisenman's somewhat ruthless stance as Institute director gaining support. Although the trustees interpreted the facts differently, and Eisenman's appointment was up for renewal, they still advocated for the Institute's continued existence. Finally, in March 1969, an agreement was reached with the Cornell team to allow the Institute to complete the semester and meet at least the minimum conditions set out in the contract. In the time remaining, the students, under Caragonne's lead, produced visualizations of a possible structural implementation of a new zoning law as a planning tool for

111 At Cornell University, for example, the 1968 students' revolt saw the mobilization of the eighty-person Afro-American Society, which occupied the Student Union building; see Charles L. Davis II, "An Appeal to Protest," *Harvard Design Magazine* 44 (2018): "Seventeen," 182–188.

112 Colin Rowe, memo to Burnham Kelly, January 4, 1969. Source: CCA Montréal, Peter Eisenman fonds: PDE-105-T.

113 Arthur Drexler, letter to Burnham Kelly, January 21, 1969; see also Burnham Kelly, letter to Arthur Drexler and letter to Peter Eisenman, January 21, 1969. Source: CCA Montréal, Peter Eisenman fonds: PDE-105-T.

New York.[114] They delivered axonometric renderings with suggestions for what vertical zoning might look like and, in particular, how subway entrances could be integrated into building volumes or how street courses might be redesigned with overlays or cul-de-sacs.[115] For the students, the semester ended when they received their diplomas. But in the spring of 1969, after less than two years, Eisenman finally declared the collaboration between the Institute and Cornell, which had made the Institute's founding possible in the first place, over due to conflicts of interest. Rowe and Caragonne had long since terminated their collaboration at that point.[116]

Orientation towards Urban Development

In the meantime, Eisenman prepared a first fundamental reorganization of activities, which was completed in April 1969. At a meeting of the Board of Trustees, he declared that the Institute was to become less of an educational institution and more of a research institution, where scholarly work was now to be produced. However, the work of the Institute director, the newly designated Fellows, and the Visiting Fellows was to consist mainly of writing proposals, for nothing less than the very existence of the Institute was at stake.[117] One realiza-

114 In the historiography of the Institute, the students' revolt has been largely ignored, while in personal accounts, the dispute between Eisenman and Rowe has been glorified. Yet the conflict-ridden events represented a crucial turning point in the Institute's history that could have recklessly sealed its fate. When Eisenman boasts retrospectively that his only act of rebellion was to have locked Rowe out of the Institute, this may be true, but it is a grossly truncated account of the first crisis the Institute endured in 1968–69, since the students' perspective played no role in this; see Colomina and Buckley, 2010, 66. In an interview in the mid-1970s, Eisenman spoke openly about the founding years of the Institute and in this context, among other things, interpreted the events of 1968–69 as an intrigue, "having gone through several 'palace revolutions' and changes of faculty," as he recounted this first messy episode in the Institute's history in reference to world history, giving greater significance to his own actions; see Eisenman, 2007, 85–86. But Eisenman's actions were not politically motivated, and strictly speaking, they were not an attempt to overthrow the ruler or superior, but rather an act of securing power. In the following, the Institute was repeatedly the site of power struggles. Eisenman not only fell out with Rowe, but also with several of his companions, often over money. Repeated reference has been made to the Oedipal relationship patterns that constituted Eisenman's psyche; see Ockman, 1995, 59. For my historiographical narrative of the Institute, I have confined myself to an analysis and critique of the mechanisms of legend-making, misinterpretation, etc.

115 After completing the studies commissioned by UDG, the Cornell team's drawings were exhibited by the New York chapter of the American Institute of Architects (AIA).

116 In 1969, following his two-year period as a Fellow at the Institute, Rowe took up a research sabbatical at the American Academy in Rome that had been planned for some time. The other members of the Cornell team benefited in part from having worked on a research project for the UDC in New York and thus having established contacts in the New York architectural world. Caragonne and two of the students, Stephen Quick and Terrance Williams, were subsequently hired by Jaquelin Robertson, who headed the Midtown Planning and Development Office from 1969 to 1972, before starting his own firm and becoming a member of the City Planning Commission.

117 In reference to the Institute's later "Program in Generative Design" study of 1971 to 1973, Lucia Allais argued that initially theory production there was merely proposal rhetoric; see Allais, 2012, 35.

tion was if the Institute was to gain more agency and increase and stabilize its budget, it would have to work only on larger projects for state or federal agencies, rather than continuing to take on smaller commissioned work for the City of New York. Burnham Kelly had explicitly warned Eisenman not to rush into this step on the road to professionalization, since the institutional structures had not yet been created, nor were there enough staff capacities available. The strategic repositioning of the Institute had been made possible by the prospect of a research project on planned and built new towns in Europe and the United States. The necessary groundwork for this was provided by Emilio Ambasz as Visiting Fellow with his work on urban systems that accommodate growth and are planned for change. The main contractor for the one-year "New Urban Settlements" study was the New York State Urban Development Corporation (UDC), headed by Edward J. Logue, which had been recently established under the Republican administration of Governor Nelson A. Rockefeller. It was granted wide-ranging powers to improve the housing situation in New York State, with a focus on New York City.[118]

The UDC appealed to the Institute because it developed large-scale housing and urban development projects that were then implemented with community participation through local Model Cities Agencies. These projects were funded by federal grants, as well as mortgages from the Federal Housing Association

118 The UDC had been established as the housing authority for New York State on April 9, 1968, in direct connection with the assassination of Dr. Martin Luther King Jr., when conservative Governor Nelson Rockefeller was able to pass a new law, the Housing and Urban Development Act. There was eager coverage of the formation of the UDC in the architectural press at the time; see "Political Progress," *Architectural Forum*, (May 1968), 37–38; see also Samuel Kaplan, "Bridging the Gap from Rhetoric to Reality. The New York State Urban Development Corporation" *Architectural Forum* (November 1969), 70–73. Regarding the political, economic, and legal aspects of the UDC's history, see Eleanor Brilliant, *The Urban Development Corporation. Private Interests and Public Authority* (Lexington: Lexington Books, 1975); see also Samuel Bleecker, *The Politics of Architecture. A Perspective on Nelson A. Rockefeller* (New York: The Rutledge Press, 1981), here 113, 126–133. Since the bill to establish the UDC, passed on the day of the African American civil rights leader's funeral, had not passed on the first ballot, Rockefeller invoked his spiritual legacy: "In tribute to Martin Luther King and to facilitate our capacity on New York State to help accomplish the things he worked for, which this legislation can do, I urge that you pass this bill, the day of his funeral as a tribute and a memory to him." cited on 132. This legislative decision made it possible to establish a quasi-public housing authority in New York State with the mandate to improve the housing situation statewide and thus guarantee a certain standard of living for all population groups, whereby private interests played a role. Nelson Rockefeller, once a multimillionaire, was obviously running low on funds due to the private financing of his election campaigns. In addition, his brother David Rockefeller, then president of Chase Manhattan Bank, was apparently behind the housing initiative and called for urban renewal to be financed by private investment and public money. To mark the thirtieth anniversary, social psychologist Susan Saegert, who had previously worked under Theodore Liebman, the UDC's chief architect, collaborated with students to organize a large-scale exhibition and symposium entitled "Policy and Design for Housing: Lessons of the Urban Development Corporation 1968–75," which documented and simultaneously critiqued the first phase of the UDC's housing and urban development policy. The exhibition was shown at the Center for Architecture in New York in 2005, at MIT's Wolk Gallery in Cambridge in 2006, and at Roger Williams University in Bristol in 2007, www.udchousing.org/ (last accessed: May 31, 2023).

and the State Housing Finance Agency.[119] In addition to the UDC, Eisenman eventually enlisted four other planning agencies as partners in their "New Urban Settlements" studies.[120] The expansion of the Institute's research activities was also accompanied by a restructuring of its teaching program. Eisenman informed the Board of Trustees that he intended to involve only senior post-graduate students, who were eager to gain practical work experience after completing their studies and displayed a certain maturity, in projects as Research Associates in the future: their duties would include project development and writing proposals as well as initial negotiations and the actual research. At the time, the Institute received applications from students at Ivy League universities and even from one student in Portugal. To advertise and recruit students as Research Associates for "New Urban Settlements," defined as "open ended systems," Ambasz designed the Institute's first poster. The poster was printed on silver Mylar foil and people could decorate and modify it themselves with various stickers that were screen-printed with text and images—DIY and participatory approaches were in vogue at the time—and thus produce their own Institute program. Informational texts about the Institute's work could be combined at will with either the logo of the Vitruvian Man or the image of an astronaut, symbolizing technological progress.[121]

In addition, Eisenman, with the assistance of Ellis, was already preparing a second major research project as a follow-up to "New Urban Settlements" in 1969. Over the summer, Eisenman and Ellis designed the outstanding research report for the NEA-funded "The Street" project in such a way that would enable them to use it as the qualifying main document in an application for an urban renewal demonstration project which they planned to submit to the U.S. Department of Housing and Urban Development (HUD). The greater part of the

119 The Model Cities Program, launched in 1966 under President Lyndon B. Johnson with the Demonstration Cities and Metropolitan Development Act, was an ambitious government support program for cities. The policy initiative was interpreted as a response to a series of problems that converged in the mid-1960s, when widespread urban violence, disillusionment with urban renewal policies, and bureaucratic difficulties led to a reform of public policy. The Model Cities Program was a new tool created by HUD to better coordinate existing urban programs. The original objective emphasized comprehensive planning that focused on new construction as well as redevelopment, social services, and citizen participation. As a result, Model Cities Agencies were created throughout the country. However, by 1969, the new administration under President Richard Nixon changed course and HUD retreated from its earlier insistence on true citizen participation. The Model Cities Program ended in 1974 and ultimately fell short of its own goals.

120 The Institute prepared the "New Urban Settlements" study on behalf of the New York State Metropolitan Transit Authority, the New York State Office of Planning Coordination, the New York State Pure Waters Authority, the New York State Urban Development Corporation, and the New York State University Construction Fund.

121 It is unclear as to what role Ambasz was granted at the Institute by Eisenman. In an official MoMA press release, he was even described as associate director of the Institute; see MoMA, Press Release no. 34, May 1976, https://www.moma.org/docs/press_archives/5382/releases/MOMA_1976_0042_34.pdf?2010 (last accessed: May 31, 2023).

document was penned by Ellis.[122] Formulated as a polemic against modernist urbanism à la Le Corbusier, their report was simply entitled "The Development of a Formal Typology and a Case Study," and drew a fundamental distinction between a spatial conception of the city and an architectural one. This included a distinction between street types, roughly delineating "positive street spaces" from streets that were purely for automobiles. In their formal typology, they favored a more traditional conception of the street that included the vertical boundary, i.e., the architectural design of the façade. As a synthesis of the first two—or as an independent third typology—they offered a more complex traffic infrastructure arranged in three dimensions, which drew on *La Città Nuova*, a series of drawings by Italian futurist Antonio Sant'Elia, although the oil economies of the postwar period meant that automobility in the United States had long since been realized. While their written documents focused on combining "physical design" with "social design," Eisenman and Ellis' proposals showed that, formally, they still wanted to define solids rather than voids. After the Board of Trustees had voted to submit the application to HUD in the fall of 1969—Armand Bartos had also been appointed a trustee—the report was attached to it to recommend the Institute for a highly endowed research proposal. Meanwhile, the Institute's attorneys had confirmed that it was legal to carry out the "Streets" project as a non-profit and receive federal funding. The project, submitted under the title "Streets as Component of the Urban Environment," was developed as a joint effort, but again Ambasz's signature was evident. Its declared goal was to approach the street not only from an architecture or planning perspective but as a complex functional and social system, as Alison and Peter Smithson had done in Great Britain. The project was planned to last several years and included research, design, and realization in three phases. The first phase would consist of various analytical studies of streets to be conducted by a team of Fellows, Visiting Fellows, and Research Associates, postgraduates in the social sciences and design, along with consultants from various disciplines (economists, traffic planners, etc.). The second phase, for which most of the budget was earmarked, would involve the development of a street prototype, while the third phase addressed the potential implementation of the prototype and an evaluation. The proposal listed Eisenman and Ellis as co-directors of the research project; Ambasz, however, was to play a major role on the project team as the designer. In addition, two of Eisenman's allies and trusted friends at the Institute—Gutman, as an architecture sociologist, and Stanford Anderson, an architectural historian and professor at MIT—were

122 IAUS, ed., *The Street. The Development of a Formal Typology and a Case Study* (New York: The Institute for Architecture and Urban Studies, June 30, 1969) Source: private archive of William Ellis. The copy of the research report to the NEA that I read included only the first part, prepared by Eisenman, Ellis, and Joseph. I am not familiar with the case study on which Caragonne and Rowe worked with the Cornell students. Apparently, the report had been backdated to meet the NEA submission deadline.

listed as external consultants, although neither of them had been involved in writing the proposal. Although the topic of streets was already a controversial one in the American architecture and planning debate of the 1960s, it was hoped that a multidisciplinary team and participatory approach would further enhance the chances of such a large research project. After all, HUD's Model Cities Program emphasized public participation.[123] The research proposal was accompanied by a letter of recommendation from Drexler, who, in his capacity as director of the Department of Architecture and Design, promised HUD an exhibition of the research project's findings at MoMA—a great incentive for the Department, which still had to rely on public relations.

With this first realignment, the Institute was now to work more as an office, and Eisenman was finally able to consolidate his own position in 1969 and secure the post of Institute director for the long term; he emerged from the disputes with Rowe and Cornell University stronger than before.[124] Not only was he henceforth solely responsible for the Institute's program and organization, his post also allowed him to invest in his own projects: Eisenman continued to develop his house designs, which he would work on from 1967 to 1977, in parallel to his research and teaching activities at the Institute and at Cooper Union. The designs were primarily for single-family or weekend homes, numbered Roman I through X, for which he would later gain international renown. Even in the early years of the Institute, with the diagrammatic, even sculptural designs for *House I* (1967), *House II* (1969), and *House III* (1970), all of which were realized and widely published, he proposed, as with his formal building analyses, a generation of forms that started from basic architectural elements and geometric operations. Eisenman did not only use the Institute as a fixed working context (it became difficult at times to separate the Institute as a project office from his own architectural practice, both in terms of space and time and in terms of work and salary), he also used it as an important PR and marketing tool to disseminate his publications and provocations and advance his career as an architect and theorist.[125] Yet the crucial factor for the Institute as Eisenman's project, namely to promote the breakthrough of a linguistic and artistic turn in American architectural culture, was that Eisenman—in the

123 A critique of the street's loss of meaning, brought about by modern, anti-urban urbanism and increasing automobile traffic, had been introduced in the United States with Jane Jacobs's *The Death and Life of Great American Cites* (1961) and established at the latest with Bernard Rudofsky's *Streets for People. A Primer for Americans* (1969).

124 The documents in CCA's IAUS fonds do not clarify whether the directorship was up for election in the summer of 1969 and whether Eisenman was subsequently elected every year, as had originally been stipulated in the Institute's by-laws. If this not the case, Eisenman would have run the Institute quasi-autocratically.

125 Later, there were also Institute projects that not only bore a strong resemblance to Eisenman's house designs, but actually were numbered as part of them.

process of restructuring and thanks to his charismatic personality, entrepreneurial spirit, and intellectual ambition—succeeded in successively expanding the inner circle of Fellows, Visiting Fellows, staff, and students as initiates, a circle that was esoteric in the epistemological sense, i.e., narrow and self-contained, so that the Institute's capital and work were increased.[126] Initially, Ellis' instrumental involvement in the preparation of the research report had qualified him for Fellowship status and allowed him to move into the office next door to Eisenman. In the 1969–70 fiscal year, Eisenman was able to draw on the Graham Foundation grant a second time and brought Kenneth Frampton, by then an associate professor at Princeton University, and Joseph Rykwert, who had been teaching at the University of Essex as a professor for art and architecture history since 1967 after completing his PhD dissertation, to the Institute for a year as Visiting Fellows. Eisenman knew both scholars from his time in Cambridge and both were highly interesting to him, bringing with them substantial academic capital, but also valuable publication experience.[127]

The Institute began the 1969–70 academic year by working exclusively on the "New Urban Settlements" study of new town planning in Great Britain, *Villes Nouvelles* in France, as comparable developments in the United States, with six Research Associates from Cooper Union, Rice University, Cornell University, and Yale University conducting research under the direction of Ambasz and Frampton. In the summer of 1969, however, Ambasz was employed as a part-time curator of design at MoMA and began attending the Institute only in the mornings, leaving Frampton primarily responsible for the analytical phase.[128] The study was to

126 The sociologist Max Weber describes charisma as a social relationship of rule: "Charisma is validated through the recognition of a personal proof by those who are ruled. This was originally effected through the performance of a miracle, bringing about a voluntary dedication to a revelation, to hero worship, to absolute trust in the leader. Where charisma is genuine, this is not, however, the for legitimation; it is instead rooted in an obligation on the part of those who have received the call to acknowledge their duty to provide personal proof. This "acknowledgement" is, psychologically, a quite personal dedication, a belief born of enthusiasm, or of despair and hope." Max Weber, "Chapter III. Types of Rule, §10: Charismatic Rule," in *Economy and Society* (Cambridge and London: Harvard University Press, [1922] 2019), 374–375.

127 Frampton, who came to the United States in 1965–66 on a Hodder Fellowship and since taught at Princeton, first as an assistant professor, and then as an associate professor, was at this time already working on his first monograph, *Modern Architecture. A Critical History* (1980), which would not appear for a while; Rykwert was working on his monograph *Adam's House. Papers on Architecture* (1972). In the report to the Graham Foundation, Eisenman highlighted both Frampton and Rykwert's publications as their academic credentials: Frampton had previously served on the editorial board of *Architectural Design* and published an essay on Pierre Chareau's *Maison de Verre* in *Perspecta*, no. 12 (1969). Rykwert wrote a column in the Italian *Domus* at the time. In addition, essays by both Frampton and Rykwert were included in the 1969 anthology *Meaning in Architecture*, edited by George Baird and Charles Jencks.

128 Ambasz already had his first exhibition at MoMA, "Paris, May 1968, Posters of the Student Revolt." In 1969 he curated Peter Wolf's exhibition "Urban Anticipations: Eugène Hénard, 1849–1923;" see MoMA, Press Release no. 106, July 31, 1969, https://www.moma.org/documents/moma_press-release_326638.pdf (last accessed: May 31, 2023). Ambasz was appointed curator of design in 1970, a position he held until 1976. Ambasz realized two major projects at

focus mainly on the growth of new town planning. Basically, the question was "whether one can develop a new approach to city building based on a concept of the urban settlement as a complex adaptive system—one capable of monitoring and regulating change and the consequences of urban design and development decisions to meet such change." The project team studied a total of six cases in relation to land use and transportation systems: two small-scale new towns, Hook (UK) and Toulouse-Le Mirail (FR), and two regional settlement patterns, Milton Keynes and South Hampshire (both in the UK), were selected for comparison with Columbia and Harvard N.C.P. (both in the U.S.). The Institute was less concerned with urban design than with the national planning policies that lay behind it. Eisenman had brought Stuart Wrede, a Yale University graduate, to the Institute to coordinate the individual studies for the research project. In 1969–70, the Institute's seminar program was tailored to the interdisciplinary nature of the research project, and the two Visiting Fellows, in particular, were also involved in teaching: Frampton commuted regularly from Princeton and gave two weekly seminars directly related to the "New Urban Settlements" study, while Rykwert flew in from England three times especially to give a total of six seminars on urban form and to hold two public events: on the city as an icon and as an institution. The program also included lectures on biological and behaviorist aspects of the environment (lecturers: Richard Chase, Raymond Studer, and Alexander Tzonis) and a four-part seminar series by Yona Friedman on infrastructure. The "New Urban Settlement" study was not completed though, and after a year Wrede was replaced by Susana Torre as the new coordinator for the research project. The Institute submitted a final report, authored by Frampton, on new town planning (main criticism: satellite towns mutate into mere bedroom communities for commuters if no jobs are created there), which became the Institute's first publication.[129] What was more important, however, was the fact that HUD was won as a new client for even larger projects immediately thereafter. In November 1970, the Institute was first commissioned to prepare a design study for a new university campus at Utica-Rome, New York, which was completed in March 1971. Although the study was not realized, it paved the way for further commissions.

MoMA: the "Universitas Project" conference (January 8 & 9, 1971 [sic!]) in January 1972; see MoMA, Press Release no. 154, n.d., https://www.moma.org/momaorg/shared/pdfs/docs/press_archives/4770/releases/MOMA_1971_0206_154.pdf (last accessed: May 31, 2023), and the exhibition "Italy, the New Domestic Landscape" during the summer of 1972 (May 26 to September 11, 1972), see MoMA, Press Release no. 26, May 26, 1972, https://www.moma.org/documents/moma_press-release_326797.pdf (last accessed: May 31, 2023).

129 IAUS, ed., *New Urban Settlements. Analytical Phase* (New York: The Institute for Architecture and Urban Studies, December 1970). The masthead listed Ambasz and Eisenman as co-directors of the study. In addition to Frampton and Torre, the project team included Robinson O. Brown, William Ellis, Gregory Gale, Lawrence Goldberg, William LaRiche, Robert Slutzky, Robert Timme, W. Stephen Wood, and Stuart Wrede. Torre had previously worked under Ambasz as an intern at MoMA. The cover design was by Robert Slutzky, who was also responsible for the layout of the text pages and the graphic design of the mapping.

While research and teaching at the Institute were still strongly influenced by the planning euphoria and criticism of the 1960s from Europe, "New Urban Settlements" as the Institute's first major project marked the beginning of a necessary professionalization and a certain professionalism as a project office. This led to the establishment of a completely new structure in terms of working and organizational forms, including the restructuring of finances and administration. In financial terms, this step was a limited success at first: despite larger contracts, the Institute recorded a deficit of US$20,000 in the fiscal year 1969–70. In the spring of 1970, Eisenman, Drexler, and Bartos therefore formed a special committee to launch a major fundraising campaign, and Bartos was appointed acting secretary of the Institute, henceforth in charge of financial affairs. Although Eisenman, as Institute director, demonstrated increasing skill in acquiring contracts and grants, the Institute's operations were subsequently shaped more and more by debt management. While the Institute's leadership assumed that the overall budget would grow steadily, it also accepted that, in the course of its further expansion, it would incur more debt. Moreover, it proved impossible to reduce the debts accumulated by the end of the 1969–70 fiscal year as quickly as planned, and the Institute was brought to the brink of bankruptcy two years later. But with the exception of the decidedly bio- and sociopolitical orientation of the "New Urban Settlements" study, the Institute did not take a position on the major issues of the time—the Cold War and racial unrest in major American cities, as well as the ongoing Vietnam War, against which not insignificant parts of the population in New York protested for years—in its programming, at least not publicly, unlike other American intellectuals and artists, architects and planners who were part of the peace movement.

Building and Expanding the Institute

The acquisition of major lucrative contracts from state and federal authorities had become attainable for the Institute under the conditions and with the human capital available at that time, but it was also necessary to secure the increasing budget. On the other hand, urban studies had the effect of attracting and engaging new Fellows to work on these group projects. A true networker, Eisenman, with the support of the Graham Foundation, was able to attract a group of aspiring architects, historians, and theorists from around the world to the Institute, most of whom would go on to pursue university careers in the New York metropolitan region. After Ambasz, Glaeser, Gutman, Slutzky (all 1968–69), Frampton and Rykwert (both 1969–70), the list of Visiting Fellows included Stanford Anderson (1970–71), Mario Gandelsonas, Anthony Vidler (both 1971–72), and finally Diana Agrest (1972–73). This migration of architects and academics to New York, doubtless attracted by its international reputation as a creative and intellectual center, was symptomatic of the strong historical and cultural connection, if not the general trend of an international exodus of the academic elite to the United States. The Institute offered its Visiting Fellows the opportunity to collaborate on large, fully funded research and design projects that both financed the Institute's operations and allowed for theoretical reflection, historical research, sociological analysis, cartographic practice, and ultimately, architectural design. The status of "Fellow" was conferred on them by the Institute's leadership after one year of dedicated group work and lasted for an initial period of three years, with the option of extension. In the process, Eisenman achieved a longer-term commitment of the Fellows' social, cultural, and intellectual capital to the Institute. This affiliation with the Institute not only made demands on the individual Fellows, it also lured them in with a wide variety of tasks and thus interesting career opportunities. At the same time, Fellows were also allowed, even encouraged, to pursue their own research, design, and publication projects. In this way, a Fellow at the Institute was assured a certain degree of freedom and could enjoy relative independence from the education and the construction industry, albeit with its inherent contradictions. Thus, institutional forms of work and organization, responsibilities, and accountabilities—participation in group projects, and attendance at Fellow meetings—were initially settled only by mutual agreement. It was not until the Institute had achieved further institutional growth that a debate about community, autonomy, and ownership emerged in the fellowship. At that time, individual Fellows received a sizeable base salary based on performance and cooperation, but their work was characterized by both self-determination and self-exploitation, by virtue of the commitment required. Ideally, they invested a large part of their time in the Institute while at the same time working as professors or architects one or even several days a week. In essence, through the powers officially conferred upon it, the Institute represented a quasi-institutional set-up that regulated the thinking of all those who participated in it and defined them as creative, entrepreneurial individuals against the backdrop of the prevailing social technologies and technologies of the self in architecture and planning. In doing so, the Institute under Eisenman's direction was in fact neither critical nor radical in the political sense, i.e., towards existing institutions. Rather it continuously probed the boundaries of autonomy and heteronomy of thought and action, tradition and avant-garde with its research and planning projects and changed the museum and the university as instances of consecration or diffusion from within, cooperating with them, but never representing a real alternative.

From a historical perspective then, if we adopt an archaeological-genealogical approach, the Institute acted as a powerhouse in the following decade, a real game changer in terms of reception and production, ultimately by "curating" individuals and projects. At a time when New York was in transition from Fordist to post-Fordist capitalism and from an industrial to service society, the metropolis became the focal point of a new architectural culture, explicitly of the new discursive formation and cultural configuration of postmodernism. Even in the early years, Eisenman was very determined in his pursuit of the goal of publishing his own journal, if not a book series. As a passionate collector of publications and paraphernalia of architectural modernism, he was aware of the strategies of the classical avant-garde and, in particular, the cultural significance of monographs and periodicals. This was particularly evident in an exhibition he curated, entitled "Modern Architecture 1910/1939: Polemics, Books, Periodicals and Ephemera from the Collection of Peter D. Eisenman," which was shown at Princeton in early 1968 and subsequently at Cornell University.[130] After failing to launch a journal as part of CASE, Eisenman tried to harken back to the heyday of modern polemics by developing corresponding formats at the Institute. As early as 1968, he repeatedly attempted to lure people with relevant experience to the Institute and to retain them for the long term. In addition to Frampton, for example, he sought to attract Wrede, who had journalistic experience as one of the editors of *Perspecta*, no. 12, and later Alexander Tzonis. *Perspecta*, published by Yale University students, served Eisenman as a model, which he acknowledged in a review in *Casabella*. After the Graham Foundation failed to approve the publication of a book series, Eisenman, with Gandelsonas as editor, planned an anthology on semiological approaches to architecture in the spring of 1971, to be published jointly by the Institute and MoMA and funded on a 50-50 basis. This was intended to be a response to Venturi's *Complexity and Contradiction in Architecture*, a manifesto in postmodern thought in architecture, whose typological studies aimed at redefining the function and place of architecture within the cityscape (and were funded by the Graham Foundation). Around the same time, Anderson and Vidler devised their own multidisciplinary journal at the Institute, entitled *Journal for Discussion and Criticism of Architecture, Planning and Urban Design* which they proposed to Michael Conelly, then head of the MIT Press, for publication in the fall of 1971. The content was to be drawn from the Fellows' research projects and the Institute's seminar offerings, although at the same time, Anderson and Vidler drew a clear distinction between their journal project and the Institute: it was to be conceived not "as the voice of the IAUS, but rather as an intellectual and communication service provided through the IAUS."[131] In their proposal, the two editors set out six thematic issues: "Architecture and Political Change," "Architecture and Conceptual Structure," "Pop Culture vs. Mass Culture: Pop Culture vs. High Art,"

130 The exhibition "Modern Architecture 1910/1939: Polemics, Books, Periodicals and Ephemera from the Collection of Peter D. Eisenman" was on display at Princeton's University Library (February 16 to April 15, 1968). Eisenman's periodical collection included the Dutch art journal *Wendingen*, which, edited by Hendricus Theodorus Wijdeveld, shaped one view of architectural modernism from 1918 to 1932. Eisenman also collected the Italian architectural magazine *Casabella* since embarking on two separate Grand Tours of Italy with Rowe in the early 1960s. He owned the volumes from 1928 to 1943 almost in their entirety; they represented his approach to architecture in fascist Italy of the 1930s and 40s.

131 Stanford Anderson and Anthony Vidler, memo to Arthur Drexler, Peter Eisenman, William Ellis, Kenneth Frampton, Mario Gandelsonas, Vincent Moore, and Peter Wolf, October 19, 1971. Source: CCA Montréal, IAUS fonds: B.1-5.

"Architecture and Social Utopias," "Meaning in Architecture," and "Low Rise/High Density." Anderson and Vidler had slated Eisenman, Ambasz, Ellis, and Frampton as guest editors for one issue each and informed them of their intentions. Ultimately, however, none of these publishing projects came to fruition. At that time, MIT Press maintained connections with individuals and projects, but not with institutions such as the Institute. And so it would be another two years before the Institute had its own journal, *Oppositions* (1973), and another eleven before it finally published its own book series, Oppositions Books (1982).

From the early 1970s, even without his own print medium, Eisenman took advantage of the freedom offered by the Institute to publish theoretical texts and speculative projects in leading journals. These included two different versions of the essay "Notes on Conceptual Architecture."[132] The first version, published in *Design Quarterly* (1970), consisted entirely of footnotes, with the numbers dotted across the white space of an otherwise empty page. With it, Eisenman not only distinguished himself as a well-read theorist (even though it is uncertain whether there is a readership for publications without a narrative or line of argument) but more importantly, his author biography names him as Institute director, thus promoting the Institute as a site for the production of theory, not necessarily architecture.[133] With the second version in *Casabella* (1971), he self-consciously inscribed his idiosyncratic notion of conceptual architecture in a theory-based, linguistic frame of reference, placing it in the tradition of American Minimalism.[134] At the same time, Eisenman also published his formal analyses of selected buildings by the Italian architect Giuseppe Terragni (1904–1943), who was also just being rediscovered in his native country, in two articles in *Casabella* (1970) and *Perspecta* (1971).[135] Using analytical drawings on formal transformation processes of individual architectural elements of the *Casa del Fascio* and the *Casa Giuliani Frigerio*, which he had instructed his students at Cooper Union to prepare, he provocatively claimed to be able to

132 Peter Eisenman, "Notes on Conceptual Architecture: Towards a Definition," *Design Quarterly*, no. 78/79 (1970): "Conceptual Architecture," 1–5; "Notes on Conceptual Architecture: Towards a Definition," *Casabella*, no. 359/360, (November/December 1971): "The City as an Artifact," 48–58

133 Anthony Grafton, historian of the footnote, referred in one of his historical essays to Gottlieb Wilhelm Rabener's *Hinkmars von Repkow: Noten ohne Text* of 1745, which assumed that learned footnotes, not eloquent texts, make authors famous, see Anthony Grafton, "The Death of the Footnote (Report on an Exaggeration)" *The Wilson Quarterly* 21, no. 1, (Winter 1997), 72–77, here 76. Whether Eisenman was familiar with Rabener is unknown.

134 Eisenman had Rosalind Krauss proofread the second version at the time; in her marginal notes, she commented on a passage in which he described every work of art as conceptual, saying, "This is bullshit!"; the annotated manuscript was on display in the exhibition "Take Note" (February to May 2010) at the CCA in Montréal, curated by Sylvia Lavin.

135 Peter Eisenman "Dall' oggetto alla relazionalità: la casa del Fascio di Terragni," *Casabella*, no. 344, (1970), 38–41; "From Object to Relationship II: Casa Giuliani Frigerio. Giuseppe Terragni," *Perspecta*, no. 13-14 (1971), 36–75.

reconstruct the design process of this proto-rationalist architecture and thus to understand its architectural language. By positing a theory of architectural form, albeit one that was incoherent and ultimately fragmentary, and calling for logical consistency in thought as well as conceptual rigor in design, Eisenman applied various approaches to the project of revalorizing, even redisciplining architecture. For example, he published the first of his series of houses, which he called "cardboard architecture," a term that he picked up and gave a positive interpretation (although, given that his houses were built out of plasterboard and rotted quickly, the term turned out to be only an honest description), first in *Five Architects* (1972), later in *Casabella* (twice, in 1973 and 1974) and in *Architecture + Urbanism* (1973).[136] Following artistic strategies of conceptual art, Eisenman sought to show the actual design process through the production and dissemination of drawings and models (although some of them were made after the fact, some even entirely without a reference building) while moving closer to modernist paper architecture.[137] Crucially, Eisenman's approach contributed to a further iteration of the autonomy of art, in the sense of the commodity character discussed and historicized by Theodor W. Adorno, opening up new opportunities as the art market was transformed, while the Institute itself became reliant on patronage.[138] In terms of historical biographical research, Eisenman's early publications offer several possible readings. Not only do they have a discursive function for self-legitimation and self-reflection, the recognition and appreciation of architecture as an ultimately commodified art form, but they also, even more than his designs or buildings, serve as a biography generator, a kind of ego document, with which he, perhaps more than any of his contemporaries, staged himself in his own perception as the most prominent representative of a new type of artist-architect.

In terms of a collective biography of the Institute, however, it must be acknowledged that Eisenman acted as purposefully and skillfully in his self-presentation and communications as Institute director as he did as an architect and author. But without Drexler and Bartos and their far-reaching and highly

136 *Five Architects* was a publication of projects by a group of emerging New York architects that Eisenman assembled around himself out of CASE, see Peter Eisenman et al. *Five Architects* (New York: Wittenborn Art Books, 1972). Previously, Eisenman had already hosted a CASE meeting of the New York subdivision at MoMA in 1969, during which Michael Graves, Charles Gwathmey, John Hejduk, Richard Meier (and William Ellis, as the sixth architect) presented projects for discussion. When the publication on the projects was published by George Wittenborn's art publishing house in 1972, as a small edition and accompanied by essays by Arthur Drexler, Colin Rowe, and Kenneth Frampton, this group became known as the "New York Five," following reviews and critique in the *New York Times* and in *Architectural Forum* and subsequently rose to international fame; see Goldberger, 1973.

137 Peter Eisenman, "Castelli di Carte: Due Opere di Peter Eisenman," *Casabella*, no. 374 (February 1973), 17–31; "Cardboard Architecture," *Architecture + Urbanism* 3, no. 35 (November 1973), 185–189; "Cardboard Architecture: castelli di carte," *Casabella*, no. 386 (February 1974), 17–31.

138 Theodor W. Adorno, *Aesthetic Theory* (Minneapolis: University of Minnesota Press, [1970] 1997).

influential connections, the Institute would not have lasted as long as it did. For it was Drexler, and thus MoMA, who gave the Institute legitimacy and visibility—if not even more. Over the years, MoMA served as an exhibition space, a conference venue, a meeting room, and as a cooperative partner, providing ideas initially for projects, exhibitions, and accompanying publications, and later for event series and other cultural productions. If the museum was generally to be regarded as a venerable instance of the consecration and legitimation of modern architecture, the Institute's leadership used the museum's capital to build an institution. To name one example, Philip Johnson, who was the founding director of the Department of Architecture and Design at MoMA in 1949 and a trustee, advisor, and decision-maker for many years, long functioned primarily as a gray eminence behind the scenes, but would later assume a central role at the Institute as a financially powerful patron and influential puller of strings.[139] Although the Institute ushered in a generational shift in New York architecture culture under Eisenman's direction, it did not dare break completely with the past and institutions as an emerging functional elite, for such power networks were too valuable, especially for architects who wanted to build in 1970s New York. Instead, MoMA helped the Institute inscribe itself into existing structures and hierarchies. After all, MoMA's status as a respected, high-culture institution and the promise of organizing and hosting major exhibitions helped the Institute to undertake two major research, planning, design, and ultimately building projects in the first half of the 1970s. The results of these projects were influenced by the fact that the discipline and profession of architecture had changed rapidly in a short period of time and that, as a result, the economic, political, and social conditions for urban renewal and for private, cooperative, and state-financed housing had changed dramatically. First, from 1970 to 1972, there was the historical-analytical and, above all, interdisciplinary research project on the function and design of the downtown street, commissioned by the HUD under U.S. Secretary of Housing and Urban Development George W. Romney after massive urban interventions had been criticized for representing slum clearance and the destruction of entire neighborhoods.[140] Second, 1972 and 1973 saw a research and design project commissioned by the Urban Development Corporation of the State of New York under Edward L. Logue, in which a prototype for low-rise yet high-density housing was to be developed and realized from 1973 to 1976 after modernist large-scale housing had come under criticism

139 Only recently there were growing calls to remove the name Philip Johnson from MoMA because of his fascist past, culminating in an open letter from the Philip Johnson Study group on January 18, 2021, see https://docs.google.com/presentation/d/e/2PACX-1vQBZHBg20U-dYfLz69NOPqPzrkz1LY97Pcgl1Pc05tBt-rYWWP6QQMqO2-yf8KGVIY1CgNQUQYlNbO88/pub?start=false&loop=false&delayms=3000&slide=id.gb660b5c816_2_0 (last accessed: May 31, 2023).

140 Francesca Ammon, *Bulldozer. Demolition and Clearance of the Postwar Landscape* (New Haven: Yale University Press, 2016).

and was being discredited by the broader public, politicians, and the housing industry. For the Institute, these two major commissions allowed it to work as an architectural office for the first and only time, as originally intended. In addition, due to the size of their contract volumes, these projects played a crucial role in enabling the Institute to develop into a significantly larger institution; for several years, they formed important cornerstones for the identification of the Fellows and the self-image of the Institute. In both cases, the Institute was forced to reorganize itself at the insistence of its clients—both in terms of the group of Fellows and the external experts that were brought in—to muster the necessary clout and expertise and to be able to bear the responsibility assigned to it. In the end, the self-imposed task of research and design projects consisted not only of scholarly and architectural work but above all of communicating both to the public.

1.2 Conducting Urban Research

Launched in January 1970, the "Streets Project" was the Institute's first major research and design project. It heralded a new decade, eclipsing all that had gone before, and set the Institute out on a new orbit. Commissioned with an Urban Renewal Grant from the U.S. Department of Housing and Urban Development (HUD), the Institute enjoyed a period of growth and stability for some time, with parts of the requested budget provided by grants from the Alfred P. Sloan Foundation and staff funding from the Graham Foundation. In its press release, HUD communicated that the high-dollar, high-stakes "Streets Project" was expected to do nothing less than develop alternative methods and techniques for urban renewal.[141] In internal parlance, the federal agency expected the Institute to produce some sort of practical guide to planning ideal street designs; it explicitly sought the publication of a research report, for which US$32,000 had been budgeted. After the contract was signed in March 1970, a new era began for the Institute, as it now appeared on the national stage as a legitimate planning consultancy; the HUD contract and, again, MoMA had made this leap toward professionalization possible. With the "Streets Project," Institute director Eisenman created jobs for a newly expanded group of Fellows, Visiting Fellows, and Research Associates, and was able to fund operations for two years (notwithstanding the larger budget and internal restructuring, however, the Institute repeatedly found itself facing insolvency in the years to come, resulting on more than one occasion in an inability to pay salaries, rent or bills). As the Institute expanded, a move seemed inevitable, and larger and more stately premises were desired. Inspired by its strategic success, Peter Eisenman was on the lookout

141 HUD News, HUD no. 70.55: "HUD Funds Demonstration Grant for Better Street Design," January 30, 1970. Source: The MoMA Archives, New York: PI II.B.579.

for new spaces for his growing Institute. The biggest change, however, involved the Institute, which had operated more or less as a single-project institution in the early years, working on several projects in parallel in the future. Key collaborators in 1970 included William Ellis, Kenneth Frampton, who increasingly preferred New York to Princeton, and Joseph Rykwert, who had since moved to New York. Along with outside consultants, research assistants, and students, the Institute's circle had now grown to twenty people—and it would continue to grow as the following academic year also saw an increase in the number of postgraduate research assistants hired to handle multiple projects simultaneously. The crucial lead came from a cousin of Rykwert's, a real estate agent: a very prestigious two-story office space on the 21st and 22nd floors of an office building on 8 West 40th Street, directly across from the New York Public Library at Bryant Park. With its central location in midtown Manhattan, conveniently situated between Grand Central Terminal and the Port Authority Bus Terminal, it was also near other institutions of social and cultural life.

Despite the exorbitant rental costs of US$43,000 per year (previously US$11,000), Eisenman was immediately convinced that he had found the right space and the right address for the next phase of the Institute. Addressing the Board of Trustees, he argued that the additional expenses could be covered by working on two research contracts for the UDC and HUD, starting the Institute's own publication series, and designing traveling exhibitions (Ezra Stoller had, according to Eisenman, shown interest). Ultimately, however, only half of the overhead costs were to be covered by the HUD budget. In May 1970, the Institute's leadership signed the lease with Jack Resnick & Sons.[142] With the move in July, the Institute had come of age and had laid the groundwork for its future work as an institution. From the elevators, one entered the central, two-story hall with a gallery, whose studio windows to the west offered a view of the American Radiator Building (1924, architects: Raymond Hood and John Howells). On the north, east, and south sides, on two floors, were numerous offices with views of the Empire State Building, the Pan Am Building, and Rockefeller Center, respectively. These offices were occupied by Eisenman and the Fellows, but some of them had to be sublet initially to cover rental costs. It was not only the fact that the Institute had its own lease that manifested a certain autonomy and independence from MoMA. Here, the Institute was also able to offer a new course program and stage public events, hold lecture series, and organize exhibitions, generating further income through their commercialization. In addition, the space helped create a sense of identity. It took on a central position in the Institute's culture and was the site of a collection of images, rituals, narratives, and codes of conduct that engendered and stimulated a sense of community among the Fellows, staff, and students who came there every day.

142 The Institute's lease agreement of May 11, 1970, was for ten years and provided for rent increases every two to three years. As a result, the Institute's leadership accepted rent debts.

A collaborative renovation of the office floor, which was in disrepair, became a community-building event. Together, the Fellows tore down walls and knocked out ceilings, put in new partitions, plastered ceilings, installed doors, repurposed light fixtures, put up shelves and tables, painted the walls, and repaired the air conditioning. More importantly, the office floor was a thoroughly historic space that had once housed the renowned Reynal and Hitchcock publishing house, which had published Le Corbusier's works for the Anglophone market. Le Corbusier, it was said, had personally gone through the proofs of the translation of *When the Cathedrals Were White* (1947) and devised the worldwide sales strategy for his publications here[143]—a legend that Eisenman was always happy to repeat, especially since it enabled him to embed the Institute in the architectural, planning, and publishing tradition of a heroic "white" modernism and to distinguish it from other contemporary trends later known as postmodernist. As daring as the decision to rent these attractive, central penthouse office floors may have been, and as constitutive for the Institute's progressive institutionalization and further development, the step-up lease meant that it was constantly in rent arrears that ultimately proved to be its undoing.

During the summer of 1970, the Institute was already working closely with federal officials (Howard Cayton, Michael Schneider, and Ralph Warburton) on the conception of the "Streets Project," since it was necessary to establish the framework and thrust before it was actually launched in the fall. One of the most important staff changes at the Institute was the appointment of Frampton as a Fellow in June 1970. Frampton, who by then had emerged as an architecture theorist and historian and had already received an offer to publish *Modern Architecture. A Critical History* with Thames and Hudson via Robin Middleton, was to work on the "New Urban Settlements" study as a Research Associate with Joachim Mantel of the ETH Zurich. His project work at the Institute eventually prompted Frampton to leave his tenured position at Princeton in 1972 and relocate to New York, where he joined Columbia University's Graduate School for Architecture and Planning (GSAP) as an assistant professor under the new dean, James Polshek. Frampton benefited from his strong loyalty to Eisenman, even though he was constantly at odds with him over his performance as Institute director and his view of architecture. The second important addition, if only for a short time, was architecture historian Stanford Anderson who, after completing his doctoral dissertation on *Peter Behrens and the New Architecture of Germany, 1900–1917* at Columbia University in 1968, was now teaching as an associate professor in the History and Theory of Architecture and Architectural Design program at MIT. After Anderson returned from a trip

143 The back cover of the original edition features a photograph of Le Corbusier on the balcony of the office floor where the Institute was located, see Le Corbusier, *When the Cathedrals Were White* (New York: Reynal and Hitchcock, 1947).

to Europe, Eisenman succeeded in convincing him to join the Institute for the 1970 fall semester to carry out the "Streets Project." A founding member and later executive secretary of CASE and a contributor to "The New City" exhibition, Anderson was to have a major impact on the Institute's research on the inner-city street.[144] The multi-unit project, which ran from the fall of 1970 to the summer er 1972, was the first and only time the Institute conducted multidisciplinary urban studies as intended. Despite the diverging interests of the Fellows and Visiting Fellows, the project's subject matter and approach differed from other research in architecture on the American city and street, for example the Las Vegas Studio, which Robert Venturi and Denise Scott Brown conducted with Steven Izenour at Yale University in 1968–69 and which formed the basis for the publication *Learning from Las Vegas* (1972), or Reyner Banham's monograph *The Architecture of Four Ecologies (*1971) and the documentary film *Reyner Banham loves Los Angeles* (1972), which was subsequently produced by the BBC. By comparison, the "Streets Project" was nowhere near as sensational and much more institutional. In 1970, in order to carry out the work at the Institute to HUD's satisfaction, Eisenman assembled a new team around Anderson, Frampton, and Rykwert, who were to work together more or less successfully for the next two years: William Ellis was appointed project lead, while Anderson and Rykwert's expertise meant that they were appointed co-directors and given Visiting Fellow status for a year. On their initiative, anthropologists, historians, sociologists, urban planners, and transportation planners were also brought in as outside consultants to the "Streets Project." One of them was Peter Wolf, an urban planner who had completed his doctorate with a dissertation on the planning approach of the French architect Eugène Hénard and had just curated a major exhibition at MoMA, and who was now added to the team for his practical experience with a traffic planning project. In addition, Elizabeth Cromley and Suzanne Frank, both of whom, like Anderson, had earned doctorates in the history of art and architecture at Columbia, were hired as Research Assistants to work with the project leaders. Cromley and Frank were thus the first women at the Institute to be hired not for a purely administrative role, but to work on content, albeit in a subordinate capacity. Initially, the Portuguese architect Duarte Cabral de Mello, Thomas Czarnowski, and Gregory Gale were also involved in the "Streets Project," all of them postgraduate students who now worked at the Institute as Research Associates.

144 In the historiography on the Institute, the "Streets Project" is often erroneously attributed to Stanford Anderson alone, probably because he was responsible for editing *On Streets* (1978). The publication, however, represents only one phase of the research project. Architecture historian John Harwood, in his text on the history of the "History, Theory and Criticism of Art, Architecture and Urban Form" doctoral program at MIT, draws a direct line from Anderson's contribution to CASE through the *Possible Futures and their Relations* to the *Man-Controlled Environment* conference to the "Streets Project;" see John Harwood, "How Useful? The Stakes of Architectural History, Theory and Criticism at MIT, 1945–1976," in Dutta, 2013, 106–143.

Project Work

During the first year of the project, the representatives of the Institute and HUD spent a long time finding a common approach and defining its content and goals. From the beginning, the project development and group dynamics were dominated by the project leaders Rykwert and Anderson's diverging ideas about what "street" meant in the first place and how it should be studied. Within just the first few weeks, it became apparent during their team meetings that their ways of thinking were irreconcilable. While Rykwert started from the functional approach of the polycentric city, Anderson, following a cognitive approach, emphasized that the people should be the focus and that it was less about efficiency and beauty than about changing the very idea of the built environment. When Eisenman submitted a work plan for the "Streets Project" to HUD in early October 1970, it was clear that it was still too broad and lacked clear lines. In terms of content, the project leaders wanted to address the use and symbolic character of the street, the demarcation between private and public, flexibility and adaptation in terms of use, and physical characteristics. In addition to theoretical work, empirical studies were also planned in different cities and at different scales. During the 1970–71 academic year, the Institute initially worked on Phases I and II of the "Street Project," i.e., an analysis of street situations and the design of a prototype. Anderson, who was studying urban structures in the United States, took a field trip to Savannah, Georgia, with a group of MIT students to analyze the historical development of the downtown street grid, which dated back to colonial urban planning in 1733, as a paradigm for the structural relationships between the development of the built environment and American society, a basic research endeavor he continued in New York.[145] Rykwert, on the other hand, was simultaneously working on a publication on the history of the street, initially approached from an art historical perspective and an etymological derivation of the word "street."[146] The publication, as the intended final product of the "Streets Project," was agreed at the Institute to be less of a practice-oriented handbook, such as what HUD was aiming for, and more of a scholarly anthology, which would include not only the Fellows' essays, but also texts from other disciplines such as anthropology, environmental psychology, and sociology.

When the Institute produced an interim report on Phases I and II of the "Streets Project" after the first year, this communicated that there was still no agreement on the methodological basis on which urban streets should be researched. The call for a "generalized approach" conflicted with the insight that only a "specific and differentiated research methodology" would lead to applicable results. One of the key sections of the report was the presentation of Anderson's research, which argued that order (street grid) always influences

145 Stanford Anderson, "Studies toward an Ecological Model of the Urban Environment," in Anderson, ed., 1978, 267–307.

146 Joseph Rykwert, "The Street. The Use of Its History," in Anderson, ed., 1978, 14–26.

structure (use). Working from a broader concept of architecture based on continuity from the individual building to the street to the city, the Institute worked with interviews and figure-ground diagrams, in addition to mapping street systems, to determine the extent to which streets were seen as positive components of urban structure and to suggest formal interventions in the existing street grid. Finally, the report proposed new processes and methods on the basis of which individual streets or entire street systems could be analyzed and become the subject of urban renewal. In line with the multidisciplinary approach of the "Streets Project," the proposals reflected socio-political, institutional, legal, and economic aspects of planning. For this work, the Institute received a first tranche of US$215,000 from HUD by the end of August 1971—a not inconsiderable amount. At the same time, however, the desired multidisciplinarity of urban studies at the Institute, despite all the good intentions of the Fellows, threatened to fail at the outset. The problem may have been that the "Streets Project" was almost exclusively carried out by Visiting Fellows and Research Associates, and Eisenman terminated the contracts of four employees at the end of the 1970–71 fiscal year, with Frank and Cromley, among others, being dismissed. In the end, however, besides the fine line that had to be navigated between authority and guidance, competition and solidarity, it was primarily personal misconduct, in addition to other commitments on the part of individual team members, that threatened the continuation of the project and the second tranche. Eventually, Ellis, who had quickly risen in the Institute's ranks, had to be removed from his role as project lead at the insistence of HUD officials, having made disparaging remarks about the contractor while walking out of earshot, before the Institute was able to continue the project and enter the design phase. The interim report, in other words, was prepared by Wolf, who by now had risen to become a full-fledged team member. While Rykwert left the Institute in the summer of 1971 due to personal differences with Eisenman over withheld wages, Anderson, who had just been unanimously appointed a Fellow in April 1971, was already departing again that fall for Cambridge, where he was involved in establishing the doctoral program in History, Theory and Criticism of Art, Architecture and Urban Form at MIT.[147] On top of that, there had been a dispute between Rykwert and Anderson that was eventually settled through the Institute's lawyers, Rubenstein, Nash & Co.

Despite all the disruption, the "Streets Project" entered Phase III in the 1971–72 academic year, when the results of the analytical and prototypical studies were to be projected onto a specific area. An area in downtown Binghamton, Upstate New York, had been selected by HUD for this purpose, which fell under the local Model Cities Program. The implementation was to be financed by grants

147 Stanford Anderson, "HTC at MIT: Architectural History in Schools of Architecture," in *Architektur weiterdenken. Werner Oechslin zum 60. Geburtstag*, eds. Sylvia Claus, Michael Gnehm, Bruno Maurer, Laurent Stalder (Zurich: gta Verlag, 2004), 330–338; see also Harwood, 2013.

that dated back to the social policy agenda of the Johnson administration, i.e., which were designed to bring together urban development and poverty reduction. The Institute's work thus took on instant realpolitik weight. With an actual design project close at hand, Eisenman declared the "Streets Project" a top priority and took over the management of the project himself, together with the architect Vincent Moore, whom he had brought to the Institute specifically for this purpose. For a year, a newly assembled team at the Institute worked on a carefully designed study. The team once again included Wolf, who was officially responsible for "legal administrative, economic planning," and first-time member Robert Gutman, who contributed "social planning." In addition, several postgraduate Research Associates worked on the "Streets Project" again, their compensation on this occasion funded by a newly established Graham Foundation Scholar Fund at Eisenman's request.[148] The Institute's goal in applying for, awarding, and executing the "Streets Project" contract was greater visibility, more expertise, and more contracts; this was evident from the value that was now placed on the professional implementation and monitoring of public relations activities. Frampton was brought in specifically for the "Streets" exhibition, to which MoMA remained committed. In October 1971, Arthur Drexler presented specific plans to the Institute's Board of Trustees for an exhibition that would focus on the role of the community and on street design as an instrument of urban renewal, following the original idea behind "The New City" exhibition. This was added to the museum's official exhibition program as #254 and scheduled for spring 1972. Meanwhile, in 1971–72, Anderson began supervising the editing of the final report and was ultimately responsible for the publication of the research findings, an anthology that would be years in the making. Most notably, the Institute had contracted with MIT Press, where Anderson sat on the editorial board, to produce the catalogue for the exhibition—this was its first collaboration with the academic publishing house, one that would later be successfully continued and expanded. Thus, all the Fellows were involved in an Institute research and design project in fiscal year 1971–72. Each team member's share in the work on the "Streets Project" was reflected in

148 In March 1971, Eisenman applied for additional grants from the Graham Foundation to establish a Graham Foundation Scholars' Fund. In his letter to John Entenza, he explained his request as follows: "This fund would be used for a variety of needs; to enable the Institute to bring people for short periods of time for special seminars; to send graduate students to other institutions for limited periods of time; to pay for unpredicted expenses for fellows at the Institute, such as making slides or incidental typing for a lecture, for attending conferences." Eisenman stated that he would be responsible for administering the grant at the Institute himself. The Graham Foundation promptly approved the application and awarded the Institute an additional US$ 10,000 for the 1971–72 academic year in addition to the Graham Foundation Visiting Fellow funding. However, Entenza made it a condition that only research assistants of the Institute should be paid directly from the Scholars' Fund to remunerate them for their work or at least to pay expenses incurred. In 1971–72, a total of eleven students received support. These included four postgraduate students at the Institute, who received the full grant amount of US$ 1,000 each: Duarte Cabral De Mello, Gregory Gale, Thomas Schumacher, and Victor Caliandro. Eisenman also billed honoraria for guest lectures and expenses for a publication through the fund. The Graham Foundation thus financed not only the Institute's human capital, but also its coffers.

the estimated budget. In fiscal year 1971–72, personnel costs were calculated as 160 daily rates for Eisenman, Frampton, and Anderson (they were to make about US$15,000 each), 100 for Moore, 80 for Wolf, and 20 for Gutman; hourly rates, however, varied by position and degree of professionalism.[149] The fact that, with the disbursement of the second tranche of US$155,000 and the request for a third tranche of another US$75,000 in the fall of 1971, it was possible to finance most of the Institute's operations through the HUD contract, was a crucial factor in securing the Institute's operations.

As a result of a hitherto inconceivable pragmatism, the "Streets Project" evolved into an intensively collaborative effort among the three partners: the Institute as contractor, HUD as client, and the City of Binghamton as testing ground for planning and design approaches to urban renewal.[150] The Institute's project team now worked closely with HUD officials, and regular meetings were held in Washington, D.C., and New York to discuss interim results and bring the extensive project to a successful conclusion. The Fellows traveled to Binghamton once a month to meet with various stakeholders in the city, such as the Urban Renewal agency, the mayor's office, the Broome County Planning Department, the Association of Business Owners, and finally the residents of the Model Cities area, to define planning goals. The project team used a wide variety of methods to collect data in the study area: formal and morphological studies of the physical shape of streets and intersections, spatial planning studies of traffic flow and density of use, and sociological studies of perceptions of the downtown and residential environments. In early 1972, the Institute also experimented with new, innovative participatory planning methods. One of them was called the "Streets Game" and it was used to simulate and prioritize planning decisions, focusing on streetscape design rather than a revision of the land use plan. After that, axonometric drawings were used to question residents about their desires and needs for street design. But unlike other forms of participation established in the United States in the late 1960s, the "Streets Project" was not about empowering underrepresented and disadvantaged populations. In the end, participation played only a subordinate, project-strategic role: residents were simply blindsided by the new methods, and the local planning agency preferred to trust in conventional methods. Ultimately, HUD was primarily concerned with the political benefits of the project and actionable outcomes. The bureaucratic burden on the "Streets Project" remained immense. The Institute, on the other hand, had readily spent much of the project budget on day-to-day operations and personnel expenses, leaving virtually nothing for the concrete realization of projects. This kind of urban renewal and Institute policy did not go uncriticized. In early

149 Organizational Chart. Demonstration Phase 3, October 1, 1971. Source: CCA Montréal, IAUS fonds: B.3-4 / ARCH401248.

150 Anderson, 1978; see also IAUS, "Demonstration Project: Streets in the Central Area of a Small American City," in Anderson, 1978, 339–375.

1971, Sarah Rubin and Steven Goldstein, two student employees who were particularly committed to social issues, submitted a written complaint addressed to Fellows, staff, and HUD representatives claiming that government funding had simply been wasted.[151] This was the second time after 1969 that students rebelled against the practices of the Institute's leadership, making their voices heard by openly criticizing the elitist position and arrogant attitude of the Fellows. But they also made constructive suggestions on how communication and cooperation could be improved, although their short stay at the Institute of only one year prevented them from having a lasting corrective effect.

Although the relationship between the two cooperation partners had become strained, in the end, Eisenman's top priority as Institute director was to achieve presentable results to deliver to HUD at the end of the project. After Eisenman was able to supplement the contract with HUD in early 1972, the Institute, equipped with an additional tranche of US$37,000, eventually delivered three theoretical models for the revitalization of the inner city. These models operated at different spatial scales—city, street, building—to answer the question of what constituted a good street. In "Model A," the project team used an analysis of the urban context to outline possibilities for spatial planning interventions at the level of the entire street system as well as individual streets and made recommendations for traffic planning and land use to optimize urban space. In "Model B," they also proposed the development of a specific street into a pedestrian zone. This planning proposal was based on conversations with local stakeholders and developed in collaboration with a course taught by Anderson at MIT. Accordingly, urban space was treated as "transactional space," based on Anderson's concept, and defined at the first-floor level as semi-private but open to the public. The urban street was to be enlivened by commercial activities. "Model C" was ultimately a concrete plan for the structural redesign of a street, with Eisenman's design team, including Victor Caliandro (MIT graduate) and Thomas Schumacher (Cornell University graduate) as Research Associates, introducing two prototypes for urban living, both with distinctive façade designs, intended to appeal primarily to the white middle class with the possibility of ownership. The design envisioned the two types of urban houses each flanking one side of the street: a four-story multi-unit building with four duplex apartments on one side, and a three-story townhouse on the other. But despite formulating three models, the Institute's "Streets Project" failed to progress beyond an intellectual exercise.

The design for the two types of houses illustrated that, in order to specifically upgrade an inner-city street into a leafy, purely residential neighborhood, the

151 Sarah Rubin and Steven Goldstein, letter to Institute Fellows, staff, and HUD representatives, January 16, 1972. Source: CCA Montréal, IAUS fonds: B.3-4; Richard Manna, proposals for a better relationship between the Institute and students, February 2, 1972. Source: CCA Montréal, IAUS fonds: B.3-4.

Institute continued to assume area-based demolition measures, and "Model C" thus did not differ very much from the approach and urban vision of the urban renewal of the previous decade, but in terms of scale and ownership. The socio-spatial context of the neighborhood was completely disregarded, and the problem of land expropriation and acquisition was not even addressed. Thus, the final report merely referred to the Institute's design testifying to a greater awareness of the contrast between private and public space. It explicitly emphasized that horizontal façade elements and the recessed building volume were intended to represent ambiguities in terms of territoriality defined by private property and residential use. Ultimately, this proposal—the Institute's first targeted building project—testified to a one-sided architectural approach rather than a multidisciplinary urban studies perspective, as it did not aim at implementing the ideals of public engagement or representative democracy to improve urban quality for diverse stakeholders, e.g., by guaranteeing mixed land use or accommodating a heterogeneous population. This limited understanding of the public sphere manifested itself in the fact that the types of houses developed relied entirely on capitalist urbanism, on the real estate market as the central mechanism for regulating urban space, and on attractive home ownership as the economic motivation for urban renewal. Thus, only exclusive functions of the street were considered. Although the City of Binghamton expressed an interest after the completion of the "Streets Project"—Mayor Alfred Libous personally lobbied for it—and the building plot was available, political, and economic considerations ultimately prevented the exemplary realization of the two prototypes in the designated Model Cities area. Even a written request from Walter Thayer, an influential MoMA trustee, to George W. Romney in his function as U.S. Secretary at HUD, to approve the budget for the building project could not change this. MoMA's "Streets" exhibition, which had been repeatedly scheduled over the years and postponed several times, was finally canceled in February 1973 on the grounds that a photo series that was to form the basis of the exhibition had not been produced yet. The publication originally planned for the exhibition eventually became an independent project of the Institute, with Anderson in his capacity as a Fellow, as the main editor. When *On Streets* was finally published by MIT Press in 1978, it was a substantial, comprehensive volume of research with numerous previously unpublished essays on the history and theory of the street. While it did have a definite influence on the architectural and planning debate, it was also slightly outdated by the time it was published.

Diversification of Activities

For a moment, with the end of the HUD contract in sight, it looked as if the Institute had finally abandoned its social goals. When Eisenman presented the Institute's future work, explicitly as an architectural think tank, at the annual meeting of the Board of Trustees in October 1971—a facility inspired by comparable institutions in the American political establishment—to further diversify research and design activities and to better incorporate the individual research interests of the Fellows, this entailed a fundamentally programmatic reorientation. To the trustees, he outlined four future research areas: 1) “Urban Components,” 2) “Urban Settlements,” 3) “Theoretical Studies. Individual,” and 4) “Theoretical Studies. Group.”[152] The research findings of the “Streets Project”—“streets as urban components,” “houses” (handwritten correction to “housing” in Eisenman’s preliminary minutes), “special building types”—were listed under the first rather than the second item, following an architectural way of thinking. The individual research projects of the Fellows (in addition to Ambasz, who was now only at the Institute on a limited basis, Anderson, and Frampton) were initially listed under the heading “Models for a Regional City;” in retrospect, Eisenman simply subsumed them under “Theoretical Studies.” Thus, for the first time, a fundamental distinction was made between the urban and the theoretical, between individual and group projects. The decisive factor was the statement that the Institute, which in the years before had basically acted as a “one project institution,” was now working on several research projects at the same time. In view of the economically strained situation, the decision to position the Institute as an extra-academic research center was also aimed at earning money in the future primarily through the production of architecture and theory. In his “Director’s Report,” Eisenman painted a thoroughly positive picture and predicted a balanced budget.[153] However, Drexler felt compelled to correct this picture and, in view of liabilities amounting to US$45,000, to point out the seriousness of the situation.

Nevertheless, Eisenman defined the Institute as a “think tank” at the meeting and presented his “Program in Generative Design” there for the first time. This was a theoretical group project for which an application for funding had been submitted to the National Institute of Mental Health (NIMH).[154] Eisenman found the ideal comrades-in-arms for his project of developing a universal theory of architecture in the Argentinean architect Mario Gandelsonas and his partner Diana Agrest, both of

152 IAUS, minutes of the meeting of the Board of Trustees (unofficial and official), October 5, 1971. Source: CCA Montréal, IAUS fonds: A.2-3 / ARCH401120 & ARCH401121; Notes on the Fellows Meeting, October 20, 1971. Source: CCA Montréal, IAUS fonds: A.2-2.

153 The financial report for 1971–72 shows that revenues of US$ 257,257 were offset by expenditures of US$ 235,335. In addition to income from contracts with HUD and UDC, the Institute had received grants from the New York State Council on the Arts (NYSCA), donations from private individuals, and grants from private foundations. The Gottesman Foundation of Celeste and Armand Bartos donated a total of US$ 40,000, Lily Auchincloss and her Van Amerigen Foundation jointly donated US$ 25,000, and the Graham Foundation gave US$ 15,000. Another US$ 11,000 was raised by renting out unused space. Debt management was not included in the budget: in fiscal year 1970–71, debts totaled US$ 46,472.80; in 1971–72, they were to increase to a total of US$ 85,370.72, and in 1972–73 to a total of US$ 130,140.77. Eisenman estimated in late February 1972 that the Institute’s debt should be all but eliminated by the end of fiscal year 1972–73 (which was not the case).

154 The National Institute of Mental Health (NIMH) was an American research center for the study of mental disorders, the largest of its kind in the world. NIMH was under the purview of the U.S. Department of Health, Education and Welfare (HEW), and its research aimed at developing a better understanding of and new treatments for mental disorders. It is not known how the Institute’s grant application was initiated or what interest the NIMH had in architecture.

whom had only moved from Paris to New York at the beginning of 1971 and were immediately accommodated at the Institute, because of their knowledge of contemporary (post)structuralist theory and French philosophy. While Eisenman officially emerged as the leader of the project, Gandelsonas was listed as co-leader and Agrest as Research Associate. The fourth member of the group was the Portuguese architect Duarte Cabral de Mello, who had earned considerable merit in the "Streets Project." The titular "Generative Design" was a direct reference to American linguist and public intellectual Noam Chomsky's theory of generative grammar, which Eisenman referred to in his theoretical texts at the time. The project aimed to use a semiotic or linguistic approach to analyze the effects of built environments on people in terms of communicative properties. Here, Eisenman ultimately conflated individual and institutional interests.[155] In a brochure published by the Institute in 1971, he stated that he had already been working privately on "Syntactic Structures. The Logic of Form in Architecture," i.e., on the application of linguistic explanatory approaches to architecture, since 1968.[156] In the fall of 1971, while Eisenman, Gandelsonas, and Agrest, were pursuing these theoretical advances, Frampton began working on a new joint research and design project that would tie in with the "New Urban Settlements" study and also, if possible, be placed at the UDC. The project, titled "Low-rise, High-density Suburban Land Settlements," involved investigating settlement patterns that explicitly referred to the suburban space; Anthony Vidler, who in the academic year 1971–72 was the second person to receive Graham Foundation Visiting Fellow funding, along with Gandelsonas, but who otherwise taught at Princeton, wrote an initial concept paper on low-rise, high-density housing, where he developed architectural and urban design alternatives to large-scale housing. However, Eisenman's move to define the Institute as a research center marked the first shift in its role as a project office away from its original intention of ideally translating each research and design study into a building project. Of course, rather than traditional architects, the Fellows at the Institute were working as a new type of academic, representatives of an emerging functional elite of knowledge workers or designers, as the boundaries between the traditional discipline and profession of architecture increasingly dissolved.

The further activities of the Institute in 1971, as presented by Eisenman in his "Director's Report" earlier that year and communicated in a first multiple-page brochure, were game-changing in that they represented the first genuine combination

155 Noam Chomsky developed his theory of transformational grammar in the 1950s. In the early 1970s, Eisenman's transferal of Chomsky's approach to architecture was primarily concerned with two questions: "The structure of form and how form generates meaning?" and "The structure of meaning and how form generates form?" IAUS, minutes of the meeting of the Board of Trustees, October 5, 1971. Source: CCA Montréal, IAUS fonds: A.2-3. Humans and the use of architecture are not considered in this transferal of a theory. Team members could not agree on Chomsky as a point of departure.

156 IAUS, brochure, 1971. Source: CCA Montréal, IAUS fonds: A.2-9.

of architectural, educational, cultural, and editorial practice in the history of the Institute. The space for this had been in place for a year, even if the main hall was only used sporadically at first. The Institute was to benefit from the opportunities offered by the special relationship between academic education, technical training, internships, and office work in the training of aspiring architects.[157] A new market for architectural education was tapped by establishing a Student Internship, for which the Institute entered into a cooperative arrangement with the Great Lakes Colleges Association (GLCA), an association of liberal arts colleges on the northeastern seaboard of the United States.[158] Within the existing Arts Program in New York, undergraduate students who did not have the option of studying architecture at their home colleges were offered the opportunity to complete a six-month internship at the Institute, where they could gain first-hand work experience by collaborating on group and individual projects. The initiative for the GLCA's cooperation with the Institute came from Richard Wengenroth, who taught in the Fine Arts Department at Ohio Wesleyan University and had established the first contact; initially, Oberlin College acted as a clearinghouse to arrange the internship.[159] In April 1971, the Institute hosted the annual meeting of colleges organized in the Association of Collegiate Schools of Architecture (ACSA). The Institute's invitation underscored its ambition to expand its internship offerings and already staked a territorial claim on the entire United States. In fiscal year 1971–72, the Institute was also awarded a US$16,000 grant by the National Endowment for the Arts (NEA) for the organization and implementation of internships.[160] Postgraduate students organized the first public series of events at the Institute in the spring of 1971: the

157 Bernard Spring gave a lecture on architectural education at City College in April 1971 that addressed this relationship.

158 In the early 1970s, the Great Lakes Colleges Association (GLCA) included twelve colleges: Denison, Antioch, Wooster, Oberlin, Ohio Wesleyan, Kenyon, DePauw, Earlham, Wabash, Hope, Kalamazoo, and Albion. As early as the 1960s, the GLCA colleges had established off-campus programs at home and abroad, including the Arts Program in New York, founded in 1968 by faculty from various fine arts departments as an experiment in alternative educational programs. As one of the co-founders and its first director from 1968 to 1973, Wengenroth was responsible for ensuring that GLCA students had a choice of opportunities to intern with an artist or at a cultural institution. The collaboration with the Institute continued this practice.

159 In the 1971–72 academic year, five interns came from Oberlin College: Le Roy "Sandy" Heck, Geoffry Koper, Frank Nicoletti, Glenn Oberlin, Julian Smith. The following year there were two: Richard Dean and Richard Wolkowitz. Beginning in the 1973–74 academic year, the Fine Arts Department at Ohio Wesleyan University cooperated with the Institute. The contact person there was Marty Kalb.

160 In the annual report of the National Endowment for the Arts (NEA), the internship program at the Institute (grant number: A-72-0-508) was summarized as follows: "Research in ways to make architectural education more stimulating—especially during the internship period between graduation and licensing—was conducted under fellowship granted to two graduate students in architecture." "Architecture and Environmental Arts, Professional Education and Development," in National Endowment for the Arts, *Annual Report Fiscal Year 1972* (Washington D.C., December 1972), 54, https://www.arts.gov/sites/default/files/NEA-Annual-Report-1972.pdf (last accessed: May 31, 2023).

"Student Fellows Lecture Series," featuring lectures by Victor Caliandro, Elizabeth Cromley, Suzanne Frank, and Susana Torre.[161] Once established, this continued as a regular lecture series with the "IAUS Spring Lectures," where locally based architects and academics could present their projects and positions.[162] In 1971, having organized CASE 8, the last meeting of the Conference of Architects for the Study of the Environment at MoMA in May, thus laying Eisenman's previous group to rest, the Institute, which in May of that year had for the first time lived up to the claim set forth in its original charter as a comprehensive educational and cultural institution—in Eisenman's words, "the Institute [was] just emerging in a creative role in education, research development and in community affairs."[163] Here it became clear that the agility of the Institute's leadership and the flexibility of its organizational and programmatic structure were now instrumental in enabling it to realize its full potential. Its ability to rapidly network with the established universities and museums in New York, various art and cultural institutions, public and private foundations, and influential and wealthy private individuals, was nothing less than epoch-making, at least in the history of the Institute, if not American architecture culture as a whole. By the early 1970s, a complex web of relationships had been established around the Institute as a "networked" actor, which not only provided its individual projects with an institutional anchor and financed their operations, but also contributed to their reputation.

Moreover, the Institute now cooperated with New York-based institutions, as well as internationally renowned ones, on major events. For example, in June 1971, the Institute opened "Art & Architecture USSR. 1917–31" (June 3 to 18, 1971), a traveling exhibition conceived by Otto Das, Gerrit Oorthuys, and Max Risselada at TU Delft, and subsequently shown at TU Berlin, Harvard University's Carpenter Center for Visual Arts, and Princeton before finally coming to the Institute. It was a first reappraisal of Russian constructivism from a Western perspective and certainly a groundbreaking exhibition for further research. At the Institute, a model of Le Corbusier's design for the Soviet Palace

161 Gale, a Cooper Union graduate, who was now responsible for organizing the "Spring Lecture Series" in 1971, had previously worked at the Institute beginning in the 1969–70 academic year for one of the post-graduate research associates, including on the "New Urban Settlements" study. For Eisenman, he also worked on the designs for *House I* and *House II* and on his Terragni study. In 1971–72 he was remunerated from the Graham Foundation Scholars' Fund.

162 In 1972, professors from Cornell University (O.M. Ungers, Werner Seligman, Fred Koetter, Colin Rowe) gave guest lectures at the Institute; in 1973, the following people gave lectures: Craig Hodgetts, R.T. Schandelbach, Stuart Cohen, Henry Wollman, James Doman, Peter Anthony Berman, Alan Chimacoff, Lance Brown, Michael Wurmfeld, Craig Whitaker, Alex Cooper, Michael Pittas. Then, in 1974, the "Spring Lecture Series" was organized for the first time by Robert Stern, who taught at Yale University and then Columbia College and was a colleague and friend of Eisenman.

163 IAUS, official minutes of the meeting of the Board of Trustees, October 5, 1971. Source: CCA Montréal, IAUS fonds: A.2-3.

(1932) had been made available from the MoMA Archives for exclusive display. Frampton reassembled the exhibits along with Mantel, and Frampton and Risselada edited a new catalogue with a cover designed by Robert Slutzky. This was published by Wittenborn Art Books.[164] The exhibition featured constructivist art and architecture of the Soviet revolutionary years that had never been seen in New York before, not only for their modernist aesthetics but explicitly in terms of social renewal in the Soviet Union. With this theme, the Institute offered a provocative challenge to the American architecture and art world, thus assuming a pioneering and mediating role in the cultural Cold War. But the exhibition also showed significant differences between the Institute and all the alternative art spaces that were founded from the vibrant New York art scene at the time as a critical counterpoint to the major museums—not only structurally and organizationally, but also in terms of professional strategies and goals, such as cultural and entrepreneurial policies. Compared to the Institute, these anti-institutional spaces, such as 112 Green Street (1970, founded by Gordon Matta-Clark), the Film Anthology Archives (1970, by Jonas Mekas), Food (1971, also by Matta-Clark), The Kitchen (1971, by Woody and Steina Vasulka) the Institute for Art and Urban Resources (1971, by Alana Heiss, almost a namesake of the Institute), and Artists' Space (1972, by Trudie Grace and Irving Sandler), were all experimental, sometimes ephemeral spaces.[165] What these rather informal art spaces had in common was that most them of received funding from one of the two major public art foundations, be it the New York State Council on the Arts (NYSCA) or the National Endowment for the Arts (NEA), i.e., precisely those two funding bodies that the Institute also targeted for the development and financing of its public events, and specifically exhibitions.

The Institute's emergence as a "think tank" was manifested in a special issue of the Italian magazine *Casabella*, for which the Institute had taken over the guest editorship, in the late fall of 1971.[166] The double issue entitled "The City as an Artifact," for which Frampton was responsible on the Institute's side,

164 IAUS, ed., *Art and Architecture. USSR. 1917–32* (New York: Wittenborn Art Books, 1971). The catalogue was primarily intended to be an exhibition guide, but also included translations of original texts, including those by El Lissitzki on the *Cloudprop* and by Moisei Ginzburg on the *Narkomfin* Communal House, and an extensive bibliography.

165 Julie Ault, *Alternative Art, New York, 1965–1985: A Cultural Politics Book for the Social Text Collective* (Minneapolis: University of Minnesota Press, 2002); see also Lauren Rosati and Mary Anne Staniszewski, eds., *Alternative Histories. New York Art Spaces 1960 to 2010* (Cambridge: MIT Press, 2012).

166 *Casabella*, no. 359/360 (November/December 1971): "The City as an Artifact." Thomas Czarnowski contributed the collage for the cover, which showed the Vitruvian man above the Manhattan street grid, while revealing a view of the city's infrastructure as its guts, from which single-family homes are excised. Stuart Wrede, inspired by Claes Oldenburg's sculptures, designed a collage of an oversized fountain for St. Peter's Square in Rome in the form of a tulip, a political sculpture, as a graphic-art contribution to the issue.

was an early form of self-reflection and self-promotion. This was the Institute's first introduction to a European readership. The Institute edited the content but also did the artwork—both the cover design and the illustrations. *Casabella* was the ideal medium for the Institute's European debut since it had had a decisive influence on the radical avant-garde architecture and planning discourse in Italy under Ernesto Nathan Rogers (1953–1965) in the early 1960s, with contributions by Aldo Rossi and Manfredo Tafuri, among others. In the editorial by then editor-in-chief Alessandro Mendini (1970–1976), *Casabella* promised its readers insights into a genuine American debate about the legitimacy of a rational but at the same time nostalgic approach to architecture. The bilingual edition, however, did not feature research and design projects at the Institute, such as the "Streets Project" or the "Program in Generative Design." Instead, it followed a twofold strategy with the editorial layout highlighting opposing positions and incorporating criticism and individual contributions giving space to very different schools and methods. Frampton divided the special issue into three sections:[167] The first part, "A Cultural Debate: The Existing Situation," was a debate between Denise Scott Brown, the only female contributor who furthermore was not associated with the Institute, and himself, in which both accused each other of populism and elitism, respectively. While clearly staged, this debate nonetheless set the tone for the American architectural discourse. While Scott Brown, in "Learning from Pop," elaborated on her central arguments for a formal analysis of landscapes shaped by consumer culture and the automobile as the basis for sign architecture, Frampton, in "America 1960–1970. Notes on Urban Images and Theory," drawing on contemporary sociological and political theory, railed vehemently against precisely this form of pop architecture, which for him not only carried the grave danger of canonizing kitsch but also had to be viewed in conjunction with the consumer and affluent society.[168] The second part, "A Dialectical Aspect. The City as an Artifact" constituted the main section, with five articles by young American architectural theorists and historians, in which Eisenman, Joseph Rykwert, William Ellis, Stanford Anderson, and Thomas Schumacher all appeared as authors associated with the Institute in one way or another. While the main body of the issue juxtaposed diverse contextual, conceptual, and largely artistic positions in architecture and urbanism, Eisenman's essay "Notes on Conceptual Architecture. Towards a Definition" was showcased as the first and thus seminal contribution. Once again blurring the lines between his role as Institute director, architect, and theorist, Eisenman spoke out against a "social or technological polemic," and by equating architecture with art, placed his practice in the tradition of American minimalism of the

167 Denise Scott Brown, "Learning from Pop" & "Reply to Kenneth Frampton," *Casabella*, 1971, 15–24 & 41–47.

168 Kenneth Frampton, "America 1960–1970. Notes on Urban Images and Theory," *Casabella*, 1971, 25–40.

1960s, explicitly conceptual art. With his characteristic rhetoric, Eisenman justified his formal approach by making explicit reference to Noam Chomsky's now classic approach to structural linguistics, specifically the distinction between surface and deep structure.[169] Picking up on the "linguistic turn" in the humanities, Eisenman was less concerned with theory than with what literary scholar Harold Bloom, in *A Map of Misreadings* (1975), called processes of constant appropriation and "creative misreading." In this regard, he proposed to explain the architectural object exclusively on the basis of its formal properties, i.e., in terms of auctorial conception, rather than individual perception. The fact that he was concerned only with the syntax and, at the very most, the semantics of architecture, i.e., grammar and morphology vis-à-vis meaning, but not pragmatics, which he left out entirely, shows that he was ultimately working from a highly reductive understanding of general linguistics. Finally, the third section, "Institutions and Artefacts for a Post-Technological Society," comprised no less than three texts by Emilio Ambasz, with which he presented his nascent "Universitas Project."[170] Having provided decisive impulses at the Institute throughout its first years, by his own account even during its founding period, Ambasz' turned his attention to a new design academy for the new society, the Universitas Project. He initially conceived this as a discursive and cultural project with the Institute as a cooperation partner, characterizing it as "post-technological" (in the vein of Alain Touraine or Daniel Bell), because technological progress was the decisive factor. A long-planned project that he ultimately carried out on his own, it finally culminated in an international, high-profile, and in the truest sense trans-disciplinary conference at MoMA in early 1972.[171] Ultimately, there were aspects that made the *Casabella* issue stand out

169 Eisenman, 1971.

170 Emilio Ambasz, "I The University of Design and Development," "II Manhattan: Capital of the Twentieth Century," "III The Designs of Freedom," *Casabella*, 1971, 87–99. Ambasz announced the Universitas Project in 1971 with text publications in both *Casabella* and *Perspecta*, no. 12/13, as well as a lecture at the "Architecture Education U.S.A." conference.

171 To discuss the possibilities of knowledge production in a post-technological society, Ambasz had invited an illustrious crowd of architects, designers, philosophers, semioticians, sociologists, etc. Louis Althusser, Jean Baudrillard, Manuel Castells, Umberto Eco, Hans Magnus Enzensberger, Roman Jakobson, Michel Foucault, Henri Lefebvre, Thomas Sebeok, Susan Sontag, and Alain Touraine, among others, all responded to his invitation, issued a written statement that they then elaborated on during the conference at MoMA, establishing a new form of intellectual work. Architecture critic Martin Pawley reported on the "Universitas Project" in *Architectural Design* with astonishment; see Martin Pawley, "Universitas. Martin Pawley Reports on the University that Never Was," *Architectural Design* (April 1972), 214–215. The stated aim was not only to express a critique of the role of institutions in society, especially universities, but also to formulate a political task for design, and thus also for architecture, against the background of changing technological, economic, and social conditions; see Emilio Ambasz, ed., *The Universitas Project. Solutions for a Post-technological Society* (New York: The Museum of Modern Art, 2006); see also Felicity Scott, "On the 'Counter-Design' of Institutions: Emilio Ambasz's Universitas Symposium at MoMA," *Grey Room*, no. 14 (Winter 2004), 46–77. It is worth noting that Eisenman had not only received a Graham Foundation grant for

and marked it as relevant for the architectural debate. The first was the editorial strategy of opposition, i.e., the confrontation between completely different approaches and contradictory positions, first devised here and later perfected by the Institute; the second was the work of the Fellows, Visiting Fellows, and Research Associates, most notably Frampton, Eisenman, and Ambasz, which was presented here in its full diversity, thus gaining international recognition, and caused a sensation.

Although the Institute did not yet have its own journal at that time, this first major publication in one of the most astute European architectural communities also marked the beginning of its publishing career. With its guest editorship of the *Casabella* issue, the Institute could finally begin to employ all those strategies of creative self-promotion and cultural valorization that had characterized heroic modernism. "The City as an Artifact" was the Institute's calling card and ultimately culminated in a three-page article about the fledgling institution, presenting the background leading up to its foundation, its principles and objectives, its general structure, and its programs and areas of research, in rather unwieldy, conspicuously institutional language. This is where the master narrative of the Institute as a "true" institution was reestablished and further disseminated.[172] The appended professional, educational, and journalistic biographies of the authors, showcasing all the social, cultural, i.e., symbolic (if not economic) capital, underscored the Institute's quasi-institutional orientation. The claim to be a serious research center, however, was countered by an ironic photo collage with the faces of the sixteen Fellows, Visiting Fellows, and Research Associates at the Institute mounted on uniform bodies, kneeling in the front row and standing in the back. This chosen form of self-presentation, similar to a soccer team photo, was simultaneously a direct or indirect reference to legendary group photos of other institutions or organizations, such as the Bauhaus, CIAM, or Team X, an attempt to carry on this venerable tradition, and a caricature of the same, and testified to the fact that work at the Institute had its share of amusing episodes. The collage, a collective form which, in addition to referencing a humanist understanding of architectural history—the Institute's self-designed logo was again emblazoned on the sweatshirts, which now promoted the ideal image of the Vitruvian man in his home country and beyond—reflected a consensus, again expressed Eisenman's fondness for professional sports. In fact, he happily declared himself to be the biggest fan of modern team sports, which historically originated in industrialized England.

a project of the same name in 1966, when Ambasz was still his student at Princeton, but also commented on concept papers in advance. Ultimately, however, the "Universitas Project" was not realized at the Institute and never to the extent originally planned.

172 Joseph Rykwert, "The Institute for Architecture and Urban Studies," *Casabella*, 1971, 100–103. The article on the Institute was reprinted without giving an author's name, which emphasizes its collective and institutional character. However, the bibliography on publications by and about Peter Eisenman compiled at CCA lists Joseph Rykwert as the author.

Despite the public image as a mixed-gender team, however, this self-portrait masked all the social relationships and intellectual differences that nevertheless prevailed among the Fellows, Research Associates, and staff, such as the hierarchical organization was certainly still present and continued to shape their gendered working conditions. And so, although the collage paints the picture of Eisenman as player and coach, the role he held at the Institute, to continue with this image, was probably that of coach and manager.

The public image of the Institute communicated by *Casabella* anticipated its future direction under Eisenman's continued leadership. The development of the Institute in the following years, with its many breaks, ruptures, turning points, and opportunities, both offered and missed, exemplified a general development in architecture culture, namely that under the changed conditions of an information and knowledge society, new forms of immaterial work increasingly came into play, which went hand in hand with new forms of organization and capital. As a new type of institution in the field of architecture, the Institute went on to shape the discursive formations and, above all, the cultural configurations that manifested themselves in the new service economy and in processes of cultural value creation. Before the end of 1971, however, the Institute's activities had already culminated in hosting the conference "Architecture Education U.S.A.: Issues, Ideas and People" (November 12 & 13, 1971) organized in conjunction with an exhibition of Cooper Union student work at MoMA—with a supporting program at the Institute.[173] Once again, the idea for the conference did not originate with the Institute but was brought to the attention of Eisenman as Institute director from outside. While all the guests who had been invited from Europe canceled, giving the event a purely American setting, Bernard Spring and Robert Geddes, the two professors who had hitherto shaped the debate on architectural education in the USA, announced their attendance. At a time when the post-1968 politicization of architecture schools was gradually being reversed and replaced by a move towards redisciplining and academization, the conference offered a powerful representation of the current state of debate simply by virtue of the abundance and quality of the young professors participating in it: in addition to the Institute's Fellows—besides Peter Eisenman, Anderson, Ambasz, Frampton, and Vidler all gave a presentation—the speakers included Jonathan Barnett, Oswald Mathias Ungers, Colin Rowe, and Denise Scott Brown, all of whom shared their pedagogical principles. In the spirit of a transdisciplinary exchange, two sociologists, Herbert Gans and Robert Gutman, were also involved. Eisenman's contribution was particularly

173 IAUS, ed., *Architecture Education U.S.A.: Issues, Ideas, and People. A Conference to Explore Current Alternatives* (New York: The Institute for Architecture and Urban Studies, 1971). Source: The MoMA Archives. See also The Cooper Union, *Education of an Architect: A Point of View* (New York: The Cooper Union, 1971).

vocal as he used the conference as a stage to pit intellectual speculation against sociological analysis, form against function, theory against practice—and ultimately to cast ridicule on any claim to reality. In his speech as host, Eisenman disputed the meaning of the conference title "Architecture Education U.S.A." and polemicized against the other participants in his paper to legitimize his own teaching: "What happens when these conflicting tendencies are presented within the confines of architectural education? They usually take the form of a debate between paired constructs as theory and practice, or form and function. These constructs are forced into an opposition or a polarity and thus given a positive or negative value only because they are seen on a scale defined in terms of 'reality,' theory and form are seen as unreal, while practice and function are considered to be super-real. In other words, these constructs are polarized and given values not because of any inherent greater validity accruing to one or the other but rather because they are made to seem so by a prevailing tendency to see them as such, within the framework of a particular bias towards reality."[174] As a speaker, he pivoted on his own axis with rhetorical deftness by suggesting that reality as a yardstick needed to be neutralized. He argued that it was the concept of reality that needed to be changed rather than that of theory. Speaking as its director, he portrayed the Institute as an alternative school of architecture that would impact reality through its education—and outlined another area of work that might be expanded in the future. As a result of this framing, later joined by all the other contributions, the conference was not so much a serious examination and discussion of didactic models and concepts as a fair of ideas. Or of vanities.

Democracy and Transparency

In 1972–73, after the conclusion of the "Streets Project," the Institute's organization underwent a significant democratization process. This continued throughout its expansion and transformation into a public institution that would go on to work on research and design, educational, cultural, and eventually publishing projects, and was accompanied by small steps towards greater professionalization and bureaucratization. Following the departure of Ambasz, who left the Institute after a dispute with Eisenman, the Fellowship was expanded to include Mario Gandelsonas and Peter Wolf who were appointed Fellows in May 1972, bringing the total up to six. Until then, the Institute as an organization had had a fairly hierarchical structure, and the Institute's leadership had been entirely tailored to Eisenman, who often single-handedly decided on the direction of the program, the appointment of Fellows, and the hiring of staff. But in 1972, building on Anderson's initiative from the year before, the six Fellows sat down to discuss their status and for the first time prepared an internal document

174 Peter Eisenman, "Preface" to "The Education of Reality," In IAUS, 1971, n.p.

that defined the rights and responsibilities of Fellows and non-Fellows, as well as the selection criteria, the election process, and the duration of a Fellowship. Eisenman himself had previously set out to define the roles of Fellows, Visiting Fellows, and Graham Foundation Visiting Fellows. A Fellowship Article, officially submitted to the Board of Trustees for approval, then stated that a potential Fellow had to first work at the Institute for one year as a Visiting Fellow, Consultant, or Research Associate, etc. on funded projects, in order to qualify for a Fellowship.[175] Accordingly, the status of a Fellow was given a tendentially neoliberal framework with an entrepreneurial imperative based on meritocratic principles, as it was grounded in the *dispositif* of autonomy and creativity, and in strategies and actions of empowerment. Since the Fellowships were now defined in terms of development, leadership, or collaboration on a project, they had not only an intellectual, cultural, and social component but also, for the institution, an economic one. Once awarded, the Fellow status was contingent on personal initiative and responsibility: only those who actively contributed to the life and well-being of the Institute over a longer period of time—be it within the framework of an Institute project, a personal project, or a special project—retained the title with all its privileges and obligations. It was specifically stated that all Fellows were provided with a workspace at the Institute, which included typing, as well as expenses for copying, telephoning, heating and lighting, and office supplies. The fundamental difference between Fellows and non-Fellows was that Fellows had a say in the election and re-election of Fellows, in the nomination of Visiting Fellows, and in the design, management, and budgeting of individual projects.

After the Fellowship Articles had been accepted, Visiting Fellows were invited to work on a specific project and received funding for one year. At Eisenman's suggestion, the election of Fellows had to be unanimous, while the election of Visiting Fellows required an absolute majority. Internally, the Fellowship formed the basis of the collective as a binary organizing principle, characterized by the interplay of hierarchy and cooperation, autonomy and leadership. The reorganization of the Fellowship (and non-Fellowship) introduced a quasi-democratic order, defined by the distribution and assumption of work and responsibility, by self-determination and continuing education, which simultaneously introduced further hierarchies and dependencies. If, for example, paid positions were now created in management, administration, research, and teaching—and this at a time when the prospects for employment as an architect, or even as an academic were becoming increasingly limited—then, from a sociology of work and organization perspective, the autonomization and responsibilization, flexibilization and precarization of work went hand in hand.

175 At the Institute, the Fellowship Article of 1972 was amended several times over the years to adapt the organizational structure to programmatic realignments, institutional transformations, and individual developments, first in July 1976 and then in July 1979, October 1979, November 1979, May 1980, July 1980 and September 1980, and June 1981.

In addition to the new transparency in Fellowship policy, further measures to reorganize the Institute's status, administration, and funding were decided at the annual meeting of the Board of Trustees in June 1972. Gibson Danes, a trustee from the very beginning, resigned, and architect Richard Meier, a cousin of Eisenman's who had already served as a trustee since October 1971, was elected as the new secretary of the Institute—family ties in the truest sense of the word. After five years, the Institute's leadership finally applied to SUNY's Board of Regents for a permanent, rather than temporary, status as an educational institution; the Institute was finally awarded the official certificate of appointment less than six months later, on January 24, 1973. In a forward-looking personnel move for the Institute, Wolf, newly elected Fellow in May 1972, was nominated for the newly created position of chairman of the Fellows just one month later. Eisenman installed Wolf as a second director alongside himself, thus relinquishing some of his power and accountability. In a letter to Wolf, Eisenman put it this way: "The intention of this appointment is that you and I, coequally, share responsibility for the management and direction of the Institute." Wolf's specific duties as chairman consisted of chairing Fellow meetings and representing the Fellows at Board meetings. He was also responsible for all financial transactions, human resources, and work coordination. While the Institute's secretary Louise Joseph had previously been the Institute's only permanent employee (next to Eisenman), further permanent positions were successively created under Wolf's direction to provide administrative continuity as the entire organization grew. These included a librarian, receptionist, managing editor, architecture education coordinator, exhibitions coordinator, and grants manager, although in keeping with the prevailing gender relations in American society in the 1970s, it was initially exclusively women who were hired. It was also Wolf who immediately championed social benefits, ensuring that Institute employees received health and life insurance coverage (Blue Cross-Blue Shield, TIAFF-CEF) and that the right to four weeks of paid vacation was now preserved. While Eisenman focused on design projects and theory production, Wolf would in the future primarily lead all the Institute's newly acquired urban planning research and design projects. The new dual directorship of the Institute subsequently embodied its two programmatic foci, "Architecture" and "Urban Studies," through their different interests and expertise. Eisenman, however, continued to appear publicly as the Institute's figurehead, and to this day has often been perceived as the sole director at the helm.

With Wolf's appointment, the Institute's administration and financial management were put on a solid footing for the first time. This institutional reform was long overdue, as Eisenman seemed increasingly overburdened with the Institute's management and administration. Nevertheless, Wolf's task turned out to be a financial suicide mission, because the Institute was already as good as insolvent. Its liabilities, which Drexler had already pointed out in 1971, now totaled US$85,000. Rent and tax arrears had to be paid as a matter of urgency.

Moreover, Eisenman had already been unable to pay wages and fees and had also had to forgo his own salary. By the end of the 1972–73 fiscal year, the Institute was even planning a total debt of over US$130,000. Wolf's main task was thus debt management, and by October 1972, the Institute urgently needed US$16,555 in capital to be liquid at all. One of Wolf's first actions was thus to introduce a new financial structure for the coming fiscal year to ensure the cash flow necessary for day-to-day operations.[176] Wolf also came up with the design of IAUS Central, a central treasury into which 40% of each of the Institute's current project's budget had to be paid to cover overhead costs; another 40% was budgeted for salaries and 20% for project costs. At that time, overhead costs were just over US$100,000, the largest item being rent and utilities, followed by personnel costs for the Institute's leadership and permanent staff. In sociological terms, the success of the two Peters as dual heads of the Institute can be explained by their being an "odd couple:" while Eisenman was the kind of employer and entrepreneur who successfully landed contracts, marketed grant applications, and raised donations, Wolf was more of a civil servant and administrator who kept track of the budget and made sure that the Institute was run on reasonably rational lines and kept as debt-free as possible. Although his power was now distributed somewhat more broadly, Eisenman's position as Institute director nevertheless remained unchallenged. To maneuver the Institute out of its predicament and reposition it as an institution, Eisenman developed various proposals for institutional services—first restructuring it into a kind of national internship and job exchange for students and architects, and then reestablishing institutions nationwide along the Institute's lines—with which he approached Bill Lacy of the NEA in June 1972. With this lofty expansion and nationalization strategy of the Institute, Eisenman was obviously also interested in enlarging its sphere of influence. To create the necessary conditions for this, he sat down in the summer of 1972 and, as he often did, drew up numerous lists: on the areas of responsibility of Fellows, staff, Visiting Fellows, and trustees, for example, the rules of procedure of the Institute's Fellow meetings, support for students, the Institute's press work, or business and the donor acquisition.[177] But starting conditions were poor: in the end, only two students came forward expressing interest and the only contacts were with journals offering internships.

During the 1972–73 fiscal year, despite several internal power struggles, the Fellows, Visiting Fellows, and Research Associates worked concurrently as a group on several research projects. For the first time, there was an explicit division of leadership across the individual work areas: Ellis headed the Library

176 Peter Wolf, financial structure, 1972–73. Sources: CCA, IAUS fonds, B.3-4.

177 Peter Eisenman, lists, n.d. Sources: CCA, IAUS fonds, B.3-4 / ARCH401264; see Karine Chemla, François Jullien, Jacqueline Pigeot, *Die Kunst, Listen zu erstellen* (Berlin: Merve, 2004).

Committee, Eisenman the External Publications Program, Gandelsonas the Internal Project Review, and Frampton the "IAUS Lecture Series," held on the theme of "Habitat & Urban Form." Starting from the fall semester of 1972, regular Fellow meetings were held every other Thursday. Synergy effects that resulted from a structural reorganization were to be used from then on, and cross-financing of the individual fields of activities was to be optimized. Diana Agrest, who was the last to receive Graham Foundation Visiting Fellow funding in 1972–73, was not assigned her own area of responsibility because she had yet to prove herself. From the categorization of the work areas, it became clear that publications, in addition to the lecture series, were seen as a public relations instrument, which Eisenman declared a top priority. While the establishment of a library had been the goal from the beginning, various concepts were now being developed, with Suzanne Frank now being designated as librarian.[178] Although research and design projects, publications, and public events were defined as separate work areas, architecture education was conspicuously absent in the 1972–73 academic year, although several undergraduate students from liberal arts colleges, most notably Oberlin, came to the Institute again. And while interns were now more intensively supervised by their mentors, and on top of that were invited to attend Fellow meetings, at least as passive listeners, however, there was no separate course offering for them this semester either. Instead, a list of all the courses taught by Fellows and Visiting Fellows at the respective universities was circulated as a substitute curriculum. In general, the Institute capitalized on the fact that Agrest was teaching at Princeton University in the fall semester of 1972, Eisenman and Wolf at Cooper Union, Ellis at Cooper Union and City College, and Frampton at Columbia University. This academic affiliation would remain one of the recipes for success at the Institute in the years that followed: most Fellows held either a professorship or, partly through Eisenman's mediation, at least a teaching position at one of the schools of architecture in the New York metropolitan area. This academic career path ensured the Fellows' livelihoods as well as cementing the Institute's considerable influence on the East Coast academic landscape, e.g., when its Fellows had a say in the development of new curricula. From the perspective of cultural sociology, aside from the research and design projects on which the Fellows worked, the Institute was already becoming a key actor in the culturalization of architecture during this first phase.

178 Frank, having already served in this role in 1971–72, was again made research associate at the Institute on February 22, 1973.

1.3 Publicly Addressing Housing

After the completion of the "Streets Project" was only partially successful for all the parties involved—in addition to the Institute, these were HUD, MoMA and, above all, the City of Binghamton—the Fellows immediately entered into the Institute's second major research and design project in 1972–73. More specifically, in addition to their own research and various publication projects (Anderson, for example, worked on the publication *On Streets* over the next few years, Eisenman on *Giuseppe Terragni*, and Frampton on *Modern Architecture. A Critical History*), they collaborated on the development of a low-rise housing scheme as an alternative to government-subsidized or public housing, on behalf of the Urban Development Corporation (UDC), their second commission. Here, the Institute again developed a prototype for modernist housing, this time under Frampton's lead, following up on the "New Urban Settlements" study, which was presented at MoMA in 1973 in the exhibition "Another Chance for Housing" and realized soon after in 1976. The designated Marcus Garvey Urban Renewal Area was in Oceanhill-Brownsville, a particularly neglected and ravaged neighborhood of New York characterized by high rates of poverty and crime after the original population had moved away. The multi-phase housing project combined research and design, and architectural and cultural production in an unprecedented way. Although it was subjected to massive criticism after its completion, it plays a major role in the Institute's history, especially since it was ultimately the only building project ever realized jointly by the Fellows. The development and planning of a prototype low-rise residential building to combat the New York housing crisis was a novelty in several respects: architecturally because the Institute was instrumental in creating an early alternative to the still widespread tabula rasa approach to large-scale construction for the state housing corporation, politico-economically because the prototype was to be implemented with government support as a public-private partnership in a variety of locations to address the housing shortage, institutionally because the Institute had entered into a unique strategic alliance with the UDC and MoMA to finally commence building, and culturally because it once again leveraged the prospect of an exclusive exhibition as a compelling argument to win a major contract. This time, however, the strategy was a success for all the parties involved: a powerful government agency, a non-profit organization working as a project office, and a world-class cultural institution. For once, the interplay between research, design, and realization, public relations, exhibition, and publication produced a measurable increase in power and influence for all three.

The idea for the project was born at the Institute in early 1972: after preliminary conceptual work by Kenneth Frampton and Anthony Vidler in the fall of 1971, Peter Eisenman brought Arthur Drexler on board. Subsequently, Drexler, as trustee of the Institute and director of the Department of Architecture and Design at MoMA, contacted Edward J. Logue, president of the UDC, and convinced him to once again enter into a promising, if not entirely voluntary, partnership with the Institute.[179] In late January 1972, Eisenman, Frampton, Vidler, and Peter Wolf met with Logue and his staff to discuss the commission. Initially, the Institute also wanted to obtain a Ford Foundation grant for the project, for which Logue again used his personal contacts, although this was an illusory undertaking. The UDC showed interest in the proposed prototype as it was undergoing a major change of direction in its housing and urban development policy at the federal level.[180] Beginning in 1968, the UDC's primary focus had been to create masses of high-quality housing for primarily middle- and low-income families. As a "super agency," it had the power to expropriate land and was exempt from taxation. It was also allowed to override existing building and zoning codes. Thus, the UDC had quickly become an attractive client for young, emerging architecture offices, with the Twin Parks large-scale housing development (1970–74) in the Bronx as its main public housing showpiece. Twin Parks was to provide 3,000 units of affordable housing and experimented with the architecture and landscape design of large-scale housing blocks, designs for public amenities, plazas, and other public spaces, such as spaces for retail, and generous floor plans that went beyond the minimum legal requirements.[181] But by 1972, the urban renewal policies of the 1960s had come under criticism, which culminated in the publication of *Defensible Space* by architect and urbanist Oscar Newman, based on data from the New York City Housing Authority (NYCHA). In his highly regarded publication, Newman, who was also presenting his research at the Institute at the time, referred to the negative social effects caused by the use of elevators

179 Frampton later described this partnership with UDC, IAUS, and MoMA as a "shotgun marriage;" see Stan Allen and Hal Foster, "A Conversation with Kenneth Frampton," *October* 106 (Fall 2003), 35–58. Drexler supported the Institute's push to the UDC with a letter of intent to exhibit the study on low-rise housing and its application at MoMA; Arthur Drexler, letter to Edward J. Logue (UDC), January 27, 1972. Source: The MoMA Archives, New York: CUR, 1037; see also Kim Förster, "The Housing Prototype of The Institute for Architecture and Urban Studies. Negotiating Housing and the Social Responsibility of Architects Within Cultural Production," *Candide*, no. 5 (March 2012), 57–92.

180 Logue had already had experience with low-rise housing in the 1960s as director of the Boston Redevelopment Authority.

181 Following a pilot study (1965–67) developed in parallel with MoMA's "The New City" exhibition by architects associated with the UDG, the UDC had designated Richard Meier, Giovanni Passanella, James S. Polshek, Prentice & Chan, Ohlhausen, Skidmore, Owings & Merrill, among others, as housing architects. See Susanne Schindler, "The Housing that Model Cities Built. Context, Community, and Capital in New York City, 1966–76," PhD diss., ETH Zurich, 2018; see also Susanne Schindler and Juliette Spertus, "A Few Days in the Bronx: From Co-op City to Twin Parks," *Urban Omnibus* (July 25, 2012), https://urbanomnibus.net/2012/07/a-few-days-in-the-bronx-from-co-op-city-to-twin-parks/ (last accessed: May 31, 2023); Nicholas Dagen Bloom and Matthew Gordon Lasner, eds., *Affordable Housing in New York. The People, Places, and Policies That Transformed a City* (Princeton: Princeton University Press, 2015). Legend has it that contracts for Twin Parks were awarded at one of Philip Johnson's garden parties at the *Glass House* in New Canaan. Besides Eisenman, John Hejduk also came away empty-handed. On the *Glass House* as the "principal base of his networking operations," see Varnelis, 2009, 120.

and too much uncontrolled public space in the discredited housing projects.[182] Previously, urban renewal had been based not only on the CIAM's urbanism principles but also on an amalgamation of hygiene and security discourse, a concatenation of obsolescence ideology and social market economy, a blend of urban planning and biopolitics, an interweaving of architecture and racism. In addition, the almost simultaneous demolition of parts of the Pruitt-Igoe public housing complex in St. Louis, Michigan, skillfully staged for maximum media attention, had marked the symbolic end of modernist large-scale housing as it had been practiced in the United States, viewed by many representatives of the world of architecture and planning, not to mention politics, as a panacea against physical decay and social ills and as a measure for growth and progress. Since then, the visually stunning demolition of one section of Pruitt-Igoe's public housing complex on March 16, 1972, which was already controversial in the run-up to the project and eventually abandoned by politicians, has served as a symbol for the "death" or failure of modern architecture and urban planning in postmodern discourse, and its transformation into a power-obsessed myth.[183] In order to test alternatives to large-scale housing and its supposed anonymity, the UDC had at the time shown increased interest in developing low-rise, and thus in its view more humane, housing typologies for the New York metropolitan area and beyond.[184] Theodore Liebman, the young chief architect of the UDC, led the charge here

182 Based on the assumption that combating and stopping signs of decay would improve neighborhoods, Newman called for the use of architectural elements as soft power instruments of passive social control—a biopolitical approach, which would shape the low-rise housing project developed jointly by UDC and the Institute; see Oscar Newman, "Defensible Space," *Progressive Architecture* (October 1972), 92–105; *Defensible Space. Crime Prevention Through Urban Design* (New York: Macmillan, 1972). The idea that architecture could remedy society's ills was contradicted in his book review in *The New York Times*; see Samuel Kaplan, "Defensible Space," *The New York Times* (April 29, 1973), 489. Architecture historian Joy Knoblauch argues that the theory of "defensible space saw vandalism and property damage, i.e., visual signs of decay, not only as an index but also as a cause of societal problems;" see Joy Knoblauch, "Defensible Space and the Open Society," *Aggregate*, Volume 2, (March 2015), http://we-aggregate.org/piece/defensible-space-and-the-open-society (last accessed: May 31, 2023). Knoblauch points to the biopolitical aspects of this criminological approach, to the fact that Newman called for architects to take an active role in strengthening civil society and fighting crime by creating such "defensible spaces" through an "environmental design" that the inhabitants themselves would control based on a "sense of ownership;" see also Knoblauch, 2012.

183 Charles Jencks, *The Language of Post-Modern Architecture* (New York: Rizzoli, 1977); see also Sabine Horlitz, "The Construction of a Blast. The 1970s Urban Crisis and the Demolition of the Pruitt-Igoe Public Housing Complex," in *Crisis, Rupture and Anxiety*, eds. Will Jackson et al. (Newcastle upon Tyne: Cambridge Scholars Publishing, 2012), 17–38; "Pruitt-Igoe: Ikone des Scheiterns? Planungsparadigmen, Lenkungsmodelle und Rezeption des US-amerikanischen Sozialwohnungsprojektes," PhD diss., FU Berlin, 2014 & "The Case of Pruitt-Igoe: On the Demolition of the US Public Housing Complex in St. Louis, 1972," *Candide*, no. 10 (2016), 61–84.

184 In 1972, *Newsweek* published a feature on Logue and the UDC's planning and policy practices, see "Housing: How Edward Logue Does It," *Newsweek* (November 6, 1972). Housing expert and planning historian Richard Plunz characterized UDC housing as a philanthropic approach, see Richard Plunz, ed., *Housing Form and Public Policy in the United States* (New York: Greenwood Publishing Group, 1980).

after completing a year-long grand tour of Europe's housing complexes, where he viewed interwar and postwar showpiece projects in the Netherlands, Switzerland, Great Britain, and other countries. Drexler's, or rather the Institute's, initiative was therefore met with a receptive ear at the UDC.

Eventually, the Institute was hired by the UDC as its architectural firm, even though the individual Fellows had hardly gathered any prior practical experience. But this new project was also a challenge for the newly appointed Liebman, who had only been responsible for selecting architects at the agency since 1971, as it was the first major building project he would oversee from planning to realization. The UDC, however, had a vested interest in gaining access to MoMA through its cooperation with the Institute, so that it could inform a broad public beyond the world of architecture, art, and culture about its housing initiative. The UDC hoped to convince as many communities in the rest of New York State as possible that state-subsidized housing with a social mix, but at the same time carried out by private developers, was possible and that low-rise housing could be used in both cities and suburbs. The Institute, which continued to seek out public authorities as clients for larger, more sensational projects, was finally given the opportunity to build and, on top of that, to project an avant-garde image for itself. And Drexler, with the prospect of an exhibition, was finally to be given the chance to exert a direct influence on building activities in New York. The first phase was to begin in 1972, with the research, design, and development of a prototype for low-rise housing. This was a sociopolitical task on the one hand, but also came with high architectural standards on the other. The Institute's low-rise housing was now a top priority at the UDC, and, like the housing authority's other building projects, it was slated for fast-track completion. The Institute was presented with a regular contract—thus operating as a "real" office for the first—and only—time in its history. The working relationship was clearly defined in an organizational chart.[185] Due to the Institute's lack of experience, however, the UDC had made it a condition that an experienced and officially registered architectural firm be brought in for construction management and execution planning.[186] Apparently, the cooperation partners agreed from the beginning that the housing prototype should not be, in Logue's words, "another theoretical exercise with a planning report and a proposal," but a real-life building project that would ideally be applied to as many locations in New York State as possible.[187] In addition to the preparation of concrete construction drawings, the actual group project at the Institute

185 Kreisler, Borg, Florman and Galay Development Corporation was brought in as the private developer.

186 Initially, Seymour Jarmal & Bernard Beizee were discussed as external architects; David Todd and Associates ultimately collaborated on the UDC/IAUS housing project.

187 Edward Logue, "Introduction," in *Another Chance for Housing, Low-Rise Alternatives*, ed. MoMA [Exhib. Cat.] (New York: The Museum of Modern Art, 1973), 4–5.

involved the analysis of potential building sites and the preparation of the exhibition and a catalog. The project team initially consisted of Eisenman as project director and Wolf as project manager; Frampton was assigned to design, and Vidler to research. As before, Frampton did most of the work and was able to prevail over Eisenman in April 1972 after internal power struggles that also concerned issues of the Institute's organizational structure, such as the status, rights, and duties of the Fellows and their working conditions. Although he had previously only worked as an architect to a limited extent in Great Britain and Israel in the 1960s, Frampton was ultimately the project architect responsible for the prototype, and Wolf took over project management.[188] As the project progressed, the Institute worked closely with a group of architects from the UDC: in addition to Liebman, these were primarily Anthony Pangaro and Michael Kirkland as project designers. In 1972, the Institute and the UDC jointly developed a four-story house type, accessed exclusively by interior stairways, as the basis for a novel form of housing that was highly dense despite its low height. In doing so, the prototype followed the UDC's "housing criteria," which defined the size, arrangement, and use of spaces in a dwelling as well as common facilities in terms of type, number, and location.[189]

In the second phase, the collaborating architects had to define fundamental architectural and planning principles for low-rise, and eventually high-density housing. Based on sociological and psychological research, these principles were designed from the outset to enable future residents to identify with the settlement and thus instill a sense of responsibility for the buildings and their neighborhood.[190] Even if this was not explicitly articulated, the design's modernist formal language still referred to the New York brownstone of the turn of the century. Distancing itself from the negative aspects of large-scale modernist housing, the reinterpretation of this typical housing typology was primarily concerned with achieving more flexibility than otherwise usual in public housing. Construction was based on a concrete structure with a brick façade, and apartment floor plans extended throughout the entire depth of the house to provide better lighting and allow for cross-ventilation. Most apartments included two separate living spaces and, where possible, bedrooms that were acoustically separated from living areas by hallways or bathrooms. In keeping with Newman's principles of "defensible space," such a low-story house type provided greater

188 Back in London, Frampton had designed the eight-story Craven Hill Gardens apartment block (with Douglas Stephens and Partners, 1964) on Leicester Square in Bayswater as an architect commissioned by the London City Council.

189 Theodore Liebman, "Learning from Experience. The Evolution of Housing Criteria," *Progressive Architecture* (November 1974), 70–77.

190 Theodore Liebman, "The UDC and the Evolution of a Housing Policy," & Anthony Pangaro and Kenneth Frampton, "Low Rise High Density: Issues and Criteria," *Another Chance for Housing, Low-Rise Alternatives*, ed. MoMA [Exhib. Cat.] (New York: The Museum of Modern Art, 1973), 12–13 & 16–17.

security and prevented vandalism by encouraging identification and maintenance by residents. The semi-private stoops in front of the houses were also a translation and update of this defining nineteenth-century architectural element, extended to include a porch, and designed to serve important functions in the neighborhood as "social condensers" in terms of livability, combined with facilitating the supervision of children playing in the street. While the UDC did not specify any style, the prototype designed at the Institute referenced numerous European examples of perimeter block developments and terraced housing, drawing on both 1920s and 30s classical modernism and post-war developments. The historical references cited included the Spangen Quarter (Michiel Brinkman, 1921) in Rotterdam, Netherlands, and Siedlung Halen (Atelier 5, 1962) near Bern, Switzerland.[191] The Swiss-French architect Le Corbusier's modernist architectural language was also used as a reference, so it was not surprising that the defining architectural element of the UDC/IAUS housing was the long window, built as a sliding window, as Le Corbusier had used this as a style-defining feature in a number of residential buildings. But while Frampton favored perimeter block developments with closed fronts abutting the street and interior courtyards that drew a clear line between private and public, Liebman insisted on the British street typology of the mews, residential buildings built inside a block for higher density with a secondary system of access routes providing access from the rear. Accordingly, Frampton, Wolf, and Eisenman, working with several interns (Randall Korman, Carl Larson, and Paul Rosen), designed not one but two simple house types at the Institute on a footprint of thirty-nine square feet with mostly two-story apartments of varying sizes, a street unit, and a mews unit over the summer of 1972. The UDC architects' plans, based on the New York street grid, involved breaking up the elongated city blocks with publicly accessible courtyards. A semi-public courtyard was envisioned as a social gathering place for residents and a separate play area for children, inaccessible to cars and visible from community facilities such as laundry rooms at the courtyard passageways, from the stoops, and from the adjacent apartments. In addition, all apartments were to have private outdoor areas as spaces for contemplation and relaxation. Gardens were envisioned for the lower ones, and balconies or terraces for the upper ones. Ultimately, the Institute and the UDC's joint design simultaneously sought to create a sense of community and responsiveness to context. While the UDC acted on the modernist belief that architectural form has a social impact, the architects at the Institute had a different interpretation of their task. Next to this architecturally conservative, and yet socially progressive design, one of the Institute's main concerns in terms of urban design, as with the "Streets Project," was the reinforcement of the streetscape.

191 Kenneth Frampton, "The Evolution of Housing Concepts: 1870–1970," in *Another Chance for Housing, Low-Rise Alternatives*, ed. MoMA [Exhib. Cat.] (New York: The Museum of Modern Art, 1973), 6–11.

In August 1972, following a presentation of the initial research and design results, the Institute proactively submitted a proposal to the UDC for implementation. Ultimately, the project, a textbook implementation of "defensible space," gained the support of Logue, who expressed enthusiasm about the two house types during a tour of the models. It was not until October that the Institute was officially commissioned to build the prototype on two radically different building sites, each in very different parts of the city. The first, a low-rise housing project, albeit fairly dense compared to the surrounding high-rise projects, was to be built in Brownsville, Brooklyn, with Frampton once again being responsible for the application of the prototype. The second was a hypothetical application of a medium-density cooperative housing scheme in Fox Hills, Staten Island, on the former site of the archdiocese, virtually a greenfield site, with a stream running through it. Eisenman was responsible for the design. The UDC deliberately juxtaposed these two applications as alternatives for both urban contexts and suburban areas. The Institute had been waiting impatiently for the contract to be awarded; after all, they were dependent on the revenue. And so, in the fall of 1972, when the UDC finally gave the go-ahead, the Institute's leadership was just able to avert insolvency, and the group of Fellows finally worked as architects over the next couple of months. In the 1972–73 fiscal year, half of the Institute's budget was made up of fees paid by the UDC and MoMA, while the other half came from other smaller commissions, research grants, funds from art and cultural foundations, and private donations. Since Frampton was now serving as both an exhibition curator and catalogue editor in the run-up to the MoMA exhibition, the Institute hired the British architect Arthur Baker for the realization of the building project. Baker, who was granted the status of Visiting Fellow, had already gained practical experience in housing construction in the 1950s when he worked for the London City Council; after moving to the United States in the 1960s, he began working at the architectural firm of Harrison & Abramovitz. At the Institute, Baker was immediately placed in charge as project architect and not only produced the working drawings and oversaw the site preparation, but also coordinated the Fellows' contributions. Here too, despite every effort to ensure professionalism, the division of labor was still quite chaotic, and while the Fellows all pursued their individual projects, somehow everyone still had a say in the housing project. At least Wolf oversaw project management for both sites. Frampton worked with Baker and Wolf, as well as with the new interns (George Snead, Richard Dean, Richard Wolkowitz), on the Brownsville project, on costing, work plans and scheduling, and on contributions for the exhibition and a concept for the catalogue. But although he had spent more time at the Institute since moving from Princeton to New York to teach at Columbia University in the fall term of 1972, he was otherwise occupied, since he was immediately intensively involved in setting up a new course on the history and theory of architecture, while the same time teaching as a Loebb Fellow at Harvard University. And Eisenman, who had been involved in all phases of the housing study up to that point, began investing most of his time in his own career in the summer of 1972, as well as in the realization of *House VI*. He had just received the commission for this from Richard and Suzanne Frank, who—Richard being his architectural photographer and Suzanne the librarian of the Institute—were fairly close to him. Again, there was very little separation between Eisenman's architectural practice and the Institute's direction, and Eisenman drew on the Institute's interns (in this case Korman, who remained for several years) as a readily available pool of labor.

In September 1972, as the Institute's Fellows were beginning work on a real-life project, Eisenman's focus turned to revising the proposal for the "Program for Generative Design," which was now his top priority, with one year of start-up funding from NIMH to the not inconsiderable tune of US$40,000.[192] Joined by the other theory-savvy people at the Institute—Gandelsonas and Agrest, as well as Cabral de Mello (and Peggy Deamer, for some time, as an intern)—Eisenman spent the following months formulating the concept for a much larger-scale study with which they proposed to establish nothing less than a theory of design creativity. When the first project proposal, however, was reviewed by the NIMH selection committee, it was criticized for four weaknesses that needed to be improved: one, "over-reliance on linguistic terminology," two, "no explicit methodology," three, "no model which was directly related to architecture," and four, "lack of definition of data." In early 1973, after three months of revision, the Institute confidently submitted an updated application, requesting the exorbitant sum of US$311,029 for a three-year project.[193] The application document differed from the first in that it was supplemented by architectural examples and explanatory diagrams, again borrowed from classical linguistics. The text of the proposal, however, retained the original structure of four individual contributions. Eisenman, Gandelsonas, Agrest, and Cabral de Mello now stated that they intended to undertake a theoretical project on two levels—analysis and design—with which they claimed to be nothing less than leaders in architectural theory. To be sure, what the four theoretical models in the proposal had in common was that they framed architecture as a process of communication, "thought of as produced by a systematic series of relationships and processes and not by things." Yet they could not agree on a common approach to creating a universal theory of architecture, because their individual approaches were ultimately not integrated—they even criticized each other. It is therefore no surprise

192 The Institute was awarded seed capital for the "Program in Generative Design" from September 1, 1972, to August 31, 1973. Eisenman budgeted an entire year's salary for himself, which made it his project; Agrest and Gandelsonas, on the other hand, were paid a salary by the Institute for six months until the end of February 1973, Duarte Cabral de Mello only for four months until the end of December 1972.

193 Peter Eisenman, grant application to U.S. Department of Health, Education, and Welfare, December 20, 1972. Source: CCA Montréal, IAUS fonds: B.6-3.

that the project did not receive funding. Ultimately, the proposal made it clear that their interests and methods were incompatible. While Eisenman professed to dispense with linguistic metaphors, the other three deliberately drew on linguistic and semiotic models to talk about the meaning of architectural forms, their design, and their effects on the environment. Eisenman pitted his formal approach against Gandelsonas's and Agrest's more cultural approaches, directly referencing his ideas of conceptual architecture and cardboard architecture.[194] And while Eisenman and Gandelsonas concentrated on the architectural scale, Agrest and Cabral De Mello focused on the urban scale. At least the fact that the supervision of the NIMH application was provided by the Center for Studies of Metropolitan Problems was anticipated in the formulation of their goals, with which they claimed to improve people's everyday lives "through radically changing how we design livable environments especially in urban areas." Accordingly, in contrast to the other approaches, Eisenman drew on structuralism as a metaphor to legitimize his formalism with verbal and diagrammatic accounts of the transformation of architectural elements in a feedback loop. He presented diagrams that drew on the design for *House IV* as the architectural object, to which his assertion of a universal theory—both context and subject-free—referred. While Gandelsonas and Agrest were reinventing themselves as architects with ideas about French philosophy, Eisenman, as usual, was intent on putting his theory into practice, "to design more controlled physical environments." He cited his own housing designs and the Institute's research and design projects, such as the low-rise housing project, the planned exhibitions and publications, and his teaching at Cooper Union in an effort to convince NIMH of the merits of the theoretical project.[195] Once again, the priorities in Eisenman's theorizing appear to be clearly set, with a strong focus on syntax, to the exclusion of meaning and use. And while he brought the application documents for NIMH up to date, the application of the low-rise prototype to the Fox Hills site had to wait.

Brownsville, Brooklyn

For the realization of the Institute's prototype in the problematic Brownsville neighborhood of Brooklyn, a report had identified a "substantial marketing and thus economic risk" in a preliminary site analysis; nevertheless, the UDC opted for this site to test the applicability of the low-rise prototype.[196] The housing

194 Ibid.

195 Ibid.

196 IAUS, "Site Alternatives and Specific Site Analysis," in *UDC Report: The Generation of Low Rise High Density Housing Criteria* (New York, The Institute for Architecture and Urban Studies, December 1972). Source: The MoMA Archives, CUR 1037; see also IAUS, "Application of the Prototype to the Marcus Garvey Park Village Urban Renewal Plan, Brownsville, New York" in *Another Chance for Housing, Low-Rise Alternatives*, ed. MoMA [Exhib. Cat.] (New York: The Museum of Modern Art, 1973), 20–27.

authority expected this to have the greatest socio-political impact, demonstrating the potential for broad-scale improvements of the poor housing conditions in precisely those neighborhoods that were economically run-down and socially marginalized. This was not only an architectural experiment, but also a social one.[197] In the early 1970s, the situation in New York was rather bleak and the "urban crisis" was showing its ugliest side. In Oceanhill-Brownsville especially, suburbanization and subsequent disinvestment led to the emergence of brownfields following vacancy, neglect, fires, and demolition.[198] The crisis affected virtually every aspect of social life. As white flight progressed, the neighborhood's former Jewish population had been replaced by an African-American community. Previously, in 1968, Brownsville had made headlines when teachers working there went on strike against the decentralization of the school system under Mayor John V. Lindsay and a racist layoff policy, as reported in the *New York Times*.[199] Subsequently, Brownsville became an inglorious symbol of "urban decay," which erupted in the 1970 "trash riots," when peaceful demonstrations that failed to produce results were followed by arson and looting by individuals. The social and political apathy in Brownsville, despite all the efforts of residents, was a direct result of New York State's misguided urban renewal policies, which had failed here across the board. These were the circumstances under which the UDC and the Institute entered the neighborhood. For after homeowners had neglected their properties or abandoned them altogether, numerous buildings had become city-owned. However, not least due to bureaucratic inefficiency, this building stock was scarcely refurbished, but for the most part had been demolished to make way for new, large-scale housing projects.[200] The planned new high-rise buildings, however, were not realized until 1972, with the result that all those residents who could afford to do so moved out to the suburbs, leaving the poorer strata of the population behind. In Oceanhill-Brownsville, then, there was plenty of city-owned land and a particularly high need for quality housing. The UDC's choice was a logical one in that the land was in the Marcus Garvey Urban Renewal Area, which had been designated an urban renewal area in 1968. It encompassed fifty-seven city blocks

197 Logue, 1973, 5. An evaluation of the housing thirty years after completion alluded to its status as a social experiment, see Kimberly Liebman, Laren Tenney, and Susan Saegert, "Good Design Alone Can't Change Society: Marcus Garvey Village (Brownsville, Brooklyn) after Thirty Years," *Planners Network* (Summer 2005); https://www.plannersnetwork.org/2005/07/good-design-alone-cant-change-society-marcus-garvey-village-brownsville-brooklyn-after-thirty-years/ (last accessed: May 31, 2023).

198 Wendell Pritchett, "A Modern Ghetto? Brownsville since 1970," in *Brownsville, Brooklyn. Blacks, Jews, and the Changing Face of the Ghetto* (Chicago: University of Chicago Press, 2003), 239–270.

199 Damon Stetson, "A Most Unusual Strike. Bread-and-Butter Issues Transcended by Educational and Racial Concerns," *The New York Times* (September 14, 1968), 19.

200 Pritchett, 2003, 345.

that had been designated by the city as the center of neighborhood revitalization efforts in the 1969 Plan for New York for Brooklyn.[201] This was the highly difficult site in which the Institute now had to prove itself.

According to the construction program, the UDC had planned the construction of approximately eight hundred residential units for the Marcus Garvey Park Village—the official name of the housing project. These were to be designed and realized by the Institute according to the specifications of the housing authority at construction costs of a maximum of US$28,000 to US$32,000 per unit.[202] The cooperation partners on site were the municipality and municipal institutions, the Model Cities Agency for Central Brooklyn as developer, and the Brownsville Community Board 16. The chosen building site—a plot of about five hectares within what is by far the largest urban renewal area, with six city blocks between Rockaway Avenue and Bristol Avenue—was suitable for testing the different qualities of the low-rise housing prototype for their urban design properties. The building site was large enough to allow for the construction of entire streetscapes and the formation of several courtyards, creating different degrees of publicness, at least to some extent, with clearly defined boundaries and thresholds, as advocated by Newman.[203] Another advantage was that large parts of the site had already been cleared, as the Institute noted in a December 1972 report: many of the existing buildings had burned out following arson attacks. But unlike other brownfield sites, site characteristics such as existing transportation infrastructure had to be considered, and existing community facilities integrated into the new development. Because some blocks still had row houses standing on them, the team of UDC and Institute architects was not able to close the block perimeter with a continuous street front in all cases. To achieve the required residential density, they therefore planned several larger-sized mews units along cul-de-sac streets instead. One complication, however, which had serious consequences, was the IRT elevated train line that cut the building site down the middle. A total distance of fifty feet (approx. 15 m) had to be maintained on either side of the line to ensure noise protection. This open space, which could not be built on, was simply designated as a parking lot so that the urban cohesion of the housing project fell by the wayside. The UDC accepted all this from the beginning. For it was certain that MGPV would

201 Kathleen Telstch, "Brownsville to get 50-Block Renewal," *The New York Times* (June 20, 1969), 1 & 75, see also CPC, *Plan for New York City. A Proposal: 3 Brooklyn* (New York: City Planning Commission, 1969). The designation as an urban renewal area was accompanied by a recommendation that educational, recreational, and childcare facilities be added to complement the new housing.

202 Lucia Allais commented on the choice of name for the housing project—Marcus Garvey was a Jamaican civil rights activist and advocate of Pan-Africanism—as being the only form of representation of African Americans in the housing project, see Allais, 2012, 34.

203 David Morton, "Low-rise, High-density. UDC/IAUS Publicly Assisted Housing," *Progressive Architecture* (December 1973), 56–63; see also Newman, 1972, 9.

go down in New York planning history as a showcase project since the new prototype was competing with the earlier generations of NYCHA housing from the 1940s and 1950s that adjoined it in the northeast. These were the six- and seven-story Brownsville Homes and the up to fourteen-story Van Dyke Houses I and II, which served as a model and a negative foil.[204] Ultimately, the Institute planned a total of 626 housing units for MGPV, which equated to a density of 50 housing units, or about 240 people, per acre. The population density thus ended up being significantly higher than initially planned to provide the housing required by the UDC. 248 housing units were apartments for large African American families and were equipped with three, four, or even five bedrooms, because according to the planning maxim of the "bedroom count," that was what mattered.[205] Disconcertingly, future residents were continually referred to as "low and middle income families," without incorporating intersectional approaches to understanding social inequalities by race and class, and the construction project was promoted by foregrounding socioeconomic aspects rather than ethnic ones. This might be due to the fact that MGPV was built on the basis of subsidies within the Section 236 program, which were granted to private construction companies under the Federal Housing Law of 1968 for the construction of federally subsidized housing units. The UDC planned to meet construction costs with New York State funds and state-supported bond sales, in accordance with standard practice at the time. Therefore, strictly speaking, the project was a public-private partnership rather than public housing, a financing option that existed only briefly in the early 1970s, because it was not owned or maintained by the public sector.

In the end, the UDC's decision to make another change on its own account, without consulting the Institute's architects, in the spring of 1973 weighed heavily. For only a few weeks before the construction started, and just as the Institute had been commissioned by Drexler to produce models and drawings for the MoMA exhibition, the UDC's director of design and construction Herbert Tessler informed the Institute that the UDC had decided to switch the first and second floors of half of the housing units on one side of the tracks, thus turning

204 Comparing the two developments in *Defensible Space*, Newman pointed out that in 1969, the crime rates and maintenance costs of the fourteen-story Van Dyke Houses in Brownsville were nearly twice as high as those of the three- to six-story Brownsville *Homes* nearby, see Newman, 1972, 39–49. See also Richard Plunz, *A History of Housing in New York City* (New York: Columbia University Press, 1990), 272–273. Today, the *Van Dyke Homes* are one of the "million dollar blocks" defined by state spending on prison inmates on a block. The Spatial Information Design Lab at Columbia University's GSAPP visualized these "million dollar blocks" in November 2006 in a research project, workshop, and exhibition on the relationship between architecture and the American legal system, focusing particularly on Brownsville because on the one hand it is home to many prison inmates, and on the other hand, undertakings were being made by developers and government agencies to resettle formerly homeless people and resettle new populations, see Spatial Information Design Lab, *Architecture and Justice* (New York: The Architectural League, 2006), www.spatialinformationdesignlab.org/MEDIA/PDF_04.pdf (last accessed: May 31, 2023).

205 Liebman, 1973, 12.

the planned arrangement upside down, with the living rooms above the bedrooms. MGPV had become a real test case, and to test not only this unconventional arrangement of the row house but also the original one, all the residential buildings in the entire construction project had to be raised. Now, only a third of the first floor could be sunk into the ground, not two thirds as initially intended, and the stoops therefore had to be built higher. They also lost their social function, as the shallow depth of the sidewalks meant that they had to be turned 90°, leaving them parallel to the street front. In addition, the gardens had to be lowered to preserve direct access to the residential areas and thus became more separated from the public space than originally intended. As a result, all the construction drawings had to be redone in the short time remaining until the groundbreaking ceremony. Overall, this change, imposed from above, caused considerable discord on the part of the Institute. More seriously, however, the commercial and social infrastructure were cut back to save money. Neither the planned community facility nor the daycare center nor even the planned playground were ultimately realized. Instead, the planners cited the existing Betsy Head Memorial Park in the neighborhood. And despite its good intentions, the Institute's design, for a variety of reasons, failed to achieve either the urban or the architectural quality that had been envisioned, which was intended to not only set the housing construction apart from other contemporary and historic projects but also to make it safer and more livable overall. For example, the planned modernist long windows à la Le Corbusier were not compatible with New York State fire codes, and considerable savings had to be made in both interior fit-outs (sheetrock instead of plastered masonry) and interior finishings (prefabricated kitchens instead of custom finishes); and finally, some architectural elements such as the garden walls and several balconies were omitted altogether to save money.

Fox Hills, Staten Island

The project application for the second site in Fox Hills on Staten Island was quite different. In August 1972, the Institute submitted a preliminary design by Eisenman to the UDC that showed how low-rise greenfield housing could function in a suburban setting.[206] This proposal was intended to preserve or, where possible, enhance the benefits of a suburban lifestyle while producing higher densities than the usual American subdivisions à la Levittown. This was relevant because Staten Island had been subject to increased suburbanization pressure since the opening of the Verrazano-Narrows Bridge in 1964, which had already led to the construction of large residential towers and, more importantly, the development of the island with single-family homes. And after all, since the East Coast, from

206 IAUS, "Application of the Prototype to Community Board 2. Fox Hills, Staten Island, New York," in *Another Chance for Housing, Low-Rise Alternatives*, ed. MoMA [Exhib. Cat.] (New York: The Museum of Modern Art, 1973), 28–37.

Boston to Washington, D.C., was by now completely urbanized with an uninterrupted development of medium-density peri- and suburban housing, the Institute's low-rise housing prototype was an architectural and planning instrument for an urgently needed upgrading and densification of the area.[207] With its two very different sites of application, it was entirely geared toward universal distribution and thus contributed to solving the prevalent housing shortage. However, in 1973—that is, at a time when Brownsville was still in the planning phase, Fox Hills had not yet begun, and the MoMA exhibition was in the making—the new political and economic developments meant that the initial situation for a state-subsidized architecture production in the USA underwent fundamental changes. This was because at the beginning of the year, on January 5, 1973, the conservative government under President Richard M. Nixon imposed a moratorium on housing subsidies as part of a far-reaching austerity program. Virtually overnight, this made new construction for low- and middle-income populations much less financially attractive to private firms nationwide.[208] After these government plans became known, the UDC searched frantically for more land in New York and across the state to submit and obtain approval for as many construction projects as possible from HUD before the amendment went into effect. Accordingly, it identified seven additional building sites and already commissioned three young architectural firms to reinterpret the prototype; the Institute was not even consulted in the process (but in the end, none of these projects were built, and it was to remain a one-time application).

In early 1973, after submitting the most urgent application for the theoretical project to NIMH, Eisenman worked with his team to gradually flesh out the prototype, although it was by now evident that an application in Staten Island would be shown in the MoMA exhibition only as a hypothetical proposal. Alongside Baker as the executive architect, Wolf as project manager, and several interns (Robert Serry, Peggy Deamer, and Randall Korman), Eisenman developed the cluster as a fundamental organizing principle. He himself made numerous drawings of the application.[209] In keeping with prevailing visions of postwar modernist architecture, the design derived from two basic elements in terms of planning, both of which corresponded to concrete specifications: the home and the automobile. Even though everyone was aware that, once the moratorium caused HUD to suspend all subsidized housing programs and issue

207 French geographer Jean Gottmann first investigated the polynuclear global city region from Boston to Washington, D.C. in his classic study *Megalopolis. The Urbanized Northeastern Seaboard of the United States* (1961); Herman Kahn and Anthony Wiener coined the neologism "BosWash" to describe this metropolitan region in 1967.

208 Charles Lamb, *Housing Segregation in Suburban America Since 1960: Presidential and Judicial Politics* (Cambridge: Cambridge University Press, 2005); see also Wendell Pritchett, "Which Urban Crisis? Regionalism, Race, and Urban Policy, 1960–1974," *Journal of Urban History* (January 2008), 266–286.

209 Eisenman developed his version of the low-rise prototype for Fox Hills around the same time as *House VII*, but in the end this project was not counted as one of his ten house designs.

strict new guidelines for urban renewal, the low-rise housing project would be a mere technical exercise, the Institute's leadership nevertheless launched its own attempts to capitalize on the draft. In a letter to Republican Senator Jacob K. Javits in March 1973, Eisenman and Wolf presented the Institute as a project office and specifically promoted the two large studies, the one on "Low-Rise Housing" and the "Streets Project;"[210] they asked him for a meeting, "to talk with [him] about how best to use this material in New York State with its social, planning and political implications." Clearly, the Institute continued to believe in their architecture and planning projects.

After much toing and froing, the UDC eventually selected a building site in Fox Hills in April 1973 and had it analyzed by the Institute, and there were further meetings with representatives of the UDC to concretize the application. At the same time, it was clear that a completely different client had to be approached. Fox Hills was an undeveloped area of about four hectares (61 acres), formerly owned by an archdiocese. Here, the building program called for the construction of 250 to 280 housing units, prompting the UDC to comment that the project would be better off not being called "high-density." As with Brownsville, Pangaro was the primarily responsible architect on the UDC side, and Matthew Cannizzaro acted as liaison at the Staten Island Housing Authority. As part of the Fox Hills study, Eisenman designed two prototype-based four-story suburban house types, which he called "cluster unit" and "stepped row unit." Like the two townhouse designs in the "Streets Project," his design was characterized primarily by carefully articulated and staggered façades that played with the dichotomy between private and public; he also devoted some, albeit less, attention to the floor plan. Eisenman's design for Fox Hills, while focused primarily on the surface, was nonetheless far more subtle and ambivalent than Frampton's Brownsville application of the low-rise prototype. Nevertheless, the study for Fox Hills also addressed the fundamental architectural and planning principles of surveillance, protection, and maintenance: the Institute's architects arranged the four-story apartment buildings, whether clustered or in rows, in such a way that the public green spaces were enclosed by buildings on at least three sides, and that ample off-street parking was provided, promising short distances to the front door, on spaces that could be viewed directly from the apartments. Anonymous, undefined, and thus unprogrammed and unsupervised space was thus reduced to a minimum. Interestingly, the green space defined by the cluster was reminiscent of the British tradition of the common, i.e., the village square. Apart from roads and a railroad line, the only features that had to be accommodated in the overall planning were natural features such as green corridors.

210 Peter Eisenman and Peter Wolf, letter to Senator Jacob K. Javits, March 14, 1973. Source: CCA Montréal, IAUS fonds: B.3-4. There is no record in the CCA archives of whether the meeting took place or whether the Institute's leadership received any response at all from the New York State Senate.

While the Institute sought to provide an alternative to the discredited public housing stock with the low-rise prototype in Brownsville, the Fox Hills study, despite little chance of success, later made a contribution to the discussion of American suburbia, which became the subject of research and teaching at the time and would continue to shape the architectural debate in the USA in the next couple of years.[211] Eisenman and his team eventually designed a total of 324 housing units for the car-oriented housing development, of which 92 apartments had one bedroom, 188 two bedrooms, and 44 three bedrooms, which roughly corresponded to the UDC's desired mix of 25, 60, and 15%. With a relatively high density for suburban areas of thirty-one dwelling units or about 120 people per acre, the goal was still to provide all the social and economic characteristics of suburban living, such as exclusivity and homogeneity, if not private ownership and capital investment. With the Fox Hill study, the Institute also made suggestions for marketing the new housing development, such as whether it would be better to build the apartments as cooperative housing or whether they should be sold or rented out individually on the real estate market. Although Eisenman wanted his rather specific architectural design to be understood as a practical test of his linguistic theory, the two housing projects in Brownsville and Fox Hills were more of a schematic juxtaposition of prototypical applications for urban neighborhoods and suburban settlements—based on simplified, generalized, rationalized, and typified notions of the lifestyles associated with each location.

This marked another, and for the time being, last time that the Institute brought itself into play as an architecture firm for private builders and developers. The dramatic changes in the social, political, and economic situation, however, made it quite clear that further orders for larger research and design projects could no longer be expected. At the same time, in June 1973, the application for the "Program in Generative Design" was finally rejected by the NIMH after all—no reasons were given. Beginning in the academic year 1973–74, it became clear that the Institute was virtually forced to reorganize itself; and although it was still eager to publish at all costs, reinventing itself as an educational and cultural institution seemed the obvious course.

211 Denise Scott Brown and Robert Venturi again set the tone on the subject of suburban America in a studio they taught on "Remedial Housing for Architects" at Yale University in 1970, in the context of which the individualization of prefabricated house types was analyzed using the example of the Levittown housing development on Long Island, New York; the studio was thus given the alternative title "Learning from Levittown." Unlike their 1968 design studio "Learning from Las Vegas," they did not publish this research and teaching project, see Colomina et al., 2022.

Another Chance for Housing

June 12, 1973, the day of the groundbreaking ceremony for the Oceanhill-Brownsville housing project, was the opening day for the MoMA exhibition "Another Chance for Housing. Low-Rise Alternatives."[212] The night before, everyone of note in New York's architectural and art community gathered in the museum's Garden Wing for the opening to witness this special high-cultural event, which showcased the Institute's housing project at a time when any other application was little more than a dream. In just a few months, the Institute had managed to transform the prototypical design into a viable building project and, in a burst of energy, to organize an exhibition and produce a catalogue. Both, part critical historiography, part political polemic, were programmatic: to bring the Institute into play as an architecture firm and service provider for private builders and developers one last time. An early exhibition review, published in the *New York Times*, noted that the two locations, linked on that very night and only a few miles apart, could not have been more different: on the one hand, the groundbreaking ceremony and "stifling reality" in one of New York's poorest neighborhoods, on the other, the fully "air-conditioned abstractions" reserved for the higher strata of society in the museum's exhibition spaces.[213] But, as MoMA had promised, the exhibition, designed by Frampton, coordinated by Barbara Littenberg, and produced by the Institute's interns, garnered international attention. In the introduction, Drexler wrote that MoMA still aimed to present low-rise housing as an alternative model to the common practice of bulldozing and redevelopment.[214] In addition to Frampton's historical research, the Institute's prototype, and its two applications in Brooklyn and on Staten Island were clearly the focus of the exhibition; they were presented in detail in numerous elevations, axonometries, floor plans, sections, and blueprints, as well as in seven architectural and urban design models and two hand-colored drawings. And in addition to the survey of historical examples of low-rise housing, another section documented the short building history of the UDC since 1968. In an unprecedented move, the exhibition and catalogue had from the outset been planned as a powerful public relations campaign by the UDC with MoMA as its advertising partner, and the Institute as author and producer, so to speak. For the exhibition, designed to address both a professional and lay audience, was also a celebration of the housing authority's fifth anniversary. Perhaps MoMA only stuck to its strategy of showing the future of housing because numerous cuts had to be made in the implementation of low-rise housing, due to legally, economically, or culturally justified architectural and planning changes;

212 MoMA, 1973. "Another Chance for Housing" replaced the "Streets" exhibition, which had been postponed indefinitely in February 1973, in MoMA's official parlance.

213 Joseph S. Fried, "Low-Rise Development Project Begun in Brownsville by U.D.C." *The New York Times* (June 12, 1973), 49.

214 Drexler, 1973, 4.

perhaps it had no other choice. In any case, the exhibition and catalogue, which basically revolved around a single building project that MoMA itself had been involved in commissioning, was vociferously promoted as a show that, in addition to innovative architectural and planning ideas, would also feature new strategies of governance and self-governance in social housing.

The Institute, in turn, portrayed itself at MoMA as the birthplace of "International Style" with its first major exhibition as an architectural firm, and a cultural producer.[215] The catalogue edited by Frampton at the same time as the exhibition, which in contrast had been fully financed by the UDC, contained introductions by Drexler and Logue as well as a historiographical essay by Frampton himself on housing concepts of the previous century, and above all extensive material on the Institute's prototype. This 40-page catalogue, which was ultimately chosen over a more comprehensive scholarly publication aimed at architects and academics, was aimed at disseminating the principles of low-rise housing underlying the prototype as widely as possible. It sold out quickly and was reprinted in a less expensive black-and-white version. Both the exhibition and the catalogue, by referring to the phenomenon of exclusion caused by urban decay on the one hand and the trends of growing suburbanization on the other, thus certainly had the potential to sell low-rise housing to New York's bourgeois, educated public as a universal solution. However, the larger cluster of problems surrounding urban development in New York, the extent, causes, and consequences of the urban transformation processes in the two vastly different boroughs of Brooklyn and Staten Island, and their economic and social demands on urban and suburban space, were not addressed. Nevertheless, "Another Chance for Housing," which as a unique pilot and demonstration project had ultimately made the Institute's only new building project possible, not only made a strong political statement that testified to the architectural will of the three partners involved—the UDC, MoMA, and the Institute—but also displayed a form of public relations that was obviously necessary to realize a project of this kind in the first place.

The exhibition generated a great deal of media coverage and brought architectural quality housing to national attention at a time when opportunities for public-private partnerships no longer existed.[216] In *The New York Times*, Ada Louise Huxtable wrote a favorable review of the exhibition, which she felt was important, "because it has caught the historical moment of change and fixed it by exhibiting an alternative proposal that could be a catalytic force in today's housing design." In general, Huxtable saw MoMA's support for selected architectural

215 MoMA's press release attributed the prototype design to Frampton and Wolf, the Brownsville application to Baker, and the Fox Hills application to Eisenman and Wolf, see MoMA, Press Release no. 47F, n.d., https://www.moma.org/momaorg/shared/pdfs/docs/press_archives/5002/releases/MOMA_1973_0073_47F.pdf (last accessed: May 31, 2023)

216 Ada Louise Huxtable, "Another Chance for Housing," *The New York Times* (June 24, 1973), 125; see also Wolf von Eckardt, "Toward New Horizons," *The Washington Post* (June 24, 1973).

positions, which helped the Institute acquire a building project thanks to its relationship with the museum, as a positive. Architecturally, she favored Eisenman's adaptation over Frampton's because "the Fox Hills project is the handsomer of the two, with a very sophisticated use of simple elements for considerable richness and surface interest. This may be because it represents the culmination of eighteen months of development work, beginning with the prototype and proceeding through the Brooklyn plan to the Staten Island scheme." The professional press, on the other hand, was not quite as kind. The Institute as exhibition organizer already had to face harsh criticism in the July/August issue of *Architectural Forum*. In an article with the blunt and telling title "It's All in the Family," which was published without naming the author (who turned out to be Suzanne Stephens), the close relationship between MoMA, the Institute, and the UDC was described as nepotism and the incestuous relationship of the three partners was condemned in the strongest terms.[217] Other negative statements followed not long after. While the exhibition was still running, James Morgen, managing editor of *Architecture Plus*, while favoring Frampton's adaption, complained that MoMA was not doing justice to its task as a leading cultural institution due to the lack of originality of the designs on display, since it was neither informative for the interested museum visitor nor for the practicing architect.[218] "To the former, the endless boards, showing rendered elevations and unit plan types are meaningless. The few models are limited to exteriors of the buildings which are generally less interesting to laymen [sic] than interior arrangements in model form. The visiting architect finds inadequate statistics and unintelligible prose. The Brownsville scheme, the more convincing of the two, presents no tabulation of unit types, while neither discloses unit sizes or costs." Morgan blamed the Institute, and implicitly Eisenman, for the triumph of a certain "cardboard esthetic" [sic], since no information about materiality was provided, and he indirectly criticized Frampton as curator of the exhibition for the fact that his historical survey barely contributed to an understanding of housing needs in the United States at the time. He also, echoing the criticism from the *Architectural Forum*, criticized the fact that only the two designs of the Institute were shown in the "Another Chance for Housing" exhibition and that there was no reference to any other contemporary housing projects, which were low-rise and high-density, but above all displayed a high degree of architectural quality, such as those by John Ciardullo in Red Hook, Brooklyn, or by Werner Seligmann in Ithaca, New York.[219] This harsh criticism was put into perspective

217 "It's all in the Family," *Architectural Forum* (July/August 1973), 25 & 27. Stephens later occasionally wrote about the Institute for *Progressive Architecture*, and in the early 1980s worked for the Institute herself as editor-in-chief of *Skyline*.

218 James Morgen, "MoMA on Housing: Nothing New," *Architecture Plus* (August 1973), 68.

219 The fact that this was a political issue, the explosive nature of which shook the entire profession, and not just attention-seeking reporting, was demonstrated by the fact that John Hejduk, the head of architecture at the Cooper Union and one of the "Five Architects," complained

by the editors, who juxtaposed Morgen's scathing review with a rather positive letter to the editor from Tom Killian, then an architect at SOM, who was full of praise for the Brownsville scheme because it represented architecturally significant housing that was finally being built in New York.[220] The December issue of *Progressive Architecture* featured an article by David Morton, managing editor of the architectural magazines, but also soon to be involved with the Institute, in which he documented the prototype extensively, but in the end presented it quite uncritically, so that the planning specifications and architectural solutions were disseminated in professional circles.[221]

From an intersectional perspective, however, it is necessary to make the caveat that none of the architectural criticism of the IAUS/UDC housing project and the MoMA exhibition adequately addressed the inscription of race or the question of class at the time. This was particularly evident in the two large-scale, watercolor perspective drawings of the prototype applications for Brownsville and Staten Island, which California architect Craig Hodgetts had been specially commissioned to produce. The two drawings, prominently displayed in the exhibition and reproduced in the catalogue (and later in the architectural press), did show the housing from a street perspective, thus making the human scale visible. But the everyday-looking street scenes not only depicted an ideal-typical use of public space; they also propagated stereotypical notions of the lifestyles of potential residents, characterized by hairstyles, clothing, habitus, social behavior, and possessions. While Brownsville's future residents were portrayed as part of urban Black America, cool, community-organized, and conforming to structural realities, those of Fox Hills were depicted as suburban couples, in intimate companionship or embracing dating culture. The problem with this colorful mode of representation was the backdrop of people, trees, and automobiles (a Cadillac in Brooklyn vs. a sport sedan with a cross on a chain dangling from the rearview mirror in Staten Island), which perhaps served as a standard of comparison, but also represented a certain image of society. There were two reasons for this: first, because it became clear that, however different the various notions of the street were at the Institute and however different the social life of the predominantly African American population in Brooklyn and the predominantly white middle class in Staten Island may have been, the urban vision was based on an energy-intensive automobile

about Morgen's fatuous review and its criticism of formalism in a letter to *Architecture Plus*; see John Hejduk, letter to *Architecture Plus*, September 5, 1973. Source: CCA Montréal, IAUS fonds: D.4-2. Hejduk defended Eisenman's work, both as Institute director and architect, out of friendship, as he himself admitted.

220 Tom Killian, "MoMA on Housing: Nothing New," *Architecture Plus* (August 1973), 68.

221 Morton, 1973. In addition to being editor of *P/A*, Morton also served as associate editor of *Oppositions* from 1973 to 1976.

culture that conformed to the ideal of the American way of life at the beginning of the 1970s, and second, because all the subjects depicted in their various constellations, both the young couples in suburban Fox Hills, who tended to belong to the Judeo-Christian culture, and the Black, more urban population in Oceanhill-Brownsville were, as was the fashion in architectural drawings at the time, completely decolorized or whitewashed, regardless of skin color. The two drawings, in which the Institute's low-rise housing was eventually relegated to the background, did not address the extent to which "Another Chance for Housing" testified to or reproduced social ambivalences in the United States—whether it was the fate of assimilation for persons of ethnic background on the one hand or the racial color blindness advocated by the civil rights movement on the other, that underlay the Institute's building project and its representation.[222]

Both the unique MoMA exhibition and the one-off application for the housing prototype ultimately failed to convince decision-makers in the municipalities of the need for a differentiated approach or to promote low-rise housing as an alternative to large housing estates—the opportunity had clearly been missed. Although this was not the Institute's fault, it did not address alternative financing models or social integration any further. The new social and cultural significance of architecture was soon to become apparent in the "postmodern turn" and the capitulation to political and economic interests.[223] For by the time the exhibition "Another Chance for Housing" opened, it was already evident that, after the change of policy in the United States, there would only be fewer state-subsidized projects for low- or middle-income, and especially African American families; moreover, in May 1973, the UDC's unrestricted position of power had been curtailed by a further amendment to the law, which gave local authorities the right to veto the housing authority's building projects. In the wake of the exhibition, the UDC had still tried everything in its power to get the project for Fox Hills completed and had, for example, given Community Board 2 representatives a tour of MoMA. In addition, the section of the exhibition on the proposed housing for Fox Hills was also to be shown at the Staten Island Museum of Arts and Science in the fall of 1973. Ultimately, however, all these initiatives failed, and the UDC finally terminated all further work on the project in August 1973. Generally speaking, paradigm shifts in American politics, society, and culture in the early 1970s were already evident here. There were apparently grants available for cultural productions, while on the other hand, no more government subsidies were being released for housing. Even the renewed and final talks about revising the prototype for another, third site in Brooklyn failed because

222 On the absence of the category "black" in the color spectrum of the North American architectural debate of the 1970s, see Mark Linder, "'Entropy Colorized:' The Gray Decades, 1966–96," *Any*, no. 16 (1996): "Whiteness," 45–49. The fact is that the Institute's low-rise housing was ultimately a research, design, and construction project by white architects; African Americans were not involved in the Institute, except for George Snead as assistant.

223 Steven Best and Douglas Kellner, *The Postmodern Turn* (New York: Gilford Press, 1997).

of a lack of funding. Thus, the MoMA exhibition, conceived as a new beginning, marked a stage in the development trajectories of the three disparate partners: for the UDC, it heralded the beginning of the end of public housing in New York State, for the Institute, the turn toward thoroughly economized knowledge and cultural production, and for MoMA, a shift toward the postmodern exhibition system of blockbuster exhibitions with postmodern content. The multiple media effects that distinguished "Another Chance for Housing" warrant special mention. The exhibition to some extent wrote the history of the UDC in the five years of its existence, bringing its vision of progressive housing to the widest possible audience, and it promoted the Institute as an architecture firm that could handle a project of this size. At the same time, it underscored MoMA's sociopolitical intentions, given that Drexler was able to launch and promote a public housing project. But when in late 1973 the UDC went on tour with "Another Chance for Housing" with the support of the Cultural Affairs Office of the U.S. Information Service, showing the low-rise housing projects as a transatlantic cultural export at the U.S. Embassy in London, it again drew criticism.[224]

Meanwhile, construction work on MGPV continued but dragged on for more than three years due to political and economic factors. When only a fraction of the apartments had been completed in the summer of 1974, Eisenman was still optimistic in his report to the Board of Trustees and expected the project to be fully completed in the spring of 1975. Following Baker's retirement from the Institute after only one year, he was replaced in his role as executive architect by Leland Taliaferro, who also worked for Eisenman. To ensure realization of the housing project, Taliaferro took over the construction supervision, signed contracts with firms, and coordinated the work of the architects, outside consultants, and contractors. Not entirely disinterestedly, Eisenman again wrote to Liebman in November 1974 offering the UDC further services on behalf of the Institute: an evaluation after completion of the housing construction, starting with the selection and profile of tenants, as there was still the opportunity to engage in urban studies. Other aspects the Institute hoped to analyze were "the performance of the building, actual use after its occupation;" "the designers themselves, you the clients, the contractors;" "characteristics of the tenants, the tenant mix, the design process, the environmental context, and the units themselves;" and "new concepts and design specifications for improving the general quality of the low rise high density housing."[225] The Institute's leadership estimated a budget of another US$240,000 for this

224 LeRoy "Sandy" Heck, a former intern at the Institute who had some insight, listed three possible interpretations of the exhibition in a review: a) "an exhibition of a particular design project," b) "a critical demonstration of how one state agency gets its job done," c) "an object lesson in dialogue between architecture and the public;" see LeRoy Heck, "Low Rise Alternatives," *Newsheet* (December 4, 1973). Source: CCA Montréal, IAUS fonds: D.4-2.

225 Peter Eisenman, letter to Theodore Liebman, November 1, 1974. Source: CCA Montréal, IAUS fonds: A.1-2.

accompanying sociological (rather than architectural) study, which was scheduled to last two years—Eisenman had once again designated sociologist Robert Gutmann as its lead in order to underscore its professionalism—and which would have addressed the processuality of the project's conception, construction, and use, which they hoped would be covered by the State of New York. But the UDC, which had to manage an internal reorganization, was currently facing new political conditions, and was under increasing economic pressure, rejected the proposal. Most importantly, the housing authority was busy elsewhere at the time, making a final push for large-scale high-rise and low-rise housing with the launch of a national public competition called the Roosevelt Island Housing Competition. The idea for this "mixed income community" for 18,000 residents on the former Welfare Island in the East River was once again based on a 1968 proposal by then-mayor John V. Lindsay, who had since been replaced.[226] The submissions, which were intended to produce feasible proposals for a new prototype of a high-rise residential building that would differ from the architectural modernism exemplified by the two modern typologies of the slab and the tower and be open to all income groups, ultimately testified to very different architectural and urban approaches.[227] The 268 participants included contributions from Fellows and Visiting Fellows of the Institute. In addition to Eisenman, who had submitted a project together with Art Net (Peter Cook) from London, Agrest and Gandelsonas also participated, as did Rem Koolhaas, who having graduated from the Architectural Association, had initially joined the Institute in 1973–74, after spending a short time at Cornell University on a scholarship, to found the Office for Metropolitan Architecture in New York (together with his partner Madelon Vriesendorp, and fellow AA graduates Elia Zenghelis and Zoe Zenghelis) in early 1975. At the Institute, the Fellows and Visiting Fellows were all able to draw on the labor of Institute interns to create their designs, although

226 Initially, a master plan had been developed for Roosevelt Island by Philip Johnson and John Burgee.

227 On the competition program, see Deborah Nevins, ed., *The Roosevelt Island Housing Competition* (New York: Wittenborn Art Book, 1975). The story of the competition was documented in the architectural press, eliciting several articles at once; see, among others, Suzanne Stephens, "This Side of Habitat," *Progressive Architecture* (July 1975), 58–63. The competition results were exhibited by The Architectural League at the McGraw-Hill Building in New York from October 15 to November 4, 1975. The Roosevelt Island Housing Competition also caused a stir internationally, with individual entries presented in *Controspazio* and in *L'Architecture d'Aujourd'hui*, see *Controspazio* 4, 1975 and *L'Architecture d'Aujourd'hui* 186 (August/September 1976). What was special about the competition announcement was that, in addition to UDC's Design Program, it also explicitly addressed the "housing issues" that the housing authority had developed with the Institute as part of the low-rise housing study and exhibited in *Another Chance for Housing* at MoMA; in this, the conceptual ideas seemed more like a shadow program of the competition, as was criticized afterwards, because Logue's goal in developing a "model mixed community" was to guarantee the same amenities in high-rise construction that had been tested in housing in Brooklyn, at a density of 110 units per acre. On Welfare Island, see Brilliant, 1975, 110–117; see also Yonah Freemark, "Roosevelt Island: Exception to a City in Crisis," *Journal of Urban History* 37, no. 3 (May 2011), 355–383.

strictly speaking they were not Institute projects.[228] But the Roosevelt Island Housing Competition, which was not only supposed to produce the UDC's largest and most symbolic building project, but also gave hope to the practicing Fellows with the promise of meaningful work and the opportunity to contribute to a federally funded housing stock aligned with local and social needs, was ultimately downgraded to a competition of ideas, and the winning projects were never realized. The competition was the UDC's swan song and gained its special significance from the enormous interest it attracted and the diversity of its entries, some of which were submitted by up-and-coming international architects. But above all, it marked a turning point in American building and social policy, symbolizing the end of competitions for large-scale housing, before the UDC as a housing authority was finally disbanded in its former form in 1976, after which it came under new management and concentrated on urban lighthouse projects dedicated to a different economy.[229]

1.4 An End to Building

By the summer of 1976, 95 percent of the Marcus Garvey Park Village (MGPV) apartments were ready for tenants to move in, and parts of the development were already occupied. At the same time, the change in federal and state policy on housing brought the first chapter of the Institute's history, during which the Fellows had conducted research and design projects primarily on behalf of public agencies, to a close. To be sure, Eisenman still spoke to the Board of Trustees in praise of the impressive architectural quality of the housing in Brownsville, Brooklyn, despite all the structural changes and the compromises that had been made, and they even submitted the building project to a national competition of the U.S. Department of Housing and Urban Development (HUD). But the Institute issued no official statement on its completion, nor was there a public presentation. In general, the Institute subsequently did little more to promote the issue of housing, let alone social responsibility for architects. Ultimately,

228 OMA participated in the competition with a project created with the help of Institute interns (Livio Dimitriu, German Martinez, Richard Perlmutter). OMA's Roosevelt Island entry was quasi as a first, if not realized project, moreover an act of Oedipal dissociation of Koolhaas from his former mentor O.M. Ungers, but also from the Institute, the place where he had originally arrived in New York and come of age, see *Architectural Design* 47, no. 5 (1977): "OMA". This dissociation, testifying to Koolhaas' irony, is suggested by the design, a mix of different urban typologies—high-rise and low-rise, towers and brownstones—reminiscent of MGPV.

229 The dissolution of the UDC—Edward J. Logue left the housing authority in early 1975, and chief architect Theodore Liebman and his entire department were terminated by the new York State administration in April 1975—marked the end of state-subsidized housing in New York State and the attempt at creating a low-rise prototype that by then was already revealing its utopian ghosts, in Suzanne Stephens' words "a model, an exemplar, a statement of what housing should be—not what it can be."

MGPV did not meet with the hoped-for response in the architecture world; the discussion was initially shaped primarily by the Institute itself.[230] Frampton, as the architect, was aware that the adaptations of the prototype to local conditions had weakened the building project in terms of the overall layout and the architecture. In "New York in White and Gray," a special issue of *L'Architecture d'Aujourd'hui* from August/September 1976 on the divides in the New York architectural scene, he himself criticized the developments and especially the policies of the UDC.[231] Shortly before completion of the construction, he used this medium as an international stage to complain about the political power and bureaucratic privileges the housing authority had had, the adjustments to the prototype that had to be made due to contextual conditions and the economic situation, and the rigid implementation of building codes by the developers. One effect was the eventual emergence of a debate about low-rise housing in the North American architectural press; in the second half of the 1970s, the editors of *Progressive Architecture* devoted no less than two issues to the topic in which, among other things, the Institute's prototype was featured.[232] *The New York Times* published a sympathetic review in May 1978, titled "The Low-Rise Solution for the Poor," albeit without naming the Institute, concluding that even in times of shortage, it would be better for developers to build many low-rise projects than none at all.[233] But while MGPV did find its way into the second edition of the 1978 *AIA Guide to New York City*, where it was discredited in a brief entry as a "pretentious experiment," it was a long time before it was subjected to serious architectural.[234] It was not until 1979 that Suzanne Stephens,

230 In 1974, the Institute's only building project was first published by an Institute Fellow, see Peter Wolf, *The Future of the City: New Directions in Urban Planning* (New York: Whitney Library of Design, 1974). In 1976, Robert Stern's design for a *Subway Suburb*, i.e., his contribution to the American section of the 1976 Venice Art Biennale curated at the Institute, questioned the urban qualities of the IAUS/UDC prototype, and what is more, he fundamentally negated housing for "low income families" as a building task by proposing a suburban single-family housing development for the middle class on the site adjacent to MGPV.

231 See Kenneth Frampton, "U.D.C. Low Rise High Density Housing Prototype," *L'Architecture d'Aujourd'hui*, no. 186 (August/September 1976), 15–21 (English version on pages XXXVII–XX). Note that the English version does not match the French translation. Frampton ended his sweeping attack with an indictment of the realignment of social policy in times of economic crisis, which he blamed for the dissolution of the UDC, and for the fact that the "housing prototype will never become the subject of further refinement, feedback and development."

232 The first special issue of *Progressive Architecture* of March 1976 on "Housing: High-rise vs. Low-rise" compared typologies; the second of October 1979 dealt with "Low-rise Housing" in general. Sharon Lee Ryder, in the introduction to the first issue, pointed to MGPV as a paradigmatic example of row houses that provided homes. Then, in the second issue, Suzanne Stephens published the first lengthy review on the Institute's only building project after its completion.

233 Josh Barbanel, "The Low-Rise Solution for the Poor," *The New York Times* (May 7, 1978), R1.

234 While1978 the social relevance of MGPV's architecture was harshly criticized in the second edition of the *AIA Guide to New York City* of—"more an architectural idea than housing for humans"—the polemic about the political impetus of the low-rise housing project in the third edition of 1988 was toned down a bit—"more a scholastic architectural thesis than a proto-

still in her capacity as an editor at *Progressive Architecture*, finally discussed the building project critically in her article "Compromised Ideal," which was unusual within the Institute's history, as well as for New York, since all the apartments had been rented in the meantime. Stephens listed in detail all the changes to the prototype and strongly criticized the fact that the construction costs of US$40,000 per unit were much higher than the UDC had initially aimed at.[235] In spite of these limitations, the article nevertheless helped the Institute's housing construction gain more attention in North America, despite all the criticism.

1973, the year in which President Nixon's moratorium on housing subsidies went into effect, was thus an incisive turning point in the Institute's history, and this harmful event had far-reaching effects on architecture culture in New York and beyond. Overall, the year represented a historical caesura in many respects and for a variety of reasons, not only in the United States and Europe, and it is now well established in architecture history that techno-aesthetic developments must be seen in their respective contexts and, above all, on a global scale. Part of the body of established narratives that follow both a socioeconomic and geopolitical line of argument is that the larger shifts manifested themselves in two ways: first, in the failure of the Bretton Woods system and a currency exchange regime that had fixed exchange rates based on the gold standard, and second, in the global oil shock and the so-called energy crisis that culminated in a combination of a production surge in the United States and the Arab oil embargo in the fall of that year. These new realities, argued economic and urban geographer David Harvey and literary and cultural critic Fredric Jameson, both of whom turned to architecture at that time, the latter even at the Institute, significantly altered architectural and cultural production in the following decades. This politico-economic line of argument has been persuasively revisited in recent historiographies of the rise of postmodern architecture, but also subjected to thorough critique and, above all, a more nuanced approach.[236] With regard to the Institute's activities, however, it was initially the very concrete events of 1973 that brought about a change in its

type for urban redevelopment," see Norval White and Elliot Willensky, eds., *The AIA Guide to New York City*, 2nd ed. (New York: Macmillan Publishers, 1978), 496; Norval White and Elliot Willensky, eds., *The AIA Guide to New York City*, 3rd ed. (New York: Macmillan Publishers, 1988), 719–721.

235 Suzanne Stephens noted in 1979 that little had been said or written about the Institute's housing project since the completion of MGPV three years earlier, see Suzanne Stephens, "Compromised Ideal: Marcus Garvey Park Village, Brooklyn, NY," *Progressive Architecture* (October 1979): "Low-rise Housing," 50–53. With her informed and reasoned critique, Stephens for the first time paid more attention to the building project, which she called a compromise, but ultimately reached a judgment that was not entirely uncritical: "The shift from ideal to real proved bumpy. Thus, while Marcus Garvey was a worthwhile experiment, it does not offer the ideal promised model for emulation so desired by those who conceived this scheme."

236 Harvey, 1989; Fredric Jameson, *Postmodernism, or, The Cultural Logic of Late Capitalism* (Durham: Duke University Press, 1991); see also Martin, 2010.

goals and self-perception, moving away from its work as a politically and practically oriented project office, if not a think tank, to embrace its original definition as an educational and cultural institution that was from now on to compete with and set itself apart from the museum and the university. The Institute's failure to maintain its focus on housing as an architectural project even after the change in policy, and to improve the prototype on the basis of the experience gained and implement it further, was compounded by another failure, namely its failure to produce an architectural theory that addressed urban ills. Indeed, in 1972–73, Eisenman financed himself for a year with his work as an architectural theorist. However, from the perspective of an institutional analysis and critique, theory production did not assume the intended role, even though in August 1973 the Center for the Study of Metropolitan Problems subsequently granted the Institute a grant of another US$37,920, which at least covered its overhead costs. When the NIMH seed funding ended, the Institute once again ran into major financial difficulties, as the Institute's leadership had been firmly counting on the income from the theoretical project.[237] In the end, Eisenman's strategy of financing the Institute through architectural *and* through theoretical production, as he had confidently formulated two years earlier, did not work out at all.

The first phase of the establishment of the Institute as a group, an organization, and an institution, which, while it ended on a rather unhappy note, was initially quite successful, since it recognized the signs of the times, was adept at drawing in architects and academics, and knew how to leverage its cooperation with authorities, foundations, associations, museums, and universities, was coming to an end. Nevertheless, Eisenman, Frampton, and Gandelsonas pulled off a coup in late 1973, when they founded the journal *Oppositions* out of the Institute. This new print medium had been in the making in one form or another for some time. Initially self-published with private, institutional, and corporate support, it enabled the most theoretically and historiographically ambitious Fellows to make a name for themselves as intellectuals by transferring their quite different knowledge, skills, and abilities, which they had previously tested and explored in research and design projects, lecturing and teaching, into academically sophisticated editorial and publishing practice.[238] From then on, *Oppositions* served

237 The theoretical models that had been formulated within the framework of the "Program in Generative Design," however premature they may have been at the time, were subsequently developed further in individual texts, as well as in the teaching and cultural productions of the Institute.

238 Allais, 2012. With Eisenman, Gandelsonas, and Agrest, three of the four protagonists of the "Program in Generative Design" published texts in the first issue of *Oppositions*; Cabral de Mello, on the other hand, was ultimately not involved. Apparently, Eisenman had originally offered the Fellows' essays to *Architectural Design* for publication. His essay on Alison and Peter Smithson was the only one to appear there, albeit in an abridged version, see Peter Eisenman, "From Golden Lane to Robin Hood Gardens; Or If You Follow the Yellow Brick Road, It May Not Lead to Golder's Green," *Architectural Design* 42, no. 9 (September 1972), 557–573, 588–592.

them in many ways as a means of raising their profile. It also featured Seligman's Ithaca Scattered Site Housing for the UDC and ultimately conveyed an approach and attitude that were truly postmodern, not least because the editors semanticized, historicized, and aestheticized developments in modern and contemporary architecture by reviving avant-garde designs, while juxtaposing different world views. What is particularly striking here is that Eisenman's assertion of autonomy, creativity, and intellectuality, which was also supported by Frampton and Gandelsonas—his fiercest critics within his own ranks—was contrasted with a reliance on commissions from the public sector or funding from national or federal foundations. This had consequences for the perception and assessment of real and theoretical projects, the transition from modern to postmodern thinking, and the detachment from principles of reason and ideals. For the conservative trends in the United States, which spelled the end of the Institute in its previous form, forced the Institute's leadership, above all Eisenman as one of the project makers of postmodernism, rather than Wolf, to open up new areas of expertise and activity and, above all, new sources of income. Once again, the Institute's future was at stake, with education and culture offering two thoroughly lucrative fields of activity in the post-industrial knowledge and service society that was gaining ground. When it finally became clear that the low-rise prototype would not be realized, and that it would not be possible to win any more major public-sector contracts, Eisenman abandoned his original goal of building with the Institute (while continuing to build institutions).

For some time, it was not clear in which direction the Institute would develop from fiscal year 1974–75, after the old working arrangements and business models had dissolved. Everyone was aware that the Institute would have to change and that, after working on housing, its projects would be completely different. The Institute worked on a number of publishing projects, including the *On Streets* anthology, a special issue of *Architecture + Urbanism* scheduled for spring 1975 that was to feature the low-rise housing in Brownsville, Brooklyn (but never materialized), and finally *Oppositions*, the Institute's own journal. There was even some brief discussion about starting an independent publishing house, but this did not seem feasible. The Institute was plagued by very concrete concerns at this time: once again, it was concerned with stabilizing funding and fighting for its financial survival. When salaries, rent, and other bills could not be paid for several months in the summer of 1974, the trustees stepped in to pay the Institute's debts and taxes. During the fiscal year, the Institute's leadership held talks with various universities and submitted applications to public and private foundations in preparation for its reinvention as an educational and cultural institution outside the university and museum. This transformation was set to begin in the fall semester of 1974. By opening the Institute to the outside world—the Institute's ongoing activities were being expanded and moving into new fields of work—its leadership sought to raise new financial capital and broaden its financial base in general.[239] Before the start of the 1974–75 academic year Wolf analyzed the administrative structure of the Institute for the first time, when the Institute extended its circle to include more interns and students than ever before.[240] At this time, even after another expansion of the inner circle of Fellows, the Institute itself comprised a total of only eight people—and almost exclusively a gentlemen's club. In addition to Eisenman and Wolf as the dual directorship, the Institute comprised Ellis, Frampton, Anderson, and Gandelsonas, as well as the newly appointed Fellows Agrest, the first and for a long time only woman to be admitted to the circle, and Taliaferro, who, in addition to completing MGPV, also worked on Eisenman's house projects.[241] And although staffing was limited, there were a number of Research Associates and Visiting Fellows around (encompassing for example young Rem Koolhaas, who was associated with the Institute, initially listed as a graduate student, while earning a living by teaching at Columbia University). As Institute director, Eisenman fostered a sense of togetherness within the group by hosting annual dinners—the "Indian Dinners" in the Institute's main hall were famous.[242]

In addition to the expansion of the Fellowship, the Institute's potential for development was ultimately demonstrated by the fact that the constitution of the Board of Trustees changed with the upcoming redesign: Drexler had already called for the appointment of new trustees at the annual meeting of the Board of Trustees in 1974, while at the same time announcing his resignation as chairman of the Board. Although he was still available, at least nominally, as a trustee, he was far less committed than before. Eventually, Armand Bartos was elected to succeed Drexler as chairman. Bartos was later to play an important role in steering the financial fortunes of the Institute. The restructuring of the Institute in 1974 thus also meant the end of its close connection with MoMA, the very institution that had helped establish it in the form in which it was to go down in architecture history in the first place. Moreover, the focus on urban studies implied by the Institute's name was over, at least for the time being, although Wolf and Ellis continued to pursue city planning

239 Peter Eisenman, "Director's Report," June 19, 1974 & IAUS, minutes of the annual meeting of the Board of Trustees, June 19, 1974. Source: CCA Montréal, IAUS fonds: A.1-2.

240 Peter Wolf, administrative structure 1974–75; Peter Wolf, "Report of the Chairman. Activities of Institute Fellows," June 19, 1974, Source: CCA Montréal, IAUS fonds: A.1-2.

241 In the 1974–75 academic year, of the Fellows, neither Anderson nor Frampton were present at the Institute, as they were each pursuing their own academic careers: Anderson continued to teach at MIT, setting up the HTC doctoral program there; Frampton was teaching at the Royal College of Art in London for two years, beginning in the fall semester of 1974, but commuted regularly to New York.

242 The Institute's inner circle in 1974 consisted of: William Ellis, Richard Wolkowitz, Peter Eisenman, Elisabeth Eisenman, Mario Gandelsonas, Madelon Vriesendorp, Rem Koolhaas, Julia Bloomfield, Randall Korman, Stuart Wrede, Andrew MacNair, Anthony Vidler, Richard Meier, an unidentifiable person, Kenneth Frampton, Diana Agrest, Caroline Sidnam, Jane Ellis, Suzanne Frank, Alexander Gorlin, see Frank, 2010, 36, figure 25. The photograph illustrated a review of Frank's memoir about her time at the Institute, see Birignani, 2011.

and even preservation and adaptive reuse projects, which at the time represented a new field of work and thus revenue, within the framework of their individual projects.[243] Individual Fellows referenced urban topics in their lectures or entire series of lectures and public events, and their teaching also included work on an urban planning study within the framework of an exhibition.[244] The departure from its former idealism, however, ultimately meant that the Institute was no longer a site of discussion on current, important political and social issues concerning urban renewal or housing, while the historians among the Fellows, Frampton and Vidler, contributed to journal issues of *Lotus International* on modern housing and industrialized cities. From then on, the Institute's common project was to practice, teach, communicate, and even celebrate architecture as an art form. As architecture in the United States became increasingly culturalized on the basis of public and private funding and philanthropy, and postmodernism asserted itself as both a discursive formation and cultural configuration on a global scale (two developments in which the Institute also played a role), the project of directly influencing the building process in New York ended in the mid-1970s, with the transition to a new accumulation regime.

243 At the annual meeting of the Board of Trustees in June 1974, Eisenman reported that Wolf and Ellis had both just completed their research projects: Wolf had been commissioned by the Manhattan Community Board 5 for a research study on the urban planning and transportation problems and potentials of the Union Square area, which included a redesign of Union Square Park, of which he had built a model with his students. IAUS, ed., *Union Square Park. Project Development: Phase 3. Report to the Manhattan Community Board* 5 (New York, June 30, 1974); Ellis, on the other hand, had prepared a showcase study on the conversion of an old mill site, the Harmony Mills in Cohoes, New York, which was subsequently transformed into a mixed-use development with attractive residential lofts. The Institute was approached, and Eisenman accepted, not because adaptive reuse and preservation were among the Institute's key competencies, but because this prototypical project covered a new subject area that was just becoming topical in the United States in the early 1970s. When the Institute received a grant from the NEA for the Cohoes project, Eisenman assigned Ellis to lead the study, with Richard Wolkowitz working as an intern. The main Harmony Mill No. 3 building, reimagined as a megastructure, had already received landmark status in 1971, and the entire site was then designated as the Harmony Mills Historic District in 1978. Because it was then one of the first projects of converting industrial buildings into housing, Ellis approached Suzanne Stephens, who published it in *Progressive Architecture.* see Suzanne Stephens, "From piano to forté. Interior Architecture," *Progressive Architecture* (February 1975), 60–67.

244 In 1976, the Institute worked on an urban study for Nicollet Island in Minneapolis, Minnesota, having been invited to an ideas competition alongside two other firms and to contribute to the exhibition "The River: Images of the Mississippi at the Walker Art Center", funded by the local City Planning Commission, see *Design Quarterly*, no. 101/102 (October 1976): "The River: Images of the Mississippi" [Exhib. Cat.]. Over the summer, a team led by Colin Rowe and Judith diMaio worked on the exhibition project at the Institute, assisted by John Hartley, Stephen Potters, Martin Kleinman, Livio Dimitriu, Bill Strawbridge, Andrew Anker, David Buege; see Colin Rowe, "Nicollet Island, Minneapolis" in *As I was Saying, Volume 3: Urbanistics* (Cambridge: MIT Press, 1996), 121–126. For the exhibition, the Institute designed a prototypical study of the revitalization of an island in a city, including the adjacent waterfront, see William Ellis, "Type and Context in Urbanism: Colin Rowe's Contextualism" *Oppositions* 18 (Fall 1979), 19ff., figure 30. The master plan, modeled on Isola Bella in Lake Maggiore, incorporated the functions of culture, recreation, utilities, and housing. The Institute's interns built the model, and individual Fellows contributed their own projects to the exhibition.

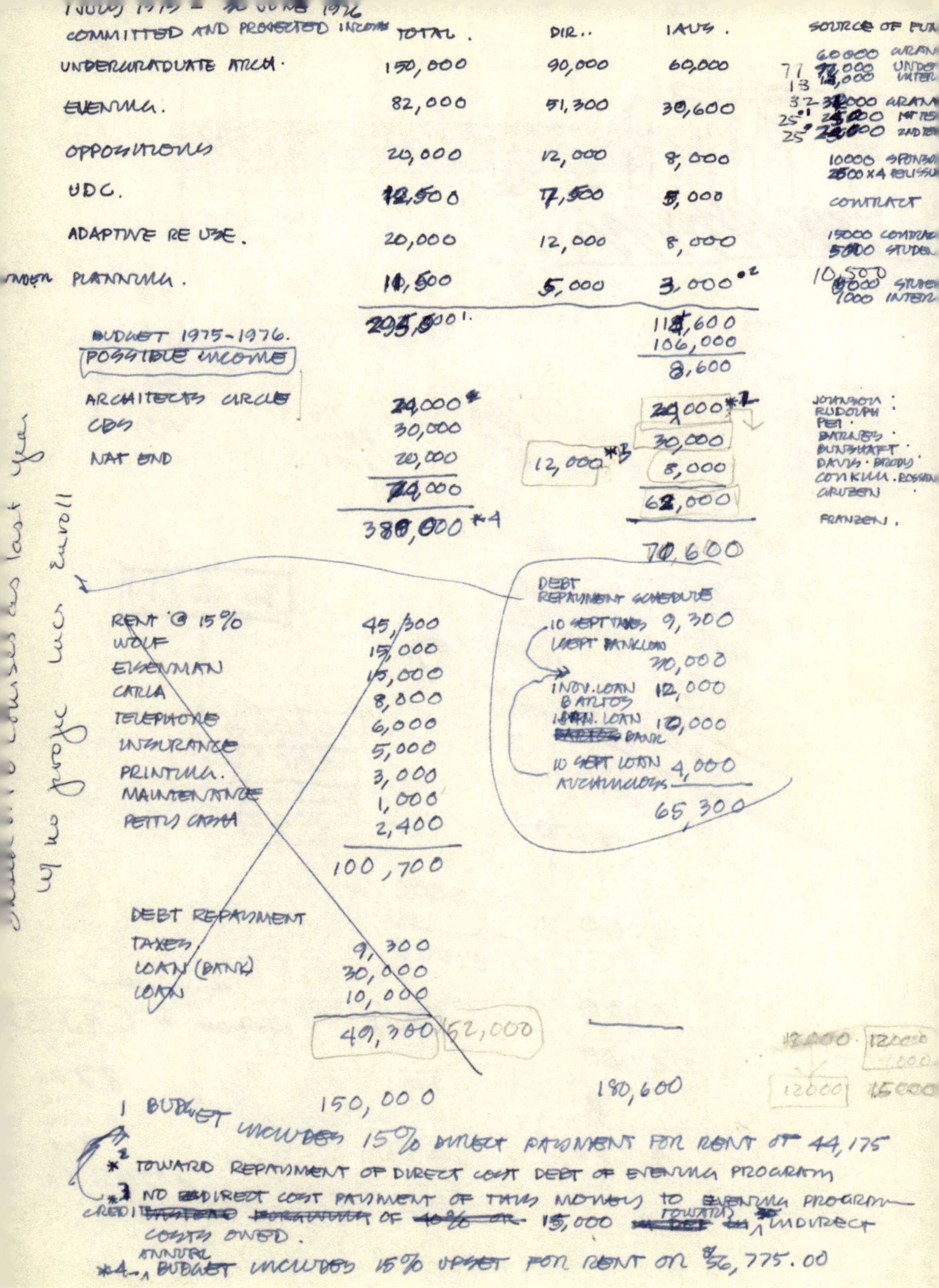

COMMITTED AND PROJECTED INCOME

	TOTAL.	DIR..	IAUS.	SOURCE OF FUN[DS]
UNDERGRADUATE ARCH.	150,000	90,000	60,000	60000 GRANT; 77,000 UNDER[GRADUATE]; 13,000 INTER[N]
EVENING.	82,000	51,300	30,600	32,000 GRANT; 25,000 1ST TERM; 25,000 2ND TERM
OPPOSITIONS	20,000	12,000	8,000	10000 SPONSOR; 2500 X 4 REISSUE
UDC.	12,500	7,500	5,000	CONTRACT
ADAPTIVE RE USE.	20,000	12,000	8,000	15000 CONTRACT; 5000 STUDENT
PLANNING.	10,500	5,000	3,000 •2	10,500: 9000 STUDENT; 1000 INTER[N]
	295,000 1.		118,600	
			106,000	
			8,600	

BUDGET 1975–1976.
POSSIBLE INCOME

ARCHITECTS CIRCLE	24,000*		20,000*2
CBS	30,000		30,000
NAT END	20,000	12,000*3	8,000
	74,000		62,000
	380,000*4		70,600

JOHNSON. RUDOLPH. PEI. BARNES. BUNSHAFT. DAVIS · BRODY. CONKLIN. ROSSANT. GRUZEN. FRANZEN.

DEBT REPAYMENT SCHEDULE

10 SEPT TAXES	9,300
1 SEPT BANK LOAN	30,000
INDV. LOAN BARNES	12,000
INDV. LOAN BANK	10,000
10 SEPT LOAN ANONYMOUS	4,000
	65,300

RENT @ 15%	45,300
WOLF	15,000
EISENMAN	15,000
CARLA	8,000
TELEPHONE	6,000
INSURANCE	5,000
PRINTING.	3,000
MAINTENANCE	1,000
PETTY CASH	2,400
	100,700

DEBT REPAYMENT

TAXES.	9,300
LOAN (BANK)	30,000
LOAN	10,000
	49,300

62,000

180,600

12000 · 12000
12000 15000

1 BUDGET 150,000 INCLUDES 15% DIRECT PAYMENT FOR RENT OF 44,175
*2 TOWARD REPAYMENT OF DIRECT COST DEBT OF EVENING PROGRAM
*3 NO DIRECT COST PAYMENT OF THIS MONEY TO EVENING PROGRAM CREDIT OF 15,000 TOWARDS INDIRECT COSTS OWED.
*4 ANNUAL BUDGET INCLUDES 15% UPSET FOR RENT ON 56,775.00

[Vertical margin note:] ... courses as last yea[r] ... if no people were enroll[ed]

Fig. 28

THE INSTITUTE FOR ARCHITECTURE AND URBAN STUDIES

The Institute for Architecture and Urban Studies
Eight West Fortieth Street, New York, N.Y. 10018

THE INSTITUTE FOR ARCHITECTURE AND URBAN STUDIES

The Institute is a non-profit educational, research and development corporation with a permanent charter from the Board of Regents of the State University of New York. It has been in operation since 1968 and has been involved in a variety of activities, including education, exhibitions, publications and the research, planning and development of prototypes.

The Institute believes that there is a critical link between architecture as a humanist discipline and professional practice. This is largely achieved through the involvement of the interns and students in the design and development of the Institute's projects. Thus in all of its activities, the Institute tries to relate its research and educational role to its involvement in practice and in this respect it seeks to sustain both education and practice as a natural continuity.

The Institute has emerged as New York's most vital and active forum for architectural discussion. It is a meeting place for architects, theoreticians and urbanists from overseas and throughout the United States. It has an active exhibitions and publications program including a highly influential journal for ideas and criticism in architecture. The Institute also offers a broad range of educational programs including an architectural intern program, undergraduate majors in architecture and planning, a work-study program in historic preservation and a special program for high school students. Of particular interest is an evening program for continuing architectural education which covers a wide scope of topics, ranging from the politics of the planning process to an analysis of current architectural developments.

At the core of the Institute is its Fellowship structure. The IAUS Fellows and Visiting Fellows are drawn from outstanding practitioners and educators from the New York area and from abroad. They have lectured widely at universities and conferences and are almost all on university faculties in the New York Metropolitan region. They serve simultaneously as the Institute's teachers, researchers and administrators.

FINANCIAL SUPPORT
National Endowment for the Arts
New York State Council on the Arts
C. B. S. Foundation, Inc.
The Duke Endowment
Graham Foundation for Advanced Studies in the Fine Arts
D. S. and R. H. Gottesman Foundation
The J. M. Kaplan Fund, Inc.
Edward John Noble Foundation
Alfred P. Sloan Foundation
Van Ameringen Foundation, Inc.

Fig. 29

EDUCATIONAL PROGRAMS

EVENING PROGRAM IN ARCHITECTURE

The Evening Program is intended to provide a range of courses in all aspects of architecture, planning and design, for a broad audience, from laymen to practitioners.
Its objective is to develop a knowledgeable constituency who have a concern for the public environment and its symbolic, social, cultural and aesthetic impact on our daily lives. Courses are offered in two ten week evening sessions, in the fall and the spring. The program at present is not degree granting, but it is possible to obtain credit with the permission of a parent school.
Admission ranges from $50 to $70, plus $5 registration fee for one credit hour (15 contact hours). Sponsorships are available and range from $175 (Individual, providing two passes good for admission to all the semester's courses) to $2000 (Corporate Sponsorship which provides a number of passes). Sponsorships are tax deductible.
Brochures for each semester are available upon request.
Director: Andrew MacNair

INTERNSHIP

The student internship program is one of the fundamental aspects of the Institute. It has provided students from many different undergraduate disciplines a break between undergraduate and graduate work. The internship is an integrated work-study program. It is an educational alternative, intended as pre-professional education for students with no architectural background.
Since the Institute is a small, close-knit center with a ratio of fellows to students of one to two, its operation depends on a cohesive spirit and unique atmosphere that can only come from this closeness of community. To this end, student initiative and willingness to assume responsibility are of the utmost importance.

UNDERGRADUATE PROGRAM IN ARCHITECTURE

A one year program, which acts as a component of an intensive non-professional architectural major for a consortium of liberal arts colleges including Amherst, Oberlin, Sarah Lawrence, Smith and Weslyan. The objective of the program is to provide a unique learning environment, while expanding the necessary and creative role of the humanities in pre-professional architectural education. Thus this program attempts to provide an alternative model for pre-professional education which can develop a cadre of students who will have an awareness of the historical and cultural aspects involved in problems of architecture and the design of the environment.
Director: Peter Eisenman

UNDERGRADUATE PROGRAM IN PLANNING

The Undergraduate Program in Planning is a one year integrated work-study program for undergraduates, which parallels the program in architecture. It is intended to supplement a general humanistic undergraduate education and at the same time to adequately prepare students for graduate work at advanced levels in the history, theory and practice of urban planning, as well as related subjects and disciplines. These two undergraduate programs are interlocking and students may choose a program which combines aspects of both.
Director: Peter Wolf

WORK-STUDY PROGRAM IN PRESERVATION AND ADAPTIVE REUSE

The first undergraduate program offering one year of specialized study in historic preservation and adaptive-reuse, open to qualified students from undergraduate schools of architecture. Its special studio classes concentrate on actual research and design problems.
The program offers a strong balance between theory, technique and technology.
Directors: William Ellis and Stuart Wrede

HIGH SCHOOL PROGRAM

The Institute has initiated a program for high school students concerned with the future of built form in the urban environment. Its purpose is to provide an elementary introduction to architecture, urban design, and planning as disciplines which deal with the problems as well as the potential of the city. It uses Manhattan as a study laboratory and the Institute as a midtown study center.
Working with architects, planners and historians, students will participate in one intensive workshop per week, combining design studios, seminars, lectures and tours.
Director: Andrew MacNair

DESIGN AND STUDY OPTIONS

A two-semester program for graduate and undergraduate architectural students. It offers intensive work in design, history and theory. The program operates under agreements with degree awarding schools, who send their students and confer credit for the work. The faculty is composed of Institute Fellows and visiting scholars and practitioners.
Director: William Ellis

RESEARCH AND DEVELOPMENT

Research and development is an activity which is central to the Institute's life. It is a necessary component of the Institute's teaching program and general philosophy.
The first two studies undertaken for the N.Y.C. Planning Commission of the Kingsbridge area in the Bronx and of street typology of Manhattan illustrate the kind of research with which the Institute is concerned. Also, the Institute was commissioned by New York State agencies to carry out a comparative analysis of new towns and has worked for HUD on a study of street forms as a strategy for the improvement of the environment. The Institute's conviction as to the relevance of aggregate built form led it to initiate, in collaboration with the State UDC, a study into low-rise housing. The results of this study were applied to the development and realization of 625 low-rise housing units in Brooklyn.

PUBLICATIONS

OPPOSITIONS, a leading international journal for ideas and criticism in architecture, published quarterly by The Institute for Architecture and Urban Studies and The M.I.T. Press.
Editors: Peter Eisenman, Kenneth Frampton, Mario Gandelsonas, Anthony Vidler.
Managing Editor: Julia Bloomfield.
Subscription annually: Students $20, Non-students $28, Institutional $35.
For further information: The M.I.T. Press, Periodicals Department, 28 Carleton Street, Cambridge, Mass 02142.

ANOTHER CHANCE FOR HOUSING: LOW-RISE ALTERNATIVES. Ed. Kenneth Frampton, The Museum of Modern Art, N.Y., 1973. An exhibition catalogue.
ART AND ARCHITECTURE – USSR 1917-32.
Exhibition and catalogue prepared by Max Risselada and Kenneth Frampton, George Wittenborn & Co., 1971.
CASABELLA. Milan, August 1971. A special issue by the IAUS.
ON STREETS. Ed. Stanford Anderson, to be published by The M.I.T. Press, Fall 1976.
A + U. The Japanese architectural journal will devote a special issue to the IAUS, Fall 1976.

EXHIBITIONS

Some of the major exhibitions organized by the Institute include:
"Art and Architecture – USSR 1917-1932" 1971
"Another Chance for Housing: Low-Rise Alternatives" (at The Museum of Modern Art) 1971
"Drawing as Architecture" 1974
"Mart Stam, Dutch Architect" 1974
"Goodbye Five, Work by Young Architects" 1975
"The Work of Massimo and Lella Vignelli" 1975
"T.M. Prentice, Sculpture" 1975
"Massimo Scolari: Drawings and Projects" 1976
"Aldo Rossi: Drawings and Projects" 1976
"A Space: A Thousand Words" 1976
The Institute has been selected to organize the American contribution in Architecture for the Venice Biennale, 1976; the exhibition is entitled: "The Suburban Alternative: 11 American Projects" 1976

AWARDS

Municipal Arts Society: Certificate of Merit, 1975
The American Institute of Architects: A. I. A. Medal, 1976, for the Institute's research and education programs.
New York Chapter, The American Institute of Architects: Special Citation, 1976, for OPPOSITIONS.

Fig. 30

FELLOWS MEETING - 25 April 1975

Present:

Julia Bloomfield
Peter Eisenman
William Ellis
Mario Gandelsonas
Rem Koolhaas
Andrew MacNair
Peter Wolf
Stuart Wrede
Leland Taliaferro

Peter Wolf summarized the present status of the Institute and noted that three new trustees had been named to the Board - Charles DeCarlo, Frank Gehry and June Larkin.

The evening program may be listed in the New School catalogue as an independent series of course offerings to be contracted independently through the Institute. Through this listing, notice of the program will reach large numbers of people and, in that way, increase enrollment.

Discussion then focussed on the nature of the interenship program. In the past, an intern was admitted to the Institute as a whole, but recently the intern has worked closely with a specific individual and become less a part of the Institute itself. The Fellows agreed that the internship should be clearly defined and structured yet retain flexibility.

It was decided that interns be associated with individual fellows, specific programs or the general Institute, depending on their different motivations for coming to IAUS. Admission will be determined by the fellow or program of the candidate's interest. Those candidates who wish to be at IAUS without any specific idea of who they wish to work with will be handled for the time being by the Program in Architecture. Each intern will then be the responsibility of the individual or program with whom he is working. On being admitted to a program (architecture, planning, adaptive reuse or other), the intern will be allocated by the program. An intern admitted to IAUS without specific ideas about what work he wishes to do will be assigned by IAUS Central as needed. At the end of the first term, the interns will be allowed to move to other assignements of their choice.

The internship will carry a $1000 fee for a one year residency at IAUS (payable half in September and half in January). The fee will cover only maintenance and administrative costs. The intern's work will be the trade-off for his learning experience. Of the $1000 fee, Institute Central will retain a 40% overhead rate. The remaining 60% will be allocated to the Fellow or program as appropriate.

Whether or not credit is received for the internship in a graduate school subsequently attended by the intern is determined solely by the graduate school.

Cooper Union students will continue to work at IAUS without fee and receive credit from that school.

Fig. 31

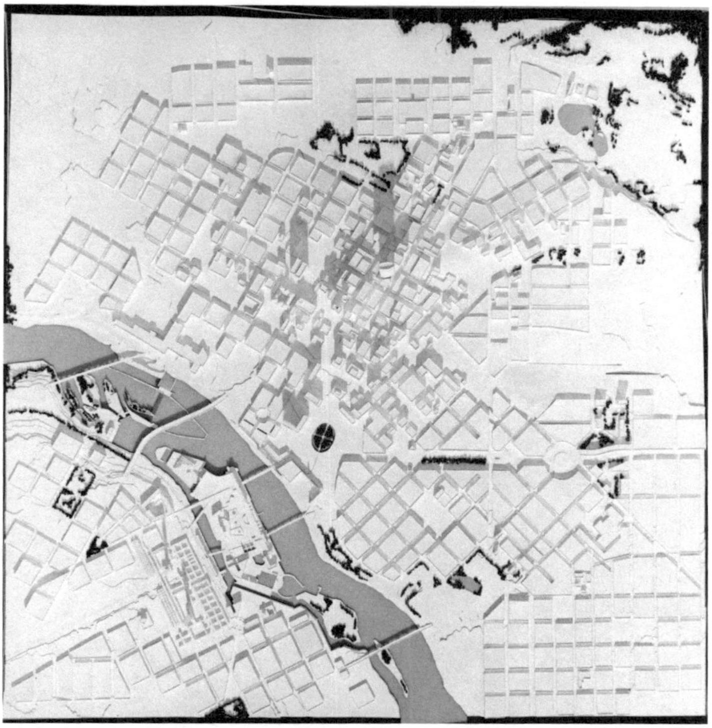

Fig. 32

Fig. 33

Fig. 34

Fig. 35

Fig. 36

ANNOUNCEMENT!
NEW SUMMER HIGH SCHOOL PROGRAM ARCHITECTURE AND PLANNING MANHATTAN: CAPITOL OF THE SEVENTIES

THE INSTITUTE FOR ARCHITECTURE AND URBAN STUDIES
Eight West Fortieth Street, New York, N.Y. 10018
212-947-0765
Tuesdays, Wednesdays, Thursdays, June 15-July 22
Grades 10 through 12
Six weeks-$400.00

The Institute announces the first semester of a new program for high school students concerned with the future of built form in the urban environment. Its purpose is to provide a format for students to become involved with architecture, urban design, and planning as disciplines which deal with the problems and potentials of the quality of life in the city. This summer program examines the forces which influence the form and development of essential elements within the city. It uses Manhattan as a study laboratory and the Institute as a midtown study center. Working with architects, planners, and historians, students will participate in three intensive worshops per week combining design studios, seminars, lectures and tours.

The Institute is located in the penthouse of an office building at the center of Manhattan, across the street from the New York Public Library and adjacent to Bryant Park. Buses, subways and trains stop within two blocks of the Institute.

MANHATTAN: A NEW VISION
Week 1: June 15,16,17, Streets & Neighborhoods: Wall Street, South Street, Mulberry Street, Nassau Street, West Broadway, Forty-Second Street
Week 2: June 22,23,24, Houses & Housing: Row Houses, Brownstones, Tenements, Apartments; from Collonade Row to the Olympic Tower
Week 3: June 29,30,July 1, Plazas & Public Places: Seagram's Plaza, Chase Manhattan Plaza, Times Square, Rockefeller Center, Grace Plaza
Week 4: July 6,7,8, Lobbies and Interior Spaces: The Ford Foundation, Bowery Savings Bank, Grand Central Station, Madison Square Garden, St. Patricks Cathedral, Waldorf-Astoria Hotel, and the Guggenheim Museum
Week 5: July 13,14,15, Parks and Natural Environment: Central Park, Riverside Park, Paley Park, the Hudson River, and Roosevelt Island
Week 6: July 20,21,22, Skyscrapers and Rooftops: The Woolworth Building, Cities Service Building, Chanin Building, Empire State Building, McGraw-Hill Building, RCA Building and Chrysler Building

DAILY PROGRAM SCHEDULE
Design Studio, Tues., Weds., Thurs., 9:00-12:
Lunch-Concert, Tues., Weds., Thurs., 12:00-1:
Lecture, Tues., Weds., Thurs., 1:00-2:00
Tour, Tues., Thurs., 2:00-4:00
Lecture (optional), Tues., 5:00-6:30*
Film (optional), Thurs., 5:00-6:30

EVENING LECTURE SERIES*
June 15: Raquel Ramati
Neighborhood Characteristics of the Street, the Sidewalk, and the Stoop
June 22: Sara Landau
The Evolution of the New York House: from the Rowhouse to the Apartment
June 29: Paul M. Friedburg
The Design of Popular Gathering Places in Outdoor New York
July 6: Paul Goldberger
Manhattan Magnificence: Lobbies, Elevators, and Interior Spaces
July 13: To be announced
The History and Ecology of New York Parks and Wetlands
July 20: To be announced
The Rise of the Skyscraper in Manhattan

FACULTY
Peter D. Eisenman, Cornell University, B.Arc Columbia University, M.S.Arch.; University o Cambridge, Ph.D.; A.I.A.; The Institute for Architecture and Urban Studies, Director; Co per Union, Visiting Lecturer.
Robert Livesey, Princeton University, A.B.; Harvard University, M.Arch.
Andrew MacNair, Princeton University, A.B.; Columbia University, M.Arch.; Evening Program in Architecture and Planning, Director; IAUS Fellow.
Stephen Potters, Cornell University, B.Arch. M.Arch & Urban Design.
Bonita Roche, Skidmore College, A.B.; Architectural Association; Mitchell/Giurgola Arch
Eugene Santomasso, Yale University, B.A.; Columbia University, M.A., Ph.D.; Brooklyn College, Assistant Professor.
Myles Weintraub, Yale University, B.A., M.Ar Harlem U.D.C., Urban Design Consultant; Urba Design Group, NYC, Principal Urban Designer;
Tod Williams, Princeton University, B.A., M. University of Cambridge, L.D.S.; Cooper Uni Adjunct Professor.
Peter M. Wolf, Yale University, B.A.; Tulane University, M.A.; New York University, Ph.D. Cooper Union, Visiting Professor; IAUS Boar Fellows, Chairman.

FOR A DETAILED POSTER AND FURTHER INFORMATI
Call or write Ruth Plawner, Assistant Direc of Summer Program, The Institute, 212-947-0

Fig. 37

he New York Times

idtown Architecture Institute 'owering as a Student Mecca

The New York Times/Jack Manning

vorking on a design project at the Institute for Architecture and Urban Studies. At right ers of student body, which now includes high-school, college and adult-education students.

By PAUL GOLDBERGER

om, a group of high-school students was how symbols of home, like chairs and te to the making of architecture. In an- , college students were grappling with of evolving a house design from three ancs. And down the hall, page proofs read for a new issue of a scholarly rchitecture.

ngs were happening at the Institute for e and Urban Studies, an eight-year-old that until recently was thought of by ects—if they thought of it at all—as the serve of a few of their rather academic, minded colleagues.

unded originally as a research organiza- d with the Museum of Modern Art, and a number of planning contracts it re- government agencies, it gained the s an organization far removed from the concerns of architectural practice.

Operates in a Loft

e last year or so the institute, which t of a casual loft space atop an office 8 West 40th Street, has suddenly moved nstream of the city's architectural life. ne largely a teaching organization that, attempt to duplicate the programs of hitectural schools, has decided instead gaps in architectural education. institute now runs an undergraduate program in architecture for 24 students from schools without architecture departments, like Oberlin, Sarah Lawrence and Wesleyan, who spend their junior year "abroad" on West 40th Street; an evening program of adult education courses in architecture that has attracted some of the nation's leading scholars and architects as lecturers; student and interne work-study programs in planning and reuse of old buildings and an experimental program in architectural education for 31 high-school students from Collegiate, Brearley, Chapin, Spence and other private day schools in the city.

Small Circulation

The institute also puts out a magazine, Oppositions, which, for the moment at least, is the nation's only actively publishing academic journal of achitecture. Its circulation is small—2,500—and it is better read in Europe than America, but it is growing steadily.

Most importantly, the institute has become the closest thing New York has to an ongoing architectural forum. It is where most overseas visitors, both scholars and practicing architects, seem to congregate when they arrive in New York, and it is the only center of architectural education anywhere where the student body ranges from the ninth-grade through postdoctoral scholars.

THURSDAY, OCTOBER 30, 1975

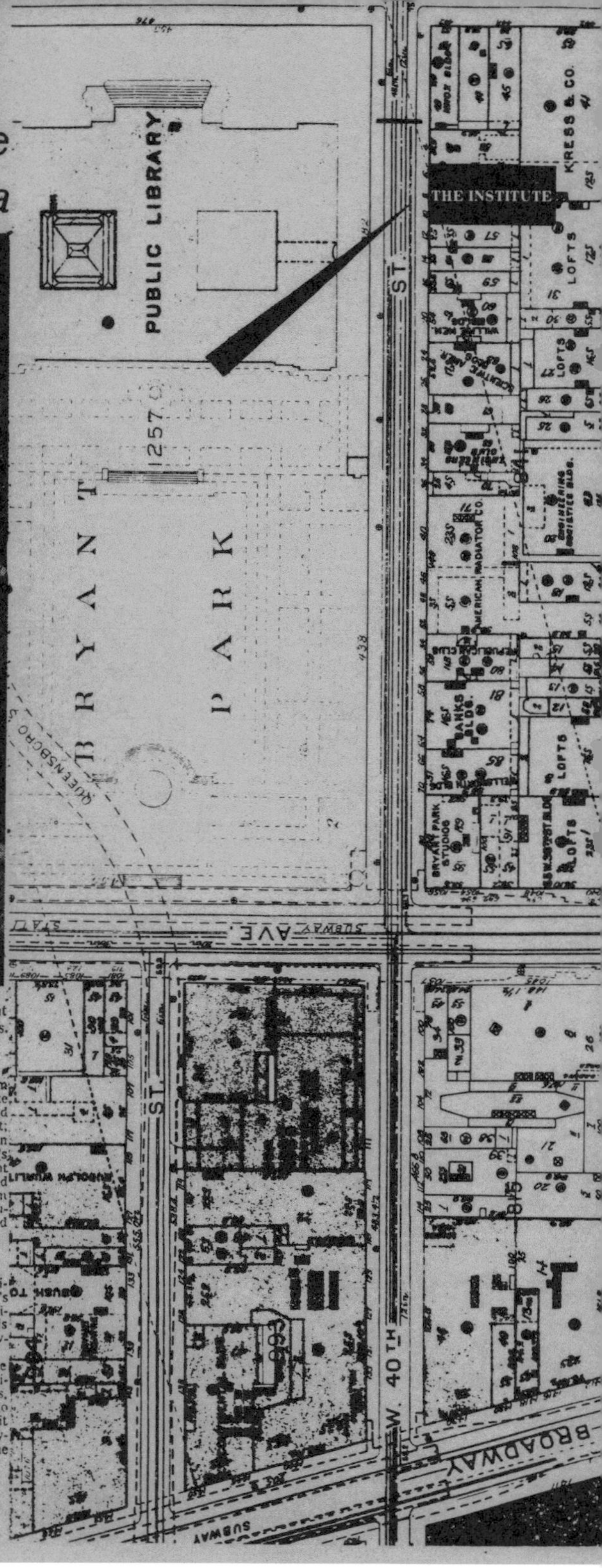

Fig. 38

FRYE

Fig. 39

Fig. 40

Fig. 41

Fig. 42

1977

28433

THE INSTITUTE FOR ARCHITECTURE AND URBAN STUDIES
Eight West Fortieth Street, New York, N.Y. 10018
Telephone: 212-398-9474

APPLICATION FOR A NEH LEARNING INSTITUTE PROGRAM
(June 1, 1977 - October 1, 1980)

Amounts Requested:	
For the Planning Period (June 1- Oct. 1, 1977).........	$ 42,000
First Program Year (Oct. 1, 1977 - Sept. 30, 1978)	115,000
Second Program Year (Oct. 1, 1978 - Sept. 30, 1979)	100,000
Third Program Year (Oct. 1979 - Sept. 30, 1980)	100,000
TOTAL AMOUNT REQUESTED	$ 357,000 ✓
Total Cost Sharing	$ 294,000
TOTAL BUDGET FOR THE PROGRAM	$ 651,000 ✓

Make check payable to: The Institute for Architecture and Urban Studies, 8 West 40th Street, New York, N.Y., 10018

Signature of the Authorizing Official:

(Peter Eisenman, Director, The Institute for Architecture and Urban Studies. 8 West 40th St., New York, N.Y. 10018)

Signature of the Program Director:

Andrew P. MacNair

(Andrew MacNair, Fellow, The Institute for Architecture and Urban Studies, 8 West 40th St., New York, N.Y. 10018)

Fig. 43

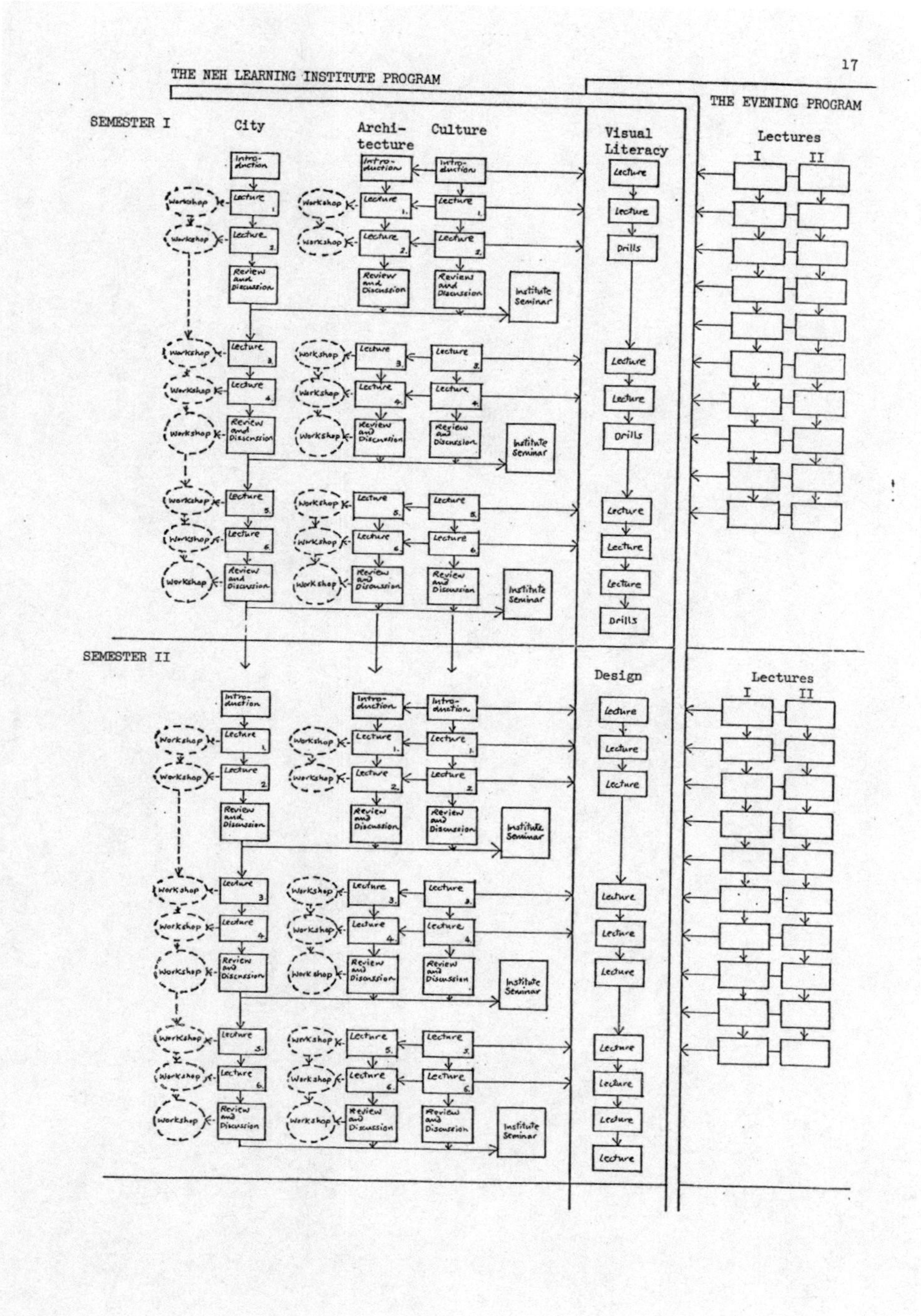
THE NEH LEARNING INSTITUTE PROGRAM
17
THE EVENING PROGRAM
SEMESTER I
City
Archi-
tecture
Culture
Visual
Literacy
Lectures
I
II
Intro-
duction
Lecture
Workshop
Review
and
Discussion
Institute
Seminar
Drills
SEMESTER II
Design

Fig. 44

Fig. 45

The Institute for Architecture and Urban Studies
8 West 40th Street, New York, N.Y. 10018
212-398-9471

THE IAUS-ADVANCED DESIGN WORKSHOP IN ARCHITECTURE AND URBAN FORM '79/'80

This program has been made possible in part with the assistance of HEW-FIPSE

The Advanced Design Workshop in Architecture and Urban Form is a one year program specifically oriented for two types of students: graduates of four year programs with non-professional degrees, and fourth year students in five and six year multiple professional degree programs.

Diana Agrest, Program Director
Jill Silverman, Program Administrator

Background

It is becoming increasingly clear that there is a need for a new form of architectural education which combines theoretical issues in architecture and urban planning with experience in an actual work situation. At the same time there is growing pressure to find more effective means for confronting the problems of the inner city. In many American Schools of Architecture, a year of study 'abroad' for selected students is recognized as a good educational idea. Students midway through a five or six year B. of Arch. or multiple degree program often need a change from their home school environment, a refresher or broadening experience in a different architectural context. Equally, graduates of four year undergraduate architecture programs likewise often feel the need to take a year off to explore various aspects of architecture before choosing a graduate school.

In the past, the Institute has offered a variety of educational programs basically directed to non-professional education. Now with the Advanced Design Workshop it has created a new program directed to the problem of relating professional education to actual work experience.

The following institutions have participated in IAUS work-study programs in the past: The Cooper Union, Cornell University, University of Cincinnati, University of Houston, Illinois University, University of Maryland, Miami University (Ohio), Syracuse University and Yale University. It is hoped that this program will expand the number of institutions involved with the IAUS work-study program.

Objective

The purpose of the Workshop is two-fold: to find new ways to make architectural education more effective and relevant to the urban situation and to find new ways for architectural students to apply theoretical concepts to existing urban problems; to serve a limited number of special students and provide an intensive and exceptional year of work and study articulating the theory and practice of design in a work situation.

Design Faculty: DIANA AGREST, Fellow IAUS
CHARLES GWATHMEY, Gwathmey Siegel Arch.
CESAR PELLI, Dean, Yale School of Architecture
ALDO ROSSI, Professor IUAV, Venice

IAUS

The location of the IAUS in New York affords an ideal situation for a one year intensive architectural program at the highest possible level.

The Program

The program offers advanced architecture students throughout the country a new and unique possibility of working directly on specific urban problems in small teams with some of the leading architects in the country. Simultaneously, it offers the planning agencies a design workshop in which certain proposals may be subjected to rigorous reassessment and where new solutions or models for limited urban development may be initiated.

Structured as an experimental laboratory exploring specific urban problems within a critical/analytical framework, the program will incorporate a two semester curriculum of work with architects in a team situation on actual problems within the context of New York City.

The Advanced Design Workshop places an emphasis on the dialectical relationship between theory and practice. It incorporates an analytical exploratory phase as part of the creative process itself. The program consists of two components, a design workshop component and a history/theory component. The Workshop will focus on both. There will be six credits awarded for the design workshop and three for theory, three for history. The course load will include special design projects for each team, and for individuals in each team as well as academic papers for the history/theory component. Each team will meet once a week with their architect, and sessions on design and planning will take place at this time with the architect/tutors. The theory/history component will develop within the frame of the Open Plan lectures and seminars.

Students

The program addresses two groups specifically: 4th year students in five year professional undergraduate programs or six year multiple degree courses and graduates from four year non-professional architecture programs. Other interested students should contact the program administrator.

Credit

The participating institutions consider the amount of credit to be given based on the work produced by the individual student. Students who are interested in participating in the program from other institutions should consult with their particular faculty to make their own arrangements or to determine final criteria for credit.

Tuition/Registration

$4,500 for students enrolled in a degree granting program ($2,500 per term)

$2,500 for students who have completed a four year non-professional degree and who are not enrolled in a degree granting program for the academic year. (Please note: that students who are not enrolled in a degree granting program must enroll for the two term sequence)

Deposit: A $250 deposit is due by June 30, 1979 for the coming academic year

Academic Year: September 6, 1979-May 17, 1980

Registration: Deadline June 30, 1979

Additional information, scholarships, application procedure contact the program administrator

Fig. 46

Fig 47

Fig. 48

2. Architecture School

In October 1975, with the start of the 1975–76 academic year, architecture critic Paul Goldberger published an article in the *New York Times* entitled "Midtown Architecture Institute Flowering as a Student Mecca." In this positive and lively portrait of the Institute for Architecture and Urban Studies, Goldberger attempted to make sense of the recent restructuring, as the Institute had totally changed within the period of a year.[245] Since being chartered as an education institution in 1967, the Institute spent most of its early years doing commissioned work for public agencies, with its labor pool comprising a handful of students operating without any proper curriculum. Thus far, the Institute's distinguishing feature as a place of learning had been providing students of various years and cohorts the opportunity to obtain practical working experience outside academia and outside the context of a conventional office job. 1973 marked a turning point in the Institute's history in many respects. Now, after many twists and turns, the Institute was increasingly presenting itself as a "true" educational institution, with a comprehensive range of offers targeting diverse groups that would qualify students for further study.[246] Although these transformations had already been implemented a year

245 Paul Goldberger, "Midtown Architecture Institute Flowering as a Student Mecca," *The New York Times* (October 30, 1975), 41 & 77.

246 Eisenman, 2007. In this conversation with Boyarsky, then director of the AA in London in its television studio, the contradiction between pedagogical aspiration and didactic reality became particularly evident when Eisenman said that the Institute "has never had a curriculum; it has never had a philosophy." This kind of rhetorical gesture was typical of Eisenman. In his characteristic ambivalence, which constantly oscillated between irony, exaggeration, provocation, and cynicism, Eisenman on the one hand negated any pedagogical task and social function of the Institute as an architecture school, but in this negation relativized his own

before the *New York Times* article, Goldberger's piece represents one of the first portraits of the Institute as architecture school. Goldberger, then a young journalist, was something of a postmodern apologist who had previously voiced criticism of the Institute, while bringing public attention to the architects who were part of its organization or worked there. In this extensive two-page article, supplemented with photographs of the students at work, Goldberger described how, over the past year, the Institute had developed into a serious "teaching organization"—a positive verdict overall. *New York Times* readers were offered a detailed description of the Institute's various education offerings which, as products designed to be purchased, were set up as self-contained modules. Prior to this, the Institute had primarily made a name for itself in the planning community as an innovative project space that dealt with public research and design commissions, and in the profession and general public thanks to a MoMA exhibition on public housing it organized in the summer of 1973. Since the fall of 1973, the Institute had garnered national and even international attention through the publication of three pilot editions of its ambitious journal *Oppositions*. Goldberger, however, made a distinct point of emphasizing how the Institute had now begun defining itself as an alternative to established schools of architecture, and how it was addressing a broader target group than it had before. The Institute was, in his words, "the only center of architectural education anywhere where the student body ranges from ninth grade to postdoctoral scholars." The article contains a quote by Peter Eisenman, the founding Institute director, that is downright baffling in light of the Institute's history: "It's true that we've become more of a service organization, and that's not what was originally intended." This confession to now viewing education in a commercial light was somewhat surprising considering that, since its early days, the Institute had presented itself as a mediator between politics and society and had primarily served public agencies. In the context of the knowledge-based society, which was changing the entire economy, it was equally surprising that Eisenman, who had been teaching at Cooper Union since 1968, was openly and in front of the broadest possible audience, describing the Institute as his place of work—even though, in professional circles, he had repeatedly emphasized that his autonomous design practice derived from his independence from universities and the architectural profession. This is especially noteworthy considering that *House VI*, the culmination of Eisenman's house designs at the time, was completed in 1975 and that Eisenman portrayed himself as a theorist and polemicist in his exhibitions, lectures, and publications. Even so, in an era of structural change—at the time, New York's financial and fiscal crisis was reaching its peak, with far-reaching negative consequences for commissions and construction activity—economic concerns ultimately trumped

teaching interests and thus created the greatest possible flexibility for the Institute in terms of curriculum design. See Kim Förster, "A Postmodern School of Architecture. Education at the Institute for Architecture and Urban Studies," in *Histories of Architecture Education in the United States*, ed. Peter L. Laurence (New York and London: Routledge, 2023), 98–117.

pedagogical arguments for Eisenman. Indeed, Goldberger's article ends with a final Eisenman quote and the observation: "Even in economically difficult times, people still want education."[247]

Educational Offerings

Reinventing the Institute as an architecture school had become unavoidable. After 1973, it was clear to the Institute's leadership that it needed to acquire new markets since it was no longer possible to acquire well-paid public sector commissions for large-scale research studies or construction projects. Neither architectural production (such as the development of a prototype for low-rise housing commissioned by the New York State Urban Development Corporation) nor theoretical production (such as the Program in Generative Design funded by a start-up grant awarded by the National Institute for Mental Health) had proven economically viable. The Institute's early recognition and capitalization of the economic value of undergraduate education in architecture is related to America's far-reaching political, social, and cultural transformations in the years after 1973. The architecture world faced a significant rupture after the conservative turn under President Richard M. Nixon and the departure from welfare state policies in the fields of housing and urban planning. No less significantly, the architecture world was also impacted by the collapse of the American building sector in the wake of the global energy and economic crisis, as well as by the unraveling of the real estate market for office buildings after the completion of the World Trade Center and the ensuing financial and fiscal crises that gripped the city and state of New York. In the Institute's early years, it had offered a small coterie of students—first, select graduate students, then postgraduate students as Research Associates, and later on larger groups of undergraduates and interns—the opportunity to collaborate on actual research and design projects, and in doing so to gather practical experience at the intersection between college and professional work (with the positive side effect of forming contacts with important architecture circles in New York). After 1974, not without a certain amount of entrepreneurial calculation, the Institute's leadership discovered how to actually *earn* money from architecture education. By introducing a one-year "Undergraduate Program" for students hailing from a network of liberal arts colleges across the Eastern Seaboard that lacked architecture programs, and by initiating an "Evening Program" of daily lectures promoted as adult education, the Institute was expanding on and developing two education initiatives that had already been in existence since 1971: first, a student internship program run in collaboration with the Great Lakes College Association (GLCA), an association of twelve northeastern colleges, and second, the IAUS Spring Lectures Series, the first public events series hosted at the Institute's premises on 8 West 40th Street. Unlike these two older programs, however, the new programs would be supported by nonprofit organizations

247 Goldberger, 1975, 77.

and sponsors, and consciously developed as education offerings with commercial promise. As a consequence, the Institute—whose small cadre of eight Fellows had previously been criticized for being overly hermetic and elitist (for example, in 1973, reviews in the professional press were critical of the MoMA exhibition on low-rise housing and again, at least indirectly, in 1974, when a feature on Italian architecture historian and critic Manfredo Tafuri cited his criticism of *Oppositions*)—was soon compelled to open up to a much more heterogeneous student body and a far broader public within an extremely short period of time.

In economic terms, the Institute's repositioning as an educational institution and brand was absolutely necessary, as it wouldn't have survived the 1974–75 fiscal year otherwise.[248] In an era of deindustrialization and deregulation, this also represented something of an "avant-garde" maneuver on the Institute's part; after all, by focusing on "education" and "culture," the Institute discovered for itself two key realms of communication and information that over the coming years would serve as core driving forces and sources of revenue—while in greater New York, the higher education landscape was in the process of expanding, cultural life was being actively promoted, and the notion of urban marketing as a neoliberal concept for reviving urban economies was being introduced.[249] Opening up the Institute along these lines served to reposition and restructure it; from this point on, the Institute would support itself primarily from the revenues from tuition fees, public and private subsidies, and donations. This multi-pronged business model succeeded in stabilizing the Institute's balance sheet while shaping its program policy and institutional, nonprofit business over the coming years. It is also important to consider the Institute's new conception of its education offerings or "educational products" against the backdrop of broader social developments in the USA: above all, the transition to a postindustrial society or service economy,[250] postwar expansion in the education field, and the post-1968 education reforms which had lingering effects on architectural pedagogy.[251] Arguably, the reason this strategic

248 Eisenman and Ellis presented the Institute's architecture education at the 1974 ACSA conference.

249 Sociologist John McHale characterized the new functions of "education" and "culture" as two central areas of the information and service society in the 1970s, see John McHale, "Education and Culture," in *The Changing Information Environment* (Boulder: Westview Press, 1976), 43–58.

250 Manuel Castells, "The Service Economy and Postindustrial Society: A Sociological Critique," *International Journal of Health Services* 6, no. 4 (1976), 595–607.

251 Aside from the transfigurations that accompanied enrollment in a specific teaching tradition, architecture education at American universities in the postwar period was characteristically determined by individuals, first of all European émigrés, e.g., Walter Gropius or Josep Lluis Sert as dean at Harvard University's Graduate School of Design in Cambridge (1938–52 and 1952–69), Ludwig Mies van der Rohe at the Illinois Institute of Technology in Chicago (1939–58), like Jean Labatut as director of graduate studies at Princeton University (1928–67), before American architects became formative, e.g., Paul Rudolph as dean at Yale University in New Haven (1958–65) or Louis Kahn at the University of Pennsylvania in Philadelphia (1957–74), see Frampton and Latour, 1980. In the

reorientation was so successful was because Institute director Eisenman managed to mobilize all the social and cultural capital the Institute had already accumulated, while simultaneously marshaling the Institute's Fellows (most of whom ran their own practices or taught at one of the other architecture schools in the metropolitan area—Princeton, Cooper Union, Columbia University, or the City University of New York—at the same time) and Visiting Fellows (many of whom had been persuaded to move to New York by Eisenman) to work together as teachers and cultural producers toward a shared goal of revalorizing architecture as an art form. In the early postwar years, architecture education in the USA had been defined by increasing specialization and compartmentalization of the disciplines, while in the 1960s the focus had been on urban studies and interdisciplinarity. Yet within a relatively short timeframe, in tandem with the emergence of postmodern service, entertainment, and attention economies, the Institute's position, role, and function underwent a dramatic transformation. What emerged from a formerly mostly self-proclaimed, yet also quite accomplished office for research and design projects, was an educational and increasingly also cultural institution that portrayed itself as "alternative" while, in the final analysis, obtaining a considerable degree of power—an organization that successfully managed to establish itself as a bridge between the college system and the universities. Marketing the new education offerings as a service, a commodity even, on an entirely separate basis from the conceptual, hitherto critical perspectives of the Institute's Fellows, was not only unusual for the time, it would also prove symptomatic of the accelerating transformation and economization of the education system occurring throughout the USA in the 1970s, as private colleges began to proliferate, and universities increasingly came to resemble factories for knowledge.[252]

1970s, there was little change in this close connection between institutions and individuals in relation to the formulation of a pedagogy. Teaching at Cornell University in Ithaca, under Colin Rowe as head of the Urban Design Studio (from 1962) and at Cooper Union in New York under John Hejduk (from 1964 as professor and from 1975 as head) proceeded from a conceptual understanding of architecture, which in the architecture debate of the time was traced back to the heroic phase of the "Texas Rangers" between 1951 and 1958 at the University of Texas School of Architecture in Austin; see Alexander Caragonne, *The Texas Rangers: Notes from the Architectural Underground* (Cambridge: MIT Press, 1995). In contrast, at Columbia University in New York, where in the 1970s several design studios for housing were offered in parallel, influenced by a modernist approach developed at Rensselaer Polytechnic Institute in Troy while all following different approaches and scales, "a certain revisionism of CIAM" could be discerned as the origin of the basic theoretical assumptions of a continuation of architectural practice in socio-political terms; see Richard Plunz, "The Four Typologies. The 'Master of Architecture' Program at Columbia University," *Lotus International*, no. 27 (1980), 106–126.

252 The academic and pedagogical context of architecture education at the Institute was the transformation of higher education in the United States, as analyzed by sociologist Alain Touraine and historian John Thelin; see Alain Touraine, *The Academic System in American Society* (New York: McGraw-Hill, 1974); see also John Thelin, "Coming of Age in America. Higher Education as a Troubled Giant, 1970 to 2000," in *A History of American Higher Education* (Baltimore: Johns Hopkins University Press, 2004), 306–326. Thelin argues that as higher education became economized—moving from "grants" to "student loans," from "gyms" to "health and fitness centers," and from "dormitories" to "apartment suites"—problems increased and all

The year 1974–75, with its diversification and commercialization of education, represented in many respects a rupture and turning point in the history of the Institute that reverberated for a long time to come—on programmatic, technical, economic, and political levels. Concurrently, this rupture shaped American schools of architecture and higher education in a broader sense. For the Institute's Fellows, the change meant that their combined teaching duties would become a central field of activity, synergistically linked to other programs in adult education and cultural production (in addition to some of their work on *Oppositions*), redefining their relationships to work and training. Their new educational programs addressed both undergraduate students and college graduates looking to complete a course of study or internship that would qualify them for a degree in architecture.[253] Yet architecture teaching at the Institute was far from a "radical" pedagogy, i.e., one that is critical and utopian, in its efforts to overturn socially constructed distinctions and destabilize norms and values that underlie designs and allow for differentiation at the intersection of race, class, and gender.[254] Indeed, the education offerings were in fact institutionalized by established colleges and developed as add-ons, extending their existing curricula. They were not necessarily critical in nature, nor did they fundamentally question or change the existing institutions. Instead, what the Institute supported, if not explicitly encouraged, was the redisciplining of architecture study, in something of a backlash or reaction to the activist-led research and teaching paradigms that had taken root after 1968—exemplified by advocacy planning and community design centers[255]—and the accompanying reorganization of university curricula vis-à-vis the contents, methods, and goals of teaching and learning. The Fellows of the pedagogically oriented Institute (in addition to Peter Eisenman these were Kenneth Frampton, Mario Gandelsonas, and Diana Agrest at the time, as well as Peter Wolf, and later Anthony Vidler) were committed to a renewed focus on history and theory, partly in a leading

institutions of higher education in the United States, despite society's countervailing perception of them as a growth industry, underwent a profound crisis in the 1970s as grant money diminished. For a critique of the transformation of universities into factories of knowledge in the course of capitalist valorization, see Gerald Raunig, *Fabriken des Wissens: Streifen und Glätten 1* (Zurich: diaphanes, 2012).

253 Reinhold Martin makes a fundamental distinction regarding education and discourse; see Martin, 2010, 66. This differentiation, however, lacks a historical dimension, as it does not take into account the legacy of the Institute's research and design projects, and, above all, does not consider the cultural and educational importance of the media and mechanisms of the Evening Program and the Exhibition Program from 1974 onward.

254 Colomina, et al., 2022. Architecture historian Beatriz Colomina conducted a research project with PhD students at Princeton University titled "Radical Pedagogies" on trends in architecture education worldwide in the second half of the 20th century; see Colomina et al., 2012; see also: Colomina et al., 2015. Yet, politically speaking, education at many institutions was anything but radical. For my earlier accounts of architecture education at the Institute, see Kim Förster, "Alternative Educational Programs in Architecture: The Institute for Architecture and Urban Studies" in Reto Geiser, ed., *Explorations in Architecture. Teaching, Design, Research* (Basel: Birkhäuser, 2008), 26–27.

255 The first and possibly most successful community design center was the Pratt Institute Center for Community and Environmental Development in Brooklyn, inaugurated in 1963.

role and partly in an advisory capacity. By developing and introducing a wide variety of education offerings in a short time, the Institute was able to establish itself in a competitive educational market with a differentiated and, above all, flexible or modular approach to architecture education—and it did so in economically challenging times, as the state of New York was undergoing a severe financial and fiscal crisis in the mid-1970s with serious consequences for the national and international economy and society. Through its educational practice, in addition to the publication of *Oppositions*, the Institute took an active role in hastening the paradigm shift to postmodernism on the East Coast of the USA and beyond, with both aesthetic and epistemological consequences.

Due to its incommensurability with the teachings of the previous years, the Institute's pedagogy now focused on imparting historical and theoretical knowledge, as well as knowledge about construction and planning, as a foundation for abstract formal exercises that, in principle, revolved around a humanistic approach and yet were informed by a formal, linguistic, and semiotic understanding of architecture, which was legitimized by recourse to architectural modernism and (post)structuralist theory. For the Institute's longstanding Fellows, this presented an opportunity to test new research and insights and bring these into circulation at a remove from their usual academic obligations, while making a name for themselves as pedagogues and intellectuals—some of them with a view to qualifying for professorships. The tuition fee-based education offerings, which students could sign up for à la carte, enabled the Institute to not only address entirely new target groups and acquire new markets, but also to redefine how architecture knowledge was produced, disseminated, and received (and habitus was created)—not just beyond the usual confines of colleges and universities, exhibitions, and periodicals, but also of the office world and construction projects.[256] At the same time, the Institute's leadership expected the new, distinctive educational focus of the Institute to first and foremost stabilize its budget.[257] Beginning in the 1974–75 academic year, after which an average of forty-five students attended the Institute per academic year, the organization and program of architecture education at the Institute went through a series of successive phases that followed an economic logic: first the introduction of new education offerings, followed by growth, consolidation, and maturation, all the way to saturation. The individual phases were fundamentally different with regard to the didactics and pedagogy of the various education offerings, the composition and expertise of the faculty, and the general relationship between teachers and students, as well as the conception and function of an architecture education that was never

256 Dana Cuff, "The Making of an Architect," In *Architecture. The Story of Practice* (Cambridge: MIT Press, 1991), 109–154.

257 Peter Eisenman, "Director's Report," June 19, 1974. Source: CCA Montréal, IAUS fonds: A.1-2.; Peter Wolf, quoted in: Richard Meier, minutes of the annual meeting of the Board of Trustees, December 10, 1974. Source: CCA Montréal, IAUS fonds: A.1-2.

fully institutionalized as the cornerstone of the Institute's complex financial apparatus. As the backbone of the Institute, the one that paid the bills and thus financed all the other activities, this apparatus retained its essential features until 1985, when after ten years as an architecture school, the Institute finally ceased its operations. In total, the Institute taught more than five hundred students, not all of whom were to become architects, and shaped their architectural approach, thinking, and practice with a distinctly postmodern habitus or social behavior.

In assessing the institutional significance of architecture education at the Institute, its social, economic, and cultural impact within the framework of an institutional history, two key aspects stand out: First, after the Institute was formally recognized as an educational institution by the Board of Regents of the State University of New York in early 1973, it no longer operated on an autonomous basis. From this point on, its authority to designate itself an architecture school was granted from the outside, i.e., it could only portray itself as an "alternative" teaching and learning institution *in relation to* established colleges and universities, even if it did come to influence these.[258] Second, the Institute managed to contribute to the liberal arts and practical education of ambitious young architecture students, some of whom would one day become part of the architectural and academic elite even though, unlike other established schools of architecture in the USA—and this is truly astonishing—the Institute *never* offered accredited degrees in architecture. This is especially striking when one considers the non-linear process of academic and professional socialization and acculturation in architecture that the Institute's students went through, i.e., the process of growing up in and assimilating into an increasingly differentiated and globally networked architecture culture.[259] In the years that followed, it would become noticeable that the Institute, with its multi-pronged, continuous teaching and learning opportunities, was indeed pursuing what was then understood as a humanistic ideal of education—yet at the same time, that this ideal turned upon a bourgeois principle of measurable output, ultimately contributing to the marketization and privatization of education in the broader architectural field.

258 Architecture historian Mary McLeod published an essay on the evolution of architecture studies from 1968 to 1990 in a comprehensive survey of the history of architectural education in the United States, see McLeod, 2012. McLeod attributes the pervasiveness of postmodernism, which she sees as "part of a larger epistemological shift" and equally as a "new stage of consumer capitalism," to the work of architecture schools as a "leading force" where she identifies a shift in "values and forms." However, she does not write about the Institute to which she herself belonged, only about accredited schools.

259 Architecture sociologist Dana Cuff refers to the architecture school and the architecture office as the two most important sites of socialization and acculturation in the education of aspiring architects, see Dana Cuff, "The Making of an Architect," in Cuff, 1991, 109–154. For an early essay arguing that the Institute, as a new educational and cultural institution outside the system of higher education in the United States in the 1970s, took on precisely these functions, see Kim Förster, "*Arch+* features 19. Die Netzwerke des Peter Eisenman," supplement to *Arch+*, no. 210 (2013).

2.1 Operating as a Teaching Facility

Launched in the fall semester of 1974, the Institute's new "Undergraduate Program in Architecture" proved to be a truly innovative and comprehensive education offering—nothing of its kind had ever existed before in the USA. Personally directed by Eisenman, the program was targeted toward undergraduates enrolled at liberal arts colleges in the Northeast that did not offer a major in architecture. The program's educational objective, as articulated by the Institute, was to teach architecture as one of the humanities—on par with art history, literature, and music—rather than as a purely artistic or technical subject. Architecture was to be embedded in a European modernist, humanist cultural ideal.[260] The undergraduates, hailing from a variety of majors at their home colleges, were offered the opportunity to spend their junior years—the semesters typically chosen for study abroad—at the Institute in New York rather than at a renowned university in Europe. At the Institute, they specialized in architecture, regardless of whether they ultimately planned to become architects or not. The Institute made a point of differentiating this one-year introduction to historical, theoretical, and aesthetic approaches to architecture, available for an initial tuition fee of US$3,000 (the fees increased successively over the years), from the undergraduate studies being offered by architecture schools whose primary focus was vocational training.[261] The Institute's offering, in other words, skillfully targeted an entirely new market, if not outright creating it. After a year, the Institute's students returned to their original colleges, without a degree accrediting them to work in the architecture field, to finish off their final year of studies. By repositioning itself within the American education landscape in this way, the Institute profiting off education reforms and an education boom taking place across the USA—a phenomenon that John Thelin, a professor of higher education and public policy, describes in his *History of American Higher Education* as "a proliferation of new degree programs and fields of study."[262]

Unlike previous years at the Institute, architecture was now taught as an integrated discipline within the didactically organized course offering. Designed as an ambitious introductory course, the Undergraduate Program comprised

260 The American educational ideal of liberal arts education favors broad fundamental knowledge over subject specialization, see Henry Crimmel, *The Liberal Arts College and the Ideal of Liberal Arts Education* (Lanham: University Press of America, 1993), here "The Principle of Liberal Education" 115ff.

261 In New York, an undergraduate major in architecture was first established at Columbia University at Columbia College under Robert Stern beginning in the 1973–74 academic year, at the same time as the establishment of the master's program at the Graduate School of Architecture and Planning (GSAP, now GSAPP, the second "p" added for preservation, a program that ran since 1964). Stern initially continued to teach the course "Elements of Architectural Design" in the "Culture: Theory" track at the Graduate School during the 1972–73 academic year and was then the architecture faculty representative at Columbia College beginning in 1974–75.

262 Thelin, 2004, 319.

five learning clusters that roughly corresponded to a liberal arts curriculum. At the center stood courses on the history and theory of architecture, planned and taught by Frampton ("The History of 20th Century Architecture") and Gandelsonas ("Elements for an Architectural Theory"). Other courses covered "Urban Development: History and Theory" (taught by Wolf) and "Structures" (Robert Silman). The curriculum was rounded off by the "Design Tutorial" taught by Fellows, Visiting Fellows, and New York-based architects, ensuring a more pluralistic approach rather than a focus on problem-solving. In the first year, in addition to Eisenman, the tutorials were taught by William Ellis, Mario Gandelsonas, Rem Koolhaas,[263] and Giovanni Pasanella.[264] As is only natural, the overall curriculum of the Undergraduate Program was conceived to match the knowledge and interests of the Fellows. The theory-laden lectures and intense discussions of texts meant that students were primarily brought into contact with the "heady world" of architectural ideas and criticism, and less with the practice and methods of building and construction. The faculty's expertise and experience, however, were paired with the didactic and methodological principles of a liberal arts education—such that Frampton's history course, for instance, focused on modernism as a case study for a temporally delineated and stylistically defined period of architectural history, while Gandelsonas's theory course was founded on close analyses of classical architecture texts. Each course was conducted as a morning lecture or seminar, followed by an afternoon design studio. First semester design studios focused on abstract tasks and went on to tackle concrete projects in the second semester. Students were assigned to a single architect in groups of five, enabling the Institute to ensure intensive supervision (and a student-to-teacher ratio of three to one). Ultimately, the Institute sought to confront the students with all the rites and rhetoric of an architecture education: the design studio as an action-oriented (and yet exploitative) form of instruction, long periods of intensive day-and-night work preparing presentations, and arguments and counterarguments as the fundamental communicative form for feedback and final reviews. And with its focus on history and theory, the Undergraduate Program also reflected the reorganization and redisciplining of architecture education that was taking place in a broader sense in the wake of 1968—as demonstrated elsewhere, for example, in the Columbia University curriculum reform, or in

263 Rem Koolhaas was at the Institute from 1973 to 1976, working on his monograph *Delirious New York*, which he researched at the Public Library at Bryant Park with the help of Institute students and interns. In the fall semester of 1974, while still assigned as a tutor in the Undergraduate Program, he was simultaneously employed as a lecturer at Columbia University, where he taught a course on "New York: An Architectural Appreciation," for which students could receive credit in the "History/Theory/Criticism" track. It was not until 1975–76 that Koolhaas was granted Visiting Fellow status at the Institute.

264 Giovanni Passanella had taught design at Columbia University since the mid-1970s, practiced independently as an architect since 1964, and was the last to be involved in the Twin Parks UDC project with three projects. Of all the Institute's design tutors, he was by far the most experienced.

the rise of new doctoral programs.[265] After the various teaching experiments of the 1960s, socio-political approaches such as "neighborhood-based advocacy planning" and "school-based community design" were quickly being rolled back across the country in favor of expanding degree programs and focusing on approaches that were centered around the arts and humanities.[266] The multidisciplinary, even "polytechnic" approach of the Institute's Undergraduate Program should therefore not obscure the fact that the curriculum was no longer centered around urban studies as a collaboration between sociologists, artists, planners, historians, etc. From 1974 onward, the architectural theory taught by the Fellows was more about a broad-strokes "Architecture" with a capital "A."

In addition to the Undergraduate Program that began in the fall semester of 1974, the Institute also launched a comprehensive series of evening lectures titled "Architecture," inspired by the course offerings of the New School for Social Research.[267] To help organize and implement the events for the "Evening Program," conceived and advertised as a continuing education offering for adults, the Institute secured funding from public art foundations—from the New York State Council on the Arts (NYSCA) in 1974, and subsequently from

265 After his transfer to Columbia University with the 1972–73 academic year under Dean James S. Polshek, Frampton was instrumental in designing the curriculum of the new master's program at GSAP and also taught there, first as an associate professor and from 1974 with tenure. He taught the theory course "Critical Comparative Analysis" and the history course "Thresholds in Modern Architecture" in the "Cultural Matrix" track for a long period of time. With the 1975–76 academic year, Frampton was involved, albeit sporadically at first, in the new housing studio led by Richard Plunz, which included a field trip to the Institute's low-rise housing development in Brownsville, Brooklyn, while it was still under construction. After returning from England in 1977, Frampton taught both history and theory courses and one of the housing design studios. The assignment he gave to his students was a refinement of the low-rise housing prototype as a perimeter block development in Manhattan.

266 Anthony Schuman, "The Pedagogy of Engagement: Some Historical Notes," *An Architektur*, no. 19–21, 2006: "Community Design: Involvement and Architecture in the US since 1963." Curricular reforms made after 1968 in the wake of student unrest in American higher education, especially due to the large role of architecture students in the political protests at schools of architecture, brought about an abolition of the Beaux Arts system and led to a focus on social and political issues such as "low-cost housing, urban revitalization, community development, social needs" and an enforcement of non-hierarchical teaching and learning methods; see McLeod, 2012. But by the end of 1973 in the United States, the broad politicization of the student body that had endured since 1970 had largely faded; see Thelin, 2004, 327.

267 The New School for Social Research (now The New School) was founded in 1919 by a group of progressive thinkers who openly criticized U.S. policies and resigned from their positions at Columbia University after an act of censorship. Modeled on the Volkshochschulen in Germany, it created a model of continuing education for adults and, beginning in 1933 as the University in Exile, provided a home for intellectuals who had been stripped of their teaching positions by fascists in Italy or were forced to flee the Nazis. With its curriculum, the New School had a significant influence on the social sciences and philosophy in the United States. In addition to pioneering courses in "African-American History and Culture" (1948, W.E.B. DuBois) and "Women's History" (1962, Gerda Lerner), the New School was also known for courses in "Creative Arts," such as courses on film history, photography, and jazz. Frank Lloyd Wright taught architecture at the New School.

the National Endowment for the Arts (NEA). Eisenman brought the young architect and Columbia graduate Andrew MacNair into the Institute's fold to coordinate the "Architecture" series; MacNair was subsequently in charge of all the Institute's public events. From 1974 to 1977, "Architecture" encompassed multiple lecture series running in parallel, offering a lecture every evening of the week throughout the semester—an unbelievable feat, involving an immense amount of work—for paying audiences and the cohort of students and interns. The Institute's "continuous education" offering was geared towards a broad audience, from young architects, and architecture students, to anyone from New York with an interest in architecture (the core target groups included artists from the East Village and SoHo and wealthy clients from the Upper East Side). Even though this form of education, like the Undergraduate Program, blurred the line between cultural product and commercial bid, "Architecture" nevertheless set itself apart as being an unusual extracurricular public education offering that was partly academic, partly popular, with a focus on the history and theory of architecture. Topics ranged from hot-button urban planning issues in New York to presentations by sought-after architects and designers. The individual lecture series, this time geared towards socializing and acculturating professionals and laypersons, i.e., potential clients, were hosted by the Institute's Fellows (including Frampton, Gandelsonas, Agrest, Vidler, and—most frequently—MacNair), Visiting Fellows, such as Rem Koolhaas, and collaborators like Colin Rowe and—most frequently—Robert Stern, all of whom brought their own individual interests to bear on the events.[268] What distinguished the Institute's adult education from existing offerings at New York art institutions, such as those by the Museum of Modern Art or the Metropolitan Museum of Art, was that the "Architecture" series focused solely on topics related to architecture and planning and—with ticket prices of US$60 per series—was slightly cheaper than its competitors. By the end of the 1974–75 academic year, thanks to the Undergraduate Program and the Evening Program (which was registered as a major public success, drawing over 400 participants for the six series offered in the first year), the Institute had managed to establish itself as a new purveyor of architectural knowledge through its innovative pedagogy, topical architectural debates, and broad-impact PR campaigns. Ultimately, this move towards education and culture spoke to the Institute's new openness, but it also testified to the postmodern sensibilities of the 1970s: the meeting of highbrow and popular

268 Stern belonged to the Mayor's Task Force on Urban Design as a young architect, encouraged by Philip Johnson, and from 1969 ran a practice with John Hagman. At the age of thirty, he published the survey volume *New Directions in American Architecture* (1969). From 1970–71 he taught at Columbia University and briefly at Yale University in 1972–73. In 1973 he positioned himself for the first time as an opponent to Eisenman, leading the "Grays" in their polemic against the "Whites." From 1973 Stern was president of The Architectural League in New York and from 1975 director of the Society of Architectural Historians. He was active at the Institute from 1974–75 and took a decisive role, although he did not hold an official position for a long time.

culture within the framework of "Architecture" as well as the fact that the Institute could cater to the alternative scene alongside a bourgeois public (for a more extensive discussion of the Institute as a cultural space, see chapter three).

Liberal Arts Education

The Institute's performance as an alternative architecture school for very different cohorts of students was only possible thanks to its institutional acceptance by established higher education institutions—the colleges that conferred a certain authority on the Institute as an educational facility in the first place, and that additionally lent it their administrative, financial, conceptual, and intellectual support. One of the most important collaborative partners in developing and running the Undergraduate Program, in both institutional and pedagogical respects, was Sarah Lawrence College in Bronxville, New York. Located north of Manhattan, a thirty-minute drive from the Institute, Sarah Lawrence was one of America's leading liberal arts colleges in the 1970s. When Eisenman first presented the idea of a one-year study program in early 1974, both institutions initially hesitated to engage more seriously with the proposition, until students launched an initiative expressing their interest. As was the case earlier in the Institute's history, personal contacts then proved decisive for institutional developments. Under the leadership of its then-president, Charles DeCarlo, Sarah Lawrence contributed significantly to the details and design of the Institute's new education program, ultimately playing a key role in facilitating its creation.[269] The preparations involved intensive consultations between representatives of the Institute and Sarah Lawrence administrators (in addition to DeCarlo, the chairman of the Curriculum Committee, Robert Wagner, was heavily involved), as the principals hashed out the fundamental principles of a humanities education and the concrete didactic content of the courses.[270] The conditions for the Institute were favorable; there had already been isolated attempts among Sarah Lawrence students and faculty to offer design courses at the college. As a result, the Sarah Lawrence directorship saw enormous opportunities in collaborating with the Institute—not only did the Institute have superlative architects and academics as teaching staff, but its upper-floor studios also afforded plenty of space for

269 In the archives of Sarah Lawrence College in Bronxville, New York, more than in any other archives, there are a number of folders of original and copied documents relating to the Institute's educational programs, from which the conceptual and administrative history of the Undergraduate Program emerges, including the *Progress Bulletin*, which reported on the current status and latest developments of the educational programs on a quarterly basis, thus providing a good understanding of the organization and program of teaching at the Institute, or, as an equivalent medium, the *Student Bulletin*, which communicated all the important information for the students participating in the Institute's educational program.

270 The formulation of the courses in history and theory as part of the Undergraduate Program at the Institute was initially rejected by some members of Sarah Lawrence's faculty as being too remote and too exemplary from a didactic point of view, while the school administration agreed in principle to cooperate, and the Curriculum Committee supported the concept.

teaching design. Beginning in May 1974, interested students could apply for a spot in the coming fall semester. It was not until then that the dean of studies at Sarah Lawrence came together with Dean Robert Wagner to formulate concrete requirements for the Institute. At the same time, DeCarlo shared his concerns and recommendations in a personal letter to Eisenman: Sarah Lawrence was to have a significant hand in shaping the Undergraduate Program's pedagogical approach. This was the starting point for the Institute to begin offering its new educational program at other liberal arts colleges in the eastern Coast United States. Financial support for the first three years of this unusual educational initiative was provided by the John Edward Noble Foundation, a private nonprofit chaired by June Larkin that specialized in working with educational programs and had previously collaborated with MoMA to support its educational offerings. In June 1974, before the Institute's leadership had even officially submitted a grant application, the Noble Foundation donated US$15,000 in seed capital.[271] This was put toward establishing administrative structures and a concrete lesson plan, making the Noble Foundation the third key institutional actor involved in the establishment and implementation of the Undergraduate Program.[272] Following this, a bulletin was drafted as the main medium of communication, which framed the offering as a humanistic education in architecture, portraying it as an alternative to existing offerings, and articulating the goals of the Institute and Sarah Lawrence with regard to the program.[273] According to the bulletin, seventeen students had already shown interest in the first year. If one examines how the Institute envisioned its new definition and role, it is telling that the document describes the Institute as an "arbiter" of various pedagogical offerings, as opposed to an "advocate" of any one political organization. In actual fact, the Institute, despite its early years as a "non-profit educational, research and development center" from 1967 to 1969, had scarcely any pedagogical experience to speak of, let alone a

271 The Institute's leadership submitted an initial, comprehensive concept paper for the Undergraduate Program in Architecture to the Noble Foundation in late June 1974, which included a precise analysis of architecture education for undergraduate students, the general objectives of the educational program, the specific structure of the courses, a budget plan for the next three years, and information about the faculty and the Institute; see IAUS, "Proposal for an Undergraduate Architectural Major," n.d., Source: CCA Montréal, IAUS fonds: AR2018.0112.001; see also Peter Eisenman, "Director's Report," June 19, 1974. Source: CCA Montréal, IAUS fonds: A.1-2. A "[t]hree year grant for a new undergraduate major in architecture among five colleges and universities, and the Institute" was approved; see "The Next Thirty Years 1970–2000" in *The Edward John Noble Foundation 1940–2000. A Report at Sixty Years* (New York: The Edward John Noble Foundation, 2000), 51.

272 In the 1970s, the Noble Foundation sponsored art education programs at major cultural institutions in Manhattan, including, since 1972, those at MoMA. Initially, grants totaling US$ 1 million were provided by the private foundation to revitalize the Educational Department there, see The Edward John Noble Foundation, 2000, 21. Mrs. June Larkin, née Noble, had once studied at Sarah Lawrence College herself, where she had graduated with a bachelor's degree in 1944. She then served as a trustee at Sarah Lawrence College from 1964 to 1973, and even chaired the Board of Trustees there during the last two years of her tenure.

273 IAUS, "Bulletin no. 1. Program in Undergraduate Architectural Education," June 24, 1974. Source: CCA Montréal, IAUS fonds: AR2018.0112.001

fleshed-out curriculum. Yet the Institute intended to profit off the intellectual and academic capital it had accumulated; here and in all its grant applications, it made a point of invoking the activities and expertise of its Fellows, who all taught at renowned schools of architecture. In doing so, it drew on an expansive definition of education, encompassing "university exhibitions, lecture series, [and] special issues of magazines." Further qualifications noted by the Institute in its application to the Noble Foundation included the housing project it was realizing in Brooklyn, as well as *Oppositions* journal, launched the year before. Over the course of the summer, Eisenman was busy promoting the Undergraduate Program, and by the fall of 1974, the Noble Foundation had upped its support to an annual sum of US$60,000. The private foundation's only condition was one of an institutional nature—namely, that any funding was contingent on the Institute being recognized by all cooperating colleges as a valid site of external studies.

The new teaching duties at the Institute in the academic year 1974–75 led to an increased professionalization of its administrative structure. Although it had not been conclusively determined at the outset whether the Undergraduate Program would function as a form of occupational training, the first academic year already saw students coming to the Institute not only from Sarah Lawrence but also from a handful of other liberal arts colleges. The Institute offered them an opportunity to specialize early, undertaking a kind of "pre-professional study" that qualified them for a future degree in architecture—a phenomenon that has been described as a "new vocationalism."[274] The Institute's immediate popularity as an architecture school can perhaps be attributed to its prior reputation, the renown of some individual Fellows, or simply the gravitational pull of New York. However, it was also bound up with the more encompassing changes that the American academic system was undergoing at the time. The greatest achievement of the Institute—and in particular of expert networker Eisenman—was that, in cooperation with Sarah Lawrence, it managed to construct an entirely new education network for the Undergraduate Program in a remarkably short time. The two initiators soon joined forces with other colleges such as Amherst, Oberlin, Smith, and Wesleyan to form a consortium that had not existed in this format before, and which was quickly expanded to include more colleges.[275] In his triple role as Institute director, director of the Undergraduate Program, and teacher, Eisenman invested much of his subsequent time in expanding his relationships with various liberal arts colleges. Once per year, Eisenman would take a "road trip" with Regina Wickham (an administrator in the Undergraduate Program), going from college to

274 Thelin, 2004, 327.

275 By the end of 1974, the Institute was already negotiating with Brown and Hampshire Colleges, as well as Stanford University. Furthermore, Bennington, Mount Holyoke, and Swathmore College were under consideration as potential partner colleges of the Institute in the Undergraduate Program; Peter Eisenman, "Director's Report," December 10, 1974. Source: CCA Montréal, IAUS fonds: A.1-2.

college as a kind of traveling architecture salesman, making contacts with potential partner schools, promoting the Institute's education offerings, and recruiting interested students (while also distributing the latest issue of *Oppositions*). Mutual agreements between the Institute and the cooperating colleges ensured that interested students would be allowed to spend two semesters at the Institute and be awarded credits for their coursework. Oberlin College acted as an interface for all the GLCA colleges that were not officially part of the consortium. Arrangements were also made for the tuition fees paid to the colleges to be forwarded directly to IAUS Central. The humanities focus of the Institute's five-part curriculum ensured that students were awarded course credits for completing a one-year course of study, even though the faculty's standards for evaluation weren't always made transparent or consistently applied. Ultimately, the Undergraduate Program enabled the Institute to tap into an important revenue stream that, over the following years, would cover the majority of its operating budget while giving its Fellows the opportunity to secure a steady income or top up their university salaries.

To further validate and implement the Undergraduate Program, the Institute set up an Advisory Committee to provide oversight and expertise. This advisory body, which met at least once per year to discuss and regulate administrative and academic issues, included Institute faculty and representatives from the cooperating colleges, as well as the president of the Noble Foundation, June Larkin, and the deans of two leading architecture schools, James Polshek of Columbia University and William Porter of the Massachusetts Institute of Technology (MIT). Institute students could also elect delegates to represent their interests and contribute their experiences. The agenda covered topics ranging from admission procedures, the centralized handling of tuition fees, the awarding of credits by individual colleges, and the teaching of individual courses, to the function of the design tutorials. More than anyone else, Polshek proved to be a passionate supporter of the Institute's education offerings, predicting early on that they would have a significant and lasting impact on architecture education in the USA. In fact, the conception and implementation of the Institute's Undergraduate Program, with its focus on history and theory, was strongly influenced by the standards at Columbia University, where an undergraduate major in architecture had been introduced under the leadership of Robert Stern in 1973–74, while in its early years, the Institute had been more strongly influenced by the instruction offered at Cornell University and Cooper Union. In early 1975, in collaboration with Stern, the Institute organized a conference on "Undergraduate Non-Professional Architectural Education" to demonstrate its newfound role. While the opening of the conference was celebrated at the Institute on January 10, 1975, the conference itself was held at Columbia's Avery Hall on January 11.[276] Organized in

276 Suzanne Stephens, "Architecture for Undergraduates," *Progressive Architecture* (March 1975), 23.

partnership with Columbia (Stern, who had proven instrumental in launching the Evening Program, had once again become an important partner for the Institute) and the Architectural League of New York, the conference was dedicated to the structures, methods, and goals of existing undergraduate programs. "The various components of such curricula," the statement of intent proclaimed, "will be discussed, including the study of history, design, technology, social sciences and the broader understanding of the relationship between architecture and culture."[277] The Institute presented its offerings as appealing to universities that ran a graduate school of architecture, but no corresponding undergraduate program as well as liberal arts colleges with no architecture major. Merely dedicating a one-day conference to the topic, with over seventy registered guests, meant that the Institute could justifiably claim to occupy a leading role in the emerging field of undergraduate architecture education, even though the conference proceedings were never published, and no survey of existing curricula was conducted. The Institute's transformation into an educational institution was completed in the summer, when Charles DeCarlo and June Larkin were named as trustees, along with the Canadian-born but California-based architect Frank Gehry. With the appointment of DeCarlo and Larkin, the Institute's new pedagogical interests would now be represented and embedded within the organizational structure; they were not only expected to contribute their pedagogical experience and institutional capital but also to serve as the Institute's envoys to the outside world.

Institutional Growth

Restructuring meant that, beginning in the fiscal year 1974–75, the Institute would finance its operations primarily via income from architecture education, with additional support coming from public grants and private donors. Most of the Fellows, Visiting Fellows, and faculty were awarded teaching salaries. The Undergraduate Program and the "Architecture" series represented a conscious decision by the Institute to open up to the outside world. In epistemological and sociological terms, the Institute's leadership and Fellows, who portrayed themselves not just as educators, but as the new elite in the architectural world, had come to realize that the Institute was a small, esoteric circle that depended on a larger, exoteric circle if it was to have any chance of survival as an educational and increasingly cultural institution. It was only thanks to this larger circle that projects like the costly and time-intensive journal *Oppositions*, as well as educational programs and public events were able to flourish. The same went for the Institute's incipient "Exhibition Program" and the labor-intensive publications that were soon to follow. All of these were of pedagogical value, in terms of content and design, but they couldn't sustain themselves or be launched independently. At the same time, the collective of architects and academics at the

277 IAUS, "Conference on Undergraduate Architectural Education. Statement of Purpose and Organization," Source: CCA Montréal, IAUS fonds: AR2018.0112.001

Institute was, thanks to its publications and combined pedagogical and cultural efforts, accruing a wide range of networks and public audiences. Whether they considered themselves to be a group, an organization, or an institution, they were presiding over a generational change—one simultaneously bound to tradition *and* sworn to destroy it, dedicated to producing, reproducing, and disseminating new historical and theoretical knowledge and creating new power structures. By investing in architectural education and developing or expanding complex networks, such as the establishment of the "Fellows" as a new form of work and organization, the Institute had managed, now that its survival was ensured, to create a context for intervention on the one hand, and on the other, to educate the next generation of architects while socializing them as future producers and consumers of the Institute's cultural products.

The Institute's success as an architecture school was initially due to the fact that Eisenman, in his capacity as Institute director, was able to incorporate nearly the entire circle of Fellows, especially Frampton, Gandelsonas, and later Vidler, into the Undergraduate Program as teachers. In return, these Fellows were given the opportunity not only to teach and publish there but also to hold and chair lectures within the framework of "Architecture," or participate in group exhibitions as part of the "Exhibition Program" launched in 1975 (also organized by MacNair). In other words, they were given the opportunity to distinguish themselves as culture producers. In this sense, architecture teaching at the Institute didn't stop at conveying basic skills and knowledge: the Fellows, Visiting Fellows, and faculty at the Institute were initiating students into a new way of thinking, as well as appearance, needs, preferences, and habits. It must have been during the academic year 1975–76 that their teaching, the lectures and exhibitions, and the Institute's publications brought about a fundamental paradigm shift for the profession and the discipline, grounded in a historically and theoretically informed reflection on architecture itself and legitimization of architecture as a self-aware practice.[278] Moreover, this was taking place at a time when "heroic modernism," after the deaths of master architects like Le Corbusier and Ludwig Mies van der Rohe in the late 1960s, was yet to be historicized (although the International Style had already sold out in service of corporate architecture). Beyond the dissemination of a specific kind of self-reflexive and self-legitimizing architectural knowledge—conceived, it is important to note, outside of the requirements of technical, social, and economic reality—the Institute's program extended to the teaching of cultural competencies

278 Both Ludwik Fleck (1935) and Thomas S. Kuhn (1962), with their epistemological approaches to the theory and history of knowledge, also posited a sociology of knowledge production and dissemination; on Fleck's theory of thought styles and thought collectives, see Ludwik Fleck, "Introduction to Thought Collectives" & "Further Observations Concerning Thought Collectives," in Fleck, [1935] 1979, 38–51 & 98–111; on Kuhn's theory of paradigm shifts, see Thomas S. Kuhn "Introduction: A Role for History" & "The Nature and Necessity of Scientific Revolutions," In Kuhn, 1962, 1–9 & 92–110.

for decoding an architectural vocabulary, and in structural and linguistic terms, the internalization of formal and semantic codes to help students comprehend architecture in all its complexity and, on that basis, formulate a new language.

In this period, Eisenman, who may not have been an equal among equals, continued to be the most vibrant and enigmatic personality at the Institute: as Institute director, he was responsible for hiring new faculty, and as director of the Undergraduate Program, he was responsible for the programming and curation with regard to pedagogical models and political strategies. In 1974–75, he ran his own one-off design studio and, even when he wasn't teaching, regularly attended the final reviews and supervised interns. Interns came to the Institute expressly to work for him, and Eisenman, whose office operated from the Institute, used them as a workforce for his own projects. What is more, not only did Eisenman share faculty and students with Cooper Union, where he had taught design since 1968, but the institutional relationship extended to a special agreement to recognize internships and waive tuition fees. When it came to personnel decisions and curriculum development, the Institute was further influenced by the teaching programs of other East Coast schools of architecture, among them Columbia University under Dean Polshek and Princeton University under Dean Geddes. Cooper Union, Columbia University, and Princeton: these were the architecture schools where some of the Fellows taught theory, history, and design, subscribing to a formal and aesthetic approach that sought to reinstate architecture as an independent practice, if not an art form, and to attach more importance to architecture—which they saw as having been lost. Eisenman and the other Fellows subscribed to and taught a critical line of thinking, one that sought to foster a better understanding of the major historical, social, and cultural contexts (less so political and economic) that had influenced architecture since the modern age.

2.2 Expanding Educational Offerings

Now that the Institute had succeeded, within the span of only a year, in establishing itself as an architecture school, positioning itself in the academic landscape with its undergraduate training and in the metropolitan culture of New York with its adult education program, after 1975, it increasingly began to target the broader American education market and expanded its offerings as an educational service provider. Next to the Undergraduate Program in Architecture and the "Architecture" series, the Institute's leadership sought out niche products to diversify its education offerings, and thus address new target groups and tap into new sources of revenue. It was certainly aided by the fact that the concept of "lifelong learning" was gaining traction in American society. The spectrum of education offerings was expanded to include the following programs launched in rapid succession in the 1975–76 academic year, most of which were rather short-lived: an integrated "Undergraduate Program in Planning" for undergraduates, slated to run in parallel to the Undergraduate Program in Architecture, the practically oriented "Work/Study Program in Preservation and Adaptive Re-Use" for undergraduate and graduate students, which did not exist in this form at any other school of architecture, and even a "High School Program" to support young talent, in a similar form, albeit different motivation, as one that had already existed at Cooper Union for some time. As a result of this expansion, the Institute was now offering educational programs for every age group (as recommended by the *Princeton Report*, a 1967 guide published by Robert Geddes and Bernhard Spring on the study of environmental design, which included architectural education).[279] Once again, the individual educational programs were institutionalized via collaborations with partner colleges. In its proposals shared with other educational institutions, in info brochures for students, and at the annual meetings of the Board of Trustees, the Institute repeated, mantra-like, that these education offerings posed no threat to established schools of architecture. Instead, they were portrayed as complementary offerings.

By 1975–76, the Undergraduate Program in Architecture was already an authority: in its second year, twenty-four students, hailing from nine different colleges, signed up for classes. Moreover, the Institute had already implemented a first reorganization of the program: having taken over Frampton's history course, Vidler now taught "Architecture in the Age of Revolutions," which now accompanied Gandelsonas's theory course.[280] Another major change was that Eisenman put Diana Agrest in charge of all the design tutorials and coordinating the tutors. The Institute hired William Ellis, Colin Glennie, Andrew MacNair, Stephen Potters, Myles Weintraub, Todd Williams, and Stuart Wrede to lead design courses alongside her. Agrest formulated the required tasks for the first semester: students would initially design a building and define its spatial program without making stylistic specifications, the second stage then encompassed the design of a residential building and the creation of a public square. Design tutorials took a postmodern approach, especially in the second semester; the large variety of individual approaches led to a prevailing air of polyphony and polysemy. Agrest's own approach, grounded in French theory, was characterized by a multi-faceted reading of public space. Her teaching shows that she conceptualized the city as an urban fabric of different structures

279 Geddes and Bernard, 1967.

280 Frampton was in London from 1975, initially on a Guggenheim Foundation fellowship, where he researched and taught on Le Corbusier for two years at the Royal College of Art. He returned briefly to the Institute in the spring semester of 1976 to offer a six-session survey lecture on Le Corbusier's work as part of the Undergraduate Program, in addition to Anthony Vidler's course. Although this series of lectures expanded the course offerings, creating elective options for the first time, Frampton's presence was viewed rather negatively on the grounds that the course distracted from the other offerings.

or as a sequence of individual objects and took this as the starting point for teaching design. Ultimately, the Institute's collective pedagogy was heavily influenced by postmodernism, understood not only discursively or aesthetically, but more generally as a cultural phenomenon and pedagogical concept.

Market Diversification

The Undergraduate Program in Planning was headed by Wolf. It was developed as an equivalent to the Undergraduate Program in Architecture but differed primarily in that it was designed to be somewhat more pragmatic, even though it similarly aimed at combining theory and practice into an integrated "work/study" program.[281] Because it was necessary for the educational program to be recognized by multiple departments at every college, administrative issues delayed the launch and institutionalization of the program by a year. Unlike other colleges, which predominantly emphasized formal aspects of urban planning, the Institute stressed a multidisciplinary approach, enabling undergraduates to better prepare themselves for graduate studies by obtaining a more practical grounding. To this end, the first module of the Undergraduate Program in Planning comprised two courses: one on the history and theory of urban planning (taught by Wolf), which explored the subject in its socio-cultural, technological, and economic aspects, and one on the social and psychological aspects of urban planning (Robert Gutman), which focused on planning public spaces and public housing. This was followed by a second module featuring case studies on twentieth-century urban planning (Craig Whitaker) and a research seminar with weekly excursions to urban infrastructures (Whitaker and Myles Weintraub). In the third module, students enrolled in an "urban design studio" (Weintraub) where they worked on independent projects or in small groups and where they were taught technical and analytical skills such as drawing, mapping, and model making.

Unlike the two undergraduate programs, the Work/Study Program in Preservation and Adaptive Re-Use, headed by Ellis and Stuart Wrede, was aimed at students from established architecture schools that did not offer this kind of program in their curricula.[282] The educational program was project-based, and the students' assignment was to spend a year preparing an inventory of buildings with landmark status. The Institute received an initial grant of US$5,000 from the Alfred P. Sloan Foundation to develop and organize the program. After the study was commissioned by the New York Landmarks Conservancy, the

281 Peter Wolf, "Program in Undergraduate Planning Education," n.d. Source: CCA Montréal, IAUS fonds: AR2018.0112.001

282 William Ellis, "Program in Adaptive Re-Use of Old Buildings," n.d. Source: CCA Montréal, IAUS fonds: A.1-2. Ellis had built a consortium of twelve architecture schools for the "Program in Preservation and Adaptive Re-Use:" Cornell University, University of Kentucky, Kent State University, Montana State University, State University of New York, Buffalo, Notre Dame University, The Pratt Institute, Rensselaer Polytechnic Institute, Rice University, University of Southern California, Syracuse University, Tulane University.

Sloan Foundation donated another US$15,000. The Institute was deemed to be qualified for the project thanks to a 1974 adaptive reuse and preservation project led by Ellis on the conversion of a mill site in Cohoes, New York, which the Institute had successfully carried out and publicized.[283] The original vision for the curriculum was quite innovative and was to include courses on history and theoretical issues, legal and financial topics, technology and landmark protection, and a design tutorial. Ultimately, however, no separate course was offered due to low enrollment—only three students and three interns signed up for the 1975–76 academic year. Instead, the program was spontaneously integrated into the Evening Program, for which Wrede chaired two lecture series on the topics "The Future of the Past: Issues in Historical Preservation" and "The New Architecture of Old Buildings: Additions, Insertions, Deletions." Although the series were primarily aimed at the general public, they also fulfilled the requirements for awarding credit to student attendees.[284]

These new additions were supplemented by the High School Program, which was introduced in the fall semester of 1975, also at Eisenman's initiative. The program was aimed at Manhattan's private school students; MacNair was initially appointed to lead it.[285] Unlike Cooper Union's "Saturday Program," which combined early training in architecture with a strategy of critical emancipation, the Institute's High School Program was conceived as an ambitious introductory course on topics relating to architecture and design—comparable to the specialized courses in art, literature, and music offered as youth summer programs by New York colleges at the time. Initially, fifteen high school juniors and seniors would meet at the Institute one afternoon per week to be taught by co-instructors Eisenman and MacNair. These students were recruited via New York Interschool, a consortium of private schools in New York.[286] But by 1976, MacNair, with the aid of a grant from the Noble Foundation, had already developed an ambitious new six-week summer course called "Manhattan: Capital of the Seventies," organized as a kind of mini-Institute for aspiring young architects. The program featured a design studio, lectures by architects and planners,

283 Stephens, 1975.

284 Wrede, who had already spent a year at the Institute in the late 1960s working on "New Urban Settlements" project and was tasked by Eisenman to design his own journal, was again a Visiting Fellow there in 1975–76 and taught design in the Undergraduate Program. Wrede left the Institute in the fall of 1976.

285 One model for the Institute's High School Program was the Saturday Program at Cooper Union, where, since 1968, students from the East Village, i.e., the adjacent neighborhood of a socially disadvantaged population, were taught architecture and the arts to give them access to private art colleges and thus open up educational opportunities; see Kim Förster, "Teaching Architecture, or, 'How to Create Spaces for Teenagers?'," in *Arts for Living*, eds. common room and Kim Förster (New York: common books, 2013), 63–108.

286 The Institute's private partner high schools in New York were Spence, Chapin, Collegiate, St. Bernards, Nightingale-Bamford, Brearly und Trinity.

and excursions, afternoon lectures, and film screenings.[287] Although the High School Program, unlike the other education offerings, did not make a significant financial contribution to the Institute's budget (or, if it did, one that was more indirect than direct), there were both institutional and personal motivations for its creation. Not only did it draw the attention of academic advisors from high schools and colleges to the Institute's new pedagogical concepts, it also served as a broader PR campaign in the local press to help brand the Institute as a serious educational institution.

Beyond diversifying its education offerings after 1975, another strategically clever move in pedagogical, more so in entrepreneurial respects on the part of the Institute's leadership was the decision to professionalize, and even commercialize the Internship Program, the Institute's longest-running educational program. Until 1974–75, internships were still integrated into the Undergraduate Program. Starting in the summer of 1975, however, they would be offered to any college graduate looking to gather first working experiences at the Institute before applying for a degree program in architecture. For a fee of initially US$1,000 (soon raised to US$1,500), Institute interns could either work directly for one of the practicing Fellows or Visiting Fellows on concrete design projects, or they could be engaged in other work at the Institute.[288] Surprisingly, the not-insignificant internship fee did not work as a deterrent. Quite to the contrary, in fact: six interns worked at the Institute in the first summer, and fourteen applied for internships in the 1975–76 academic year. This can be attributed, on the one hand, to the Institute's power of attraction, its reputation and prestige that it had built up and consolidated over the course of time; on the other hand, it can also be attributed to the fact that American society in the 1970s generally accepted that "good education" had its price, as "internships, field experiences, study abroad and numerous other innovations gradually came to be accepted components of the bachelor's degree experience" in architecture education as well.[289] A one-year stay "abroad" at the Institute promised interns an in-depth engagement with, initiation into, and development of a specific habitus and thought styles they sought to learn from Eisenman and other Fellows. At the Institute, interns were confronted with the most important theoretical and historiographical debates of the day while coming into direct contact with an Institute Fellow (possibly one of their personal heroes)—even if their education was more about conceptual skills and abilities than practical experience. In exchange for the interns' collaboration on individual or collective projects, the Institute offered intensive supervision in the

287 IAUS, "High School Program" leaflet, 1976. Source: Walker Arts Center

288 The following Fellows offered internships: Diana Agrest, William Ellis and Stuart Wrede, Mario Gandelsonas, Andrew MacNair, Stephen Potters and Todd Williams, Myles Weintraub, and Peter Wolf.

289 Thelin, 2004, 330.

preparation of a portfolio—which was crucially important for students applying for a coveted spot in a master's program in architecture. The student bulletin expressly stated that the Institute would assist them in the selection of the right architecture school: "It is the intention of this program to work closely with the graduate school to develop a mutually beneficial and interlocking relationship."[290] Crucially, the Institute operated in close collaboration with college staffers responsible for handling applications and acceptances, and Fellows provided sought-after letters of recommendation to students at the end of their one-year studies. By advising interns on their choice of university, the Institute played a key role in their education, if not their career. What the Institute had effectively introduced, in other words, was a market in recommendation letters.

School Routine

Within a very short time, the Institute, formerly more interested in its own research and design projects than in mentoring students, had managed to fundamentally transform itself, as described by Goldberger. With its education offerings and public events, it had morphed into a new educational center of basic and further training for future architects and planners, an unusual site for communitization and communication not only between Fellows, Visiting Fellows, staff, students, and interns but also the New York public in general. During the day, the Institute was enlivened by its busy educational programs, while crowds flocked to the evening lectures every night of the week during the semester. Starting in 1975, exhibitions were on constant rotation in the two-storied main hall. The creativity and vitality displayed at the Institute at this point, as evidenced by brochures and posters, had an enormous appeal. In the 1975–76 academic year, just one year after its reinvention, the Institute boasted nearly forty students, the consortium had been expanded to include ten colleges, and the faculty had grown to seventeen members.[291] A total of four hundred people attended the evening lectures in the "Architecture" series. In the context of the Institute's history, it is crucial to note that the focus on education, increased public interest, support from public and private foundations, and networking with academia ensured that the Institute, which in the years before had repeatedly faced insolvency, was finally on solid financial footing. And it wasn't just the Institute that profited from this transformation; students and interns also benefited from the intellectual debates and passionate exchange of ideas taking place there, even if they weren't obtaining a vocational degree. Many of the college graduates who came to work at the Institute came specifically to work for Eisenman, whose *Houses* had been widely published and seen as demonstrating

290 IAUS, "Student Bulletin Two. Program in Undergraduate Architectural Education," February 1975. Source: CCA Montréal, IAUS fonds: AR2018.0112.001

291 In the 1975–76 academic year, most students came from Brown, Hampshire, Oberlin, and Sarah Lawrence Colleges.

a radical attitude toward design. His interns finished a small series of models or drawings for him, which he then displayed in museums and galleries or sold to collectors and archives; only a few interns gained any actual experience in building.[292] Other interns worked for the other Fellows, who were themselves practicing architects, such as Diana Agrest and Mario Gandelsonas, who entered local and international competitions in the mid-1970s from the Institute and planned building projects in Argentina.[293] In addition, interns worked for Visiting Fellows like Rem Koolhaas and helped him research a number of topics for his upcoming monograph *Delirious New York*, which was developed at the Institute and had already been announced in its publications, lectures, and exhibitions. Interns also helped finish drawings for Koolhaas's Manhattan Projects and, after the establishment of his Office for Metropolitan Architecture in early 1975, contributed to his submission to the Roosevelt Island Housing Competition organized by the Urban Development Corporation.[294] Those interns who didn't work directly under one of the Fellows or Visiting Fellows were assigned to one of the Institute's more general productions, such as *Oppositions*, managed by

292 Interns who have worked for Eisenman include Sam Anderson, David Buege, Read Ferguson, Randall Korman, John Leeper, Jay Measley, and Caroline Sidnam. For *House VI*, which Eisenman designed from the fall of 1972, commissioned by Richard and Suzanne Frank, and which eventually was built, he had hired Randall Korman as executive architect. In 1975, when *House VI* was just completed, Eisenman had interns retrospectively make series of drawings of the transformations in the design process based on his sketches, in addition to Randall Korman, Read Ferguso, and Caroline Sidnam. The drawings of transformations of *House VI* were shown at MoMA in the spring of 1975 as part of the exhibition "Architectural Studies and Projects"; they were also scheduled for a May 1975 exhibition in Naples, which was ultimately cancelled. Sidnam, who was one of Eisenman's interns in 1974–75, produced drawings for *House III* and *House IV*. A model of *House II* produced by interns was exhibited at the Institute in November 1976 in the exhibition "Idea as Model". Joan Ockman referred to Eisenman's habit of only theorizing his designs after the fact as "ex post facto diagrams of the design process of his own houses;" see Ockman, 1995, 59. However, she failed to point out that the authorship of the drawings and models was shared, since the interns were involved in the theorizing, and that they were produced to become art objects on the art market. In his monograph *Houses of Cards*, Eisenman mentioned the interns by name, but reduced their contribution to making the drawings or building the models; see Peter Eisenman, *Houses of Cards* (New York: Oxford University Press, 1987), VI. Some names were missing here though, for example David Buege, who built the model of *House II*.

293 Interns who have worked for Diana Agrest and Mario Gandlesonas include Stan Allen, Andrew Anker, Peggy Deamer, Livio Dimitriu, Christian Hubert, Joan Ockman, Miguel Oks, and Pat Sapinsley. They worked on projects for Roosevelt Island, New York (1975), the French Ministry of Housing (1975), La Villette, Paris (1976), and Nicollet Island, Minneapolis (1976). Miguel Oks, who was an Agrest/Gandelsonas intern from 1975 to 1977, prepared drawings for three apartment tower blocks (1977) in Buenos Aires, which were subsequently built.

294 On one of the last pages of *Delirious New York*, Rem Koolhaas states: "Between 1972 and 1976 much of the work on the Manhattan Projects was produced at the Institute for Architecture and Urban Studies in New York, with the assistance of its interns and students." Rem Koolhaas, *Delirious New York. A Retroactive Manifesto for Manhattan* (New York: Oxford University Press, 1978), 255. The students and interns who worked on the Manhattan Projects were: German Martinez, Richard Perlmutter, and Derrick Snare. For OMA's entry to UDC's housing competition for Roosevelt Island, the Institute seconded Livio Dimitriu, German Martinez, and Richard Perlmutter.

Julia Bloomfield, or the library that Suzanne Frank had stepped in to run, both providing access to content and knowledge as an educational resource and educational facility.[295]

Soon the Undergraduate Program teaching staff was recruited from outside the Institute's circle of Fellows and New York architects. Eisenman had always used his frequent visits to Europe to persuade young and ambitious architects and academics to move from Europe to New York—something that added to the allure of the education offered at the Institute.[296] The Institute promised them a creative working environment, plus the opportunity to exhibit, lecture as part of "Architecture," and publish in *Oppositions*. Eisenman even managed to hire some of them to teach at the Institute as Visiting Fellows. In early 1975, Archigram came from England to give a workshop. In early 1976, Bernard Tschumi and Grahame Shane—like Koolhaas previously, who had now left New York and thus the Institute to finish *Delirious New York* in London—both came from the Architectural Association to the Institute, where they taught in the Undergraduate Program as design tutors on temporary contracts. But neither Tschumi nor Shane remained in their intended roles for long, only lasting one semester. Tschumi, however, continued to use the Institute to write for *Oppositions*, to work on his own projects, and to organize the exhibition "A Space: A Thousand Words," while Shane immediately turned his back on the Institute. For both, the Institute proved to be a springboard (as it had for Koolhaas) to establishing a footing in New York and launch their international careers.[297] Despite the constantly changing workforce, the Institute's ever-widening social, educational, cultural, and financial networks were an

295 One of the Institute's interns at the time was Ockman, who, after initially being assigned to Diana Agrest, worked for the *Oppositions* editorial staff as an editorial consultant on the basis of her bachelor's degree in Comparative Literature, where she was paid for her copywork, contrary to standard practice, before going on to study architecture at Cooper Union. Following the internship, Ockman was active at the Institute for a long time: first, from 1976 to 1980, working alongside her studies for the *Oppositions* editorial staff, where she rose to associate editor, and in the early 1980s, after graduation, also as executive editor of Oppositions Books and as editorial consultant to several exhibition catalogues. Finally, in 1981, Ockman was made a Fellow, published in *Oppositions*, and was involved in the Young Architects' Circle.

296 Eisenman made several trips to Europe in 1975 and 1976 for different purposes: a contribution to the "Conceptual Architecture" exhibition (January 1975) at Peter Cook's Art Net with an accompanying conference at the AA, in the context of which he was interviewed by Alvin Boyarsky in the TV studio of the AA; a meeting of *Arquitecturas Bis*, *Lotus International* and *Oppositions* editors in Cadaqués near Barcelona (November 1975); and the curation of and participation in the American contribution to the architecture section of the Venice Art Biennale (July 1976).

297 Bernard Tschumi, who as an architect was interested in a performative approach, quickly turned away from the Institute, as he did not feel at home there, although he positioned himself in relation to the New York architectural scene. His diverse projects, some of which he worked on at the Institute, the Architectural Manifestos, the Manhattan Transcripts, and the various Follies, and which he also published in *Skyline* at the time, were not exhibited at the Institute, but at Artists' Space (1978), the Architecture Room at P.S.1 (1979), Max Protetch Gallery (1981), and Leo Castelli Gallery (1983).

enormous benefit to students. For example, in the summer of 1976, students in the Undergraduate Program had the opportunity (as an addition to their regular curriculum) to work on a design and exhibition project headed by Colin Rowe and Judith DiMaio to redesign Nicollet Island, a now derelict former industrial zone in inner-city Minneapolis.[298] As part of the Bicentennial exhibition "The River: Images of the Mississippi," highlighting the significance of the river for the city, the Institute was commissioned by the Walker Art Center to conduct a design study to be exhibited in the fall of 1976 (October 2, 1976 to January 9, 1977); which was immediately turned into a research and design project, and subsequently an exhibition project, with a US$10,000 endowment.[299] Working on the project—further evidence of how architecture education at the Institute had evolved from the urban studies projects of the early years (in which Rowe had also been involved before leaving in protest)—nevertheless gave students the opportunity to gather experience in planning and staging exhibitions and more generally engaging in cultural production. For the urban design project, students produced architectural drawings and an urban model under the supervision of DiMaio and Rowe. Their proposal called for low-rise buildings on the island's shore, postmodern urban forms (terraces, labyrinths, gardens), and water features (pools, fountains). For the urban districts just north and south of Nicolette Island, the students also designed low-rise residential buildings, plazas, and a shopping district, as well as skyscraper towers and apartment blocks.

The Institute generally saw many of its students and interns spend a formative phase of socialization and acculturation in architecture, thanks in particular to the Undergraduate Program and the Internship Program, where their understanding of the profession and discipline was strongly influenced by the (predominantly male, with a few exceptions) faculty and mentors assembled there. One pedagogical effect of studying at the Institute was to not only shape and define the aspiring

298 Colin Rowe, "Nicollet Island, Minneapolis" in *As I was Saying, Volume 3: Urbanistics* (Cambridge: MIT Press, 1996), 121–126. At the Walker Art Center in Minneapolis, the archives contain some documents on the exhibition and also on the Institute.

299 *Design Quarterly*, no. 101/102 (October 1976): "The River: Images of the Mississippi" [Exhib. Cat.]; see also Martin Friedman, "The River: Images of the Mississippi, Walker Art Center, Minneapolis, United States of America," *Museum International* 32, no. 1/2 (1980), 15–21. The exhibition "The River" included commissioned artworks as well as 19th-century classics, boat tours, film screenings, tours, and seminars, and three architectural projects in the form of design studies: next to the Institute, The Hodne-Stageberg Partners from Minneapolis and Craig Hodgetts were also invited, along with Charles Moore from Los Angeles. Source: Walker Arts Center. In preparation for the Institute's contribution, consisting of a model and several drawings, the Undergraduate Program students took time off from the Design Tutorial and instead worked exclusively on the project for a month to ensure that the exhibits were ready in time for the exhibition opening. They then started their regular classes slightly behind schedule. The original plan was for the three architectural projects to also be displayed at the Minneapolis City Planning Office following the exhibit, but this did not happen. Thus, the design study ultimately had no political effect.

young architecture students' approach to design but also to formulate a new habitus and thought style specific to architecture for them. Grounded in theoretical concepts and historical knowledge, this approach led students to adopt a more reflected way of writing and speaking about architecture and the city, and about their own creative practice and cultural values in general. At the Institute, the students acquired cultural skills related to postmodern practice—drawing and model making, exhibiting and lecturing, writing and editing—that would prove essential to their future careers. There is growing evidence that outside of their everyday studies and work, students and interns also participated in the social and cultural life of the Institute. After setting up seating and helping with ticket sales for the "Architecture" series, for example, they were able to sit in on the evening lectures from the gallery. Beginning in 1976, they also increasingly had the opportunity to witness a variety of exhibitions at the Institute of architectural drawings and contemporary projects by rising European architects who were largely unknown in the United States at the time, e.g., Aldo Rossi and Massimo Scolari, who not only exhibited their works in connection with the lecture series "A New Wave of European Architecture" in 1976 but also held seminars for the Institute's students and interns.[300] In this golden era of the Institute, the list of architects and academics holding lectures and exhibiting their work was both immense and illustrious. For example, students had the opportunity to listen to Manfredo Tafuri on one of his rare trips to the United States. While the Evening Program increasingly came to feature designers and artists, humanities scholars, progressive thinkers, and important figures from public life, the "Exhibition Program" quickly grew to include not only historical exhibitions, but also work by sought-after architects from Europe, the US, Japan, Austria, and South America. Some students and interns were involved in installing the exhibitions or helped with the poster design and merchandizing. They were also involved in the preparations for the group exhibition, first "Good-bye Five" (1975), an exhibition of architectural drawings by young architects, including Fellows and Visiting Fellows, most of them from New York and the United States, and then "Idea as Model" (1976), an exhibition in which long-term Fellows, friends of the Institute, and a few envoys from the New York Five presented architectural models as conceptual works, and even works of art.[301] Generally speaking, the Undergraduate Program

300 Following the "European New Wave," Rossi spent a few days after holding his lecture at the Institute, where he was to exhibit his drawings for the first time. Apparently, rather than spend his spare time at the Institute, Rossi preferred to go on excursions across New York with a group of students and interns. Theses daily excursions were an experience for everyone involved and took them to Coney Island and Central Park, among other places.

301 A few students and interns who helped with the installation on the night before the opening of "Idea as Model" even witnessed (and thereby became the exclusive audience for) Gordon Matta-Clark's notorious "Window Blow-Out," see Philip Ursprung, "Blinde Flecken der 1970er Jahre: Gordon Matta-Clarks 'Window Blow-out'," in *Reibungspunkte, Ordnung und Umbruch in Architektur und Kunst. Festschrift für Hubertus Günther*, eds. Hanns Hubach, Barbara von Orelli-Messerli, Tadej Tassini (Petersberg: Michael Imhof Verlag, 2008), 293–300. Legend has it that Matta-Clark was originally invited to carry out one of his famous cuttings on a seminar

and the Internship Program in particular helped assimilate these students into the organization, gender roles, and culture of a competitive, and ultimately precarious architectural practice. Moreover, the Institute was initiating not only its students and interns, but also its Fellows, Visiting Fellows, and faculty into a series of educational, cultural, publishing, and ultimately professional practices that were typical of the New York architecture scene at the time. The Institute exposed them to all the ambivalences, contradictions, and complexities inherent to the New York architectural community—confronting them, in other words, with the very micro-politics and micro-economies that, in this early phase of postmodernism, were paving the way for new forms of research and education and new architectural jobs.

School Profile

In the mid-1970s, a period of transition, both for architecture culture in general and the Institute in particular, between symptom and cause, the Institute's leadership attached central importance to education in institutional, financial, social, and cultural respects. From 1974 to 1977, the Institute's education offerings were not only its most reliable source of income, but they also exerted a multi-pronged multiplier effect, magnifying the resonance of the Institute's production and reproduction as well as the dissemination of new theoretical and historiographical knowledge. In "economically difficult times," as Eisenman had put it, the Institute's continued visibility was more important than ever and was achieved through its teaching activities, outreach to colleges and universities, as well as through its own public relations efforts to promote its education offerings, lecture series, and exhibitions, and through coverage in the press. For this reason, Goldberger's laudatory article in the *New York Times* on the revival of the Institute as an architecture school, published on October 30, 1975, came at exactly the right moment. Of all people, it was Goldberger—one of Eisenman's harshest critics and a champion of Robert Stern's "Grays"—who gave the Institute exactly the kind of non-polemical publicity that it (its occasional penchant for polemics notwithstanding) so desperately needed to recruit an even larger number of new students and attract new paying audiences. This was repeated when, in the summer of 1976, Goldberger again provided the Institute with media exposure with another feature article in the *New York Times* in which he wrote exclusively about the new High School Program.[302] The network of the Institute as an educational institution, however, also encompassed professional organizations such as the American Institute of Architects,

room at the Institute, which in this context would have referred less to the buildings shaped by suburbanization, deindustrialization, and real estate speculation, as their departure from valorization, but rather to the specific power structures that underlay architecture education at the Institute.

302 Paul Goldberger, "Young Summer Class Students Learn Architecture by 'Building'," *The New York Times* (July 27, 1976), 33.

i.e., the national association of practicing architects in the United States, which in 1976, at the suggestion of James Rossant of AIA's New York chapter, awarded the Institute a Gold Medal.[303] Praised as an "unusual and innovative educational institution," the Institute was recognized for becoming "a controversial center of discussion and debate on architecture and planning. Through dedication to excellence in education, research, and publication, it has extended its sphere of influence beyond its New York home."[304] Although this was only an institutionalized honor, it brought further attention to the Institute, and the Institute's leadership capitalized on this in its subsequent publicity and advertising campaigns. In the following years, the Institute came to increasingly rely on professional public relations and outreach for student recruitment and fundraising and hired Frederieke Taylor as the first director of development. In 1976, the Institute published another prospectus designed by its own Massimo Vignelli to promote its range of education programs.[305] Next to listing the two Undergraduate Programs and the High School Program, the printed brochure, in accordance with the graphic identity of the Institute, announced that, in the following academic year, the program "Design and Study Options" would be expanded beyond the six existing courses of study. In doing so, the Institute hoped to place a greater emphasis on design and recruit more students from schools of architecture. Parallel to this, it pursued a strategy of enlisting other colleges to help institutionalize the education offerings—for example, by offering the "Architecture" series as an on-site complement to the course catalogues of other colleges, such as the Parsons School of Design (today The New School of Design), the Pratt Institute, or the New Jersey Institute of Technology. In addition to undergraduates and graduates, the Institute was now reaching out to a new target group of "non-traditional students," a group that had in recent years "worked its way into admission offices and student affairs centers."[306] During the Institute's period of prosperity as an architecture school, students from private universities and art and design schools in the surrounding region were able to take advantage of its adult education offerings, with their flexible

303 IAUS, portfolio for an application to the 1976 AIA medal, n.d. Source: AIA Archives. The Institute had applied for an award from the American Institute of Architects (AIA). The Institute's application folder, which includes documentation of the 1975–76 educational and cultural programs, the two Undergraduate Programs in Architecture and in Planning, the Program in Preservation and Adaptive Re-Use, announcements of public programs, exhibitions, lecture series, and closed events, a public relations overview, and a compilation of letters of recommendation, is documented in the AIA archives.

304 American Institute of Architects, letter of December 23, 1975. Source: AIA Archives.

305 IAUS, prospectus, 1976. Source: CCA Montréal, IAUS fonds: A.2-7. Taylor had come to the Institute in the spring of 1976 to fill the new post of director of development, established for two years at the initiative of Armand Bartos, chairman of the Board of Trustees, with a grant from the Gottesman Foundation. Her primary tasks were to prepare financial reports for the Institute for the first time, apply for grants, and improve public relations.

306 Thelin, 2004, 326–327.

modular structure, so characteristic of goods and services in the information and knowledge society, to learn about architectural history and theory, urban planning, and contemporary debates in architecture, design, and the arts—all in a central location in Midtown Manhattan, in a space embedded in the local architecture and design scene. The Institute had turned into a kind of night school of architecture, allowing students to further specialize while obtaining the credits necessary for their studies.

Apart from critical size, and even business continuity, there was one more major difference between the Institute and the leading international schools of architecture to which it compared itself, through collaboration or competition, especially the AA in London, Cooper Union in New York, and the Istituto Universitario di Architettura di Venezia (IUAV). This was the fact that it never held the status of an officially accredited university. There were both structural and strategic reasons for this. The 1976 edition of *Peterson's Guide to Architecture Schools in North America* noted that the Institute required no entrance exam, but also that students could not attain any kind of recognized degree.[307] Yet this circumstance did not diminish the economic success or cultural influence of the Institute as an architecture school. The lack of admission requirements was attractive to students, even though, as an architecture school, the Institute was not as anarchic, creative, or critical as its role models. For some time, however, the Institute did discuss collaborating with Sarah Lawrence College to develop a professional master's degree based on its Undergraduate Program that would have qualified graduates to practice architecture. On Charles DeCarlo's initiative, a concept paper was even drawn up, outlining a joint degree program with a focus on the public role of architecture and urban planning.[308] Arthur Drexler, then still on the Board of Trustees, supported the push, arguing that the Institute needed to take a leadership role vis-à-vis other institutions of higher education. In the past few years, he argued, the consortium of colleges had shown an interest in cooperating with the Institute: "because the universities believe[d] the Institute ha[d] a clear idea where architecture [wa]s heading."[309] A committee was set up that, besides DeCarlo and Drexler, included Richard Meier, Peter Wolf, and Peter Eisenman. The strategy they formulated for architectural education at the Institute was based on the Internship Program and the Evening Program and was called a "Terminal Master Program." According to this concept paper, the Institute's master's program would have been aimed primarily at

307 Karen Collier Hegener and David Clarke, eds., *Architecture Schools in North America* (Princeton: Peterson's Guides, 1976).

308 The establishment of a master's degree program was on the agenda of the 1976 annual meeting of the Board of Trustees, see Peter Eisenman, "Director's Report," June 10, 1976. Source: CCA Montréal, IAUS fonds: A.1-2.

309 Ibid.

graduates from liberal arts colleges and at professionals looking for further qualifications, as well as at young mothers planning an entry into professional life. It was assumed that, after their completing degrees, master's students would be seeking positions in management, education, industry, or governance—i.e., not in architectural practice per se. For this reason, the courses on offer were to cover the public sector, public administration, construction supervision, taxation, development planning, urban development planning, land use planning, urban history, and so forth, all of which problematized the various ways influence could be exerted upon the built environment. But in the end, the Institute never did establish a master's program. It was Eisenman who opposed the establishment of a fully-fledged degree program from the very beginning, and for good reason. The Institute lacked the necessary financial and human resources, space for teaching, and other facilities, such as a fully equipped library, to make such a step towards professionalizing its education program possible. Moreover, the fact that the Institute, with its education offerings and especially its Undergraduate Program, had only assumed duties from its partner colleges that it had the capacity to fulfill, also proved to be a strategic advantage. Ultimately, when it came to formulating course content and requirements, the Institute was accountable to nobody. The flexibility that lay at the heart of the Institute's education offerings with their modular design made it possible for the Institute's leadership to redefine the program each successive year, depending on the budget and staff available, and modify it as needed. This made it possible to build up surpluses and maximize profits. From a pedagogical point of view, the spectrum of education programs on offer each academic year was ultimately determined by supply and demand, rather than the need to meet educational objectives or humanistic ideals. After all, one of the fundamental prerequisites for creating and maintaining an educational program at the Institute was that it would pay for itself. When, after a one-year test phase, an education offering failed to demonstrate financial viability or couldn't garner a minimum number of students, it was unceremoniously canceled or, at best, replaced by a new offering.

Teaching and Learning Success

In terms of the relevance of the architecture education offered at the Institute, the Undergraduate Program and the Internship Program would ultimately, at least indirectly, influence the next generation of architecture students, while the Fellows played a role in shaping the broader architectural curricula. The innovations relevant in this context were that the Undergraduate Program functioned as a kind of preliminary education opportunity for ambitious liberal arts students and that the Internship Program operated as a pre-sorting mechanism for people looking to apply for one of the coveted spots at an Ivy League university. What is more, the programs enabled the Institute and its Fellows to expand their academic and intellectual capital by forming ties to all the major institutions on the East Coast. In doing so, they did not explicitly position themselves in competition with the established schools of architecture (predominantly those of the Ivy League on the East Coast, as well as those in California), but instead operated as an interface between colleges and universities. The Institute thus came to occupy a unique position within the academic and higher education landscape in America, especially by the second half of the 1970s. After three years of being supported by the Noble Foundation, the Institute's leadership had come to measure its success as an architecture school not so much qualitatively, in terms of having achieved certain teaching and learning goals, but rather quantitatively vis-à-vis the institutional network, in terms of the admission rates of former students to graduate schools. The university acceptance details were published in the annual *Progress Bulletin.*[310] The significance attached to architecture education when it came to the education not only of prospective architects but also of responsible citizens and future developers, is evident in a memo Eisenman once wrote characterizing the Institute's basic educational strategy as that of a "trojan horse" within the architectural field.[311] Eisenman was also alluding to the new position of power occupied by the Institute, in its role as a major but non-institutionalized architecture school. The Institute's power expanded with each new academic year, as students and interns educated at the Institute infiltrated the universities and the professional world, disseminating, in settings both educational and practical, the postmodern paradigm, thinking, and practice they had learned at the Institute.

Ultimately, the Institute's success was also due to the fact that its restructuring as an educational institution, and its strategy of flexible adaptation to the market, was profitable in a variety of ways. In its first two or three years as an architecture school, as student enrollment rose to thirty-eight, the Institute immediately began earning significantly more income from architectural education. In the 1975–76 academic year, the Institute already began running out of space; a new seminar room and a provisional library were hastily erected on the mezzanine floor. Suzanne Frank, whose work at the Institute was initially confined to copying texts for Undergraduate Program students for free, now worked in an honorary capacity as Institute librarian.[312] What made the expansion of the Institute's education offerings in 1975–76 even more surprising was the fact that, in the same academic year, nationwide college attendance had declined for the first time since the 1944 G.I. Bill of Rights, and public and nonprofit institutions in general were undergoing a period of crisis.[313] In this context, one possible determinant for the Institute's success was the fact that, in 1970s America, people were newly

310 As schools of architecture, Princeton, Columbia University, Cooper Union, and MIT were particularly popular, followed by Yale University, Berkley University, and UCLA.

311 "Progress Bulletin no. 3," January 27, 1975. Source: Sarah Lawrence College Archives.

312 Frank, 2010, 6. The installation of a library as the basis for teaching architecture had been an unresolved problem at the Institute for years at that time and was to remain an issue.

313 Thelin, 2004, 321 & 323. The GI Bill of Rights aimed to promote the reintegration of U.S. soldiers into the workforce and thus into American society.

conscious of the need for lifelong learning, and students, in general, were adopting an "academic consumerism"—a development that took hold across the field of architecture in a broader sense and was expedited by the Institute. Actual professionalization of the Institute's architecture education was finally achieved in the 1976–77 academic year, when, with the appointment of Carla Skodinski as the new Undergraduate Program coordinator, the Institute began to actively set about improving its administration. In concrete terms, this meant that in the period from 1974 to 1977, the Institute was largely able to finance its operations and pay the rent for its representative penthouse premises through its work as an architecture school. Although the tuition figures declined slightly in the 1976–77 academic year (after optimistic predictions of thirty students and ten interns), the offerings were nevertheless profitable, primarily due to the increase in tuition fees and further economization, i.e., the decision to charge more for internships. In the 1976–77 fiscal year, the Institute earned nearly US$150,000 from the Undergraduate Program alone, having attracted "only" nineteen students from twelve colleges, plus fourteen interns.[314] Despite the various issues facing the administration and public relations team, the Institute's consortium now comprised seven colleges: Amherst, Brown, Franklin & Marshall, Hampshire, Hobart and William Smith, Oberlin, and Sarah Lawrence. Advertising for the Undergraduate Program, however, was subpar on both sides. Not only did some of the colleges neglect to list the classes in their course catalog, but the Institute failed to print a poster advertising the education offerings for three years running. Nevertheless, the Institute's income continued to rise (even as the expenses for teachers and tutors increased): after years of debt management, the Institute now, for the first time in its history, was operating on a balanced budget.[315]

In the mid-1970s, at a time when there were scarcely any jobs available for New York architects due to financial and fiscal crises, the Institute under Institute director Eisenman was transformed into a new kind of workplace, one that allowed Eisenman, the other Fellows, and external instructors to support

314 IAUS, "Summary Report: Budget Summary 1974–1977," n.d. Source: CCA Montréal, IAUS fonds: A.1-3.

315 Lucia Allais describes the years 1974 to 1977 at the Institute as "stable educational years," see Allais, 2010, 35. She does not however differentiate between the individual programs and their respective funding sources to support this statement. To confer a financial, and therefore institutional dimension on architecture education at the Institute, revenues from tuition must be considered separately from other revenues and always in relation to total revenues. Grant funds were used specifically to help set up an educational program and therefore must also be attributed to this institutional role of the Institute. For example, in 1975–76, the Undergraduate Program in Architecture was the most financially successful, with revenues of US$ 140,000, compared to the other newly launched educational programs. The Undergraduate Program in Planning had tuition revenues of just over US$ 12,000. The Program in Preservation and Adaptive Re-Use brought in an additional US$ 20,000. Nevertheless, the Undergraduate Program in Architecture incurred debts in its first two years. In terms of the 40 percent overhead, the Institute's income from all educational programs in 1975–76 still contributed just under US$ 65,000 to a total budget of just over US$ 108,000. Architecture education revenues were slightly less than the projected total.

themselves by teaching while still raising their profile by participating in the "Architecture" series, exhibiting at the Institute, and writing for *Oppositions*. After visits by Koolhaas, Tschumi, and Shane, the Institute's relationship with the AA cooled off slightly. Eisenman maintained his networks, e.g., by inviting Spanish architect Rafael Moneo, a professor of architectural theory at the Escola Tècnica Superior d'Arquitectura in Barcelona, to the Institute for the 1976–77 academic year. He had met Moneo in 1967 at the Aspen Design Conference along with Oriol Bohigas, Frederico Correa, Ignasi de Solà-Morales, and Nuno Portas, and again in 1975 at a meeting of the editorial staffs of *Arquitecturas Bis*, *Lotus*, and *Oppositions* in Cadaquès, Spain. While Moneo, who remained in New York for one-and-a-half years, was teaching at the Institute, he was also conducting his own research at Cooper Union, which he published in *Oppositions*. Curating and taking part in the American contribution to the Venice Art Biennale in the summer of 1976, and editing and publishing *Oppositions* 5, the so-called "Italian Issue," in October 1976, Eisenman expanded the Institute's relationship to the IUAV, in particular through his intensified intellectual and artistic dialogue with Manfredo Tafuri and Aldo Rossi, who had both been guests of the Institute before, in order to develop collaborative education programs, and later research and publishing projects. Even if these discursive and institutional encounters did not result in a program or project, they helped Eisenman establish what later became called the "New York–Venice axis," which, proceeding from Tafuri's neo-Marxist critique of contemporary postmodern architectural practice, was intended to enliven architectural debate in North America in the second half of the 1970s.[316] By offering Fellows, Visiting Fellows, and external scholars the opportunity to teach, publish, lecture, and exhibit at the Institute—that is to say, in a broader sense, to work as cultural producers in both material and immaterial respects—the Institute was able to activate its collective social, cultural, and symbolic capital and—through education *and* culture—transform it into economic capital. In the process, work at the Institute took on the shape of a collective learning experience, with fluid boundaries between teachers and students. In 1976–77, after various attempts to establish an in-house reference library had failed in the face of financial and organizational limitations, Frank was now officially entrusted with the task of reviving the library project at the Institute as a Visiting Fellow.[317] She estimated a budget of US$14,500 for a core collection of two hundred books, emphasizing that the Institute's book collection and archive would play a vital role in the students' educational experience. As cities in general

316 On the intellectual and institutional exchange between Eisenman and Tafuri and the relationship between the IUAV and the IAUS, see Ockman, 1995; On Tafuri's reception in the United States, see Diane Ghirardo, "Manfredo Tafuri and the Architectural Theory in the U.S., 1970–2000," *Perspecta*, no. 33 (2002): "Mining Autonomy," 38–47; on Tafuri's biography, see Leach, 2005.

317 Suzanne Frank, "A Proposal for Reviving the Library," n.d. Source: CCA Montréal, IAUS fonds: ARCH401150.

and architecture, in particular, became culturalized, the Institute was subsequently designed and managed to ensure that both the educational *and* cultural programs contributed to the production, reproduction, and dissemination of a wide variety of architectural knowledge.[318] In the final analysis, the interaction between various fields of activities served to evoke a multiplicity of internal synergies and external networking effects. This had repercussions in the pedagogical and discursive spheres, but also on an economic and political level. For many Fellows, however, leisure time and working time were barely distinguishable.

Alternative Education

The opportunities and challenges of the Institute's work as an architecture school—which resulted from the fact that the modular education offerings could be designed flexibly and altered at short notice as needed, thanks to rapid decision-making processes and an administration geared around Eisenman—became apparent when, in the 1976–77 academic year, contrary to what had been advertised in the brochure, the Institute was unable to maintain its full catalogue of education programs. In its third year as an architecture school, after a phase of growth and differentiation, the Institute was entering its first consolidation phase. The Undergraduate Program in Planning was discontinued after only one year, as Wolf hadn't been able to obtain the necessary funding. On top of this, the Institute also faced bureaucratic problems in obtaining recognition from the relevant departments of the individual colleges for the work performed by students in the program. Nevertheless, the Urban Design Tutorial (one of the modules in the Undergraduate Program in Planning) was retained under the leadership of Weintraub in order to remain true to the humanistic ideal of a "broad education." The internship in urban planning at the Institute was also retained. Meanwhile, the Work/Study Program in Preservation and Adaptive Re-Use under Ellis's direction was extended for another year thanks to the renewed sponsorship of the J. M. Kaplan Fund, although it continued to feature no course offerings. Another program created under Ellis's direction was the "Design and Study Options," offered in 1976–77 following a short preparation period. Unlike the Undergraduate Program, these options would primarily be aimed at architecture students who, as was common at the time, were completing a five- or six-year bachelor's program at one of America's schools of architecture and would typically spend one year of their degree studying in Europe.[319] This education offering was tantamount to a public proclamation: the Institute was now going to increasingly focus on obtaining new market

318 Andreas Reckwitz, "Creative Cities: Die Kulturalisierung der Stadt," in *Die Erfindung der Kreativität. Zum Prozess gesellschaftlicher Ästhetisierung* (Frankfurt: Suhrkamp, 2012), 269–312.

319 In 1976–77 the Institute attracted a total of six schools of architecture as partners for the Design and Study Options: University of Notre Dame, University of Illinois, University of Miami (Ohio), Syracuse University, University of Virginia, University of Cincinnati.

segments and student groups and aggressively solicit students at established schools of architecture by focusing on "integrated design" as a pedagogical selling point. To set itself apart from its European competitors, the Institute planned to simply fly in European faculty and hire internationally renowned architects from New York's architectural scene as tutors. During the conceptual phase, the following names came up as dream candidates: from England, James Stirling, Peter Smithson, Leon Krier, Rem Koolhaas, and Elia Zenghelis (from the AA in London), from Italy, Aldo Rossi, Massimo Scolari, and Manfredo Tafuri (from the IUAV in Venice), from Germany, O. M. Ungers (Cornell, UCLA), from Spain, Rafael Moneo (Universidad Politèchnica de Catalunya de Barcelona), and from the USA, John Hejduk, Richard Meier, and Giovanni Pasanella. Meanwhile, the Design and Study Options would appeal to American architecture students because, unlike a one-year study abroad in Europe, it was easy to have their coursework recognized, thanks to prior agreements with their home universities. In addition to this, prospective students could take advantage of the courses already on offer in the Undergraduate Program, which now held a central position in communicating the Institute's perspectives, topics, and approaches.[320] At this point, it had become clear that the Institute's significance as an architecture school, despite its international aspirations, lay primarily in the USA. Over time, it had managed to build up a solid national reputation.

At the end of its three years of funded activities, the Institute's reinvention as an architecture school was deemed a success. The final report to the Noble Foundation in January 1977 made a particular point of highlighting the progress that had been made, especially with the Undergraduate Program and the Internship Program (although the report did discuss the administrative and academic issues involved in expanding the consortium and tackling bureaucratic obstacles at the individual colleges).[321] After taking stock, the Institute's leadership concluded that its best strategy lay in supplementing the liberal arts education offered at colleges, and thereby offering an "alternative" to undergraduate studies at schools of architecture. Once again, the admissions figures at graduate schools were cited as an indicator of the Institute's success. For instance, eighty percent of Undergraduate Program students who applied for architecture degrees were accepted to their university of preference. At the same time, the Institute's leadership admitted that they hadn't managed to integrate theory and history courses into the Undergraduate Program's design studio as successfully

320 Interest in the Design Study Options was limited. The Institute was unable to expand the consortium to the targeted number of twenty colleges even in the following year. In 1977–78, only five students were enrolled. Only five schools of architecture cooperated with the Institute: University of Cincinnati, Syracuse University, University of Miami (Ohio), University of Illinois, Kentucky State University.

321 IAUS, "Final Report to the Noble Foundation," n.d. Source: CCA Montréal, IAUS fonds: A.1-3.

as they had hoped. The Internship Program, meanwhile, was chalked up as another success; it had offered college graduates a variety of ways to prepare themselves for an architecture degree beyond a conventional internship at an architecture firm. The report specifically noted that feedback from former Institute students was largely positive. These students attributed their academic success to the fact that they had more experience in translating ideas into forms than other students in their cohorts. In the end, the Institute's final report to the Noble Foundation, more self-referential than self-reflective, stated that the end result of its three-year education initiative was precisely the goal that had been formulated three years earlier, amounting to something of a self-fulfilling prophecy: "As we have stated from the beginning our intention is to educate a unique cadre of people with the notion of architecture as a fundamental cultural resource as a mirror and a repository of man's [sic] hopes and aspirations."[322] The project undertaken by the Institute (as an architecture school), Peter Eisenman (in his triple function as Institute director, director of the Undergraduate Program, and intern mentor), and the teaching Fellows (as a new architectural and academic elite) was to "nurture youth" in two senses. On the one hand, they were nurturing youth in general, by promoting a new generation of producers and consumers in the field of architecture who would put their trust in the Institute and have faith in its cultural production. On the other hand, in a much more concrete sense, the Institute was grooming its own academic progeny, insofar as many students and interns either remained attached to the Institute—distinguishing themselves by working as teaching faculty or editorial staff, or in certain cases even rising into the hierarchy of Fellows—or else followed careers later in life as academics attached to American schools of architecture, where various Fellows also worked as professors of history, theory, and design.

Given that, in early 1977, not only the future of the Undergraduate Program was at stake but also that of the Institute, the final report to the Noble Foundation can be read as both a programmatic document and an instrumental one. This explains why Eisenman outlined that, over the following three years, the Institute planned to continue expanding its education offerings—to implement architectural education more consistently through tighter integration of theory and history courses and closer coordination among the course directors. Pedagogical approaches would be differentiated, in didactic terms, by introducing "team teaching," course offerings would be diversified by focusing more on art (explicitly film), painting, and sculpture as well as the sociology of the built environment, to meet the requirements of the educational ideal of a "liberal arts education" in terms of both content and methods. Of course, the Institute's leadership hoped that the Noble Foundation's patronage would continue but did not express satisfaction with the scope or influence of the Institute as an architecture school. Therefore,

322 Ibid.

the Institute made a systematic effort to minimize its dependence on funding from the foundation by continually expanding its consortium and thus educating more students in total. In the end, despite its initial promises to the contrary, the Noble Foundation did not maintain its support. On top of this, the close cooperation between the Institute and Sarah Lawrence also came to an end after three successful years. In the 1977–78 academic year, Sarah Lawrence did not send a single student to the Institute, and the numbers were limited in the following year as well. From 1977 to 1980, the number of students enrolled at the Institute certainly increased, but enrollment ultimately remained far below the optimistic predictions of fifty students in the Undergraduate Program and thirty interns per year. Without further funding, the Institute's expected income of US$350,000 from architecture education alone remained unrealistic. From this point on, the Institute could only depend on revenues from tuition fees, even as the significance of the pedagogy practiced at the Institute clearly began to recede behind its cultural activities.

2.3 Entering into a Phase of Consolidation

In 1977, after three years of growth and experimentation following its relaunch as an educational institution, the Institute began to reconfigure its work in all areas, including architectural education. However, as the Institute celebrated its tenth anniversary in the autumn of 1977, Institute director Eisenman began to prioritize other programs over education, by formalizing and professionalizing public programs even more and expanding publication programs. This was because, in January of 1977, anticipating its upcoming anniversary, Eisenman made a proposal that the Institute undertake another strategic reorientation. In his "Director's Memo" to the trustees, which he titled "Definition of the Institute: The Next 10 Years," and which sought to reposition the Institute again, he wrote extensively about the Institute's past programs and the challenges it had faced in recent years.[323] In his characteristically polemical tone, Eisenman argued that, for the first ten years of its existence, the Institute had lacked a proper objective, being more of an "ad hoc collection of programs based on fund raising." Not only did he dismiss the successes already achieved by the Institute, both working on real-life research and design projects in its original capacity as a link between office and academia and later as an architecture school, but he also dismissed its original ambitions. In the context of the history of the Institute, it was crucial that the lack of a purpose, according to Eisenman, could only be changed by increasingly viewing it as a "cultural resource," while linking it more intensively to international architectural debates (see chapter three). Eisenman made no secret of the fact that he

323 Peter Eisenman, "Director's Memo: Definition of the Institute: The Next 10 Years," January 11, 1977, Source: CCA Montréal, IAUS fonds: A.1-3 / ARCH401031.

was chiefly interested in the Institute's two journals: primarily *Oppositions*, which since 1976 had been published by MIT Press and was already a staple of university curricula and library collections, but also the art criticism journal *October*, launched in the same year (see chapter four). At the same time, however, he firmly defended the idea of architectural education as a business model. To focus human resources when it came to teaching activities, he recommended a politics of consolidation—suggesting, for example, that, in the future, the Institute should prioritize its most successful education offerings: "the Undergraduate Program in Architecture with both its Undergraduate and Internship components; the Evening Program of Continuing Education; and the Summer High School Program."[324] However, he argued in favor of canceling the less successful ones, for example, the Work/Study Program in Preservation and Adaptive Re-Use, claiming that it was impossible to maintain a sufficiently high standard of quality. Ultimately, Eisenman's memorandum, bridging the past, present, and future of the Institute, argued for continuity. But soon after, it became apparent that the Institute's leadership intended to apply for larger grants, while maintaining the education offerings, in order to expand the "Architecture" series and to professionalize it as a center for adult education. The Institute's work over the coming years, in addition to its various grant applications, testify to the extent that American architecture education and culture as a complex were increasingly being permeated by an economic approach, an attention economy, and a desire for power. Over the years, the Institute's continued existence as an institution was ultimately contingent on personal and professional networks, even as it sought to compete with other institutions, universities—especially the Ivy League schools of architecture—and museums. In the end, these networks were political in nature, not only in terms of Institute policy but educational policy as well. With the launch of new services in the educational and cultural sphere, the social dimension of these networks—the contacts and connections—became even more economically important.

Learning Institute

Only shortly thereafter, as the Noble Foundation and NEA grants expired, architectural education was given a new role when the Institute's leadership capitalized on the opportunity that arose and applied for a large grant from the National Endowment for the Humanities (NEH) under their "Cultural Institution Program."[325] The comprehensive proposal, complied by Taylor, that the Institute submitted to the NEH in April 1977 was preceded by about a

324 One personal motivation for keeping the High School Program might have been that Eisenman brought his children to the Institute on weekends, where they attended sessions of the High School Program.

325 IAUS, application to the NEH for a "cultural institution grant" (EH-28433-77-547). Source: NEH Archives.

year of discussions with the National Council on the Humanities. In this proposal, departing somewhat from the management rhetoric that Eisenman had established in his "Director's Memo," the Institute portrayed itself as serving a dual function as both a humanistic educational institution and an international research center for architecture. Originally, this fundraising initiative was part of a larger plan, referred to here as the "NEH Learning Institute Program," which was conceived as an ambitious two-semester adult education program that would complement existing educational offerings at the Institute. The proposal, co-authored by Kenneth Frampton, Mario Gandelsonas, and Anthony Vidler, followed a didactic concept, and identified three subject areas that were to form the core of the curriculum in teaching architecture as one of the humanities, and which now addressed the public: "the American city and the development of cities, the nature and problems of contemporary artistic culture, and the development of modern architecture from the turn of the century." The concept built on the Institute Fellows' expertise, encompassing three core courses on the themes of "The City," "Architecture," and "Culture," which were to be supplemented by Institute Seminars.[326] With this methodological approach, the Institute proposed to promote interdisciplinarity and to provide insights on a variety of themes. The "NEH Learning Institute Program," which, had it been realized as planned, would have included workshops, lectures, reviews, and discussions, was to run concurrently with a continuation of the Evening Program and was to be supplemented by a fourth course. This fourth course, which was to be taught by MacNair, would have focused on "Visual Literacy" in the first semester and "Design" in the second. However, with this complex educational structure and the renewed blending of education and culture in the same vein as "Architecture," the Institute's main objective was not so much the humanities as such but continued to be to popularize architecture and bring both the discipline and the profession to the attention of New York metropolitan society.

The Institute's proposal presented the "NEH Learning Institute Program" as a key component of a wider restructuring of the Institute, according to which the Fellows would focus primarily on two areas of activities, teaching, and research, involving cultural production and publishing, over the next five years. Accordingly, the Institute would have consisted of a two-part structure: a Center for Public Education, with the existing educational programs and the yet-to-be-established "NEH Learning Institute Program" assigned to it, and an International Study Center, which would have encompassed in particular the Fellows' individual research projects, the library project, and various publication formats. Three years of experience in the field of adult education and the

326 In its NEH application, the Institute repeatedly emphasized, as a statement of intent, that the long-term Fellows were to serve as facilitators. Attached to the application were course plans by Vidler ("The City"), Frampton ("Architecture"), Krauss ("Arts"), plus a fourth on "Visual Literacy" (by MacNair). There was a slight mismatch between arts and culture.

Fellows' affiliations with various universities were cited to support the expansion of the Institute into a public educational facility, while the Institute's international institutional network was advertised in support of its expansion into a research center, explicitly including a future cooperation with the IUAV and international guests as Visiting Fellows. Given the flirtation with the humanities, it became clear that the Institute's understanding of research had changed considerably since the early years, a fact that the Fellows repeatedly noted with regret. For the NEH proposal no longer cited projects for the City Planning Commission, the U.S. Department of Housing and Urban Development, or the Urban Development Corporation of New York State as qualifications, but instead listed the Institute's publications, including new formats that were planned but not yet realized, e.g., the publication of historical documents or a book series, all of which were interwoven in the proposal.[327] Other documentation on research and publication plans mentioned a major study on New York that was to be produced in cooperation with IUAV professors and a conference on formalism in architecture.

The NEH proposal was far more professionally devised, more comprehensively framed, and more thoughtfully formulated than any of the Institute's previous proposals. Strikingly, the Institute's application for a Cultural Institution Grant was nothing less than a claim to leadership in architecture education directed at the public at large. The proposal text proclaimed that the Institute intended to produce additional teaching and learning materials as part of its adult education, which ambitiously included the production of course notes and course books, a new book series of selected lectures, and the publication and sale of slide series and audio recordings of each lecture series. With its rhetoric of a humanistic architecture education for the general public, its promise of an interdisciplinary study of architecture, and the expert opinions enlisted from well-known architects and university professors of the humanities, the proposal finally convinced the National Council on the Humanities to such an extent that, in June 1977, the Institute was granted the requested funding amount in full. Apparently, the federal agency in Washington, D.C. expected the Institute to make a significant contribution to the study of architecture as one of the humanities in the United States.

Continuous Education

When the Institute received the US$357,000 NEH grant in the summer of 1977, the largest project to date in financial terms, the NEH Learning Institute was not implemented, for whatever reason, and many of the ideas advertised were also not to be realized. But the Institute's leadership, now presiding over an enormous budget and the associated planning security for the next three

327 In this context, the grant application also mentioned for the first time an English translation of Aldo Rossi's seminal monograph *L'Architettura della Citta* planned with MIT Press.

years, launched "Open Plan," a continuation of its adult education offerings. The program included both lecture courses and more in-depth seminars that could be booked for US$60 and an additional US$45 and, like "Architecture," addressed a metropolitan audience. For the price of US$200, you could also become a "friend" of "Open Plan" and lend your financial support. A limited number of scholarships were also available. The promotional material made sure to mention that tuition fees were tax deductible. As with the Evening Program before, the ultimate effect of the new education offering was to forge an even closer tie between social, cultural, and symbolic capital, architectural debate, and an economy of attention. One effect of "Open Plan" on the Institute was that, by acquiring such a large source of funding and raising its budget by over one hundred percent, the lecture and event series necessitated further professionalization, even bureaucratization of the Institute's operations, a development that was not untypical for the educational landscape of the 1970s.[328] The shift in emphasis from education toward culture and the resulting organizational and programmatic developments already anticipated by Eisenman in his "Director's Memo"—especially the choice to refocus on public events, exhibitions, and publications—would prove to have a major impact on the pedagogy of the Institute's other architectural education offerings (especially the Undergraduate Program, more so than the Internship Program), since the development of the curriculum and the composition of the faculty were no longer the chief priority and were instead contingent on the availability of various Fellows and Visiting Fellows. Again, the profound changes happening at the Institute were also reflected in the composition of the Board of Trustees: in the summer of 1977, Charles DeCarlo and June Larkin, who were both instrumental in the introduction of the Undergraduate Program and the Institute's performance as an architectural school, resigned from their posts as trustees. While the Board of Trustees was now to more closely reflect the public interests of the Institute as a cultural institution, including representatives of the architecture world and New York patronage, the interests of architecture education were to be safeguarded by the inclusion of William Porter (dean of the MIT School of Architecture) and Colin Campbell (president of Ohio Wesleyan University).

In 1977, the year in which the Institute celebrated its tenth-year anniversary, it was at the peak of its activity and vitality, due to the complexity of its programs, the range of offerings and products, and the skillful leveraging of the opportunities presented. A popular architecture school and a trendy cultural space, with a publishing imprint that was soon to expand, it was now competing for public funds, audience, and attention with New York's largest institutions—with the city's museums, theaters, libraries, and universities. *Oppositions* was not only shaping debates in the field of architecture but also influencing the history and

328 Thelin, 2004.

theory education being offered at schools of architecture. In 1976–77, alongside prefaces and articles from its four editors (Eisenman, Frampton, Gandelsonas, and recent addition Vidler), it also featured writings by external authors such as Rem Koolhaas on Ivan Leonidov and Coney Island, Rafael Moneo on Aldo Rossi, James Stirling on typologies, and Manfredo Tafuri on a transatlantic comparison of postmodern approaches, the historical context behind the Five Architects, and the work of Giuseppe Terragni, etc. Meanwhile, with "Open Plan," featuring "the most prominent practitioners and thinkers in the field," the Institute was helping to embed contemporary architectural thinking in New York society, at a time when the construction industry was booming again, and raise public awareness. In the fall, "Open Plan 77" covered such topics as "Style and Meaning in American Architecture," "The Anglo-American Suburb," "Cities within Cities," "The Modernist Vision," "The Metropolitan Vision of New York and Paris," "The Languages of Design," and "The Interior Landscape." As Fellows like Frampton and Vidler entered the ranks of New York intellectuals by hosting lecture and event series and publishing their own work, the Institute as a group, organization, and institution became, functionally speaking, the new architectural elite. As the design of the public program made clear, this was happening in both discursive and institutional respects: by assuming the power of interpretation, by defining who belonged and who didn't, the Institute was drifting further and further into the establishment. This development also manifested itself in the architecture education offered at the Institute as a cultural and educational center. Like the knowledge and cultural economy, education at the Institute underwent a phase of consolidation and maturation from 1977 to 1980.

Simultaneously, the Institute's directors were planning, designing, and launching new formats for communicating contemporary examples of architectural practice, often drawings or models, followed by all kinds of cultural production in the guise of education offerings. As work at the Institute became increasingly attuned to the higher education market, this became more of a focus. This also meant that, from 1977–78, the Institute was able to continue expanding its most successful education offerings: the Undergraduate Program, the Internship Program, and initially also the Design and Study Options. This process was aided by the fact that the necessary structures already existed. The enormous significance that was still attached to architecture education at the Institute was demonstrated when Carla Skodinski, coordinator of the Undergraduate Program, was appointed a Fellow in 1977, while Eisenman was officially still directing it. Yet by offering architecture education outside of the Institute, and by opening itself up to the arts and the humanities, the Institute was increasingly becoming an institutional authority, itself tasked with consecrating and legitimizing those who taught, lectured, exhibited, or published there, and ultimately with authorizing the dissemination of postmodern architectural styles and architectural thought. In the winter of 1977, the Institute's exhibition "Princeton's Beaux-Arts and Its New Academicism: From

Labatut to the Program of Geddes" featured student works by former Princeton University students under the last two deans, the most prominent of whom were perhaps Robert Venturi and Charles Moore (but Mary McLeod's thesis, who had just graduated, was also shown, five years before she became involved in architecture education). Eisenman, in a letter to Robert Geddes, emphasized the great influence of Donald Drew Egbert, the architectural historian who had once taught there.[329] The full spectrum of architecture education continued to be a strategic cornerstone of the Institute's work, yet, in terms of the economic, legal, and ideal interests of the Institute, individual Fellows, staff, and students, foundations and sponsors, such work was no longer accorded the significance it had been before. After a brief decline in enrollment in the 1977–78 academic year, the Institute once again quickly reached its full capacity as an architecture school. In 1978–79, the consortium for the Undergraduate Program was expanded to include thirteen liberal arts colleges; the number of registered students had risen to fifty-two.[330] Although the Noble Foundation's support was gone, revenue from education programs rose to approximately US$260,000 after a slump in the 1977–78 fiscal year.[331] This meant that, notwithstanding considerable funding from the NEH, the share of revenue from tuition fees still amounted to nearly forty percent of the total budget of US$660,000.

Institutionalizing postmodernism through education and culture also meant that the Institute was able to grow and expand its fields of activities and diversify its range of products which, in addition to cultural value, also had an educational value. The public programs—be they lecture series, exhibitions, or a variety of new publication formats—not only served a discursive function within the field of architecture but also proved to have institutional dimensions. In 1978, at the same time as the establishment of "Open Plan," the "Exhibition Program" was professionalized with the aid of funding from the NEA. The exhibitions, be they historiographic or contemporary, thematic or monographic, would prove to influence debates such as the historiography of modernism or the full range of postmodern positions. Meanwhile, in addition to *Oppositions* and *October*, which were changing the way people discussed architecture and art, the Institute was simultaneously developing new publications and formats. These included: *Skyline* (launched in 1978), an architecture newspaper that featured reviews and interviews alongside announcements for the Institute's own

329 Stanford Anderson, "Architectural History in Schools of Architecture," *Journal of the Society of Architectural Historians* 58, no. 3 (September 1999), 282–290.

330 The consortium of the Undergraduate Program included thirteen colleges: Amherst, Brown, Colgate, Connecticut, Franklin and Marshall, Hobart and William Smith, Middlebury, Oberlin, Sarah Lawrence, Skidmore, Smith, Swarthmore, Wesleyan. The bulk of the Design and Study Option students came from Syracuse University.

331 Berlin and Kolin, "Accountant Report," 1977–78, June 30, 1978; "Accountant Report," 1978–79 & "Accountant Report," 1979–80, May 30, 1980. Source: CCA Montréal, IAUS fonds: A.2-10.

events and education offerings, IAUS Exhibition Catalogues (1978) that included not only plans, sections, and drawings, but also critical essays, and Oppositions Books (1981), which aimed at changing the architectural canon by publishing famous authors (eventually including two books by Aldo Rossi, and one each by Alan Colquhoun, Adolf Loos, and Moisei Ginzburg). In 1978, the "National Architecture Exchange" was also promoted as a new platform for communicating architecture. Using the "National Tour," as this key mechanism of architecture education came to be called, institutions across North America—universities and museums—could book traveling exhibitions and lecture series conceived at the Institute and order exhibition catalogues and slide series as teaching materials—much more diverse, albeit less professionalized than Monica Pidgeon's audio-visual series which she started a year later at the RIBA in the UK. In addition, the "New Wave" series, which showcased architectural scenes in different countries (first after Europe came Japan, then Austria), was redesigned as a lecture tour by architects, which included a traveling exhibition featuring their projects. Over the course of this phase, the Institute evolved into a postmodern salon for all kinds of people—not only practicing architects, critics, and historians, but also academics from other disciplines, authors, artists, curators, publishers, and gallery owners from New York and all across the USA, even from Europe, Japan, and Latin America. In addition to Rafael Moneo, Aldo Rossi, and Arata Isozaki, people like Gerrit Oorthuys (TU Delft), Giorgio Ciucci, and Massimo Scolari (both from the IUAV) came to the Institute as visiting faculty in the late 1970s. As the decade came to a close, the Institute's leadership sought out an institutional collaboration with the IUAV, having formed personal relationships with individual professors—with Manfredo Tafuri in particular.[332] The relationship with what in the Anglophone world became referred to as the "Venice School," however, seemed somewhat unusual, as the two institutions differed in their interpretations not only of history and theory but also of their pedagogical project in general, despite a shared interest in the classical modernist avant-garde and the postmodernist neo-avant-garde. However, they agreed to organize a joint workshop in Venice scheduled for the summer of 1979—and the Institute even advertised it. In the end, it was canceled, like so many other things before. Other pedagogical projects planned by Institute leadership in

332 The Istituto Universitario di Architettura di Venezia (IUAV), founded in 1926 as an independent school of architecture and officially recognized as a university, had succeeded in repositioning itself in the early 1970s with its historical research projects on the American, Soviet, and European city that resulted, among other things, in the publication *The American City* (1979), edited by Giorgio Ciucci, Francesco Dal Co, Mario Manieri-Elia and Manfredo Tafuri. In the mid-1970s, under the direction of Manfredo Tafuri, the "Venice School" broke new ground. A preoccupation with Michel Foucault and a transference of discourse analysis to architecture took place in 1978, when the French philosopher, at Tafuri's invitation, participated in a series of discussions in Venice with the professors teaching there. These were published in the volume *Il Dispositivo Foucault* (1977); On the positions of the "Venice School," see Hilde Heynen, *Architecture and Modernity: A Critique* (Cambridge: MIT Press, 1999), 128–147; see also Leach, 2014.

this period failed to get beyond the conception phase. For example, the idea of preparing teaching materials to accompany "Open Plan" events proved to be far too time-consuming for the course instructors and program leadership.

Core Curriculum

By the end of the 1970s, a decade in which the higher education system in the USA—architecture in particular—had undergone major reforms, and education in general was becoming increasingly marketized, the Institute was no longer pursuing the humanistic ideal it had originally set for itself for its pedagogical project as consistently as before. In 1978, as part of its fundraising and development efforts, the Institute mailed out a promotional brochure introducing its organization and program, its history, and its mission, with architecture education playing only a secondary role.[333] Overall, the brochure highlighted the extent to which the Institute's education offerings were no longer as committed to a humanistic ideal of education—the notion of educating a human holistically in the arts and sciences, in keeping with the standards imposed by the ideal of a general education—as well as the extent to which architecture was no longer taught as an integrative approach to problem-solving, but rather as an intellectual and artistic practice. Instead, the brochure emphasized historical references and the idea of creative activity from a humanistic tradition—i.e., on the basis of cultural if not religious heritage. This backdrop served to demonstratively frame the discussion of contemporary structuralist and post-structuralist approaches. The Institute's increased focus on history and theory was accompanied by an increased pedagogical push to address specific architectural problems and principles via recourse to formalism and modernism. In the teaching of design, for example, a conceptual approach was encouraged, and this had impacted how architecture was conceived and discussed at American schools of architecture. At a later meeting of the Fellows, some complained that architecture education at the Institute stopped being sufficiently practical after the "Undergraduate Program in Planning" and the "Program in Preservation and Adaptive Re-Use" were discontinued.[334] Ultimately, the interdisciplinary methods inherent in the teaching of urban studies, a subject that had once given the Institute half its name (instantiated at the Institute through the concrete research and design projects carried out under Peter Wolf and William Ellis until the mid-1970s), were increasingly replaced by drawing boards and textbooks. In the end, "urban studies" only appeared on the Institute's curriculum in the context of "Open Plan," with lecture series like "Forum on New York: The Place of Urban Design" (fall 1978) and "Housing Versus the City" (fall 1979). In the meantime, the Undergraduate Program had evolved in such a way that the primary aim in

333 IAUS brochure, 1978. Source: CCA Montréal, IAUS fonds: C.3-2 / C.3-3

334 Marguerite McGoldrick, official minutes of Fellows meeting, October 8, 1981. Source: CCA Montréal, IAUS fonds: A.2-9; Marguerite McGoldrick, unofficial minutes of Fellows meeting, October 8, 1981. Source: CCA Montréal, IAUS fonds: A.2-9.

developing the curriculum and selecting faculty members was to juxtapose different postmodern philosophical and architectural approaches, even though the five-unit requirements remained essentially the same: courses in history, theory, urban planning, structure, and a design tutorial. This is unsurprising when you consider that the Institute's contribution to the marketization of basic architectural education, training, and continuing education represented a postmodern phenomenon par excellence. The Institute's history and theory courses remained set in stone and continued to be taught by Eisenman, Frampton, Gandelsonas, and Vidler, who here, similar to *Oppositions*, also laid claim to what might be termed "interpretive sovereignty." However, in the years between 1977 and 1980, course offerings repeatedly had to be rearranged to accommodate the individual obligations and interests of the teaching Fellows—not least because they were now in higher demand as teachers or else involved in expanding other programs at the Institute. Frampton, after returning from London in 1977, went back to the Graduate School of Architecture and Planning at Columbia University, where he taught history, theory, and design, and prepared his first monograph, *Modern Architecture: A Critical History*. Vidler, who had always remained attached to Princeton, worked there as chairman of the dissertation committee and headed the "Program in European Cultural Studies" together with cultural historian Carl Schorske. Therefore Eisenman, in his role as director of the Undergraduate Program, took over the history course, which was usually taught by Frampton and Vidler, for the 1978–79 academic year. He put a creative spin on the curriculum, inspired by Colin Rowe's humanistic approach, which he had experienced firsthand.[335] In the first semester, he problematized the relationship between history and theory by examining the sixteenth-century Italian Renaissance, particularly the architecture of Andrea Palladio and Vincenzo Scamozzi. In his lectures and seminars, he focused on imparting a visual understanding of architecture as a historical text. In doing so, he was interested in the structural qualities of individual buildings rather than their textual qualities—in syntax rather than semantics. He treated historical examples as a theoretical tool for thinking analytically and architecturally, i.e., for focusing on formal and structural issues. Students were tasked with drafting plans and axonometric drawings, constructing models, and presenting their analyses of individual structures in the form of diagrams. For Eisenman, it was not about "historical accuracy or thoroughness" but about conceptual precision. While this understanding of architecture and history was somewhat one-dimensional, it allowed him to conclude the course with

335 Eisenman was once a traveling companion of Colin Rowe on his Grand Tours of Italy. For the course description and reading list, see Peter Eisenman, "History," Undergraduate Program in Architecture. Course Outlines and Reading Lists, Fall 1978. Source: private archive of Patrick Pinnell. Unlike Frampton and Vidler, Eisenman was also primarily interested in teaching formal and structural analysis as a basis for design. Both socio-economic and political issues, as well as the urban scale were deliberately left out, as were the technological or practical problems that govern architectural practice. It should be noted that Eisenman did not design the course alone, but together with Giorgio Ciucci.

an analysis of Giuseppe Terragni's *Casa del Fascio* (1932–36). In the second semester, as a follow-up, Eisenman offered a course on "Architectural Analysis: Image and Text." This was essentially a theory course centered around a formal analysis of canonical buildings, similar to what he had taught at Cooper Union since 1968. The case studies were selected buildings by Terragni and Le Corbusier.[336] The list of required reading covered the established historiography of architectural modernism: Sigfried Giedion's *Space Time and Architecture* (1941) and *Mechanization Takes Command* (1948), Rudolf Wittkower's *Architectural Principles in the Age of Humanism* (1949), and Françoise Choay's *The Modern City: Planning in the 19th Century City* (1969). But beyond reading these classics, students at the Institute were given no further instruction in the historiographical method, as would have been required in a liberal arts college.

The 1978–79 theory courses, on the other hand, were divided between Gandelsonas and Vidler. While the fundamental approach of Gandelsonas's fall semester course ("Elements for an Architectural Theory") was pragmatic, exploring how architectural knowledge is produced and how contemporary practice can use it as an instrument for critique and transformation, Vidler's spring semester course ("Ledoux, or the Formation of Modernism") focused solely on the biography of a single architect, hoping to use Claude-Nicolas Ledoux as a case study to reveal the contradictions of modern architecture.[337] Another time, Gandelsonas, whose thinking was informed by post-Marxist and post-structuralist approaches, semiotics, and linguistics, organized his theory course as a reading course of Leon Battista Alberti's *Ten Books of Architecture* and Le Corbusier's *Towards a New Architecture* to teach the differences between normative and interpretive theories and discuss their structure and function in terms of practice. He taught interpretive theories in relation to specific themes: the role of architectural history, the merits and limitations of formal and typological analysis, and history as a prediction of the future or memory of the past. In epistemological terms, he was concerned with the fundamental possibility of establishing an objective theory. The reading list included essays by Agrest and Gandelsonas well as Louis Althusser and Claude Levi-Strauss, John Lyons, and John Searle. Vidler, on the other hand, worked biographically and followed the exemplary principle in his theory course, conceptualizing it around the contradictory figure of Ledoux, whose life and work exemplified for him the transition from the classical to the modern era. Based on a reading of Ledoux's texts, he discussed various textual formats in relation to the question "What are the Limits of the Text?" and the education of new architects in relation to the question "What is Architectural Education?" but also addressed issues of utopia and politics. For Vidler, the intensive discussion

336 Peter Eisenman, "Architectural Analysis: Image and Text," Undergraduate Program in Architecture. Course Outlines and Reading Lists, Fall 1978. Source: private archive of Patrick Pinnell.

337 Mario Gandelsonas, "Elements for an Architectural Theory" & Anthony Vidler, "Ledoux, or the Formation of Modernism," Undergraduate Program in Architecture. Course Outlines and Reading Lists, Fall 1978. Source: private archive of Patrick Pinnell.

of Ledoux as one of the main representatives of revolutionary architecture, whom he positioned between Palladio and Le Corbusier, was about questions of modernism and postmodernism in a genealogical sense.

For both Gandelsonas and Vidler, the introduction of "French theory" to the United States offered a means of underpinning formal analysis and formal design as well as historiography and theory production, regardless of whether the space being referenced was conceived at an architectural or a more urban scale.[338] The one constant in the Undergraduate Program, and in architecture education at the Institute in a broader sense, however, was the "Structures" course taught by Robert Silman, which he had offered since the beginning in 1974. This course borrowed heavily from Henry Cowan and the teaching approach of Mario Salvadori, an engineer at Columbia University. The course in "Urban Development" had to be scrapped, since Wolf had taken a temporary break from teaching at the Institute. But Eisenman as director of the Undergraduate Program managed once again to attract international guests to the Institute to teach this unit instead. In the fall semester of 1978, Italian architectural historian Giorgio Ciucci, who was also teaching at the Rhode Island Institute of Design (RISD), taught a course entitled "Representation of Space, Space of Representation," which he had prepared at the IUAV, and now only had to recycle it. Ciucci also supervised interns at the Institute. In the spring semester of 1979, the French architectural historian Antoine Grumbach, who was also teaching at Princeton University, held a course entitled "Urban History: Paris as an Urban Form; The City as Collage." Another constant was the design tutorial led by Diana Agrest; the team of tutors was reassembled from scratch every year, however, leading to a rather eclectic design pedagogy.[339] The development of the Institute's curriculum and faculty over the years clearly showed that different perspectives could be juxtaposed and brought into contact with each other, even if its publications and public events, in addition to the required learning materials for students of the Undergraduate Program, at least hinted at a postmodern understanding of history and theory, which would have highlighted metafiction and deconstruction, rather than objectivity and facts, based on a critical, academic understanding of research.[340] Naturally, this stemmed from the fact that the staff imported their own stylistic preferences and methodological approaches to their teaching.

338 However, contemporary French philosophy, both structuralist and post-structuralist approaches, while referenced in *Oppositions*, and published in *October* and *Skyline*, were rarely taught at the Institute.

339 Stephen Harris and Stephen Potters were tutors in the fall 1978 semester, as well as Italian architect Massimo Scolari, who was also a guest at the Institute along with Giorgio Ciucci; in the spring 1979 semester, Harris and Potters were joined by Patrick Pinnell and Myles Weintraub.

340 However, it has been pointed out there is no clear definition of the term postmodernism. French philosophers who are generally considered postmodern, such as Michel Foucault and Jacques Derrida, but also the American historian Hayden White, strictly refuse to be identified with postmodernism, see White, 1973.

As an unconventional architecture school, the Institute enabled its Fellows to build careers as academics and architects, including by showcasing their skills as pedagogues, historians, and theorists, or otherwise financing their work. The best illustrations of this are Agrest and Gandelsonas, both of whom were unknown before coming to the USA in 1971 and joining the Institute. The Institute provided a framework in which they could make a name for themselves, whether with articles in *Oppositions*, lectures within the context of the "Architecture" or "Open Plan" series, or contributions to the "Exhibition Program." As the decade progressed, Agrest would become an assistant professor at Princeton University in 1972 and a professor at Cooper Union in 1976; Gandelsonas worked various teaching jobs, first at Sarah Lawrence and then at Columbia University.[341] Towards the end of the 1970s, they were the primary shapers of architectural education at the Institute, where they had both risen to leading positions on the faculty. In addition to heading the theory course, Gandelsonas also became the coordinator of the entire Internship Program when intern numbers began to rise.[342] In the meantime, it had become mandatory for all interns to spend two semesters at the Institute. During the first semester, they were required to complete one of the Undergraduate Program's design tutorials, and in the second semester, they worked on personal projects with individual Fellows. They also had to participate in two "Open Plan" courses per semester. Another consequence of the Internship Program's administrative reorganization was the rule that interns were recognized for all of their work and shared in the profits, at least in cases where Fellows made money off their projects. Eisenman, for example, compensated his interns with drawings and models that they had produced according to his specifications. As assets, these objects were likely to appreciate on the art market, or in museum archives. To maintain its credibility as an educational institution, the Institute limited letters of recommendation to two per person.[343] Although the interns could list two preferred architecture schools each, it was ultimately the Fellows who decided among themselves how to allocate the letters, thus exerting a key influence on the interns' choice of the most suitable architecture school. The Institute predominantly sent graduates of its Internship Program and Undergraduate Program to Columbia University or Yale, but a few also ended up at Princeton, Harvard's Graduate School of Design, or MIT. The function and purpose of the internships were a controversial issue at the Institute. The Fellows all agreed that the Internship Program

341 Agrest became a professor at Princeton University as early as 1981, Gandelsonas not until 1991.

342 Mario Gandelsonas & Giorgio Ciucci, "Internship Design Tutorial," Undergraduate Program in Architecture. Course Outlines and Reading Lists, Fall 1978. Source: private archive of Patrick Pinnell.

343 The new "Graduate School Admission Policy" was on the agenda at the beginning of 1979; Advisory Committee, meeting minutes, February 8, 1979. Source: Sarah Lawrence College Archives.

explicitly functioned as a form of career guidance. The decision to formalize the internships was justified by the fact that interns were now working exclusively for one of the Fellows; this meant that the goals were different from before, when interns worked alongside the Fellows on urban research and design projects. As director of the Internship Program, Gandelsonas argued it should also be the individual Fellows' responsibility to train interns in basic skills and abilities—after all, they were being used for their labor at the Institute. He argued this because, unlike students in the Undergraduate Program, interns often came to the Institute with no previous training. But the idea of an eight-week crash course was never implemented. The role of interns at the Institute and their relationship to the Fellows was neatly summarized by Agrest, who described their status as that of "apprentices," suggesting a master–apprentice relationship. What the internships amounted to was a pre-modern, artisanal approach to teaching and learning, based on modeling and imitation, rather than a modern scientific or even postmodern problem-oriented approach.

Compared to its other offerings, the Institute's High School Program was exceptional, something of a luxury, in that it displayed quite a high degree of continuity over the years. The reasons why the Institute maintained the program after 1977, once it had repositioned itself as a high culture educational institution, were both institutional and personal. While contributing little to the funding of the Institute's operations, it was extremely important as a public relations tool. Perhaps the main reason the Institute preserved the program was its interest in maintaining the self-image it had cultivated, as a site that offered a "continuous education" encompassing all age groups.[344] In the summer of 1977, after Eisenman and MacNair had successfully launched the program, two junior New York architects, Lawrence Kutnicki and Deborah Berke, were appointed co-directors, MacNair having been assigned to other tasks. Eisenman entrusted the two to lead the program, even though they were both only recent graduates and had scarcely any teaching experience.[345] By the fall of 1977, Kutnicki and Berke had transformed the High School Program into a

344 Compared to offerings at other New York institutions, a pedagogical history and didactic analysis shows that the Institute's High School Program was committed to teaching architecture as art according to a humanities ideal; see Förster, 2013. Each semester, the Institute produced its own poster, designed in-house on the basis of its graphic identity. Source: private archive of Lawrence Kutnicki. In addition, the High School Program found a major supporter in Guy Trebay, a journalist with *The Village Voice*. To target students, Kutnicki and Berke wrote to private schools in the metropolitan region, where they worked with art teachers, but also to some public high schools. In November 1978, they presented the program at the National Institute of Architectural Education's (now the Van Alen Institute) Career Day, which was graced by no other than Philip Johnson.

345 Kutnicki's qualifications included a bachelor's degree in architecture from Cooper Union and a master's in urban planning from City College; Berke had a bachelor's degree in architecture and art from the Rhode Island School of Design.

comprehensive, carefully thought-out education offering, operating on a shoestring budget with relative independence from the Institute's usual activities. They quickly assumed full autonomy, teaching Saturday courses during the fall and spring semesters, although the core of the program was still the multi-week course offered during the summer holidays. With Eisenman's support, they were able to use all the Institute's facilities and resources to develop a sophisticated program for talented architectural youth. Since the High School Program, unlike the Undergraduate Program, wasn't tied to any institutional requirements or learning outcomes, Berke and Kutnicki had a free hand in taking an experimental approach to designing the courses, both in terms of content and methodology. From 1977 to 1980, every course (which they redesigned each semester) was dedicated to an overarching theme, alternating between architecture and urban planning. In "Architecture and the Arts," they delved into architecture's position in relation to the other arts like poetry, drawing, painting, sculpture, photography, film, dance, music, and literature. In "Five Architects," they focused on specific heroes of architectural modernism (Frank Lloyd Wright, Le Corbusier, Ludwig Mies van der Rohe, Alvar Aalto, Louis Kahn) and projects from contemporary architects. In "Mapping Manhattan," they focused on the architecture of New York and the development of specific urban structures and building typologies. Initially, Eisenman and some of the Fellows participated in the High School Program, but over time the program directors increasingly sought to hire artists from their circle of acquaintances and friends, sometimes even former students, as faculty. The fundamentals of architecture were conveyed to the students from a theoretical and historical perspective, and course topics ranged from architectural principles to historical figures, from construction to symbolism. The quality of the courses was extremely high, as evidenced by the intellectually demanding texts that were on the required reading lists. The High School Program design studio, meanwhile, followed an approach that was both formal and artistic, comparable with the education offered by Cooper Union. In this case, however, John Hejduk's design principles were adapted to the language and life experience of teenagers. In addition to this, the directors organized regular excursions to contemporary exhibitions or selected buildings, during which students were asked to make on-site drawings. Because Kutnicki and Berke invested a great deal in mentoring students, they achieved their declared goal of introducing them to abstract concepts and spatial thinking.[346] Ultimately, the High School Program was meant to convey an impression of what it meant to work as an architect, critic, and historian. Even though

346 Kutnicki and Berke demonstrated their work as educators each year with books of students' works compiled by them, which next to copied records of drawings, models, and collages included written reflections and course evaluations by the students. Source: private archives of Lawrence Kutnicki and Suzanne Frank. Based on these books, students can be said to have achieved not only cognitive goals, but instrumental ones of learning to see and draw, and affective ones of developing a critical awareness of architecture and the built environment.

it was a relatively small program, featuring ten to fifteen students per course, both formats, "Summer Architecture" and "Saturday Architecture," were very popular, and some students returned in the following years. There is evidence of students who went on to study architecture.

Design Education

Ultimately, by offering the Undergraduate Program, the Internship Program, and the High School Program, the Institute performed a pedagogical function that hadn't existed in this form in the American education system, an initiating function as much as a gatekeeping one—despite constant changes in curriculum development, certain major personnel changes in the faculty, and the renunciation of responsibility for a traditional degree qualification. But the educational programs always had an institutional function as well, for while the Institute had built a philanthropic network of architects and builders through its offerings in adult education and had increasingly come to serve the architectural establishment and building industry, the architecture education it provided contributed to the formation of a new architectural elite in the United States. Near the end of the decade, the Institute launched another (very expensive) teaching format: the "IAUS Advanced Design Workshop in Architecture and Urban Form." Offered for the first time in the 1979–80 academic year, the program was led by Diana Agrest and represented an exclusive environment for conveying a postmodern attitude towards design. The course, conceived as an experimental design studio, was a one-year program predominantly aimed at advanced architecture students. Applicants were required to have already finished a four-year bachelor's degree or be enrolled in a five- or six-year bachelor's degree program. In this regard, the Advanced Design Workshop was a successor to the Design and Study Options, which served Agrest as a cornerstone in both administrative and institutional respects.[347] Once again, the Institute teamed up with a consortium made up of state schools primarily aimed at vocational training (meaning, in this case, technical colleges), in addition to Cooper Union and the schools of architecture at Cornell University and Yale.[348] By using the qualifier "advanced" in the title, the Institute purposefully set the program apart from conventional architectural programs, no longer placing itself in a supplementary role, but for the first time positioning itself as a site of further education. Equipped with a state education grant from the Fund for the Improvement of Postsecondary Education of the U.S. Department of Health, Education, and

347 Agrest for her part was not only able to take over the institutional network of Design and Study Options that Ellis had built, but also use the remaining credit; see Frank, 2010, 135. The administration was initially held by Jill Silverman, and in the second year it was taken over by Berke.

348 The ADW consortium included: Cooper Union, Cornell University, University of Cincinnati, University of Houston, Illinois University, University of Maryland, Miami University (Ohio), Syracuse University, Yale University. Students came in large part from Syracuse University and Tulane University.

Welfare (HEW-FIPSE), Agrest was able to aim higher with the new education offering than with the former Design and Study Options. She managed to hire acclaimed architects as teachers for the program, most of whom held professorships at renowned universities. In the first year, guest professors included Charles Gwathmey, Cesar Pelli (dean of the Yale School of Architecture), and Aldo Rossi (IUAV)—big names who were tied to Eisenman and the Institute in various ways. Their likenesses were also used on the posters specially designed by Vignelli to advertise the Advanced Design Workshop in a nationwide campaign. One element that set it apart from other schools of architecture was the high tuition fees: when the program first began, the cost was US$4,500 per year or US$2,500 per semester; in the following academic year, these figures rose to US$5,500 and US$3,500. In addition to the Undergraduate Program and the Internship Program, the Advanced Design Workshop was a commercial offering with which the Institute sought to position itself in the new decade as the "Center for the Study of Architecture and Urban Design," as it called itself in a new funding application to the NEH.[349] However, in institutional, pedagogical, financial, intellectual, didactic, and personnel respects, the Institute remained dependent on the Ivy League schools of architecture.

While the Institute positioned the Undergraduate Program as a foundational course and the Internship Program as an in-depth training program within the changing discipline of architecture, it made no claim towards the Advanced Design Workshop, a form of expanded college study, amounting to a vocational training program.[350] In addition to its claim to professionalism and exclusivity, the Advanced Design Workshop's distinguishing feature was that it reached a balance between theory and practice—that is to say, by integrating history and theory into design. To achieve this, Agrest stressed that all design projects must have a relationship to the urban context. On the posters advertising the program, interested students read the following dual objective: "to find new ways to make architectural education more effective and relevant to the urban situation and to find new ways for architectural students to apply theoretical concepts to existing urban problems; to serve a limited number of special students and provide an intensive and exceptional year of work and study articulating the theory and practice of design in a work situation." For potential participants, the poster presents New York (even though it speaks in general terms) as an urban setting with a dual function: as a learning environment and a case study. Didactically, the course was split into a workshop component and a theory and history component. The program was too small, however, for there to be a course offering in history and theory led by Agrest. Instead, Advanced Design Workshop students were invited to the "Open Plan" lectures, which in the fall semester of 1979 were

349 Frederike Taylor, "NEH Proposal," n.d. Source: CCA Montréal, IAUS fonds: A.2-10.

350 Ibid.

quite eclectic, with courses on "Piranesi/Le Corbusier" (led by Gandelsonas and Vidler), "The American Monument" (Patrick Pinnell), "Housing Versus the City" (Frampton), and "Architecture in the 1980s" (MacNair). Additionally, Advanced Design Workshop students were referred to the Undergraduate Program's curriculum, specifically the courses taught by Eisenman, Frampton, Gandelsonas, and Vidler. Beyond its institutional synergies, the modularity of architecture education at the Institute now had a pedagogical component, as leadership anticipated that the older, ambitious graduate students would exert a positive influence on the undergraduates and interns at the Institute.

In terms of service orientation, when it came to hiring a teaching staff, the Institute continued to profit from its networks, friendships, cooperations, and affiliations that had been carefully curated and designed by Eisenman. For those involved, one asset was that the Institute was finally able to secure Rossi as a design tutor for the Advanced Design Workshop course in the 1979–80 academic year. Affiliated with the IUAV, Rossi had been a regular guest at the Institute since 1976 and frequently taught at Cooper Union;[351] he was to teach at the Institute for two years, even after rising to international stardom with his appearance at the first Architecture Biennale in Venice in 1980. At the time, the Institute was involved in translating his *L'architettura della città* and preparing the first edition of *A Scientific Autobiography*, both of which were to take Rossi's reception in the Anglophone world to a new level. There were some students who came to New York and attended the Advanced Design Workshop exclusively because of Rossi and his famous postmodern, contemplative, and melancholy approach to design (some of this fame was no doubt attributable to his essays in *Oppositions* and exhibitions at the Institute).[352] Rossi's approach—visible in his writing, photography, drawings, and projects—was characterized on the one hand by the historical and geographical contextualization of urban architecture, and on the other, by the typological, artistic, and pictorial approach of analogous architecture. The innovation and transformative power of the Advanced Design Workshop, however, cannot be ascribed solely to the personality of Rossi, some credit is due to the format itself. Broadly speaking, the distinguished and diverse faculty stood out from the faculties of other universities in the metropolitan area. The composition of the faculty demonstrated the extent to which Eisenman had over the years deliberately assembled a variety of perspectives in teaching at the Institute. But it also reflected a postmodern tipping point in American architecture education.

351 Eisenman had charmed Rossi and promoted him to the utmost, the Institute granting him a solo exhibition twice after the spring of 1976 and once again in the fall of 1979 (only Scolari had the opportunity to do this).

352 One student was Kyong Park, who had come to New York especially to study with Rossi, but soon turned his back on the Institute in disappointment, only to start the Storefront for Art and Architecture in 1982; see Joseph Grima, José Esparza, Chong Cuy, Charles Sneath, Suzannah Bohlke, Cesar Cotta, Pernilla Ohrstedt, and Danny Wills, eds., *Storefront Newsprints 1982–2009* (New York: Storefront for Art and Architecture, 2009).

Prominent design tutors like Rossi stood for postmodern fashions, dispositions, thinking styles, and behaviors, employing self-referential, sometimes even self-satisfied language and diction. This kind of habitus was echoed in figures like Agrest, Eisenman, Gandelsonas, and Vidler, all of whom maintained friendly or otherwise close relationships. The fact that, in the 1980–81 academic year, Cesar Pelli was replaced by Robert Stern and O. M. Ungers, two architects who had once stood for quite different approaches to design, was emblematic of how postmodernism was institutionalized at the Institute. In the early 1980s, when the Institute sought to capitalize on the emerging "star system" in architecture that it had helped to create, architecture education played a significant part. Ultimately, the Advanced Design Workshop, as a new type of education offering with which the Institute competed with other educational institutions in the academic landscape, was only effective for a short period of time, namely the length of time it had a sufficiently large budget to be able to afford experienced, well-regarded teachers. When funding from HEW-FIPSE ended after two years, and the Institute, as a result, curtailed its public relations campaigns for the 1981–82 and 1982–83 academic years, the composition and performance of the Advanced Design Workshop faculty fell considerably short of its high expectations, with enrollment dropping from twenty-one to nine. The shelf life of Agrest's unique education offering was brief. Even expanding the target group to include master's students was of no avail, and the program was ultimately canceled.

Nevertheless, the Advanced Design Workshop is symptomatic of the changes in American architecture education after the revolts and reforms of 1968 and the general tendency toward redisciplining—that is, the return to a more intellectual and artistic understanding of architecture. As the Institute became more institutionalized as an architecture school over the years, it contributed to this process with its various education offerings, as well as its conferences, exhibitions, events, and publications—beginning with the conference it organized at MoMA in 1971, "Architecture Education U.S.A.: Issues, Ideas and People," as well as the "Universitas Project" conceived by Emilio Ambasz in 1972.[353] Eisenman regularly sought to position the Institute in the tradition of the Bauhaus, the modernist school par excellence, which exerted its influence on the American architecture education system via personal continuity, even though the opportunities of 1970s New York were entirely different. In his polemics, he also compared the Institute to other contemporary architecture schools like the AA, Cooper Union, or even the IUAV. Yet the Institute never (or only to a limited extent) engaged in a critical, historical, or even theoretical

353 On the 1970s shift of architecture education and the triumph of formalism, see McLeod, 2012. McLeod, and more generally the *Architecture School* anthology edited by Ockman, only address the role and function of the Institute and the Fellows in the American higher education system sporadically, for whatever reason.

reflection of the role and function of architectural education in a time of social change. The sole exception was Kenneth Frampton, who published an article entitled "Apropos Ulm: Curriculum and Critical Theory" in *Oppositions* 3 about the Ulm School of Design, founded in the aftermath of World War II, which was known for its modern conception of design and its efforts to spearhead a democratic reorganization of living conditions in postwar Germany.[354] Frampton also contributed to a 1980 issue of *Lotus International* dedicated to "Architecture in the American University," with selected examples of contemporary architecture education (in addition to his various editorial tasks at the Institute, he was part of the Executive Council of the Italian journal, which he had joined in the fall of 1976 with issue 12).[355] The issue compared the doctrines of three American schools of architecture, Cornell, Cooper Union, and Columbia, and thus had to be selective in its approach. It opened with an introductory essay co-authored by Frampton with Alessandra Latour, outlining the historic development of architecture education in the United States throughout the twentieth century. It then went on to profile, in extensive detail, Cornell University, and Colin Rowe's Urban Design Studio in particular (written by David Middleton), Cooper Union (written by Rafael Moneo and Robert Slutzky, who taught design there), and Columbia University (written by Richard Plunz, who was chairman of the Architecture Division there). Rowe himself contributed a text revisiting a talk on the utility of education he had given in 1971 at the conference "Architecture Education USA," which was organized by the Institute but never published and has now been erroneously reduced to Rowe and his case for style.[356] The editorial expressed the hope that an analysis of the structure of teaching at the different colleges would provide insights into the self-perception of American architecture. Accordingly, Cornell University (a collaborator with the Institute in its early days) stood for the attempt "to reconcile the Beaux Arts tradition with the modern movement," Cooper Union under John Hejduk (who had hired Eisenman and maintained close relations to the Institute) stood for "the entrance of the artistic avant-grade into university teaching in the wake of the Bauhaus," and Columbia University (where Frampton himself taught) stood for "a pragmatic tradition that has found its field of application in the impact with the social problems of the metropolitan city."[357]

Yet the enormous changes that took place in higher education and the architectural world of 1970s America were barely the subject of historical comment,

354 Kenneth Frampton, "Apropos Ulm: Curriculum and Critical Theory," *Oppositions* 3 (May 1974), 17–36.

355 *Lotus International*, no. 27 (1980): "Architecture in the American University."

356 Colin Rowe "Architectural Education in the USA: Issues, Ideas, and People. A Conference to Explore Current Alternatives," *Lotus International*, no. 27 (1980), 42–46.

357 Frampton and Latour, 1980.

critical reflection, or public debate. The *Lotus International* editors paid no attention to some of the East Coast's most important architecture schools—not just the University of Pennsylvania, Yale, and Princeton, but also Harvard and MIT, to say nothing of the schools of note on the West Coast, including Berkeley, UCLA, or the recently founded Southern California Institute of Architecture SciArc in Los Angeles. However, even though such an approach yielded only a rough sketch of architectural debate and education in the USA, the juxtaposition of teaching approaches at Cornell, Cooper Union, and Columbia revealed the shadowy outlines of the Institute as an architecture school in this renegotiated space of transatlantic, modernist approaches to architecture. The Institute was always in evidence here, even if only implicitly, as a negative foil. Peter Eisenman was even discussed in the *Lotus* editorial, albeit as an architect, not as the Institute director. He was mentioned in the same breath as American architectural historian Vincent Scully, in order to illustrate two opposing positions held by American architecture vis-à-vis Europe. Standing on one side was Scully, an avowed advocate of "genuine" American architecture, with Eisenman, a militant interpreter of Europe's architectural modernism, on the other. Eisenman's characteristic rhetorical gestures were described here as "a complex play of transatlantic influences and exchanges," which the editors claimed was in turn exerting an influence, in a truncated form, on European architecture. Ultimately, in the perception of the *Lotus* editors, it was the traditional schools of architecture, some of which were backed by over a hundred years of tradition, rather than the Institute, that continued to set the standard for architecture education. The Institute did, however, exert an influence on education, whether directly or indirectly, with its Fellows' teaching, its cultural activities, and the teaching material it provided. For it was through its relational, complex, and differentiated work—through the interplay of its educational, cultural, and publishing practices—that it came to alter the architecture culture, in the USA and internationally, in critical and decisive ways. Arguably, the Institute was the most postmodern of all America's architecture schools. What was exemplified at the Institute was a deeply engaged, highly ambitious pedagogy that aimed beyond merely keeping the Institute alive. Despite the faculty's commitment and skills, academic display, and professional advancement eventually came to trump the needs of a quality architecture education. Such an approach, if sustained, would not only have prepared students of architecture for diverse living and working environments by imparting core competencies in architectural, ecological, organizational, methodological, and socio-economic fields but would also have fostered social responsibility in the profession and discipline.

2.4 Commercially Exploiting Learning

At the end of the 1979–80 academic year, when the NEH grant for Open Plan expired and was not renewed, the chapter of adult education at the Institute was closed once and for all. Yet as America stood at the brink of a major conservative turn, affecting the federal endowments, the Institute's leadership had no choice but to undertake yet another reinvention in tandem with a restructuring of its finances for the 1980–81 fiscal year. Despite the recent shift in emphasis toward a much larger portfolio of publications and exhibitions as main fields of activities, architecture education continued as a successful business model, even though at this point the Undergraduate Program, the Internship Program, and the Advanced Design Workshop made up only a fourth of the total budget, which had risen to nearly US$877,000. While teaching was still a central focus, the figures for the Institute's education offerings had actually even begun to dip into the red: a total deficit of US$45,000 was anticipated for the 1980–81 fiscal year, and only part of the debts were paid off in the following year.[358] The complexity, however, is reflected in the fact that the Institute reached peak enrollment in the 1980–81 academic year, with a total of seventy students enrolled in the three commercially successful education offerings, not least because it had conducted a successful marketing campaign with new posters and brochures.[359] As an educational institution, and especially in terms of personnel, the Institute had once again reached its limits. The neoliberalization of Institute operations, now financed by a challenge grant from the NEH, which had to be matched by private donations at a 2:1 ratio, therefore manifested itself in the restructuring of the Board of Trustees. This began in the 1980–81 fiscal year. In addition to Philip Johnson, more architects, developers, and businessmen were added to the board, and in the future, they would decide the Institute's fate. New management positions were being created to be filled by people from the business world. For example, Hamid-Reza Nouri, an auditor, was nominated associate director, and Lynn Holstein became director of development after the departure of Frederieke Taylor, both of them being tasked with making individual programs profitable.

In "Education Programs," one of the four pillars of the new decade, next to "Publication Programs," "Public Programs," and "Development Programs," however, the biggest challenge was the composition of the faculty, as veteran Fellows took on new projects and new people had to be brought to the Institute. In fact, the biggest change in architecture education at the Institute revolved around Peter Eisenman who relinquished his position as director of the Undergraduate

358 IAUS, "Undergraduate Program/Internship Program," 1980–81, May 23, 1980, Source: CCA Montréal, IAUS fonds: A.2- 10; IAUS, funding requirements of each program, 1980–81, Source: CCA Montréal, IAUS fonds: A.2- 11.

359 IAUS, "Student lists," 1980–81. Source: CCA Montréal, IAUS fonds: A.1-4.

Program, which he had occupied for six years, for the fall semester of 1980, after having founded the firm Eisenman/Robertson with Jaquelin Robertson on January 1, 1980.[360] In 1980, Diana Agrest and Mario Gandelsonas did the same, founding Agrest and Gandelsonas Architects, headquartered in New York.[361] Although Eisenman encouraged young faculty members, the Fellows' primary complaint was that his successor wasn't satisfactorily vetted. Initially, Lars Lerup from the University of California, Berkeley, who did not make an appearance at the Institute beforehand, took on the administration of both the Undergraduate Program and the Internship Program, as a Visiting Fellow. He was assisted by Deborah Berke as administrator, while Lawrence Kutnicki and the painter Robert Slutzky, returning to the Institute after many years, supervised the interns. The Undergraduate Program's history and theory courses continued to be taught by the long-tenured Fellows—Frampton, Gandelsonas, and Vidler, as well as Robert Silman—at least ensuring their continuity and quality. Replacing Agrest, who was concentrating on the Advanced Design Workshop, Gandelsonas took over the administration of the design tutorial, which he taught in collaboration with Deborah Berke and Christian Hubert as tutors. Ignasi de Solà-Morales, the second Visiting Fellow at the Institute in 1980–81, was also brought back to teach and headed a course on "Urban History." Another major issue was that, beginning in 1980–81, the Institute was on the constant lookout for new premises, as already announced by Eisenman in 1977, both because it lacked space and because the stepped rent for the 8 West 40th Street penthouse had become a problem. The Institute's leadership visited a variety of properties that met the requirements for an architecture school, but in 1981 the Institute was able to expand its spaces once again, renting additional studio rooms one floor below, on the building's twentieth floor, since the former design studios on the upper mezzanine floor had in the meantime all been converted to offices for the editorial staff of the various publications. At the outset of the 1981–82 academic year, addressing a meeting of the Board of Trustees, Eisenman appeared happy with these developments: "[E]ducation programs were doing well, for the first time we have adequate quarters."[362] Under the new political auspices—after 1981, with President Ronald Reagan taking office, America was becoming a "nation of the rich" and occupying a central place in the rise of global neoliberalism—the Institute had transformed into what was primarily a training ground for architects of the post-industrial knowledge and information society since education now represented an investment in the future.

360 To mark the founding of the office, Eisenman published early projects; see *Architecture + Urbanism*, no. 112, (January 1980): "Peter Eisenman."

361 Agrest and Gandelsonas followed; see *Architecture + Urbanism*, no. 114 (March 1980): "Diana Agrest, Mario Gandelsonas."

362 IAUS, minutes of the meeting of the Board of Trustees, October 6, 1981. Source: CCA Montréal, IAUS fonds: A.1-5.

Throughout 1981, the Fellows held a series of meetings to develop plans for the future and draw up new scenarios that would allow the Institute to continue to fulfill its function as an architecture school, especially with junior teachers as faculty. The proposals submitted in these meetings made it clear that architecture education had an economic dimension that was just as important as the pedagogical one.[363] First of all, for the first time since 1976, the agenda included the idea of establishing a degree-granting program. The Fellows discussed the details of adding a graduate program to their education offerings and, eventually, offering students the opportunity to obtain an accredited degree at the Institute. One prerequisite the Institute would have to meet before becoming an officially recognized school of architecture was that such an education offering would need to be approved by the U.S. Department of Education. However, establishing a fully-fledged degree program would require an estimated start-up capital of US$500,000, which the Institute could not raise. Another option would have been for the Institute to partner with another educational institution that already had state recognition, as was discussed with Sarah Lawrence College in 1976. But this was only possible if the arrangement offered something to both parties. In any case, no degree program was possible without the existence of a research library, which would need to be constructed around the existing inventory of Eisenman's library, but doing so would cost an estimated US$1.5 million. The discussions clearly revealed that the Fellows were divided over whether the Institute had the time, energy, and money to invest in developing such a program. The Fellows had quite differing ideas about the Institute's direction. Frampton, for instance—following up on a suggestion by Lerup, who envisioned the Institute as a research center—even suggested incorporating a doctoral program in art history, since art historians were underrepresented when it came to teaching history and theory at schools of architecture.[364] As Frampton saw it, such a move would place the Institute in an ideal position to enter a consortium with NYU, Harvard, and Columbia University to offer postgraduate academic architecture education in place of a master's degree. Eisenman, on the other hand, was toying with the idea of founding a new institution along the lines of the Canadian Centre for Architecture in Montréal, founded in 1979 by Phyllis Lambert, yet without the corresponding spaces.[365] The Institute pursued this idea seriously in the 1981–82 fiscal year, briefly drafting plans for what was called "The Philip Johnson Center for Architecture."[366] In its existing form, the Institute

363 Silvia Kolbowski, minutes of Fellows meeting, March 10, 1981. Source: CCA Montréal, IAUS fonds: A.2-9. / ARCH401178

364 Doctoral programs were now offered at Princeton, MIT, and Berkeley; see McLeod, 2012.

365 Kolbowski, 1981.

366 IAUS, project description for "The Philip Johnson Center for Architecture," 1981, Source: CCA Montréal, IAUS fonds: A.1-13 / ARCH263662.

would have been continued, albeit merged into this new institution, creating something unprecedented in New York: part archive, part museum, part research library, and part study center. The plans extended to founding a new school of architecture that would finally have offered fully fledged, accredited architecture degrees, but the project ultimately fell through due to a lack of funds.

The Institute's success from 1974–75 derived from the fact that, on the basis of architecture education, with government funding and in the course of its professionalization and bureaucratization, it had become a powerful institution that knit two areas of society, education and culture, tightly together while also functioning as a publishing imprint (or, in certain cases, as a co-publisher).[367] Nevertheless, its financial future remained uncertain. The Institute hoped to find a programmatic solution to this uncertainty when, over a few days in the fall of 1981, it called a series of meetings of the Fellows to "departmentalize" itself. The idea was to reorganize the Fellows' activities into four different working areas with separate responsibilities while drafting a strategy for development and priority-setting over the next five years. The result was a five-year plan that divided the Fellows' work into the following categories: "Publication Programs," "Education Programs," "Public Programs," and "Development Programs."[368] By then at the latest, development—i.e., fundraising and public relations, as well as grant applications and income-generating project planning—was the Institute's driving force, to which education offerings, public events, and publications all catered to. The unofficial minutes of these meetings recorded the course of the discussions between the Fellows present (unlike the official minutes, which had been revised and only summarized the content of the discussions), highlighting the problems facing the individual education offerings.[369] As testimonies, they reveal that certain Fellows had come to see the Institute's architecture education project as fundamentally doomed to failure. They did acknowledge that the Institute had achieved a certain measure of renown as an alternative architecture school, attracting students from liberal arts colleges and schools of architecture not only throughout the USA but now from around the world. Students of the Advanced Design Workshop, for example, came from places as diverse as the AA in London, Ireland, and South Africa. Yet enrollment figures were already declining to the extent that the Institute's leadership felt it had no choice but to increase tuition fees. In the 1981–82 academic year, fees for the Advanced Design Workshop were raised from US$5,500 to US$7,500, while fees for the

367 Ockman, 1988.

368 CCA's IAUS fonds contains a folder of unofficial and official minutes of Fellows meetings from the fall of 1981. Source: CCA Montréal, IAUS fonds: A.2-9.

369 Marguerite McGoldrick, official minutes of Fellows meeting," October 8, 1981. Source: CCA Montréal, IAUS fonds: A.2-9 / ARCH401164; Marguerite McGoldrick, unofficial minutes of Fellows meeting, October 8, 1981. Source: CCA Montréal, IAUS fonds: A.2-9 / ARCH401167.

Undergraduate Program were raised from US$5,700 to US$6,000, and then to US$7,800 in 1982–83. The Institute's leadership defended this drastic increase in the fees paid by undergraduates—by nearly a third within only two years—by claiming that the Institute's tuition fees needed to match those of competing liberal arts colleges for it to survive on the education market. Some of the Fellows had already expressed grave concerns about this development, even though its main intention was to ensure that architecture education didn't turn into a loss-making venture.[370] Deborah Berke argued that education offerings were no longer worthwhile, pointing out that tuition fees of US$7,000 were effectively "outpricing the market." The Fellows also discussed the fact that higher tuition fees would present certain "social issues" affecting the composition of the student body, as students from less well-off families would be effectively excluded. Such debates over tuition fees, target groups, pedagogy, and market viability testified to broader trends in American higher education. Since at least the 1974–75 academic year, if not since its foundation, the Institute had found itself in the middle of an overall trend towards an academicization of architecture education, reflected in the widespread growth of new master's programs, undergraduate degrees, and doctoral programs in architecture. The diversification, intellectualization, internationalization, and even commercialization of the architecture education landscape was a steamrolling trend—one to which the Institute had ultimately contributed but with which it eventually fell in line.

In the end, the Institute's education offerings influenced a rather small, but select contingent of students over the years, even if the Institute only contributed to one stage of their induction and incorporation into the New York architectural community. Some nonetheless went on to pursue careers as architects and even academics. The Institute's influence was even greater, perhaps, when it came to the production of teaching materials. Even though they weren't explicitly conceived as such, many of the publications edited and published at the Institute—ranging from *Oppositions*, Oppositions Books, to the IAUS Exhibition Catalogues—were quickly incorporated into architectural curricula. Yet even if only temporary, the decline in the Institute's enrollment figures, especially in the Advanced Design Workshop, presented a major financial problem that endangered the health of the Institute. Some of the Fellows blamed this on insufficient recruitment and lack of leadership.[371] In any case, by the 1981–82 academic year, the Institute was no longer at full enrollment. One obvious explanation was that Eisenman was no longer recruiting students as actively as he had in years past. Additionally, some of the Institute's contracts with cooperating universities expired without anyone doing anything to renew them. The diminished interest in the Advanced Design Workshop was

370 Kolbowski, 1981.

371 McGoldrick, 1981.

blamed on controversial hirings, lack of funds to print posters, and poor word-of-mouth advertising. Once again, events demonstrated that good public relations and outreach work had a major influence on the success of architecture education at the Institute. Only the Internship Program continued to see unabated popularity, largely thanks to the reputation of practicing Fellows. There were so many applicants for internships, in fact, that the Institute began selecting interns according to their past academic achievements. Meanwhile, the Fellows discussed whether the Internship Program had now grown disproportionately large in relation to the Institute's activities and whether interns were still getting sufficient insight into the profession. Cooperating with architecture firms was ruled out, however, for practical reasons. Instead, the Fellows decided to continue educating the interns at the Institute, in certain cases under individual Fellows, so that in their second semester, they could focus on producing a portfolio that included model making. What this demonstrates is that, regardless of which educational program is under consideration, the student-teacher relationship at the Institute had clearly changed: students no longer pursued their own education on an equal footing with the teachers, as advocated in the progressive pedagogical debates of the 1970s. At the Institute, the relationship was conceived quite traditionally, and students and interns were subordinated to the Fellows. In extreme cases, this could manifest in a certain kind of paternalism. Still, in a time of institutional upheaval, Eisenman tried to precipitate a generational change by transforming the Fellowship, bringing in a new culture and pedagogical approach. At the dawn of the 1980s, the Institute promoted junior faculty. As the old guard, consisting of Eisenman, Frampton, Gandelsonas, and Vidler, as well as Agrest, who had shaped architecture education in history, theory, and design at the Institute since 1974–75, were busy with other things, increasingly pursuing careers as architects by founding firms and taking leading academic positions, the Institute appointed Deborah Berke (1980), Larry Kutnicki, and eventually Robert Silman (both in 1981) as Fellows. For example, in the 1981–82 academic year, the urban planning course taught by Vidler was turned into an "Urban History" course taught in conjunction with David Mohney, who had studied under Vidler at Princeton, as a teaching assistant. The new course called "Architecture and the City," which contrasted utopian urban designs with ones that had actually been realized, played a fundamental role in reintroducing a humanities orientation to the liberal arts curriculum at the Institute.[372] In 1982–83, architecture education at the Institute was widely advertised as an alternative, mainly with posters in the by now typical Vignelli design, which was also used for all cultural events. The High School Program was communicated separately. The Undergraduate Program's history and theory courses, taught by Frampton and Gandelsonas, experimented with forms

372 Anthony Vidler, "Architecture and the City," Undergraduate Program in Architecture. Course Outline, fall semester 1981. Source: private archives of Patrick Pinnell.

of "team teaching," and tutors in the design studio were now composed primarily of Princeton graduates. While the veteran Fellows continued to monitor the quality of education, these late arrivals brought new approaches to methodology and content—thinking less dogmatically and operating more openly. Yet ultimately, Eisenman's resignation as Institute director—in the summer of 1982, after fifteen years, he withdrew from his post, while still remaining at the Institute in some capacity as a trustee—represented an enormous rupture in the Institute's history, with far-reaching consequences for the Institute and the fellowships, and above all the leading role of education from this time on. Despite Eisenman's departure, architectural education thrived, although students were unaware of the changes at the Institute and complained that they were not being taught by him.[373] To maintain day-to-day operations, Frampton had taken on the role of director of programs in June 1982, at least in the short term, but he too would resign from all his posts before the end of the year—for good reasons. While the Institute, having moved to its new premises on Union Square in 1983, tried to maintain operations, balancing organization and programming, architecture education was placed at the fore and Gandelsonas, appointed the new director of education, took on temporary leadership of the education offerings. In the 1982–83 academic year, the Institute was to embark on a new beginning.

373 Although he was no longer Institute director, nor a Fellow at the Institute, Eisenman again supervised interns in the academic year 1982–83.

Adopted 13 January 1977
Board Meeting

Statement of Trustees' Authority

The Board of Trustees shall have the following capacities with respect to the operation of the Institute:

1. General Institute policy: The Trustees establish general Institute policy. Program initiatives rest with the Fellows.

2. Prior review: The Trustees exercise their responsibility through a continuing review of the quality of the Institute's leadership in the maintenance of high quality programs and by a periodic check of the integrity and efficiency of the programs. Major changes in policy and any substantial new claims on funds will be brought to the Trustees for review before final decisions or commitments are made. The Trustees thus exercise a prior and general review in such matters as the allocation of a significant proportion of the Institute's resources, the setting of priorities for development, changes in programs of broad bearing for the institution, the determination of tuition and fees, plans calling for new construction, the establishment or abolition of departments or schools, changes in admissions policies affecting sizeable categories of potential students, and changes in relations with outside educational and social institutions and governmental agencies.

3. Authority directly exercised: In matters concerning financial health and physical properties, the Trustees participate directly in the formulation of policy and the conduct of the business of the Institute. The Trustee Executive and Finance Committee directs the investment of Institute funds. The Trustees establish fund raising policies, approve major development programs, help to identify important sources of potential financial support and raise funds. The Trustees actively supervise long-range physical and economic planning.

In addition, the Board may contribute advice and criticism on the shaping of programs and the conduct of affairs of the Institute. If the Board is to assess general policies wisely, it must be fully and currently informed and be alert and sensitive to particular conditions and requirements. Members of the Board often have experience and competence that can be helpful to the Institute in its dealing with specific problems, and their advice is most valuable in the early consideration of new policies.

Fig. 49

Fig. 50

Fig. 51

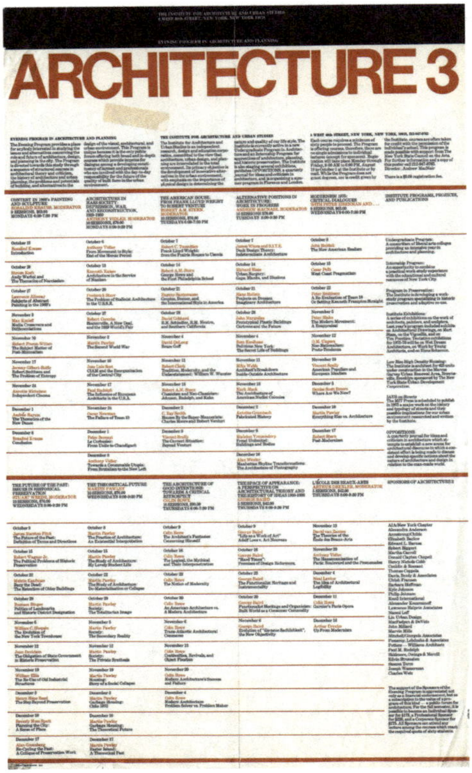

Fig. 52

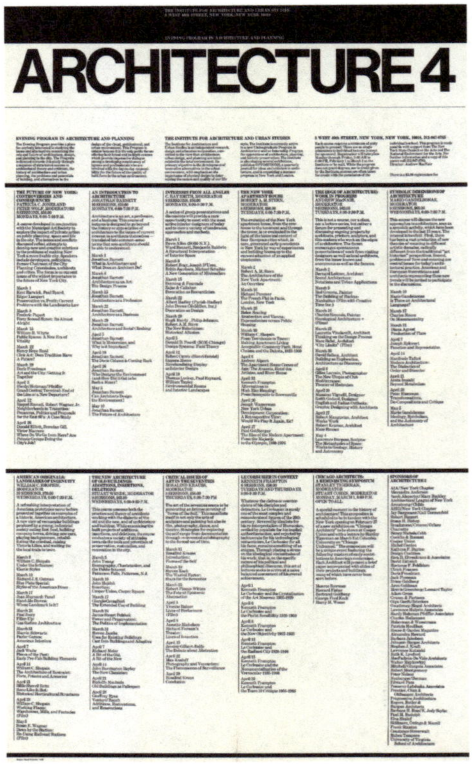

Fig. 53

Fig. 54

Fig. 55

Fig. 56

Fig. 57

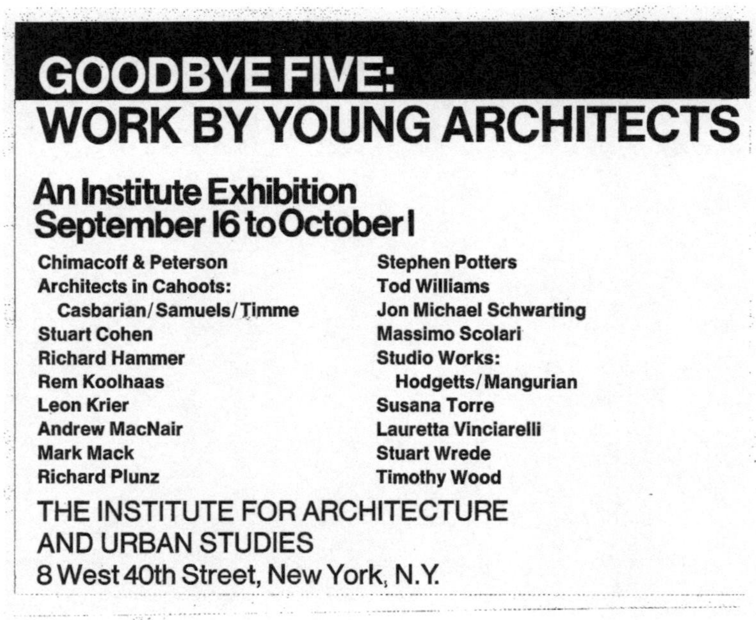

Fig. 58

Idea as Model
THE INSTITUTE FOR
ARCHITECTURE AND
URBAN STUDIES
8 WEST 40th STREET, NEW YORK,
NEW YORK
December 16, 1976 to January 14, 1977
This program is made possible with public funds from the New York State Council of The Arts.

Fig. 59

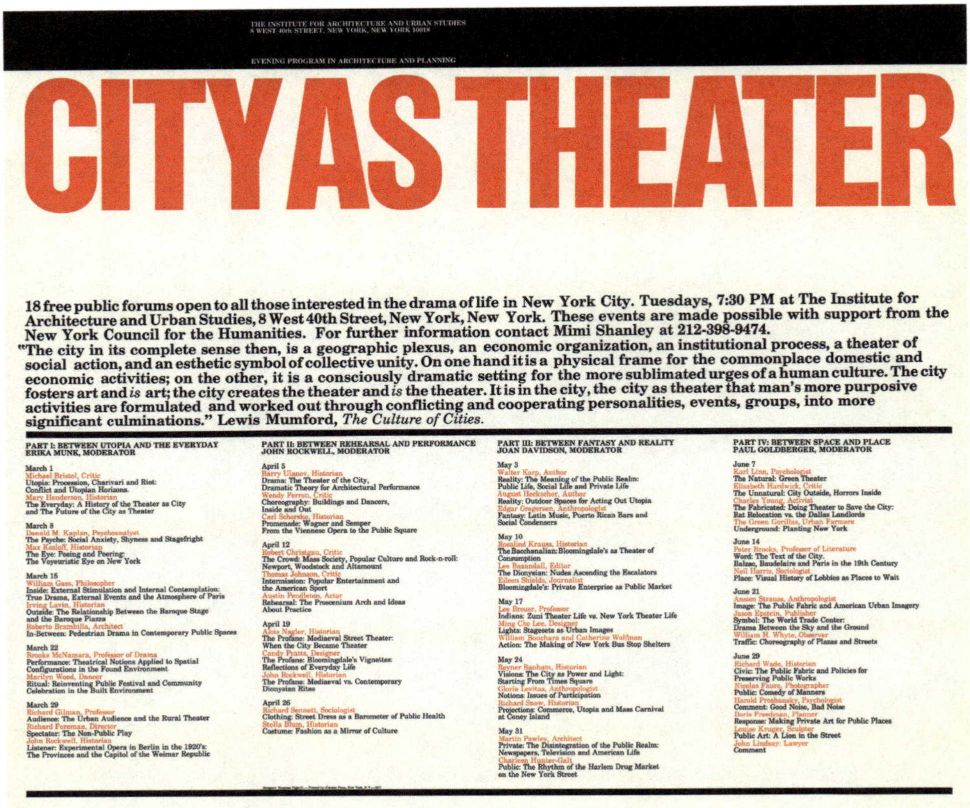

THE INSTITUTE FOR ARCHITECTURE AND URBAN STUDIES
8 WEST 40th STREET, NEW YORK, NEW YORK 10018
EVENING PROGRAM IN ARCHITECTURE AND PLANNING
CITY AS THEATER
18 free public forums open to all those interested in the drama of life in New York City. Tuesdays, 7:30 PM at The Institute for Architecture and Urban Studies, 8 West 40th Street, New York, New York. These events are made possible with support from the New York Council for the Humanities. For further information contact Mimi Shanley at 212-398-9474.
"The city in its complete sense then, is a geographic plexus, an economic organization, an institutional process, a theater of social action, and an esthetic symbol of collective unity. On one hand it is a physical frame for the commonplace domestic and economic activities; on the other, it is a consciously dramatic setting for the more sublimated urges of a human culture. The city fosters art and is art; the city creates the theater and is the theater. It is in the city, the city as theater that man's more purposive activities are formulated and worked out through conflicting and cooperating personalities, events, groups, into more significant culminations." Lewis Mumford, The Culture of Cities.
PART I: BETWEEN UTOPIA AND THE EVERYDAY
ERIKA MUNK, MODERATOR
March 1
Michael Bristol, Critic
Utopia: Procession, Charivari and Riot: Conflict and Utopian Horizons.
Mary Henderson, Historian
The Everyday: A History of the Theater as City and The Future of the City as Theater
March 8
Donald M. Kaplan, Psychoanalyst
The Psyche: Social Anxiety, Shyness and Stagefright
Max Kozloff, Historian
The Eye: Posing and Peering: The Voyeuristic Eye on New York
March 15
William Gass, Philosopher
Inside: External Stimulation and Internal Contemplation: True Drama, External Events and the Atmosphere of Paris
Irving Lavin, Historian
Outside: The Relationship Between the Baroque Stage and the Baroque Piazza
Roberto Brambilla, Architect
In-Between: Pedestrian Drama in Contemporary Public Spaces
March 22
Brooks McNamara, Professor of Drama
Performance: Theatrical Notions Applied to Spatial Configurations in the Found Environment
Marilyn Wood, Dancer
Ritual: Reinventing Public Festival and Community Celebration in the Built Environment
March 29
Richard Gilman, Professor
Audience: The Urban Audience and the Rural Theater
Richard Foreman, Director
Spectator: The Non-Public Play
John Rockwell, Historian
Listener: Experimental Opera in Berlin in the 1920's: The Provinces and the Capitol of the Weimar Republic
PART II: BETWEEN REHEARSAL AND PERFORMANCE
JOHN ROCKWELL, MODERATOR
April 5
Barry Ulanov, Historian
Drama: The Theater of the City, Dramatic Theory for Architectural Performance
Wendy Perron, Critic
Choreography: Buildings and Dancers, Inside and Out
Carl Schorske, Historian
Promenade: Wagner and Semper From the Viennese Opera to the Public Square
April 12
Robert Christgau, Critic
The Crowd: Mass Society, Popular Culture and Rock-n-roll: Newport, Woodstock and Altamount
Thomas Johnson, Critic
Intermission: Popular Entertainment and the American Sport
Austin Pendleton, Actor
Rehearsal: The Proscenium Arch and Ideas About Practice
April 19
Alois Nagler, Historian
The Profane: Mediaeval Street Theater: When the City Became Theater
Candy Pratts, Designer
The Profane: Bloomingdale's Vignettes: Reflections of Everyday Life
John Rockwell, Historian
The Profane: Mediaeval vs. Contemporary Dionysian Rites
April 26
Richard Sennett, Sociologist
Clothing: Street Dress as a Barometer of Public Health
Stella Blum, Historian
Costume: Fashion as a Mirror of Culture
PART III: BETWEEN FANTASY AND REALITY
JOAN DAVIDSON, MODERATOR
May 3
Walter Karp, Author
Reality: The Meaning of the Public Realm: Public Life, Social Life and Private Life
August Heckscher, Author
Reality: Outdoor Spaces for Acting Out Utopia
Edgar Gregersen, Anthropologist
Fantasy: Latin Music, Puerto Rican Bars and Social Condensers
May 10
Rosalind Krauss, Historian
The Bacchanalian: Bloomingdale's as Theater of Consumption
Lee Baxandall, Editor
The Dionysian: Nudes Ascending the Escalators
Eileen Shields, Journalist
Bloomingdale's: Private Enterprise as Public Market
May 17
Lee Breuer, Professor
Indians: Zuni Theater Life vs. New York Theater Life
Ming Cho Lee, Designer
Lights: Stagesets as Urban Images
William Bouchara and Catherine Wolfman
Action: The Making of New York Bus Stop Shelters
May 24
Reyner Banham, Historian
Visions: The City as Power and Light: Starting From Times Square
Gloria Levitas, Anthropologist
Notions: Issues of Participation
Richard Snow, Historian
Projections: Commerce, Utopia and Mass Carnival at Coney Island
May 31
Martin Pawley, Architect
Private: The Disintegration of the Public Realm: Newspapers, Television and American Life
Charleen Hunter-Galt
Public: The Rhythm of the Harlem Drug Market on the New York Street
PART IV: BETWEEN SPACE AND PLACE
PAUL GOLDBERGER, MODERATOR
June 7
Karl Linn, Psychologist
The Natural: Green Theater
Elizabeth Hardwick, Critic
The Unnatural: City Outside, Horrors Inside
Charles Young, Activist
The Fabricated: Doing Theater to Save the City: Rat Relocation vs. the Dallas Landlords
The Green Gorillas, Urban Farmers
Underground: Planting New York
June 14
Peter Brooks, Professor of Literature
Word: The Text of the City. Balzac, Baudelaire and Paris in the 19th Century
Neil Harris, Sociologist
Place: Visual History of Lobbies as Places to Wait
June 21
Ansom Strauss, Anthropologist
Image: The Public Fabric and American Urban Imagery
Jason Epstein, Publisher
Symbol: The World Trade Center: Drama Between the Sky and the Ground
William H. Whyte, Observer
Traffic: Choreography of Plazas and Streets
June 29
Richard Wade, Historian
Civic: The Public Fabric and Policies for Preserving Public Works
Nicolas Faure, Photographer
Public: Comedy of Manners
Harold Proshansky, Psychologist
Comment: Good Noise, Bad Noise
Doris Freedman, Planner
Response: Making Private Art for Public Places
Louise Kruger, Sculpter
Public Art: A Lion in the Street
John Lindsay: Lawyer
Comment

Fig. 60

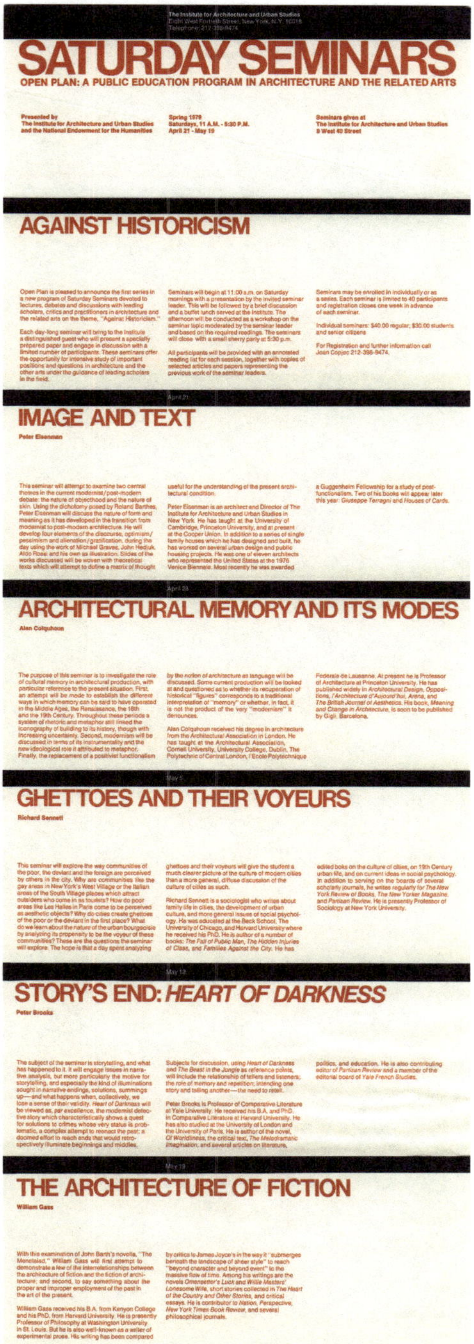

The Institute for Architecture and Urban Studies
Eight West Fortieth Street, New York, N.Y. 10018
Telephone: 212-398-9474

SATURDAY SEMINARS

OPEN PLAN: A PUBLIC EDUCATION PROGRAM IN ARCHITECTURE AND THE RELATED ARTS

Presented by
The Institute for Architecture and Urban Studies and the National Endowment for the Humanities

Spring 1979
Saturdays, 11 A.M. - 5:30 P.M.
April 21 - May 19

Seminars given at
The Institute for Architecture and Urban Studies
8 West 40 Street

AGAINST HISTORICISM

Open Plan is pleased to announce the first series in a new program of Saturday Seminars devoted to lectures, debates and discussions with leading scholars, critics and practitioners in architecture and the related arts on the theme, "Against Historicism."

Each day-long seminar will bring to the Institute a distinguished guest who will present a specially prepared paper and engage in discussion with a limited number of participants. These seminars offer the opportunity for intensive study of important positions and questions in architecture and the other arts under the guidance of leading scholars in the field.

Seminars will begin at 11:00 a.m. on Saturday mornings with a presentation by the invited seminar leader. This will be followed by a brief discussion and a buffet lunch served at the Institute. The afternoon will be conducted as a workshop on the seminar topic moderated by the seminar leader and based on the required readings. The seminars will close with a small sherry party at 5:30 p.m.

All participants will be provided with an annotated reading list for each session, together with copies of selected articles and papers representing the previous work of the seminar leaders.

Seminars may be enrolled in individually or as a series. Each seminar is limited to 40 participants and registration closes one week in advance of each seminar.

Individual seminars: $40.00 regular; $30.00 students and senior citizens

For Registration and further information call Joan Copjec 212-398-9474.

April 21

IMAGE AND TEXT

Peter Eisenman

This seminar will attempt to examine two central themes in the current modernist/post-modern debate: the nature of objecthood and the nature of skin. Using the dichotomy posed by Roland Barthes, Peter Eisenman will discuss the nature of form and meaning as it has developed in the transition from modernist to post-modern architecture. He will develop four elements of the discourse: optimism/pessimism and alienation/gratification, during the day using the work of Michael Graves, John Hejduk, Aldo Rossi and his own as illustration. Slides of the works discussed will be woven with theoretical texts which will attempt to define a matrix of thought useful for the understanding of the present architectural condition.

Peter Eisenman is an architect and Director of The Institute for Architecture and Urban Studies in New York. He has taught at the University of Cambridge, Princeton University, and at present at the Cooper Union. In addition to a series of single family houses which he has designed and built, he has worked on several urban design and public housing projects. He was one of eleven architects who represented the United States at the 1976 Venice Biennale. Most recently he was awarded a Guggenheim Fellowship for a study of post-functionalism. Two of his books will appear later this year: *Giuseppe Terragni* and *Houses of Cards.*

April 28

ARCHITECTURAL MEMORY AND ITS MODES

Alan Colquhoun

The purpose of this seminar is to investigate the role of cultural memory in architectural production, with particular reference to the present situation. First, an attempt will be made to establish the different ways in which memory can be said to have operated in the Middle Ages, the Renaissance, the 18th and the 19th Century. Throughout these periods a system of rhetoric and metaphor still linked the iconography of building to its history, though with increasing uncertainty. Second, modernism will be discussed in terms of its instrumentality and the new ideological role it attributed to metaphor. Finally, the replacement of a positivist functionalism by the notion of architecture as language will be discussed. Some current production will be looked at and questioned as to whether its recuperation of historical "figures" corresponds to a traditional interpretation of "memory" or whether, in fact, it is not the product of the very "modernism" it denounces.

Alan Colquhoun received his degree in architecture from the Architectural Association in London. He has taught at the Architectural Association, Cornell University, University College, Dublin, The Polytechnic of Central London, l'Ecole Polytechnique Federale de Lausanne. At present he is Professor of Architecture at Princeton University. He has published widely in *Architectural Design*, *Oppositions*, *l'Architecture d'Aujourd'hui*, *Arena*, and *The British Journal of Aesthetics*. His book, *Meaning and Change in Architecture*, is soon to be published by Gigli, Barcelona.

May 5

GHETTOES AND THEIR VOYEURS

Richard Sennett

This seminar will explore the way communities of the poor, the deviant and the foreign are perceived by others in the city. Why are communities like the gay areas in New York's West Village or the Italian areas of the South Village places which attract outsiders who come in as tourists? How do poor areas like Les Halles in Paris come to be perceived as aesthetic objects? Why do cities create ghettoes of the poor or the deviant in the first place? What do we learn about the nature of the urban bourgeoisie by analyzing its propensity to be the voyeur of these communities? These are the questions the seminar will explore. The hope is that a day spent analyzing ghettoes and their voyeurs will give the student a much clearer picture of the culture of modern cities than a more general, diffuse discussion of the culture of cities as such.

Richard Sennett is a sociologist who writes about family life in cities, the development of urban culture, and more general issues of social psychology. He was educated at the Beck School, The University of Chicago, and Harvard University where he received his PhD. He is author of a number of books: *The Fall of Public Man*, *The Hidden Injuries of Class*, and *Families Against the City*. He has edited boks on the culture of cities, on 19th Century urban life, and on current ideas in social psychology. In addition to serving on the boards of several scholarly journals, he writes regularly for *The New York Review of Books*, *The New Yorker Magazine*, and *Partisan Review*. He is presently Professor of Sociology at New York University.

May 12

STORY'S END: *HEART OF DARKNESS*

Peter Brooks

The subject of the seminar is storytelling, and what has happened to it. It will engage issues in narrative analysis, but more particularly the motive for storytelling, and especially the kind of illuminations sought in narrative endings, solutions, summings up—and what happens when, collectively, we lose a sense of their validity. *Heart of Darkness* will be viewed as, *par excellence*, the modernist detective story which characteristically shows a quest for solutions to crimes whose very status is problematic, a complex attempt to reenact the past; a doomed effort to reach ends that would retrospectively illuminate beginnings and middles.

Subjects for discussion, using *Heart of Darkness* and *The Beast in the Jungle* as reference points, will include the relationship of tellers and listeners; the role of memory and repetition; intending one story and telling another—the need to retell.

Peter Brooks is Professor of Comparative Literature at Yale University. He received his B.A. and PhD. in Comparative Literature at Harvard University. He has also studied at the University of London and the University of Paris. He is author of the novel, *Of Worldliness*, the critical text, *The Melodramatic Imagination*, and several articles on literature, politics, and education. He is also contributing editor of *Partisan Review* and a member of the editorial board of *Yale French Studies*.

May 19

THE ARCHITECTURE OF FICTION

William Gass

With this examination of John Barth's novella, "The Menelaiad," William Gass will first attempt to demonstrate a few of the interrelationships between the architecture of fiction and the fiction of architecture; and second, to say something about the proper and improper employment of the past in the art of the present.

William Gass received his B.A. from Kenyon College and his PhD. from Harvard University. He is presently Professor of Philosophy at Washington University in St. Louis. But he is also well-known as a writer of experimental prose. His writing has been compared by critics to James Joyce's in the way it "submerges beneath the landscape of sheer style" to reach "beyond character and beyond event" to the massive flow of time. Among his writings are the novels *Omensetter's Luck* and *Willie Masters' Lonesome Wife*, short stories collected in *The Heart of the Country and Other Stories*, and critical essays. He is contributor to *Nation*, *Perspective*, *New York Times Book Review*, and several philosophical journals.

Fig. 61

Peter Wolf
Chairman

Eight West Fortieth Street, New York, N.Y. 10018
Telephone: 212 947 0765

9 June 1977

To: M. Shanley
F. Taylor
A. MacNair
P. Eisenman
T. Vidler

From: P. Wolf

Re: Proposal

DEVELOPMENT

A. MacNair (3 days wk.)	F.Taylor (2 days wk.)
	-Trustee Liason
	-Architects' Circle Liason
	-National (Federal) Gov't. Applications
	-State Gov't. Applications
National Media Contact------------	-National Media Contact
Local Media Contact---------------	-Local Media Contact
Sponsors (Exhibits) (NEH) (Evening)	
Contributions (Individuals) (Corporations) (Business)	
The National Program (NEA)---------	-NEA
The Exhibitions Program(NYSCA)----	-NYSCA
Others----	-Others

NEH

A. MacNair (2 days wk.)	F. Taylor (3 days/wk.)
	-Organization
	-Budget
	-Books-Accounts
	-Schools - Liason
	-Students
	-Course Material Budget liason
	-Course Directors liason
Evening Operations	
Sponsors Contributions	
4th Course Director	

Fig. 62

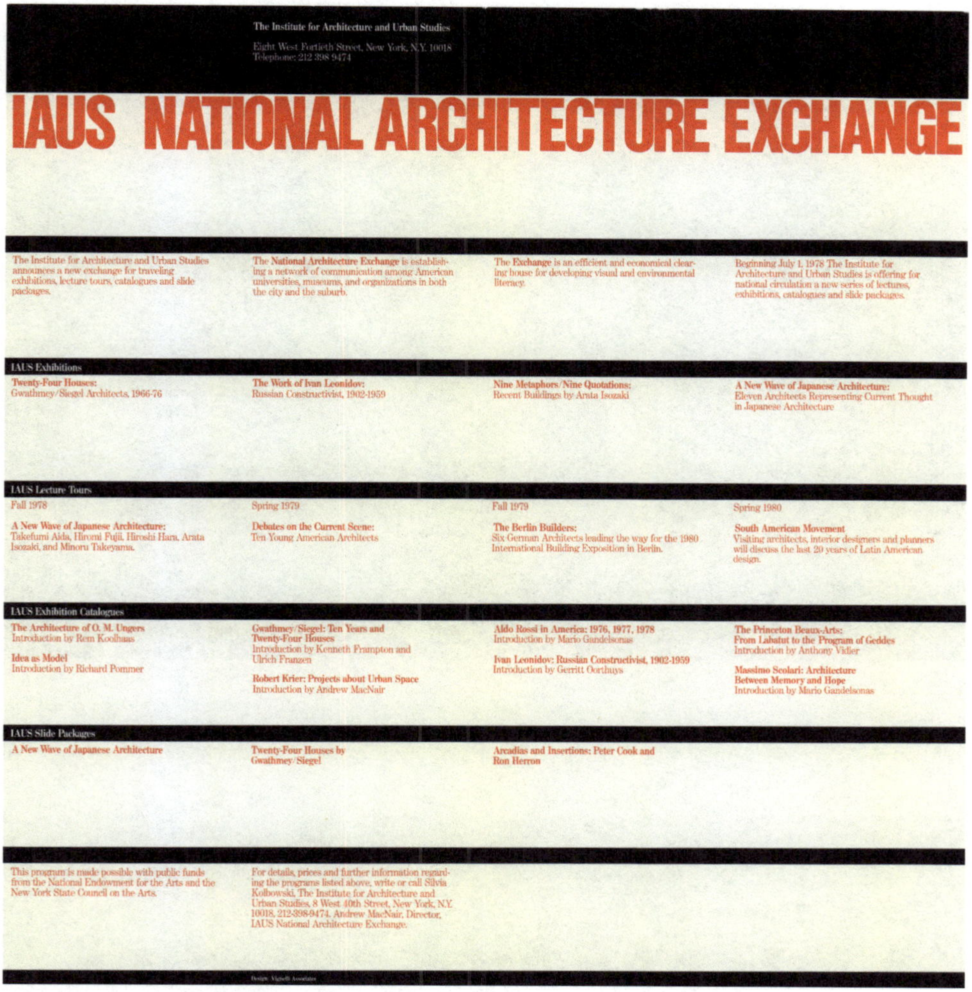

The Institute for Architecture and Urban Studies
Eight West Fortieth Street, New York, N.Y. 10018
Telephone: 212 398 9474
IAUS NATIONAL ARCHITECTURE EXCHANGE
The Institute for Architecture and Urban Studies announces a new exchange for traveling exhibitions, lecture tours, catalogues and slide packages.
The National Architecture Exchange is establishing a network of communication among American universities, museums, and organizations in both the city and the suburb.
The Exchange is an efficient and economical clearing house for developing visual and environmental literacy.
Beginning July 1, 1978 The Institute for Architecture and Urban Studies is offering for national circulation a new series of lectures, exhibitions, catalogues and slide packages.
IAUS Exhibitions
Twenty-Four Houses: Gwathmey/Siegel Architects, 1966-76
The Work of Ivan Leonidov: Russian Constructivist, 1902-1959
Nine Metaphors/Nine Quotations: Recent Buildings by Arata Isozaki
A New Wave of Japanese Architecture: Eleven Architects Representing Current Thought in Japanese Architecture
IAUS Lecture Tours
Fall 1978
A New Wave of Japanese Architecture: Takefumi Aida, Hiromi Fujii, Hiroshi Hara, Arata Isozaki, and Minoru Takeyama.
Spring 1979
Debates on the Current Scene: Ten Young American Architects
Fall 1979
The Berlin Builders: Six German Architects leading the way for the 1980 International Building Exposition in Berlin.
Spring 1980
South American Movement
Visiting architects, interior designers and planners will discuss the last 20 years of Latin American design.
IAUS Exhibition Catalogues
The Architecture of O. M. Ungers
Introduction by Rem Koolhaas
Idea as Model
Introduction by Richard Pommer
Gwathmey/Siegel: Ten Years and Twenty-Four Houses
Introduction by Kenneth Frampton and Ulrich Franzen
Robert Krier: Projects about Urban Space
Introduction by Andrew MacNair
Aldo Rossi in America: 1976, 1977, 1978
Introduction by Mario Gandelsonas
Ivan Leonidov: Russian Constructivist, 1902-1959
Introduction by Gerritt Oorthuys
The Princeton Beaux-Arts: From Labatut to the Program of Geddes
Introduction by Anthony Vidler
Massimo Scolari: Architecture Between Memory and Hope
Introduction by Mario Gandelsonas
IAUS Slide Packages
A New Wave of Japanese Architecture
Twenty-Four Houses by Gwathmey/Siegel
Arcadias and Insertions: Peter Cook and Ron Herron
This program is made possible with public funds from the National Endowment for the Arts and the New York State Council on the Arts.
For details, prices and further information regarding the programs listed above, write or call Silvia Kolbowski, The Institute for Architecture and Urban Studies, 8 West 40th Street, New York, N.Y. 10018, 212-398-9474. Andrew MacNair, Director, IAUS National Architecture Exchange.
Design: Vignelli Associates

Fig. 63

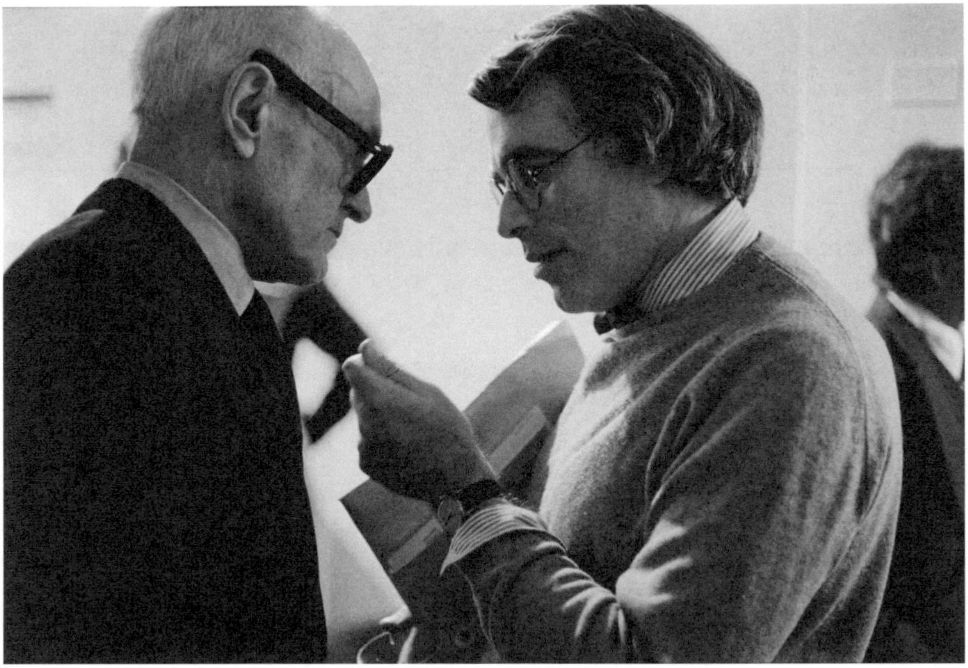

Fig. 64

Fig. 65

The Institute for Architecture and Urban Studies
XI1167IAUS
XI1177NYC

Fig. 66

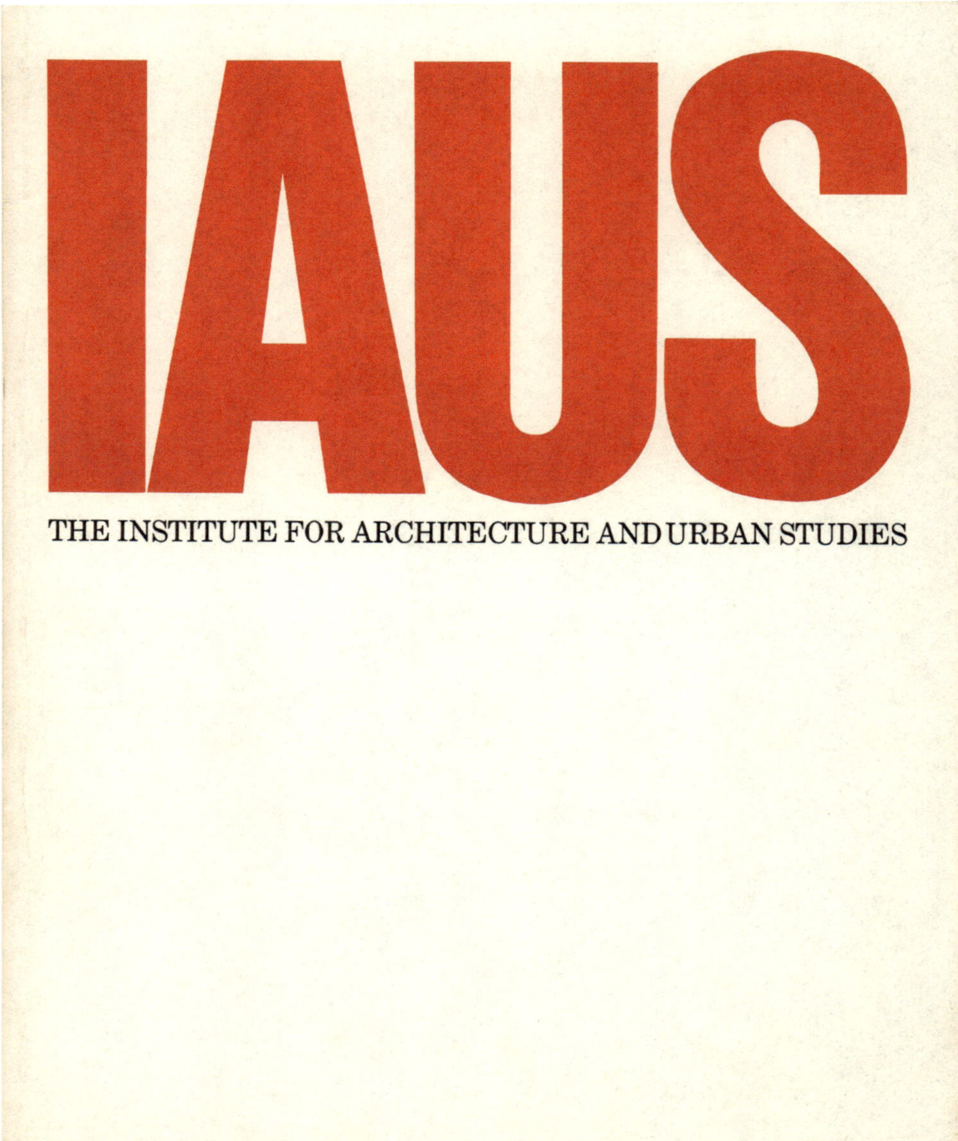

Fig. 67

Fellows 1978–1979

Diana Agrest
Stanford Anderson
Julia Bloomfield
Peter Eisenman
William Ellis
Kenneth Frampton
Suzanne Frank
Mario Gandelsonas
Andrew MacNair
Stephen Potters
Carla Skodinski
Leland Taliaferro
Frederieke Taylor
Anthony Vidler
Myles Weintraub
Peter Wolf

Fig. 68

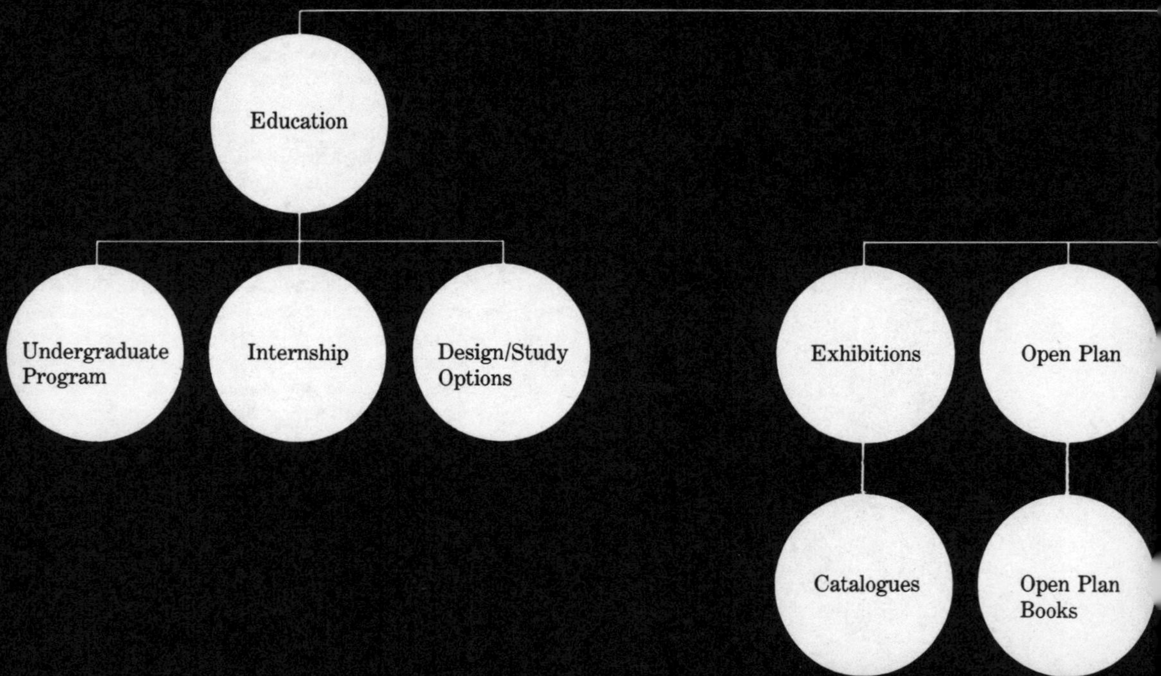
Education
Undergraduate Program
Internship
Design/Study Options
Exhibitions
Open Plan
Catalogues
Open Plan Books

Fig. 69

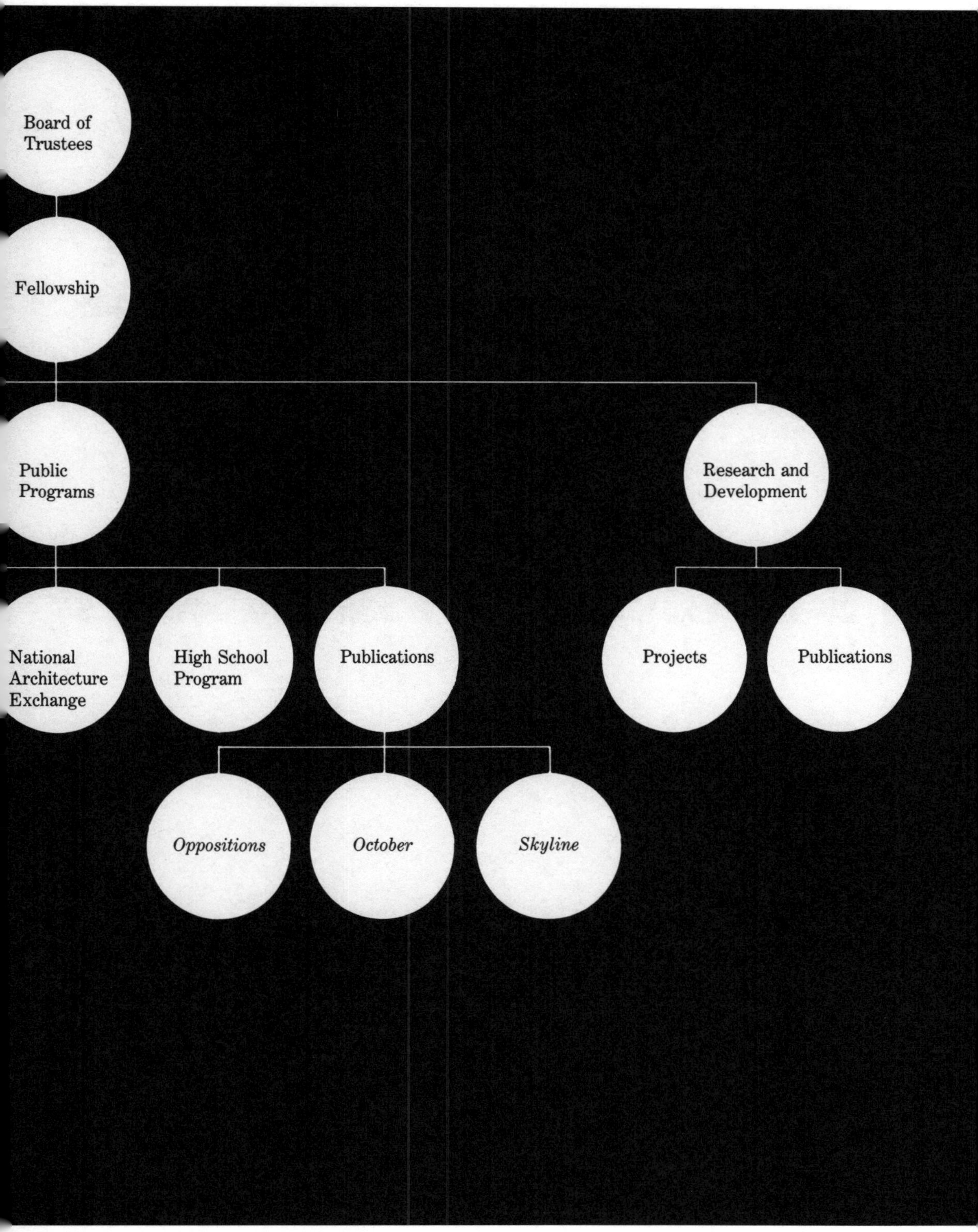
Board of Trustees
Fellowship
Public Programs
Research and Development
National Architecture Exchange
High School Program
Publications
Projects
Publications
Oppositions
October
Skyline

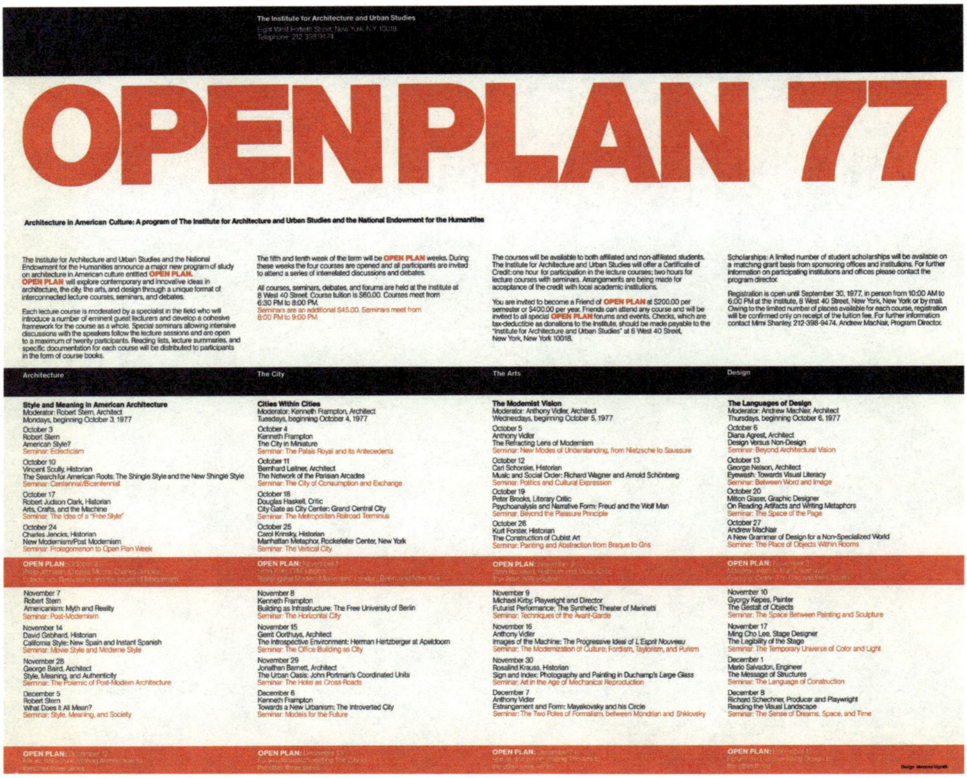

The Institute for Architecture and Urban Studies
OPEN PLAN 77
Architecture in American Culture: A program of The Institute for Architecture and Urban Studies and the National Endowment for the Humanities
The Institute for Architecture and Urban Studies and the National Endowment for the Humanities announce a major new program of study on architecture in American culture entitled OPEN PLAN.
OPEN PLAN will explore contemporary and innovative ideas in architecture, the city, the arts, and design through a unique format of interconnected lecture courses, seminars, and debates.
Each lecture course is moderated by a specialist in the field who will introduce a number of eminent guest lecturers and develop a cohesive framework for the course as a whole. Special seminars allowing intensive discussions with the speakers follow the lecture sessions and are open to a maximum of twenty participants. Reading lists, lecture summaries, and specific documentation for each course will be distributed to participants in the form of course books.
The fifth and tenth week of the term will be OPEN PLAN weeks. During these weeks the four courses are opened and all participants are invited to attend a series of interrelated discussions and debates.
All courses, seminars, debates, and forums are held at the institute at 8 West 40 Street. Course tuition is $60.00. Courses meet from 6:30 PM to 8:00 PM.
Seminars are an additional $45.00. Seminars meet from 8:00 PM to 9:00 PM.
The courses will be available to both affiliated and non-affiliated students. The Institute for Architecture and Urban Studies will offer a Certificate of Credit: one hour for participation in the lecture courses; two hours for lecture courses with seminars. Arrangements are being made for acceptance of the credit with local academic institutions.
You are invited to become a Friend of OPEN PLAN at $200.00 per semester or $400.00 per year. Friends can attend any course and will be invited to all special OPEN PLAN forums and events. Checks, which are tax-deductible as donations to the Institute, should be made payable to the "Institute for Architecture and Urban Studies" at 8 West 40 Street, New York, New York 10018.
Scholarships: A limited number of student scholarships will be available on a matching grant basis from sponsoring offices and institutions. For further information on participating institutions and offices please contact the program director.
Registration is open until September 30, 1977, in person from 10:00 AM to 6:00 PM at the Institute, 8 West 40 Street, New York, New York or by mail. Owing to the limited number of places available for each course, registration will be confirmed only on receipt of the tuition fee. For further information contact Mimi Shanley, 212-398-9474. Andrew MacNair, Program Director.
Architecture
Style and Meaning in American Architecture
Moderator: Robert Stern, Architect
Mondays, beginning October 3, 1977
October 3
Robert Stern
American Style?
Seminar: Eclecticism
October 10
Vincent Scully, Historian
The Search for American Roots: The Shingle Style and the New Shingle Style
Seminar: Centennial/Bicentennial
October 17
Robert Judson Clark, Historian
Arts, Crafts, and the Machine
Seminar: The Idea of a "Free Style"
October 24
Charles Jencks, Historian
New Modernism/Post Modernism
Seminar: Prolegomenon to Open Plan Week
OPEN PLAN:
November 7
Robert Stern
Americanism: Myth and Reality
Seminar: Post-Modernism
November 14
David Gebhard, Historian
California Style: New Spain and Instant Spanish
Seminar: Movie Style and Moderne Style
November 28
George Baird, Architect
Style, Meaning, and Authenticity
Seminar: The Polemic of Post-Modern Architecture
December 5
Robert Stern
What Does It All Mean?
Seminar: Style, Meaning, and Society
OPEN PLAN:
The City
Cities Within Cities
Moderator: Kenneth Frampton, Architect
Tuesdays, beginning October 4, 1977
October 4
Kenneth Frampton
The City in Miniature
Seminar: The Palais Royal and its Antecedents
October 11
Bernhard Leitner, Architect
The Network of the Parisian Arcades
Seminar: The City of Consumption and Exchange
October 18
Douglas Haskell, Critic
City Gate as City Center: Grand Central City
Seminar: The Metropolitan Railroad Terminus
October 25
Carol Krinsky, Historian
Manhattan Metaphor, Rockefeller Center, New York
Seminar: The Vertical City
OPEN PLAN:
November 8
Kenneth Frampton
Building as Infrastructure: The Free University of Berlin
Seminar: The Horizontal City
November 15
Gerrit Oorthuys, Architect
The Introspective Environment: Herman Hertzberger at Apeldoorn
Seminar: The Office Building as City
November 29
Jonathan Barnett, Architect
The Urban Oasis: John Portman's Coordinated Units
Seminar: The Hotel as Cross-Roads
December 6
Kenneth Frampton
Towards a New Urbanism: The Introverted City
Seminar: Models for the Future
OPEN PLAN:
The Arts
The Modernist Vision
Moderator: Anthony Vidler, Architect
Wednesdays, beginning October 5, 1977
October 5
Anthony Vidler
The Refracting Lens of Modernism
Seminar: New Modes of Understanding, from Nietzsche to Saussure
October 12
Carl Schorske, Historian
Music and Social Order: Richard Wagner and Arnold Schönberg
Seminar: Politics and Cultural Expression
October 19
Peter Brooks, Literary Critic
Psychoanalysis and Narrative Form: Freud and the Wolf Man
Seminar: Beyond the Pleasure Principle
October 26
Kurt Forster, Historian
The Construction of Cubist Art
Seminar: Painting and Abstraction from Braque to Gris
OPEN PLAN:
November 9
Michael Kirby, Playwright and Director
Futurist Performance: The Synthetic Theater of Marinetti
Seminar: Techniques of the Avant-Garde
November 16
Anthony Vidler
Images of the Machine: The Progressive Ideal of *L'Esprit Nouveau*
Seminar: The Modernization of Culture; Fordism, Taylorism, and Purism
November 30
Rosalind Krauss, Historian
Sign and Index: Photography and Painting in Duchamp's *Large Glass*
Seminar: Art in the Age of Mechanical Reproduction
December 7
Anthony Vidler
Estrangement and Form: Mayakovsky and his Circle
Seminar: The Two Poles of Formalism, between Mondrian and Shklovsky
OPEN PLAN:
Design
The Languages of Design
Moderator: Andrew MacNair, Architect
Thursdays, beginning October 6, 1977
October 6
Diana Agrest, Architect
Design Versus Non-Design
Seminar: Beyond Architectural Vision
October 13
George Nelson, Architect
Eyewash: Towards Visual Literacy
Seminar: Between Word and Image
October 20
Milton Glaser, Graphic Designer
On Reading Artifacts and Writing Metaphors
Seminar: The Space of the Page
October 27
Andrew MacNair
A New Grammar of Design for a Non-Specialized World
Seminar: The Place of Objects Within Rooms
OPEN PLAN:
November 10
Gyorgy Kepes, Painter
The Gestalt of Objects
Seminar: The Space Between Painting and Sculpture
November 17
Ming Cho Lee, Stage Designer
The Legibility of the Stage
Seminar: The Temporary Universe of Color and Light
December 1
Mario Salvadori, Engineer
The Message of Structures
Seminar: The Language of Construction
December 8
Richard Schechner, Producer and Playwright
Reading the Visual Landscape
Seminar: The Sense of Dreams, Space, and Time
OPEN PLAN:

Fig. 70

Fig. 71

Fig. 72

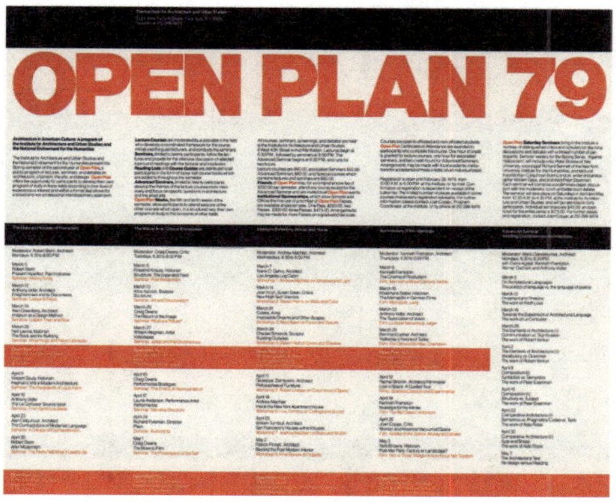

Fig. 73

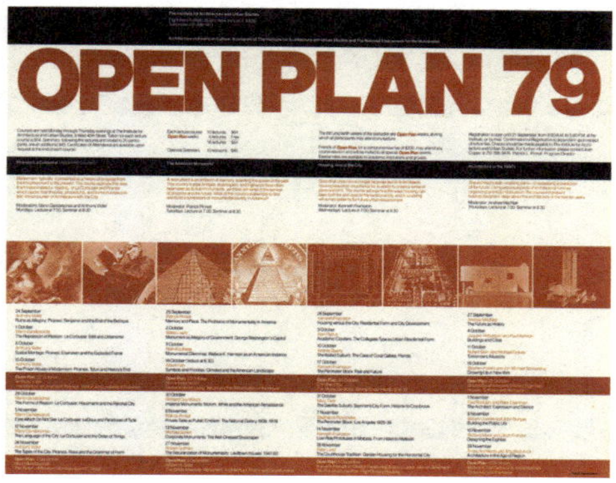

Fig. 74

Fig. 75

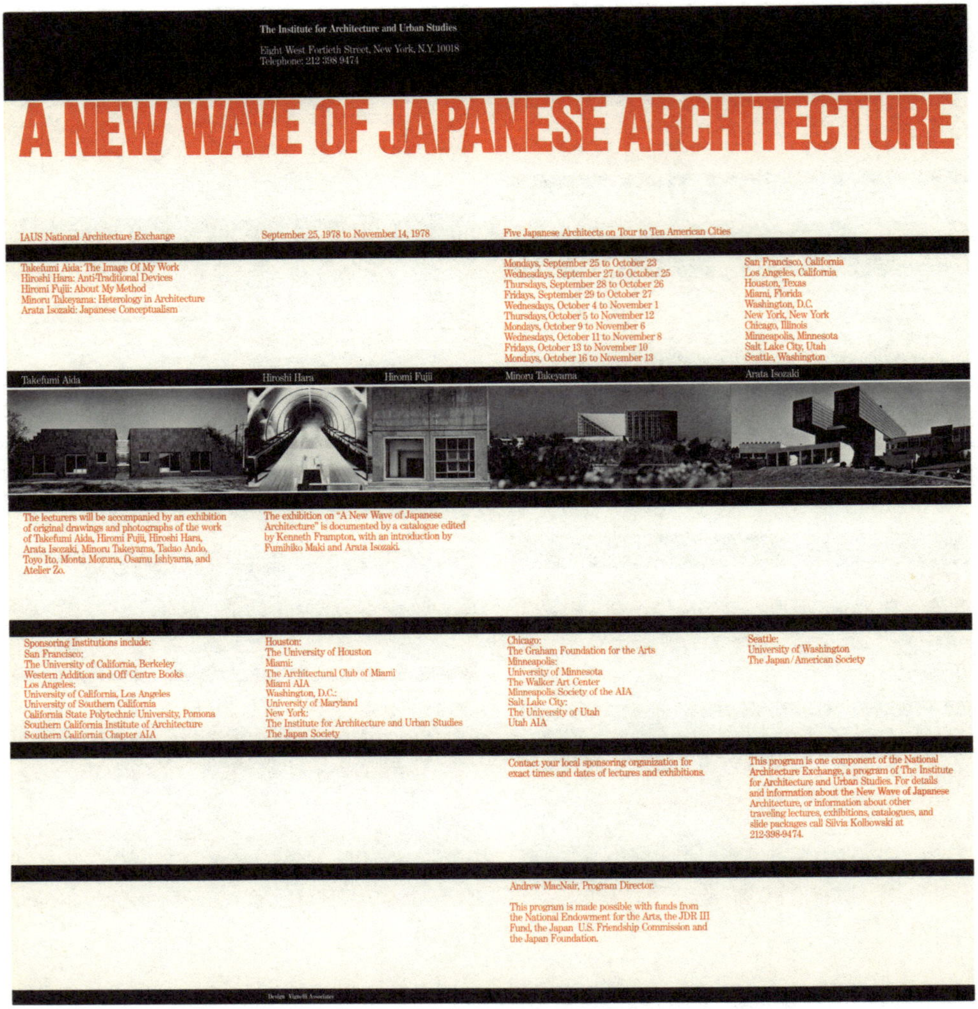

The Institute for Architecture and Urban Studies
Eight West Fortieth Street, New York, N.Y. 10018
Telephone: 212 398 9474
A NEW WAVE OF JAPANESE ARCHITECTURE
IAUS National Architecture Exchange
September 25, 1978 to November 14, 1978
Five Japanese Architects on Tour to Ten American Cities
Takefumi Aida: The Image Of My Work
Hiroshi Hara: Anti-Traditional Devices
Hiromi Fujii: About My Method
Minoru Takeyama: Heterology in Architecture
Arata Isozaki: Japanese Conceptualism
Mondays, September 25 to October 23
Wednesdays, September 27 to October 25
Thursdays, September 28 to October 26
Fridays, September 29 to October 27
Wednesdays, October 4 to November 1
Thursdays, October 5 to November 12
Mondays, October 9 to November 6
Wednesdays, October 11 to November 8
Fridays, October 13 to November 10
Mondays, October 16 to November 13
San Francisco, California
Los Angeles, California
Houston, Texas
Miami, Florida
Washington, D.C.
New York, New York
Chicago, Illinois
Minneapolis, Minnesota
Salt Lake City, Utah
Seattle, Washington
Takefumi Aida
Hiroshi Hara
Hiromi Fujii
Minoru Takeyama
Arata Isozaki
The lecturers will be accompanied by an exhibition of original drawings and photographs of the work of Takefumi Aida, Hiromi Fujii, Hiroshi Hara, Arata Isozaki, Minoru Takeyama, Tadao Ando, Toyo Ito, Monta Mozuna, Osamu Ishiyama, and Atelier Zo.
The exhibition on "A New Wave of Japanese Architecture" is documented by a catalogue edited by Kenneth Frampton, with an introduction by Fumihiko Maki and Arata Isozaki.
Sponsoring Institutions include:
San Francisco:
The University of California, Berkeley
Western Addition and Off Centre Books
Los Angeles:
University of California, Los Angeles
University of Southern California
California State Polytechnic University, Pomona
Southern California Institute of Architecture
Southern California Chapter AIA
Houston:
The University of Houston
Miami:
The Architectural Club of Miami
Miami AIA
Washington, D.C.:
University of Maryland
New York:
The Institute for Architecture and Urban Studies
The Japan Society
Chicago:
The Graham Foundation for the Arts
Minneapolis:
University of Minnesota
The Walker Art Center
Minneapolis Society of the AIA
Salt Lake City:
The University of Utah
Utah AIA
Seattle:
University of Washington
The Japan/American Society
Contact your local sponsoring organization for exact times and dates of lectures and exhibitions.
This program is one component of the National Architecture Exchange, a program of The Institute for Architecture and Urban Studies. For details and information about the New Wave of Japanese Architecture, or information about other traveling lectures, exhibitions, catalogues, and slide packages call Silvia Kolbowski at 212-398-9474.
Andrew MacNair, Program Director.
This program is made possible with funds from the National Endowment for the Arts, the JDR III Fund, the Japan U.S. Friendship Commission and the Japan Foundation.
Design Vignelli Associates

Fig. 76

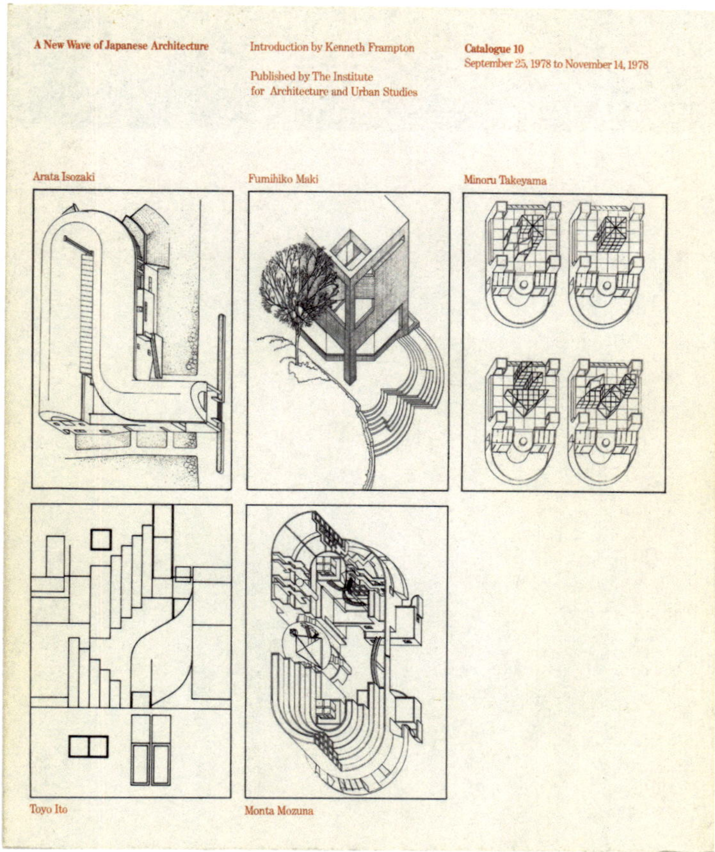

Fig. 77

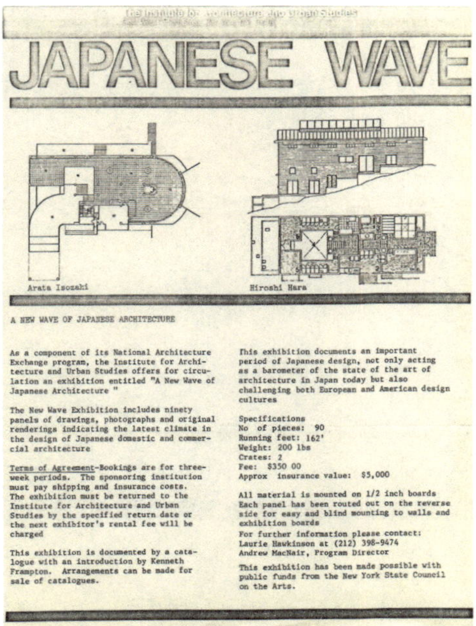

JAPANESE WAVE

Arata Isozaki

Hiroshi Hara

A NEW WAVE OF JAPANESE ARCHITECTURE

As a component of its National Architecture Exchange program, the Institute for Architecture and Urban Studies offers for circulation an exhibition entitled "A New Wave of Japanese Architecture "

The New Wave Exhibition includes ninety panels of drawings, photographs and original renderings indicating the latest climate in the design of Japanese domestic and commercial architecture

Terms of Agreement-Bookings are for three-week periods. The sponsoring institution must pay shipping and insurance costs. The exhibition must be returned to the Institute for Architecture and Urban Studies by the specified return date or the next exhibitor's rental fee will be charged

This exhibition is documented by a catalogue with an introduction by Kenneth Frampton. Arrangements can be made for sale of catalogues.

This exhibition documents an important period of Japanese design, not only acting as a barometer of the state of the art of architecture in Japan today but also challenging both European and American design cultures

Specifications
No of pieces: 90
Running feet: 162'
Weight: 200 lbs
Crates: 2
Fee: $350 00
Approx insurance value: $5,000

All material is mounted on 1/2 inch boards Each panel has been routed out on the reverse side for easy and blind mounting to walls and exhibition boards
For further information please contact:
Laurie Hawkinson at (212) 398-9474
Andrew MacNair, Program Director

This exhibition has been made possible with public funds from the New York State Council on the Arts.

Fig. 78

The Institute for Architecture and Urban Studies
Eight West Fortieth Street, New York, N.Y. 10018
Telephone: 212 398 9474
PHILIP JOHNSON
Processes
This exhibition is made possible
with public funds from the
New York State Council on the Arts
September 12, 1978 to October 31, 1978
Design: Vignelli Associates

Fig. 79

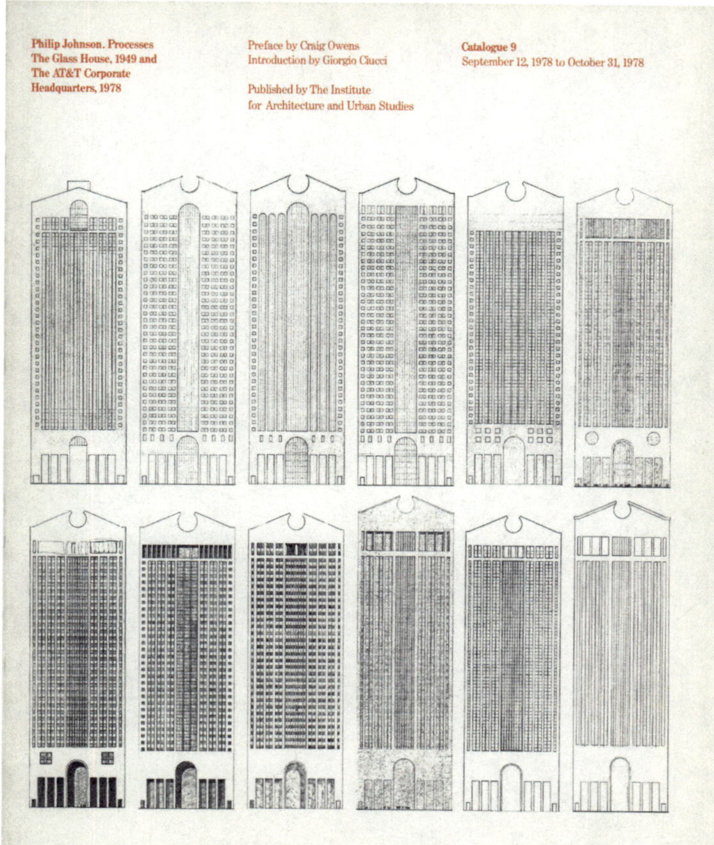

Fig. 80

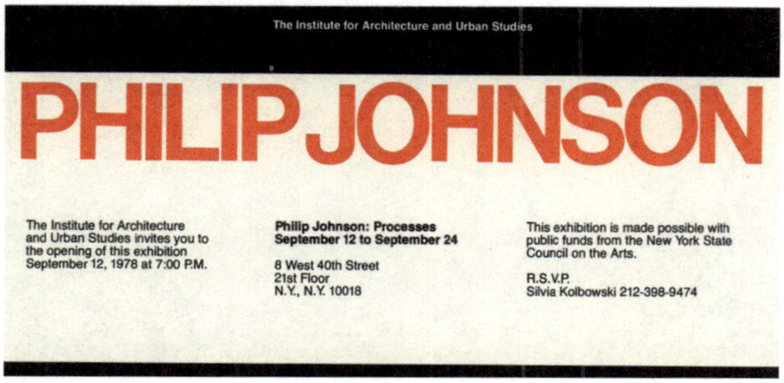

Fig. 81

Fig. 82

Fig. 83

Fig. 84

Fig. 85

Fig. 86

Fig. 87

The Institute for Architecture and Urban Studies
Eight West Fortieth Street, New York, N.Y. 10018
Telephone: 212 398 9474
OFFICE FOR METROPOLITAN ARCHITECTURE
Toward a modern (re)construction of the European city:
Four Housing Projects
An exhibition of recent projects by Rem Koolhaas and Elia Zenghelis
This exhibition was made possible in part with funds from the New York State Council on the Arts.
March 12—May 28, 1982
Monday through Friday, 9 to 5

Fig. 88

3.
Cultural Space

On January 20, 1975, Peter Eisenman was interviewed by Alvin Boyarsky, Director of the Architectural Association in London, about the Institute for Architecture and Urban Studies, which recently had distinguished itself through various research and design projects, educational programs, and cultural productions. This conversation, conducted and recorded in the television studio of the AA, which was at the time an internationally renowned school, testified to its friendly recognition of its American counterpart and a mutual interest on the part of both institutions.[374] Boyarsky began by introducing Eisenman as a "compere emcee," i.e., master of ceremonies or an announcer, and praised him for stimulating a debate through the formation of groups, as "someone who puts together many packages, involving many people in many places."[375] Eisenman spoke candidly and at length about the Institute's early years, its composition and funding, and paid tribute to the British architecture culture he had encountered in his student days at Cambridge, which to him was largely about cultured debate, rather than just the design of buildings. He returned Boyarsky's kindness by pointing to the AA as "some sort of a hybrid," even a role model for the Institute, given the Institute's recent work as an educational institution since the fall semester of 1974, with its offerings for liberal arts college students and its novel adult education program, specifically noting its collaboration with

374 "Peter Eisenman–in conversation with Alvin Boyarsky," AA School of Architecture, January 20, 1975, www.youtube.com/watch?v=rhQLaM0Q11g (last accessed: May 31, 2023); the conversation was recorded and later published as an almost verbatim transcript; see Eisenman, 2007, 83–87.

375 Ibid, 83.

Archigram for a workshop held shortly thereafter in New York.[376] The interview was significant not only in terms of its talking points, as both interviewees pointed out that the Institute was now active on an international level thanks to the ambitious journal *Oppositions*, which began publication in late 1973, and other productions that earned it its reputation as a center of debate and drew the attention of the architecture community, extending far beyond the East Coast of the United States, but also in terms of media, as the interview was broadcast on AATV, the school's own television channel. It is this combination of form and content, the emphasis on the importance of public debate, and the highlighting of the power of curating people, that foreshadowed the advancing medialization of architecture, the culturalization of the social, and the economization of the cultural: a development in which the Institute and the AA were both instrumental.

For parallel to its reinvention as an architecture school after the turning point in 1973, which brought about a massive redesign and restructuring and ultimately radically changed the market for architecture education in the United States, the Institute after 1974–75, under Eisenman's leadership, also increasingly made its mark as a new kind of cultural space, both an event space and an exhibition space.[377] From then on, cultural production and cultural products at the Institute, constantly oscillating between bourgeois and countercultural forces, took on different functions, both discursive and institutional. Strikingly, after funding for urban renewal and public housing projects had ceased, New York was no longer of concern and urban studies now had to be reinterpreted. Against the backdrop of societal transformation in the United States, the Institute set out to act as an educational and cultural service provider to the architecture community, academia, and the world of arts and culture, and to finance operations not only through revenue from student and internship fees, but also from grants from the National Endowment for the Arts (NEA) and the National Endowment for the Humanities (NEH), private foundations and—increasingly—private donations for public events, series of lectures, and exhibitions. This complex business model would shape the Institute's politics and economics for years, until its demise in 1985. This reorientation also entailed new ways of working for Fellows and Visiting Fellows, new organizational

376 Ibid, 85.

377 Kim Förster, "Institutionalizing Postmodernism: Reconceiving the Journal and the Exhibition at the Institute for Architecture and Urban Studies in 1976," in *Mediated Messages: Periodicals, Exhibitions, and the Shaping Postmodern Architecture*, eds. Véronique Patteeuw and Léa-Catherine Szacka (London: Bloomsbury, 2018), 213–229. In this essay on institutions of postmodernization, I argue that the Institute can be seen as a paradigmatic example of the "well-defined cultural spaces," that Manfredo Tafuri wrote about in his critique of the New York architectural scene as being "entrusted with the task of pleasurably entertaining a highly selected public;" see Tafuri, 1976, 53–72, here 53; Tafuri, 1987; here, Tafuri merely alluded to the Institute, and instead discussed individual architects, the usual suspects, and their individual positions.

structures, and new kinds of programming in the context of broader trends in architecture. It was thus inevitable that, in the wake of the reorganization that accompanied the economization of the "Undergraduate Program" and the "Internship Program," the Institute opened up even more to an increasingly broad target audience. This turn towards the cultural sphere, which must be seen in the context of the transformation of American culture and which aimed to open up historical and theoretical approaches in architecture to both a professional and general public with new trends in liberal arts education, adult education, alternative art spaces, and architecture publications, took place against the backdrop of the democratization of higher education, as well as the promotion of the arts and the humanities under a welfare-state policy that, while having been in place since the mid-1960s, took on new proportions under President Jimmy Carter in the wake of the United States Bicentennial in 1976.

Education and Culture

By holding a series of public lectures and setting up exhibitions after the move to its new premises in 1971, the Institute had already established two new formats and acquired a wealth of expertise, which it continued to develop and expand from the fall of 1974. In doing so, the Institute, as both a socio-cultural and epistemic actor, accompanied—or even pioneered—a postmodern turn in American architecture culture. For the public events served a dual purpose: the presentation of projects and discussion of positions on the one hand, and on the other, the acquisition of public grants from the arts and culture sector and of private funds by drawing on the American tradition of philanthropy and cultural sponsorship. With the 1974–75 academic year, the previous IAUS Spring Lectures series, which had served as a platform for Research Associates, Fellows, and friends in the past four years, was transformed into a curated year-round program: under the simple title "Architecture," evening lectures were now organized in the fall and spring semesters. The "Evening Program," originally conceived and advertised as adult education, similar to what had once been postulated for American architecture education in the *Princeton Report* of 1967, served to simultaneously academize and popularize the debate. It was aimed at a diverse audience, even if it ultimately consisted mainly of architects and designers speaking to their peers. At the same time, the Institute's foray into holding its own exhibitions was revived with a premiere in 1971 and transformed into an independent "Exhibition Program." At first, these shows were quickly-made, rather eclectic group and solo exhibitions of drawings and models, later supplemented by retrospectives of forgotten protagonists, sometimes heroes, of architectural modernism from Europe and America, who now served as references for a new theory production and modernist historiography.

It was no coincidence that with the architectural *dispositif* of autonomy and creativity—comparable to developments in art and culture—the Institute invested in a culturalization of architecture at precisely the same time when

New York, as an international financial center, was particularly affected by the economic downturn as a result of progressive deindustrialization and the onset of globalization, with the collapse of the Bretton Woods system and the transition from Fordism to post-Fordism as the defining economic form.[378] Urban studies have highlighted that in 1974–75 the metropolis was on the verge of insolvency, having been simultaneously plunged into a financial and fiscal crisis by the actions of banks and the absence of tax revenues resulting from the suburbanization of large segments of the population. Lacking the opportunity, the Institute was no longer concerned with making a contribution to society by, for example, regenerating inner cities as places to live, spend leisure time, or work, or even by organizing or regulating housing, albeit in a technocratic approach, as with the townhouse design commissioned by the U.S. Department of Housing and Urban Development as part of the "Streets Project" (1970–72) or the prototype for low-rise alternatives (from 1972) commissioned by the New York State Urban Development Corporation—for both urban and suburban sites—with its shift toward public-private partnership or even ownership. From a sociology of culture perspective, the introduction of new mechanisms for communicating architecture and tools for marketing, public relations, and acquisition, meant that, in the course of a semiotization, historicization, and aestheticization of not just the urban but the architectural, there was an increasing focus on architecture as a work of art and thus the architect as an artist.[379] With education and culture, two key features of capitalist ideology and thus of social life in the post-industrial knowledge, information, and service society now became the focus of the Institute's work. This combination created new financially lucrative forms of labor, employment, and work in architecture, with its gendered division and morality of competition, individualism, and meritocracy, that became attractive, alongside training a new generation of architects. The architectural project of the Institute thus changed abruptly from a contribution to architectural production to its management, i.e., the administration, dissemination, and reproduction of architectural knowledge, through the processing and control of information.[380] In line with the rules of an immaterial, symbolic economy that exists alongside the goods of a classical, material economy, activities at the Institute such as lecturing, debating, and exhibiting foregrounded circular processes of re-evaluation and self-legitimation of architecture as a

378 On the transformations of New York in the wake of the globalization of telematics and economic transactions, see Saskia Sassen, *The Global City. New York, London, Tokyo* (New York: Princeton University Press, 1992); on the transformation of New York from an industrial metropolis to a global city, see Moody, 2007.

379 Reckwitz, 2012. Cultural sociologist Andreas Reckwitz discussed the culturalization of cities, with New York in the 1970s and 80s as an example, in his study on the invention of creativity.

380 McHale, 1976.

discipline.[381] The intellectual and artistic practices of the Fellows who exhibited there went hand in hand with the professionalization and economization of cultural production in architecture—a development at the Institute that certainly paralleled the emerging transformation of universities into "factories of knowledge" and of cities into the sites of a cultural and creative economy.[382] Globally speaking, New York was a pioneer and model in this respect, as can be seen, for example, in the emergence of alternative art spaces, the launch of new journals, the proliferation of conferences, and the prominence of research centers, and at the same time the transformation of the museum, increasingly driven by blockbuster exhibitions, as well as a burgeoning gallery sector and art market.[383]

In the mid-1970s, as a neoliberal trend and economic revitalization took hold in the United States, characterized by government de-investment, privatization, deregulation, and a belief in market self-regulation, the Institute with its educational offerings and public events, from the perspective of institutional critique, quickly became a forum or meeting place beyond the inner circle of Fellows, staff, students, interns, and extended circle of friends, colleagues, and architects on a national and international level. In an earlier essay about Eisenman's house designs, entitled "L'Architecture dans le Boudoir" and published in *Oppositions* in 1974, the architecture historian and critic Manfredo Tafuri compared the culture of the Institute and its journal, at least indirectly, to the activities in the boudoirs of eighteenth-century France, i.e., small and secluded interior spaces in country houses and city mansions where literature and art were consumed by the aristocracy or bourgeoisie.[384] The Institute had similarly transitioned into the coquetry space of architecture, where a new scene emerged, one that initiated an epistemological shift toward postmodernism, as evidenced by various historical sources of self-representation and external perception. From the perspective of the architecture humanities, it pioneered innovation, variation, differentiation, diversification, and ultimately commodification—not only of education but also culture—providing a blueprint for an institution of architecture in a globalized, postmodern society that always incorporated entrepreneurial and governmental dimensions as well. Moreover, education and culture at the Institute were subject to what was later termed an "economy of attention." This involved interlocking, self-amplifying networks and played out at different scales of groups, organizations, and institutions and in the process transformed cultural,

381 Bourdieu, [1971] 1983, Bourdieu, 1983.

382 Raunig, 2012; see also Gerald Raunig, *Industrien der Kreativität: Streifen und Glätten 2* (Zurich: diaphanes, 2012).

383 Ault, 2002.

384 Tafuri, 1974.

social, and symbolic capital into economic capital.[385] While the Institute provided a stage for architects, designers, scholars, and artists to position and make a name for themselves, the work of Fellows and Visiting Fellows, now acting as cultural producers and consumers, can best be seen as a hybrid form of material and immaterial labor based on flexible, precarious relations.[386]

The Institute made history, for parallel to its efforts beginning in the mid-1970s to distinguish itself as a serious educational institution with a broad range of offerings, it cultivated a name for itself as a cultural institution, eventually evolving into a "fashionable," if not postmodern, salon in New York.[387] With their Evening Program and Exhibition Program, the Fellows henceforth worked to ensure that architecture and art, education and entertainment, culture and consumption were closely intertwined. At the Institute, cultural production was approached by way of pluralism, which became the condition of postmodern discourse in its social and aesthetic, creative and intellectual assumptions. The project of foregrounding both the design and the tools of design, drawings and models, and inviting other architects, artists, critics, scholars, and writers to reflect on these cultural techniques, architectural knowledge, and modes of representation and perception, which were discussed and displayed as autonomous and creative acts in lectures and exhibitions, was not without interest, and the Institute managed to captivate other audiences. With the Institute's growth, funding from the major state and national foundations, and the expansion of its reach to the American, if not global stage, the Fellows not only sought to professionalize and eventually bureaucratize management and curation, but also the design of programs and products—even if they succeeded only for a few years. The turn to the architecture establishment in New York, which combined philanthropy and cultural sponsorship, was historically significant. After all, the Institute, as a new, self-created and self-sustaining group, organization, or institution in architecture, was producing a new generation of architects and academics. While education paid the bills, the convergence of lectures, exhibitions, and finally publications produced stars—fostering, if not creating a celebrity culture. Finally, the Institute promoted the practicing architects among the Fellows and Visiting Fellows who succumbed to the lure of the art and architecture, if not the real estate market, and pioneered the fusion of architecture and sculpture that was advanced in the 1980s under the label of deconstructivism.

385 Franck, 1998; Franck, 2000; see also Lapassade, [1967] 1972; Pierre Bourdieu, "The Forms of Capital," in *Handbook of Theory and Research for the Sociology of Education*, ed. John G. Richardson (New York: Greenwood Press, 1986), 241–258.

386 Lazzarato, 1996.

387 Ockman, 1988, here 198–199. The Institute's evolution toward what Ockman once aptly described as a nexus of "its bureaucratization, its cultivation as a fashionable salon and power base in New York, and its solicitation of mainstream patronage" began as early as the mid-1970s; see Tafuri, 1976; Tafuri, 1987.

3.1 Providing Adult Education

In the fall semester of 1974–75, the Institute set about fundamentally redefining the relationship between new cultural producers and consumers in architecture on the one hand, and the ambiguity of production and reception of architectural knowledge on the other by reinventing and honing its education and culture profile—first as an alternative to universities and museums as the classic venues of legitimation and dissemination of knowledge, and later in competition with them. The pedagogy that the Institute now developed as an architecture school, swiftly launching alternative educational offerings targeting diverse cohorts (see chapter two), not only guaranteed its ability to operate after years of impending bankruptcy—now staved off with the income from tuition and internship fees—but also made its emergence as one of the "well-defined cultural spaces," to adopt and expand on Tafuri's words, possible in the first place.[388] With the start of the fall semester, a new, comprehensive lecture series with the catchy and apt title "Architecture" replaced the IAUS Spring Lectures series that had been running since 1971. The series had been modeled on the New School for Social Research, with a variety of course offerings in the evenings to appeal to the widest possible audience of professionals and laypersons. For the Evening Program, which focused almost exclusively on architecture and to a lesser degree on urban issues, to be financially self-sustaining, the Institute applied for funding from the New York Council on the Arts (NYSCA), a state foundation of very great importance for art and culture projects. In the ambitious application text, jointly written by Kenneth Frampton, the originator of the idea, and Eisenman in his capacity as Institute director, "Architecture" was advertised in November 1973 as a "continuing education" program that was to be larger, above all more professional, and better marketed than the previous series of lectures.[389] The rationale was that there was an existing demand for a public debate on architectural topics. The event format, which contributed to the Institute's survival with additional income from course fees, was ultimately an instrument of both self-marketing and identification with the Institute.

With "Architecture," the Institute was following the example of the larger museums in New York, where comparable offerings already existed, especially since an awareness of lifelong learning had been gradually gaining traction since the American higher education reform of 1968. Qualifications cited included both the Institute's past activities and the Fellows' individual work, as well as the institutional network of funding bodies, cultural institutions, and American

388 Tafuri, 1976, 53; Tafuri, 1987, 293.

389 IAUS, "Request for Assistance for the Institute for Architecture and Urban Studies. A Program for Continuing Education in Architecture and the Urban Environment." Source: Yale University, Robert A.M. Stern Archives.

universities.[390] In the proposal text, the authors presented an extensive program spanning several years and including one or even multiple lectures every weekday evening during the semesters from October to December and from March to May—an incredibly high number. They also sketched out concepts for possible lecture courses on the history and theory of architecture to prove that they had sufficient expertise and the networks to be able to implement an event series of this magnitude. Thus, in addition to Frampton, a number of architecture historians from the most prestigious American universities, notably Colin Rowe as host and Vincent Scully as lecturer, were to present their current research to a broader, and in particular urban audience. There were also courses on the history of architecture theory and the semiology of architecture, with Mario Gandelsonas as a scheduled speaker. "Architecture," with a capital "A," was henceforth the Institute's unique selling point vis-à-vis other institutions, and the Evening Program found an audience not only among architects—young professionals and architecture students from master's programs were targeted—but also in the Manhattan community. The educational focus was also evident from the fact that the offering was explicitly presented as a supplement to the established curricula, not just of schools of architecture: students of all disciplines were to receive credit for participation from their home universities. The threefold objective articulated in this program testified to the Institute's high aspirations: "1.) to stimulate and strengthen the overall approach of the profession to environmental design and to establish a more profound common cultural base from which to practice design; 2.) to ultimately raise the general level of design performance within the New York region; 3.) to demonstrate the model of an independent extra-mural educational institution which may eventually come to be inaugurated in other urban centers in the United States."[391] Accordingly, the Institute proposed nothing less than to provide guidance and direction to the profession, to influence architectural events through theoretical and historiographical considerations, and to extend its own sphere of influence on a national level. These three goals were repeatedly modified and reformulated in the years that followed, with various cultural productions using slightly different wording.

With "Architecture," the Institute had designed an event format that was unprecedented on this scale. Even as he prepared the application to NYSCA, Eisenman had already pulled strings, activated his networks, and written to a number of architects and heads of architecture schools asking for comments and criticism on the planned program, and above all for letters of recommendation.

390 Qualifications cited included affiliation with MoMA, support from the Graham Foundation, collaboration with Cornell University, Cooper Union, Rice University, Rutgers University, institutionalization of the internship through that of the Great Lakes College Association, sponsorship by NYSCA and the NEA, and collaboration with the UDC.

391 IAUS, "Request for Assistance for the Institute for Architecture and Urban Studies," n.p. Source: Yale University, Robert A.M. Stern Archive.

These letters, contributed by Ulrich Franzen, James Polshek (dean of the Graduate School of Architecture and Planning at Columbia University), John Hejduk (dean of the School of Architecture at Cooper Union), Robert Stern (president of the Architectural League), and Tim Prentice (president of the New York chapter of the American Institute of Architects), were enclosed with the grant application. This not only served to establish an institutionally grounded support network but also ensured that the project was announced in advance in architecture circles, thus already laying the foundation. In this context, Eisenman had also written to Robert Stern in November 1973. Stern, in addition to his architectural practice, was also teaching at Yale University and especially at Columbia College; as president of the Architectural League in New York, where he had started out as Program Director, he was a leading figure in East Coast architecture despite his youth.[392] Since then, Eisenman and Stern cultivated a friendly rivalry for a while, especially since the two stood for different camps in the architecture scene—the much-discussed "Whites vs. Grays" debate was already buzzing with myth—although they were not so different socially, in terms of habits, skills, dispositions, etc. despite their different aesthetic preferences.[393] On the contrary, by positioning himself as an opponent in the battle for attention—a framing that introduced the neoliberal paradigm according to which the market is the most efficient coordinating mechanism for society and thus for culture, into the architecture debate—Eisenman made common cause with Stern. In a letter to Constance Eiseman, the director of NYSCA, Stern explicitly expressed his support for the Institute's plan to organize "Architecture." The unlikely alliance between Eisenman and Stern was crucial to the success of the grant application and, by extension, to the "Architecture" series coming to fruition. For with this distinguished series of lectures, the Institute was actually competing with the League and its public programs. Now, however, synergies were being exploited instead: Stern was even originally slated to direct the Evening Program and was also appointed Visiting Fellow in 1974, having already organized the IAUS Spring Lectures series that spring. But although he ultimately declined, and the collaboration with the League thus failed to materialize, Stern nevertheless invested considerable time and energy in the Institute in the years that followed, acting as a presenter and lecturer, crowd puller and campaigner in the service of a particular variant of architectural postmodernism, and ultimately, in the early 1980s, playing a not insignificant role as a source of ideas and advice regarding public events.

392 Stern, et al., 1995.

393 Stern was first invited to participate in the Institute in 1972 before supporting *Oppositions* in 1973 and was previously asked for advice on grant acquisitions and public relations strategy. In 1973, he ignited the debate by curating a series of articles "Five on Five" in the May issue of *Architectural Forum* with contributions by himself, Jaquelin Robertson, Charles Moore, Alan Greenberg, and Romaldo Giurgola; see Robert Stern, "Stompin at the Savoye," *Architectural Forum* (May 1973): "Five on Five," 46–48.

Adult Education

In June 1974, the Institute received US$19,000 in start-up funding for "Architecture" from NYSCA, the very foundation that otherwise supported alternative art spaces and art projects across all boroughs of New York, and yet also recognized adult education in Midtown Manhattan as being worthy of support. However, the Institute had bigger plans: "Architecture" finally made it possible for the Institute to establish itself as a new force for intellectual, often polemical debate in the New York architecture scene. Eisenman found a suitable replacement for Stern in young Andrew MacNair, whom he appointed to coordinate the Evening Program. Having just finished his studies in architecture, first at Princeton and then at Columbia, where he had already gained experience in organizing small exhibitions and self-publishing his own newspaper, McNair was approached on the street one day by Eisenman in a seemingly preordained encounter.[394] With the active support of Eisenman, Frampton, and Stern, MacNair was tasked with quickly putting together an extensive program of events. This was to take place in the Institute's penthouse on 8 West 40th Street, which required a new structure and organization. The spacious two-storied main hall provided nearly ideal conditions for holding public events, especially a series of lectures since it was large enough to accommodate about one hundred people. Nevertheless, the Institute had to be remodeled, primarily to incorporate the daily operations of "Architecture" as a public and at the same time commercial event. Partition walls were installed to allow the audience members who took one of the elevators up to the twentieth floor to be easily redirected so that admission could be charged centrally in one corner of the space. However, there was no infrastructure on site yet to host an event of this scale: furniture (seating and lectern) and technology (carousel projectors) had yet to be purchased or rented. The first step was to renovate the event space: new carpets and shades for the windows were installed, so that the first lectures could begin before dark. MacNair may not have been as intellectually interested as the long-time Fellows and always remained a junior partner, but he was thoroughly creative, filling the role of mover and shaker perfectly, and, despite his punk attitude, became a central figure who soon took on all of the Institute's public events and public relations, significantly shaping its public image, while simultaneously demonstrating street credibility.[395] For "Architecture," he was aided by William Eitner, who served as his technical assistant. In addition, he worked

394 MacNair, who studied under Vidler and Frampton, among others, was initially assigned Stern as his thesis advisor at Columbia, but he turned him down and instead chose Raimund Abraham as his supervisor and Robert Smithson as his reviewer; he completed his thesis on the redesign of the Central Park Zoo under Frampton. After graduation, MacNair first worked for Haus-Rucker-Co. on a project on rooftops in New York.

395 By his own admission, MacNair was all over the New York art and music scene in downtown and midtown Manhattan: there he frequented the hip clubs, attended live performances by the bands of the hour, listened to punk, new wave, and disco, and befriended young architects and artists, who were invested in land art, post-minimalism, and performance art.

closely with a series of secretaries, who reported to him. In the first two years, these were Regina Wickham, who was also charged with coordinating architecture education in the Undergraduate Program, followed by Ruth Plawner in 1976, as his own assistant, and finally Mimi Shanley in 1977.

"Architecture" figuratively and literally provided a stage for Fellows and friends, both the inner and outer circle, emerging and already acclaimed architects and academics, who made appearances there as presenters and speakers, to reach an audience beyond the universities and museums, as well as publications. The first events in the Evening Program began in September 1974 and were scheduled to run for ten weeks. The program featured academic lectures on current approaches to the history and theory of architecture, as well as to urban planning and policy, alongside presentations of the positions of up-and-coming architects and designers. Massimo Vignelli, who had previously contributed the graphic design for *Oppositions*, specially designed a square poster announcing a total of six lecture courses on various days of the week, Mondays through Thursdays, sometimes two lectures in a row: "Public Places in New York" (presented by Michael Kirkland), "Introduction to the History of Modern Architecture, 1900–1920" (Kenneth Frampton), "The Background Work, and Influence of Louis I. Kahn" (Robert Stern), "An Introduction to Urban Design" (Jonathan Barnett), "The Architecture of the Italians: 16th Century" (Colin Rowe), and "Human Versus Natural Environment" (Andrew MacNair). "Architecture" thus encompassed different types of knowledge production and consumer tastes, both introductory and advanced, and appealed to a diverse audience with a programmatic novelty and complexity that was the secret to its success. The choice of presenters and lecturers deliberately broke down distinctions between highbrow and popular culture, and the themes depicted on the poster appealed to both the middle classes and an alternative clientele. This testified to the Institute's freshness and openness to the architecture profession, as well as towards the general public, but also to the new postmodern sensibility that had come to characterize the art and culture scene in 1970s New York.

In a short space of time, the Institute was able to establish itself as a unique, sophisticated venue: a community space for young architects, designers, intellectuals, and creatives—predominantly from New York, the East Coast of the USA, and the rest of North America, but increasingly from other parts of the world as well—and an event space for an educated yet consumerist public that could afford the admission fees. This receptiveness once again strengthened the internal cohesion of the group, as the Fellows were now more strongly integrated into complex discursive and institutional networks. "Architecture" provided a forum of national standing for staging controversial debates among peers, for engaging in a practice of communication and interaction that was typical of the architecture discipline, and for demonstrating individuality and distinctiveness in public. In this way, conversations about architecture increasingly found their way into academic and metropolitan culture at the Institute, which in some ways took on a pioneering role.

Institutional Identity

At the Institute, Vignelli's graphic design assumed a central function in culture production, in processes of group formation and identification, and in the attention economy.[396] Vignelli, having first worked with Unimark International to design the corporate identity of large American companies and thus of corporate culture in general in the second half of the 1960s, and then going on to independently develop the graphic identity for New York and several national institutions (for example, the information system for the Botanical Garden in Brooklyn and the brochures and information boards of the National Park Service of the USA) in the 1970s, in addition to redesigning the signaling system and subway map of the New York Metropolitan Transportation Authority, now brought his experience and practice to the Institute.[397] After designing the cover and layout for *Oppositions*, Vignelli developed an entirely new graphic language—modernist in style but certainly contemporary—for the Institute, which he applied to its overall institutional identity. In addition to all the Institute's publications, Vignelli went on to design almost all of its printed materials, including stationery, envelopes, invitation cards, posters, leaflets, flyers, and brochures, etc. With this contribution to the Institute, its meaning and purpose, he shaped a sense of belonging, if not community, i.e., a living environment and experiential space. The design for "Architecture" was modeled on the layout and logotype for *Oppositions*, with its attention-grabbing elements, sans-serif font, capital-letter title, and constructivist color scheme. In addition, with the black bar at the top and the red lettering, the poster introduced two graphic elements that would become identifying features and trademarks of the Institute as a cultural space, replacing the old logo, the Vitruvian Man. The "Architecture" series was also, with reference to the sociologist Pierre Bourdieu and applying his theory of capital and the economy of cultural production to the field of architecture, coded according to the mechanisms of culturation and dissemination in the art, literature, and culture sector.[398] And while the fourth issue of *Oppositions*, scheduled for the fall of 1974, had to wait for funding, "Architecture" became an important flagship for the Institute—in no small part because of this poster and the new institutional identity. Building on

396 Vignelli's archive, which contains many of the graphic designs for the Institute, is housed at the Vignelli Center for Design Studies at Rochester Institute of Technology; CCA's IAUS fonds also has a large selection of Vignelli's designs.

397 With Unimark International, Vignelli played a decisive role in the dissemination of Helvetica in the United States; see *Helvetia* (2007, director: Gary Hustwit), https://www.hustwit.com/helvetica (last accessed: May 31, 2023); see also Jan Conradi, *Unimark International: The Design of Business and the Business of Design* (Baden: Lars Müller, 2009); see Emilio Ambasz, *Design Vignelli* (New York: Rizzoli International, 1981); Eric Larrabee and Massimo Vignelli, *Knoll Design* (New York: Harry N. Abrams, Inc., 1981); Massimo Vignelli, "On Rational Design," (interview with Steven Heller and Elinor Pettit) in *Design Dialogues*, eds. Steven Heller and Elinor Pettit (New York: Allworth Press, 1998), 3–7.

398 Pierre Bourdieu, "Social Space and Symbolic Power," *Sociological Theory* 7, no. 1 (Spring 1989), 14–25.

Oppositions, Vignelli used the poster to position and showcase the Institute as a genuine institution and a strong brand with a high level of recognition. Printed in a run of several thousand copies and mailed to architectural firms and schools of architecture across the country, the poster provided information about the conception of the series of lectures, as well as the standing of its presenters and speakers. It also provided quick and easy access and insight into the Institute's discursive and institutional networks: merely listing the names and large number of illustrious guests secured and disseminated the Institute's reputation as a center of debate. Each semester, a large print of the new poster was displayed on the bulletin boards of universities and colleges or in the kitchenettes of offices, while smaller versions were used in apartments, adorning refrigerators and bedroom doors. Part artwork and part advertising, the poster for "Architecture" immediately gave the Institute greater visibility than *Oppositions*; a staggering US$6,000, one-third of the NYSCA grant, was spent on printing and mailing.

In the mid-1970s, when there were hardly any building contracts for architects in New York due to the economic downturn, the Institute was able to transform itself into a new working context. By offering a select group of people the opportunity to lecture, exhibit, and publish there, i.e., to work as cultural producers, to operate and realize their potential as entrepreneurs in their own right, the Institute motivated the assembled individuals to profitably mobilize their cultural and social capital and transform it into economic capital in the medium term. Eisenman, like other longstanding Fellows and Visiting Fellows who attended for a year, financed his work by holding teaching positions at the Institute and at one of the local schools of architecture, and by receiving additional funding for cultural production. Crucially, the interplay of individual activities and one of the Institute's programs produced a variety of internal synergies and external networking effects. The Institute, now a constitutive part of New York architecture culture, positioned itself as a self-regulating system that adopted and continued certain traditions in order to shape the future of the profession and discipline. "Architecture" initially provided a framework for architects to present their historiographical or theoretical research, or their design projects, and to position themselves as public intellectuals. The Fellows who took advantage of this in the first two years were Frampton ("Introduction to the History of Modern Architecture," "Architecture 1," "History of Modern Architecture 1920–1940," "Architecture 2;" Le Corbusier in Context, "Architecture 4")[399] and Gandelsonas ("Symbolic Dimensions of Architecture," "Architecture 4") as *Oppositions* editors, alongside Visiting Fellow Anthony Vidler

399 Throughout the 1970s Frampton drew on lecture courses for his work on his monograph *Modern Architecture. A Critical History* (1980); in the eight-part lecture course "Le Corbusier in Context" he elaborated and presented the first results of his research project "Le Corbusier and the Evolution of the Purist Sensibility, 1898–1928," for which he received a grant from the John Simon Guggenheim Memorial Foundation in 1974–75. Frampton published his studies on Le Corbusier in *Oppositions* 15/16 and *Oppositions* 19/20 in the early 1980s.

("Architecture in Mass Society: Depression War and Reconstruction, 1929–1959," "Architecture 3"), and Peter Wolf ("The Future of New York: Controversies and Consequences," and, with Patricia C. Jones, "Architecture 4") as chairman of the Fellows,[400] each of whom presented a series of lectures. As Program Director, MacNair also regularly assembled a lecture course with peer-to-peer presentations from contemporary architects, artists, and designers, some of them highly renowned. At its best, with "Architecture" the Institute demonstrated its ability to serve as a cultural space in terms of the joint project of reevaluating and redisciplining architecture, as well as a collective teaching and learning process with boundaries between faculty and students, lecturers and audience, exhibitors and visitors, authors and readers becoming increasingly fluid. With lectures being held every night, work and leisure at the Institute were inextricably linked, which went hand in hand with the emergence of immaterial labor among architects and academics. While the individual Fellows' contributions were largely based on self-exploitation, the Institute distinguished itself by establishing new centralized and distributed networks for the production, dissemination, and reception of architectural knowledge, within and beyond the profession and the discipline. The Fellowship itself was based on the liberal principle of merit and gratitude and, above all, as its further development shows, nevertheless shaped and reflected by the then prevailing power (and gender) hierarchies in society, other labor remained flexibly organized and precarious.

Within this neoliberal institution of an emerging postmodern architecture culture, Rem Koolhaas was to play a special role in the Evening Program and the Exhibition Program, initially as a presenter and lecturer in the lecture course "Who is New York? The Future of the Metropolis," which MacNair organized with him as part of "Architecture 2" in the spring semester of 1975.[401] Koolhaas, who was officially a Visiting Fellow at the Institute in 1974–75, began work there on *Delirious New York* by giving lectures on "Manhattanism," the specific architecture and culture of New York. He researched this topic during his time at the Institute, which provided him with students and interns for this purpose, and published articles about it in *Oppositions*, and later in *Architectural Design*, before the study was

400 Wolf had just published the monograph *The Future of the City* (1974), with funding from the Ford Foundation.

401 Koolhaas was listed on the posters for the "Architecture" series several times, with lectures on "Coney Island," November 1974, "Delirious New York: Manhattan as Laboratory," March 1975, "Recent Projects and New Zoning 1976," May 1975, "Delirious New York: The Secret Life of Buildings," November 1975, and "Dali, Le Corbusier, and New York," March 1977. These topics were all included in *Delirious New York*. In 1976, after two years at the Institute, Koolhaas went back to the AA in London, where he held lectures and completed his monograph, but he was to return to New York and the Institute repeatedly, contributing to exhibitions and publications, and later presenting his retroactive manifesto there in the fall of 1978.

published in book form.[402] The Institute provided him not only with a workspace but also an income, for in 1974–75 he taught as a tutor in the Undergraduate Program, in parallel with a course he taught at Columbia University, which allowed him to co-found OMA. Yet Koolhaas was never made a Fellow: he once described his role at the Institute and his contributions to "Architecture" as that of the proverbial "fly on the wall," seeing and hearing events as they occurred.[403] Early in his international career, Koolhaas benefitted greatly from the Institute, which saw him reach maturity as an architect. Not only was he able to publish certain findings from the archive early on in *Oppositions*, thus making his book project officially known, but his appearances in the "Architecture" program, for which larger premises were rented specially and which attracted local architecture critics, allowed him to set the agenda and establish his own voice vis-à-vis the New York architecture scene. Koolhaas and MacNair, with whom he was friends and with whom he also cooperated, were the "young savages" compared to the more serious Eisenman, Frampton, Gandelsonas, and Agrest, or Vidler. And yet the Institute benefitted greatly as well, as it could enlist the support of the traveling wayfarer on his peregrinations between the old world and the new.

Architects, Intellectuals and Critics

Within a short period of time, the Institute produced a new type of architect: not just an artist, but an intellectual, a public figure who performed in front of an audience, similar to developments in philosophy and the humanities, and the public events, "Architecture" and the "Forum" release event for *Oppositions*, from 1974 onwards, were instrumental in this.[404] Several Fellows were able to make a name for themselves as historians or theorists by presenting lecture courses or contributing content. Meanwhile, Eisenman assumed the role of the host in "Architecture," not infrequently emerging from his office, where he had used the time to work, shortly before the start of the event to greet the audience and, as Institute director, to announce the evening's speakers—in other words, he was the "master of ceremonies," as Boyarsky rightly noted,

402 Rem Koolhaas, "'Life in the Metropolis' or 'The Culture of Congestion'," *Architectural Design* 47, no. 5 (1977): "OMA," 319–325; see also Kenneth Frampton, "Two or Three Things I Know About Them: A Note on Manhattanism," *Architectural Design* 47, no. 5 (1977): "OMA," 315–318, George Baird, "Les Extremes. Qui se Touchant," *Architectural Design* 47, no. 5 (1977): "OMA," 326–327, and Demetrios Porphyrios, "Pandora's Box. An Essay on Metropolitan Portraits," *Architectural Design* 47, no. 5 (1977): "OMA," 357–362.; see Rem Koolhaas, "Why I wrote Delirious New York and Other Textual Strategies," *Any*, no. 0 (May/June 1993), 42–43.

403 In other writings, I have used the figure of the harlequin to characterize Koolhaas, as his ironic attitude led him to perform veritable jumps, somersaults, and cartwheels in design, aesthetics, writing, and rhetorics; see Kim Förster, "From Remment to Rem. A Quite Literary Story of Someone Who Made It in New York," *Clog* (June 2014), 32–33.

404 Michel Foucault and Gilles Deleuze, "Intellectuals and Power," in *Language, Counter-Memory, Practice. Selected Essays and Interviews*, ed. Donald F. Bouchard (Ithaca: Cornell University Press, 1977), 205–217.

always in the spotlight.[405] Eisenman himself gave lectures too. For example, he expounded on the historical references for his formalist stance ("Giuseppe Terragni and Italian Rationalism," May 1975), presented his rationalist designs ("Transformation, De-Composition and Critique," April 1976), and discussed his theoretical approaches ("Post-Functionalism. A Continuing Modernism," November 1976), thus legitimizing himself and publicly promoting his work as both an architect and a theorist. Despite a relatively small output of buildings, texts, and other contributions, the Institute at that time was Eisenman's project, and he was masterful in building his public persona as an architecture intellectual, in part by skillfully enlisting Fellows and friends, architects and critics, to his cause and imposing his own thinking style on his colleagues.[406] One characteristic aspect of the Institute's artistic and intellectual work in "Architecture," however, was that each lecture course was aimed at very different audiences endowed with intellectual or financial capital: local and national architects, but also a broader, culturally minded public. Eisenman and the other Fellows spoke to their colleagues, mostly converts, with whom they competed for attention, if not commissions, but they also preached to lay audiences, so that they had to adapt their content to the interests of the general public and popularize, or, to use Pierre Bourdieu's terms, trivialize, even vulgarize it.

The intellectualization and popularization of architectural knowledge, history, and theory in the context of adult education, the Institute's public relations efforts, and the marketing of its educational offerings in the arts and humanities as a commodity were accomplished through a number of promotional strategies, improved advertising, and the involvement of critics. MacNair placed more ads in the local press, in the liberal daily *The New York Times*, the popular weekly *New York Magazine*, and the left-wing neighborhood newspaper *The Village Voice*. Despite declining to take on the role of program director, Stern played a crucial role in shaping the program of "Architecture" by presenting his own lecture course every semester, thus supporting the Institute in its public outreach mission, for example by holding overview lectures on popular and yet sophisticated topics ("The American House. From Jefferson to Wright," "Architecture 2," "The American House: From Frank Lloyd Wright to Robert Venturi," "Architecture 3," "The New York Apartment House," and "Architecture 4"). Stern, more than any other presenter, succeeded in appealing to an Upper East Side clientele that was used to comparable formats at the Metropolitan Museum or the Museum of Modern Art, and thanks to his efforts "Architecture" became a huge success with

405 Eisenman, 2007, 83.

406 Ghirardo, 1994, 71. Architecture historian Diane Ghirardo characterized Eisenman's "masterful public relations work" in terms of his constant gamesmanship and criticized his talent as a self-promoter: "With a canny talent for showmanship more akin to P.T. Barnum than to Walt Disney, Eisenman in the early 1970s managed to parlay a miniscule design portfolio and a wide range of acquaintances into the New York-based Institute for Architecture and Urban Studies." In other words, Eisenman was successful not so much for his curiosities, but for his public relations.

the audience. Stern also appeared regularly on discussion panels and, as spokesman for the "Grays" in the perceived stronghold of the "Whites," took on the role of the eloquent *advocatus diaboli*, challenging Eisenman time and again, so that the paying audience was party to a spectacular debate. Stern was soon joined by other presenters who enjoyed a high standing and were recognized outside the world of architecture. Paul Goldberger, for example, contributed to "Architecture 2" in the spring of 1975, along with Ada Louise Huxtable, the second, younger architecture critic at the *New York Times*, who was responsible for more popular topics and contemporary, primarily conservative postmodern architects. The lecture course presented by Goldberger, "The American Architectural Establishment. A Critical Reassessment," in which a representative of the younger generation of architects addressed a project by a successful architectural firm, was extremely popular. Stern featured Johnson/Burgee Architects, for example, who were successful despite the economic downturn and had just completed Pennzoil Place in Houston, Texas. This comprised two skyscrapers that introduced a postmodernist formal language with their trapezoidal plan as a variation on the modernist glass box and had been much lauded and made famous by Huxtable, who dubbed them the "building of the decade," although the firm itself was not well regarded in New York architecture circles due to the commercial nature of its corporate architecture. Eisenman, who also contributed to this lecture series, chose Paul Rudolph, the architect of the Brutalist art and architecture building at Yale University in New Haven, Connecticut, where he also served as dean, and whose work was thus historicized and elevated at the Institute at a time when the Brutalist style was gradually slipping out of fashion, despite the advocacy of Reyner Banham, and Rudolph was increasingly pursuing projects in other countries.[407]

In its internal report for the 1974–75 fiscal year, Institute director Eisenman noted that the Evening Program, as a newly introduced format of educational, cultural, social, and symbolic value, had in its first semester already recorded a total of more than four hundred paying participants who attended one of the lecture courses for a tuition fee of sixty dollars. Compared to the adult education offerings at other cultural institutions in New York, the program was relatively inexpensive. The Institute had gone into debt to set up "Architecture;" this was due to the high demands of the program but also to the inexperience of everyone involved in event management and accounting. In the second semester, the Evening Program experimented with tiered ticket prices and used cultural sponsorship as an additional source of revenue, as it had done when *Oppositions* was first launched. Occasionally, private individuals were enlisted as donors, who in return received free admission to all events and whose names were listed on the

407 Other architectural firms whose projects were presented and discussed included: Skidmore, Owings & Merrill, John Portman, Davis Brody & Associates, Bertrand Goldberg and Harry Weese, I.M. Pei & Partners, Gunnar Birkerts, Kevin Roche, and John Dinkeloo as students of Eero Saarinen, Cesar Pelli, and Edward L. Barnes.

poster. In the mid-year report to NYSCA, MacNair cited quantitative attendance figures as a measure of the attention paid to the Institute as a cultural institution. In addition, an evaluation was conducted at the end of each semester to gather qualitative information about participants. This allowed a new programming and curation focus to be selected each semester according to audience demand and interest. Although initial revenues of US$33,000 were offset by expenses of US$45,000, after some teething problems, the Evening Program as a commercial offering in its second year contributed a not insignificant 30% to the funding of the Institute's operations. Then, in fiscal year 1975–76, US$82,000 of revenue was collected for the Evening Program, slightly more than half of the Undergraduate Program, which brought in US$150,000. The fact that, despite the Institute's increasing annual budget, the fees of individual presenters and lecturers repeatedly went unpaid, with Fellows and friends of the Institute even foregoing the remuneration to which they were entitled, was indicative of the continued tenuous financial situation and the precarious nature of cultural production. Although jobs were created at the Institute, the working conditions remained questionable due to poor payment practices and haphazard financial management.

Exhibition Activity

After years of no further initiatives, an effort was made to organize exhibitions at the Institute again and for it to start exhibition operations on its own. The internal historiography of the Institute mentions an exhibition in 1974, entitled "Drawing as Architecture"—an equation that would have elevated representational technologies to the rank of architecture—with contributions by Raimund Abraham, Diana Agrest, Peter Eisenman, Mario Gandelsonas, John Hejduk, and Robert Stern.[408] The initiative to hold exhibitions on a regular basis originated with MacNair who, alongside his time-consuming and labor-intensive task of coordinating "Architecture," started out by organizing a number of smaller, informal exhibitions.[409] A newspaper advertisement announced six exhibitions for 1975, beginning with a historical exhibition titled "Mart Stam: Dutch Architect 1920–1965" (January 14 to January 31, 1975), which MacNair quickly assembled

408 Suzanne Frank in her memoirs of the Institute refers to "Drawing as Architecture" that is said to have taken place; see Frank, 2010, 167; see also Taylor, 2010, 317; Architecture historian Jordan Kauffman casts doubt on this; see Jordan Kauffman, *Drawing on Architecture: The Object of Lines, 1970–1990* (Cambridge: MIT Press, 2018), 94f. However, there are no references to be found in internal documents (lists, memos, notes, reports) in CCA's IAUS fonds. The 1978 IAUS brochure, on the other hand, already lists this exhibition, without naming the participants, which could be interpreted as an early claim to authorship of an idea, as was so often the case, and even an early case of valorizing architectural drawings. Kaufman alludes to the rift that had opened up on an institutional level between MoMA and the Institute by then, and on a personal level between Drexler and Eisenman, not to mention Ambasz, which is a longer story.

409 Taylor, 1990, 315–322. Taylor, who worked as director of development at the Institute before founding her own gallery, had written this essay about the Institute's Exhibition Program for a seminar course taught by Mary McLeod at Columbia University.

from exhibits from the Institute's 1971 exhibition "Art & Architecture USSR 1917–32," since these were still stored in crates at the Institute and had to be constantly moved back and forth to free up space for the Evening Program to run smoothly.[410] The first half of 1975 saw further exhibitions by members of the inner circle and friends of the Institute: the postcard collections of Rem Koolhaas and Madelon Vriesendorp, designs by Lella and Massimo Vignelli, photographs by Dorothy Alexander, who documented the Institute's 1973 "Low-Rise Alternatives" exhibition at MoMA, designs for Central Park by MacNair himself, the kinetic sculptures of Tim Prentice, a friend of Eisenman's, formerly an architect and now a practicing artist, and the environmental and land art of Alan Sonfist, whose pioneering role was amply illustrated by "Time Landscape," an urban forest in New York. These exhibitions were always scheduled for a brief run (a week or two, sometimes a month) and made quickly; the works were simply hung on the wall as the central main hall was needed for the nightly events. At first, most of the exhibitions were devoted to art; there were hardly any architecture exhibitions. MacNair's curation did not seem to follow any approach other than the available opportunities and offerings and his own interests and tastes; the exhibition design was rather self-made and amateurish, which had a charm of its own. In addition, MacNair's programming and curation of "Architecture" was notable for its introduction of new formats designed to tap into new markets and reach new audiences. For one thing, the Institute supplemented its adult education program in the spring of 1975 by adding an extraordinary workshop series by the British group Archigram, quite in keeping with the times, on the subject of "Toward the Urban Suburbia." This included six workshops, for which, along with Peter Cook, Ron Herron, and Mike Webb, the protagonists of London's former pop avant-garde were brought to the Institute in their first-ever visit to New York. In addition, the Institute offered educational trips to Florence and London in the summer of 1975, billed as a "Summer Program," although in the end they were not carried out. But there was huge potential and eventually further steps were taken to expand and develop it into a viable and competitive cultural institution. Despite the diversification of its offerings and commercialization of its activities, however, the Institute never really participated in the development that Max Horkheimer and Theodor W. Adorno criticized in their analysis of the "culture industry" in *Dialectic of the Enlightenment*, notwithstanding the Marxist concept of culture which is diametrically opposed to that of Bourdieu. Compared to the developments in film, radio, and print in the first half of twentieth-century America, and television in the second half, the Institute's cultural production—the evening lectures, exhibitions, and other products, especially print products—were still largely publicly

410 IAUS, "Portfolio for an Application to the 1976 AIA medal," n.d. Source: AIA Archives. As so often, chronologies should be contested, for the exhibition dates given in the adverts are not only themselves erroneous, but also differ from those listed later in promotional materials and historiography; see Taylor, 1990, 321.

funded and not just business, education, and culture in the broader sense, and on top of that, they were far too uncertain and volatile.[411]

In fiscal year 1975–76, after a successful start, the Institute's leadership focused on growth with all cultural activities. Not only was the NYSCA funding renewed, but the Institute also received grants from the Gottesman Foundation (of Adam and Celeste Bartos) and the Duke Foundation. These funds allowed Eisenman and MacNair to expand the thematic focus of the Evening Program from architecture to contemporary trends in the arts, which was particularly consistent with Eisenman's understanding of the architectural object as the result of a formal transformation. In addition, the Institute now established exhibition activities that would, within a year, make it a player in the field of architecture exhibitions. One reinforcing moment in this development was the exhibition "Architectural Studies and Projects" (March 13 to May 15, 1975) at MoMA, curated by Emilio Ambasz in the museum's penthouse on the sixth floor, which housed the lounge and cafeteria, on the initiative of Barbara Jakobson: It displayed drawings by twenty-two architects.[412] In MoMA's press release, Ambasz first defined architectural drawings, which were perceived as critical, if not visionary, as an art form in their own right: "Paper Projects have in many instances influenced architecture's history as forcefully as those committed to stone. Whether their intent is aesthetic, evocative, ironic, polemical, methodological, ideological, or conjectural, their strength has always resided in their poetic content." In keeping with the times, with the complications of the macroeconomic environment that accompanied the 1973 oil crisis, the individual exhibits were for the most part fictitious architectural drawings that did not necessarily correspond to a concrete building project; some of them were even unbuildable.[413] "Architectural Studies and Projects," which was critically received with two reviews in the *New York Times*, was also trendsetting and quite strategic in other respects, as it paved the way for New York's architecture

411 Adorno and Horkheimer, [1944] 1972.

412 MoMA, "Architectural Studies and Projects," Press Release no. 14, March 13, 1975, https://www.moma.org/documents/moma_press-release_332895.pdf (last accessed: May 31, 2023). Among the architects exhibited were: Raimund Abraham, Peter Cook, Peter Eisenman, John Hejduk, Rem Koolhaas, Richard Meier, Gaetano Pesce, Cedric Price, Ettore Sottsass, Friedrich St. Florian, and Superstudio; see Barbara Jakobson (interview with Sharon Zane), (New York: MoMA, October 29, 1997), 22; https://www.moma.org/momaorg/shared/pdfs/docs/learn/archives/transcript_jakobson.pdf (last accessed: May 31, 2023). See also Kauffman, 2018, 78ff. Kauffman points out that the authorship of the idea was disputed.

413 Through MoMA, architectural drawings have entered the history of art and architecture as visionary projects, due in part to an early group exhibition titled "Visionary Architecture" (1960), a number of solo exhibitions in the early 1970s, and the establishment of the Howard Gilman Collection curated by Pierre Apraxine, who was head of the Art Lending Service from 1970 to1973. This collection was later celebrated and historicized in another group exhibition titled "The Changing of the Avant-Garde: Visionary Architectural Drawings" (October 24, 2002, to January 6, 2003); see Terence Riley, ed., *The Changing of the Avant-Garde: Visionary Architectural Drawings from the Howard Gilman Collection* (New York: MoMA, 2002).

culture to become more commercial.[414] It was held as an art, promotional, and sales event of the Art Lending Service, a subdivision of the Junior Council under Jakobson's direction that lent works of modern art donated by galleries to individuals and corporations to display in their homes and offices, and thus supported MoMA. "Architectural Studies and Projects" contributed to the development of a market for architectural representations by putting a price tag on young architects' drawings and selling them straight off the wall, virtually as a capital investment, making them both fashionable and profitable. The highest selling prices were fetched by variants of drawings that were later published in *Delirious New York*, Koolhaas' "Manhattan Projects," and especially drawings by Elia and Zoe Zenghelis.[415] From among the Fellows, Eisenman was featured, whose series of drawings "Transformation No. 14" for his recently completed design for *House VI* commissioned by the Franks and produced by interns at the Institute, found no buyers. However, this development resonated at the Institute, although it represented an intellectual counterpart (not least because of its non-profit status), and yet it was implicated through a symbolic economy.

While public interest in architectural drawings was fostered with exhibitions such as those at MoMA, the Institute began organizing regular architecture exhibitions as part of its exhibition activities under MacNair's banner in the fall of 1975. As aggregations of individual architects from the Institute's circles, these were initially group exhibitions of works by Fellows and friends, forming groups of like-minded individuals and providing frameworks for interpretation, to be later supplemented by exhibitions of works by contemporary postmodern American and European architects. The first group show was "Goodbye Five. Work by Young Architects" (September 16 to October 1, 1975), an exhibition of projects by eighteen young architects assembled by MacNair. Even though the publication *Five Architects*, towards which the title referred, was only a few years old, this represented a provocative proclamation of a generational change in the New York architecture scene, as a continuation of the existing polemic about Eisenman and the New York Five.[416] The exhibition showed a cross-section of a younger American architecture scene that was not really doctrinaire or ideological, nor necessarily

414 Paul Goldberger, "Architectural Drawings at the Modern," *The New York Times* (March 14, 1975), 24; Ada Louise Huxtable, "Poetic Visions of Design for the Future," *The New York Times* (April 27, 1975), 142.

415 Of the sixty-six drawings on display, fifty-three were for sale. According to a list of works, the highest purchase prices were achieved by two drawings by OMA, "The Square of the Captive Globe" [sic!] with US$ 770, which Rem Koolhaus [sic!] had made with Zoe Zenghelis, and "The Egg of Columbus Center" with US$ 780, drawn by Elia and Zoe Zenghelis. None of Madelon Vriesendorp's drawings were sold.

416 The exhibition "Good-bye Five: Work by Young Architects" included works by Alan Chimacoff & Steven Peterson, Architects in Cahoots (John Casbarian, Danny Samuels, Robert Timme), Stuart Cohen, Richard Hammer, Rem Koolhaas, Leon Krier, Andrew MacNair, Mark Mack, Richard Plunz, Stephen Potters, Tod William, Jon Michael Schwarting, Massimo Scolari, Studio Works (Craig Hodgetts, Robert Mangurian), Susana Torre, Lauretta Vinciarelli, Stuart Wrede, and Timony Wood.

committed to a single stance but thrived equally on formal playfulness and figurative allusions, as in the works of Architects in Cahoots from Houston or Studio Works from Los Angeles, and with Susana Torre and Lauretta Vinciarelli also included two female architects.[417] In addition to MacNair as Fellow and Koolhaas as Visiting Fellow, other Institute staff members were also featured (Stephen Potters, Michael Schwarting, and Stuart Wrede). Two European architects, Leon Krier and Massimo Scolari, who stood for postmodernism and were later featured in solo shows, made their first appearance at the Institute in this context. MacNair had buttons that read "Goodbye Five" produced for the opening and distributed them among the audience to deliberately popularize the first group show at the Institute which, like its historically legitimized and now famous predecessor, served primarily as provocation and self-promotion. Simply by wearing this emblematic button, as was common in punk or alternative culture at the time, that evening's exhibition visitors were making a statement, consciously or unconsciously, against the elitist habitus and discursive supremacy of the New York Five. The exhibition was the first at the Institute to attract attention in the New York dailies and architecture press, with critical reviews appearing in the *New York Times* and *Progressive Architecture*.[418] Reviewers complained that the installations were disorganized and unprofessional and lacked labels. Most importantly, they again drew attention to the retreat of architects into artistic practice, which they believed, as had been evident at the Institute in the mid-1970s, to be primarily due to the country's poor economic situation. The Institute's exhibitions, the critics pointed out, and its public events in general were evidence of this: the sociopolitical approaches, participatory design, and environmental issues that had been all the rage in the field at the beginning of the decade, after 1968, were now apparently no longer in vogue among this new generation of architects.

3.2 Hosting Evening Entertainment

With the fall semester of 1975, the Evening Program became much more extensive and attractive, at times including up to ten, sometimes even eleven lecture courses per semester, with two lectures per day, at 6 p.m. and 8 p.m. "Architecture 3" already covered a fairly broad range of topics. In addition to the

417 In 1973, Susana Torre co-founded the Archive of Women in Architecture at the Architectural League of New York, on which basis she curated the 1977 exhibition "Women in American Architecture: A Historic and Contemporary Perspective;" see Susana Torre, *Women in American Architecture: A Historic and Contemporary Perspective* (New York: Whitney Library of Design, 1977), see also https://archleague.org/women-in-american-architecture-1977-and-today/ (last accessed: May 31, 2023).

418 Paul Goldberger, "Young Architects Display Recent Visionary Designs," *The New York Times* (September 19, 1975), 25; Peter Papademetriou, "Good-bye Five: Work by Young Architects," *Progressive Architecture* (November 1975), 20–21.

Fellows' interests and areas of expertise, the lecture courses now included courses facilitated by outside presenters on topics such as historic preservation and the conversion of buildings, humanism in architecture, the history of ideas in nineteenth-century architecture, an introduction to issues in architecture, positions in interior design, the architecture of Chicago, and topics in contemporary art, painting, sculpture, and film. In its dual role as a bridge between education and culture, the Institute cooperated with various local universities and targeted individual lecture courses towards so-called "non-traditional students," emphasizing the focus on architecture as a unique selling point in the course catalogues. For two semesters in a row, art critic and theorist Rosalind Krauss, who had come to the Institute in 1975 at the invitation of Eisenman along with her colleague Annette Michelson after both had quit as editors at *Art Forum*, offered lecture courses on "Content in 1960s Painting and Sculpture" and "Critical Issues of Art in the 1970s." The following spring, Krauss and Michelson founded the art criticism and theory journal *October*, with the Institute as its initial publisher.[419] In a different radical twist, Arthur Drexler presented a lecture course on "L'École des Beaux Arts" in the fall semester of 1975 as a special format designed to coincide with the exhibition he curated at MoMA, "The Architecture of the École des Beaux Arts" (October 1975 to January 1976), even though he had fallen out with Eisenman and had in the meantime resigned as chairman of the Board of Trustees of the Institute. The Beaux-Arts exhibition, which foregrounded the Beaux-Art system that was still prevalent when the Bauhaus made its mark in the 1920s, made a case for postmodernism—as a departure from modernism and as a stylistic movement rather than a social development, which was now being revisited at the Institute, and ultimately both institutions benefited from the debate and the attention.[420] Drexler was viewed rather critically among the Fellows for this MoMA exhibition, which showed original Beaux-Arts drawings and photographs of Beaux-Arts buildings in the USA, and which came across as visually powerful and thus crowd-pleasing. Subsequently, MoMA's recent reorientation was hotly debated at the Institute in the context of a "Forum" for *Oppositions* 4 in January 1976, under the title "The Architecture of the École of Beaux Art-Exhibition," announcing a paradigm shift in architecture culture that had finally arrived in New York.[421]

419 Not only Krauss, but also Michelson contributed to the Institute's public events in return for being hosted there with *October*; Michelson for example offered a lecture course on "Soviet Film 1925–1935. A Study in Revolutionary Cinema" as part of the "Architecture" series, conceived as a film seminar in the Rizzoli Publishing House Screening Room, and sponsored by the Rizzoli Bookstore nearby, on 5th Avenue between 55th and 56th Street. Later, *Oppositions* featured writing on Sergei Mikhailovich Eisenstein; see Manfredo Tafuri, "The Dialectics of the Avant-Garde: Piranesi and Eisenstein," trans. Marlène Barsoum and Liviu Dimitriu, *Oppositions* 11 (Winter 1977), 72–80.

420 MoMA, "The Architecture of the Ecole des Beaux Arts," Press Release no. 59, August 8, 1975, https://www.moma.org/momaorg/shared/pdfs/docs/press_archives/5289/releases/MOMA_1975_0074_59.pdf (last accessed: May 31, 2023); see Kauffman, 2018, 14ff.

421 William Ellis, "Beaux," *Oppositions* 5 (Summer 1976), 131–134.

The curation of "Architecture" as an ongoing series of lectures involving a variety of architects, designers, and artists, academics, and critics in New York and turning them into content providers and cultural producers at the Institute was aimed at amplifying the intricacy, if not the complexity of the debate, and helping the discipline flourish outside of the university lecture hall, the museum/gallery, and beyond of the pages of books and journals. In economic terms, beyond the cultural transformation, the expansion and differentiation of the Institute's teaching and cultural products went hand in hand with the increasing focus of "Architecture" on the philanthropic model in 1975–76, when three types of cultural sponsors were introduced: individuals (US$175), professional sponsors (US$220), and corporate sponsors (US$275). In the spring of 1976, the Institute also received institutional sponsoring for the Evening Program for the first time. At the same time, this establishment of cultural production at the Institute, of social processes involved in the generation and circulation of cultural forms, practices, and values, which was supported by state arts funding, an outwardly open but ultimately closed system of opinion that revealed shared understandings (and misunderstandings) within the New York architecture community, demonstrated to the public that the Fellows were no longer bound by a common research and design project. Nonetheless, in epistemological terms, the Evening Program and Exhibition Program showed that the Fellows, Visiting Fellows, lecturers, and audience formed a collective of thought that was part of a larger community of thought, sharing a common interest in an architectural project: namely, to introduce a new postmodern thinking style and, in this way, to re-culturalize architecture, i.e., to re-semantize, re-historicize, and re-aestheticize it. In an article about the Evening Program in the *New York Times* in October 1975, Goldberger gave the Institute additional support, a good argument against its somewhat elitist image: "But probably the farthest from the ivory tower is the Institute's Evening Program, a New School-like potpourri of courses designed to bring the gospel of architecture to the average man on the street."[422] While Western society at the time, explicitly American society, was from an urban sociological perspective deemed to have lost the capacity for impersonal cooperation and thus for political action, Goldberger, as an architecture critic who was himself involved in and benefited from the reorientation and repositioning of the Institute, with its opening to the public, became an accomplice to this culturalization of architecture.[423]

The immense significance attached at the Institute to the Evening Program, not only as an institutional and discursive format, but above all as a socio-cultural event, was also underscored by the fact that Eisenman himself now presented

422 Goldberger, 1975, 41 & 77.

423 Richard Sennett, *The Fall of Public Man. On the Social Psychology of Capitalism* (New York: Alfred A. Knopf, 1976).

a lecture course entitled "Modernism 1975" for the first and only time in the fall semester of 1975 as part of "Architecture 3," inviting prestigious guests, architects, historians, and critics to debate current trends in architecture and the status quo of contemporary architecture in general, which he provocatively defined as a problem of style. The "Critical Dialogues," as they were captioned, were not a classical lecture course, but rather public events featuring conversations in a talk show format. His guests were John Hejduk, Cesar Pelli, Peter Smithson, Peter Blake, O.M. Ungers, Vincent Scully, Denise Scott Brown, Martin Pawley, and once again Robert Stern, an illustrious circle of intimates and colleagues from various backgrounds who stood for a variety of ideological, conceptual, and architectural approaches. The topics of conversation were the various prevailing styles of the time: realism, pragmatism, modernism, neo-rationalism, populism, idealism, and post-modernism. The talk show format was explicitly modeled on the successful PBS television program "Firing Line" (since 1966) with William F. Buckley Jr. in order to, in Eisenman's words "engender a spirit of participation," i.e., to make the program more viewer-friendly, and perhaps also television-friendly.[424] A stage had even been set up in the Institute's main hall for the first time at Eisenman's particular request. In addition, Eisenman, along with MacNair, had briefly considered recording the events and broadcasting them on New York City Cable TV, then a fledgling medium based on the idea of public access. Although the plan for such a media offensive was not pursued further, Eisenman once again took center stage with the "Critical Dialogues" by conceiving and hosting these events.

A critical historiography of the "Critical Dialogues" based on psychoanalysis and deconstruction might focus on Eisenman's constant pursuit of controversy at the Institute; not only the editorship of *Oppositions* but also the organization of "Architecture" was a means for him to emphasize the Institute's central role and its extraordinary position of power in the American architecture world. The title "Modernism 1975" was self-explanatory, and so the events needed no further explanation beyond their aim, which was to propagate the continuing relevance of a modernist rather than postmodernist architectural language at the dawn of a postmodern pluralism, a culture of ambiguity and indifference. Nor did it require a didactic or communicative concept beyond a simple juxtaposition of eminent architects, historians, theorists, and critics. With forty-nine paying spectators in the audience, the "Critical Dialogues" were comparatively successful, although Eisenman did not come close to the popular appeal of his counterpart Stern. Above all, however, it became clear that the Institute was now attempting to fulfill its educational mandate as broadly as possible with

424 New School, "Course Catalog," 1975–76, 65. The New School course catalogue for the fall semester of 1975, directly below Eisenman's, announced as another lecture course of the "Architecture" series that urban planning scholar and Civil Rights activist Paul Davidoff was to present at the Institute on "Community Participation in the Environmental Design of Neighborhoods: Programs for Improvement." But the education offering on the history, politics, and strategies of planning from below was, for unknown reasons, not offered at all.

"Architecture," and that the Evening Program admirably served the needs of the New York cultural audience as evening entertainment, comparable to television. While the education and culture programs pushed for professionalization and economization, the Institute itself in 1975–76 became more of a stage than a classroom, with controversial positions being discussed and counter-positions taken. This practice echoed that of the "Whites" vs. "Grays" debate, in that it suggested a choice, but ultimately represented no real alternative.[425]

The emergence of postmodernism, conceived as both a discursive formation and an institutional network, ultimately as a cultural phenomenon in architecture, became evident at the Institute in "Architecture 3," where intellectual debate took a back seat to quality of performance. What is notable about Eisenman's greatest contribution to adult education is that he called these public conversations "critical." His choice of guests suggests that he meant this as more of a formal or rhetorical attribution than a philosophical or socio-political one. For not only the debate conducted with *Oppositions*, but also the public events, series of lectures, and exhibitions were ultimately additive and collective, characterized by competition, solidarity, enmity, and friendship. The definition of the position adopted with "Modernism 1975" already testified to the fact that—even before Charles Jencks finally made the pluralism of varieties of post-modernism socially acceptable, presentable, and intelligible with *The Language of Post-Modern Architecture* (1977)—the narratives of modernism had been exhausted and emptied of all meaning.[426] Attention was now focused on the staging of a plurality, which was always also socially specific, gendered, and generationally contingent, which represented a distraction from architecture, housing, and urban planning as a formerly social task. Despite all the claims of "criticality," aside from its popular format, "Modernism 1975" was in institutional terms an elitist event, characterized by the fact that it no longer made any difference whether the issues under discussion were "real" or "bogus."[427]

With regard to the Institute's architectural project—disciplinary and professional—which was directly related to the self-realization and self-marketing both of the Fellows as individuals and the Institute as a group, Eisenman found a congenial partner in Stern, who had initially claimed "post-modernism" for himself, his thinking, and his activities in New York architecture circles, and not

425 Watson, 2005; Martin, 2010, 29.

426 See especially the "Architecture" chapter in Martin, 2010, 174–179.

427 In epistemological terms, the distinction between "real problems" and "bogus problems" is central, not only to modern science, but to any exclusive community; see Fleck, [1935] 1979, 104; in architecture history, the accusation of the falseness of postmodernism, explicitly formulated against Eisenman's networks, has been repeatedly made and fiercely debated; see Ghirardo, 1994. The Institute's original claim to stand for the real, i.e., for "real facts" and for "real problems," has since faded.

only for the "Critical Dialogues."[428] Eisenman and Stern joined forces by staging a controversial debate that was performed in public, thus apparently producing architectural truths. Their shared interest in the power of controversy and an economy of attention was first manifested in their joint publication of a special issue of the Japanese journal *Architecture + Urbanism* on "White and Gray: Eleven Modern Architects" in April 1975.[429] With their appearances in each other's lecture courses as part of "Architecture," Eisenman and Stern continued the "Whites" vs. "Grays" debate that had previously been generated and conducted in journals, i.e., *Architectural Forum* and *Oppositions*, at the Institute, literally "in-house," showcasing themselves and their projects in an open confrontation.[430] In 1976, just when the Institute was escaping the narrow national focus of the debate and was able to position itself internationally, for example by signing a contract with MIT Press for *Oppositions* (see chapter four) and by being commissioned to curate the American contribution to the 1976 Venice

428 Architecture historian Kazys Varnelis points out that Eisenman and Stern were friends and documents the extent to which both were aware of their roles and expressed that they needed each other as opponents: "As Stern described it, his friendship with Peter Eisenman was based on 'the very oppositeness of his nature from mine ... [he] is my perfect alter-ego: If I didn't invent Peter Eisenman who would have?'" Eisenman, for his part, once claimed, "If Stern had not existed, I would have had to invent him, and vice versa." See Kazys Varnelis, "Philip Johnson's Empire: Network Power and the AT&T Building," in *Philip Johnson: The Constancy of Change*, ed. Emmanuel Petit (New Haven: Yale University Press, 2009), 123.

429 *Architecture + Urbanism*, no. 52 (April 1975): "White and Gray: Eleven Modern Architects". This issue featured projects by twelve East Coast architects (all men) who roughly fell into the two camps of the "Whites" and the "Grays": Peter Eisenman, Robert Stern (with John Hagmann), Charles Moore, Richard Meier, Michael Graves, Jaquelin Robertson, Werner Seligmann, Charles Gwathmey (with Robert Siegel), Richard Weinstein, Tim Prentice with Lo-Yi Chan and Rolf Ohlhausen, Giovanni Pasanella, and John Hejduk. The projects were framed by two introductory texts by Vincent Scully ("The Shingle Style Today or the Historians Revenge") and Colin Rowe ("Collage City"). The occasion for this renewed grouping, which resumed the polemics of the early 1970s, was a discussion meeting at the University of California in May 1974, hosted by Cesar Pelli, Anthony Lumsden, Tim Vreeland, Craig Hodgetts, and Paul Kennon, who called themselves the "Silvers." In the second half of the 1970s, the juxtaposition was then superseded by the emergence of debates on architectural postmodernism vs. the neo-avant-garde.

430 In *Utopia's Ghosts*, Martin points out that the dichotomous "Gray/White" camp formation overlapped with the launch of *Oppositions* and other activities at the Institute in 1973, when Stern edited the special feature "Five on Five" for *Architectural Forum*: "Thus, also in 1973, elements of this apparatus interacted with those of another apparatus speaking its own dialect in the Gray/White debate, an in-house power struggle of five against five, for which architecture-as-language was a foregone conclusion. It made no difference that one side spoke of semantics while the other spoke of syntactics, because these two levels ultimately converged—again, quite pragmatically—in architecture's new home within an ecology and an economy of signs." See Martin, 2010, 66. Elsewhere, Martin compares the staging of the debate to a televised boxing match, as a media event in which the point is to take sides with one of the two opponents; see Martin, 2010, 29. When he writes that the debate was held "in-house," he seemed to be referring to the fact that it was hermetic. It should be noted, however, that the debate was not only mediatized in journals, but also took place live at the Institute, and that the competition, due to the concerted efforts and prearranged agenda, resembled a wrestling match rather than a boxing match; less of a sporting contest than one that was always about high ratings and good entertainment; see Förster, 2018, 215, 226, footnote 5.

Biennale, the programming and curation of "Architecture 4" provided concrete proof that the two supposedly opposing but ultimately quite similar schools of thought represented by Eisenman and Stern may have differed in terms of their preferred architectural language, but not in their media or forms of communication, despite their individual habituses and cultural differences.[431]

International Stage

After audience figures for the Evening Program stagnated, albeit at a high level, in 1975–76, the Institute was able to obtain additional funding from the NEA by including art theory in the spring of 1976. The broader scope of "Architecture" alone enabled the Institute to demonstrate openness and inclusiveness, lay claim to specific topics, and involve a wide range of people. It was no less original to legitimize architectural theory by including art theory, analogous to the steps taken by Eisenman with his house designs earlier in the decade, and thus to assert the autonomy of a postmodern, creative, and intellectual architectural practice on a broad basis. Notwithstanding, with "Architecture 5" in the fall semester of 1976, when the economic downturn was still definitely noticeable after the height of the crisis, the Institute offered more established architects and designers than ever a major platform on the international stage to present and discuss their current projects: practicing architects from the Institute's broader network and friends from Eisenman's circles (John Hejduk, Charles Gwathmey, Ulrich Franzen, Philip Johnson), international guests (Arata Isozaki, Rafael Moneo, James Stirling, O.M. Ungers) and New York and other American architects (Samuel Brody & Lewis Davis, William Conklin, Charles Moore, Paul Rudolph). MacNair also managed to engage renowned graphic and fashion designers (George Nelson, Candy Pratts, Ward Bennett, Mary Joan Glynn, Milton Glaser, Massimo Vignelli, Ivan Chermayeff) for the Evening Program. The highlight of "Architecture 5," however, was Manfredo Tafuri, who was enlisted as an internationally renowned scholar to make an appearance at the Institute with a lecture on "Modern Architecture: The Dialectics of Order and Disorder." Tafuri was Eisenman's and the Institute's harshest and most vehement critic and had previously torn apart the self-proclaimed avant-garde, or postmodern "neo-avant-garde," on the basis of a historically grounded reading, a combination of structuralist linguistics and political economic analysis.

431 By the time Tafuri unmasked the demarcation between the different camps in American architecture circles as arbitrary in his [1976] 1987 essay "The Ashes of Jefferson," the PR strategies of the "Whites" and the "Grays" had already been transformed into a debate about the legitimacy of an architectural postmodernism. The Five Architects were finally, with *Oppositions* 5 (Summer 1976), declared dead. A collage of portraits of the five protagonists, as part of Vignelli's graphic design, which preceded Tafuri's contribution "American Graffiti. Five x Five = Twenty-five" provided another self-image, through which this short-lived group was simultaneously historicized and heroized. Eisenman, with his polemical text "Post-Functionalism," which was finally published as an editorial to *Oppositions* 6 (Fall 1976), claimed to have outlined a basis for a general theory of architecture.

In view of this comprehensive and impressive program, which was once again communicated via the poster and fully justified the funding from the NEA, it hardly mattered that individual speakers who had been initially announced did not appear in the end and in some cases, entire lecture courses had to be canceled due to lack of audience interest. What became apparent, however, was that the two art courses planned by Krauss (Lawrence Alloway, Max Kozloff, Jeremy Gilbert-Rolfe, Steven Koch, Robert Pincus-Witten, and Yvonne Rainer were scheduled as external guests), for example, were ultimately not held due to lack of interest.

In the mid-1970s it became evident that the Institute, with all its public events, "Architecture" as a series of lectures, and the exhibitions, increasingly functioned as what in social research is called a tastemaker, as well as a gatekeeper, since it not only mediated access to the globalizing architecture world of New York, but also represented a reference system in that it did not act and network as a functional elite in the service of a power elite, but instead shaped at least the American, if not international architecture debates and education with new patterns of interpretation, structures of meaning, and identities. In the spring of 1976, new trends from Europe were introduced at the Institute when a new series of lectures called "New Wave of European Architecture," aimed at professionals and students, brought a number of young European architects to New York for the first time as part of "Architecture 5." Among the invitees were O.M. Ungers (representing Berlin), Rem Koolhaas (Amsterdam), Elia Zenghelis (London), Massimo Scolari (Milan), Carlo Aymonino (Venice), Robert Krier (Vienna), Leon Krier (London), Jürgen Sawade (Berlin), Aldo Rossi (Milan) and, at the end, O.M. Ungers again (this time representing Cologne). For ten weeks, they met on Monday afternoons to discuss their current projects in the context of the respective local architecture scenes. While the title of the series was a reference to the French Nouvelle Vague—the auteur films of young cineastes produced in the 1950s and 60s that opposed established commercial cinema, its imagery, and narrative flow—the international new generation of architects was, at the suggestion of MacNair, announced as the new architectural avant-garde in the subcultural *Village Voice.*[432] And while a central role was assigned to Ungers, Koolhaas was singled out as the secret star. The distinctive hallmark of the "European New Wave" as another public event was that a group of young European auteur architects, who represented unconventional, even dissenting approaches while simultaneously espousing a shared postmodern aesthetic, were brought together under one label and marketed by the Institute. Unlike the avant-garde movements historicized in *Oppositions*

432 The tone of the advert in the *Village Voice* was rather suggestive: "Facades: From the urban centers of Europe—London, Berlin, Vienna, Milan, etc.—come nine of the architectural avant-garde to tell us their theories about the endless quest 'to make a new architecture.'" See "Options" *Village Voice* (January 19, 1976).

and communicated by "Architecture," it was not so much about the marriage of artistic expression and social progress or even radicalism in the face of changing socio-economic and political conditions. The "European New Wave" was primarily significant in conceptual and methodological terms; in New York, it marked the beginning of a globalized, postmodern architecture culture based on a glamorized culture of celebrity. For the Italian architect Aldo Rossi, who in 1976 began commuting regularly between Milan and the United States, where he taught as Mellon Professor at Cornell University and Visiting Professor at Cooper Union, it was the first, but not the last, appearance at the Institute; he was absorbed and appropriated by Eisenman for his own purposes. The Krier brothers were also largely unknown in the United States at the time. Ultimately, all of the Institute's programs and products aimed at both peers and a general public had a cultural, social, and symbolic added value in addition to their educational value. With the "New Wave" format, which was not well received by the New York architecture public at first, but was later continued and expanded, the Institute campaigned for a young generation of up-and-coming architects from Europe at an early stage, quasi a postmodern variant of what MoMA had previously done with its International Program. It identified who was popular internationally and subsequently, in combination with its publications, shaped the taste of a generation of architects in the United States.

In addition, monographic exhibitions of architectural designs began to be regularly shown at the Institute. Characteristically for the architectural innovation of this period, it was mainly drawings and models, i.e., the classic instruments of design, that were on display—preferably of visionary or theoretical projects, and less of realized or even projected projects. In the spring of 1976, for example, Scolari and Rossi each exhibited their fictional "Drawings and Projects" in solo exhibitions at the Institute after their "New Wave" lectures. Again, exhibitions and even openings at the Institute were sparsely attended, although MacNair, now supported by Silvia Kolbowski, produced specially designed advertisements and posters based on Vignelli's graphic identity concept. Because the main hall was always available except for in the evenings and the times when design studio crits were held for the educational programs, MacNair was able to produce exhibitions in a very short time and on a low budget. In addition, in his typical DIY style, he produced fanzine-like booklets to accompany the exhibitions. These contained photocopied materials and were quickly stapled together. In curating the Exhibition Program, he then primarily implemented the ideas of Eisenman and other Fellows and worked closely with the exhibiting architects. The short lines of communication and comparatively unbureaucratic structures at the Institute made it possible to make quick decisions and thus respond to needs as they arose. The exhibits, still presented in a makeshift manner, were often simply hung on the walls, and most of the time no special exhibition design was developed at all. Even if a certain plurality was celebrated within "Architecture" as the highest virtue at the Institute, what the various cultural productions, the positions represented in the series of lectures, and the aesthetics conveyed by the exhibitions had in common was that they testified to the verbally and visually powerful assertion of artistic and intellectual autonomy. In this sense, the Institute acted more as a non-commercial art space of architecture, at least formally, while the works did not have a local angle (which gained in complexity as art and commerce became increasingly intermingled).

If the 1970s saw the permeation of capitalism into all areas of social and cultural life, then the Institute, with its repositioning and restructuring, stands as an example of how, against the backdrop of the triumphant advance of the culture, media, and advertising industries in the United States, the fine dividing lines between education and entertainment, culture and commerce became increasingly diffuse—including in the field of architecture.[433] By the mid-1970s, the Institute was able to establish itself as a new architecture institution with an extremely ambitious product and program policy, in terms of form and content, in very differently regulated knowledge markets. In addition to architecture education with the "Undergraduate Program," the "Internship Program," and the "Design and Study Options," and cultural production with "Architecture," the Exhibitions Program, and new formats such as the "New Wave" series, a more comprehensive publication practice emerged as a new, third field of activity for Fellows and editorial staff (see chapter four). With its complex reprogramming and curation, the Institute enriched the educational and cultural landscape of the New York metropolitan region while asserting its role as a leading architecture center on the East Coast of the United States, if not the international stage, thanks to its broader cultural aspirations. By engaging with very different forms of capital, the Institute now established itself as a powerful instance of consecration and legitimation on the lines of Bourdieu's analysis in "Sociology of the Market of Symbolic Goods," thus exerting influence on the reconstitution of the architecture field and the reinterpretation of an architectural habitus.[434] After its repositioning, the Institute's leadership did everything within its power to ensure that it acted as an institution itself and was legitimized by other cultural and professional institutions. Having already been honored by the Municipal Art Society in 1975, the Institute was awarded the Gold Medal of the American Institute of Architects (AIA) in early May 1976. The Institute's operations, its

433 On the mechanisms of culture, media, and advertising industries, see Adorno and Horkheimer, [1969] 2009. The Institute's contribution to the transformation of architecture culture lay less in new technologies of information transmission or telecommunication that were gradually gaining acceptance in metropolitan society, as well as architecture firms; personal computers did not play a role yet, and cable television was not used. At the Institute, the copy machine was the most important device. It was used by the Institute's leadership, editors, and faculty to copy documents, texts, and teaching materials; writing and editing was done using the "cut and paste" technique. Eisenman also made copies of many internal documents, which he apparently took home with him and later gave to the CCA as the Institute's (un-)official archive, so that they form the source basis of this publication.

434 Bourdieu, [1971] 1983.

policies, and economics were now clearly aimed at becoming an authority on education and culture and a serious competitor to other New York institutions.

Inner Circle and Extended Circle

The crux and major challenge facing the institutionalization of new cultural forms, practices, and values at the Institute, as well as its further expansion as a network, were once again Eisenman's charismatic, entrepreneurial, and ambitious behavior, bearing, and presence, which determined the Institute's Fellowship and personnel policies. Recent organizational, structural, and functional changes in 1975–76 were reflected in the expansion of the Fellowship and new structures of actors and relationships, underscoring the growth and maturity of architecture education, the diversification of cultural production, and the introduction of public relations activities. The Institute now included ten Fellows; in addition to Eisenman, as the established, unchallenged Institute director and Wolf as chairman of the Fellows, they were, in alphabetical order: Diana Agrest, Stanford Anderson, Julia Bloomfield, William Ellis, Kenneth Frampton, Mario Gandelsonas, Andrew MacNair, and Leland Taliaferro.[435] The Fellowship, despite its members' obligation to pay dues to the Institute while retaining their status, was a flexible structure that accommodated the particular conditions and needs of the Institute's cultural work. This was illustrated by the modification of the Fellowship Article, undertaken in 1975–76 for the first time to reflect the Institute's newly defined orientation as an educational and cultural entity. Not only was the status of Visiting Fellows redefined and regulated, but it was also established that the Institute, even if it continued to grow, would have a maximum of only twenty Fellows, ensuring the exclusivity of their status. After Koolhaas and Wrede had been Visiting Fellows for the first half of the year, both of them contributing to individual lecture courses and deriving an income from teaching, Eisenman succeeded in bringing two more architects from Europe: Bernard Tschumi and Grahame Shane. They both came to the Institute from the Architectural Association in the spring of 1976, after Koolhaas had left to return to London, and were on the Institute's faculty in the second half of the year.[436] One of Eisenman's successes, a milestone for the Institute,

435 The Fellows, however, were not all equally involved in cultural and educational activities, let alone stationed at the Institute at all: Anderson was editor in charge of the Institute's *On Streets* publication, to be published by MIT Press, but was hardly ever present in New York, Frampton taught at the RCA in London from 1975 to 1977, occasionally visiting New York to hold lectures at the Institute and working primarily as editor of *Oppositions*, and Taliaferro worked as executive architect on the MGPV housing project in Brownsville, Brooklyn until 1976, and then as Eisenman's executive partner on *House X* outside of the Institute.

436 In contrast to Shane, who left the Institute after only one semester, Tschumi stayed a little longer, showing "A Space: A Thousand Words," the exhibition which he curated, and first exhibited at the RCA in London in April 1975, for a second time, while also working on his postcard series "Advertisements for Architecture" and writing his manifesto-like text "Architecture and Transgression;" see Sandra Kaji-O'Grady, "The London Conceptualists: Architecture and

its future activities, and its reputation was a contract signed with MIT Press in April 1976 for the continued publication of *Oppositions*. This transformed what was supposed to be a "little magazine" into a scholarly journal, strengthening the format, and the content was now provided by the Institute's international circles. In the spring of 1976, the Institute added a second publication: the art theory and criticism journal *October*, edited by Krauss and Michelson, and initially self-published. This meant that, in addition to the Evening Program and the Exhibition Program, two formats whose reach was primarily local and regional, the Institute now had two ambitious publications with an international reach. Thus, at least temporarily, the Fellows secured a near-monopoly position in the architecture and art debate and participated in the international debate about postmodern theory, historiography, and aesthetics which, in the case of *October* (more than *Oppositions*), reflected the rise of poststructuralism, and was particularly influenced by French philosophers.[437] Eisenman succeeded in bringing more people, who were responsible for individual productions, to the Institute as Visiting Fellows for 1976–77, including Vidler as the new editor of *Oppositions* and Krauss as editor of *October*. Finally, the Institute evolved into a lively and popular hub for architects from the United States and abroad during this period, cultivating a variety of relationships as more and more architects from Europe, Japan, and Argentina stopped by the Institute on their journey to the East Coast of the USA via New York (in no small part due to its convenient location between Grand Central Terminal and Penn Station/the Port Authority Bus Station). It thus became an incubator for architectural ideas, the commercialization of drawings and models, and new projects. For example, in addition to regular visits from Rossi, Rafael Moneo, among others, spent time at the Institute at Eisenman's invitation beginning in 1976, and Arata Isozaki was also a frequent guest. Curating content and building complex relationships with individuals who participated as lecturers, exhibitors, authors, teachers, etc., the Institute was not only a platform and showcase for its distinguished Fellows but also a gateway for high-profile and soon-to-be prominent international guests to the American architecture and art world.

Another step toward professionalizing the Institute's cultural work was the creation of the post of director of development in the spring of 1976 at the initiative of Armand Bartos, a long-time chairman of the Institute's Board of Trustees, which was to become the central point of contact for the development and implementation of fundraising, cultural sponsorships, and third-party

Performance in the 1970s," *Journal of Architectural Education* 61, no. 4 (2008), 43–51. Ultimately, Tschumi also turned his back on the Institute, not least because he felt less connected to the New York architecture scene than to the local art scene. His subsequent exhibitions, "Architectural Manifestos" and "Manhattan Transcripts" were both shown at other New York art venues, at Artists' Space and at P.S.1.

437 The journal *Semiotext(e)* had organized the conference "Schizo-Culture" at Columbia University in late 1975 to introduce post-1968 French radical philosophy to New York.

financing. From a pool of thirty applicants, Wolf ultimately selected Frederieke Taylor, who had previous experience in public relations in England and had also served on MoMA's Junior Council.[438] Upon her appointment as director of development, Taylor, who was initially given Visiting Fellow status in 1976–77 before being elected Fellow in 1977, filled one of the few full-time positions at the Institute paid for by the Bartos' Gottesman Foundation. Building on the American tradition of philanthropy and existing grant programs, her primary responsibilities were to improve outreach, solicit private donations, and submit applications to the major arts and cultural foundations at the state and federal levels. In addition, she was responsible for the Institute's relations with its trustees and its broader network. As a marketing tool, the Institute commissioned a new prospectus which was mailed out in the spring of 1976 and for the first time listed all seven of the Institute's educational offerings and its past research and design projects, various publications, exhibitions, and awards.[439] Obviously, the establishment of the new post restructured the Institute's entire operations, which had a positive effect on the financial planning of its leadership and made its business practices both more professional and more transparent. Once in office, one of Taylor's first official acts was to organize a fundraiser for all of the past donors of the "Architecture" series; in June 1976, she invited representatives of the business and finance communities to a reception at the Union League Club, where they heard a talk about the importance of adult education in architecture, which was then one of the Institute's flagships. As early as the second half of the 1970s, her duties also included establishing a circle of friends of the Institute as a philanthropic network, the newly launched Architects' Circle, cultivating relationships so that the organization and program of all the public events would be supported by a network of patrons in the future. As "friends," Taylor initially courted successful architectural firms that had already presented lectures as part of "Architecture," and thus were beholden to the Institute. Philip Johnson, who had officially sponsored *Oppositions* since 1974 but had otherwise remained in the background, was one of the first donors of the Evening Program. Thanks to Johnson's recommendations, Taylor was subsequently able to successively expand the Institute's Architects' Circle, organizing exclusive events for members in return for an annual contribution

438 Taylor's father, Pieter Sanders, a Dutch art collector who specialized in Piet Mondrian, became interested in architectural models and drawings at the time and bought a model of Eisenman's *House X*.

439 The Institute's 1976 prospectus advertised not only the names of both trustees and Fellows, but also Visiting Fellows at the time. It listed all the public and private grants awarded to the Institute since 1974–75 as a form of recognition and to solicit funding: NEA, NYSCA, C.B.S. Foundation Inc, The Duke Endowment, Graham Foundation, D.S. and R.H. Gottesman Foundation, J.M. Kaplan Fund, Inc, Edward John Noble Foundation, Alfred P. Sloan Foundation, and Van Ameringen Foundation, Inc.

of US$1500.[440] In addition, the Institute listed all its cultural sponsors in promotional materials, and they also received a subscription to *Oppositions*. The professionalization of cultural production in architecture, following in the footsteps of established institutions, was ultimately made possible by the fact that the Institute's operations were now sustained by a complex economic model which, while still being based on tuition fees and reliant on grants from various foundations, took an increasingly professionalized approach to cultural sponsorship as a third source of revenue, i.e., a reciprocal venture. This was quite common in the United States for 501 (c) not-for-profit organizations. Crucially, however, larger sums were available as a result of the federal funding policy during this period. It was explicitly the NEA and NEH under President Jimmy Carter in the wake of the bicentennial celebration in 1976 that enabled the Institute to exist and evolve as an agenda-setting cultural space, a social space for work, life, and experience, and a style-setting architectural firm. In its as-yet new dual role as an institution of education and culture, the Institute flourished by intensifying and differentiating its publishing activities. It shaped material and non-material architecture culture in the United States and beyond, informing postmodern thinking styles, transforming aesthetic norms and attitudes, and introducing new cultural ideas, values, and beliefs—a truly epistemic space that produced bodies of knowledge, power structures, and creative and intellectual routines through exchanges with other disciplines, the arts, and the humanities.

Transatlantic Dialogue

The Institute's focus on culture, i.e., the assertion and establishment of a complex web of meaning, and especially its focus on curation, i.e., in Eisenman's case specifically the compilation of lists and assemblage of groups, was again evident in the summer of 1976, when it was officially invited by the U.S. Committee on International Exhibitions of the American Federation of Arts to curate the American contribution to the XXXVIII Biennale di Venezia. For the first time, the famed art biennial, traditionally framed as a competition between nations, included an architecture section called "Europe/America: architettura urbana, alternative suburbane" in the Magazzini del Sale under the curatorial direction of Vittorio Gregotti in addition to the national pavilions.[441] The architecture focus at the 1976 Venice Biennale was thus the dichotomous confrontation and exchange between European and American positions on the relation of architecture to the city, which was played out as a conflict of interests and generations between representatives

440 The Institute's Architects' Circle in fiscal year 1977–78 involved: Edward L. Barnes; Davis, Brody & Associates; Conklin and Rossant; Ulrich Franzen; Philip Johnson & John Burgee; Richard Meier; Mitchell/Giurgola; I.M. Pei; Paul Rudolph; Skidmore, Owings & Merrill; and Robert Stern.

441 "Europe–America. Historical Centre–Suburbia," in *Environment, Participation, Cultural Structures: General Catalogue*. Second Volume, ed. La Biennale di Venezia (Venice: Alfieri Edizioni d'arte, 1976), 235–264.

of modernism (members of Team X such as Aldo van Eyck, Giancarlo De Carlo, Alison and Peter Smithson) and postmodernism (Manfredo Tafuri, Aldo Rossi, and several American architects invited by the Institute).[442] For the curation of the American projects, the Institute received funding in advance from the NEA and the Graham Foundation; in addition, various other American universities (Cooper Union, Columbia University, UCLA, Pratt Institute) participated in the exhibition—both financially and through student labor.[443] As Institute director, Eisenman made the curation of the American contribution his own project and invited Stern to serve as co-curator. Under the title "Suburban Alternatives. Eleven American Projects," the two assembled several architects and firms whose designs for suburban homes were meant to represent a genuinely American architecture, to the exclusion of others.[444] At the Institute, the selection of participants, whose contributions and exhibits were coordinated by Taylor, i.e., a kind of public relations exercise, became a veritable power play, sometimes involving violent reactions, angry phone calls, and bruised egos. In his opening speech to the panel discussion that he held as curator in Venice on August 1, 1976, Eisenman declared that for him the fundamental problem was one "of building urbanity, of suburbia, or society," only to assert immediately afterward that it was the Americans who were now bringing in new ideas because of their experience with suburbanization.[445] By, above all, resisting the resumption of the debate on modernity, he was deliberately preparing the way for postmodernism. Adopting the role of architecture historian, he acknowledged "the change of the modern movement," before summarily

442 Léa-Catherine Szacka, "Debates on Display at the 1976 Venice Biennale," in *Place and Displacement: Exhibiting Architecture*, eds. Thordis Arrhenius, Mari Lending, Wallis Miller, Jérémie Michael McGowan (Zurich: Lars Müller, 2014), 97–112. It would, however, be short-sighted to limit the debate to the generational conflict alone. While architecture was given its own subsection at the 1976 Venice Biennale with the emergence of postmodernism, an independent architecture biennale was staged for the first time in 1980 under the curatorial direction of Paolo Portoghesi; see Léa-Catherine Szacka, *Exhibiting the Postmodern. The 1980s Venice Biennale* (Venice: Marsilio, 2016).

443 During the bicentennial celebration of the independence of the United States, the funding volume of the federal foundations in the art and the humanities was significantly increased, explicitly with regard to art education (and thus also architecture education) programs; see Donna Binkiewicz, *Federalizing the Muse: United States Policy and the National Endowment for the Arts, 1965–1980* (Chapel Hill: University of North Carolina, 2004), 184ff.

444 At the 1976 Venice Biennale, "Suburban Alternatives: Eleven American Projects" featured: Raimund Abraham, Emilio Ambasz, Peter Eisenman, John Hejduk, Craig Hodgetts, Richard Meier, Charles Moore, Cesar Pelli, Robert Stern, Stanley Tigerman, Denise Scott Brown with Robert Venturi (of the New York Five, Michael Graves, Charles Gwathmey & Robert Siegel were not involved; of the Fellows, Diana Agrest and Mario Gandelsonas). In 1977, the projects of four architects who had exhibited in Venice were presented in New York in the exhibition "Abraham / Eisenman / Hejduk / Rossi" at Cooper Union.

445 IAUS, "Transcript of a Conference at Venice Biennale," August 1, 1976. Source: CCA Montréal, IAUS fonds: ARCH153618. The participants were: Carlo Aymonino, Peter Eisenman, James Stirling, Raimund Abraham, Aldo Rossi, John Hejduk, Denise Scott Brown, Hans Hollein, Robert Stern, Peter Smithson, Emilio Ambasz, Giancarlo De Carlo, Oriol Bohigas, Aldo Van Eyck, Manfredo Tafuri, and Alvaro Siza.

dismissing functionalism in a multitude of respects: political, economic, sociological, and aesthetic. Eisenman concluded his speech by quoting Michel Foucault and referring to a new episteme of "post-functionalist sensibility," which he nevertheless pointedly set apart from a "positivist sensibility." The two American curators, a real dream team, brought practical ideas to the exhibition rather than theoretical ones, and used the biennial to their own ends: Eisenman showcased designs of *House X*, and Stern presented his project for a subway suburb.[446] In sum, the American contribution to this twentieth-century debate on urbanism/suburbanism as a way of life thus once again drew attention to the ways in which author architecture and the real estate industry were intertwined. With its plea for urban development largely based on automobility, home ownership, and mortgages, the exhibition curated at the Institute dovetailed effortlessly with the agenda of architectural postmodernism, which in Venice was pitted against the representatives of a now-defunct European structuralist approach to modernist urbanization.[447] If participation in the exhibition meant media attention on the very big international stage for the American architects who took part, especially for Eisenman, the Institute was also able to make a name for itself and gain recognition as a new cultural actor and network on an international level.[448]

446 Peter Eisenman, "Five Easy Pieces: Dialectical Fragments Toward the Decomposition and Reintegration of Suburbia," in *Environment, Participation, Cultural Structures: General Catalogue*. Second Volume, ed. La Biennale di Venezia (Venice: Alfieri Edizioni d'arte, 1976), 256; see also Fredric Jameson, "Modernity versus Postmodernity in Peter Eisenman," in *Cities of Artificial Excavation. The Work of Peter Eisenman, 1978–1988*, ed. Jean François Bédard (New York, Rizzoli International, 1994), 27–37. Stern's design indirectly criticized the Institute's only realized building project by contrasting the low-rise housing complex in Brownsville, Brooklyn, commissioned by the Urban Development Corporation, financed in public-private partnership, and completed in the summer of 1976 after three years of construction, with a counter model based on small-scale, private ownership and located in the immediate neighborhood.

447 At the 1976 Venice Biennale, the now transatlantic dispute over the interpretation of architecture—polarizing between generations, styles, continents, and politics—erupted on August 1, 1976, during a panel between Aldo van Eyck and Manfredo Tafuri. Eisenman, commenting on Oriol Bohigas' *Oppositions* essay on "Aldo van Eyck and the New Amsterdam School," described the situation with the following words: "However, unlike the usual transition between generations, this was not just the passing of an age or the changing of a style. Instead, it revealed the existence of a profound schism between the architects of the fifties and sixties and those of the seventies. This split is marked by a galaxy of complex liaisons and alliances, as well as by the conflicts represented by the architects of the Biennale—United States vs. Europe, Team Ten vs. Tendenza, near Left vs. far Left, populism vs. elitism, realism vs. formalism—all of which are too fraught with subtle nuances to be easily condensed in this context." See Peter Eisenman, "Commentary," *Oppositions* 9 (Summer 1977), 19–20.

448 It emerged from the assembly of participants that by 1976 the demarcation between the two camps of the "Whites" and the "Grays" had become obsolete. In his essay "Les cendres de Jefferson," written in the summer of 1976, Tafuri pointed out that the two camps were ultimately not that different; see Tafuri, 1976. Reinhold Martin took up Tafuri's criticism of the indistinguishability of the architectural positions in his history of postmodernism: "By 1976, the date of Tafuri's 'Ashes of Jefferson' text, the debate had played itself out with both sides fully identified with postmodernism." See Martin, 2010, 30.

On the other side of the Atlantic, the Institute benefitted from the contacts it had made, emerging in the second half of the 1970s as an important hub and nexus for a transatlantic dialogue, contributing its share to the globalizing postmodern European and North American cultural sphere. This was largely due to Eisenman's esteem for both intellectual and creative positions from Europe and his efforts not only to invite criticism but to make the critics great. After the 1976 Venice Biennale, Eisenman continued to expand his contacts in Italy, especially with the architects, historians, and theorists teaching at the Istituto Universitario di Architettura di Venezia (IUAV). It was, as so often with Eisenman, a personal, social network superimposed over the institutional one that was cultivated, presented, and exhibited here. After Rossi's and Scolari's first appearances at the Institute in the wake of the "European New Wave" earlier that year, Eisenman planned to engage them as faculty in the newly created "Design and Study Options" (which ultimately never came to fruition), as he did with Tafuri, who headed the IUAV in 1976, positioning it at the intersection of criticism and historiography and thus charting a course for action, both for himself and for the discipline.[449] Even if this plan to integrate some of the most prominent European professors of architecture and architecture history and persuade them to commit to the Institute's larger cultural project failed, a "New York—Venice" axis was invoked, at least from the Institute's vantage point, which was reflected in the cultural production manifested on the pages of *Oppositions*. This axis was later translated into further exhibitions by Rossi and Scolari and was then reflected in publications (both planned and realized) by Rossi, Tafuri, and others. This was somewhat surprising since the Venice School was committed to highlighting the conditionality and thus fragility of historical certainties, rather than simply reproducing architectural ideologies, as was often the case at the Institute.

In terms of both architecture and media history, it is noteworthy that as the Institute's relations with Europe intensified in the summer of 1976, the European architecture press also became more critical of the transformations in American architecture culture. This was most evident in the French *L'Architecture d'Aujourd'hui*, edited by Bernard Huet, whose August/September 1976 issue, a special issue on New York's building culture called "New York in White and Gray" was devised as a commentary on the unfolding urban crisis of the American metropolis and resulting developments in the profession and discipline of architecture.[450] With contributions from Huet as editor-in-chief and Brian Brace Taylor as correspondent editor, who was intimately

449 For Tafuri, writing mainly as a historian, historiographical research represented a form of resistance after the project of modernity had become fully enmeshed in the capitalist system; see Leach, 2014.

450 *L'Architecture d'Aujourd'hui*, no. 186 (August/September 1976): "New York in White and Gray."

familiar with the American architecture world and was responsible for this issue, "New York in White and Gray" addressed two parallel, reinforcing developments, later exemplified by the image of the demolition of the Pruitt-Igoe housing complex: the emergence of an architectural postmodernism versus the end of public housing. The issue exclusively presented individual positions from the Institute's inner circle: Agrest (with Alessandra Latour) and Frampton each contributed an article, documenting a housing competition announced by the Urban Development Corporation for Roosevelt Island in 1975 while also discussing the subsequent demise of the housing authority and its consequences for the Institute's only building project, published here as a sort of post-mortem. In addition, *L'Architecture d'Aujourd'hui* published an article by Stern (but not by Eisenman) which was presented as the first theory of postmodern architecture. The thematic issue also included articles by Taylor and Tafuri, both of whom provided detailed descriptions and reflections based on participant observations, offering key texts that, from a historiographical, theoretical, and—ultimately—narrative perspective, provided initial insights into what an institutional critique of architecture might look like, i.e., how the work of architecture institutions in general and the Institute, in particular, might be addressed, and while not necessarily representing a disruption, nevertheless addressed questions of injustice, or even oppression.

Taylor's article "Self Service Skyline" stood out as the first text to explicitly address the Institute as such, and offered an outside perspective, emphasizing that from its inception the Fellowship had been much more culture-focused by design, as compared to two other, quintessentially socio-politically defined New York institutions in architecture and planning, which were founded around the same time and operated at municipal and state level, and with whom the Institute had successfully collaborated on research and design projects in its early days: the Urban Design Group (UDG) and the Urban Development Corporation (UDC), i.e., the subdivision of the City Planning Commission on the one hand, and the state housing authority on the other.[451] While the title unmistakably alluded to the self-centeredness of New York's architectural production, Taylor criticized the debate between the "Whites" and the "Grays" for its aesthetics, which he saw as being detached from reality, saying that its lack of congruence was once elevated to the new design paradigm, resulting in architecture as a system becoming self-constituting, self-referential, and self-reproducing. Taylor's verdict was devastating, for he attested that at this point since the UDG and the UDC had ceased to exist, at least in their original form, all three players would have held only a marginal position in relation to the American architecture and construction world. The Institute, whose funding, collaborations, programming, and curation he discussed in detail, did not, in his view,

451 Brian Brace Taylor, "Self Service Skyline," *L'Architecture d'Aujourd'hui*, no. 186 (August/September 1976), 42–46.

represent any position more radical than those of the established schools of architecture, although he did recognize a number of critical approaches. For him, the Institute was distinguished from the established institutions by its function and role within an international network: "In point of fact, the IAUS's principal attribute has been its capacity to bring together for limited periods the representatives of a non-American intelligentsia and to offer them an open forum for ideological debate. This is something that neither MoMA nor a university faculty was able to provide." Taylor acknowledged that the economic crisis in the United States meant that priorities had to be set differently and, as a result, architecture education and cultural production at the Institute were paramount. In the Marxist tradition, he concluded that the practices of architects operating within this closed system of close relationships had done little to change modes of production or forms of labor; however, he did not analyze the immaterialization and precarization of the Fellow's activities as a new reality in greater detail. Written for a French readership, the article noted that no new cultural values had emerged, nor had the role of architects been redefined in relation to mass culture. For Taylor, those involved accepted that the gap between rich and poor would continue to grow, or even that architecture helped to drive this socio-economic process.

The special issue also included Tafuri's article "Les cendres de Jefferson," in which the Italian historian and critic expressed himself (initially in French) similarly to Taylor on New York's architecture circles.[452] Writing about contemporary culture in the Marxist tradition, Tafuri, having visited and gained first-hand knowledge of the Institute that same year, for the first time focused on the new "organizational structures of intellectual work" as a historically significant event. While his text mostly focused on the different positions of selected American architects, Tafuri opened by criticizing the snobbery and self-referentiality of the architecture scene there.[453] At the outset, he briefly remarked on the scarcity of jobs for architects on the US East Coast in general, after the UDC, one of the largest employers, had been dissolved the year before. Regarding the culturalization of architecture, he expressed outrage that "once high levels of comprehensive integration have been achieved, it [has become] possible to maintain well-defined cultural spaces, entrusted with the task of pleasurably entertaining a highly selected public," and that "in such a way, new circuits of production and use do come to

452 Tafuri, 1976; Tafuri, 1987. It should be noted that the collection of essays was initially scheduled to appear in the Oppositions Books series.

453 Again, Tafuri came up with the New York Five: Eisenman, Graves, Gwathmey, Hejduk, and Meier, as distinct from Paul Rudolph and I.M. Pei, Philip Johnson, Robert Venturi, John Portman, and Kevin Roche; but he also cited the representatives of the other camp: Robert Stern and John Hagmann, Diana Agrest, Mario Gandelsonas, Rodolfo Machado and Jorge Silvetti, and Emilio Ambasz. Apart from Agrest, these architects were again all male.

be created."[454] In keeping with the theories of the time, Tafuri argued from the perspective of psychoanalysis and literature, architecture and sociology, that the code of behavior in New York was dominated by "vanity" and "comedy," and that "formalism" and "a systems of solitudes" were among the basic requirements for sheer survival. Although he did not explicitly mention the Institute in this context, and elsewhere in the article merely referred to it as Eisenman's current group, after the demise of other groups founded by Eisenman, the Five Architects, and the Conference of Architects for the Study of the Environment (CASE). His main critique of the Institute as a cultural space, being an "instrument of self-promotion and identification," however, was more profound and addressed the Institute's recent cultural efforts, particularly its public events: the Evening Program, the Exhibition Program, and especially the "New Wave" series. He drew on French cinema, noting that "architecture [came] to be exhibited in its own *cinémas d'essai*," i.e., art-house cinemas that were run independently of the industry and existed only because they received financial support from the government. Tafuri thus repeated the criticism of architecture's assertion of autonomy that he had voiced earlier in his 1974 *Oppositions* essay "L'Architecture dans le Boudoir," in which he had denounced the "formalism" of the New York Five and compared it to Marquis de Sade's method of depicting sexual transgression and sexual gratification, in that "maximum freedom springs forth from maximum terror;" a criticism he revisited and amplified two years later in "Les cendres de Jefferson" with the phrase "the formal terrorism of Eisenman."[455] He was indirectly referring to the fact that contemporary architects like Eisenman discussed their supposedly avant-garde projects, which became entangled in endless language games, behind closed doors, preferring to fete themselves rather than accept their social responsibility. This time, however, Tafuri's attack was no longer directed only at building, but now also referred to cultural production in general, i.e., talking, exhibiting, and writing about architectural practice.

Shortly thereafter, in her article "Fiddling While New York Burns" in the October 1976 issue of the British *Architectural Design*, architect Camilla Ween echoed the serious claims made in *L'Architecture d'Aujourd'hui* and elaborated on the criticism for the publication's English-speaking readership.[456] In particular, Ween found fault with what she saw as the complacent actions of the New York architecture scene, which she attempted to capture with the metaphor "the architectural avant-garde fiddles with increased intensity," and in the same vein, the work of "cultural institutions" like the Institute—she was actually the first to use the term, which she used to include professional groups like the UDG and

454 Tafuri, 1987, 293.

455 Tafuri, 1987, 300; see Ockman, 1995, 71, footnote 71.

456 Camilla Ween, "Fiddling While New York Burns," *Architectural Design* 46, no. 10 (October 1976), 630.

educational institutions like the Cooper Union as well—for creating an incestuous environment. With the title of her article, Ween was referring not only to the designs of a new generation of "cult architects," thus anticipating a critique of stardom or star architecture, but also and primarily to the architecture, education, and culture produced by institutions in light of the grave socio-economic situation in New York, which in 1975–76 was verging on bankruptcy, while in some neighborhoods, explicitly in the Bronx, abandoned houses burned down. Ween's perspective from across the Atlantic was therefore instructive, in conceptual and methodological terms, in that it classified and situated cultural change in architecture at the transition to neoliberalism, deregulation, and privatization, without having to name these abstract processes. For by adopting the ancient idiom "fiddling while Rome burns," she was referring to the burning of Rome in 64 B.C., and thus to Emperor Nero, who was not only accused of arson but was also alleged to have sought even greater self-aggrandizement by rebuilding the city. This was a more than harsh criticism of the elitist habitus and self-important actions of Eisenman, Stern & co. Koolhaas' postmodern, humorous take on architecture and the city, "once imaginary, continually hysterical and finally delirious," was the only approach she regarded favorably, wishing he had infected the Institute with it. Like Taylor and Tafuri in *L'Architecture d'Aujourd'hui*, Ween explained historical developments in architecture culture not only from their context, which was not self-evident but on the basis of their political and economic conditions and constraints, while also attributing a certain share of responsibility for the situation in neglected neighborhoods to the architects. For her, the comparison to Nero was a simple yet striking illustration of how the architecture world could be blamed for having closed itself off from reality, and how its disengagement had only widened the "gulf between the ruling classes and the newly urbanized poor;" an accusation that was repeatedly leveled toward the Institute.

Clearly, a large part of the architecture community on the East Coast of the USA, not just at the Institute, had turned to showcasing knowledge, defending beauty, and feting itself in a hedonistic vein, rather than continue to examine sociopolitical phenomena and serve as a corrective. The informed, even eloquent criticism in *L'Architecture d'Aujourd'hui* and *Architectural Design*, two leading European architecture journals, highlighted the extent to which voices in the European architecture community were dissociating themselves from developments in New York, explicitly from the paradigm shift toward a postmodern rhetoric and aesthetics, mediated through cultural production, which at the Institute was increasingly characterized by an internationally oriented entrepreneurial attitude and philosophy.[457] For the Fellows, on the other hand, it became

457 In the German-speaking world, the paradigm shift in architecture from a social project to a cultural project was neglected and hardly ever (if at all) addressed in the more theoretical journals: in Switzerland, *Archithese* reported favorably on the specific metropolis of architecture in 1976 with three themed issues on New York, see *Archithese*, no. 17, 18, 20; in Germany, *Arch+* took a

clear that, with their creative and intellectual work, the architects and academics assembled there, through the interplay of a wide variety of production contexts, were never truly autonomous in the face of an increasingly globalizing architecture culture, but also made themselves dependent on social power hubs and economic, political, and cultural decision-makers, i.e., the power elite. One key development in New York and especially at the Institute since the mid-1970s was that architects and academics, while consciously arguing against and thus, at least rhetorically and aesthetically, attempting to detach themselves from the structural conditions of architectural production, were nevertheless creating new dependencies on the building industry and the real estate market. However, the budgeting of education and culture as the predominant forms of architecture mediation, i.e., the reproduction of architecture in neoliberal times—of which the Institute is a paradigmatic example—made it clear, via the economization of architecture education on the one hand and increasingly privately funded cultural production on the other, that the lecture series, exhibitions, and publications were more than just a mechanism in the market of the symbolic economy, but rather their primary aim was to open up new markets.

Attention Economy

The extent to which the Institute's cultural events which, despite being a commercial offering, were partly a public forum and partly an elite salon held behind closed doors, not only provided a cultural and social frame of reference but also had economic value, was evident from the fact that the intellectual and above all symbolic capital of the assembled elite was put to profitable use. The evening lectures, like the exhibition openings, were a veritable who's who of seeing and being seen. Under the title "Forum," the publication of the latest issue of *Oppositions* was now increasingly celebrated at the Institute with special release events that might include thematic lectures or panel discussions and usually culminated in well-attended cocktail parties. In late October 1976, the publication of *Oppositions* 5, dubbed the "Italian Issue" at the Institute, which featured a review of Aldo Rossi's San Cataldo Cemetery in Modena by Rafael Moneo and a concluding critique of the New York Five by Tafuri, the first issue published by MIT Press, was duly celebrated at the Institute with a debate on Rossi's architectural drawings. Yet ultimately, the Latin name "Forum" with its allusion to an outdoor public space in Rome, or any other city (Latin: civitas), used as a

look across the Atlantic with a special issue on "Der Tod der Architektur" in 1978, but only published a translation of Tafuri's avant-garde critique of the New York Five (not of the cultural spaces of the architecture scene) and a Marxist critique of urban developments in New York lamenting the death of architecture; see Manfredo Tafuri, "Die Kritik der Architektursprache und die Sprache der Architekturkritik," trans. Michael Haase, Marc Fester, Nicolaus Kuhnert, *Arch+*, no. 37 (April 1978): "Der Tod der Architektur," 4–16 (translation: Michael Haase, Marc Fester, Nicolaus Kuhnert); on the Americanization of cities and the turn in domestic and social policy underlying the financial crisis, see also Francis F. Privens and Richard A. Cloward, "Die Krise der Stadt am Beispiel New Yorks," *Arch+*, no. 38 (May 1978): "'Amerikanisierung' der Städte?" 24–27.

marketplace for the sale of goods, was at least ambiguous, if not misleading. For while the "Forum" served an economy, it was in fact an exclusive social occasion reserved for Fellows and Visiting Fellows, and beyond that for the journal's donors and other invited guests from the world of building and construction, as a way of expressing gratitude for their financial and philanthropic support. Here, the new civility at the Institute overrode the supposed radicalism of its architects and academics. The events were photographed by Dorothy Alexander, who had previously exhibited at the Institute and rose to become the Institute's in-house photographer in the years that followed.[458] While in the beginning the panels of the release events, and thus the curation and intellectual capital, were shown, later often two or three people from the inner and outer circles of the Institute are photographed—deep in conversation, sometimes even with a cocktail glass in hand. These photos, which depicted a largely male-dominated social class in which women were merely allowed to do the legwork in subordinate roles, were also published, captioned by name(s), in each upcoming issue of *Oppositions*. The Institute thus absolutely reveled in its own celebrations, giving *Oppositions* the character of a popular illustrated society magazine and the Institute a prestigious reputation and, arguably, a bit of glamour. The diagnosis formulated by sociologist Richard Sennett at about the same time in his 1976 publication *The Fall of Public Man*, with the subheading *On the Social Psychology of Capitalism*, with regard to developments in American society and economy at large, can thus be applied to the New York architecture community as a specific, yet representative segment of metropolitan society: an "erosion of public life in the cities" and the reduction of all actions in public to a form of "playacting," i.e., a performance or spectacle that is primarily concerned with acclaim and prestige.[459] This shift in emphasis at the Institute, which was certainly contingent on socio-political and economic constraints, but nonetheless readily carried out, from welfare-oriented urban renewal and housing projects and a humanistic educational ideal to an increasingly publicity-oriented cultural institution, competing on the market of adult education, art, and exhibitions, highlighted the extent to which attention had now become a commodity and currency in architecture culture as well, as had the organization of new job assignments, scopes of work, and divisions of labor.[460] With the cultural production, indeed profit orientation of all formats,

458 Dorothy Alexander's photographs (negatives) can be found in the Beinecke Library of Yale University.

459 Sennett, 1976.

460 From as early as the mid-1970s, the Institute's stance and operations can be understood as a paradigmatic example of an attention economy, as described by philosopher and architecture theorist Georg Franck in close reference to Marx's theory of capital (the structure of both publications shows extreme similarities) and Bourdieu's concept of capital, which he developed with regard to the deconstructivist architecture of the 1980s; see Franck, 1998 and Franck, 2000; see also Karl Marx, *Das Kapital. Kritik der politischen* Ökonomie (London: Faber & Faber, [1867] 2007); and Bourdieu, 1986.

i.e., the Evening Program and the Exhibition Program, and now the "Forum," architecture culture, advertising, outreach, and public relations at the Institute had become indistinguishable with regard to the interplay of all forms of capital; what was featured was who or what was currently in fashion.

In 1976–77, in the context of "Architecture," the Institute focused on further popularizing and semanticizing, historicizing, and aestheticizing architecture in New York. For "Architecture 5," it secured funding from the NEA, the J.M. Kaplan Fund, and the Mary Duke Biddle Foundation. Two respected journalists who wrote about society, art, and culture and who were known to their New York audiences from the local cultural press were secured as presenters: Brendan Gill (editor at *The New Yorker*) for the lecture course "The Preoccupations of Critics and Architects 1976," and Grace Glueck (*The Village Voice*) for "Design without Architects." The lecturers in these two courses were almost exclusively big names from the American architecture and design communities, to whom this kind of publicity brought further attention. Gill's course featured Philip Johnson, William Conklin, Roy Allen of SOM, Samuel Brody and Lewis Davis, Ulrich Franzen, Charles Gwathmey, and Paul Rudolph, almost all members of the Architects' Circle, plus Arthur Drexler and Charles Moore; Glueck's course featured George Nelson, Ward Bennett, Mario Salvadori, Milton Glaser, Massimo Vignelli, Ivan Chermayeff, and two women, Candy Pratts and Mary Joan Glynn. Stern, on the other hand, who was working on a book series as a Visiting Fellow in 1976–77, made it his mission to introduce postmodern architecture to the New York public as a hitherto barely theorized or historicized style with his lecture course on "New Modernism/Postmodernism." He promoted a new historicism, classicism, and eclecticism with various representatives and apologists of architectural postmodernism as lecturers, e.g., Alan Greenberg, Hugh Hardy, and Rodolfo Machado. Eisenman's renewed appearance on Stern's course testified once again to their shared interests and strategies on behalf of a truly postmodern culturalization of architecture. In his lecture, "Post-Functionalism," Eisenman presented his most recent outlines of a general theory of architecture, which he then published as an editorial to *Oppositions* 6, where he used a post-humanist approach to lay the foundation for his deconstructivist sculptural practice.[461]

Along with the blossoming of postmodernism as a stylistic form and practice, a postmodernization of architecture culture became apparent in all fields of the Institute's work, in addition to a pluralization of ideas and knowledge, encompassing an attention economy, i.e., affective, attention-seeking performances, and speech acts that commodified the cultural, the social, and the symbolic: in the overall pedagogy and concrete didactics of its teaching, the product range of its

461 Peter Eisenman, "Post-Functionalism," *Oppositions* 6 (Fall 1976), n.p. In modernism, as it breaks with the past, Eisenman recognized a "non-humanist attitude toward the relationship of an individual to his physical environment."

cultural production, the content of its publications, first and foremost *Oppositions* and *October*, as well as in every other format conceived, produced, or published there from 1977 onward. This includes the tabloid architecture newspaper *Skyline* (from 1978), the IAUS Exhibition Catalogues (1978), and the Oppositions Books series (1981), (see chapter four). An operationalization of history and theory, their division in postmodern terms, that served to underscore the assertion of the autonomy of architecture, if not its criticality, was evident in the editorials of *Oppositions* 4 through 7. These were written individually by the editors between 1976 and 1977 and published as personal manifestos, a development that was critically reflected in Tafuri's contributions. Since *Oppositions* had been elevated to the status of an academic publication by MIT Press, according to Eisenman's "Director's Report" of the summer of 1976, another key priority during fiscal year 1976–77 was to develop the exhibitions into a regular and vital part of the program in their own right.[462] Discursively and institutionally, this paradigm shift meant that heroes of modernism who had fallen into obscurity were shown alongside representatives of postmodern sensibility who were now surging to the fore. With a larger budget, a more professional organization, and a promise to appeal to the public, the Institute's exhibitions sought to attract funding from the major art and humanities foundations and participate in the global art establishment and business. After the American art scenes had been generating numerous alternative art practices (performance art, conceptual art, minimal art, etc.) and with them new cultural subjects since the 1960s, the Institute's exhibiting and curating operations were to benefit from the vibrant art community and booming art market in New York. The fact that the art world, which had emerged and survived not least thanks to federal funding, was also subject to drastic changes in the mid-1970s in terms of institutional structures became evident, among other things, in the increasing marketization of the creative *dispositif* and the progressive replacement of the "art community" by a "financial community."[463] While the format of the spectacular blockbuster exhibition was consolidated in the summer of 1976 with the "Treasures of Tutankhamen" exhibition and its lavish display at the Met, the triumph of institutional critique was simultaneously reflected in the establishment of the P.S.1 Contemporary Art Center in a vacant school building in Long Island City, Queens, which had been purchased by Alanna Heiss, the founder of the Institute for Art and Urban Resources.[464] In the context of the culturalization of New York,

462 Peter Eisenman, "Director's Report," June 10, 1976. Source: CCA Montréal, IAUS fonds: A.1-2.

463 Reckwitz, 2012.

464 For "Rooms" (June 9 to June 26, 1976), the inaugural exhibition at P.S.1 curated by the Institute for Art and Urban Resources, a large group of artists, among them Gordon Matta-Clark, was invited to work on the building in an act of institutional critique; see Hal Foster, "1976," in *Art Since 1900. Modernism, Antimodernism, Postmodernism*, eds. Hal Foster, Rosalind Krauss, Yve-Alain Bois, Benjamin H. D. Buchloh, David Joselit (London: Thames & Hudson, 2006), 620–623; see also *Art Forum* (October 1976).

the institutionalization of the local alternative art scene was further aided by the fact that until the mid-1970s, the budgets for art spaces were largely funded by the NEA, primarily from the "Visual Art Program" funds under the direction of Brian O'Doherty. Following "Architecture," the Institute's Exhibition Program benefited from this development. More crucially, however, the Institute was implicated when architectural projects, drawings, and models were exhibited there, and soon after shown, sold, and promoted in commercial galleries for the first time, first at the Leo Castelli Gallery (beginning in 1977), which at the time was beginning to specialize in architecture, and soon after at the Max Protetch Gallery (1978).[465] The Institute, both directly and indirectly, became involved in the business of art, which had a significant impact on architecture production and culture.

Ideas as Models

This newly launched Exhibition Program began with "Idea as Model" (December 16, 1976, to January 14, 1977), a much larger group exhibition than "Good-bye Five" and other earlier DIY exhibitions that MacNair had organized and curated almost single-handedly.[466] At Eisenman's suggestion, MacNair had written to selected architects, many of whom were already associated with the Institute in one way or another, in the summer of 1976, specifically asking them to contribute scale models this time rather than drawings in order to showcase their communicative qualities and conceptual diversity. Initially titled "Ideas as Models" (plural), the aim of this exhibition was "to present ideas and problems of architecture as investigated in model form."[467] Eisenman wanted the generic scale model to be understood not merely as an instrument of design, but as a medium of knowledge at the intersection of architecture and conceptual art. While he had already engaged with this idea in his theoretical texts and house designs, the exhibition now officially bore the subheading "Investigation about Architecture." In this way, the Institute presented itself to the professional world as an innovative exhibition space while at the same time distancing itself again from the most recent developments at MoMA, where the major exhibition of architectural drawings from the Ecole des Beaux Arts, curated by Drexler, had been on view the year before. When "Idea As Model" (singular), finally opened at the end of the year, after a three-month delay, the Institute displayed a total of twenty-four models of very different make and quality, architecture

465 Kauffman, 2018, "The Changing Nature of Architectural Drawings," 134–221 & "Normalized Practice: Architecture in the Galleries," 222–272.

466 Martin Hartung, "Idea as Model," in *Exhibit A: Exhibitions That Transformed Architecture 1948–2000*, ed. in Eeva-Liisa Pelkonen (London: Phaidon, 2018), 196–201; preparations for the exhibition "Idea as Model" began as early as July 1976, and it was originally scheduled to open on September 9, 1976; see also Stefaan Vervoort, "Scale Models and Postmodernism: Revisiting Idea as Model (1976–81)," *Architectural Theory Review* 24, no. 3 (2020), 224–240.

467 Andrew MacNair, letter to Robert Stern, July 28, 1976. Source: Yale University: Robert A.M. Stern Archive; see Richard Pommer, "The Idea of 'Idea as Model'," in *Idea as Model: 22 Architects 1976/1980*, Catalogue 3, ed. IAUS (New York: Rizzoli International, 1981), 3–9, here 3.

and art, mostly by New York architects (featuring representatives of both camps, the "Whites" as well as the "Grays," with Michael Graves, Charles Gwathmey, John Hejduk, Charles Moore, Jaquelin Robertson, and Robert Stern) and once again by architects from Europe (O.M. Ungers, Massimo Scolari, and Leon Krier, as well as Rafael Moneo and Stuart Wrede, who were Visiting Fellows), and by all practicing Fellows (in addition to Eisenman, these were Diana Agrest and Mario Gandelsonas, as well as William Ellis). In keeping with the economy and aesthetics of the time, the exhibition poster, hand-crafted from a design by Graves, now appeared in a strikingly postmodern design: a three-dimensional collage of materials (wood, cardboard, and paper) that Graves' students at Princeton had produced at great expense, applied to the very template designed by Vignelli (silkscreen, chalk).[468] In an act of both creativity and commercialization, the handmade posters, produced in a small series of 100 copies and hand-signed by Graves, were sold during the opening and in the exhibition afterward, in part to cross-fund the exhibition and cover the costs of labor and organization.

The most distinctive aspect of the group exhibition was that it showed scale models as both an independent intellectual achievement and an artistic work. Conversely, it bore witness to the derealization of postmodernism, blurring the boundaries between reality and fiction. For the exhibits presented very different approaches, comparable to model making in the daily routine of architecture firms or in architecture studies, where very different techniques are used for concept and design models, models for testing materials or construction, and presentation and exhibition models. Thus, in the context of "Idea as Model," some models referred to a single idea, while others told more of a story. Ultimately, however, many of the models on display did not meet the criteria established beforehand, because they did not generate new ideas that led to design decisions.[469] Even the Eisenman model of *House II* exhibited here, which had been made especially for the exhibition by David Buege, one of his interns, did not meet the requirements, since the design idea underlying *House II* had already been formulated in working drawings and realized in the *Falk House* (1970). Contrary to what was postulated, the model made of colored Plexiglass plates intersecting at right angles was not strictly speaking a "study of a hypothesis, a problem, or an idea of architecture," since the formal transformations of the architectural structure were merely illustrated retrospectively for the purpose of the exhibition, and not fundamentally explored.[470] In other words, it was ultimately a presenta-

468 The poster for "Idea as Model" is neither to be found in CCA's IAUS fond, nor at the Vignelli Center for Design Studies. However, a copy is archived at the Museum für Gestaltung at ZHdK in Zurich (inventory no: 3DK-0003).

469 Hartung, 2018, 198f.

470 Buege subsequently reproduced the plexiglass model of *House II* in an edition of three identical copies; in 1980, Eisenman sold one of the models to the Deutsches Architekturmuseum (DAM) in Frankfurt for DM 3,000 (inventory no: DAM: 066-001-052).

tion model, not a working model. Nevertheless, as curator of the exhibition and principal author of the catalogue that followed years later, the model of *House II* portrayed Eisenman as the most intellectually and architecturally consistent of the architects on display, demonstrating the conceptual finesse and artistic willpower of his house designs by emphasizing the steps taken rather than the results achieved by the formal transformations.[471] Although the official catalogue originally planned for the exhibition could not be realized on schedule, "Idea as Model" was revisited by Paul Goldberger in the *New York Times*, where he reviewed the Institute's second group exhibition under the heading "How Architects Develop Ideas."[472] Here, Goldberger criticized the curation in particular, saying that the models on display had nothing to do with each other and that the selection did not follow any particular concept. Internally, however, "Idea as Model" was considered a success due to the enormous number of visitors, up to 150 people per day by their own account: a powerful argument, and one that was communicated in a letter of thanks from MacNair to the participating architects after the exhibition.

In the context of the Institute's history, and even more so from the perspective of sociology and architecture culture, the postmodern obsession with power exhibited in the scale model of "Idea as Model" made it a significant historical event, one that was symptomatic of the assertion of autonomy communicated by Eisenman in his 1976 editorial for *Oppositions* "Post-Functionalism." This represented a departure from earlier social and political commitments.[473] At the same time, it marked the conclusion of an early phase in terms of the market for contemporary architectural objects, to which the 1975 MoMA exhibition "Architectural Studies and Projects," especially the works of Barbara Jakobson and Emilio Ambasz, contributed.[474] The Institute's exhibitions, starting with "Idea as Model," were both symptom and cause for a new phenomenon in the globalizing architecture culture of postmodernism: on the one hand, the assertion that architecture, the architectural drawing and model were proposed as an autonomous art form, and their changing nature, as a once purpose-bound instrument of design, with their incorporation into the already globalized art market on the other. As a result, the Institute, with its Exhibition Program and in synergy with its other cultural productions, was instrumental in making the architectural object a form of capital investment on par with built architecture, thus rendering it a fetish in the Marxist sense.[475] By that time, it was already

471 Peter Eisenman, "House II Transformations," In IAUS, 1981, 34–35.

472 Paul Goldberger, "How Architects Develop Ideas," *The New York Times* (December 27, 1976), 58.

473 Vervoort, 2020, 235.

474 Kauffman, 2018, 78ff.

475 Franziska Stein, "Peter Eisenman: House II (Falk House)," in *Das Architekturmodell. Werkzeug, Fetisch, kleine Utopie*, eds. Oliver Elser, and Peter Cachola Schmal (Zurich: Scheidegger und Spiess, 2012), 250–254.

evident that the Institute was anything but naive and that architects had already lost their innocence.[476]

This development, later criticized for its hypocrisy, had already manifested itself at the Institute in the run-up to "Idea as Model." A conflict emerged following an incident on the eve of the opening, involving a performance by New York artist Gordon Matta-Clark.[477] Himself a trained architect, having studied architecture at Cornell University in the late 1960s and then taking part in the SoHo art scene since the early 1970s instead of going into practice, Matta-Clark had been invited by MacNair to contribute because of his conceptual rigor. It was not only Matta-Clark's early work under the moniker "Anarchitecture" that became one of the great founding myths of radical art and architecture of the decade.[478] In fact, he made his name with his architectural dissections (or literal deconstructions), including "Splitting" (1974), a suburban single-family house that could no longer be sold on the market, "Day's End" (1975), a pier that had fallen victim to deindustrialization, and also internationally, "Conical Intersect" (1975), a Parisian apartment building that had to make way for the urban renewal during the construction of Centre Pompidou.[479] MacNair had invited Matta-Clark to execute one of his "cuttings" in a seminar room as a site-specific artwork that would have revealed the financial workings underlying the Institute as an architecture institution itself.[480] But things turned out differently, and a crucial aspect of this chapter of the institution's history—as well as this form of institutional critique—is that Matta-Clark's contribution to "Idea as Model" exists mostly in the memories of those who attended it and, in the absence of evidence, can hardly be substantiated. Legend has it that Matta-Clark, whose relationship with the New York architecture scene was ambivalent, showed up at the Institute late at night during the final preparations for the exhibition, apparently under the influence of alcohol and accompanied by his partner, Jane Crawford. According to eyewitness accounts, he hung a series of photographs of destroyed windows of houses in Harlem and the Bronx, which he had brought with him, between the windows in the main hall, which were

476 Richard Pommer, "Post-script to a Post-mortem," in IAUS, 1981, 10–15. The "Idea as Model" exhibition was then contextualized in the catalog, which was not published until 1981, as an independent publication.

477 The explosive power of the exhibition was not apparent to its immediate surroundings and was only acquired retrospectively through its reception.

478 Mark Wigley, *Cutting Matta-Clark. The Anarchitecture Investigation* (Zurich: Lars Müller, 2018).

479 Stephen Walker, *Gordon Matta-Clark: Art, Architecture and the Attack on Modernism* (London: I.B. Tauris, 2009).

480 Rosalyn Deutsche, "The Threshold of Democracy," in *Urban Mythologies. The Bronx Represented since the 1960s*, ed. John Alan Farmer (New York: Bronx Museum of the Arts, 1999), 94–101, here 95.

twice as high, and then shot the panes with a BB, that gun he had borrowed from Denis Oppenheim, a friend and colleague; a performance witnessed by only a few people that nevertheless went down in architecture and art history as "Window Blow-Out."[481] Oral history further relates that when Eisenman entered the Institute the next morning and noticed the broken windowpanes, he felt compelled to act immediately, thus completing the performance in a sense: he had Matta-Clark's photographs removed and the broken panes unceremoniously replaced by a glazier before the guests appeared for the opening that night. The photographs were wrapped up and left for the artist to pick up. Whether consciously or unconsciously, it was Eisenman's reaction and action, later likened to an act of censorship, that completed the artwork and gave it that mythical, indeed legendary aura.[482] Whether he was acting as Institute director, i.e., because of the insurance or the lease, or for other minor, even personal reasons, and thus intervening in curatorial practice, was not recorded. The real fascination of "Window Blow-Out," however, lies in the fact that, unlike Matta-Clark's other performances, this work of art lived on only as a narrative, since he apparently did not document it in photographs or film, as was his usual practice.[483]

481 Besides Matta-Clark and Crawford, the only people present that evening were MacNair himself and a few of the Institute's interns and students who helped set up the exhibition; see Jane Crawford, transcript of an interview with Jürgen Harten, March 27, 1979, Source: CCA Montréal, Gordon Matta-Clark collection: PHCON2002: 0016:006:124. Although Crawford confused the Institute with the Architectural League, gave the wrong date and address of the exhibition, and omitted the title, this is an early reception, only two years after the event.

482 Andrew MacNair, in Mary Jane Jacob, *Gordon Matta-Clark: A Retrospective* (Chicago: Museum of Contemporary Art 1985), 96. It has however been frequently pointed out that Eisenman may have overreacted in this situation. Crawford mentioned that the press at the time did not cover the censorship of "Window-Blow Out" at all. MacNair, however, in an early interview stressed that Eisenman felt reminded of the Kristallnacht. Art historian Thomas Crow reiterated this statement, without discussing the comparison that seems exaggerated and inappropriate; see Thomas Crow, "Site-Specific Art: The Strong and the Weak," in *Modern Art in Common Culture* (New Haven: Yale University Press, 1996), 134. It remains unresolved whether Eisenman voiced this comparison. For to compare an artistic performance, however aggressive it may have appeared, with the centrally organized and state directed violence against the Jewish population across the Third Reich on the night of November 9, 1938, would ultimately testify to a misjudgment of the extent and brutality of the anti-Semitism of the Nazi regime; a provocation at any cost.

483 Philip Ursprung, "Blinde Flecken der 1970er Jahre: Gordon Matta-Clarks 'Window Blow-out'," in *Reibungspunkte, Ordnung und Umbruch in Architektur und Kunst*, Festschrift für Hubertus Günther, eds. Hanns Hubach, Barbara von Orelli-Messerli, Tadej Tassini (Petersberg: Michael Imhof Verlag, 2008), 293–300. In his essay on "Window Blow-Out," art historian Philip Ursprung argued that no records exist, neither in CCA's Gordon Matta-Clark collection nor in its IAUS fonds. To my knowledge, neither an invoice from the glazier nor a complaint to the police are archived there. In an oral history interview, I was given to understand that one of the Institute's interns apparently photographed the performance, but further research for documentation yielded no tangible results.

In the years that followed, the story of Matta-Clark's performance at the Institute was regularly reiterated and reinterpreted.[484] While it has hardly played a role in architecture history, "Window Blow-Out" has always been relevant to art history. Echoing sentiments expressed about the Institute in the architecture press in 1976, the site-specific performance was seen on the one hand as a rebellious act against the profit drive of the architecture establishment, and on the other as a symbolic confrontation with the unacceptable housing conditions of an impoverished strata of society—both being criticisms that had been raised at the time.[485] The first review, an essay by Richard Pommer, which was supposed to be published in the Institute's catalogue but did not appear until 1981, already highlighted the different social responsibilities assumed by art and architecture and the generational conflict that was revealed here: "The late Gordon Matta-Clark wanted to show photographs of vandalized New York windows against panes broken for the occasion at the Institute, but at last minute, with the cold air coming in, his exhibit was pulled. A pity, whatever the reasons: it would have called attention to the rival conceptions of younger artists, who often seem less afraid of social statements than these architects do."[486] The roles were clearly assigned: while according to this interpretation, Matta-Clark stood for a socially engaged art practice that opposed the revitalization and beautification of urban space, the avant-garde architecture practice at the Institute reproduced the hierarchical organization of the segregated city. In the end, both parties benefited from this confrontation, which was elevated to a battle. Through the rebellious performative act at the heart of architecture culture, which was not spontaneous but planned, and the outraged action that followed, an ultimately political reaction to this powerful allegory, Matta-Clark and Eisenman defended and even grew their standing and reputation in their respective scenes. Only the Institute, which had previously always portrayed

484 To address the reception of Matta-Clark in recent art history, a veritable fascination with his person and work that informs historiography, Ursprung reflected on the conditions, possibilities, and limits of oral history. To this end, he has given individual protagonists and witnesses of "Window Blow-Out" the opportunity to have their say, while highlighting inconsistencies in subjective accounts of the events by contrasting their statements with the reception history in essays by art historians and a biography of Matta-Clark. In my oral history, Andrew Anker, who helped build the exhibit as one of the Institute's interns, indicated even that it might not have been Matta-Clark at all who shot the windowpanes, but someone else.

485 Matta-Clark's preference for alternative art spaces was already noted by Crawford in 1979, only one year after the artist's early death from cancer, in an interview, when the first large retrospective at the Städtische Kunsthalle Düsseldorf marked the beginning of Matta-Clark's reception, see *Gordon Matta-Clark: One for All – All for One* (Düsseldorf: Städtische Kunsthalle Düsseldorf, 1979). Here Crawford spoke about Matta-Clark's training as an architect at Cornell University and individual projects. She also touched on "Window-Blow Out," which addressed the abandonment and decay of housing, in her eyes a "very strong, very powerful piece, since this is what was relevant, a major problem;" for further reception of "Window-Blow Out," see Crow, 1996; see also Deutsche, 1999.

486 Pommer, 1981, 6.

itself as being progressive and now came across as quite conservative, came off badly in the process, and subsequently proved to be the venue of the architecture establishment.

The Ten-Year Anniversary

The steady growth of the Institute and the associated "growing pains" were evident in 1977, the year of its tenth anniversary, when it hosted a conference with the evocative title "After Modern Architecture" organized by the *Oppositions* staff.[487] The editors of *Oppositions* had invited colleagues from like-minded journals in Europe—*Arquitecturas Bis* (from Spain), *A.M.C.* (France), *Controspazio* and *Lotus International* (both from Italy)—to present an overview of postwar architecture and its coverage. At this "little magazine" conference, sponsored by MIT Press and exclusively covered by Ada Louise Huxtable in the *New York Times*, Eisenman, Frampton, Gandelsonas and Vidler introduced their personal manifestos from *Oppositions* 4, 5, 6, and 7 for discussion, followed by thematic workshops, thus asserting their claim to leadership in the historicization and theorization of postmodernism among their peers—again behind closed doors, by invitation only, and with only a few guests, with the difference that this time, instead of art and architecture, the topic of conversation was textual, editorial, and publishing work, and thus cultural production.

At the same time, the Institute exhibited a more open and accessible approach than ever before, with "Architecture 6," funded once again by the NEA, in the spring semester of 1977. This time, it offered a comprehensive program of nine lecture courses, including a workshop on street photography, and an educational trip to the Netherlands and England.[488] At the conclusion of the three-year Evening Program, as it had been organized since 1974, academic approaches no longer played a role due to the fact that the long-term Fellows did not present lecture courses. Instead, Stern, the one constant over the years, continued his investigation into a postmodern turn, inviting notable architecture critics from the conservative camp (Charles Jencks, Brent Brolin, C. Ray Smith, Paul Goldberger, John Morris Dixon, Vincent Scully, Peter Blake) to debate the polemically phrased question "What is Happening to Modern Architecture?" with him under the heading "Critics Speak." In addition to thematic lecture courses on such topics as "The Interior Room," "The Making of the Natural Landscape," "Human Behavior and the Physical Environment," as well as "Frank Lloyd Wright

487 Ada Louise Huxtable, "Architecture View: A Sense of Crisis About the Art of Architecture, Architecture in Crisis," *The New York Times* (February 20, 1977), 99. The status of the conference at the Institute was not clear, and the question of which division should bear the costs was an in-house problem, since neither *Oppositions*, nor public programs, nor IAUS Central felt responsible.

488 The Institute's educational tours were to be led by Frederike Taylor and Julia Bloomfield, who were both newly elected Institute Fellows, but were most probably not carried out due to lack of interest.

and the Rise of European Modernism," one highlight of "Architecture" was the relaunch of the "New Wave of European Architecture" series, now limited to six participants, Peter Cook, Rem Koolhaas, Robert and Leon Krier, Elia Zenghelis and Massimo Scolari, all of whom returned to the Institute. Compared to the previous year, the series had dropped its avant-garde appeal and was now advertised more widely. The Institute took on a more glamorous image, treating participating architects like pop stars by flying them by helicopter from JFK Airport to the heliport on the Pam Am Building, not far from the Institute.[489] Gandelsonas took over the contextual framing this time with his lecture "Rowing Upstream. An American View of A New Wave," while Suzanne Stephens, an architecture critic at *Progressive Architecture*, provided a summary and commentary with "European Transfer." One innovation, which was to transform cultural production at the Institute, was that after their lectures there, the "European New Wave" participants were sent as a bookable package on a lecture tour to thirteen architecture schools across the United States. This was made possible by an extremely affordable round-trip ticket from Pan American World Airways, and MacNair was also able to secure the airline as a sponsor of "Architecture." This redesign of the "New Wave" series proved a game changer for the Institute's market positioning as it expanded its sphere of influence as a cultural institution across the North American continent and significantly raised its profile nationwide. The Institute brought European architects into the national public eye by simple and direct means and, as they began to take on their first teaching positions at American universities, established a monopoly on the reception and overarching institutional framing of contemporary European architects in the United States.

A unique event in the history of the Institute, if not of architecture culture in New York occurred concurrently with "Architecture 6," when MacNair and Shanley organized "City as Theater," an independent series of lectures that represented a discursive and institutional innovation precisely because it was made possible by a grant from the New York Council for the Humanities. In doing so, the Institute took advantage of the opportunity that presented itself and adopted the underlying idea from the newly established humanities funding body which, as it happens, was located across the road on the other side of Bryant Park on 42nd Street. Contact had already been established by the end of 1975, when the Council's director Ronald Florence, a regular in the Institute's "Architecture" series, approached the Institute's leadership with a proposal for a thematic event focusing on the intersection between architecture and the humanities. The young Richard Sennett, who at the time was a professor of sociology at New York University and also directed the Center for Humanistic Studies there, was initially mooted as a cooperation partner for the Institute.

489 In the spring of 1977, the midtown heliport reopened for a short time until a tragic accident ended this commercial chapter of the city's transportation infrastructure, depriving the Institute of some of the glamour of the jet-setting age.

Sennett's hypotheses from his forthcoming book *The Fall of Public Man* were already reflected in an initial concept for the series from 1976, which brought together contemporary urban issues and performative approaches from the arts.[490] Since Sennett himself was at the time in the process of founding his own institution with historian Thomas Bender, the New York Institute for the Humanities, which opened at NYU later in 1977, he played a smaller role in "City as Theater" than originally planned.[491] The Institute's grant application to the New York Council for the Humanities, ultimately penned by Vidler, revealed his influence and focused on public space, criteria for publicness, and measures to improve the quality of urban life. It was successful. Vidler, in turn, enlisted Carl Schorske, a historian who specialized in Central European intellectual and cultural history and who, like him, taught at Princeton University and directed the "Program in European Cultural Studies" there, as an outside consultant for "City as Theater" to lend academic legitimacy to the one-off series.

What was special about "City as Theater" was that the Institute was able to offer it free of charge as a non-commercial event, thanks to funding from the humanities. Incorporated into "Architecture 6," it included four lecture courses, for which once again well-known personalities, mainly journalists and other public figures, were enlisted as presenters: Erika Munk (editor of *Drama Review*, who also wrote the "Cross Left" column in the *Village Voice*), John Rockwell (music and dance critic for the *New York Times*), Joan Davidson (chair of the J.M. Kaplan Fund, which also sponsored the Institute), and Paul Goldberger (architecture critic for the *New York Times*). Drawing directly on "What is a City?" an essay by Lewis Mumford from *Architectural Record*, and explicitly on his 1938 classic work *The Culture of Cities*, "City as Theater" focused on an expanded, transdisciplinary concept of architecture and the city and on a multi-layered analysis of modern life. On the poster, also designed by Vignelli (this time in landscape format and in bold constructivist colors, black and red, the Institute's signature colors), the Institute announced the concept with a lengthy quote from Mumford, printed in large letters: "The city in its complete sense, then, is a geographic plexus, an economic organization, an institutional process, a theater of social action, and an esthetic symbol of collective unity. On one hand it is a physical frame for the commonplace domestic and economic activities; on the other, it is a consciously dramatic setting for the more sublimated urges of a human culture. The city fosters art and is art; the city creates theater

490 Sennett's objective in *The Fall of Public Man* was to analyze the link between metropolitan culture and theatrical performance against the backdrop of the formation of a capitalist mode of production; see Sennett, 1976.

491 Apart from the concurrence evident in the naming of both the Institute for Architecture and Urban Studies and the New York Institute for the Humanities, it is curious that the two once briefly converged, when the Institute flirted with the humanities, only to clash shortly thereafter due to very different understandings of architecture, urban studies, institutions, the humanities, and criticism.

and is the theater. It is in the city, the city as theater, that man's more purposive activities are formulated and worked out through conflicting and cooperating personalities, events, groups, into more significant culminations."[492] While the performing arts had to be relegated to the background, the city in its multiple meanings was brought back to the foreground again, if only for a moment.

As part of "City as Theater," a program of eighteen panel discussions was held at the Institute every Tuesday at 7:30 p.m. from March to June 1977, with a total of fifty-two lecturers: prominent figures from cultural and intellectual life, professionals, established academics, journalists, city politicians, and up-and-coming activists. The topics were the city and the theater in the broadest sense, both literally and figuratively.[493] Ultimately, "City as Theater" became one of the few Institute events to present a truly public forum for anyone interested in the spectacle of urban life. Strikingly, not one of the long-term Fellows gave a lecture; from the Institute's circles, only Krauss contributed. In general, with a few notable exceptions (Renyer Banham, Martin Pawley), architects and planners were hardly represented in this series—for the purposes of the humanities, the Institute was thus entering into completely new territory. In contrast to its involvement in city planning and public housing projects in the early years and its interdisciplinary theorizing, the Institute was for the first time opening up to a more humanities approach to urban studies. Individual lectures addressed topics in urban sociology (William H. Whyte), urban psychology (Donald Kaplan, Karl Linn, Harold Proshansky), urban planning (Doris Freedman, the first director of cultural affairs in New York) and urban politics (John Lindsay, the mayor of New York from 1966 to 1973). MacNair arranged for "City as Theater" to be advertised on one of the first digital displays in Times Square. Because a larger crowd was expected for at least some of the panels, he booked a City University auditorium in the immediate vicinity of the Institute. Ultimately, "City as Theater" might ideally have stimulated a

492 Lewis Mumford, "What is a City?" *Architectural Record* 82 (November 1937), 92–96.

493 For example, there was a panel with contributions by writer and philosophy professor William Gass on "Inside: External Stimulation and Internal Contemplation: True Drama, External Events and the Atmosphere of Paris," by art historian Irving Lavin on "Outside: The Relationship Between the Baroque Stage and the Baroque Piazza," and by New York architect Roberto Brambilla on "In-Between. Pedestrian Drama in Contemporary Public Spaces." Sennett (on "Clothing: Street Dress as Barometer of Public Health") and Schorske (on "Promenade: Otto Wagner and Gottfried Semper") were joined by Max Kotzloff (executive editor of *Artforum*), Jason Epstein (co-founder of the *New York Review of Books*), Brooks McNamara (professor of theater studies), and Richard Foreman (theater director and founder of Ontological-Hysteric Theater). Other panels focused on various popular culture formats (folk and rock music festivals, the entertainment program in professional sports, Latin American music and Puerto Rican bars, department stores as temples of consumption, newspapers, and television as mass media), or the design of public space (sidewalks, bus shelters, lobbies, plazas). Other contributions addressed specific sites and buildings in New York City (Bloomingdales, Times Square, Coney Island, World Trade Center). It was striking that the quota of women was quite high compared to other series of lectures at the Institute, including contributions by Charlayne Hunter-Gault ("Talk of the Town" column in *The New Yorker*) or Liz Christy (founder of the first community garden, here representing the Green Guerrillas).

multi-, if not transdisciplinary exchange of knowledge, not just among the Fellows; architects, professionals, and students, who normally made up the Institute's audience, were given the opportunity to hear and meet with literary scholars, theater professionals, journalists, artists, and so on.[494] All this, however, can obscure the fact that "City as Theater" ultimately proved to be a strategic move on the part of the Institute's leadership to secure new funding opportunities.

In the Institute's anniversary year, the two management fields "Development" and "Communication" became very important, and MacNair was not only given the task of securing media contacts at the local and national levels but also, in addition to Taylor's work, looking after the sponsors of the two public programs, as well as the individual donors.[495] In contrast to the Evening Program, much less effort was expended on the Exhibition Program at the Institute in the first half of 1977. The exhibition "Princeton's Beaux-Arts and Its New Academicism," with student projects from Princeton under Dean Robert Geddes, was followed by monographic exhibitions on European architects who participated in the "New Wave" series, this time by Rob Krier and O.M. Ungers, with posters again being sold in small editions, like those for "Idea as Model." In this context, the copying machine played a central role as a contemporary reproduction technique: in the mid-1970s, Xerox launched a color copier as a technical innovation for the broader market. Combining mass production techniques and manual labor, the Institute's interns reproduced the motifs for the exhibition posters that year with a photocopier, which were then partially recolored by hand by the exhibiting architects and pasted onto the template designed by Vignelli to retain the graphic identity and identify them as a product of the Institute. Printed in an edition of a few hundred and valorized by customization, the posters were sent out to donors and sold during the exhibition. To foster relationships with donors, especially with members of the Architects' Circle, it was announced that they would not only receive a personal invitation to special events and openings but also an original poster as a collector's item.

A New Self-Image

It was no coincidence that at this point in time changes in architecture culture and the changing role of architects in society became a major topic of discussion in the American architecture community, and the Fellows and friends of the Institute contributed to this discussion as well. This was reflected in the May 1977 issue of

494 However, no conclusions can be drawn about the response, since there is no documentation of the number of visitors; apart from an extensive interview with Taylor in the *Soho Weekly News*, possibly for public relations, if not outreach, the series was not reviewed in either the local or the architecture press; nor was "City as Theater" as the one-off event mentioned in any particular way in the Institute's historiography.

495 Peter Wolf, "Proposal for Development," June 9, 1977. Source: CCA Montréal: IAUS fonds, A.2-7 / ARCH401152.

Progressive Architecture, which asked fundamental questions about the future of the profession and a recasting of the function and social role of architects. This multi-layered coverage also reflected the new cultural paradigm in architecture, especially (but not only) historiography and the production of theory as practice. According to the editorial, the aim was to draw new insights from the "changing cast of characters" in order to present a possible role model for young architects for discussion or to suggest alternatives.[496] In accordance with the discursive and institutional debate of the time, the stage metaphor was deliberately selected to address the public appearance of architects as actors of themselves, regardless of whether they slipped into one of the existing roles, or took completely new paths (the term "role-model" was used here in a double sense, combining the exemplary and the performative)—after all, for architects it was primarily about being in the limelight. This issue featured an article about the recent evolution of the profession by Robert Gutman, a sociologist of architecture who had been familiar with the Institute from its founding, having himself served on the faculty in its early years, and followed the New York architecture scene closely, as he was repeatedly called in as a sociological consultant on various projects over the years.[497] In his article "Architecture: The Entrepreneurial Profession," Gutman posited that architects in the 1970s held outdated ideas of themselves and an exaggerated self-image. Drawing on this diagnosis, based on quantitative research, and starting from four distinctive characteristics of the profession—first, that there is no urgent need for architecture, second, that architects share design activities with other professions, third, that their work is subjective in nature, and fourth, that demand for architecture is contingent on the economic cycles of the real estate market—Gutman argued that architecture had become entrepreneurial and that architects must take the initiative and create demand themselves. He acknowledged that architects were then offering other services and involved in other media, exhibitions, and graphic design. While he noted that the profession had increasingly shifted to large, industrial-type offices with specialized subdivisions, he did not believe that the alternative was for architects to limit themselves by viewing architecture as art and asserting themselves as artists, in imitation of avant-garde practices. Committed to scholarly objectivity, Gutman obviously held a mirror up to the Institute and explicitly to his old companion Eisenman by criticizing the ideology, doctrine, and system of autonomous architecture. Instead, he demanded that architects take their competencies and responsibilities seriously and create not only forms but buildings that must also meet the requirements of their inhabitants and users.

496 In a 1972 discussion of the role of intellectuals in society, Michel Foucault and Gilles Deleuze pointed out that theory "does not express, translate, or serve to apply practice: it is practice." Discourse is thus always directed against power, a "counter discourse," see Foucault and Deleuze, 1977.

497 Gutman, 1977; see also Dana Cuff and John Wriedt, eds., *Architecture from the Outside in. Selected Essays by Robert Gutman* (New York: Princeton Architectural Press, 2010), 32–42.

The bulk of the issue consisted of a multi-page feature, conceived and partly written by Suzanne Stephens as senior editor of *P/A*, dedicated to the "Multiple Protagonists" of American architecture, the different types of firms and architects, which can be read both as an architecture debate on the understanding of architects' roles in the present and in the history of the Institute and as a compelling case study and insightful analysis of the identity crisis of the New York architecture community.[498] In addition to the roles of "the individual" (exemplified by Richard Meier), "the corporate architect" (Paul Kennon), and "the gamesman" (Jaquelin Robertson), which were prevalent in the mid-1970s, Stephens also analyzed the relatively new role model of the "the polemicist-theorist" (Stern and Eisenman).[499] While the feature suggested a clear distinction, the examples make it clear that they all in some way emphasized the new entrepreneurial spirit, thought, and activities within the architecture profession, while the Institute was a place where the associated ideas, values, and practices converged in cultural, social, and symbolic forms. In Stephens' view, however, the "polemicist-theorist" as a distinctive type was in the first case to be characterized by staging a debate—"stirring up controversy, debate, excitement"—and achieving an enormous outreach—"to students, magazines, foreign architects, and even (to some degree) mass media."[500] "And while each together or apart was considered elitist, exclusivist, and clubby, what they had done was turn the theorist-polemicist image around. They had taken it out of its anti-hero, outsider role and put it center stage, made it a star." With this, Stephen described what was crucial for the formation of a scene and forms of communization, as well as the beginnings of what came to be called "starchitecture:" the multiple mechanisms of differentiation and demarcation, the adoption of artistic practices, and the play with elitist strategies of inclusion and exclusion. Stephens ultimately presented her feature as a kind of history of intellectual work at the Institute, whose network of relationships was in her opinion characterized not only by discursivity and criticality, but above all by institutionality and exclusivity; as it was a closed circle, she used the term "coterie-ism" in reference to the context of networks of communication, evaluation, and emotion. It is noteworthy that Stephens was one of the few to discuss the complementary pairing of Stern and Eisenman, which was later alluded to in architecture historiography, and which she described as the greatest "coup du theatre."[501] In her view, they were both in their own way "impresarios of exotica,"

498 Suzanne Stephens et al., "Role-Models. Multiple Protagonists," *Progressive Architecture* (May 1977), 59–71.

499 Stephens points out that several other roles of architects were not addressed in the feature, e.g., the "architect-developer," the "architect-researcher," or the "architect-who-has-chosen-to-leave-the-field," i.e., the professional dropout.

500 Suzanne Stephens, "Polemicist-theorist," *Progressive Architecture* (May 1977), 68.

501 Martin, 2010, 66.

but with "different orientations as well as design attitudes"—a media masterstroke, yet one in which she—in her role as *P/A* editor—was not entirely uninvolved. Although Stephens focused on male role models, the issue of *P/A* was an exercise in introspection at a time of increasing differentiation and representation that prompted further analyses of the social networks, relationships, and dependencies in the New York architecture community. While well observed, the journalistic presentation of architectural types—itself a form of embedded architecture journalism due to its access to the underlying evidence base, insider knowledge, information strategies, and human interests—nevertheless leaves readers with the bitter taste of publicity-mongering and self-promotion.

This became pronounced again in the next issue of *P/A*. For Eisenman himself had, at about the same time, compiled three separate reviews of *House VI* published in the June 1977 issue under the title "Critique of Weekend House by Philosopher, Sociologist and Architect Himself," which discussed the underlying design process, meaning, and use of this cottage house of the Franks in Connecticut, with the furniture in the photographs curated by Vignelli. In addition to a critique from a literary and philosophical perspective, for which he enlisted the writer William Gass, and one from a sociological perspective by Robert Gutman, it also featured a text of his own from an architecture perspective, in which Eisenman portrayed himself as an interpreter of his own designs.[502] This form of orchestrated architecture criticism had become almost a trademark, since over the years Eisenman had repeatedly succeeded in enlisting renowned theorists and historians from among the Fellows and friends to write reviews of one of his ten experimental house designs, *House I* to *House* X, 1967–77, for national and international journals.[503] After Gandelsonas and Frampton had reviewed the first projects and while Gass and Gutman were critiquing *House VI*, he was already able to persuade top-ranking art and architecture critics Rosalind Krauss and Manfredo Tafuri to review his first book project *Houses of Cards*, which would not appear until ten years later.[504] In his

502 Peter Eisenman, William Gass, Robert Gutman, "House VI. Residence. Critique of Weekend House by Philosopher, Sociologist and Architect Himself," *Progressive Architecture* (June 1977), 57–67.

503 The idea of producing ten house designs in ten years apparently originated with John Hejduk, who designed according to his Nine Square Grid method. Hejduk himself first exhibited his Diamond series (1962–1967) in The Architectural League in November 1967; see Michael Jasper, "Working It Out: On John Hejduk's Diamond Configurations," *Architectural Histories* 2, no. 1, (2014), https://journal.eahn.org/articles/10.5334/ah.cb/ (last accessed: May 31, 2023). Regarding his house designs, Eisenman, like the postmodern project maker he was, took ideas and made them big, often overdoing it: while others started with an idea and then abandoned it, Eisenman often started without a plan and then ended up draping a concept over his projects.

504 Mario Gandelsonas, "On Reading Architecture," *Progressive Architecture* (March 1972), 68–88; "On Reading Architecture," *Architecture + Urbanism* 2, no. 9 (September 1972), 51–69; "Linguistics in Architecture" & "Due opere di Peter Eisenman: Castelli di carte," *Casabella*, no. 374 (February 1973), 17–31; "On Reading Architecture II [House IV]," *Architecture + Urbanism* 4,

choice of critics, Eisenman always acted according to the principle that negative attention is also a form of attention, as long as it is intellectually stimulating, with both sides ultimately gaining in reputation and prestige. Thus, in the 1970s, Eisenman—through the seductive power of the ironic iconoclasm displayed by his projects and the persuasive appeal of his masterful public relations strategy—succeeded in drawing attention both to his design approach and to himself as Institute director and project maker. In addition to his teaching position at Cooper Union and his fees as an architect, the Institute was an essential working context and source of income for Eisenman, enabling him to see himself as an autonomous and critical architect, who could project an image of himself as independent and free from vested interests. In this context, Gutman wrote in his *P/A* article that architects often chase the image of the "romantic loner," free and independent, "cultivating personal relationships with an understanding and appreciative client," while the majority were actually wage-earners.[505] Not only did the focus on the architectural autonomy of the artist-architect, communicated and maintained in cultural production, mask the shift of the construction industry that created a global flexiblized workforce in the 1970s, the understanding of the architect's role soon underwent a change as architectural drawings and models were assigned a monetary value on the art market, in addition to their artistic value. This was exemplified by the exhibition "Architecture I" at the Leo Castelli Gallery, which opened in the fall of 1977—the most important and profitable point in time in the gallery world—once again on the initiative of Jakobson in collaboration with Ambasz like the sales show "Architectural Studies and Projects" at MoMA two years before.[506]

no. 39 (March 1974), 89–100; see Kenneth Frampton, "Criticism: Eisenman's House I," *Architecture + Urbanism* 3, no. 11 (November 1973), 190–192; "Five Architects," *Lotus International*, no. 9 (1976), 136–151 (English original 231–233); "Maison VI," *L'Architecture d'Aujourd'hui*, no. 186 (August/September 1976), 63–66; see Rosalind Krauss, "Death of a Hermeneutic Phantom. Materialization of the Sign in the Work of Peter Eisenman," in *Houses of Cards*, ed. Peter Eisenman (New York: Oxford University Press, 1987), 166–184; see Manfredo Tafuri, "Peter Eisenman: The Meditations of an Icarus," in *Houses of Cards*, ed. Peter Eisenman (New York: Oxford University Press, 1987), 167–187.

505 Cuff and Wriedt, 2010, 37.

506 Leo Castelli Gallery showed contemporary architects in a cycle of three exhibitions every three years, "Architecture I" (October 22 to November 12, 1977), being the first, followed by "Architecture II: Houses for Sale" (October 18, to November 22, 1980) and "Architecture III: Follies: Architectures for the Late-Twentieth Century Landscape" (October 22 to November 15, 1983); see "Architecture I" In Eeva-Liisa Pelkonen, ed., *Exhibit A: Exhibitions That Transformed Architecture 1948-2000* (London: Phaidon, 2018), 196; the architects featured in "Architecture I" were: Raimund Abraham, Emilio Ambasz, Richard Meier, Walter Pichler, Aldo Rossi, James Stirling, Roberert Venturi, and John Rauch. See Paul Goldberger, "Architectural Drawings Raised to an Art" *The New York Times* (December 12, 1977), 50; Eisenman then exhibited in "Architecture II," together with Emilio Ambasz, Vittorio Gregotti, Arata Isozaki, Charles Moore, Cesar Pelli, Cedric Price, and O.M. Ungers; and again in "Architecture III".

3.3 Representing the Institutional Establishment

Despite its recent successes, the growth of the Institute was deceptive. With *Oppositions* 5 and the "Idea as Model" exhibition, the academic year 1976–77 had indeed begun with a bang, and the Institute had achieved great things in the previous three years—not just as an educational, but as a cultural institution, organizing lectures every night, regularly showing exhibitions, and publishing two accompanying journals. However, in its anniversary year, the Institute was once again facing organizational and programmatic changes. For at the end of fiscal year 1976–77, funding from both the John Edward Noble Foundation for the "Undergraduate Program" and the NEA for the "Evening Program" expired and was not renewed. The previously successful tripartite business model of tuition fees, private and public grants, and individual sponsorship was proving vulnerable. Against this background, Eisenman took the initiative in early January 1977, addressing the Board of Trustees with a memorandum, several pages long. As Institute director, he was accountable to them, aware that it was they who defined Institute policy. With his "Director's Memo," he not only took stock of the Institute's activities and tasks to date and described the present situation but, building on this, outlined future goals for the next decade.[507] In framing the Institute's position and specifying a path of consolidation, Eisenman, in order to redefine the Institute for the next ten years, undertook a detailed consideration of "its position in the specific community and in the society at large." The "Director's Memo" was a strategy paper and was thus based on conceptual and financial, political and economic considerations. In launching this initiative, he was, of course, also pursuing his own institutional agenda, boldly claiming to give the Institute "a definition and a limitation" and "in short a sense of structured purpose" for the first time. Subsequently, Eisenman once again pitched the Institute, much as in the first version of the 1967 by-laws, as a "unique cultural institution" that could not be compared to other academic institutions or professional practices. Leading the way for the ambitious project of further institutionalizing the Institute—here, Eisenman emphasized the "careful choice of the term 'Institute' in its title"—were all those institutions he cited as models for a possible new orientation. He had in mind an architecture institution that was comparable to "a policy group such as the Brookings Institute in Washington, or a think tank such as the Institute for Advanced Studies in Princeton." These references and comparisons, however much they may have differed in function and operation and however pretentious they may have seemed, highlight that Eisenman was obviously concerned with the future scope of the Institute's research, educational, and cultural activities, and not least with maintaining or even expanding his power in his capacity as Institute director.

507 Peter Eisenman, "Director's Memo," January 11, 1977.

Eisenman had so far been opposed to institutionalizing the individual fields of work, or at least institutionalizing them completely, and had thus, for example, not founded an Institute publishing house or introduced a degree program, preferring instead to rely on networks. The situation for the public programs was somewhat different. In his "Director's Memo," however, he made it clear that, as a charismatic leader, he continued to believe in the cultural and ultimately architectural project that had been built at the Institute over the past three years. Consequently, he set very high goals for its future development. The Institute was to be transformed into a "cultural resource" with a twofold structure, "concerned with the creation of information about architecture and the public environment, the nature of design and the design of the public environment," and "concerned with the dissemination of this information through publication, exhibition, and educational programs." According to this realignment, architecture education would remain a strategic cornerstone but was to be consolidated and refined. However, by defining the Institute as a site for the production, reproduction, and dissemination of architectural knowledge, Eisenman laid claim to intellectual hegemony: "Such an agency has the capacity to become an international center for research, design and discussion which will place the Institute at the center of future thinking on the nature, design, and maintenance of this country's major undefined resource: the public environment." The "public environment," then, was Eisenman's new buzzword, which he used abundantly in his "Director's Memo." The inclusion of the socially relevant and nationally fashionable debate about the relationship between public space and the built environment in the Institute's program, realized with "City as Theater," can perhaps be explained by the fact that the memorandum was the basis for a "major capital development fund raising drive" to apply for major grants from America's two federal funding agencies, the National Endowment for the Arts (NEA) and the National Endowment for the Humanities (NEH), both established under President Lyndon B. Johnson with the Arts and Humanities Bill on September 29, 1965. In the mid-1970s, it was the allocation of greater amounts of funding that made the work of museums and smaller institutions possible.

Eisenman's initiative was rhetorically adept for it was once again a matter of rallying the trustees behind the Institute's continued existence. He believed that it was fundamentally important for the Fellows, as the Institute's most important resource, to be retained for the longer term, e.g., by finally compensating them adequately for all their work. But Eisenman was also concerned with the diverse social, cultural, and other capital of the architects at the Institute. Above all, he emphasized their intellectual work and academic affiliation as a special feature of the Institute—in his eyes an architecture elite—and once again, as he had done in 1971, stressed its "think tank component."[508] Unlike in the Institute's early years,

508 Ibid, 3.

the Fellows' contribution was now defined as that of scholars and educators, built environment experts and policy consultants, architecture intellectuals and cultural producers, and not as practicing architects. In keeping with the emerging information society, the focus was now on the creation of "qualitative and software" information "as a basis for the future development of a new cultural policy toward the public environment." The Institute was thus to be understood as: "1. A place to conceptualize basic issues of design related to the public environment in its most fundamental issues: its iconography, its history, its design, and its use," "2. A center for discussion groups, conferences, and lectures, concerning the work on these basic issues," "3. A cadre of leaders from schools, and the profession," and "4. A focus for the development of the discipline of architecture." This multi-faceted definition of the Institute and the architects networked within it encompassed content and format, discursive and institutional aspects, structures, and functions. From a sociology of culture perspective, Eisenman thus defined a clear picture of the Institute's role as a sanctifying and disseminating authority, comparable to the classical academies, the salons of the nineteenth century, or the universities, museums, and publishing houses of the twentieth century.

Eisenman concluded his "Director's Memo" with a call to take advantage of the opportunities presented by the NEA's invitation to apply for a Challenge Grant and the NEH's funding under their "Cultural Institutes Program" [sic!].[509] These grants, he believed, could help expand the existing program in adult education into a comprehensive and even more audience-oriented program as "a public cultural facility;" for fiscal year 1977–78, the Institute's leadership anticipated an additional US$600,000 in revenue from these two funding sources alone. On the one hand, Eisenman's interest lay in continuing the three successful education offerings as instruments for the dissemination of knowledge and securing the financing for the Institute's operations. On the other hand, Eisenman was particularly concerned with the publications *Oppositions*, now academically legitimized, and *October*, while plans were already in place for new publication projects: he was negotiating with Roger Conover of MIT Press for a book series in which his Terragni monograph would be the first. Above all, however, in this "Director's Memo," he informed the Board of Trustees for the first time that they were now looking for an "expanded, and possibly rent-free, centrally located building or space" in order to finally be able to accommodate their own library in addition to providing adequate office spaces and exhibition spaces. The 1970 stepped rent for the 8 West 40th Street penthouse had in the meantime become a real financial burden: in 1976–77, the rental costs of US$44,562.50 accounted for about one fifth of the budget. Ultimately, the "Director's Memo" did not contain many new ideas, a realistic plan, or a genuine vision for the Institute, but rather a description of the status quo mingled with

509 Ibid, 5.

inflated expectations. At best, it could be explained by the opportunities offered by the still broad funding landscape in the United States. However, developments already hinted at a conflict that was to become even more important: the imminent professionalization and bureaucratization of all fields of work and the financing of a "Building Project" as a basis for further institutionalization.

On the same day that Eisenman submitted his "Director's Memo," he was confirmed in office by the Board of Trustees as Institute director. Likewise, all other board positions were confirmed: Armand Bartos as chairman, Arthur Drexler as treasurer, and Richard Meier as secretary. Thus, the organizational foundations for the Institute's work in the years to come were laid. At the same time, the dual leadership of the Institute was awarded a salary increase for 1977, so that both received compensation of US$18,000 each annually: Eisenman for his job as Institute director and Wolf in his capacity as chairman of Board of Fellows, with both continuing to share responsibility. With a workload of half a position each, their duties included "responsibilities for budgetary matters, program development, financial assessment, liaison between Trustees and the Fellows, annual reports and meetings." In addition, Eisenman continued to draw a salary as director of the "Undergraduate Program in Architecture" for the next three years, while his work as editor of *Oppositions* remained unpaid. But apparently, he had overshot the mark with his uncoordinated initiative, for less than two days later the trustees responded to his "Director's Memo" with a written statement, asserting their authority and, more importantly, their primacy over the Institute's leadership.[510] In their statement, they emphasized that it was they, and not the Institute director, who would set the Institute's medium and long-term policy; the administration and day-to-day work, on the other hand, would be the responsibility of the Institute's leadership and the Fellows. Moreover, they spelled out that it was they who assessed the Institute's work and ensured that high quality was maintained by reviewing the integrity and efficiency of individual programs in addition to other measures. They stressed that interventions in setting priorities could only be made in consultation with them and that they had the final say in all major decisions: "The trustees thus exercise a prior and general review in such matters as the allocation of a significant proportion of the Institute's resources, the setting of priorities for development, changes in programs of broad bearing for the institution, the determination of tuition or fees, plans calling for new construction, the establishment or abolition of new departments or schools, changes in admission policies affecting sizable categories of potential students, and changes in relations with outside educational and social institutions and government agencies." The trustees also asserted their authority over financial planning and property ownership through the formulation of policies and business administration. In addition, there was to be a governance committee and a finance committee to guide the Institute's business: "The trustees

510 Board of Trustees, "Statement of Trustees' Authority," January 13, 1977. Source: CCA Montréal, IAUS fonds: A.1-13.

establish fund raising policies, approve major development programs, help to identify important sources of potential financial support, and raise funds." With this statement, an important document about the structure and constitution of the Institute as it had actually existed since its founding, the trustees communicated their relationship to the Institute and its leadership; accordingly, they were willing to give advice and offer criticism but demanded to be informed about new conditions and requirements in a timely manner. Whatever the trigger may have been, whether it was the "Director's Memo," which was far more demanding than earlier "Director's Reports," or a growing dissatisfaction with Eisenman's tendency to overstep the freedom he had been given in previous years, the trustees responded to his perhaps somewhat premature and precipitous actions and impressively demonstrated their leadership by reinstating the hierarchy between them and the Fellows as originally set forth in the 1967 by-laws. In this dispute, a power struggle was emerging that would increasingly preoccupy the Institute and ultimately wear it down in the years to come.

Cultural Institution Grant

After the spring of 1977 had, in view of the Institute's upcoming anniversary, seen the acquisition and establishment of contacts with the worlds of industry and business, globally active corporations, the realms of architecture and construction in general, various foundations, and academia, it became apparent in the summer of 1977 that Eisenman's solitary push to define the Institute as a "cultural resource" had not been in vain, and would eventually pave the way for his greatest coup to date. No sooner had the last "Architecture" and "City as Theater" events been held, than the Institute was awarded a Cultural Institution Grant from the NEH totaling US$357,000, having originally proposed the introduction of a "NEH Leaning Institute Program" in addition to the existing Evening Program as well as the transformation of the Institute into a Center for Public Education and an International Study Center: an immense sum for such a project.[511] Not only was this the largest grant in the Institute's history, it was also the largest grant the NEH had ever awarded to an architecture institution. This grant, awarded by a federal agency, elevated the Institute to the rank of other cultural institutions, having previously qualified for a grant from the New York Council for the Humanities for "City as Theater." This was because the Institute, as host and organizer, had demonstrated interest in the humanities—not only in the arts, but in cultural history, sociology, and anthropology, while at the same time impressively demonstrating that it could play an important, if not leading role as a cultural space in the broader academic and intellectual culture of New York.

511 IAUS, application to the NEH for a "Cultural Institution Grant" (EH-28433-77-547). The peer reviews, much more than the required four, were submitted by Philip Johnson, John White, William Turnbull, Norbert Birnbaum, Paul Rudolph, Lee Copeland, Thomas Hess, Barry Ulanov, Carl Schorske, Edward Logue, and Charles Moore.

Finally, with the NEH grant, the Institute's leadership launched "Open Plan" in the fall of 1977, an interdisciplinary, more tightly curated and structured series of lectures and additional events, as a successor to "Architecture." Although the ambitious plans from the NEH proposal were implemented only rudimentarily at the time, with the planned restructuring not realized and the educational claim watered down, the immense sum provided the Institute with exceptional planning security for the following three years. "Open Plan" once again blurred the distinction between education and culture, while operations, from an organizational sociology perspective, continued largely as usual. Nevertheless, cultural production was selectively expanded and intensified, and high-profile programs were professionalized, which was reflected in the quantity and quality of overall output. Finally, the redefinition of the Institute as a "cultural resource" and the consecration and legitimization through the NEH brought about institutional change, as the Institute needed to network differently. In the summer of 1977, the composition of the Board of Trustees changed, with Charles DeCarlo and June Larkin stepping down and representatives of both the universities (William Porter, Colin Campbell) and the profession (Edward Logue, Tim Prentice, Charles Gwathmey, Ulrich Franzen) stepping in to replace them. In addition, two wealthy influential representatives of New York society, Christophe de Menil and Marietta Tree, both philanthropists and socialites, were added to the board, and Carl Schorske represented the humanities. Massimo Vignelli, the Institute's long-term graphic designer responsible for its graphic identity, brand image, and institutional reputation, was also rewarded for his commitment to a position as a trustee while continuing to work unpaid on all printed materials and new publication formats. At the same time, the Fellowship was also expanded: with the 1977–78 addition of new Fellows Carla Skodinski, Frederieke Taylor, and Anthony Vidler, who had contributed to the Institute as coordinator of the undergraduate program, director of development, and editor of *Oppositions*, the number of Fellows grew to thirteen. The Institute's history took a decisive turn in the course of the professionalization necessitated by the high level of funding, as women, in particular, were subsequently appointed to leading positions, although from a feminist perspective, it should be emphasized that these were mostly still subordinate positions, ranking below the more veteran Fellows, and thus perpetuated the established patriarchal and hierarchical structure of the Institute.

With federal funding, the professionalization of the management of adult education and the Institute's work as a cultural institution in general were imperative, as was the greater bureaucratization of the Institute's operations, since it was now held more accountable than before.[512] Immediately after the

512 Ockman, 1988, 199. Ockman, in her history of the journal *Oppositions*, pointed out the connection between the Institute's departure from its original purposes and the increasing bureaucratization that accompanied its institutionalization. However, she locates this development far later, although it was evident by 1977–78 at the latest, ironically with the opening up to the humanities.

grant was announced in the summer of 1977, Terry Krieger, the Institute's contact person at the NEH, announced that he would collaborate closely with the Institute in conceptual and administrative matters, which also meant close control over the design and implementation of the planned program. And although programmatic changes were apparently not an issue, the Institute initially had to completely revise its budget because the NEH finance department was surprised by the extremely high projected overhead costs. The NEH grant, paid in three annual installments of US$127,000, US$125,000, and US$105,000, eventually covered the rising operating costs and salary expenses of the Fellows and staff. During the grant period, the Institute was required to submit six-monthly reports on the progress and success of the program. This meant that, beginning in fiscal year 1977–78, for the first time since the Institute's inception, financial reports were prepared annually by official auditors demonstrating the budgeting of all programs and the proper use of funds—a practice that was maintained for the next three years.[513] Additional staff had to be hired or contracted to bring administration and accounting up to the requisite standards. The NEH grant thus brought greater transparency to the Institute's operations. The Institute's hopes in this regard rested on both Taylor and MacNair, both of whom were seen as playing a critical role in terms of development and communication. Wolf had drafted a two-tier job description with the NEH application, under which Taylor would henceforth be primarily responsible for the administration of adult education and the NEH grant, further fundraising, communicating with the trustees, overseeing the Architects' Circle, which was to be expanded, and increasing local and national outreach. MacNair, on the other hand, was to be primarily responsible for the implementation and coordination of all public programs, communications with foundations, the NEA, and NYSCA, as well as sponsor relations and press contacts. This new division of responsibilities reflected the fact that public relations, along with external acquisition, philanthropy, and cultural sponsorship, had become an important area of action and business activity for the Institute, combining self-presentation and external perception, and helping to ensure the Institute's health and financial growth. Coincidentally, communication had become a much-vaunted panacea in the very year that New York itself was elevated to a brand with the "I ♥ NY" image campaign designed by Milton Glaser and implemented on July 15, 1977: urban branding was intended to boost urban tourism, and the Institute, like the metropolis, arguably became a symbol of neoliberalism.[514]

513 Berlin and Kolin, "Accountant Report," 1977–78, June 30, 1978; "Accountant Report," 1978–79 & "Accountant Report," 1979–80, May 30, 1980. Source: CCA Montréal, IAUS fonds: A.2-10.

514 Greenberg, 2008.

"Open Plan"

Following the four courses described in the original NEH proposal to provide adult education on a larger scale and with more exposure for an additional three years, the fall semester of 1977 saw the launch of "Open Plan." Initially coordinated by Andrew MacNair and administered by Frederieke Taylor, with Mimi Shanley assisting them, the name translated the modern architectural principle of floor plan organization into a didactic approach, although the humanistic "NEH Learning Institute Program" originally outlined in the grant application was only implemented in a greatly scaled down and eventually commercialized version. Like "Architecture," "Open Plan" served as a communication tool. It had a clearly structured poster in a typical Vignelli design, printed in a circulation of 20,000 copies and mailed to addresses taken from the databases kindly provided free of charge by established New York institutions and the editors of periodicals. Accordingly, "Open Plan 77" included four courses on "Architecture," "The City," "The Arts," and "Design," each of which ran on one weekday, from Mondays through Thursdays. As presenters, Frampton, Vidler, and MacNair offered courses on the history of large-scale architectural forms ("Cities within Cities"), modernism in the various arts ("The Modernist Vision"), and the contemporary practice of architects and designers ("The Languages of Design"), respectively. The fourth instructor was once again Robert Stern (instead of Krauss), who still did not hold Fellowship status but once again drew a large audience with his course on current trends in American architecture ("Style and Meaning in American Architecture"). The "Open Plan" offering thus once again combined academic and industry knowledge production, covering high-brow and popular culture, now dubbed humanities research, but ultimately not so different from its predecessor and repeating its recipe for success. Each course comprised ten dates with eight lectures from 6:30 to 8:00 p.m., with a fee of US$60 per course. As an add-on, course participants could attend eight accompanying seminars following each lecture, for which an additional fee of US$45 was charged, making the educational package more intense for learners and teachers, and more lucrative for the Institute. The special feature of "Open Plan," however, was a so-called "Open Plan Week" inserted into the fifth and tenth weeks: a special format that, e.g., offered participants the opportunity to hold a panel discussion on the main topic of each course. The course participants benefitted from this because they could attend all the events in these weeks and thus shop for ideas for the next course. The programming and tiered price structure made it clear that "Open Plan" was once again a cultural, educational, and ultimately commercial format. And although prices were still lower than those charged by other institutions for comparable offerings, the public programs guaranteed the Institute additional revenue from admission fees in addition to the NEH grant.

Apart from the fact that the lecture series was better structured, the content, methods, and objectives were fairly well coordinated, and there was a recognizable overall concept, the main difference between "Open Plan" and

"Architecture" was that there was much more money involved. This time, course instructors received a salary of US$3500 and, in return, took on more responsibility, contractually agreeing to design courses and produce teaching and learning materials, on top of delivering a total of three sessions themselves, including the introduction and facilitation of the "Open Plan Weeks." At the beginning of the program, participants received course notes that included a schedule, a description, and a bibliography. Presenters were paid a fee of US$200. Despite the overarching humanities framing, "Open Plan" turned out to be a far more architectural program than advertised, if only because architects, historians, and theorists had been commissioned as instructors, as well as lecturers. Unlike "City as Theater," there were hardly any speakers from other disciplines or professions; the focus on culture was either dropped, or architecture was interpreted as a cultural asset that interacts with other arts and design, and materializes within urban space. In addition, there was an even stronger focus on course instructors. Frampton and Vidler, who had progressed their careers at Columbia and Princeton University, respectively, were to make a name for themselves as architecture historians. The "Open Plan Weeks," which were intended to stimulate interdisciplinarity dialogue or even scholarly exchange, ultimately also functioned more as promotional events—both for the signing up for the next course in the following semester and for the participating architects. From an institutional point of view, they were instrumentalized by inviting members of the Architects' Circle to the discussions in order to stage topical debates in front of a live audience. These included, for example, Philip Johnson, Charles Moore, and Charles Jencks on "Eclecticism, Revivalism, and the Issues of Modernism," Leon Krier and O.M. Ungers on "Revising the Modern Movement: London, Berlin, and New York," and Massimo Vignelli and Ivan Chermayeff on "Forms of Order: The Grid and the Column." With "Open Plan," the Institute then continued to benefit from market and management-oriented strategies rather than the public-focused strategies that had previously been introduced and maintained with "Architecture." From 1977, the generous humanities funding from the NEH enabled the Institute to make all of its efforts bigger, louder, and more profitable, and to present itself to a New York audience as what was later criticized for being "a fashionable salon and power base" of the architecture intelligentsia.[515] Under these circumstances, with this mix of actors, interests, networks, and stakeholders, "Open Plan" eventually institutionalized the debate on forms, ideas, and values of architecture, a historiography of modernism, including post-structuralist approaches drawn from French theory, in response to the varieties of architectural postmodernism that, following philosophical postmodernism, came to dominate the debate and education at the time.

515 Ockman, 1988, 199.

Society Events

Throughout the year, preparations were underway for the anniversary celebration in the fall of 1977. In January, on Eisenman's initiative, several committees were formed to prepare, among other things, a benefit dinner and—quite classically, but perhaps a bit prematurely—a *Festschrift*, all of which was combined with fundraising activities.[516] But when the Institute celebrated its tenth jubilee, there was no sign of this originally planned comprehensive program. Nor did the hoped-for US$1 million in donations that would have financed a library and six fellowships from 1977 to 1980 materialize. Other activities that year, including *Oppositions*, seemed to have exhausted all the Institute's capacities. Nevertheless, at the height of its power in 1977–78, the Institute reached its creative climax, due in no small part to the successive expansion of its public programs and funding from the NEH grant. The anniversary was duly celebrated on November 11, 1977, with a grand ceremony, a string quartet, and in evening attire.[517] In addition to the invitation, Vignelli designed a poster for the event, very pared down, with a black bar and red lettering on a gold background, and only the dates of the Institute's founding and anniversary printed in Roman numerals and the abbreviations IAUS and NYC in capital letters. Flags were hung in the Institute's stately main hall for the ceremony, and Vignelli was responsible for designing a banner, on which IAUS—now used more and more frequently for the Institute's branding—was emblazoned in large capital letters. In addition, the Fellows, Visiting Fellows, students, and interns of the Institute were invited to design individual posters to contribute to the festivities, which were displayed that day. A meeting of the Board of Trustees was held in the morning, and notables from the field of architecture were welcomed in the evening. The character of the celebration, with all its pomp and circumstance, underscored the fact that the Institute was more firmly rooted in mainstream, middle-class culture than it had purported to be with its supposedly radical, autonomous, and critical stance symbolized by its banner in Russian constructivist shades of black and red. The showcasing of the Institute and all its stances was not meant to be ironic but corresponded with the often elitist, rather conservative ideas of architecture that figured prominently at the Institute. The extent to which the Institute had by then become a fixture in New York society became obvious when Brendan Gill later reported on its big anniversary as one of the major social events in New York in his society column "Talk of Town" in *The New Yorker*, which usually combined local reportage and political commentary, but on this occasion bore the headline

516 IAUS, "Structure of Working Committees for IAUS 10th Anniversary," February 1, 1977. Source: CCA Montréal, IAUS fonds: A.2-7. In keeping with the thematic focus indicated in his "Director's Memo," Eisenman also planned a major conference on "The Design of the Public Environment as it Affects the Public Interests."

517 Because of his penchant for number symbolism, Eisenman scheduled the anniversary for November 11, 1977, even though the Institute had been charted on September 29, 1967.

"Partygoing." Indeed, the guest list displayed here was a veritable who's who of the New York architecture community at the time.[518]

In fact, communication with society and the media became increasingly important at the Institute. The Institute took advantage of its good contacts, not only with architecture critics in the trade press, but especially with journalists at the major New York dailies and weeklies, such as the *New York Times*, *The New Yorker*, and even *The Village Voice*. The reports about the Institute and especially the reviews of its exhibitions raised its profile and attracted new audiences to its public events. For example, Paul Goldberger, who had long been acquainted with the Institute, included the NEH grant in his annual review of major architecture events in the *New York Times* in 1977.[519] In it, he placed the Institute's new sponsorship alongside the opening of the Citicorp Center (architect: Hugh Stubbins), the expansion of the Frick Collection (Harry van Dyke and John Barrington Bayley), and the construction of the new Bronx Development Center (Richard Meier). In the context of these major building projects that marked the end of the economic downturn in New York, Goldberger described the growth of the Institute as follows: "The Institute for Architecture and Urban Studies, which started out as a rather cliquish think tank, had evolved by its 10th anniversary last autumn into a lively diverse center of architecture exhibitions, seminars, lectures and classes for the general public. The NEH recognized the Institute's role last year with a 350,000 grant and given the Institute's remarkable ability to bring architectural ideas to a wide public, that grant has to rank as one of the major architecture events of 1977." Clearly, Goldberger welcomed the new developments at the Institute. To single out the highly endowed funding for "Open Plan" and thus for education and culture at the Institute as an important architecture event, comparable to a building, was as surprising as his argument that the Institute would excel particularly in its ability to bring architectural issues to the general public. So while Goldberger was once again drawn to promote the Institute, public relations, and press relations were used ever more effectively to compete for national and international attention. The Institute, which had long positioned itself as a complement to existing offerings, now entered into direct competition with the major art and cultural institutions in New York: with museums, theaters, libraries, but also universities, when it came to grant money, audience favor, and reputation.

518 Brendan Gill, "The Talk of the Town. Partygoing," *The New Yorker* (November 28, 1977), 45.

519 Paul Goldberger, "How the Cityscape fared in 1977," *The New York Times* (January 5, 1977), *The Home Section*, C1 & C11. In his review of the year, Goldberger also explicitly addressed the Department for Housing, Preservation and Development's new J-51 tax incentive, which facilitated property redevelopment and, in his view, had a significant impact on the cityscape for the first time in 1977, justifying the conversion of commercial space into luxury lofts. He cited the landmark status for Grand Central Terminal, which was thus saved from demolition, with relief. He also felt that new restaurants could now be expected to offer good interior design, mentioning the River Cafe in Brooklyn as an example.

"Open Plan" Contd.

From 1977 to 1980, in tandem with its consolidation and maturation as an educational and cultural institution, the Institute's "Open Plan" played a key role in fostering public debate about architecture, urban history, art, and design in New York, with themes like American architecture and suburbanization clearly taking center stage at first. However, its role in reinventing the symbolic economy in New York architecture culture and serving an emerging consumer culture in American architecture was perhaps even more crucial. In the second year of the series, Frampton continued to teach the "Architecture" course and Vidler the "The Arts" course. MacNair stepped down from directing "Open Plan," which Vidler took over on an interim basis in 1978 (first with Silvia Kolbowski, then with Joan Copjec as coordinator) in order to devote himself to other duties at the Institute, most notably the publication of the Institute's own newspaper format, *Skyline*, which featured a calendar of cultural events in New York, including the Institute's Evening Program and Exhibition Program, which were thus placed in this context. Nevertheless, he continued to lead the "Design" course. The program of events was well filled with notable architects, designers, and artists: Georges Nelson, Milton Glaser, Ivan Chermayeff, Gyorgy Kepes, Mario Salvadori, Frank Gehry, and Michael Graves, to name but a few, made an appearance as part of "Open Plan 78." And after Stern held the "City" course for the last time in the spring of 1978 with a course on "The Anglo-American Suburb: Village, House, Garden," this was followed in the fall of 1978 by a course on "Forum on New York. The Place of Urban Design" (instructed by Craig Whitaker), which once again recalled the former focus on urban studies, albeit for the last time. After that, "Open Plan" was supplemented by other, additional formats such as seminars and film screenings, eventually eliminating the rigid arrangement of four thematically defined and discipline-bound courses altogether.[520] With Craig Owens ("Visual Arts: Critical Encounters," spring 1979) and Patrick Pinnell ("The American Monument," fall 1979), both of whom were working as editors for *Skyline*, "Open Plan 79" finally offered the Institute's own junior staff the opportunity to make a public appearance as lecturers in their own courses.

"Open Plan 78" already had a strong historiographical focus, especially with the "Architecture" and "The Arts" courses led by Frampton and Vidler, but also provided space for engaging with current theoretical debates in the other courses. Twice, in the fall of 1978 and in the spring of 1979, a so-called "Advanced Seminar" was offered under the supervision of Gandelsonas, designed specifically as an in-depth study for participants of earlier courses or graduate students in architecture. This

520 Copjec, who had previously studied film at the Slade School in London, learned about the position of Kolbowski through a mutual friend with whom she shared a reading group on feminism and psychoanalysis. She was interviewed by Eisenman and Vidler and immediately offered the job on the strength of her knowledge of French theory. Copjec received her PhD in Cinema Studies from NYU under Michelson and then worked as her assistant. From 1981 she worked on the *October* editorial staff, taught theory courses at the Institute (1982–1984) and worked as a ghost-writer for Gandelsonas.

was a unique offering in architecture theory, with the goal of providing an introduction to current structuralist and poststructuralist philosophy, which in the English-speaking world often goes by the name "French Theory," and a related analysis of contemporary architectural projects. While Gandelsonas gave the general introduction to classical and modern architecture, other Fellows, especially the *Oppositions* editors, led the individual sessions. Frampton introduced Sigfried Giedion's *Space, Time and Architecture* and Vidler Manfredo Tafuri's *Architecture and Utopia: Design and Capitalist Development*—examples of different ways of writing the history of architectural modernism that can be described as operative or critical. Eisenman presented his formal-aesthetic approach using *House X* as an example, again taking the opportunity to introduce his own projects to a wider audience through the Institute.[521] Agrest drew upon her text "Design vs. Non-Design" (from *Oppositions* 6, Fall 1976) for her reading of architecture and the city. Finally, Gandelsonas discussed Michel Foucault's *The Order of Things* as a new model for theory production in architecture. The two international Visiting Fellows, Giorgio Ciucci and Massimo Scolari, both of whom were professors at the IUAV in Venice, were also involved in the seminar, each with their own sessions on "Modes of Representation." With sophisticated courses such as the "Advanced Seminar," the Institute proved that it had become a stronghold of the architecture intelligentsia and a guardian of intellectual debate. Once again, it was evident that there was a common interest in history and theory among the Fellows and that they still shared an intellectual, if not architectural project, even if individual attitudes, for example regarding the question of the autonomy of architecture or the modernist dichotomy of form and function, were quite divergent. The "Advanced Seminar" was offered a second time, but this time the focus was no longer on guided readings of historiographical classics or theoretical writing, but on the idiosyncratic interpretation of three contemporary architects. Essentially centering on the contemporary debate on style in the spirit of postmodernism, following an introduction "On Architectural Languages," two seminar sessions were each devoted to the elements in the work of Robert Venturi, composition in the work of Peter Eisenman, and comparison in the work of Aldo Rossi; Gandelsonas was joined by Agrest, Frampton, Vidler, and Swiss architecture historian Werner Oechslin as speakers. Finally, Gandelsonas himself spoke on semiotic and linguistic aspects in a session on "The Architectural Text." Through Gandelsonas' framework, seminar participants were schooled in a poststructuralist, at times post-Marxist theoretical discourse. However, by presenting them with contemporary projects as arguments for theory, the Institute also engaged in fame-making in terms of a discursive, autonomous, and critical architecture.[522]

521 Eisenman, who first presented *House X* at the 1976 Venice Biennale, subsequently lost the commission; see Peter Eisenman, *House X* (New York: Rizzoli International, 1982).

522 Following the "Advanced Seminar," Gandelsonas wrote his essay "From Structure to Subject" in December 1978, in which he acknowledged the autonomy of form in principle but criticized

In the spring of 1979, "Open Plan" provided the Institute with a platform for another format, the "Saturday Seminars." The concept of this offering, "Against Historicism," was that on five Saturdays in April and May, seminar participants were guided by renowned lecturers from various disciplines through a full-day program that included a thematic lecture followed by a discussion and an afternoon workshop. These "Saturday Seminars" were less of an academic seminar, and more of an exclusive event aimed at an educated middle-class audience that could afford to take the time on weekends to attend such educational offerings. Again, the Institute recruited distinguished scholars, critics, and architects to speak to a select audience; Alan Colquhoun, Richard Sennett, Peter Brooks, and William Gass were invited as keynote speakers. The seminar, however, was introduced by none other than Eisenman himself with a session on "Image and Text," though he was otherwise absent from "Open Plan." Drawing on the literary theory and criticism of structuralist/poststructuralist Roland Barthes and using the example of the postmodern architecture of Michael Graves, John Hejduk, and Aldo Rossi, he addressed key contemporary approaches such as textual analysis and close reading rather than biographical or contextual interpretation.[523] With the success of the "Saturday Seminars," the Institute proved that adult education in architecture was a viable option for the premium segment as well. The discussion of architecture-related, social, cultural, and humanities topics, as framed by Eisenman, could apparently also be sold as an intellectual treat over lunch; however, this one-time offer was to be the only one.

During the period from 1974 to 1980, the Institute held public events almost exclusively on topics intrinsic to architecture—"City as Theater" being the major exception. Criticism of relevant urban planning or broader socio-political issues, on the other hand, was hardly ever voiced. The major contemporary issues of the decade with implications for the profession and the discipline, such as the housing crisis in the context of the shifts in economic and social policy in the United States, global issues such as the energy crisis or the realization that our resources are finite, or local issues such as the economic crisis or urban development policy were not debated—at least not publicly or in front of an audience. When the neoliberal restructuring of federal and urban policies in the wake of the financial and

Eisenman for not taking the subject into account in his designs. He did not reintroduce the subject until his 1976 *Oppositions* editorial "Post-Functionalism," after which he put it into practice with *House X*; see Mario Gandelsonas, "From Structure to Subject. The Formation of an Architectural Language," *Oppositions* 17 (Summer 1979), 6–29.

523 While poststructuralist theory was debating the author during the decade in which Eisenman produced his designs for *House I* through *House X*, Eisenman himself was instrumental in establishing an American version of the figure of the artist-architect that would rise to stardom in the wake of cultural and economic globalization; see Roland Barthes, "Death of the Author," *Aspen Magazine*, no. 5–6 (1967); republished in *Image-Music-Text* (New York: Hill and Wang, 1977), 142–148; Michel Foucault, "What is an Author?" in *Language, Counter-memory, Practice Selected Essays and Interviews*, ed. Donald F. Bouchard (Ithaca: Cornell University Press, 1977), 113–138.

fiscal crisis in New York became a trial run for transformations on a global scale, when privatization and deregulation of all areas of the economy and society, including architecture and the construction industry, set in and the state withdrew, the New York architecture community was forced to realize that its place in the postindustrial or information society, marked by the transition to a service-based economy, i.e., a networked and knowledge society was now fundamentally different. Eisenman, for example, presented an image of himself as an architect-intellectual primarily concerned with design-related decision-making, Frampton and Vidler pursued their own social and intellectual agendas with their historiography of questions of style and epochs, historic periodization, and critical genealogies of a 1970s cryptomodernism, while Gandelsonas and Agrest, on the other hand, were influenced to some extent by Barthes, Foucault, and others on whom they based their postmodern practice of architecture and urban design, etc. The Institute's example, however, shows that not all architects were content to limit themselves to interior design or corporate work, but wanted to break new ground.[524] By focusing on a supposedly autonomous creative practice and emphasizing the formal and contextual, the Institute designed images, roles, and functions of the profession and discipline that were artistically conceptualized, if not necessarily critically reflected. However, as a contribution to postmodernism as a broader cultural phenomenon, the Institute's activities were also symptomatic of "the cultural logic of late capitalism" discussed in literary and cultural criticism at the time and that involved both architectural and cultural production, affecting both material and immaterial culture, images and text.[525] The history of the Institute shows how, by the end of the decade, it had evolved into a major and dominant player in education, culture, and publishing, and had forgotten its origins. This is reflected, among other things, in the fact that the Institute's development and communication now increasingly turned to the establishment and the real estate industry, not only as sponsors but as collaborators. Instead of practicing institutional critique itself, and criticizing the museum and university with their constant assertions of avant-gardism, the Institute, despite maintaining close contact to and friendly relations with major institutions, now tended to confirm and reinforce the status quo, promoting a postmodernization in the educational and cultural spheres through its powerful and pivotal position between the college and the university and its close collaboration with the world of museums and galleries and the economization of everything.

524 Marxist urban and economic geographer David Harvey discussed objects of postmodern architecture, explicitly Philip Johnson's AT&T Building, as a cultural expression of globalized financial capitalism; see Harvey, 1989, 292.

525 The Marxist literary and cultural critic Fredric Jameson, writing along similar lines, saw postmodern architecture as the most visible expression of changes in aesthetic production; see Fredric Jameson, "Postmodernism, or, The Cultural Logic of Late Capitalism," *New Left Review*, no. 146 (July-August 1984), 59–92. Indeed, Jameson even explains that his own conception of postmodernism was born primarily out of an engagement with architecture, which was initiated at the Institute.

Comprehensive Product Range

From 1978, the Institute's exhibition activities were further expanded and professionalized, coordinated by Andrew MacNair, at first primarily with funding for arts and culture. Cross-financed by other programs and, in some cases, even financed by the exhibiting architects themselves, the exhibitions later became largely self-supporting, although low-budget productions still continued. Now managed as an Exhibition Program in its own right, the exhibition activities also served to showcase architects from the Institute's immediate circle—as a way of expressing appreciation or thanks for their contribution and support, as it were. In 1978, for example, it featured solo exhibitions by Charles Gwathmey and Philip Johnson shortly before they were appointed as trustees. The exhibition "Gwathmey Siegel Architects. Twenty-four Residences" (December 15, 1977, to January 15, 1978), which featured axonometric drawings and additional documentation of the firm's 1966–67 residences, was the first for which the Institute received a grant from the New York Council on the Arts. Subsequently, Gwathmey even took on a leading role at the Institute as president. Generally speaking, the Exhibition Program had become more hegemonic and exclusive, and so it was a great honor and distinction for architects to exhibit at the Institute. Contemporary European, Asian, and American practices such as Rob Krier, O.M. Ungers, Leon Krier, Arata Isozaki, Ron Herron and Peter Cook, Lauretta Vinciarelli, Gaetano Pesce, Aldo Rossi, John Hejduk, Mark Treib, or Massimo Scolari were presented to an American audience, in some cases for the first time, but group exhibitions also remained a fixture, now with an overarching theme, such as "The Image of Home: Giuliano Firenzoli, Nancy Goldring, Michael Webb" or "Beyond Historicism." Retrospectives were introduced as a further category, featuring protagonists of European and American modernism, who had previously played no or only a minor role in architecture historiography, and external curators were often commissioned, as for example for the exhibition "Ivan Leonidov: Russian Visionary Architect, 1902–1959" (February 1–20, 1978), which Gerrit Oorthuys of TU Delft presented as Visiting Fellow together with Rem Koolhaas, with photographs the two had taken on their individual research trips through the USSR, and which was a success with both the public and the press. In 1978, a total of eight exhibitions were shown at the Institute, more than ever before in a single year, which led to MacNair being assigned a coordinator, the architect Laurie Hawkinson, who soon became a curator as well.

Not surprisingly, it was the conceptually inclined architects who benefited from the Institute's new exhibition opportunities beyond MoMA or one of the commercial galleries. Following the activities of the Junior Council and the Art Lending Service, however, the dynamics of the art market, and thus the architecture market, changed dramatically in the second half of the 1970s, and architectural drawings and models were now increasingly valorized and marketed as works of art. This became evident at the latest in the fall of 1978, when the Max Protetch Gallery, an established marketplace for minimal and conceptual art founded in

1969 in Washington, D.C., moved to New York, opening a space on 37 West 57 Street that focused on various architectural representations and techniques alongside works of art.[526] For its inaugural exhibition (September to October 1978)—which in addition to Eisenman featured Michael Graves as the other architect, as well as artists Siah Armajani, Richard Fleischner, and Denise Green—Eisenman contributed an anamorphic model of *House X*, shown only at Cooper Union and Princeton University, with which he sought to prove that his experimental house designs, theoretical texts, and autonomous artworks were less concerned with subjectivizing the formalist approach than with individualizing architectural practice. The attention-grabbing strategies displayed by Eisenman, both as a practicing theorist and writing architect, made it clear that he obviously placed more emphasis on the media presence of his projects than on their structural solidity: in 1978, for example, he designed *House El Even Odd* (11a) for a competition called by *Progressive Architecture* on behalf of Kurt Forster, an axonometric model that was never intended as a building but always had a sculptural quality.[527] With a view to enforcing commercial values and norms, the transformation in the New York gallery system set new standards in the field of architecture as well. At the same time, a new exhibition landscape developed beyond the museum: in 1978, for example, P.S.1 opened the Architecture Room, a non-commercial exhibition space initially curated by Lindsay Stamm Shapiro and devoted exclusively to hip North American architects.[528] The Institute, at the latest with its collaboration with Max Protetch, would eventually contribute to an increasing commercialization of architecture culture.[529]

Interestingly enough, 1978 was the year that AIA ended the long-standing advertising ban for architects, an issue that Paul Goldberger had reported on repeatedly.[530] In the same year, also coordinated by MacNair, the Institute launched the "National Architecture Exchange," a platform designed to create synergies between adult education, exhibition activities, and publishing, providing clients with a comprehensive cultural offering and expanding the Institute's

526 Kauffman, 2018, 224ff. According to Kauffman, Max Protetch regularly met with Eisenman and Hejduk during this time to discuss whom to show and what might constitute an architecture gallery in the first place. Protetch is quoted here as saying that he had in mind something similar to the Institute, only commercial. Eisenman later designed the interior of his gallery for him.

527 "Citation: Architectural Design (House 11a)," *Progressive Architecture* (January 1979), 84–85.

528 The Architecture Room at P.S.1 run by Lindsey Stamm Shapiro featured exhibitions on Frank Gehry, Diana Agrest and Mario Gandelsonas, Stanley Tigerman, Melvin Charney, Friday Architects, Studio Works, and Bernard Tschumi. Stamm Shapiro then worked for the Institute, serving in the position of a curator and editor.

529 Kauffman, 2018, 236. The Institute and the Max Protetch Gallery showed four coordinated exhibition pairs: Aldo Rossi (1979), John Hejduk (1980), Massimo Scolari (1980), and later OMA (1982).

530 Paul Goldberger, "Architects Will End Ban on Advertising," *The New York Times* (May 25, 1978), A20; "Institute of Architects Keeps Bans on Advertising and Contracting," *The New York Times* (June 9, 1977), 45.

sphere of influence, although it was only short-lived. As of July 1, 1978, institutions across the United States and Canada were offered the opportunity to book the Institute's lecture series and exhibitions, resulting in a kind of nationalization of a "best-of" the Evening Program and Exhibition Program. For this service orientation, which built on and expanded the "New Wave" series, the Institute again received substantial funding from the NEA and NYSCA. Four exhibitions and two series of lectures were slated to tour each year as part of the "National Architecture Exchange." According to the text on the accompanying poster, the new cultural and educational platform had two main goals: first, building a network, or more precisely, "establishing a network of communication among American universities, museums, and organizations in the city and the suburb," and second, opening up a new market, "offering for national circulation a new series of lectures, exhibitions, catalogues and slide packages." By providing teaching and learning materials, already advertised as a central mechanism in the NEH proposal for the Cultural Institution Grant, the Institute now sought to benefit nationally from the production, circulation, and dissemination of architectural knowledge. Had the platform been operating as planned, all of the Institute's cultural activities, including its publications, with the IAUS Exhibition Catalogues as a new product, would have been subject to a capitalist commodity-type market, rather than a humanist educational mission. The production of a catalogue series, which distributed documents on all the important exhibitions and enabled the Institute to apply for further grants from art and cultural foundations, was meant to cross-finance the organization and realization of exhibitions. The publishing portfolio was thus to be supplemented by the catalogues as a further, ultimately independent, and above all autonomous print product. In this context, the poster for the "National Architecture Exchange" already referred to eight catalogues that were planned retrospectively for the exhibitions of 1976, 1977, and 1978.

Furthermore, the Institute's leadership even envisioned new types of Institute satellites in other North American metropolises. These were advertised to the NEH as part of a decentralized network for regionally adapted content and cooperation with international institutions with a view to developing a program of lectures, seminars, and exhibitions. This plan ultimately failed to find funding.[531] Nonetheless, the Institute established a presence across North America, initially with "A New Wave of Japanese Architecture," a continuation of the "New Wave" series that had already been promoted as part of the 1978 "National Architecture Exchange," this time with a traveling exhibition and slide series.[532] Having focused on the local scenes in a number of European

531 Frederieke Taylor, NEH proposal, 1979. Source: CCA Montréal, IAUS fonds: A.2-10.

532 The following further series of lectures were conceived: "Debates on the Current Scene (Ten Young American Architects)" for spring 1979, "The Berlin Builders (Six German Architects Leading the Way for the 1980 International Building Exposition in Berlin)" for fall 1979, and "South

metropolises before presenting a distinct European practice in its first two iterations, the "New Wave" concept now packaged nationally defined architectural trends from selected countries: first architects from Japan (winter 1978–79), bringing Arata Isozaki, Hiromi Fujii, Toyo Ito, Monta Mozouna, and Osamu Ishiyama to New York for a series of lectures that subsequently toured to nine more cities in the United States and received two favorable reviews in the *New York Times*.[533] The poster exhibition, however, was not on view at the Institute until later (December 20, 1978, through January 30, 1979). In each case, the Institute collaborated with local architects as external curators in designing the exhibitions and accompanying catalogues, but clearly put its own stamp on the cultural productions by prescribing its own specifications for the design of the poster and publications. Here, even more than before, the Institute acted as both cultural entrepreneur and artistic director, a central clearinghouse for maximum publicity. The smaller, sometimes provincial partner institutions, most of which were unable to raise a large production budget of their own, became mere purchasers of cultural products. By offering other institutions the opportunity to efficiently and cheaply book lecture series and traveling exhibitions, trademarked by the Institute, they became involved in the self-organized, flexibilized, and precarized form of cultural production at the Institute as part of a centralized network, in keeping with a cultural critique of postmodernism. And although it is impossible to speak of mass production for a mass audience in a way that would bear comparison with the culture industry, the Institute's advertising for its niche products, which were (self-)produced in small batches, tends to reveal aspects that have now become characteristic of flexible production systems.[534] Ultimately, both the "National Tour," as the "National Architecture Exchange" was affectionately called, and the "New Wave" series proved to be a commercial failure; they were very labor-intensive to produce and were quickly abolished.[535]

American Movement (Visiting Architects, Interior Designers and Planners will Discuss the Last 20 Years of Latin American Design)" for spring 1980 (ultimately, none of these came to fruition). Further slide series included: "Twenty-Four Houses by Gwathmey/Siegel" and "Arcadias and Insertions: Peter Cook and Ron Heron" (it is not clear whether these were ever produced).

533 Paul Goldberger, "An Overview of Japanese Architecture," *The New York Times* (December 22, 1978), C26; Ada Louise Huxtable, "The Japanese New Wave," *The New York Times* (January 14, 1979), D27. Huxtable's verdict was clear: "Stunning and provocative.... If there is an active avant-garde today, this is it."

534 Baird, 2001, 11.

535 Further "New Wave" series on contemporary architectural trends from Switzerland, Spain, and South America had already been planned and announced by the Institute but did not happen due to lack of financial support.

3.4 A Lack of Follow-up Financing

The trouble with the large NEH grant was that plans had to be made for follow-on funding while it was still in place, in order to be able to even sustain operations at this size, especially since the Institute was not built on an endowment. While expansion was already in the cards for the publishing portfolio of *Oppositions*, *October*, and *Skyline* in 1978, with the IAUS Exhibition Catalogues and Oppositions Books as new formats for channeling earlier efforts for a book series, all the Institute's educational, cultural, and publishing activities were increasingly driven by fundraising, public relations, and marketing, with the "cultural resource" being part of the sales pitch. Branding became key, and while Vignelli's strong, distinctive graphic design for all printed materials, not just publications, and especially for the posters, provided an institutional identity, the acronym "IAUS" was now more frequently used as a brand name for the Institute. The impending termination of "Open Plan" ushered in the next hurdle and the Institute's redesign eclipsed everything in the 1978–79 fiscal year. In light of the reconfiguration of government funding for the arts and humanities, a foreshadowing of the neoliberal turn in federal grant policy-making, the task of increasing patronage tied in with previous development work; it first manifested itself at the Institute in late 1978 with the production of a new brochure, compiled by Taylor with Abigail Moseley under a grant from the Charles E. Culpeper Foundation, which provided an updated overview of the Institute's history, its current structure, and program organization, as it became more commercially oriented.[536]

The Institute's brochure was not only a means of communication but can also be read as another institutional document. For the forty-four-page, richly illustrated publication presented the Institute, past projects, current programs, and even future products from an institutional point of view that applied to not one particular person: a retrospective and statement of intent in its best light. To provide clarity, it opened with an organizational chart designed by Vignelli with an overview of the Institute's various fields of work; curiously, the "Public Programs," under which "Exhibitions" and "Open Plan," the "National Architecture Exchange," but also the "High School Program" and all "Publications" were grouped, took center stage here. In 1978, "Research and Development" and "Education" were relegated to the sidelines in the interests of outreach and public relations. While information on the housing projects and educational offerings was placed in the front section of the brochure with a double-page spread, cultural production was portrayed by merely listing details of, for example, all the exhibitions shown to date, the titles of lecture courses, and the names of everyone involved, especially the presenters, but also the lecturers, highlighted as the Institute's network. Just as importantly,

536 Ockman, 1988, 199. Ockman referred to this as the "solicitation of mainstream patronage."

the back section of the brochure listed all the names of the Institute's trustees, Fellows, staff, and faculty at the time. In this well-designed "we"-construct of the Institute, the photographic depiction of the 1978–79 Fellowship now framed the sixteen Fellows as individuals rather than a collective with portraits and roundels, proof of individuation and differentiation.[537] As all areas of the Institute grew and became even more professionalized and bureaucratized, the proportion of female Fellows increased significantly to nearly one-third, albeit in service roles, since women were brought in to perform administrative tasks. The brochure can thus be viewed as a kind of collective biography, a comprehensive documentation of the life and career of the Institute that eliminated all the inconsistencies through seemingly neutral information and networks described in the lists of names. Ultimately, this form of presentation obscured the fact that, in contrast to the cliquish, hermetic group of the early years, the organization of the Institute in the late 1970s now consisted of an inner and an extended circle, was organized hierarchically, and highly stratified in terms of race, class, and gender, and that, starting with a conservative, male representative at its center, it was not simply a reflection of societal conditions, but a social construct in its own right.

On the other hand, the brochure was also an extremely revealing historical document that concealed gaps, breaks, turning points, and ruptures and, with the number and variety of activities, testified to the Institute's current position and marketability in entrepreneurial processes, while still being listed as a nonprofit organization. Page by page, it listed the names of individual supporters, the members of the Architect's Circle, the sponsors of "Open Plan" and *Oppositions*, and the public and private foundations that had financially supported individual areas of work. By the time Taylor began mailing the brochure, it was clear that after the Cultural Institution Grant, the financial base would need to be diversified further. In the future, the Institute would hope for an NEH's Challenge Grant. This first required the collection of private donations and public grants, which would then be matched by government funding—a practice widely used in North America for cultural funding. Beginning in 1979, the Architects' Circle, which Taylor had now expanded and formalized as administrative director, played a crucial role through which the Institute's leadership hoped to find common ground with established architectural firms and successful builders. This circle of friends now included: Edward L. Barnes, Davis, Brody & Associates, Conklin and Rossant, Ulrich Franzen, Philip Johnson & John Burgee, Richard Meier, Mitchell/Giurgola, I.M. Pei, Paul Rudolph, Skidmore, Owings & Merrill, and Robert Stern. In fiscal year 1979–80,

537 The Fellows of the Institute in 1978/79 were: Diana Agrest, Stanford Anderson, Julia Bloomfield, William Ellis, Peter Eisenman, Kenneth Frampton, Suzanne Frank, Mario Gandelsonas, Andrew MacNair, Stephen Potters, Carla Skodinski, Leland Taliaferro, Frederieke Taylor, Anthony Vidler, Myles Weintraub, and Peter Wolf.

one year before the existing NEH grant for "Open Plan" expired, the Institute first entered into negotiations with the National Council on the Humanities in Washington, D.C., for this purpose.

Philip Johnson and the Institute

The 1979 IAUS brochure featured a close-up of Philip Johnson as a representative of the Architects' Circle, arguably the most prominent and provocative exponent of his guild at the time, who was to play a key role at the Institute as a benefactor.[538] Johnson had featured repeatedly in the years before as a donor of *Oppositions* and "Architecture," but had otherwise remained in the background as more of an *éminence grise*. Since Johnson's influence in the world of architecture and building in New York was as great as ever, he advanced to become the most powerful patron of the Institute and was to become closely tied to its fate. Now, as illustrated by the brochure, he was increasingly feted by Eisenman, in a postmodern hagiography, and included in the Institute's various cultural activities.[539] Previously, Johnson had been courted by the Institute's cultural productions in a variety of ways: the tenth issue of *Oppositions* in 1977 was devoted almost entirely to Johnson, exploring his writing and hailing his *Glass House* as a masterpiece,[540] and in the May 1978 issue of *Skyline*, Johnson was given a lot of space to justify his new postmodern stance in an in-depth interview (the first interview in the tradition of Andy Warhol's *Interview Magazine*),[541] and finally, in the fall of 1978, the Institute mounted an exhibition on the AT&T Building.[542] The tremendous attention paid to Johnson at the time by all areas of the Institute—a genuine media hype—was largely due to Eisenman's strategy as Institute director; others were much more critical. The Institute's tribute to Johnson, whose reputation as a corporate and postmodernist architect had been damaged, exemplified the complex mechanism of heightened attention, as public events and publications were used to orchestrate targeted media exposure and

538 In the 1970s, Johnson held the position of power broker in the New York architectural world. In interviews, Eisenman indicated that Johnson regularly helped him by writing checks. No evidence of Johnson's financial support of the Institute before the early 1980s, however, is found in the CCA's IAUS fonds.

539 With the demise of the Institute, Michael Sorkin in The Village Voice retrospectively criticized Eisenman for engaging in hagiography with his publications about Johnson; see Michael Sorkin, "Reforming the Institute," *The Village Voice* (April 30, 1985), 102; republished in *Exquisite Corpse: Writing on Buildings* (New York: Verso, 1991), 110–113.

540 Peter Eisenman, "Behind the Mirror: On the Writings of Philip Johnson," *Oppositions* 10 (Fall 1977), 1–13; Robert Stern, "The Evolution of Philip Johnson's Glass House, 1947–1948," *Oppositions* 10 (Fall 1977), 56–67.

541 Philip Johnson (interview w/ Martha Carroll and Craig Owens), "Skylights: Philip Johnson on Philip Johnson," *Skyline* (May 1978), 7–8.

542 IAUS, ed., *Philip Johnson: Processes. The Glass House, 1949 and The AT&T Headquarters, 1978*, Catalogue 9 (New York: The Institute for Architecture and Urban Studies, 1979). The Institute visited the completed AT&T Building in 1984.

create a longer-term media presence. As part of the power structure, Eisenman made no secret of the fact that he was well aware of Johnson's central importance in the New York architecture community. In a mixture of whitewashing and caricature, he himself had written a short article for *Skyline*, albeit under a pseudonym, about the premier league of New York architects around Johnson.[543] Under the title "The Philip Johnson All Stars" and the pseudonym of Ernesto di Casarotta, he had composed a piece in the jargon of a sports report, providing insights into the web of relationships in the architecture community. While there was talk of "long-standing jealousies and rivalries," Eisenman unabashedly flattered Johnson under the protection of his pseudonym by awarding him a pivotal position. Obviously, there was no way around Johnson at that time if one wanted to become part of the American architecture establishment.[544] Typically, Eisenman also inserted himself into this narrative, virtually a sociogram of the relationships between those who considered themselves the most important New York architects, as one of the protagonists. The fact that he could take the liberty of publishing such an odd piece in *Skyline*, which combined human interest, hype, and gossip, underscored his unique position of power at the Institute, which he refers to here, in passing, offensively as "Istituto nero" [sic!]. Johnson, in turn, benefited from this publicity at the Institute, which he approvingly nicknamed "The Eisenman Institute," while reinforcing its credentials, partially rehabilitating his reputation in the architecture scene.

The exhibition "Philip Johnson: Processes" (September 12 to October 31, 1978), in which the design of Johnson/Burgee Architects for their AT&T Building at 550 Madison Avenue in Midtown Manhattan, i.e., the skyscraper that would become an icon of postmodern architecture upon its completion in 1984, was presented comprehensively for the first time and published in a catalogue, played a decisive role in this respect both for Johnson's profile and for the position of the Institute.[545] The plans for the new, prestigious headquarters of the market-dominating American communications company had just been made public in the spring, whereupon the high-rise, which differed significantly from the Seagram Building, for example, immediately attracted attention because of its postmodern design. This recalled the Roman and Florentine Renaissance, not least because of its striking interpretation of the tripartite structure: an extra-high loggia at the base, a luxurious marble façade for the office floors, and a pediment that immediately evoked associations of broken Chippendale furniture. The Institute itself, where the exhibition was not entirely uncontroversial and provoked strong reactions from some of the Fellows, who rejected the design

543 Peter Eisenman [Ernesto di Casarotta, pseud.], "Quarta Roma: Report from Rome" *Skyline* (August 1978), 6.

544 Plunz and Kaplan, 1984.

545 IAUS, 1979.

out of hand because of its historical eclecticism, came under fire at the time. In his exhibition review in *The Village Voice*, for example, young architecture critic Michael Sorkin sharply attacked the incestuous conditions that specifically underlay the exhibition's conception, before revealing himself to be a harsh critic of the Institute's work in general; the crucial role played by Johnson for the Institute at the time, as "both the prime benefactor and éminence grise," was thus well known in New York.[546] Criticism of architectural production and cultural production, and an analysis of social relations and institutional networks, arose time and again.[547] The major publishers, in turn, supported Eisenman in his efforts to reinstall Johnson as a central figure in American architecture. The volume *Philip Johnson: Writings*, which he edited with Stern, once more a congenial partner, and which was published by the New York office of Oxford University Press in 1979, with a preface by Eisenman and an introduction by Vincent Scully, was a celebration of Johnson the author. Like Stern, Eisenman thus secured a special position in Johnson's entourage and was high up on his list, before Richard Meier, Michael Graves, and Frank Gehry.

For Eisenman, this alliance with Johnson came to play a key role in his search for further financial backing and corporate patronage. In his quest for power and fame, and to secure the Institute's continued operations, Eisenman got involved with Johnson, even though it was fairly well-known in the American architecture world that Johnson had sympathized with fascist ideology in the 1930s and had even, as a correspondent for the German Reich, published anti-Semitic texts in the leading American newspapers *Examiner*, *Social Justice*, and *Today's Challenge*.[548] Eisenman, like many others, was apparently not that concerned; on the contrary, he continued to court Johnson and tried to use this knowledge to his own advantage. Over the years, a close yet complicated relationship developed between Johnson and Eisenman, one of mutual esteem and dependence, and by supporting him, he made the Institute dependent on

546 Michael Sorkin, "Philip Johnson: The Master Builder as a Self-Made Man," *Village Voice* (October 30, 1978), 61–62; republished in *Exquisite Corpse: Writing on Buildings* (New York: Verso, 1991), 7–14.

547 Kazys Varnelis, "The Spectacle of the Innocent Eye: Vision, Cynical Reason, and the Discipline of Architecture in Postwar America," PhD diss., Cornell University, 1994.

548 Franz Schulze, Johnson's biographer, was the first to thoroughly review Johnson's fascist past; see Franz Schulze, *Philip Johnson. Life and Work* (New York: Alfred A. Knopf, 1994). On Johnson's political activities in the 1930s, see also Kazys Varnelis, "'We Cannot Not Know History:' Philip Johnson's Politics and Cynical Survival," *Journal of Architectural Education* 49, no. 2 (November 1995), 92–104. Varnelis argued that Johnson's entire career, indeed his entire life, must be seen against this backdrop and compared his reappraisal of Johnson's right-wing past to discussions of Paul de Man and Martin Heidegger. There was little discussion of how Eisenman and the Institute dealt with this knowledge.

Johnson's favor, even turning it into his metaphorical backyard.[549] Eisenman may have been acting out of responsibility towards the Institute, and perhaps towards the Fellowship, but he was entering dangerous territory, turning a blind eye to reality and pursuing a policy of double standards—and double-speak. Institutionally and discursively, the Institute became Johnson's ally or accomplice and was also to benefit from the attention economy, as the latter rose to superstardom in the American architecture world.[550] In January 1979, Johnson was featured on the cover of *Time* magazine, holding a striking model of the AT&T Building, which announced an in-depth feature on "U.S. Architects. Doing their own thing."[551] At the latest, this cover, which turned the architectural model into a media event, made it clear that architecture was entering into a new relationship with the market in the wake of geo-economic restructuring, which made it interesting for the Institute again.

In the process, from the perspective of a Marxist critique of urban development and the profession, architects were now increasingly assuming a merely decorative role for the new global accumulation regime, for which they provided the enticing images: "fiction, fragmentation, collage and eclecticism, all suffused with a sense of ephemerality and chaos," these were the strategies adopted by postmodern and deconstructivist architecture and urban design at this time in order to attract even more investment.[552] Changes in architecture and media politics in general, like the coverage of notable architects, already indicated that in the coming decade, some would achieve celebrity status, turning into commodities themselves. Johnson knew how to play this particular game like no other. In May 1979, the seventy-two-year-old architect was awarded the Pritzker Prize for lifetime achievement. Already compared to the Nobel Prize in the announcement, this was the first annual Pritzker Prize in Architecture to ever be awarded, and thus a particularly symbolic one. Johnson received it, according to the jury, "for 50 years of imagination and vitality embodied in a myriad of museums, theaters, libraries, houses, gardens, and corporate

549 Eisenman's dependence on Johnson defined the Institute until its closure in 1985, and finally culminated in 1988 in the "Deconstructivist Architecture" exhibition at the Museum of Modern Art (June 23 to August 30, 1988); see MoMA, "Deconstructivist Architecture," Press Release no. 29 (March 1988), https://www.moma.org/momaorg/shared/pdfs/docs/press_archives/6526/releases/MOMA_1988_0029_29.pdf (last accessed: May 31, 2023); see also Philip Johnson and Mark Wigley, eds., *Deconstructivist Architecture* (New York: Museum of Modern Art, 1988).

550 Varnelis used the example of the contract awarded to Johnson/Burgee for the AT&T Building to discuss the underlying networks, see Varnelis, 2009.

551 *Time Magazine* (January 8, 1979): "U.S. Architects. Doing their Own Thing." Varnelis also pointed out that the photo of Johnson in the pose depicted there is reminiscent of Moses and the tablet with the Ten Commandments, see Varnelis, 2009.

552 Harvey, 1989, 66ff., especially 98.

structures."[553] In his extraordinary way, Johnson dedicated the award to the art of architecture, ennobling himself in the process. In an attempt to tie the newly crowned superstar to the Institute as closely as possible, and thus intensify contacts in the building and real estate world, Johnson was appointed a trustee on February 1, 1980, which gave him more influence on Institute policy than before—a conscious decision that in the medium term led to Institute business now also being negotiated at the Century Association, an exclusive New York social club headquartered on 43rd Street, not far from the Institute, i.e., in the heart of New York high society, where Johnson held court.[554]

The Beginnings of "Starchitecture"

Through a thorough analysis of its cultural productions and its cultural politics, a critique of the Institute as one of the "cultural spaces" of the New York architecture community will, by undertaking a close examination of the interplay between education, culture, and publishing, ultimately help us understand the emerging phenomenon that was subsequently described with the neologism "starchitecture."[555] From a sociology of culture perspective, the Institute offers important insights into how the genesis of the star system obscured or normalized the interconnections between architecture and the market, i.e., the economic mechanisms of a capitalist construction and real estate economy.[556] The Institute illustrates not only how architecture became intertwined with other arts, the humanities, and the cultural sphere in general, but also how, not least due to the growing sectors of the creative and cultural industries, an economization of all forms and formats of cultural production took place in the 1970s. The Institute's remarkable list of lecture series and exhibitions demonstrates that even then, before the key events of postmodernism in the 1980s, a celebrity culture and eventually a global star system were established, which received an additional boost when architects became brands themselves and increasingly competed as actors for the realization of iconic projects in a globalized architecture world. Through the program and organization of "Architecture," "Open

553 Philip Johnson, 1979 Laureate, www.pritzkerprize.com/laureates/1979 (last accessed: May 31, 2023).

554 The Century Association was frequented by architects; in addition to Johnson, who had belonged to the club since 1968, Stern (since 1976) and Eisenman (since 1977) also became members, as did many others. Among the Institute trustees: Armand Bartos (since 1978), John Burgee (1979), Colin G. Cambell (1978), Henry N. Cobb (1974), Charles DeCarlo (1974), Gibson Danes (1960), George A. Dudley (1971), Ulrich Franzen (1983), Edward Logue (1972), Richard Meier (1976), Cesar Pelli (1983), T. Merrill Prentice (1966), Jaquelin Robertson (1974), Edward L. Saxe (1984), Frank Stanton (1948), John F. White (1964), and Peter Wolf, (1976); see *The Century Yearbook* (New York: Century Association), see also Plunz and Kaplan, 1984.

555 Davide Ponzini and Michele Nastasi, *Starchitecture. Scenes, Actors, and Spectacles in Contemporary Cities* (New York: The Monacelli Press, 2016).

556 Andreas Reckwitz, "Die Genese des Starsystems," in *Die Erfindung der Kreativität. Zum Prozess gesellschaftlicher Ästhetisierung* (Frankfurt: Suhrkamp, 2012), 239–268.

Plan," the "New Wave" series, and the "National Architecture Exchange," as well as various exhibitions, the Institute furthered the popularity of various architects—emerging architects from Europe such as Rem Koolhaas, Elia Zenghelis, Robert and Leon Krier, Aldo Rossi, and Massimo Scolari, alongside established firms from the United States, as well as architecture historians—and was thus partly responsible for the economization of creativity and criticism. In the second half of the 1970s, reviews by the two regular architecture critics writing for the *New York Times*, Ada Louise Huxtable and Paul Goldberger, now appeared more frequently and reported favorably on the Institute. This is indicative of how and how much the Institute, personally championed by Eisenman, shaped Rossi's reception in North America. Under Laurie Hawkinson as program director, "Aldo Rossi in America. Città Analoga Drawings" (September 19 to October 20, 1979), Rossi's second solo exhibition at the Institute after 1976 and the first to be coordinated with the Max Protetch Gallery, tried to satisfy both intellectual interests and commercial ones, though this distinction made was not clear or unequivocal. While the Institute showed drawings by Rossi made during his previous stays in the United States, the Max Protetch Gallery, with "Aldo Rossi: Architectural Projects," (September 18 to October 13, 1979) offered drawings of realized and unrealized projects for sale.[557] The two exhibitions opened on two consecutive evenings, with Max Protetch first, highlighting the commercial interests.[558] The symbiotic nature of the arrangement along the culture/commerce axis was reflected in the exhibition catalogue produced by the Institute with an edition of 1500 copies at a retail price of five dollars; in exchange for a certain number of free copies and the placement of ads in *Skyline*, the Max Protetch Gallery covered the costs of shipping and framing the images. Then, in the October 1979 issue of *Skyline*, Rossi was thoroughly hyped with an exclusive but oddly edited interview by Diana Agrest, and his drawings were featured on the cover and inside the paper.[559] At the time, reviews in both the *New York Times* and *Progressive Architecture* were uniformly positive.[560] This enormous exposure in New York greatly increased Rossi's popularity and reputation in the United States, in part because the Institute subsequently sent the exhibition on tour, stopping off at ten North American cities as part of the National Tour. In addition, Eisenman had previously paved the way at the Institute for two monographs: the long overdue English-language translation of *L'Architettura della Città* and the valuable first publication of *A Scientific Autobiography*, for

557 Kauffman, 2018, 236, 264. Legend has it that even some of drawings by Rossi exhibited at the Institute may have been sold, with Eisenman acting as a facilitator.

558 "Skylights: Rossi Opening Crowds" *Skyline* (October 1979), 15.

559 Diana Agrest, "The Architecture of the City: An Interview with Aldo Rossi," *Skyline* (September 1979), 4–5.

560 Ada Louise Huxtable, "The Austere World of Rossi," *The New York Times* (October 1979), D31; see also "Aldo Rossi: Two Exhibits," *Progressive Architecture* (October 1979), 21, 23.

which negotiations with MIT Press had begun. By acting as an intermediary in the sale of Rossi's drawings off the wall, Eisenman overstepped his competencies as Institute director; likewise, he overstepped his authority as a critic by contributing his creative misinterpretations of Rossi's understanding of autonomy and architecture in his preface to the exhibition catalogue, or, later on, in his introduction to one of the books.[561] More than that: Eisenman used this cultural management in his usual manner to disseminate his own ideas of culture, sociology, and art.

At the beginning of the new decade, the Institute held its first major retrospective of an American architect, titled "Wallace K. Harrison: New York Architect" (December 18, 1979, to January 12, 1980). This was apparently the first-ever major retrospective of Wallace Harrison, who was best known for his contribution to Rockefeller Center, the design of the UN Headquarters, and Lincoln Center, all prominent New York modernist buildings.[562] Once again Koolhaas, who had become something of a star himself with the publication of *Delirious New York* in 1978, an exhibition at the Guggenheim Museum that same year, as well as a portrait in *Interview Magazine*, was invited to curate the exhibition—a new chapter in the Institute's historiographical endeavors.[563] Koolhaas used the same tactics as when researching for his monograph that brought Harrison out of obscurity: having gained access to Harrison's private archive, he selected sketches, plans, and photographs of realized and unrealized projects, both well-known and unknown, to celebrate the American corporate architect for his metropolitan architecture as well. Koolhaas also knew how to use exhibition design as a provocative device and built a curved wall as a special feature, copying Harrison's formal language, creating more wall space, and, to top it off, mocking the orthodox Le Corbusier reception of the Institute's Fellows. Reviews in the *New York Times* were favorable.[564] Immediately thereafter, the Institute was to partner with the Max Protetch Gallery on two more occasions. First with a coordinated exhibition on Hejduk in the winter of 1980, with "John Hejduk: Seven Houses" (January 22 to February 16, 1980) on display at the Institute, and Max Protetch showing "The Works of John Hejduk" (January to February 16,

561 Peter Eisenman, "The House of the Dead as the City of Survival," in *Aldo Rossi in America: 1976–1979*, Catalogue 2, ed. IAUS (Cambridge: MIT Press, 1979), 4–15; Peter Eisenman, "The Houses of Memory: The Texts of Analogue," in Aldo Rossi, *The Architecture of the City* (Cambridge: MIT Press, 1982), 2–11.

562 Kim Förster, "The Institute for Architecture and Urban Studies: Wallace K. Harrison. New York Architect (Press Release)," in *Architecture Itself*, ed. Sylvia Lavin (Leipzig: Spector Books, 2019), 121–124.

563 Rem Koolhaas, *Delirious New York. A Retroactive Manifesto for Manhattan* (New York: Oxford University Press, 1978).

564 Paul Goldberger, "Architecture: Harrison Retrospective," *The New York Times* (January 2, 1980), C 15; Ada Louise Huxtable, "Reexamining Wallace Harrison," *The New York Times* (January 6, 1980), D23.

1980)—both the series "The Thirteen Watchtowers of Cannaregio" and a selection of house designs. This was all the more remarkable because Hejduk had been excluded from the MoMA exhibition "Transformation in Modern Architecture" shortly beforehand in 1979 because of Drexler's curatorial approach, which had dictated that only photographs of realized projects be shown there,[565] a provocative statement that was highly contested at the Institute, which in 1980 also published Hejduk's book of poems, *The Silent Witness and Other Poems*.[566] The Hejduk exhibition was followed by a coordinated exhibition on Massimo Scolari in the spring of 1980, with "Massimo Scolari: Architecture. Between Memory and Hope" (May 6 to June 20, 1980) at the Institute, and Max Protetch showing "Massimo Scolari: Drawings and Watercolors" (May 13 to June 7, 1980). For the time being, however, no further cooperation with Max Protetch was to take place. The early 1980s saw exhibitions on a less regular basis, when the opportunity arose: "Mark Treib: Some Posters on the Theme of Architecture" and "A New Wave of Austrian Architecture," coordinated by Missing Link, the last iteration of the "New Wave" series.

The End of Plenty

With the inevitable expiration of the NEH Cultural Institutions Grant at the end of fiscal year 1979–80, the Institute's publicly funded output as a cultural space within a carefully balanced range of education offerings, public programs, and publications began to shift. The roles and responsibilities for the Evening Program and the Exhibition Program had been switched around repeatedly in previous years, leaving little continuity. For the final year of "Open Plan," Patrick Pinnell had been entrusted with directing the program, with Vidler as his advisor. Nearly all of the veteran Fellows contributed to the series of lectures in one way or the other, however, to justify the NEH funding and to endorse the Institute's application for follow-on funding. Gandelsonas and Vidler co-taught a course on "Piranesi/Le Corbusier," with lectures on the modern reception of the Italian artist and the modernist urbanism of the Swiss-French architect, and Frampton presented "Housing versus the City," with lectures on building and settlement types in Europe and North America—explicitly on the perimeter block, the estate, and the suburb, and for the last time on the Institute's prototype of low-rise, high-density housing. By the late 1970s, New York architecture culture had certainly changed, and the Institute was implicated. As his substantive contribution, Pinnell now offered a course called "The American Monument," which presented

565 Paul Goldberger, "Architecture: Houses Designed by John Hejduk," *The New York Times* (January 32, 1980), C15; Ada Louise Huxtable, "John Hejduk: A Mystic and Poet," *The New York Times* (February 3, 1980), D25.

566 John Hejduk, *The Silent Witness and Other Poems* (New York: The Institute for Architecture and Urban Studies, 1980). Pieter Sanders collected works by John Hejduk in particular, and the two became friends. He also financed the production of Hejduk's book of poems with US$ 5000, an investment he called "big business."

fairly national positions on the current monumentality debate, with lectures on the Capitol in Washington, D.C., and the National Gallery, on Wallace K. Harrison, on Frederick Law Olmstead, on Frank Lloyd Wright, on Levittown, and on the skyscraper as a building type in general. Staying true to his earlier approach, MacNair presented the course "Architecture in the 1980s," anticipating developments of the new decade, which would be all about postmodernism. In it, he paired two architects at a time to exchange ideas, including some of the Institute's trustees and friends: Charles Gwathmey and Stanley Tigerman, Richard Meier and Ulrich Franzen, and, as newcomers to the scene, Friday Architects and Arquitectonica. Even Wallace Harrison made a personal guest appearance during "Open Plan Week."

But in the second semester, "Open Plan," as it had existed since 1977 with its humanistic structure of courses on "Architecture," "The City," and "The Arts," as well as "Design" ended. In its last iteration in the spring of 1980, the conservative traits that dominated the transformation of the American economy and culture of the 1980s and ultimately led to a drastic restructuring and polarization that culminated in the culture wars, were already becoming palpable. Thus, "Open Plan," now with the subheading "Architecture in American Culture," was transformed into a more profitable and popular format, with eight smaller courses organized in two blocks, and featured a wider range of topics, some of which were quite a delicate affair. In the first block, Frampton and Gandelsonas each presented a course with American content: "Louis Kahn. Modernism as Tradition" and "America vs. Europe: Symbolic Exchanges and Transformations." Given the Institute's history and social standing, the invitation of Michael Sorkin—architecture critic for *The Village Voice* and troublemaker—who had previously followed the Institute closely, to present a course was a new departure. Titled "The Family: Sources of the Architectural Status Quo," he took the opportunity to focus on the powerful networks in American architecture culture and in particular the role of educators, clients, curators, gallery owners, critics, and architects: the clearest attempt yet at an institutional analysis, even as a critical, conceptual, and curatorial approach. In addition, Mary McLeod, the first female presenter at "Open Plan," offered a course on "Architecture and the Social Order: Style, Politics, and Regeneration," which not only focused on political interdependencies and reformist approaches but also outlined a feminist critique of architecture and the city. The second block showed that the debate surrounding postmodernism was only just gaining momentum, reproducing the prevailing ideologies of tradition and innovation, author and work, to reflect and reinforce the origins and lines of development of the new American architecture culture. Pinnell presented a course on "Frank Lloyd Wright: Tradition as Modernism" and William Howard Adams on "Architecture and the Ideology of Nature: Gardens as Ideal Forms." Vidler, however, offered a course with the title "Shadowboxing: Modern and Postmodern in the 1980s," which focused on the analysis and meaning of new positions and approaches, the discourse and culture of postmodernism from a humanities perspective. With his own contribution, a lecture titled "Beyond the Isms: The Question of Architecture Itself," and lectures by Eisenman ("Ghosts in the Stadium. Players and Programs"), Rosalind Krauss ("Is there Culture without Style?"), John Hejduk ("Is there a Fascia in Mies?"), and Alan Colquhoun ("Newspeak? Architecture Parlante in the Eighties"), Vidler's course, unlike MacNair's, dealt with the big unresolved questions of the new, all-important decade of architecture as a brand, culture, politics, and style.

According to the agreement between the Institute and the NEH, the original plan was to organize another event to conclude the cultural promotion of architecture as part of the humanities in 1980. At that time, there were serious plans to organize a conference on "The Architect and the Developer" (with, among others, Tafuri as the most prominent speaker, who was to contribute a capitalist critique of real estate), which promised not only to furnish a genuine debate but also at the same time facilitate bridge-building with the construction and finance industries. By 1980–81, negotiations were already underway with Jonathan Barnett, formerly a partner in the City Planning Commission, who had presented a course in the framework of the Evening Program on architecture as an art, profession, and business, and who was now developing and revising a concept for a conference on the topic of "Architecture, Development and the New Investment Pattern: Can They Co-Exist?" already scheduled for September 25, 1981 (it is not clear, however, whether the conference actually took place).[567] Ultimately, it was Vidler who, after helping to write the concept in 1977 but then remaining largely in the background, submitted a final report on "Open Plan" to the NEH in which he explicitly and confidently attested to the institutional and discursive success of the program.[568] At the end of the report, Vidler retrospectively concluded that the adult education offering had indeed not reached as much of the "wide and general non-professional audience" outside of architecture and had not contributed as much to "the study, discussion and understanding of architecture in contemporary culture" as hoped, openly admitting that the main goal had not been achieved. But despite these concessions, he concluded that "Open Plan" had nevertheless succeeded, at least in the first two years, in stimulating a dialogue between a wide variety of disciplines "ranging from history, city planning, aesthetics, cultural studies, interior design, urban design, to architecture." In his opinion, it had been extremely successful in promoting "a discussion of the relations among the different arts and the humanities" (though he did not say that he himself was instrumental in this with the "The Arts" course) and, to a certain extent, in demonstrating "the central role of architecture as a humanistic discipline." According to Vidler's

567 Jonathan Barnett, proposal for a conference, October 20, 1980, November 10, 1980, November 11, 1980, April 6, 1981, April 10, 1981. Source: CCA Montréal, IAUS fonds: A 5-11.

568 Anthony Vidler, "Report on Open Plan for the NEH," July 22, 1981. Source: CCA Montréal, IAUS fonds: A.1-6.

self-evaluation, the Institute had made an important contribution to fostering a greater awareness of "architecture and its public role" in metropolitan society and the mass media. In its response, the NEH acknowledged the Institute's work with "Open Plan" and the quality of the individual courses but criticized the Institute for not holding the long-planned conference on the relationship between architecture and the humanities as the culmination and conclusion of the Cultural Institution Grant.

The end of the generous funding in the summer of 1980 marked the beginning of a new era. The Institute lost one of the most important cornerstones of its programming and funding model. This not only led to considerable destabilization and uncertainty; above all, with the end of the "Open Plan," it lost the public character it had been cultivating since 1974 with its educational and cultural work, holding lectures every evening during the semesters over a period of six years—first with "Architecture" and then with "Open Plan." When this collapsed, the Institute was left with no high-profile forum for debate in the New York architecture community and no central mechanism for generating attention. Besides the education offerings, only the Exhibition Program continued after 1980 and, along with the IAUS Exhibitions Catalogues, became an important component. When Lindsey Stamm Shapiro took over as program director in 1981, the production of exhibitions became larger and more professional, powerfully networked, and on several occasions built on collaboration: "Le Corbusier's Saint-Pierre de Firminy, Early Drawings: 1961–1962," curated by José Oubrerie and shown as a double exhibition at Cooper Union as well, "Clorindo Testa: Architecture and Personal Mythology," curated by Jorge Glunsberg, "Kazuo Shinohara: 11 Houses (1971–1976)," with Frampton as the driving force behind it (and which later travelled to UQAM in Montréal, sparking regionalist debates), "Raymond Hood," curated by Robert Stern, "Office for Metropolitan Architecture. Toward a modern (re)construction of the European city: Four Housing Projects," conceived by the office and coordinated with the Max Protetch Gallery (March to April 3, 1982), "William Lescaze," curated by Barbie Campbell Cole, with the support of Syracuse University's School of Architecture, and "New Symbolism: The West Coast Architects," coordinated by Frank Gehry and Mark Mack. In many cases, the NEA and NYSCA funding and further corporate and private sponsorships once again revealed the entrepreneurial side of cultural production at work alongside the curatorial one. Apart from "Idea as Model" in the early days, none of the Institute's practicing Fellows had been granted a solo exhibition: Eisenman's Terragni show did not materialize, and an exhibition on projects by Agrest Gandelsonas that was still under discussion in 1981 was eventually overturned by Eisenman.

By 1980, the Institute's Fellows had turned primarily to publishing as a cultural practice. The various formats were to become the Institute's primary mission, based on the contracts with MIT Press for *Oppositions*, the IAUS Exhibition Catalogues, and, most recently, the Oppositions Books. Implementing this and making it at least somewhat profitable required further restructuring and reinvention of the Institute, which would continue to operate as an architecture school, working primarily as an editorial office, but would function differently from the cultural space when it came to creating synergies between the various areas of activity. The Institute's contribution to architecture culture, as a network of networks, was to have enabled and promoted a mode of cultural production that created an attention economy that was not only symbolic but also political. More than that, by producing and reproducing knowledge, information, and communication, by talking about individual positions and approaches, and by exhibiting the *dispositif* of creativity and autonomy, it had demonstratively staged a new canon of theoretically and historically considered practices. By facilitating and encouraging architects, historians, and theorists to work as cultural producers in the 1970s—when jobs in New York were scarce—especially since funding for the arts and the humanities had been available for a time as an important source of income, the Institute, as cultivator, contributed decisively to the assertion of postmodern architecture as a fashionable style and to postmodernism as a cultural phenomenon.

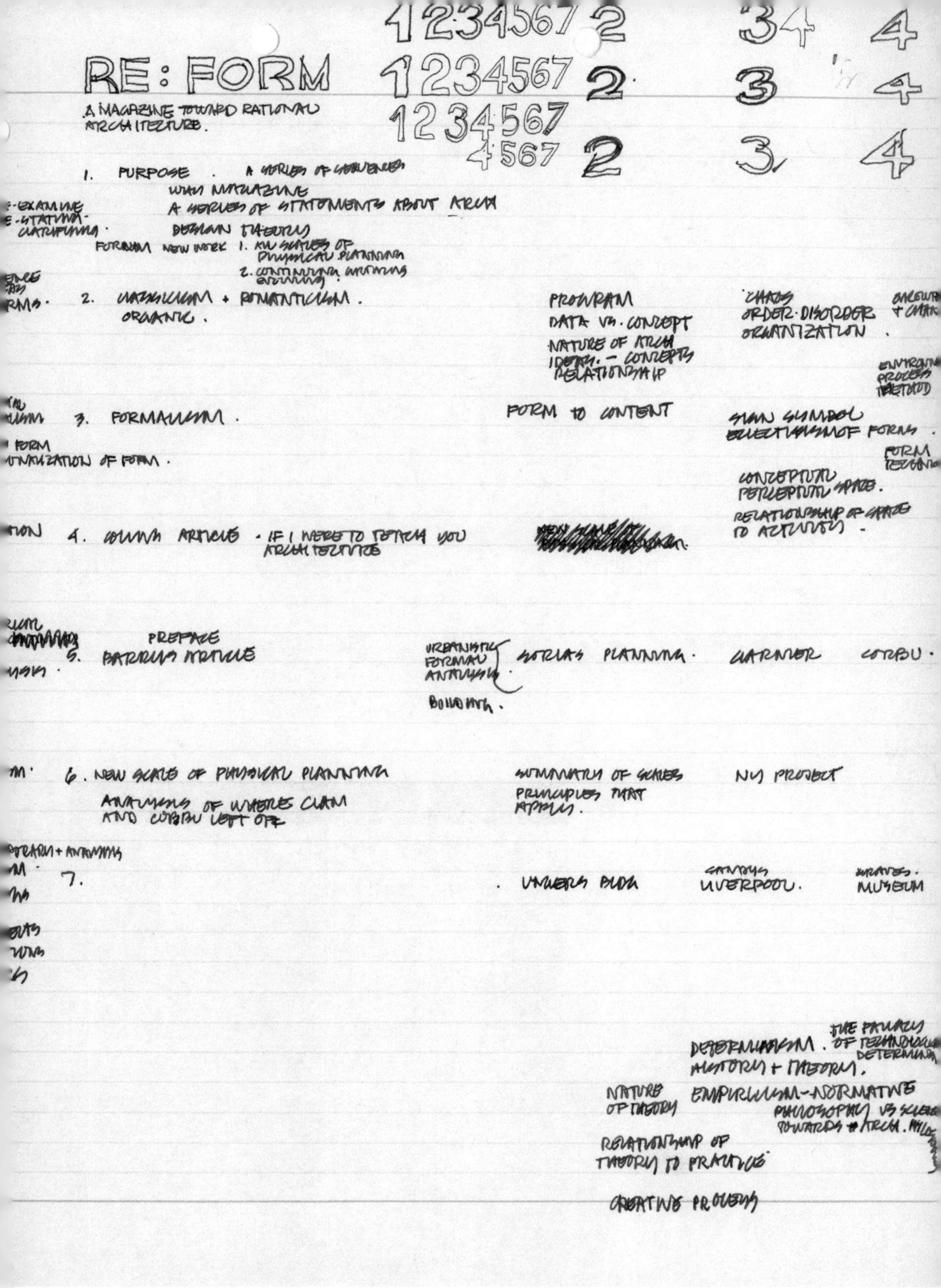

RE: FORM

1234567 2 3 4

1234567 2 3 4

1234567

4567 2 3 4

A MAGAZINE TOWARD RATIONAL ARCHITECTURE.

1. PURPOSE . A SERIES OF SEQUENCES
WITH MAGAZINE
A SERIES OF STATEMENTS ABOUT ARCH
DESIGN THEORY
FORUM NEW WORK 1. NEW SCALES OF PHYSICAL PLANNING
2. CONTINUING WRITINGS

2. CLASSICISM + ROMANTICISM .
ORGANIC .

PROGRAM
DATA VS. CONCEPT
NATURE OF ARCH
IDEAS. – CONCEPTS
RELATIONSHIP

CHAOS
ORDER · DISORDER
ORGANIZATION .

3. FORMALISM .

FORM TO CONTENT

SIGN SYMBOL
ELECTICISM OF FORMS .

CONCEPTUAL
PERCEPTUAL SPACE .

RELATIONSHIP OF SPACE
TO ACTIVITY -

4. COLUMN ARTICLE · IF I WERE TO TEACH YOU ARCHITECTURE

[redacted]

PREFACE
5. BARRIS ARTICLE

URBANISTIC
FORMAL
ANALYSIS .
BUILDING .

SORIAS PLANNING . GARNIER CORBU .

6. NEW SCALE OF PHYSICAL PLANNING
ANALYSIS OF WHERE CIAM
AND CORBU LEFT OFF

SUMMARY OF SCALES
PRINCIPLES THAT
APPLY .

NY PROJECT

7. . LEICESTER BLDG   STIRLING LIVERPOOL .   GRAVES . MUSEUM

DETERMINISM . THE FALLACY OF TECHNOLOGICAL DETERMINISM
HISTORY + THEORY .

NATURE OF THEORY   EMPIRICISM – NORMATIVE
PHILOSOPHY VS SCIENCE
TOWARDS A ARCH. ...

RELATIONSHIP OF
THEORY TO PRACTICE

CREATIVE PROCESS

Fig. 89

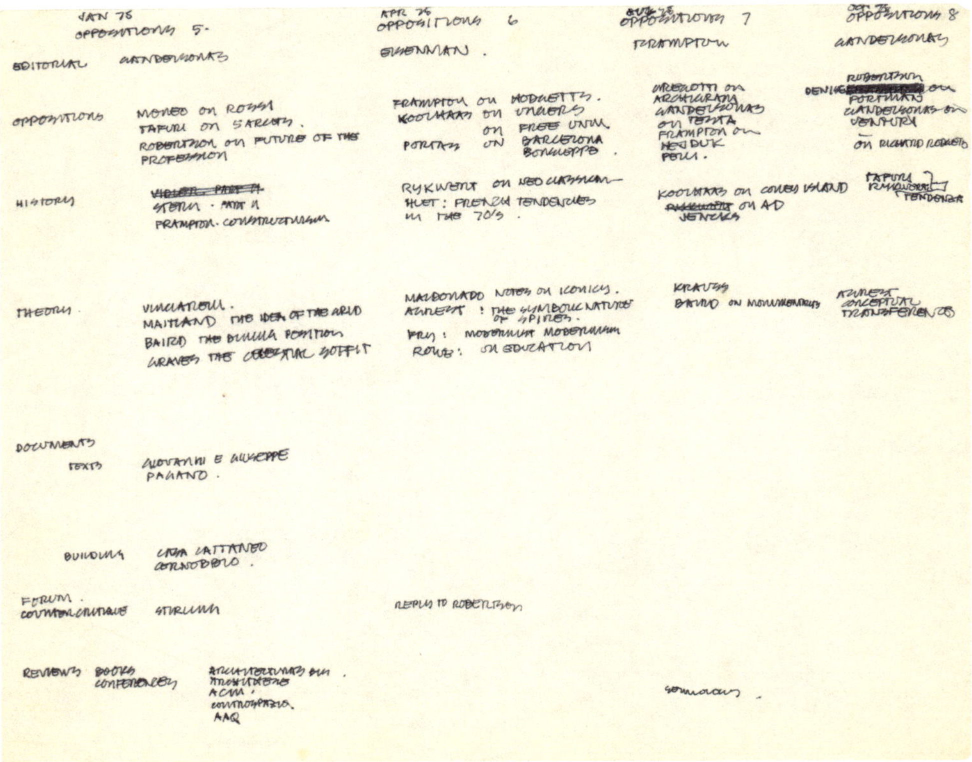
JAN 75
OPPOSITIONS 5.
APR 75
OPPOSITIONS 6
OPPOSITIONS 7
OPPOSITIONS 8
EDITORIAL
OPPOSITIONS
HISTORY
THEORY
DOCUMENTS
TEXTS
BUILDING
REVIEWS BOOKS
CONFERENCES

Fig. 90

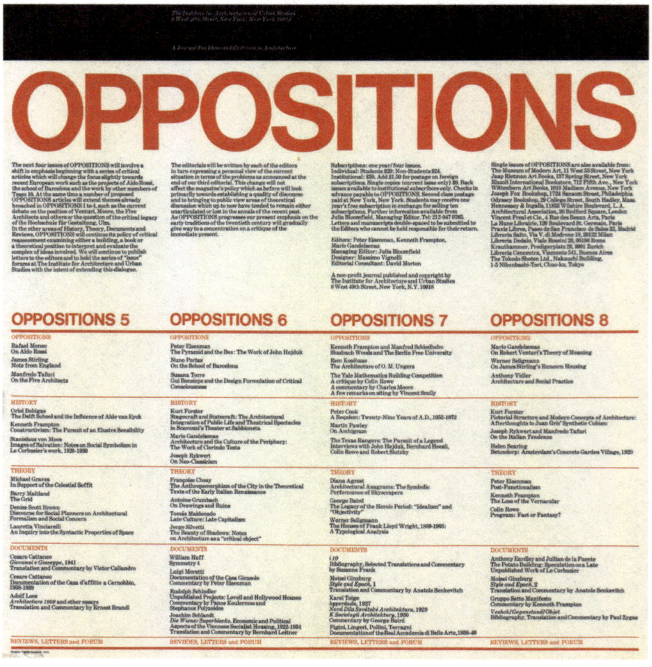
OPPOSITIONS
OPPOSITIONS 5
OPPOSITIONS 6
OPPOSITIONS 7
OPPOSITIONS 8

Fig. 91

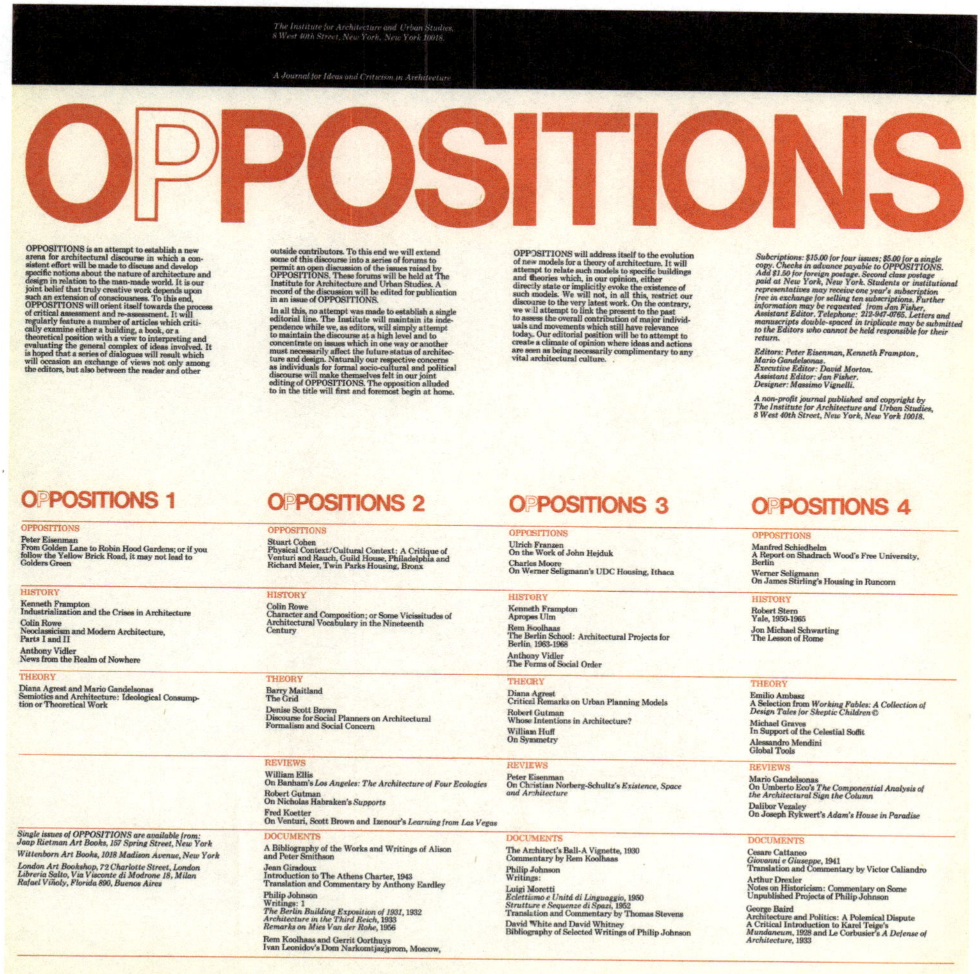

The Institute for Architecture and Urban Studies, 8 West 40th Street, New York, New York 10018.

A Journal for Ideas and Criticism in Architecture

OPPOSITIONS

OPPOSITIONS is an attempt to establish a new arena for architectural discourse in which a consistent effort will be made to discuss and develop specific notions about the nature of architecture and design in relation to the man-made world. It is our joint belief that truly creative work depends upon such an extension of consciousness. To this end, OPPOSITIONS will orient itself towards the process of critical assessment and re-assessment. It will regularly feature a number of articles which critically examine either a building, a book, or a theoretical position with a view to interpreting and evaluating the general complex of ideas involved. It is hoped that a series of dialogues will result which will occasion an exchange of views not only among the editors, but also between the reader and other outside contributors. To this end we will extend some of this discourse into a series of forums to permit an open discussion of the issues raised by OPPOSITIONS. These forums will be held at The Institute for Architecture and Urban Studies. A record of the discussion will be edited for publication in an issue of OPPOSITIONS.

In all this, no attempt was made to establish a single editorial line. The Institute will maintain its independence while we, as editors, will simply attempt to maintain the discourse at a high level and to concentrate on issues which in one way or another must necessarily affect the future status of architecture and design. Naturally our respective concerns as individuals for formal socio-cultural and political discourse will make themselves felt in our joint editing of OPPOSITIONS. The opposition alluded to in the title will first and foremost begin at home.

OPPOSITIONS will address itself to the evolution of new models for a theory of architecture. It will attempt to relate such models to specific buildings and theories which, in our opinion, either directly state or implicitly evoke the existence of such models. We will not, in all this, restrict our discourse to the very latest work. On the contrary, we will attempt to link the present to the past to assess the overall contribution of major individuals and movements which still have relevance today. Our editorial position will be to attempt to create a climate of opinion where ideas and actions are seen as being necessarily complimentary to any vital architectural culture.

Subcriptions: $15.00 for four issues; $5.00 for a single copy. Checks in advance payable to OPPOSITIONS. Add $1.50 for foreign postage. Second class postage paid at New York, New York. Students or institutional representatives may receive one year's subscription free in exchange for selling ten subscriptions. Further information may be requested from Jan Fisher, Assistant Editor. Telephone: 212-947-0765. Letters and manuscripts double-spaced in triplicate may be submitted to the Editors who cannot be held responsible for their return.

Editors: Peter Eisenman, Kenneth Frampton, Mario Gandelsonas.
Executive Editor: David Morton.
Assistant Editor: Jan Fisher.
Designer: Massimo Vignelli.

A non-profit journal published and copyright by The Institute for Architecture and Urban Studies, 8 West 40th Street, New York, New York 10018.

OPPOSITIONS 1

OPPOSITIONS
Peter Eisenman
From Golden Lane to Robin Hood Gardens; or if you follow the Yellow Brick Road, it may not lead to Golders Green

HISTORY
Kenneth Frampton
Industrialization and the Crises in Architecture
Colin Rowe
Neoclassicism and Modern Architecture, Parts I and II
Anthony Vidler
News from the Realm of Nowhere

THEORY
Diana Agrest and Mario Gandelsonas
Semiotics and Architecture: Ideological Consumption or Theoretical Work

Single issues of OPPOSITIONS are available from:
Jaap Rietman Art Books, 157 Spring Street, New York
Wittenborn Art Books, 1018 Madison Avenue, New York
London Art Bookshop, 72 Charlotte Street, London
Libreria Salto, Via Visconte di Modrone 18, Milan
Rafael Viñoly, Florida 890, Buenos Aires

OPPOSITIONS 2

OPPOSITIONS
Stuart Cohen
Physical Context/Cultural Context: A Critique of Venturi and Rauch, Guild House, Philadelphia and Richard Meier, Twin Parks Housing, Bronx

HISTORY
Colin Rowe
Character and Composition; or Some Vicissitudes of Architectural Vocabulary in the Nineteenth Century

THEORY
Barry Maitland
The Grid
Denise Scott Brown
Discourse for Social Planners on Architectural Formalism and Social Concern

REVIEWS
William Ellis
On Banham's *Los Angeles: The Architecture of Four Ecologies*
Robert Gutman
On Nicholas Habraken's *Supports*
Fred Koetter
On Venturi, Scott Brown and Izenour's *Learning from Las Vegas*

DOCUMENTS
A Bibliography of the Works and Writings of Alison and Peter Smithson
Jean Giradoux
Introduction to The Athens Charter, 1943
Translation and Commentary by Anthony Eardley
Philip Johnson
Writings: 1
The Berlin Building Exposition of 1931, 1932
Architecture in the Third Reich, 1933
Remarks on Mies Van der Rohe, 1956
Rem Koolhaas and Gerrit Oorthuys
Ivan Leonidov's Dom Narkomtjazjprom, Moscow,

OPPOSITIONS 3

OPPOSITIONS
Ulrich Franzen
On the Work of John Hejduk
Charles Moore
On Werner Seligmann's UDC Housing, Ithaca

HISTORY
Kenneth Frampton
Apropos Ulm
Rem Koolhaas
The Berlin School: Architectural Projects for Berlin, 1963-1968
Anthony Vidler
The Forms of Social Order

THEORY
Diana Agrest
Critical Remarks on Urban Planning Models
Robert Gutman
Whose Intentions in Architecture?
William Huff
On Symmetry

REVIEWS
Peter Eisenman
On Christian Norberg-Schultz's *Existence, Space and Architecture*

DOCUMENTS
The Architect's Ball-A Vignette, 1930
Commentary by Rem Koolhaas
Philip Johnson
Writings:
Luigi Moretti
Eclettismo e Unitá di Linguaggio, 1950
Strutture e Sequenze di Spazi, 1952
Translation and Commentary by Thomas Stevens
David White and David Whitney
Bibliography of Selected Writings of Philip Johnson

OPPOSITIONS 4

OPPOSITIONS
Manfred Schiedhelm
A Report on Shadrach Wood's Free University, Berlin
Werner Seligmann
On James Stirling's Housing in Runcorn

HISTORY
Robert Stern
Yale, 1950-1965
Jon Michael Schwarting
The Lesson of Rome

THEORY
Emilio Ambasz
A Selection from *Working Fables: A Collection of Design Tales for Skeptic Children©*
Michael Graves
In Support of the Celestial Soffit
Alessandro Mendini
Global Tools

REVIEWS
Mario Gandelsonas
On Umberto Eco's *The Componential Analysis of the Architectural Sign the Column*
Dalibor Vezaley
On Joseph Rykwert's *Adam's House in Paradise*

DOCUMENTS
Cesare Cattaneo
Giovanni e Giuseppe, 1941
Translation and Commentary by Victor Caliandro
Arthur Drexler
Notes on Historicism: Commentary on Some Unpublished Projects of Philip Johnson
George Baird
Architecture and Politics: A Polemical Dispute
A Critical Introduction to Karel Teige's *Mundaneum*, 1928 and Le Corbusier's *A Defense of Architecture*, 1933

Fig. 92

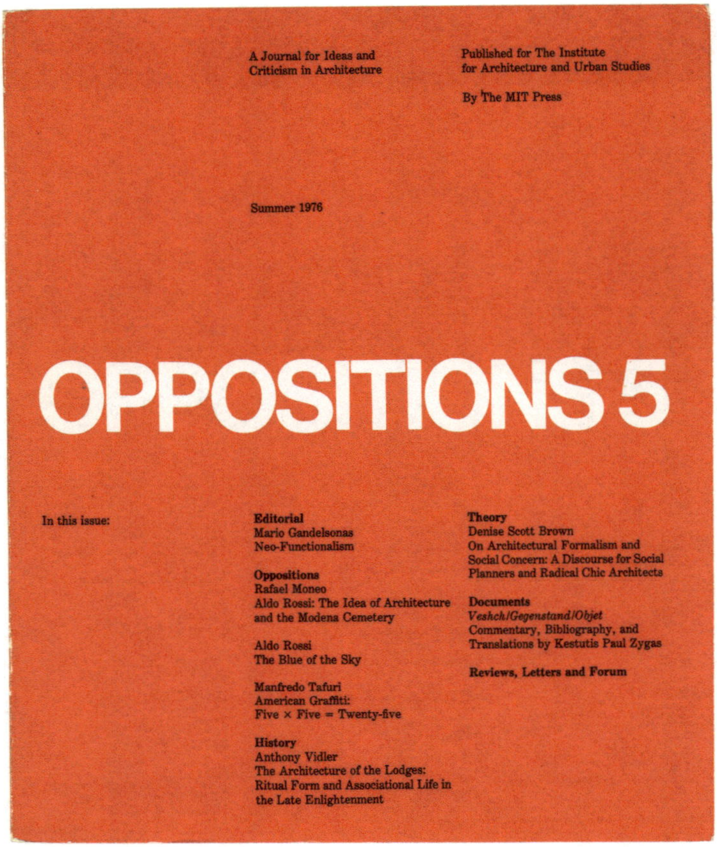

A Journal for Ideas and
Criticism in Architecture

Published for The Institute
for Architecture and Urban Studies

By The MIT Press

Summer 1976

OPPOSITIONS 5

In this issue:

Editorial
Mario Gandelsonas
Neo-Functionalism

Oppositions
Rafael Moneo
Aldo Rossi: The Idea of Architecture
and the Modena Cemetery

Aldo Rossi
The Blue of the Sky

Manfredo Tafuri
American Graffiti:
Five × Five = Twenty-five

History
Anthony Vidler
The Architecture of the Lodges:
Ritual Form and Associational Life in
the Late Enlightenment

Theory
Denise Scott Brown
On Architectural Formalism and
Social Concern: A Discourse for Social
Planners and Radical Chic Architects

Documents
Veshch/Gegenstand/Objet
Commentary, Bibliography, and
Translations by Kestutis Paul Zygas

Reviews, Letters and Forum

Fig. 93

Fig. 94

The Institute for Architecture and Urban Studies

Eight West Fortieth Street, New York, N.Y. 10018
Telephone: 212 947 0765

OPPOSITIONS
A Journal for Ideas and Criticism in Architecture

Peter Eisenman
Kenneth Frampton
Mario Gandelsonas
Anthony Vidler
Editors

Julia Bloomfield
Managing Editor

Massimo Vignelli
Designer

David Morton
Editorial Consultant

October 15, 1976

Mr.Peter Eisenman
c/o IAUS

Dear Peter,
By now you have received Oppositions 5, and thus know that The MIT Press has assumed the publication and distribution of the journal. For this we are all grateful. But more importantly, it means that Oppositions is coming out regularly every three months and that the dialogue we have begun will be more continuous and thus more meaningful.

We would like to thank you now for your past support of Oppositions which made possible the survival and growth of the journal. Your initial support was particularly appreciated because it came at a time when the stability and future of the journal was uncertain.

Your sponsorship of Oppositions actually expired with Oppositions 5. We hope you will renew your support for the next four issues, Oppositions 5, 6, 7, and 8. Due to increased prices, we have been forced to raise the sponsorship contribution to $130 for the next four Oppositions. Sponsorship of the journal is extremely important as it enables us to maintain the high quality of the journal and to subsidize the students' subscription rate. But remember, without your support there can be no journal, MIT notwithstanding.

An invitation accompanies this letter to invite you to the Forum on the occasion of the publication of Oppositions 6. We are repeating the format of the successful Beaux Arts forum of having practicing architects and educators both from Europe and America make brief presentations. It should be a good evening. This will be open to sponsors only and we are hopeful that you will continue your most generous support.

We look forward to seeing you.

Yours sincerely,

Peter

Peter Eisenman

Fig. 95

Fig. 96

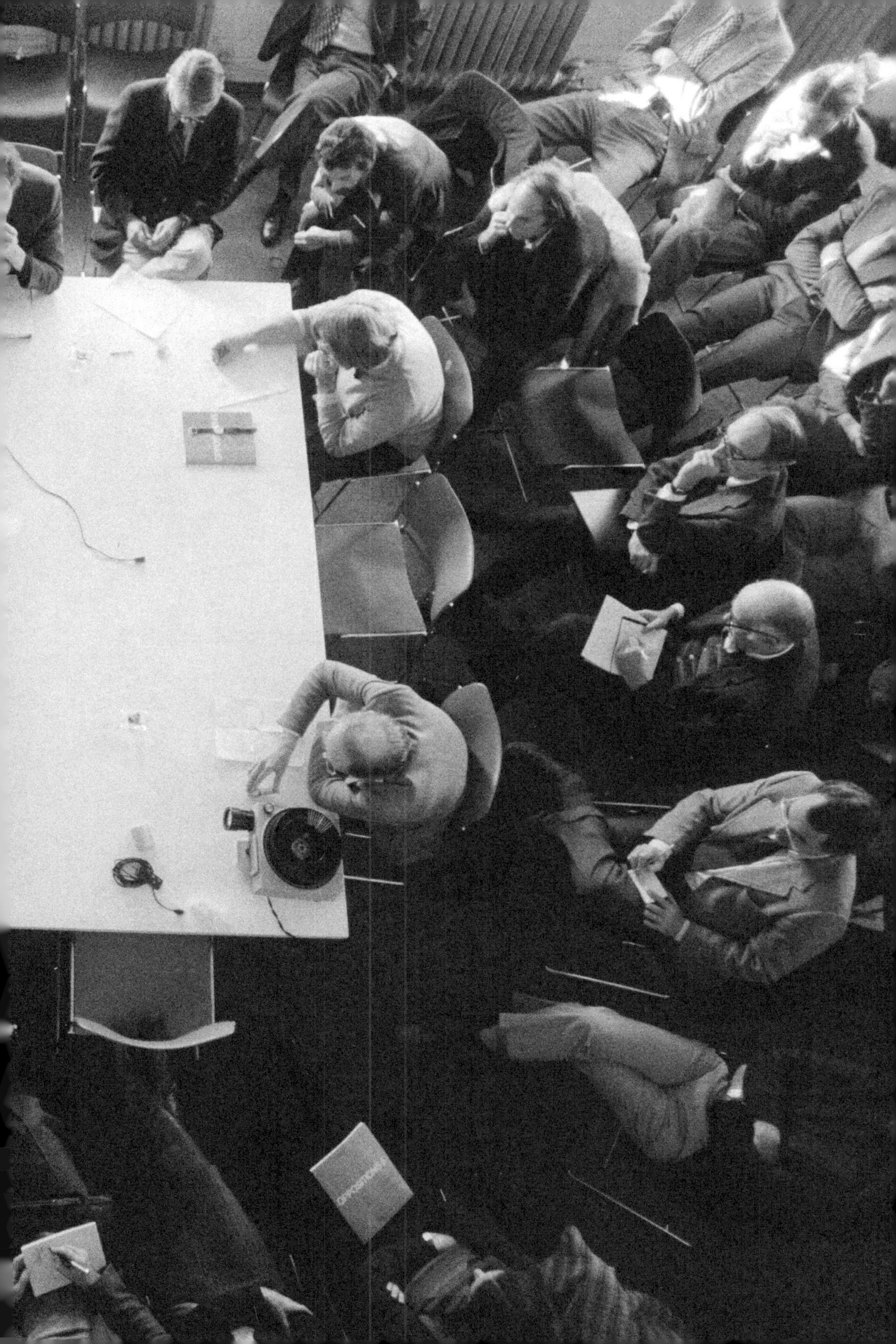

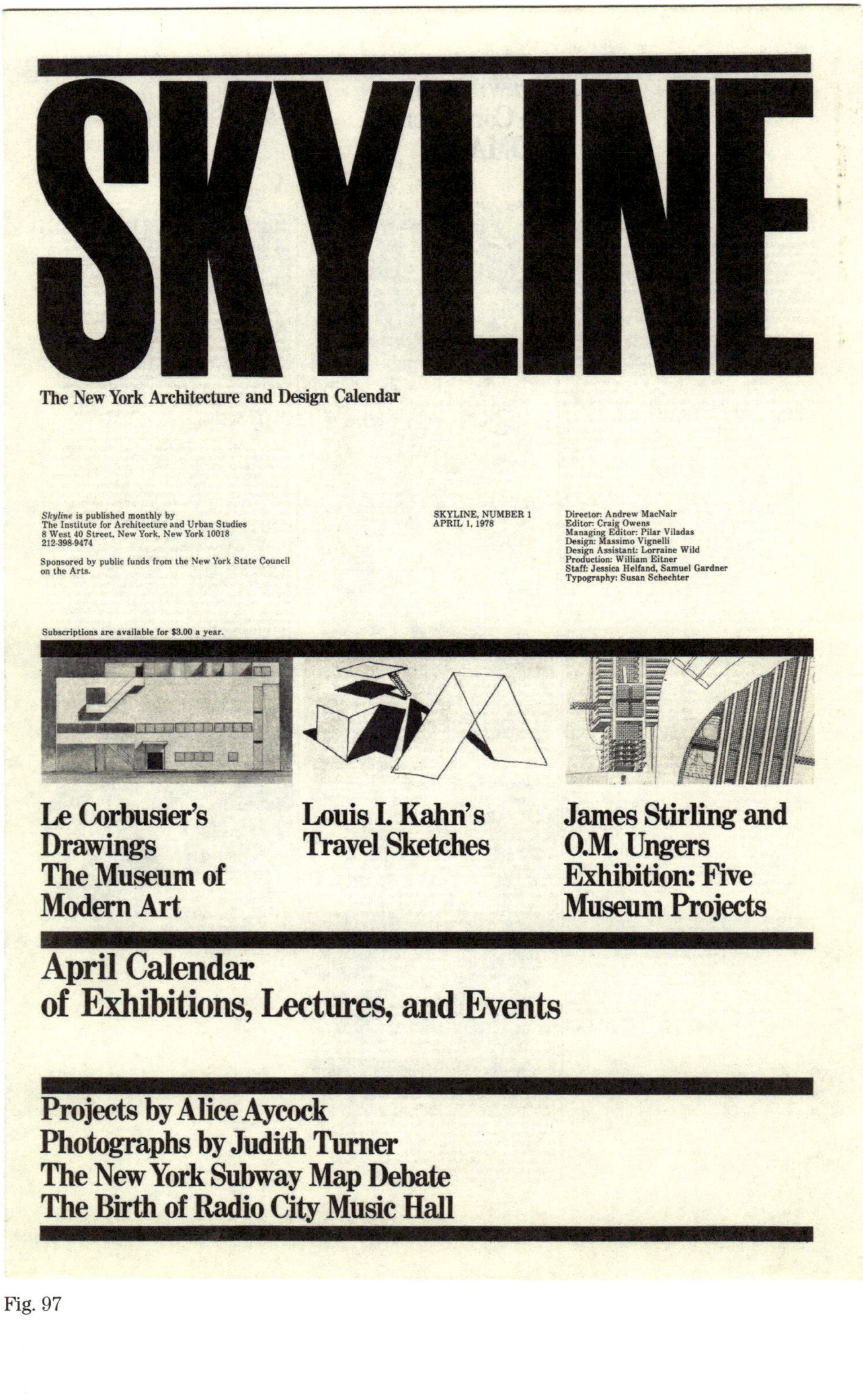

SKYLINE

The New York Architecture and Design Calendar

Skyline is published monthly by
The Institute for Architecture and Urban Studies
8 West 40 Street, New York, New York 10018
212-398-9474

Sponsored by public funds from the New York State Council on the Arts.

SKYLINE, NUMBER 1
APRIL 1, 1978

Director: Andrew MacNair
Editor: Craig Owens
Managing Editor: Pilar Viladas
Design: Massimo Vignelli
Design Assistant: Lorraine Wild
Production: William Eitner
Staff: Jessica Helfand, Samuel Gardner
Typography: Susan Schechter

Subscriptions are available for $3.00 a year.

Le Corbusier's Drawings The Museum of Modern Art

Louis I. Kahn's Travel Sketches

James Stirling and O.M. Ungers Exhibition: Five Museum Projects

April Calendar of Exhibitions, Lectures, and Events

Projects by Alice Aycock
Photographs by Judith Turner
The New York Subway Map Debate
The Birth of Radio City Music Hall

Fig. 97

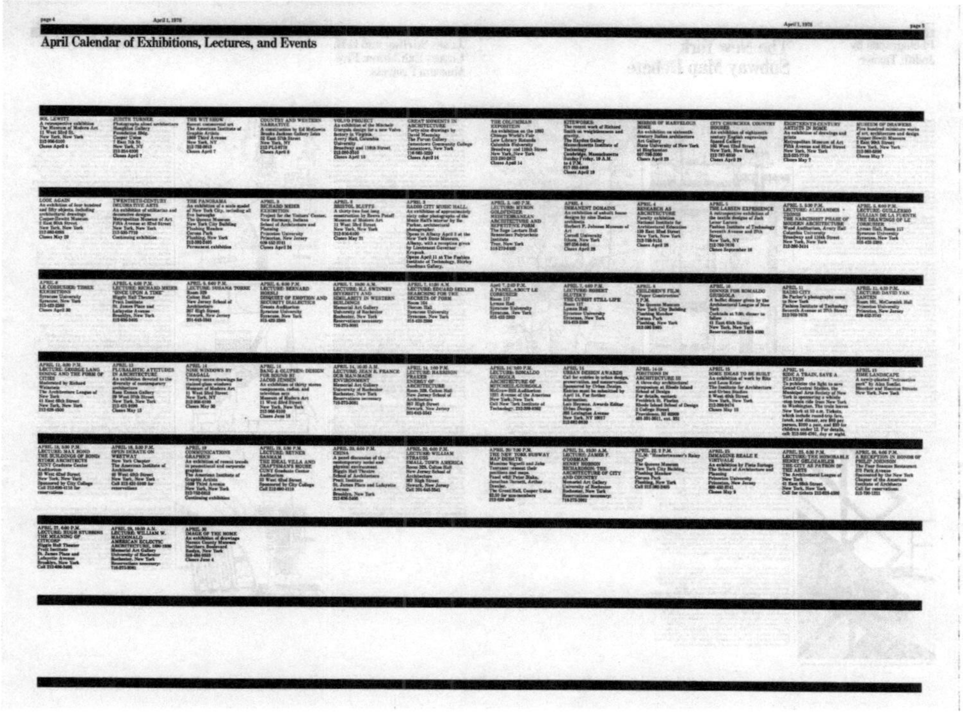

Fig. 98

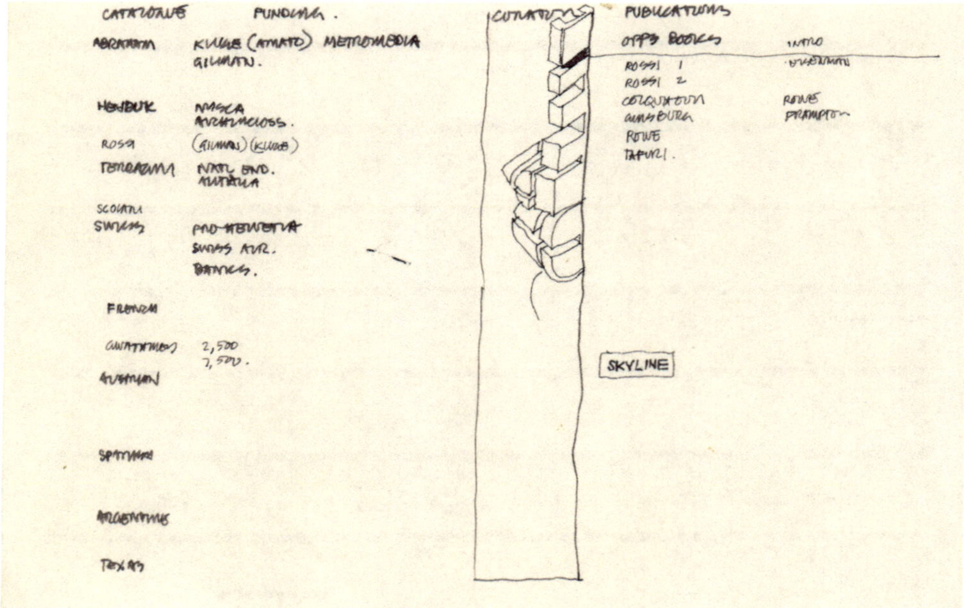
CATALOGUE
FUNDING.
PUBLICATIONS
OPPS BOOKS
ROSSI 1
ROSSI 2
ROWE
TAFURI.
METROMEDIA
SWISS AIR.
BANKS.
2,500
7,500.
SKYLINE
TEXAS

Fig. 99

The Institute for Architecture and Urban Studies
Eight West Fortieth Street, New York, N.Y. 10018
IAUS EXHIBITION CATALOGUES
IAUS Catalogue 1
Massimo Scolari: Architecture Between Memory and Hope
Introduction by Manfredo Tafuri
$12.00/copy, $1.50 shipping
All orders must be prepaid.
Add $3.00 for each foreign order ($15.00)
Available May 1980. ☐ No. of copies ____.
IAUS Catalogue 2
Aldo Rossi in America 1976 to 1979
Introduction by Peter Eisenman
Texts by Aldo Rossi
$10.00/copy, $1.50 shipping
All orders must be prepaid.
Add $3.00 for each foreign order ($13.00)
Available now. ☐ No. of copies ____.
IAUS Catalogue 7
Gwathmey/Siegel Architects: Five Houses
Introduction by Kenneth Frampton
Preface by Ulrich Franzen
$12.00/copy, $1.50 shipping
All orders must be prepaid.
Add $3.00 for each foreign order ($15.00)
Available March 1980. ☐ No. of copies ____.
IAUS Catalogue 9
Philip Johnson: Processes
The Glass House, 1940 and AT&T, 1978
Preface by Craig Owens
Introduction by Giorgio Ciucci
$10.00/copy, $1.50 shipping
All orders must be prepaid.
Add $3.00 for each foreign order ($13.00)
Available now. ☐ No. of copies ____.
IAUS Catalogue 10
A New Wave of Japanese Architecture
Introduction by Kenneth Frampton
Texts by 10 Japanese Architects
$12.00/copy, $1.50 shipping
All orders must be prepaid.
Add $3.00 for each foreign order ($15.00)
Available now. ☐ No. of copies ____.
IAUS Catalogue 11
Wallace Harrison: Fifty Years of Architecture
Introduction by Rem Koolhaas
$12.00/copy, $1.50 shipping
All orders must be prepaid.
Add $3.00 for each foreign order ($15.00)
Available April 1980. ☐ No. of copies ____.
IAUS Catalogue 12
John Hejduk: 7 Houses (1955-1962)
Introduction by Peter Eisenman
Texts by John Hejduk
$12.00/copy, $1.50 shipping
All orders must be prepaid.
Add $3.00 for each foreign order ($15.00)
Available now. ☐ No. of copies ____.
IAUS Catalogue 13
A New Wave of Austrian Architecture
Introduction by Kenneth Frampton
Texts by 5 Austrian architects
$12.00/copy, $1.50 shipping
All orders must be prepaid.
Add $3.00 for each foreign order ($15.00)
Available May 1980. ☐ No. of copies ____.

Fig. 100

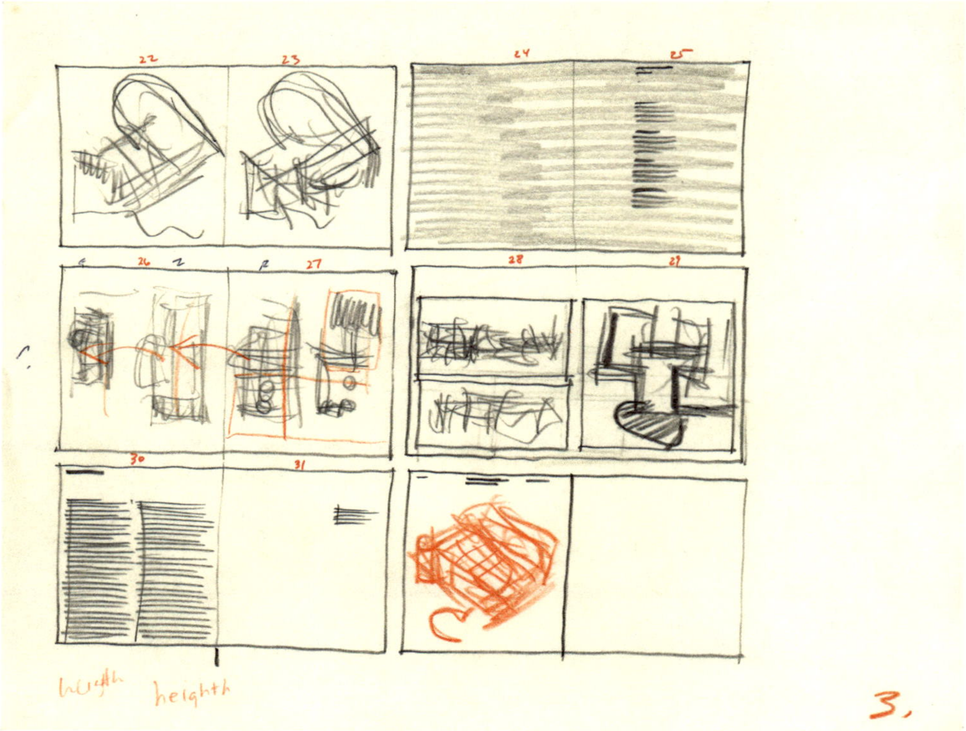

Fig. 101

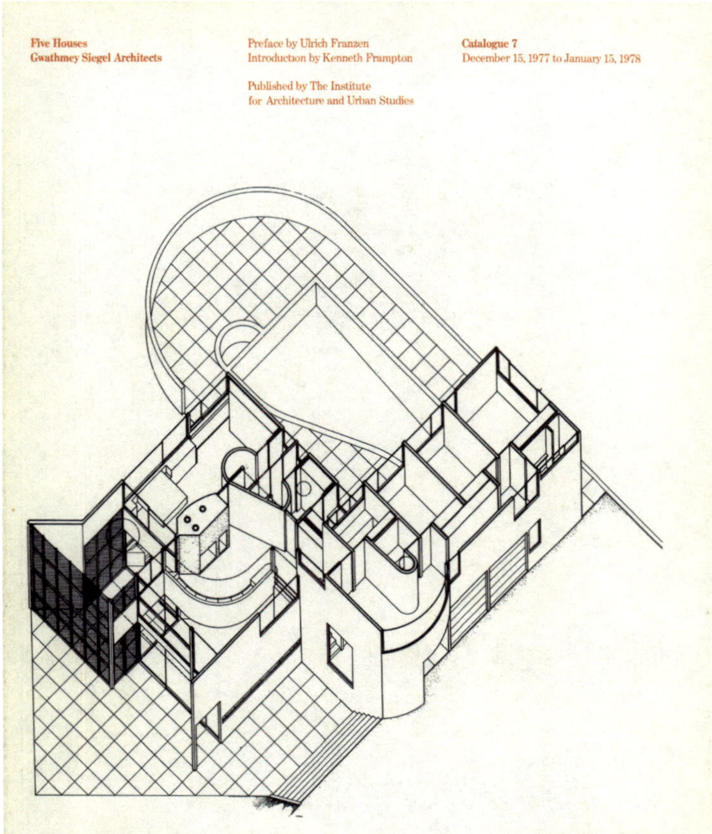
Five Houses
Gwathmey Siegel Architects
Preface by Ulrich Franzen
Introduction by Kenneth Frampton
Published by The Institute
for Architecture and Urban Studies
Catalogue 7
December 15, 1977 to January 15, 1978

Fig. 102

Fig. 103

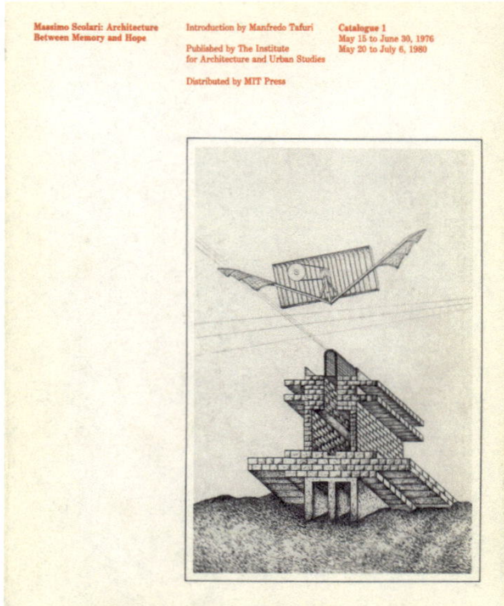

Fig. 104

Fig. 105

Fig. 106

Fig. 107

Fig. 108

Fig. 109

Fig. 110

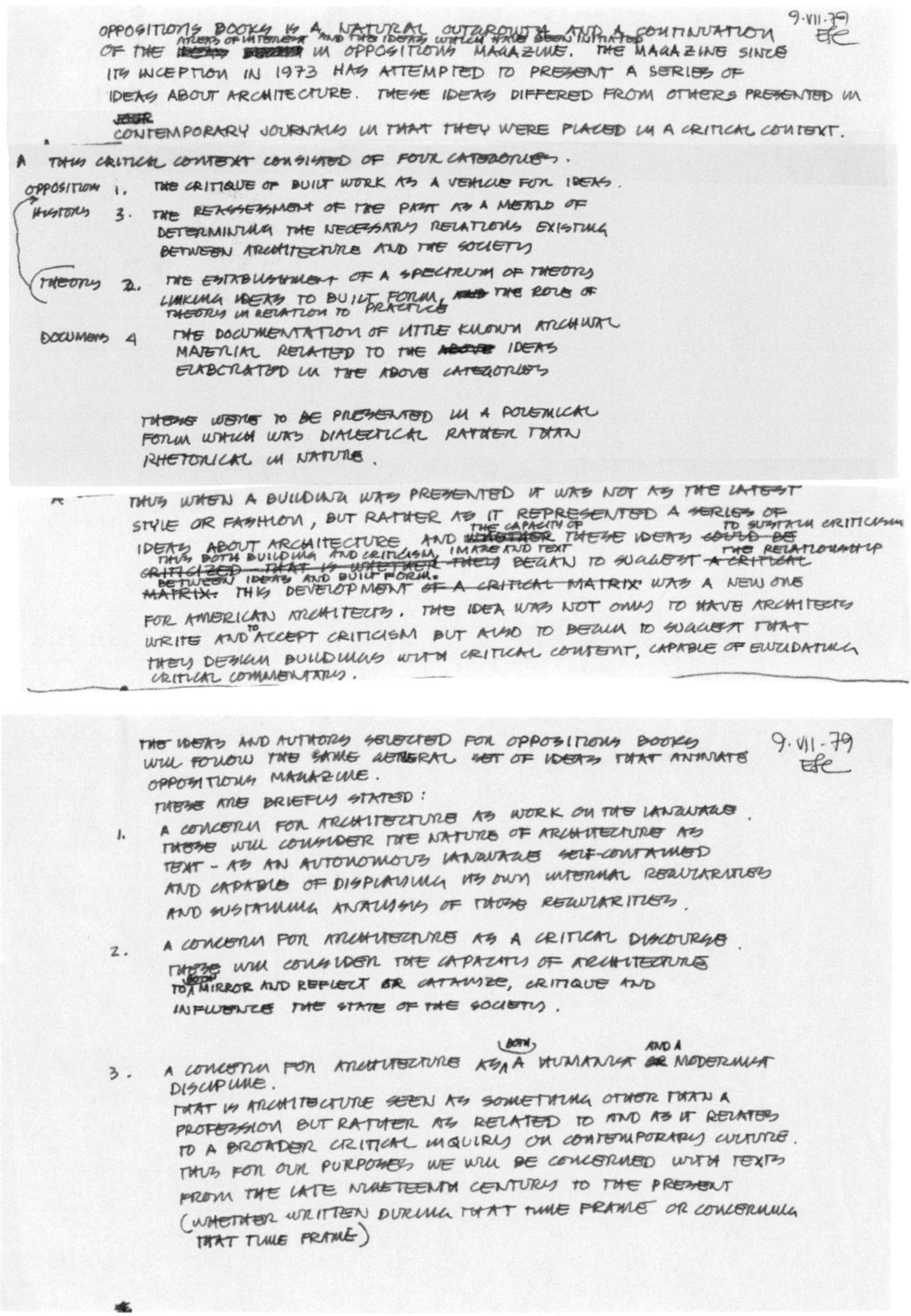
9.VII.79 EFE
OPPOSITIONS BOOKS IS A NATURAL OUTGROWTH AND A CONTINUATION OF THE IDEAS OF INTEREST AND THE IDEAS WHICH HAVE BEEN INITIATED IN OPPOSITIONS MAGAZINE. THE MAGAZINE SINCE ITS INCEPTION IN 1973 HAS ATTEMPTED TO PRESENT A SERIES OF IDEAS ABOUT ARCHITECTURE. THESE IDEAS DIFFERED FROM OTHERS PRESENTED IN ~~OUR~~ CONTEMPORARY JOURNALS IN THAT THEY WERE PLACED IN A CRITICAL CONTEXT.
A THIS CRITICAL CONTEXT CONSISTED OF FOUR CATEGORIES.
OPPOSITION 1. THE CRITIQUE OF BUILT WORK AS A VEHICLE FOR IDEAS.
HISTORY 3. THE REASSESSMENT OF THE PAST AS A METHOD OF DETERMINING THE NECESSARY RELATIONS EXISTING BETWEEN ARCHITECTURE AND THE SOCIETY
THEORY 2. THE ESTABLISHMENT OF A SPECTRUM OF THEORY LINKING IDEAS TO BUILT FORM, ~~AND~~ THE ROLE OF THEORY IN RELATION TO PRACTICE
DOCUMENTS 4 THE DOCUMENTATION OF LITTLE KNOWN ARCHIVAL MATERIAL RELATED TO THE ~~ABOVE~~ IDEAS ELABORATED IN THE ABOVE CATEGORIES
THESE WERE TO BE PRESENTED IN A POLEMICAL FORM WHICH WAS DIALECTICAL RATHER THAN RHETORICAL IN NATURE.
A THUS WHEN A BUILDING WAS PRESENTED IT WAS NOT AS THE LATEST STYLE OR FASHION, BUT RATHER AS IT REPRESENTED A ~~SERIES OF~~ IDEAS ABOUT ARCHITECTURE AND ~~WHETHER~~ THE CAPACITY OF THESE IDEAS ~~COULD BE CRITICIZED - THAT IS WHETHER THEY~~ TO SUSTAIN CRITICISM. THUS BOTH BUILDING AND CRITICISM, IMAGE AND TEXT BEGAN TO SUGGEST ~~A CRITICAL MATRIX.~~ THE RELATIONSHIP BETWEEN IDEAS AND BUILT FORM. THIS DEVELOPMENT ~~OF A CRITICAL MATRIX~~ WAS A NEW ONE FOR AMERICAN ARCHITECTS. THE IDEA WAS NOT ONLY TO HAVE ARCHITECTS WRITE AND TO ACCEPT CRITICISM BUT ALSO TO BEGIN TO SUGGEST THAT THEY DESIGN BUILDINGS WITH CRITICAL CONTENT, CAPABLE OF ELUCIDATING CRITICAL COMMENTARY.
9.VII.79 EFE
THE IDEAS AND AUTHORS SELECTED FOR OPPOSITIONS BOOKS WILL FOLLOW THE SAME GENERAL SET OF IDEAS THAT ANIMATE OPPOSITIONS MAGAZINE.
THESE ARE BRIEFLY STATED:
1. A CONCERN FOR ARCHITECTURE AS WORK ON THE LANGUAGE. THESE WILL CONSIDER THE NATURE OF ARCHITECTURE AS TEXT - AS AN AUTONOMOUS LANGUAGE SELF-CONTAINED AND CAPABLE OF DISPLAYING ITS OWN INTERNAL REGULARITIES AND SUSTAINING ANALYSIS OF THOSE REGULARITIES.
2. A CONCERN FOR ARCHITECTURE AS A CRITICAL DISCOURSE. ~~THESE~~ WILL CONSIDER THE CAPACITY OF ARCHITECTURE TO BOTH MIRROR AND REFLECT ~~OR~~ CATALYZE, CRITIQUE AND INFLUENCE THE STATE OF THE SOCIETY.
3. A CONCERN FOR ARCHITECTURE AS BOTH A HUMANIST ~~OR~~ AND A MODERNIST DISCIPLINE.
THAT IS ARCHITECTURE SEEN AS SOMETHING OTHER THAN A PROFESSION BUT RATHER AS RELATED TO AND AS IT RELATES TO A BROADER CRITICAL INQUIRY ON CONTEMPORARY CULTURE. THUS FOR OUR PURPOSES WE WILL BE CONCERNED WITH TEXTS FROM THE LATE NINETEENTH CENTURY TO THE PRESENT (WHETHER WRITTEN DURING THAT TIME FRAME OR CONCERNING THAT TIME FRAME)

Fig. 111

Prospective Series of Oppositions Books

Translations:

1. The Architecture of the City by Aldo Rossi
Translated by Diane Ghirardo. Introduction by Peter Eisenman.

2. Style and Epoch by Moisei Ginzburg.
Translated by Anatole Senkevitch. Introduction by Kenneth Frampton.

Collected Essays:

3. The Collected Writings of Alan Colquhoun.
Introduction by Kenneth Frampton.

4. The Collected Writings of Adolf Loos
Introduction by ~~Kenneth Frampton~~ ALDO ROSSI AND CARL SCHORSKE

5. The Collected Writings of Theo van Doesburg.
Introduction by Sergio Polano.

6. The Selected Writings of Arata Isozaki.
Introduction by Peter Eisenman.

7. The Selected Writings of Massimo Cacciari.
Introduction by Anthony Vidler.

8. The Collected Writings of Eduardo Persico.
Translation and introduction by Diane Ghirardo.

9. The Collected Writings of Hans Sedlmayr.
Translation, editing and critical commentary by Werner Oechslin.

10. The Selected Writings of Colin Rowe.
Introduction by Peter Eisenman.

11. The Critical Writings of Manfredo Tafuri.

12. Avanguardia e razionalita by Tomas Maldonado.
Introduction by Kenneth Frampton.

New Texts:

13. Autobiographica Scientifica by Aldo Rossi.
Introduction by Peter Eisenman.

14. Architecture and Industrialized Society by Kenneth Frampton.

Fig. 112

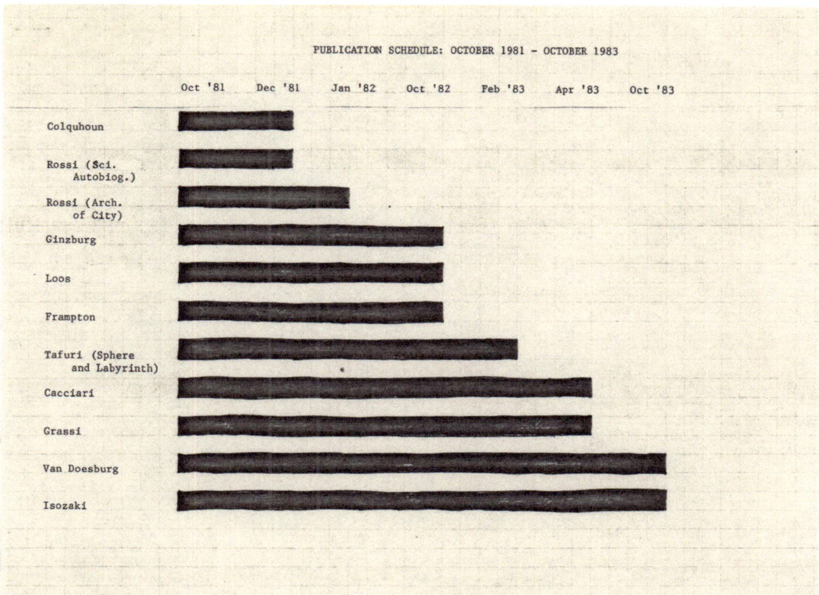

Fig. 113

KENNETH FRAMPTON PROPOSAL FOR AN OPPOSITIONS BOOK
December 3, 1981

Title: LABOR, WORK, AND ARCHITECTURE
Essays in Architectural Criticism
INTRODUCTION : K. FORSTER.
INTRODUCTION : K. FRAMPTON
Contents:
Part I
"The Status of Man and the Status of His Objects" reworked with "Labor, Work, and Architecture," including amplification on Frankfurt School (1966/1979)
"Industrialization and the Crisis in Architecture" with reworked definition of architecture and more specific background on Ecole des Ponts et Chaussees (Oppositions 1)
"Constructivism: The Pursuit of an Elusive Sensibility," with expanded section on Vertov (Oppositions 6)
Part II
"The Work and Influence of El Lissitzky" (Architects' Yearbook 12, 1968)
"The Humanist Versus the Utilitarian Ideal" (A.D. 38, 1968)
"The City of Dialectic" (A.D. 39, 1969)
"The Vicissitudes of Ideology" (L'architecture d'aujourd'hui 177, 1975)
"Apropos Ulm" (Oppositions 3)
Part III
"Notes on American Architecture and Urbanism" (Casabella XXXV, 1971)
"Maison de Verre~~s~~ and Duchamp's Machine Celibataire" (unwritten)
an essay on Loos (unwritten)
"Louis Kahn and the French Connection," reworking Sullivan-Wright-Kahn relationship as suggested by Botero

Fig. 114

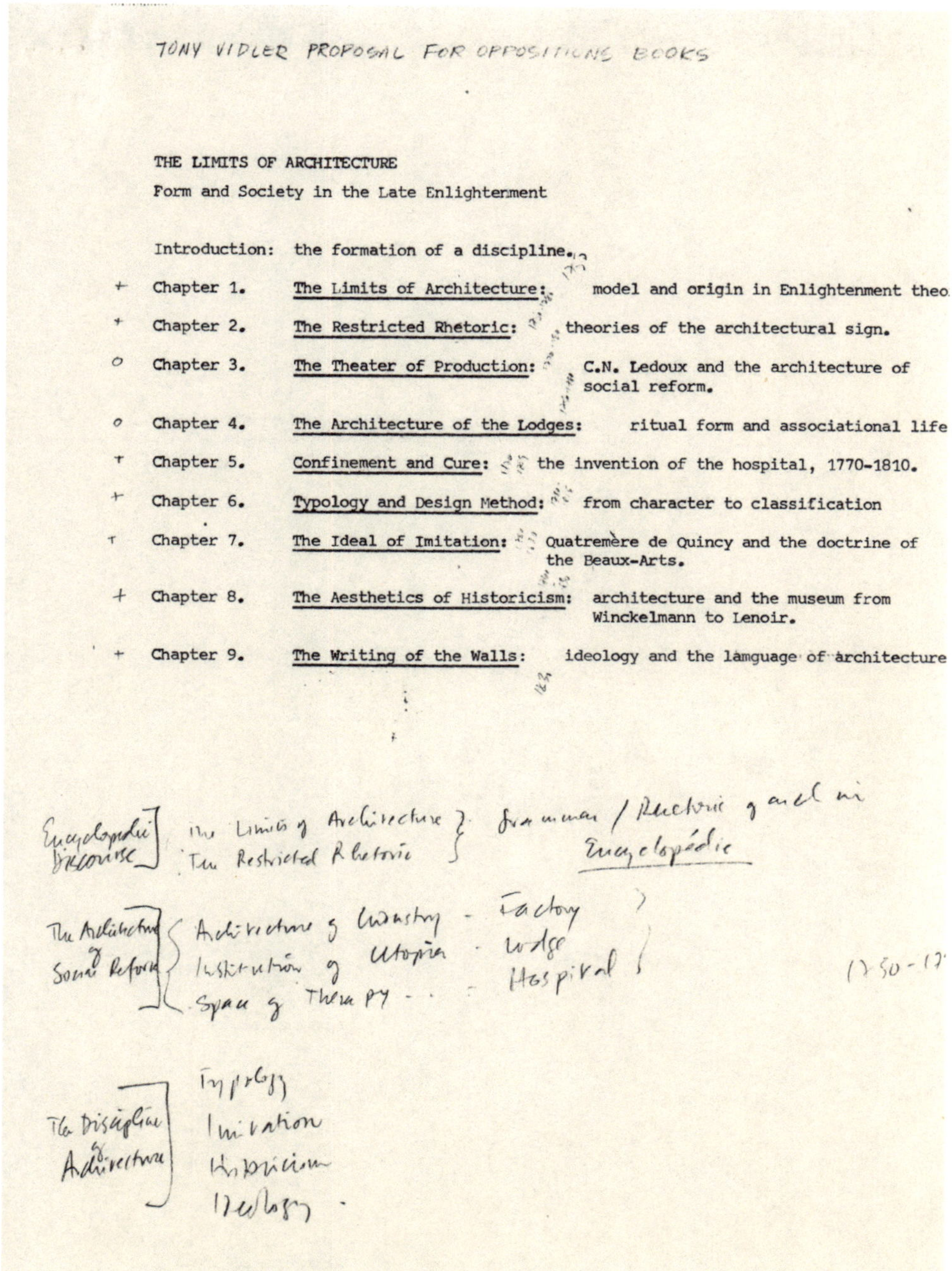

TONY VIDLER PROPOSAL FOR OPPOSITIONS BOOKS

THE LIMITS OF ARCHITECTURE

Form and Society in the Late Enlightenment

Introduction: the formation of a discipline.

+ Chapter 1. The Limits of Architecture: model and origin in Enlightenment theo

+ Chapter 2. The Restricted Rhetoric: theories of the architectural sign.

o Chapter 3. The Theater of Production: C.N. Ledoux and the architecture of social reform.

o Chapter 4. The Architecture of the Lodges: ritual form and associational life

+ Chapter 5. Confinement and Cure: the invention of the hospital, 1770-1810.

+ Chapter 6. Typology and Design Method: from character to classification

+ Chapter 7. The Ideal of Imitation: Quatremère de Quincy and the doctrine of the Beaux-Arts.

+ Chapter 8. The Aesthetics of Historicism: architecture and the museum from Winckelmann to Lenoir.

+ Chapter 9. The Writing of the Walls: ideology and the lamguage of architecture

Encyclopédie Discourse — The Limits of Architecture / The Restricted Rhetoric — Grammar / Rhetoric of and in Encyclopédie

The Architecture of Social Reform — Architecture of Industry - Factory / Institution of Utopia - Lodge / Space of Therapy - Hospital — 1750-17

The Discipline of Architecture — Typology / Imitation / Historicism / Ideology.

Fig. 115

The Institute for Architecture and Urban Studies
Eight West Fortieth Street, New York, N.Y. 10018
Telephone: 212 398 9474

20 July, 1981

Dinner for Sponsors of Skyline given by Philip Johnson

Philip Johnson
John Burgee
Edward Barnes
Harry Cobb
Lew Davis
Paul Rudolph
Der Scutt
Paul Kennon
Gordon Wildermuth
Raul de Armas
Arthur Davis
Peter Eisenman
Suzanne Stephens
Charles Gwathmey
Hamid Nouri
GIAN FRANCO MONACELLI
JORDAN GRUZEN
ANTHONY VIDLER
15 to attend

Peter,

Hamid mentioned that Monacelli had provided a photographer for tonight. Also, is there anything else that you might need?

Fig. 116

Fig. 117

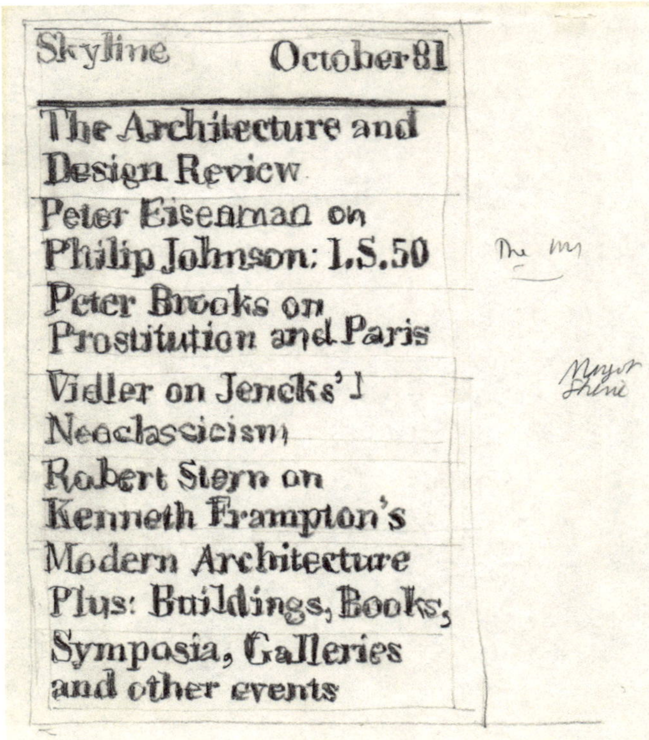

Fig. 118

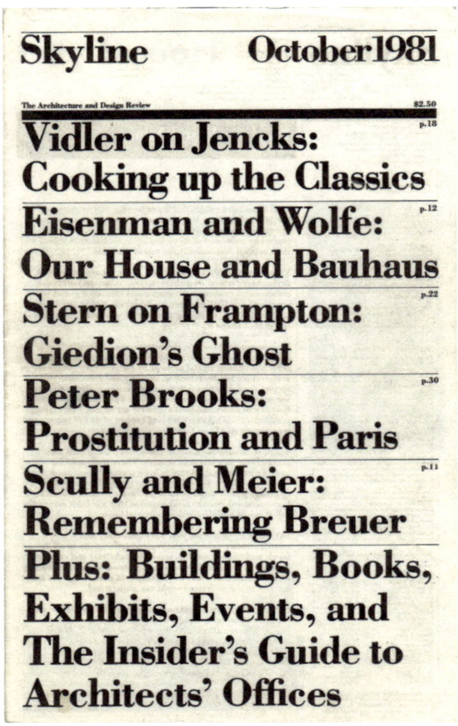

Fig. 119

The Institute for Architecture
and Urban Studies
and Rizzoli Communications, Inc
Skyline
invite you to celebrate the return of
Skyline
Returns!
Tuesday, September 29, 1981
from 6pm until 8pm
Sept. 29, 6-8
IAUS 8 W. 40
at the IAUS
8 West 40 Street, New York City

Fig. 120

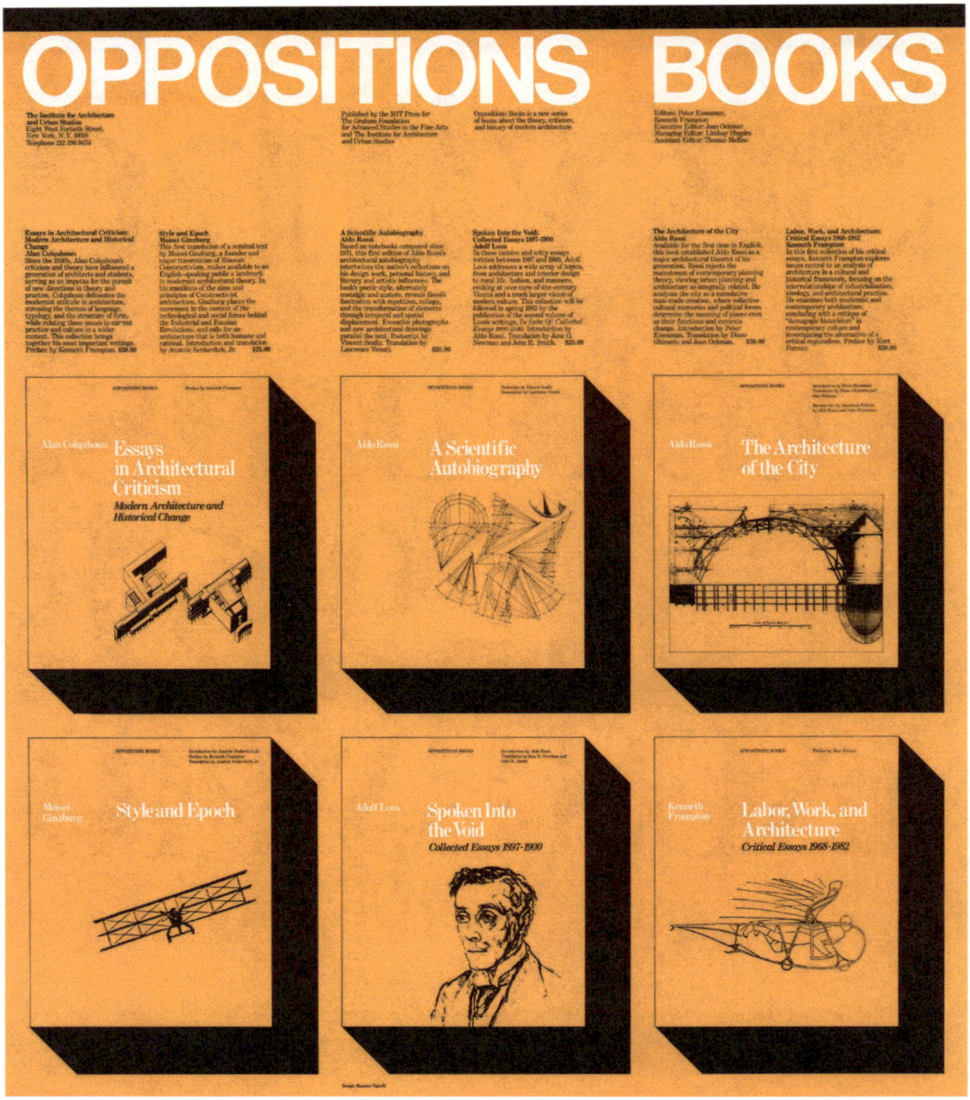

OPPOSITIONS BOOKS
The Institute for Architecture and Urban Studies
Eight West Fortieth Street,
New York, N.Y. 10018
Telephone 212 398 9474
Published by the MIT Press for The Graham Foundation for Advanced Studies in the Fine Arts and The Institute for Architecture and Urban Studies
Oppositions Books is a new series of books about the theory, criticism, and history of modern architecture
Editors: Peter Eisenman, Kenneth Frampton
Executive Editor: Joan Ockman
Managing Editor: Lindsay Shapiro
Assistant Editor: Thomas Mellins
Essays in Architectural Criticism: Modern Architecture and Historical Change
Alan Colquhoun
Since the 1950's, Alan Colquhoun's criticism and theory have influenced a generation of architects and students, serving as an impetus for the pursuit of new directions in theory and practice. Colquhoun delineates the modernist attitude in architecture, stressing the themes of language, typology, and the structure of form, while relating these issues to current practice and culture in a wider context. This collection brings together his most important writings. Preface by Kenneth Frampton. $30.00
Style and Epoch
Moisei Ginzburg
This first translation of a seminal text by Moisei Ginzburg, a founder and major theoretician of Russian Constructivism, makes available to an English-speaking public a landmark in modernist architectural theory. In his manifesto of the aims and principles of Constructivist architecture, Ginzburg places the movement in the context of the technological and social forces behind the Industrial and Russian Revolutions, and calls for an architecture that is both humane and rational. Introduction and translation by Anatole Senkevitch, Jr. $25.00
A Scientific Autobiography
Aldo Rossi
Based on notebooks composed since 1971, this first edition of Aldo Rossi's architectural autobiography intertwines the author's reflections on his design work, personal history, and literary and artistic influences. The book's poetic style, alternately nostalgic and austere, reveals Rossi's fascination with repetition, collage, and the transformation of elements through temporal and spatial displacement. Evocative photographs and new architectural drawings parallel the text. Postscript by Vincent Scully. Translation by Lawrence Venuti. $20.00
Spoken Into the Void: Collected Essays 1897-1900
Adolf Loos
In these incisive and witty essays written between 1897 and 1900, Adolf Loos addresses a wide array of topics, from architecture and interior design to rural life, fashion, and manners, evoking at once turn-of-the-century Vienna and a much larger vision of modern culture. This collection will be followed in spring 1983 by the publication of the second volume of Loos's writings, In Spite Of: Collected Essays 1900-1930. Introduction by Aldo Rossi. Translation by Jane O. Newman and John H. Smith. $20.00
The Architecture of the City
Aldo Rossi
Available for the first time in English, this book established Aldo Rossi as a major architectural theorist of his generation. Rossi rejects the mainstream of contemporary planning theory, viewing urban planning and architecture as integrally related. He analyzes the city as a cumulative man-made creation, where collective cultural memories and political forces determine the meaning of places even as their functions and contexts change. Introduction by Peter Eisenman. Translation by Diane Ghirardo and Joan Ockman. $30.00
Labor, Work, and Architecture: Critical Essays 1968-1982
Kenneth Frampton
In this first collection of his critical essays, Kenneth Frampton explores issues central to an analysis of architecture in a cultural and historical framework, focusing on the interrelationships of industrialization, ideology, and architectural practice. He examines both modernist and contemporary architecture, concluding with a critique of "demagogic historicism" in contemporary culture and investigating the alternative of a critical regionalism. Preface by Kurt Forster. $20.00
Alan Colquhoun
Essays in Architectural Criticism
Modern Architecture and Historical Change
Aldo Rossi
A Scientific Autobiography
Aldo Rossi
The Architecture of the City
Moisei Ginzburg
Style and Epoch
Adolf Loos
Spoken Into the Void
Collected Essays 1897-1900
Kenneth Frampton
Labor, Work, and Architecture
Critical Essays 1968-1982

Fig. 121

Fig. 122

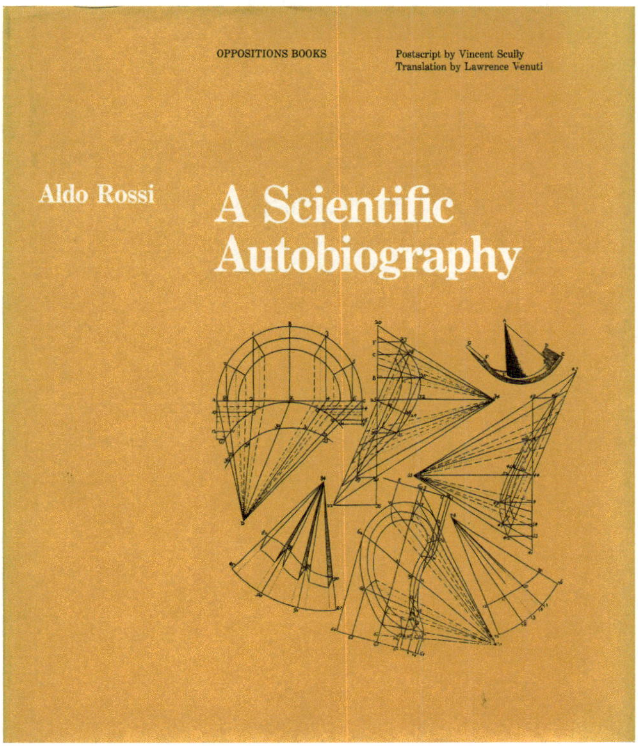

Fig. 123

C

<u>Oppositions 27</u> Winter 82 (DEADLINE END SEPT 81)

SPECIAL ISSUE ON INSTITUTIONS POWER AND ARCHITECTURE
KENNETH FRAMPTON AND ANTHONY VIDLER

Teyssot/Architecture for the EMployed
Budensigg/ Berlin Britz (KFo)
Murard/Neat and Sane
J.L. Cohen/Lurcat and the French Popular Front (KF)
Scarpa/Cultural Policy and Rationalization

<u>Oppositions 28</u> Spring 82 (DEADLINE NOV 81)

SPECIAL ISSUE ON POST MODERNISM/maybe Reg issue to help its progress
PETER EISENMAN

Frampton/The Resistance of Architecture
Rose/Nature of Physical Reality at the Quantum Level (PDE?)

<u>Oppositions 29</u> Summer 82 (DEADLINE FEB 82)

SPECIAL ISSUE ON HISTORY AND PRACTICE
ANTHONY VIDLER

<u>Oppositions 30</u> Fall 82 (DEADLINE May 82)

SPECIAL ISSUE ON NIETSCHE AND ARCHITECTURE
~~ANTHONY VIDLER~~ FORSTER + BUDDENSIEK

Fig. 124

4. Publishing Imprint

In the fall of 1976, Peter Eisenman received a letter dated October 15 informing him of the publication of *Oppositions* 5 by MIT Press.[569] What is remarkable about this document, printed on the journal's stationery, is that the sender was Peter Eisenman himself, who as editor of *Oppositions* had signed and also mailed the letter to his own address—a truly postmodern expression of self-referentiality. It is evidence that *Oppositions*, which had provided the Institute with a base of loyal readers, subscribers, authors, and sponsors, was currently in the process of repositioning itself in the marketplace. Sent to all of the journal's sponsors, it informed them of the recent signing of a contract with MIT Press, after long negotiations, and the promise of what would now be a regular quarterly publication. Eisenman, always the provocateur and publicist, not only promised the continuation of a dialogue but combined this with an appeal for financial support for the forthcoming volume, a donation of US$130 for the issues *Oppositions* 5, 6, 7, and 8—a rhetorically clever, if transparent, move. This appeal for donations was not just another promotional tool of the Institute after the new issue had already been sent to its erstwhile sponsors, but ultimately a written document, one that historians would call an ego-document: a source of insight into how Eisenman perceived and represented himself at the smallest intersection of the circle of editors and the circle of sponsors. *Oppositions* had already been on the market for three years, and in the meantime, not least due to Eisenman's constant advertising—whether after lectures or during interviews—it had earned a reputation as a sophisticated journal. While donations

569 Peter Eisenman, letter to Peter Eisenman, October 15, 1976. Source: CCA Montréal, IAUS fonds: D.4-7.

had once been the prerequisite for launching the journal as a self-published initiative, the letter announced that now, after the successful pilot issues, the next step had been taken to guarantee the publication of *Oppositions* in the longer term. But despite the conclusion of the contract, Eisenman, the publisher, said they were still dependent on donations. For this, along with all the authors and essays, editing, and production, was the capital that the Institute brought to the collaboration with MIT Press. But the letter does not only testify to a politics of journal-making, a combination of architecture journalism and cultural management. Since the sponsors were simultaneously invited by Institute director Eisenman to one of the "Forum" events celebrating the publication of *Oppositions* 6, a closed event at the Institute exclusively reserved for the sponsors and dedicated to the last major MoMA exhibition titled "Beaux-Arts," the letter testified, above all, to an economy of culture that was practiced there, more than to a belief in discourse or the interplay of ideas and criticism, and as a document of philanthropy as practice, set the future course of financing the publication of the Institute's own journal through cultural sponsorship. Along with ensuring the survival of *Oppositions*, Eisenman's announcement of the conclusion of the MIT Press contract flattered the sponsors—for by including himself in the list of addresses he put himself on par with the other sponsors, be they private individuals, institutions, or corporations. Moreover, the letter testified that the Institute was now distancing itself from plans to start its own publishing house. The price for this, however, was that the editors had to bury the myth of the journal as merely a "little magazine."

4.1 Investing in Academic Journals

For when in 1973, a circle of Fellows at the Institute once again set out to found their own journal to stimulate architecture discourse, this time successfully, the main questions, apart from the appointment and composition of its editorial board, concerned the content and financing of the first issues, i.e., the traditional tasks of a publishing house: production, marketing, and distribution. Publications had always played an important role for Peter Eisenman, and the launch of the Institute's own journal had thus been particularly important to him since its founding, as he was well aware of its historical role in establishing interpretative sovereignty. Not only was Eisenman a passionate collector of avant-garde periodicals of European modernism, but in 1968 he even exhibited his private collection at Princeton University.[570] At the time, he also

570 Eisenman exhibited his private collection at the Princeton University Library under the title "Modern Architecture 1910/1939: Polemics, Books, Periodicals and Ephemera from the Collection of Peter D. Eisenman" (February 16 to April 15, 1968). Tafuri highlighted Eisenman's passion: "Not to be overlooked is the fact that Eisenman is an avid collector of magazines and

published essays, reviews, and theoretical articles in international magazines such as the Italian *Casabella* and the British *Architectural Design*. Otherwise, he favored *Perspecta*, edited by students at Yale University, *Design Quarterly*, published by the Walker Arts Center, and the short-lived *Architectural Forum*. For him, these were the only serious architecture journals and magazines in the United States, in contrast to the major American architecture press such as *Architectural Record* and *Progressive Architecture*. At an early stage, Eisenman therefore gathered people around him who had experience in publishing, such as Kenneth Frampton (as a Fellow) and Stuart Wrede (as a Research Associate), from whom he hoped to acquire the necessary knowledge and skills. In the early 1970s, Eisenman and Mario Gandelsonas planned a series of books on architecture theory, edited by the Institute, in cooperation with MoMA, and with support from the Graham Foundation, as a response to Robert Venturi's publication *Complexity and Contradiction in Architecture* (1966). This never materialized. There were in fact contacts with MIT Press, which at that time was already considered the best publishing house in the field, via Stanford Anderson. But even after five years of the Institute's existence, it took a long time for any major publication projects to materialize, whenever the opportunity to publish presented itself—only *New Urban Settlements* (1970), a comprehensive research report on British and American New Towns and French *Villes nouvelles*, the two exhibition catalogues *Art & Architecture USSR. 1917–31* (1971) and *Another Chance for Housing. Low-Rise Alternatives* (1973), and "The City as an Artifact," a special issue of *Casabella* (1971), for which the Institute had taken over the guest editorship, had been published. By 1973, several attempts to launch a journal had already been made, including by Anderson and Anthony Vidler, among others. Now, in the spring of 1973, when the question of publishing was revisited with *Oppositions*, there was even internal discussion on Gandelsonas's initiative for the Institute to found its own publishing house—the proposed names were "The IAUS Publishing Corporation" and "IAUS Publications, Inc."—i.e., to define an entity with legal capacity, to which certain rights and, above all, limited responsibilities would have been attached. Above all, however, these considerations regarding the business model also concerned the economic intentions and safeguards associated with the planned publications.

However, the Institute did not start a publishing house operating on its own account, neither at that time nor at a later stage. Following the failure of the joint attempt by Eisenman, Gandelsonas, and Diana Agrest to capitalize on theory production by applying for impressive grants, *Oppositions* was launched

documents of the avant-garde. The spirit of the collector is not that of the bricoleur, but presupposes a process of selection." See Tafuri, 1976, here 49. Oddly enough, the essay is titled "European [sic!] Graffiti. Five x Five = Twenty-five" in the journal, which adds to the confusion as to who is appropriating whom.

in November 1973, two months late, and self-produced as a supposedly "little magazine." The first issues were financed by private funds and donations from a network of private, institutional, and corporate sponsors. Despite recurring financial difficulties during this period, the Institute, as a societally and culturally well-networked organization, provided the framework that made this journal possible. From then on, publishing *Oppositions* offered Eisenman and the long-serving Fellows, as well as selected external authors, the opportunity to develop their own ideas as essays, to contribute them to the larger, more widespread debate, and confer on them the weight of a publication. Writing, i.e., historicizing, theorizing, and critiquing, provided them with the opportunity to make a name for themselves on a national and soon international level. When the first three issues of *Oppositions* were produced between 1973 and 1975 as pilot issues alongside the Fellows' other teaching and cultural production, they were successful in raising the Institute's profile beyond New York and the East Coast of the USA, first in architecture circles, and later in other circles as well. After that, the Institute was not only frequently equated with *Oppositions* from an outside perspective, but Eisenman's reputation in the profession as a "publisher" and "collector of many fetishes" soon preceded him.[571] In an interview that he gave to Alvin Boyarsky, the head of the Architectural Association in London, in their television studio at the beginning of 1975, he talked about the Institute as a site of theory production with reference to *Oppositions*, thus elevating it to an almost mythical site of architecture: "And then we have a magazine, which we are using to try and develop a level of discourse internationally about ideas, and to see architecture as a critical vehicle." While Eisenman referred to the different values, motivations, goals, and intentions of the editors and external, in some cases international authors in this context, he once again did not clarify what exactly he meant by "discourse" or "critical."[572] At the same time, it was precisely the unresolved publishing situation that meant that, in winter 1975, the continuity of this ambitious project was anything but assured, and *Oppositions* 4, the issue to be published by the New York publisher Wittenborn Art Books, was thus delayed.

571 Eisenman, 1975.

572 Ibid. Eisenman was using a rather broad concept of "discourse" here, for especially in architecture the term colloquially denotes any form of debate. In contrast, public intellectuals in the 1970s increasingly used the concept of "discourse" to refer to the approaches of post-structuralist philosophy and linguistics in the wake of the theoretical turn, above all by Michel Foucault, who theorized his historical-genealogical approach in *The Archaeology of Knowledge* ([1969] 1972) and delivered his 1970 inaugural lecture at the Collège du France on *The Order of Discourse*. In a 1972 conversation with Gilles Deleuze, Foucault emphasized that for him, discourse analysis was always directed against power, as a "counter-discourse;" see Foucault and Deleuze, 1977. In addition, Eisenman first elaborated on what exactly he meant by "critical" in a lecture he gave at Cooper Union in the fall of 1986, see Eisenman, 1988, 190–193.

The crucial factor for the Institute's self-perception as a publishing house, regardless of its legal status, was that the role and importance of its publications changed fundamentally in the spring of 1976 when Eisenman managed to negotiate a contract with MIT Press for the publication of *Oppositions*. This led to the expansion of the editorial team. Next to Eisenman, Frampton and Gandelsonas were initially responsible for *Oppositions* in their dual function as editors and publishers, with Julia Bloomfield soon taking over as managing editor. The circle of editors was soon expanded to include Vidler, who was later joined by Kurt Forster and eventually Agrest. Yet *Oppositions* was not to remain the Institute's only publication, for as a result of its repositioning in terms of teaching and cultural production, the Institute's publishing activities were also expanded and extended to include other formats. Thus, although the Institute was never an independently operating publishing business, even in the medium term, it was subsequently also run—quite efficiently—as an editorial department, and at least some of the Fellows and an increasing number of editorial staff practically served as a writing and editing workshop, constantly devising new formats and content. Following the example of *Oppositions*, which had to provide for sections such as "History," "Theory," "Criticism," and "Documents," *October*, a quarterly journal for art theory and criticism, was launched in 1976, and then, after the Institute's 10th anniversary, *Skyline* (starting in April 1978), a monthly tabloid aimed at the New York architecture, art, and culture community.[573] This was followed by the series of IAUS Exhibition Catalogues (from 1979 onwards), which began with a documentation of the Institute's "Exhibition Program," and finally by the canon-changing series Oppositions Books (from 1982). This development was aided by further collaboration with MIT Press as an academic publisher, at least for *October*, the catalogue and the book series, and later with Rizzoli International as a commercial publisher—both providing

573 The publication of *Oppositions* as possibly the Institute's prime print product has been historicized extensively, the first account coming from Joan Ockman, herself a former member of the editorial staff, who nevertheless maintained a historical distance. As part of the inner circle, she first noted—in relation to *Oppositions*' history and to the relationships within the editorial staff—that over the years, in the twenty-six issues produced between 1973 and 1984, there had been a shift from theory to historiography and that, on balance, less architecture criticism was published than initially anticipated; see Ockman, 1988. This dichotomy was reproduced later on, with the *Oppositions* revival on the occasion of the publication of the *Oppositions Reader* (1999); see Hays, 1998. Since then, much emphasis has been placed on the beginnings of *Oppositions* with regard to the emergence of a theoretical debate in North America, the initial idea of founding a journal, the cultural technique of journal-making, and the interplay of "the real and the theoretical," but without clarifying the extent to which the theory, history, and criticism of architecture intersected with institutional, educational, and cultural politics. Drawing on Ockman's essay, Louis Martin elaborated on the prehistory of *Oppositions*; see Martin, 2008; Beatriz Colomina, together with PhD candidates at Princeton, compared the practice of journal-making in the 1960s and 1970s; see Colomina and Buckley, 2010; Lucia Allais linked theoretical research at the Institute to the rhetoric of grant proposals; see Allais, 2010. However, the fixation on *Oppositions* failed to recognize that the Institute became a legitimating and consecrating institution precisely because of its synergetic effects.

professional production, publicity, and distribution for prestige projects. At the time, while the Fellows' editorial work was mostly either unpaid or offset by other sources of income, editorial staff and production were cross-financed by foundation grants and donations.[574]

The Institute's shift in emphasis toward publishing is representative of, contributed to, and performed pioneering work for general growth in the journal and book market in architecture and urban studies in the United States, and had a symbolic significance for the increasing professionalization of the Institute's work. With regard to the textual and editorial practices of the Fellows, new insights into the postmodern turn in North American architecture culture can be gained by combining the histories of production and reception and reading and analyzing the individual formats. Once again, the combined study of the real social and discursive formations, while also taking into account the everyday work of editing and publishing, the way the editorial offices and editorial boards were organized in each case, and the associated institutional economy, will highlight the paradigm shift toward redefining the professional image of the architect and celebrating the figure of the architect as artist. This is not to question the very large significance attached to *Oppositions* by the editors and other Fellows, and by authors and readers alike, in terms of the novelty of the approaches and topics presented there, nor the strong identification of the Institute with the journal. Nevertheless, studying the conditions and constraints under which *Oppositions* was produced also helps to clarify the extent to which the Institute's knowledge production at the transition from Fordism to post-Fordism, in terms of the emergence of what was understood as a neo-avant-garde discourse on concepts such as "autonomy" and "criticality" in architecture or in terms of the creativity and intellectuality involved, was based on the enforcement of flexibilized, precarious labor: ultimately the expectation of dedication and, accordingly, self-exploitation.[575]

By publishing *Oppositions* and through the establishment of a complex and interlocking textual and editorial apparatus for *October*, *Skyline*, the IAUS Exhibition Catalogues, and Oppositions Books, the Institute fostered a

574 The work of the editorial staff for all other publication formats—the architecture newspaper, the exhibition catalogues, and the book series—has not yet been critically examined and reviewed, except in a few cases, e.g., Aldo Rossi's two monographs, *A Scientific Autobiography* and *The Architecture of the City* (both 1982). Architecture historian Mary Louise Lobsinger analyzed these two Oppositions Books by Rossi as prominent publications of the American architecture debate in the 1980s for the specific textual format chosen, as autobiography and urban theory, respectively; see Mary Louise Lobsinger, "That Obscure Object of Desire: Autobiography and Repetition in the Work of Aldo Rossi," *Grey Room*, no. 8 (Summer 2002), 38–61; Mary Louise Lobsinger, "The New Urban Scale in Italy: On Aldo Rossi's L'architettura della città," *Journal of Architectural Education* 59, no. 3 (February 2006), 28–38.

575 Somol, 1998; *Perspecta*, no. 33 (2002): "Mining Autonomy."

transatlantic dialogue between a genuinely North American postmodernist architecture debate and a truly European one, thus helping to found, if not significantly shape a publishing practice in architecture that can be understood as both a discursive formation and a cultural configuration. Not unlike MoMA before it, with its exhibitions on modern architecture, the Institute promoted a certain sense of global architecture culture, albeit viewed from New York. *Oppositions* in particular published the next generation of Japanese and Latin American authors and architects, next to European (especially Italian) ones, and was accordingly disseminated abroad. The Institute's publications also helped to establish a new kind of postmodern textual and editorial production across all publishing formats that combined both scholarly and popular, critical and autobiographical writing and included: theoretical and historiographical essays, programmatic, sometimes polemical editorials in *Oppositions*; architecture reviews, book and exhibition reviews, event listings, popular culture interviews, shopping and reading tips, obituaries, insider reports, reportages in *Skyline*; monographic texts, and forewords, prefaces, and articles, which increased or demanded credibility, in Oppositions Books. Through publishing, the Institute, as the self-proclaimed architectural avant-garde in North America (or at least the East Coast) ultimately projected a self-image and legitimized itself with regard to narrative structures, plot, and setting—precisely because of the general thrust of neo-avant-garde formalism or modernism against other backward-looking postmodern styles such as historicism, classicism, and eclecticism—with reference to architectural and artistic strategies, but not necessarily the social role of the political avant-garde movements of the twentieth century. The architecture debate at the Institute, according to Hayden White's 1973 classification of historiography, sometimes took on the form of "drama" or "comedy," or, more specifically, the kind of situation comedy that was particularly popular in North America at the time, and the publications broadcast this to the world.[576]

Contrary to outside perceptions, which have also been reproduced by architecture historians, the Institute and *Oppositions*, although not congruent, were not simply responsible for "teaching" and "discourse" vis-a-vis the discipline, although the two respective groups at the Institute—the teaching staff of the various education offerings and the editorial staff of the journal—took on both pedagogical and discursive tasks in their day-to-day work.[577] Moreover,

576 In the 1970s, the American historian and literary scholar Hayden White, borrowing from French post-structuralism, developed his approach of meta-history; see Hayden White, *Metahistory. The Historical Imagination in Nineteenth-Century Europe* (Baltimore: Johns Hopkins University Press, 1973). Translating historiography to architecture history, it can be interpreted as a discursive strategy of inscription in history, what White calls "emplotment;" he distinguishes four forms: "romance," "tragedy," "comedy," and "satire," which are accompanied by different "tropes," "modes," "arguments," and "ideologies."

577 Martin, 2010, 66.

the Fellows' teaching and cultural production served to finance not only the Institute's publishing activities but also its overall operations, since both the textual and editorial practices and the educational operations at the Institute also played a significant role in producing the next generation of architects and academics. The work on the two academic journals, as well as the exhibition catalogues and the book series, was primarily concerned with disseminating ideas and criticism, changing both architecture and art discourse—with lasting effects. Then again, by publishing *Skyline*, which appeared alongside the other more respectable formats only to eventually outdo them, the Institute, with all the media formats developed and produced there, also represented a market for attention from a sociological perspective, if not of vanities from a psychological one.[578] And although the Institute was never a real publishing house according to economic standards, i.e., with professional marketing and distribution structures, it was more than just an institutional framework for the *Oppositions* editors who, in addition to the professorships they held at New York universities and colleges, increasingly portrayed themselves there as architects, theorists, or historians, often with other publication projects up their sleeves.

Again, Pierre Bourdieu's sociology of cultural production, art, and literature can serve as a lens to better understand the history of the Institute's publications and its publishing networks, as it would not have been able to successfully manage the individual productions without the collaborative efforts of Fellows, Research Associates, assistants, students, and interns, as well as the numerous others involved. Kenneth Frampton, who took on a central role in the newly created position of director of publications towards the end of the 1970s, was responsible for all publication formats of the Institute. Working from the assumption of an interrelationship between society, architecture, and other fields of cultural production, the focus here is on the extent to which the interplay between the fields of activity at the Institute, including education and cultural production, was fundamental for writing, editing, and publishing, not only in terms of the cross-fertilization of ideas but also cross-financing and cross-promotion.[579] Yet the Institute's contribution to the new discursive formation of architectural postmodernism can only be understood by examining its collaborations with the multitude of external authors, with Massimo Vignelli as the Institute's longstanding in-house graphic designer (later replaced by Michael Bierut, Vignelli's erstwhile employee), with the publishers of choice—MIT Press represented by Roger Conover and Rizzolli represented by Gianfranco Monacelli—and with the editors of other publications on the book and journal market. Once again, the Institute's overall publishing

578 Bourdieu, 1983a; Franck, 1998 & 2000; Tzonis and Lefaivre, 1978.

579 Bourdieu, 1983b.

apparatus, its ambition, and its ability were grounded in the interplay of architecture, knowledge, and power. At the start, it was about the sovereignty of interpretation over two argumentative contexts that were characteristic of postmodernism, namely the production of theory inspired by French theory and a revisionist, yet mostly operative historiography of modernity, and if it was critical, then it was so in the sense propagated by the Frankfurt School.[580] From the mid-1970s on, it was a matter of hegemony in terms of the production and dissemination of knowledge relating to the built environment, not only in North America but in the entire English-speaking world of architecture, academia, and culture. The aim was to exert an influence on architecture debate and education in both the short and long term, globally speaking, through the scientific, graphic, didactic, and cultural quality, visibility, and longevity of its journals, exhibition catalogues, and book series.

Pilot Issues

When, after some initial difficulties, the first issue of *Oppositions* came out in November 1973, Peter Eisenman finally had his own journal—or more precisely: "A Journal for Ideas and Criticism in Architecture," as the subheader read.[581] As one of the three editors, alongside Kenneth Frampton and Mario Gandelsonas, who had equal rights, he was actively supported by other editorial staff in this ambitious and demanding publishing project: David Morton, who otherwise served as editor of *Progressive Architecture*, contributed input on editorial questions as editorial consultant and Suzanne Frank was initially assigned to provide editorial support for the first three issues, along with two interns, Jan Fischer and Susan Carter.[582] From a sociological perspective, *Oppositions*, which emphasized the relevance of writing and reading in the newly emerging architecture culture, initially tied to the East Coast, can be understood as an auto-poetic network, i.e., one that was self-constituting, self-referential,

580 Martin Jay, *The Dialectical Imagination. A History of the Frankfurt School and the Institute of Social Research, 1923–1950* (Berkeley: University of California Press, 1973); see also Francois Cusset, *French Theory: How Foucault, Derrida, Deleuze, & Co. Transformed the Intellectual Life of the United States* (Minneapolis: University of Minnesota Press, 2008).

581 Martin, 2008. The publication date of *Oppositions* 1 indicated on the cover is September 1973, but it was not published until two months later. The first issue was thus already behind schedule, but this back-dating practice was not an isolated case. In the eleven-year publication history, none of the total of twenty-six issues appeared on time.

582 As part of Princeton's *Clip Stamp Fold* research, exhibition, and publication project, Eisenman, Frampton, Gandelsonas, and Vidler commented on their respective contributions to editorial work in a public conversation with Beatriz Colomina, Urtzi Grau, and Daniel Lopez-Peres; see "Small Talks: Oppositions, Architeqturas Bis, Lotus" Storefront for Art and Architecture, New York, January 23, 2007, www.vimeo.com/user1360843 (last accessed: May 31, 2023); see Colomina and Buckley, 2010. Not present was Julia Bloomfield as the long-time managing editor of *Oppositions*, or Kurt Forster and Diana Agrest, who joined the editorial staff later.

and self-reproducing.[583] Yet the editors' different theoretical, historiographical, and ultimately creative approaches and concerns—in their own words, their "respective concerns for formal, socio-cultural and political discourse"—meant that they were engaged in productive competition with one another.[584] At the Institute, Eisenman, Frampton, and Gandelsonas were to play a key role because of their shared journalistic and editorial practices. From an epistemological perspective, their journal, by virtue of its formal, substantive, and institutional seclusion, shaped and cemented the Institute's inner circle. Over time, other Fellows and friends of the Institute in addition to the three editors were involved as authors and editorial board members, students and interns as additional editorial staff, and numerous outside individuals, graphic designers and their assistants, editors, and translators were involved in producing the journals. As a result, through *Oppositions*, the Institute established, developed, and strengthened far-reaching networks with the New York architecture scene, schools of architecture, and the cultural and publishing world.

The title *Oppositions* clearly signaled the postmodern, poststructuralist qualities and features of the journal. Rhetorically, it expressed contradiction, linguistically, opposition, and politically, resistance. But the provocation that lay in this nomenclature went even further. In Eisenman's design for the journal's logo, the first "P" was drawn as an outline so that the title could be read both as "positions" and as "zero positions," i.e., both in the plural and as a negation or dissolution of any stance at all.[585] Here his predilection for language games was clearly in evidence. In addition to a linguistic-discursive plane of reference, the ambiguities also had a formal-aesthetic one; the format and graphics of the journal could be read as a historical quotation in several respects. The cover and layout of *Oppositions* were developed in collaboration with the Italian New York-based graphic designer Massimo Vignelli, who had designed the corporate identities of American Airlines, Bloomingdales, Heller, and Knoll International, among others, and at the same time provided his services to

583 In their joint essay on the networks of artworks, in which they compare various approaches to the sociology of art, the architecture theorist Niels Albertsen and the sociologist and philosopher Bülent Diken argue that although these must be understood as an autopoetic system, it is precisely a matter of analyzing them in terms of their underlying networks in order to anticipate their role as mediators; see Albertsen and Diken, 2004, 35–58. Accordingly, *Oppositions* as a cultural product also performed social work and thus had social relevance.

584 Peter Eisenman, Kenneth Frampton, Mario Gandelsonas, "Editorial," *Oppositions* 1 (September 1973), n.p.

585 In official correspondence, *Oppositions* was initially referred to with as "Positions/Oppositions." Ockman distinguished the three ways of reading the chosen title, referring to Roland Barthes's 1953 publication *Le degré zéro de l'écriture* (English: *Writing Degree Zero*); see Ockman, 1988, 182. Suzanne Frank pointed out that the ambiguous logotype of *Oppositions* was based on a drawing by Duarte Cabral de Mello, then a Research Associate at the Institute, and that Gandelsonas had originally suggested the title, see Frank, 2010, 41–42.

nonprofit institutions such as the Institute. The layout reflected a modernist rationale, i.e., the will to order and organize all content in a structuralist grid. Only two fonts were used—Helvetica as a *sans serif* typeface for the logo and Century Schoolbook as a serif typeface for all other text—and formed the basis for the entire institutional identity of the Institute from 1973 on. The Pantone color Super Warm Red was chosen for the journal's cover, a catchy signal color that made issues of *Oppositions* an instant eye-catcher in bookstores, libraries, offices, and on private bookshelves.[586] The high-gloss finish and full cover flap gave *Oppositions* the appearance of a high-quality print product that could nevertheless be treated as a "little magazine" by the editorial team and produced in accordance with the principles of independence and cost reduction.[587]

To finance the journal, Eisenman attempted to raise US$100 each from a total of 100 sponsors in the run-up to publication in early 1973 to cover printing costs.[588] The three editors and Diana Agrest, who did not become an editor until much later, also subsidized the production by contributing US$3000 each as start-up capital.[589] In the following, the Institute's textual and editorial practice was seen as an independent one in its own right. *Oppositions* was initially

586 Strikingly, the cover design and page layout of *Oppositions* bore a strong resemblance to graphic design from Switzerland that was dominant in the 1950s and 60s, e.g., of the design journal *Neue Grafik / New Graphic Design / Graphisme actuel (*1959–1965). Vignelli had been strongly influenced by the Basel School around Josef Müller-Brockmann since his architecture studies in Milan; see Kerry William Purcell, *Josef Müller-Brockmann* (London: Phaidon, 2006). For the Institute, he incorporated numerous graphic elements of the Basel School into his repertoire. For example, the color of the *Oppositions* cover, "Oppositions red," as it became known, played a major role in Swiss graphic design of the 1960s; see Josef Müller-Brockmann, *Gestaltungsprobleme des Grafikers* (Heiden: Arthur Niggli, 1961), *Raster Systeme* für die *visuelle Gestaltung* (Heiden: Arthur Niggli 1981); see also Lars Müller, *Josef Müller-Brockmann. Pioneer of Swiss Graphic Design* (Baden: Lars Müller, 2001).

587 Massimo Vignelli, *Grids. Their Meaning and Use for Federal Designers* (U.S. Government Printing Office: Federal Design Library, December 1978).

588 *Oppositions* 1 was largely financed by individual sponsors; Eisenman had managed to win a total of ninety-nine sponsors. The sponsors also included two schools of architecture: UCLA and the University of Kentucky. With *Oppositions* 2, 124 private individuals, seventeen institutions, and three corporations were named as sponsors, including all the major schools of architecture at the Ivy League universities on the East Coast of the United States. The institutional sponsors of *Oppositions* were Boston Architectural Center, Carnegie-Mellon University, Columbia University, Cooper Union, Cornell University, Harvard University, Massachusetts Institute of Technology, MoMA, the New York division of the AIA, Pratt Institute, Princeton University, SUNY Buffalo, UCLA, University of Kentucky, University of Manitoba, University of Puerto Rico, University of Texas, Yale University. The neo-avant-garde ambitions and neo-Marxist attitudes of the editors and authors notwithstanding, it is striking that the three corporate sponsors who gave US$ 100,000 each were all large American oil companies (including Exxon), after the 1973 oil crisis.

589 Initially, three issues of *Oppositions* were planned; see the list of articles for *Oppositions* 2 and 3. Source: Columbia University, Shadrach Woods Collection. The list comprised articles by Diana Agrest, Stuart Cohen, Peter Eisenman, William Ellis, Kenneth Frampton, Mario Gandelsonas, Robert Gutman, William Huff, Frederick Koetter, Colin Rowe, Denise Scott Brown, Robert Stern, Shadrach Woods.

marketed as a non-commercial journal, which meant that mailing costs were cheaper. But the list of sponsors on the back cover made it clear from the outset that the journal was not disinterested—a total of twenty-six issues was produced at the Institute over the years. Through *Oppositions*, the Institute gradually built a philanthropic patronage within the architecture, academic, institutional, and corporate culture that was already firmly established in the North American arts and culture sector. *Oppositions* offers early evidence that Philip Johnson, an influential architect and powerful broker, once a curator and a trustee at MoMA, played a crucial role, not only as a wealthy patron of the journal but also as a closet supporter of the Institute working behind the scenes.[590] As a reward for this collective form of philanthropy, the publication of *Oppositions* 2 in late April 1974 brought not only public attention and a complimentary copy of each new issue, but also invitations to exclusive release events, lectures, and discussions held at the Institute under the title "Forum," where architects and academics debated the topics of the hour, ironically behind closed doors.[591] With the publication of each issue, the Institute began to establish its preeminence as a "postmodern salon" in American architecture culture.

While historiographies of *Oppositions* have so far mostly highlighted the opposition between architecture theory and history as a key characteristic, thus reproducing intradisciplinary lines of conflict between the editors, a closer look at the actual contents of the journal indicates that from a cultural studies perspective, it was a genuinely postmodern journal, as can be seen from the montage of set pieces in conjunction with the very contemporary form of sponsorship.[592] The first

590 Despite his fascist leanings becoming known to the architecture audience, Johnson was given a forum with the release of *Oppositions* 2; see Philip Johnson, "Rejected Architects: The Berlin Building Exposition of 1931, Architecture of the Third Reich," *Oppositions* 2 (January 1974), 81–94.

591 The first "Forum" celebrating the release of *Oppositions* 2 was held on April 29, 1974, and was about Ludwig Mies van der Rohe, with Ludwig Glaeser (presenter), Arthur Drexler, Philip Johnson, and Colin Rowe as guests.

592 Ockman wrote about the history of *Oppositions* in 1988, shortly after the journal ceased publication, detailing the interests and roles of the editors, editorial strategies, and significance to architecture debates; see Ockman, 1988. Ockman also elaborated on the history of the Institute's reception of Tafuri, explicitly by *Oppositions* editors; see Ockman, 1995. The 1999 *Oppositions Reader*, featuring a selection of essays edited by K. Michael Hays, was published by Princeton Architectural Press. Following the publication, various authors in the late 1990s commented on the journal's conception and organization, the different positions of its editors, the relationship between theory and history, and its significance for architecture debates and architecture education; see Hays, 1998; Vincent Pecora, "Towers of Babel," in *Out of Site. A Social Criticism of Architecture*, ed. Diane Ghirardo (Seattle: Bay Press, 1991), 46–76; Schwartzer, 1999; Sherer, 1999; Stern, 1999. Louis Martin began an oral history at CCA in the early 2000s with Eisenman, Frampton, and Forster, among others, and subsequently published an essay on the prehistory of *Oppositions*; see Martin, 2008. Meanwhile, Bloomfield and Frank, two other individuals associated with *Oppositions* and the Institute, respectively, published writings about the journal; see Bloomfield, 2010; Frank, 2010.

issue of *Oppositions* already offered space for the editors and some other Fellows to present their approaches in thematic essays. Agrest (along with her partner), Colin Rowe, and Vidler were featured as authors here, alongside contributions by Eisenman, Frampton, and Gandelsonas. With the second issue, the journal was then divided into five sections, in keeping with the new disciplinary logic that was just being developed in academia. "Oppositions" (later renamed "Criticism"), "History," "Theory," "Documents," and "Reviews, Letters, etc." focused on a critical, theoretically considered, and historically grounded examination of selected building projects, practicing (and demonstrating) discursivity, providing archival and historical texts, and forming opinions through the publication of book reviews and letters to the editor. The "Forum" column—which already resembled the society column "Talk of the Town" in the weekly *The New Yorker*—was, after all, a kind of glimpse behind the scenes, explicitly reporting on the preceding release event at the Institute.[593] What's more, the articles, written by Elis in sometimes scathingly satirical language, were illustrated with photographs of the cocktail parties that followed. These photographs placed the Institute in a glamorous light, elevated the Fellows and guests to celebrity status, and aroused the envy of those architects who had not been invited. As an early voice, *Oppositions* thus not only represented the constitution of the emerging New York architecture scene that met at the Institute in the mid-1970s, but also played a constituent role in terms of its networks. One effect was to provide interested readers with insights into the Institute's complex social and institutional fabric without their being invited to the party.

In the editorials initially co-authored by Peter Eisenman, Kenneth Frampton, and Mario Gandelsonas, the three editors communicated that they were well aware that, as with any journalistic work, they were dealing with knowledge and power.[594] In *Oppositions* 1, they expressed a common interest in influencing contemporary architecture through theory and history.[595] In *Oppositions* 2, they deliberately positioned their journal in the tradition of modernist publications, such as the art magazines *De Stijl* (1917–1928) and *L'Esprit Nouveau* (1920–1925), only to immediately distance themselves from a glorified image of the avant-garde and any intention to revive a polemical discourse.[596] In *Oppositions* 3, Eisenman,

593 Initially, informed book reviews along the lines of *The New York Review of Books* were envisioned, so that new publications would be reviewed from a variety of disciplinary perspectives. Starting with *Oppositions* 3, William Ellis was in charge of the "Forum" section.

594 Peter Eisenman, "Post-Functionalism" *Oppositions* 6 (Fall 1976), n.p.; Kenneth Frampton, "On Reading Heidegger" *Oppositions* 4 (October 1974), n.p.; Mario Gandelsonas, "Neo-Functionalism," *Oppositions* 5 (Summer 1976), n.p.

595 Eisenman, Frampton, Gandelsonas, 1973.

596 Peter Eisenman, Kenneth Frampton, Mario Gandelsonas, "Editorial," *Oppositions* 2 (January 1974), n.p.

Frampton, and Gandelsonas, then still in the additional role of publishers, lamented, in what was for the time being their last joint effort in providing an editorial line, their ineffectiveness, since the topics they were interested in, their meaning, and their significance occupied only a marginal position in the world of architecture and building.[597] They interpreted their resignation or retreat into the realm of signs as political action. However, it became clear that their ideas of the social function of history, theory, and criticism in architecture diverged widely. Eisenman liked to flirt with his apolitical stance and repeatedly invoked the myth of the autonomy of architecture, for example in April 1974 at a roundtable on "Theory" at Princeton University with Lionel March, Manfredo Tafuri, Rodolfo Machado, and Mario Gandelsonas as participants. In doing so, he hoped to shift the focus back onto the architectural object, as distinct from the corporate architecture of the 1960s. He also repeatedly championed this formal approach on the pages of *Oppositions*. The authorship of the three editors was clearly discernible in individual parts of the editorials, but their joint signature presented a united front to the outside world: "Whatever our differences, *Oppositions* continues to assert our belief in the importance of theory as the critical basis of significant practice."[598] In the end, it was precisely this positioning, the flirtation with or celebration of ambiguities and contradictions, theoretical and historical approaches, avant-garde and nostalgic attitudes, and self-confident or self-reflexive behavior, that ultimately made *Oppositions* an exciting read. The journal was read by practicing architects, students, and professors alike. Eisenman took it upon himself to personally hand out each new issue from his suitcase following his public appearances—like a traveling salesman distributing discursive abilities and skills. Despite being touted as a "little magazine," an epithet that was readily received and disseminated by contributors and outsiders alike, *Oppositions* was at best a simulation of an avant-garde magazine, since the Institute hardly saw itself as the vanguard of a social movement and instead set itself apart in an elitist fashion; after the prototype housing project was never realized, its focus shifted to education and culture and lost sight of the problems of the times.[599]

Aside from the fact that the pilot issues of *Oppositions* only appeared irregularly, the production had to be cross-financed by the Institute during the restructuring period, in addition to the donations, even though the editors were exempt

597 Peter Eisenman, Kenneth Frampton, Mario Gandelsonas, "Editorial," *Oppositions* 3 (May 1974), n.p.

598 Ibid.

599 What has hardly been mentioned in the history of *Oppositions* but plays an essential role for an institutional analysis and critique, was to what extent the networks and their conditions in the 1970s and 1980s differed from those of the 1920s and 1930s—not to mention the different intentions and ambitions among architects conditioned by political, economic, and social developments.

from overhead costs of all the projects at IAUS Central. However, in contrast to the few renowned magazines on the North American market, the quasi-academic journal legitimized the Institute's academic networks, offering the editors the opportunity to set their own priorities through the choice of subject matter and approach. Eisenman initially brought to the first issues his predilection for architects from England, especially Alison and Peter Smithson and James Stirling, whose projects he appropriated with his own formalist interpretations.[600] Frampton, on the other hand, formulated a socio-political critique of architecture, based largely on his reading of Hannah Arendt's *The Human Condition*, while writing about Russian Constructivism and the pedagogy at the HfG Ulm.[601] Gandelsonas and Agrest were concerned with a semiology of architecture influenced by post-Marxist and post-structuralist theory.[602] Further topics of the pilot issues were: in the "Oppositions" section, contextualism in projects by Venturi and Rauch and Richard Meier, respectively, and Werner Seligmann's housing for the Urban Development Corporation, in the "History" section, an essay by Colin Rowe on the vocabulary used in British architecture to describe composition, and in the "Theory" section, an essay by Rosalind Krauss on intention in Minimal Art, and one by Manfredo Tafuri on the language of architectural postmodernism in Italy and the United States and the possibilities of an architecture critique. As editors, Eisenman, Frampton, and Gandelsonas thus complemented each other very well in terms of their interests and the projects, architects, texts, and authors they selected for publication and formed a well-rounded team; even if they did not agree in their research questions, methods, and outcomes, their thinking styles at least had in common that they were all interested in the legacy of architectural modernism in their architectural practice, theory production, and historiography and were thus oriented toward Europe. At the same time, these were the cornerstones of the new order, which were being constructed and communicated with *Oppositions* in a reversal of a post-war transatlantic dialogue. As the publication's spin doctor, Eisenman challenged the other two editors to take a stand on certain issues in shorter articles, and himself pitted theory production against historiography. Based on a closer reading of all the editors' editorials, essays, introductions, and commentaries, which functioned as post-scripts, they theorized about the characteristics of architecture and historicized avant-garde practices as precedents for a postmodern architecture.

600 Peter Eisenman, "From Golden Lane to Robin Hood Gardens; Or If You Follow the Yellow Brick Road, It May Not Lead to Golder's Green," *Oppositions* 1 (September 1973), 27–56; "Real and English: Destruction of the Box I," *Oppositions* 4 (October 1974), 5–34.

601 Kenneth Frampton, "Industrialization and the Crisis of Architecture," *Oppositions* 1 (September 1973), 57–82; "Apropos Ulm: Curriculum and Critical Theory," *Oppositions* 3 (May 1974), 17–36; "On Charles Jencks and Nathan Silver's Adhocism: The Case for Improvisation," *Oppositions* 3 (May 1974), 104–105; "George Wittenborn. 1905–1974," *Oppositions* 4 (October 1974), 14.

602 Mario Gandelsonas and Diana Agrest, "Semiotics and Architecture: Ideological Consumption of Theoretical Work," *Oppositions* 1 (September 1973), 93–100.

Many contributions directly or indirectly referenced key texts in linguistics and semiotics, authors from the Frankfurt School, and contemporary philosophers from France, thus injecting them into American architecture debate via their own idiosyncratic interpretations. A glance at the references and footnotes cited in *Oppositions* suffices to trace the extent to which new architectural thinking and discursive terminologies were introduced here, based on the interdisciplinary references to other fields of knowledge. Ultimately, the journal, in keeping with Roland Barthes's aphorisms, testified to the editors' desire to disseminate and debate their own ideas, but it also always served as a powerful instrument of self-aggrandizement and self-representation, as well as management, i.e., the administration of architectural knowledge, through the handling and control of information. Moreover, even though polemical, at times cynical, and critical tones sometimes crept in, Eisenman, Frampton, and Gandelsonas all believed in the power of the text. *Oppositions* thus portrayed the Institute as a place of intellectual debate, which here took the form of postmodern, rather than modern views and thinking in its historiography and theory production.

From the outside, *Oppositions* was perceived early on as the governing body of the Institute, although texts from the immediate or extended circle of the Institute, e.g., Rosalind Krauss, Colin Rowe, Emilio Ambasz, Robert Stern, etc., were published in the first three issues, in addition to essays by the editors and other Fellows. Obviously, it would be inappropriate to equate those views and attitudes expressed by individual authors and exchanged through *Oppositions* with those of the Institute as a whole.[603] After all, not all the Fellows and Visiting Fellows were represented in the journal over the years, and the younger generation in particular was barely granted access.[604] And yet *Oppositions* set out to be the journal of history and theory and—probably out of strategic considerations and in order to address the disagreement between supposedly modern and postmodern positions—a dichotomous confrontation was set up in the pilot issues through the choice of architects (besides Stirling, Venturi, Meier, and Seligmann, these included Peter Eisenman, Michael Graves, and Aldo Rossi) and authors (Stuart Cohen, Charles Moore) featured there, thus providing a platform for the ideological battle between the Whites and the Grays. By contrast, very little was published about

603 Rosalind Krauss, "The Fountainhead," *Oppositions* 2 (January 1974), 61–70; Colin Rowe, "Character and Composition, or: Some Vicissitudes of Architectural Vocabulary in the Nineteenth Century," *Oppositions* 2 (January 1974), 41–60; Emilio Ambasz, "A Selection of Working Fables," *Oppositions* 4 (October 1974), 65–74; Robert Stern, "Yale 1950–1965," *Oppositions* 4 (October 1974), 35–62.

604 One exception was Rem, or rather Remment Koolhaas, as he was then known under his full name, who was invited to contribute to *Oppositions* twice in 1973–74, when he was visiting at the Institute; see Rem Koolhaas and Gerrit Oorthuys, "Ivan Leonidov's Dom Narkomtjazjprom, Moscow," *Oppositions* 2, (January 1974), 95–103; Rem Koolhaas, "The Architects' Ball – A Vignette, 1931," *Oppositions* 3 (May 1974), 91–96.

current affairs—apart from housing, with essays about Twin Parks in the Bronx and another UDC project, Elm Street Housing in Ithaca.[605]

With the emergence of postmodernism as a new discursive formation as per Michel Foucault, two editorial lines could be discerned throughout the pages of *Oppositions* from the very beginning, at a time when education, culture, and publishing shared similar patterns of concerns, perspectives, concepts, and themes:[606] on the one hand, an examination of the new, self-proclaimed architectural avant-garde dedicated to critical theory, i.e., to a new, Western European Marxism, and on the other, a historiography of architectural modernism from Europe that, in contrast to classics such as the works of Sigfried Giedion, displayed revisionist streaks not only by linking architecture, technology, and urbanization but also by giving a voice to the architects themselves as protagonists. When it appeared in the fall of 1974, *Oppositions* 3—the previously ghostly letter "P" in the title had been filled in by now, giving the journal an even more combative stance—set new trends with regard to the internationalization and intellectualization of the American architecture debate and education. In this issue, the *Oppositions* editors published an article by the Italian architecture historian and critic Manfredo Tafuri for the first time. This had the effect of contributing to the creation of a whole new translation culture, however awkward and stilted some phrases and wordings may have sounded as a result of linguistic interference and theoretical terminology.[607] Another factor was that *Oppositions* initiated an intellectual exchange, grounded in non-discursive formations, between the Institute for Architecture and Urban Studies and the Istituto Universitario di Architettura di Venezia (IUAV), where Tafuri had taught since 1968 and where he was head of the history department.[608] In "L'Architecture dans le Boudoir," he offered a critical reading of the formalism, or rather language games, of postmodernist tendencies on both sides of the Atlantic that incorporated both architectural and theoretical works by Robert Venturi, James Stirling, Peter Eisenman,

605 There was no mention, for example, of the solar homes built at this time by Douglas Kelbaugh, a student of Eisenman.

606 Michael Foucault, "Discursive Formations," in Foucault, ([1969], 1972), 31–39.

607 Translations were made by Victor Caliandro, Marlène Barsoum, and Liviu Dimitriu.

608 For an intellectual biography of Manfredo Tafuri, see Andrew Leach, *Manfredo Tafuri. Choosing History* (Gent: A&S/books, 2007). The exchange between and networks of the IAUS and the IUAV have hardly been studied under discursive and institutional aspects. Following Tafuri's death in 1995, Ockman focused on the construction of the Venice-New York axis and, among other things, also elaborated on the relationship between Eisenman and Tafuri; see Ockman, 1995. Ockman's essay was republished in German under the one-dimensional title "'Boudoir Architecture' als Anschauungsmaterial: Manfredo Tafuri und New York," in the Swiss journal *werk, bauen + wohnen* (September 1995), and yet the transatlantic dialogue worked both ways. Italian architect Ernesto Ramon Rispoli initially stated that his dissertation at the Politecnico di Torino would specifically highlight the performance of Italian architects and academics in the United States but fell short of this goal; cf. Ernesto Ramon Rispoli, *Ponti sull'Atlantico. L'Institute for architecture and urban studies e le relazioni Italia-America (1967–1985)* (Quodlibet: Macerata, 2012).

Michael Graves, and Aldo Rossi.[609] In light of the transformation of the capitalist system, Tafuri, a committed Marxist, criticized the self-referentiality of contemporary architectural practice, particularly in the United States, its retreat into the realm of signs, and its disassociation from the production process. Although he did not explicitly mention the Institute here—he visited it for the first time in May 1974 on a trip to the USA that took him primarily to Princeton—he nevertheless addressed some of the Fellows, especially Eisenman, very directly in his critique of the architects' claim to power and the myth of autonomy (and criticality). In view of his diagnosis that contemporary architecture was only discussed in the "boudoir" (French for back room) as an abstract work of art, Tafuri concluded by referencing Walter Benjamin's classic essay "The Author as Producer" and outlining that the only way out for architects was to look for alternative possibilities of action within the existing relations of production; that is to say, for productive intellectual work that has an effect on the relations of production.[610] Not only did Tafuri later re-engage with the New York architecture scene and the new spaces of cultural production that opened up there, but *Oppositions* afforded him the opportunity on several occasions to publish his critique of the globalized neo-avant-garde and an operationalized historiography for an English-speaking readership and to define an autonomous role of the architecture historian or critic. By mediating this exchange, the Institute was instrumental in the production, distribution, and reception of an "American Tafuri," as the Italian theoretician was subsequently labeled.[611]

609 Tafuri, 1974. Tafuri's essay "L'Architecture dans le Boudoir" was based on a lecture he had previously given at Princeton in April 1974 at Agrest's invitation as part of the lecture series "Practice, Theory and Politics in Architecture"; see "Introduction," in *Oppositions* 3, 1974, 37; see also Ockman 1995, 67, footnote 4. In the panel discussion the following day, moderated by Gandelsonas, Tafuri met Eisenman, Rodolfo Machado, and Vidler, and again voiced his criticism of the architectural language of a self-proclaimed avant-garde; see audio recording of panel discussion, no date. Source: Princeton University, School of Architecture Archive. Here, Tafuri emphasized that he was interested in Eisenman, Graves et al. precisely because their architecture had had no political meaning for him and had simply been useless. In his examination of the New York architecture scene, he subsequently drew on findings from field work and participant observation. In April 1974, on the last day of his three-day stay in the USA, he was visiting the Institute, and took part in an editorial meeting of *Oppositions*. In the introduction to "L'Architecture dans le Boudoir," the editors then announced the future reception of contemporary architecture as well as history, theory, and criticism from Italy.

610 Walter Benjamin, "The Author as Producer," *New Left Review* 62, no. 1 (July-August 1970), 83–96. Obviously, the term "boudoir" alluded to Marquis de Sade's classic *La Philosophie dans le* boudoir of 1795.

611 Tafuri was first published in English in 1971 in the catalogue for the MoMA exhibition "The New Italian Landscape" that was curated by Emilio Ambasz; see Manfredo Tafuri, "Design and Technological Utopia," in *Italy: The New Domestic Landscape* ed. Emilio Ambasz (New York: The Museum of Modern Art, 1972), 388–404. Subsequently, monographs, essays in edited volumes, and other texts by Tafuri were published by MIT Press and the Institute. On Tafuri's reception in American higher education, the so-called "Venice School" and the "American Tafuri," i.e., Tafuri as adapted by American architecture circles, see *Any*, no. 25/26 (2000): "Being Manfredo Tafuri: Wickedness, Anxiety, Disenchantment;" see also Ghirardo, 2002.

The third issue of *Oppositions* enabled the Institute to expand and strengthen its reputation, its services, and its market position far beyond New York for the first time; the number of sponsors and subscribers increased continuously.[612] Also, as a result of the epistemological insights and effects that the expansion of the circle of authors and the incorporation of criticism had, it was able to establish itself as a new actor in the production and dissemination of knowledge, for educational and cultural purposes, outside of the traditional world of academia and academic publishing. Despite, or perhaps because of, its comparatively small circulation, the journal's editorial focus on theory and history was met with open arms and advanced another form of mediatization in the United States beyond the typical architecture press, e.g., *Architectural Forum*, *Architectural Record*, or *Progressive Architecture*, or more academically minded journals, e.g., *Perspecta*. With regard to a sociology of knowledge, culture, and media, *Oppositions* can be seen as the written manifestation of a thought collective specific to East Coast architecture in the 1970s, and even as a constitutive part of a transatlantic dialogue. One change that became crucial to the journal's development was the involvement of Julia Bloomfield as managing editor with *Oppositions* 3 in 1974.[613] As one of the few permanent staff members, Bloomfield oversaw all phases of production from text acquisition to print approval until 1982, was responsible for the editors' time and work management, communication with authors, text and image editing, coordination of graphics and typesetting, fundraising and sponsorship, communications with the publisher or printer, preparation and correction of galley proofs, printing support, and so on; after one year she was elected Fellow, which underscored her importance and usefulness at the Institute. Despite the successful establishment of *Oppositions*, however, the journal was never able to support itself financially; debts had already been incurred with the pilot issues, which was mainly due to the high production and personnel costs totaling about US$15,000 per issue. As a result, the editorial team intensified its search for a professional publisher. At a time when a reinvention of the Institute as an architecture school and a cultural space was on the horizon, the Institute's management could not afford to continue the journal

612 Institutionally speaking, Eisenman measured the success of *Oppositions* by the number of sponsors and subscribers: "Oppositions already has three corporate, seventeen institutional, and one hundred and ten individual sponsors. It has over 400 subscriptions and is beginning to expand its distribution to Europe and western United States." Peter Eisenman, "Director's Report," June 19, 1974. Source: CCA Montréal, IAUS fonds: A.1-2. By the end of 1974, with the publication of *Oppositions* 3, there were already 576 subscribers, including 496 individuals and eighty institutions, and a total of 151 sponsors, including 129 individuals, nineteen institutions, and three corporations.

613 Bloomfield's contact with the Institute originally came about via Frampton, whom she knew from England. After Bloomfield moved to New York, she contacted Eisenman in 1973 looking for work. At the Institute, she initially worked at the front desk. She contributed the bibliography on the Smithsons to *Oppositions* 2. Bloomfield was one of the non-architect women at the Institute who worked there as permanent staff. Her salary was initially set at US$ 2,500 per issue; with four issues per year, this amounted to US $200 per week, which was paid when the budget allowed.

as a loss-making project. In addition, the aim was to make *Oppositions* available internationally in selected bookstores, and such distribution at home and abroad was hardly feasible for the Institute on its own. For *Oppositions* 4, the Institute had been able to reach a one-time agreement with George Wittenborn (who passed away shortly after) in 1974, so that this issue was the first (and only) to be published by Wittenborn Art Books and distributed through Wittenborn's art and architecture bookstore in Manhattan; yet editorial and financial complications meant that the production of this very issue dragged on for nearly a year and a half.[614]

It goes without saying that by publishing *Oppositions* with the help of sponsors, the Institute assumed a new role and responsibility in American architecture culture without becoming dependent on any single person, institution, or corporation. The subscription structure and pricing policy established a differentiated readership and guaranteed consistent sales as special offers and promotional efforts appealed to architecture students and professors alike.[615] The institutional sponsorship secured the conceptual and financial commitment of architecture schools and other institutions and, as a positive side effect, enabled the journal to find its way into the most important libraries nationwide. As a quarterly journal, each issue provided comprehensive teaching material for history and theory courses in master's and even bachelor's degree programs that were being added to the curriculum at schools of architecture, as well as course material for the new doctoral programs in architecture that were just being developed at prestigious Ivy League universities. Partly because almost all of the editors and Fellows taught as university professors themselves and thus acted as multipliers through their extra-academic activities, *Oppositions* also introduced a new postmodern thinking style to American academia. Many articles in the journal offered guidance in the perception and appreciation of both modernist architecture as a historical period and architectural postmodernism as a contemporary architectural style. In terms of theory production and historiography, the Institute set out to promote the emergence of postmodernism in the field of architecture in general. For the historical circumstances and special characteristics of the newly interpreted publication format meant that *Oppositions*—as a fictitious or, to use Jean Baudrillard's terminology, "hyperreal," "fake" or "artificial" thought collective on which Eisenman imposed his thinking—offered manifold possibilities for the production and dissemination of knowledge, for the definition of real and apparent problems, for the reception of methods and concepts, for intellectualized and yet depoliticized reflections, and ultimately for eclipsing socio-economic and socio-political issues.

614 *Oppositions* 4, dated October 1974, did not appear until January 1976. While editing the issue, Eisenman was already working on the conceptual design of *Oppositions* 5.

615 In 1975, an annual subscription was US$ 20 for students, US$ 24 for non-students, and US$ 30 for institutions. One of the sales strategies was to engage students by giving them a year's subscription for free in exchange for taking out ten subscriptions.

Professional Journal-Making

Winning MIT Press as a collaborative partner in the mid-1970s, after long and tough negotiations, was of great importance for the Institute, enabling it to not only continue *Oppositions*, but also to establish another journal, *October*, and successfully reposition, restructure, and realign itself as group, organization, and institution. What was to become the "Publication Program" at the Institute was thus decisively strengthened and even secured for the coming years. In the run-up to the contract negotiations for *Oppositions*, the Institute printed a poster for backlog issues and a flyer for upcoming issues which, in addition to the first public announcement of its association with the publisher, also named potential authors and topics for contributions for an entire issue.[616] Previously, Eisenman had prepared a list of potential topics for issues 5 to 8, thus setting out the editorial line of the journal, incorporating shared and individual interests of the editors while at the same time committing them to contribute.[617] The poster was immediately sent to all subscribers to engage them in the publicity drive for the coming issues. In addition to institutional purposes, however, the poster also served discursive purposes, since two new editorial strategies were communicated here in an info text: on the one hand, the editorial team planned to increasingly direct the focus of *Oppositions* across the Atlantic to contemporary positions in Italy and Spain (with less of a focus on architecture from Great Britain) in addition to continuing the high-profile dispute between the "Whites" and the "Grays," on the other hand, they also intended to advance the architecture debate with individually written editorials. The *Oppositions* poster as cultural product and medium, similar to those for the "Evening Program" and the educational programs, served promotional purposes—demonstrating that the journal was to become an even more powerful instrument of knowledge, its production, reproduction, and dissemination, while the editors positioned themselves internationally as theorists or historians. By pre-selecting authors and themes and promoting a transatlantic dialogue, the Institute cast itself as an authority of legitimacy and consecration, impressively underscoring its self-appointed role as gatekeeper for the American architecture scene. Although ultimately only a fraction of the articles listed on the poster were to be published in *Oppositions*, the poster nevertheless communicated its approaches and ambitions.

With the signing of the contract for *Oppositions* on April 1, 1976, the "little magazine" became an academic publication. The in-house production with a smaller budget and a smaller print run, which had meant freedom of content and allowed for irregularities in the publication, had to subsequently be transformed into a more professional production which would benefit both

616 *Oppositions* (poster), ca. 1975. Source: CCA Montréal, IAUS fonds: C.3-7 / ARCH250449.

617 Peter Eisenman, notes on the content for *Oppositions* 5 to 8, n.d. Source: CCA Montréal, IAUS fonds: C.1-2 / ARCH401325.

contractual partners.[618] The deal ensured the continuation of *Oppositions* in the medium term and promised to provide the Institute with international exposure. MIT Press, which not only guaranteed scholarly quality but also promoted artistic innovation, had the necessary structures to ensure distribution at home and abroad, and to handle advertising and publicity. On top of this, the publishing house also assumed a large part of the production costs of *Oppositions* while the Institute committed itself to the regular production of four issues a year while benefitting from the international reputation of the university press. It was agreed that Vignelli, together with the editors and staff, would retain control over the graphic design, layout, and printing of the journal, which was established as a brand. One of the publisher's conditions, however, was that the Institute would continue to acquire donations from individuals, institutions, and corporations so that it could contribute its financial share. Any debts incurred were to be shared between the two contracting parties. On this basis, issues 5 through 24 of *Oppositions* were produced from 1976 to 1982, before the Institute switched to Rizzoli International in 1982. Essentially, MIT Press took care of the journal's business development, leaving the editors to concentrate entirely on content and, through journal-making, contribute to debates and education in architecture while redisciplining and intellectualizing it. Both the theoretical and historiographical approaches kept up the appearance of disinterested involvement. But this is only the first impression, for although *Oppositions* never really contributed to economic revenue, i.e., to the financing of the Institute's operating costs or to the Fellows' income, its contribution can nevertheless be measured in symbolic gains. The new collaboration with the Institute enabled MIT Press to raise its profile in the longer term, not only in the journal segment but in the book segment as well, by establishing an architecture segment, thereby strengthening its market position here alongside its segments in science and the arts. The university press became an important partner for the Institute, not least thanks to Roger Conover, who was appointed acquisitions editor at MIT Press in 1976 and was now responsible for the architecture segment there. In this role, Conover showed a strong interest in building on the relationship with *Oppositions* by establishing further contacts with New York architecture circles and expanding existing ones. For him, importing intellectually ambitious authors and publishing a new sophisticated body of texts were both quite attractive.

618 MIT Press, contract between the Institute and MIT Press, appendices and tables, April 1, 1976. Source: CCA Montréal, IAUS fonds: C.1-1 / ARCH401320. Previously, the Institute had been in talks with MIT Press regarding the publication of other print products, an exhibition catalogue for *The Streets* exhibition, a book series, and also a journal. While being dependent on the collaboration, they disputed the correct designation of the collaboration. The Institute's leadership succeeded in defining the relationship of the university press to the Institute as a subservient one. The imprint eventually stated, "Oppositions is a journal published for The Institute for Architecture and Urban Studies by The MIT Press." The copyright for the journal was held by the Institute.

The Institute also saw a shift in its work. By the time the contract with MIT Press was signed, *Oppositions* represented only one of the Fellows' activities (albeit a particularly labor-intensive and high-profile one) among multiple others. At the same time, the three editors Eisenman, Frampton, and Gandelsonas were still involved in the "Undergraduate Program," the flourishing "Evening Program," and the expanding "Exhibition Program." Nevertheless, another journal was already being developed and produced at the Institute at the same time: *October*, a new journal of art theory and criticism, edited by Rosalind Krauss and Annette Michelson. This second regular publication, which was to occupy a special position at the Institute, both institutionally and intellectually, was an excellent addition to the portfolio in terms of Eisenman's aspirations and was produced according to the same procedure. Nevertheless, *Oppositions* would remain the more important driving force, as the textual work on and controversial discussions around the journal served to stabilize the thought collective of Fellows and authors. With regard to the new epistemology of architecture debate and education, pitting a modern way of thinking, designing, and implementing against a postmodern one, two strategies and successes played a seminal role in the journal's continued authority and reputation: first, with the MIT Press deal, *Oppositions* was distributed to schools of architecture via special subscription offers for institutions and was thus represented in libraries nationwide, so that within a very short time the journal advanced to become a teaching and learning resource. Second, due in part to its text-heavy design—plans, drawings, and photographs were used rather sparingly—it supported the Institute's reputation as a center of architecture intelligentsia on both a national and international scale. A close reading of *Oppositions*, not only the first four issues but also the new edition under MIT Press, reveals that the journal was also a medium for reinventing the architect's role as intellectual or artist. By producing and distributing new architectural knowledge and featuring theorists, historians, and critics as authors whose texts would prove groundbreaking and pioneering, the editors had a strong influence on postmodern discourse, at best in terms of a critical-reflexive understanding of theory that allowed for differences, and in terms of a genealogical-archaeological understanding of history that functioned beyond established models, precedents, and references. Moreover, by providing new perceptual and evaluative criteria for contemporary and modern architecture, history, and theory from America and Europe, a central mechanism was created in a market of symbolic goods.

In 1976, Anthony Vidler was added as a fourth permanent editor to ensure professionalism in the editorial work while Frampton was mostly absent from the Institute over the next few years. Vidler, who was initially given Visiting Fellow status for a year, strengthened the architecture history focus of *Oppositions*; he was the first to work on a thematic issue on nineteenth-century Parisian urbanism in L'École des Beaux Arts. In June 1976, at a meeting of the Board of

Trustees, Eisenman reported the conclusion of the contract with MIT Press and announced an increase in productivity: according to his report, *Oppositions* 5 was already in print in the summer of 1976, *Oppositions* 6 was being typeset, and the first manuscript for *Oppositions* 7 had already been written.[619] In order to cope with this new productivity, Bloomfield, as managing editor responsible for ensuring the increased editorial work and production, began working with the Institute's interns on each issue to manage the extra workload.[620] One of these interns was Joan Ockman who, by virtue of a BA in Comparative Literature and her experience as an editorial assistant at *The New Yorker*, took over editing duties starting with *Oppositions* 7.[621] In addition, David Morton, who usually served as a senior editor at *Progressive Architecture*, now advised the Institute as an editorial consultant. In the second half of the 1970s, while Eisenman (and in other ways Frampton) continued to feed the myth of modern architecture, *Oppositions* went on to become a significant medium in disseminating, legitimizing, and consecrating the postmodern architecture debate, which continued to spread throughout the globe, initially through publications. Compared to leading European journals, such as *Architectural Design* (from the UK), *Casabella* (Italy), *L'Architecture d'Aujourd'hui* (France), *Archithese* (Switzerland), and even *Arch+* (Germany), which occasionally featured architecture from the United States, *Oppositions* was now much more internationally oriented and had a much stronger focus on intellectual discourse than on the mere discussion or presentation of individual new buildings.

"The Italian Issue"

When *Oppositions* 5 finally appeared in October 1976, the issue manifested the dual ambition of its editors and the Institute to not only cover the American debate but also to link it to an international or transatlantic dialogue. It was Eisenman himself, having scribbled handwritten notes on the selection of authors and topics on a concept paper for *Oppositions* 4 as early as the summer of 1975, who was responsible for the issue.[622] One historical factor that must be considered is that this new issue marked the onset of an internationalization of the American architecture debate at the very moment when the foundations of architectural practice were being radically altered by new neoliberal

619 Peter Eisenman, "Director's Report," June 10, 1976. Source: Sarah Lawrence College Archives.

620 Andrew Bartle had done the editing for *Oppositions* 5 and 6. In addition, Bloomfield later worked with Raleigh Perkins and Jay Johnson.

621 Ockman came to the Institute in early 1976, initially working as an intern for Agrest, but was soon assigned by Eisenman to edit texts for *Oppositions* as well as his own publications. Eisenman was working on two publications at the time, on *Giuseppe Terragni* and on *House X*. Even after Ockman began studying architecture at Cooper Union in the fall of 1976, she remained with the editorial staff, first as editor consultant and later, from *Oppositions* 11 onward, rising to associate editor.

622 Peter Eisenman, *Oppositions* 4, n.d. Source: CCA Montréal, IAUS fonds: C.1-2 / ARCH401321.

politics and the urban crisis in New York. In contrast to this historical context, *Oppositions* 5 primarily featured architects and authors from Italy, which is why the issue became known at the Institute as "The Italian Issue." The "Oppositions" section, reserved for architecture reviews, featured two articles, the first being "Aldo Rossi: The Idea of Architecture and the Modena Cemetery," a rather positive review of Rossi's 1966 monograph *L'Architettura della Città* and his 1971 award-winning project for the San Cataldo Cemetery in Modena, Italy, by Rafael Moneo.[623] This detailed contribution was complemented and enhanced by exclusive drawings by Rossi, printed on glossy black paper, as well as the reprint of a translation of Rossi's project text "The Blue of the Sky."[624] The second contribution in this section was a text by architecture critic Manfredo Tafuri, who wrote about individual, small-scale works by the New York Five from 1965 to 1970 under the title "American Graffiti: Five x Five = Twenty-five." Here, he resumed his linguistic-semiotic critique of postmodern language games in the United States first outlined in "L'Architecture dans le Boudoir" and provided another personal, yet theoretically justified take on contemporary architectural practice, which the editors illustrated with a specially made heroic collage of the protagonists, a farewell, so to speak, to the "Whites."[625] Interestingly, Tafuri's text explicitly addresses Eisenman's involvement with two groups, each of which resulted in exhibitions at the Museum of Modern Art: first, his role at CASE, whose urban renewal project for Harlem was shown at "The New City" in 1967, and secondly the Institute, whose design of a housing prototype was shown at "Another Chance for Housing" in 1973. Again, Tafuri censured both designs for not being based on socio-political interests, but rather on exclusively formal-aesthetic ones. Ultimately, the Italian author, with reference to contemporary French philosophers such as Jean Baudrillard or Michel Foucault who, speaking as critics and historians of the present, had begun to use terms such as "simulacra" or even "coercion," criticized the architecture intelligentsia of the 1970s for moving far away from the original tradition of the modern avant-garde in Europe. Thus, in a feedback loop, *Oppositions* itself provided one of the key texts of the self-observation and self-description of the architectural project, whose shifts were renegotiated under terms such as "neo-avant-garde" or "post-modernism."

623 Rossi and Tafuri were guests at the Institute in the spring of 1976: Rossi presented his latest projects in March 1976 as part of the "European New Wave" series, after which he exhibited his architectural drawings and stayed for a few days. Tafuri, following a stay at MIT, visited the Institute a second time in April 1976 and gave a lecture on "Modern Architecture: The Dialectics of Order and Disorder" as part of the "Architecture" series.

624 Rafael Moneo, "Aldo Rossi: The Idea of Architecture and the Modena Cemetery," trans. Angela Giral, *Oppositions* 5 (Summer 1976), 2–21; Aldo Rossi, "The Blue of the Sky," trans. Marlène Barsoum, Livio Dimitriu, *Oppositions* 5 (Summer 1976), 31–34.

625 Tafuri, 1976; Tafuri, 1974.

These contributions in the "Oppositions" section of *Oppositions* 5 demonstrated strategies that were to become characteristic of the discursive, editorial, and journalistic practice at the Institute: on the one hand, trends that were already in vogue were addressed by presenting, for example, Rossi, one of the most dazzling actors of European postmodernism, who was, however, still largely unknown in North America; on the other hand, critical voices such as those of Tafuri were presented right away and thus appropriated for the Institute itself to a certain extent. Eisenman, theoretically well-read and rhetorically gifted, knew how to use both strategies for his own purposes. His introduction to Moneo's text on Rossi is another case of "creative misreading."[626] While referring to the transformation of the architectural field over the past decade, and not only situating Rossi's approach, evident in *L'architettura della Città* and San Cataldo Cemetery, within the neo-rationalism of the Italian Tendenza but also contextualizing Moneo's 1973 article, Eisenman presented "autonomous architecture" as the only possible concept—without, however, discussing the transatlantic differences. Eisenman saw autonomy, which conceptualizes architecture as an independent art form in contrast to the city, not only in the Tendenza, but in "the metaphysical Scolari, the romantic Krier brothers, the delirious Koolhaas" and thus not only connected these disparate figures, but classified his own approach at the same time.[627] His introductory text, which he ended with the sentence "And who will dare cry in the face of all this-Formalism!" was a battle cry, and he used the opportunity to paint a picture of himself as an eloquent and polemical architect and theorist, in order to distance himself from his critics.[628] He built Tafuri up as an adversary to legitimize his formalist approach and thus repeatedly used him as a fame-maker, similar to what he had done earlier with Frampton and Gandelsonas. Both profited from this: Eisenman was able to legitimize and enhance his own position through Tafuri's criticism, negative though it was, and Tafuri used the opportunity to publish his texts in English and thus reach an international readership.[629]

The release of *Oppositions* 5, published in a run of 3,000 copies and at a new price of US$6, was duly celebrated in October 1976 with a "Forum" on Aldo Rossi. As a good host, Eisenman could not resist personally inviting all

626 Harold Bloom, *A Map of Misreadings* (New York: Oxford University Press, 1975).

627 Peter Eisenman, "Introduction" to Rafael Moneo's "Aldo Rossi: The Idea of Architecture and the Modena Cemetery," *Oppositions* 5 (Summer 1976), 1.

628 Ibid.

629 A total of five texts by Tafuri were published in *Oppositions*, following "L'Architecture dans le Boudoir" (1973) and "American Graffiti" (1976) were "The Dialectic Of The Avant-Garde" (1977), "Giuseppe Terragni: Subject And 'Mask'" (1977), and "The Historical Project" (1980). In addition, Tafuri was published in the IAUS Exhibition Catalogues and in *Skyline*. Next to the Institute, MIT Press played a major role in the creation of the "American Tafuri," as the university publisher published *Architecture and Utopia* (1976), *The Amercian City* (1979), and eventually *The Sphere and the Labyrinth* (1987).

the subscribers—and himself—cleverly using the occasion to remind them all to renew their subscriptions, as the contract with MIT Press demanded. The "Forum," with which the conclusion of the contract was once again solemnly celebrated at the Institute six months later, was intended as an event for subscribers, sponsors, and friends, but also offered a well-attended panel discussion in which Rossi's architectural projects, i.e., his drawings were interpreted by Fellows and invited speakers—in this way, Rossi did after all become the hot topic of the day in the New York architecture scene in the fall of 1976.[630] In contrast to the previous "Forum" section, this one did not just illustrate the podium, but rather the well-attended cocktail party that followed, which was extensively documented by Dorothy Alexander as the Institute's new in-house photographer and published in the next issue. A spotlight for the first time was cast on Philip Johnson. *Oppositions* now also had self-reporting in the style of high society, which proved to all readers at a glance that the Institute was able to attract the who's who of the New York architecture scene, who celebrated there in style, as befitting their social status.

For the Institute, *Oppositions* 5 thus meant a new beginning and a new orientation in many respects, not only because of the academic publisher behind it. The journal subsequently served less to set up a new genuinely American theory, as originally claimed, but instead expressed itself primarily in the popularization of a rather provincial architecture debate in the first half of the 1970s, centered on the two East Coast axes "New York—Cornell" and "Yale—Penn." Instead of engaging on a more intellectual level, *Oppositions* sought to raise its international profile. But "The Italian Issue" also showed that the polemics and division of the American architecture scene into the "Whites" and the "Grays" had been exhausted, as Manfredo Tafuri had already aptly noted in his essay "Les cendres de Jefferson," which first appeared in French in *L'Architecture d'Aujourd'hui* in 1976.[631] The editors' new focus took them across the Atlantic, or even across the Pacific. Along with this internationalization, Peter Eisenman had already guest-edited the issue of "White and Gray: Eleven Modern Architects" in the Japanese magazine *Architecture + Urbanism* in April 1975, together with his closest collaborator and rhetorical counterpart Robert Stern, who was very well-networked in New York architecture circles as president of The Architectural League.[632] However, journal issues did not only serve the dissemination of postmodern

630 IAUS, invitation card to *Oppositions* "Forum 6," October 26, 1976. Source: Vignelli Design Center, RIT. The exhibition of Rossi's works at the Institute in the spring of 1976 as part of the "European New Wave" series had not yet generated much of an audience.

631 Tafuri, "Les cendres de Jefferson," 1976.

632 *Architecture + Urbanism*, no. 52 (April 1975): "White and Gray: Eleven Modern Architects."

design and thought.[633] When Institute director Eisenman was commissioned to curate the American contribution to the Venice Art Biennale in the summer of 1976, he also cooperated with Stern and was able to use and further expand his contacts in Italy, especially with the IUAV. These networks were reflected not only in the selection of authors and topics featured in *Oppositions*, but also in the public events, exhibitions, lecture series, and teaching, and later in other publications of the Institute, especially in the conception of their own book series. The transatlantic dialogue along the new "Venice—New York" axis was based on mutual interests, networks and friendships, promises and commitments.[634]

One decisive factor in the new discursive dispositions and relations in the 1970s was the fact that, following the "Italian Issue," Manfredo Tafuri was built up by the *Oppositions* editors—Eisenman was certainly the driving force—to become the journal's most published author. Tafuri placed three more essays in the following issues as well as two texts for publications that Eisenman was planning on Terragni and his own projects.[635] Although his book *Architecture and Utopia. Design and Capitalist Development* had already been published by MIT Press in January 1976 (Conover's first publication as acquisitions editor), the Institute was instrumental in Tafuri's reception in the English-speaking architecture world, and thus in international debate and research. One great merit of *Oppositions* was that, by importing and translating Tafuri's texts, it valorized and simultaneously vulgarized a critical approach in the style of a historiographical metafiction. In particular, the younger generation of Fellows clearly adopted Tafuri's approaches in their sociopolitical readings of architecture. Interestingly, while the incomprehensibility of his texts and the poor quality of their translations were criticized in letters to the editors, this did not detract from this development. On the contrary, Tafuri has since become an integral part of architectural scholarship in North America (as opposed to Europe) and of the curriculum of American universities. In addition, next to Tafuri, Eisenman was vehemently committed to publicizing and popularizing the texts and drawings of Aldo Rossi, who taught at Cooper Union and became a regular at the Institute from 1976 during his trips to the USA, through the Institute's public events and publications.

633 Patteeuw and Szacka, 2018. In Europe, this task was assigned to the British *Architectural Design*, or the Swiss *Archithese*, from the mid-1970s, and later the German *Arch+*, among others.

634 Ockman, 1995.

635 Tafuri was commissioned by Eisenman to write two essays for his monographs: "Giuseppe Terragni: Subject and 'Mask'" (1977) for *Giuseppe Terragni* and "The Meditations of Icarus" (1980) for *Houses of Cards*.

Editorial Policy

The postmodern thinking style, linked with an academic habitus, on the one hand turning away from the project of modernity and on the other hand operationalizing theory production and historiography at the Institute, was also evident in the editorials of *Oppositions*, which the editors began signing individually in *Oppositions* 4 to 7 between 1976 and 1977.[636] With these short texts, all of them personal manifestos, Eisenman, Frampton, Gandelsonas, and Vidler communicated their different takes on the discipline and its autonomy at the transition from the late modern to the postmodern period, showcasing their influences and references, and simultaneously referencing each other and distancing themselves from one another.[637] In this lineup, they each asserted authority over contemporary architecture. Frampton had already made a start in the spring of 1976 in *Oppositions* 4 with "On Reading Heidegger." Writing as a historian and starting from Martin Heidegger's thesis, "That language, far from being a servant of man, is all too often his master," he rejected any rhetoric of autonomy since this mystified the—economic, social, and political—conditions of architecture rather than revealing them.[638] In *Oppositions* 5, Gandelsonas continued with "Neo-Functionalism." Under this neologism, following Tafuri's semiotic/linguistic analysis, he combined the symbolic meaning of postmodernism with the modernist architectural doctrine of functionalism, aiming to overcome the division of the contemporary architecture world into the antagonistic camps of "neo-rationalism" and "neo-realism," both of which he criticized as being "anti-functionalist."[639] With "Post-Functionalism" in *Oppositions* 6, Eisenman then attempted to develop the foundations of his theory of modernism.[640] He self-consciously distanced himself from two trends of the 1970s, rationalism and postmodernism, both of which he believed were still indebted to a humanist approach to architecture. According to his "non-humanist attitude," he defined modernism "as a sensibility based on the fundamental displacement of man," and instead, by excluding the subject and denying buildings any use-value—which he outlined as an "ethical positivism of form and function"—called for the recognition of an autonomous architecture based on the transformation of geometric

636 Kenneth Frampton, "On Reading Heidegger," *Oppositions* 4 (October 1974), n.p.; Mario Gandelsonas, "Neo-Functionalism," *Oppositions* 5 (Summer 1976), n.p.; Peter Eisenman, "Post-Functionalism," *Oppositions* 6 (Fall 1976), n.p.; Anthony Vidler, "The Third Typology," *Oppositions* 7 (Winter 1976), 1–4.

637 Ockman, 1988, 196–197.

638 Frampton, 1974.

639 Gandelsonas, 1976.

640 Eisenman, 1976. Ullrich Schwarz interpreted Eisenman's editorial as the "first sketchy formulation" of a theory of modernism, see Schwarz, 1995, 17.

bodies or the generation of form out of itself.[641] Although Eisenman ultimately did not succeed in writing a general theory of architecture, even though he had read Foucault and used the concept of the *épistémè* from *The Order of Things* (English translation of 1970), he did manage to expand the "Peter Eisenman brand" in line with the IAUS brand.[642] What is astonishing is that with the publication of some rather sketchy fragments of a theory, he engendered keywords for the debate on a "critical" or rather "post-critical" architecture and in doing so also proved to be the founder of an entirely new discourse based on idiosyncratic appropriations, theoretical set pieces, and historical interpretations. After various versions of his "Notes on Conceptual Architecture," i.e., his analyses and drawings of the transformation processes in the designs of the Italian rationalist Giuseppe Terragni, and after his contributions to *Oppositions*, e.g., his analysis of the architecture of the Smithsons and James Stirling, Eisenman's "Post-Functionalism" had an inaugural value for a new, post-modernized form of architecture theory.[643] Finally, Vidler also contributed an editorial with "The Third Typology" in *Oppositions* 7.[644] Turning away from abstract nature (Laugier's primitive hut) and technological utopia (Le Corbusier's machine aesthetic), he proposed—in line with the urbanism of Aldo Rossi and Leon Krier and as a critique of formalism—the traditional European city as a third typology. Thus, with a neo-rationalist typology that was both self-referential and self-reproducing, he was concerned not with isolated buildings but with the city and public space; he explicitly pointed out that the polis had always been political by its very nature. Taken individually and above all together, the contributions of the four editors thus impressively demonstrated that postmodernization, i.e., the derealization of architecture culture, was also reflected in the topics and method of *Oppositions*.

641 Eisenman, 1976, n.p. Eisenman himself referred to his programmatic text "Post-Functionalism," as he called his reflections, in distinction to Gandelsonas' editorial, "existing fragments of thought." The poster for *Oppositions* 5 to 8 still announced that he had planned to publish a longer essay under the same title in the "Theory" section. Nevertheless, Eisenman's editorial can be seen as theoretical base for the sculptural approach that was to materialize in *House VI*. Eisenman was subsequently more active as a writing architect than as a practicing theorist. In 1976 he was working on the book on *House X*, for which he had just lost the commission, but which he had just exhibited: first at the 1976 Venice Art Biennale, and later at Cooper Union and Princeton.

642 Foucault, [1966] 1970.

643 Eisenman, 1970 and 1971.

644 Vidler, 1976. Vidler's editorial was a version of the lecture he gave at the *Little Magazine Conference* at the Institute in February 1977. The text was subsequently published in several versions. The version printed in *Oppositions* was a heavily revised paper: the remarks on the first two typologies had been condensed. Vidler's emphasis on public space was watered down, a section on the decomposition and recomposition of fragments was added, and the critique of examples of contemporary practice was toned down; see Anthony Vidler (interview with Beatriz Colomina and Daniel Lopez-Perez), in Colomina and Buckley, 2010, 537–541.

In the end, it was very clear that each editorial was a polemic, with the authors referencing different historical and theoretical considerations to reflect on the conditions, meanings, histories, theories, concepts, methods, built examples, and textual references of contemporary practice. Taken together, these texts announced the Institute's program. While comparing different and quite contradictory positions and in the intellectual competition between the editors, they nevertheless formed a self-contained, self-sufficient, and even self-serving discussion group. Furthermore, the editorials were written according to the principle of juxtaposed positions and competition between the editors. But this was precisely because of their very special relationships with each other—their sympathies, dependencies, agreements, and disputes. As textual documents, the four editorials, while demonstrating the sophistication of their authors, testified to the extent to which postmodernism—although new poststructuralist and postmodernist theorems and concepts were only rudimentarily valorized and appropriated, if at all—had made its way into the Institute as a discursive formation, with *Oppositions* as the medium for presenting an image of oneself as an intellectual of architecture, if not beyond. Frampton, for example, as a now-recognized historian of modernism, repeatedly made his voice heard as the harshest critic of postmodern architecture, and here, with his focus on "the socially experienced quality of place," he was already arguing for a central aspect of his approach, which he later formulated as an alternative postmodernism under the banner of Critical Regionalism.[645] Gandelsonas, on the other hand, as an architect and theorist, was the most explicit advocate of postmodern thinking and design among the *Oppositions* editors. While Eisenman used the term postmodernism as a polemical concept and sought to expose architectural postmodernism as a media construct, Vidler, as a historian, countered the accusation of being an apologist, here and in other editorials, with the argument that he was less interested in a new architectural doctrine. The separation of architectural style and mindset was a workaround, albeit an inadequate one, that allowed him to address "the city and typology," for him in a

645 Frampton, 1974. Frampton would not publish his version of "Critical Regionalism," modelled on Alexander Tzoni and Liane Lefevre, until 1983, but then twice, in two different textual formats: a philosophical essay in Hal Foster's *The Anti*-Aesthetic, later translated into German, French, and Spanish, and a project-based architecture critique in *Perspecta*; see Kenneth Frampton, "Six Points Towards and Architecture of Resistance. Towards a Critical Regionalism," in *The Anti-Aesthetic. Essays on Postmodern Culture*, ed. Hal Foster (Seattle: Bay Press, 1983), 16–31; "Prospects for a Critical Regionalism," *Perspecta*, no. 20, (1983), 147–162. Frampton developed his fundamental idea on place as a central category in relation to urban development as early as the 1960s in his essay "Labour, Work and Architecture," published in George Baird and Charles Jencks, eds., *Meaning in Architecture* (New York: Braziller, 1969), 150–168. In the 1970s and 1980s Frampton published a piece on "Production, Place and Reality" and celebrated regional alternatives to the globalized American postmodernism with publications and conference speeches, including at UQAM in Montréal in 1983. His book project, which would have collected examples of "Critical Regionalism" from around the world, did not materialize, but Frampton subsequently discussed these in his revisions of *Modern Architecture* and prefaces.

sense a continuation of the modern movement, "as the only possible bases for the restoration of the critical role of architecture;" he was thus arguing for the "public nature" of all architecture and against the "private vision of romantic individualists."[646] The *Oppositions* editors were united, however, in their criticism of the emerging architectural postmodernism, that is to say, of a mix of classicism, eclecticism, and historicism popular among architects, historians, readers, curators, museum visitors, and private and corporate clients.[647] For a long time, the editors had even planned to devote two issues of *Oppositions* to debates about the heritage of modernism or a critique of postmodernism, but these never appeared. All editorials, however, were subsequently republished in other contexts, publications, and languages—in France, Spain, and Germany—which brought the individual positions to international attention beyond the typical readership of *Oppositions*.[648]

At the time, the Institute was opening up to the arts, humanities, social sciences, and cultural studies in order to benefit from government funding in the years to come. Yet as a postmodern salon, in terms of its funding, programming, and production, it could not escape the conservative trend and privatization wave in the country, as also evidenced by the journal. While the editors, with the backing of MIT Press, entered into a competition for the best references, since the narrative

646 Vidler, 1976, 4. Vidler's position has been criticized as an apologia for postmodern architecture. Eisenman thus continued to polemicize that *Oppositions* turned to historicism; see Peter Eisenman (interview with Beatriz Colomina and Urtzi Grau), in Colomina and Buckley, 2010, 261–264. It remains unclear whether he is referring to his personal relationship with Leon Krier here. Krier first exhibited at the Institute in 1975 and was a guest there two more times as part of the "European New Wave" series. In 1977 they met at Princeton University, where Eisenman exhibited and Krier contributed a portrait of Eisenman for the poster. They kept up the dialogue; see Peter Eisenman, "Interview. Leon Krier and Peter Eisenman," *Skyline* (February 1983), 12–15; see also Cynthia Davidson, *Eisenman/Krier: Two Ideologies* (New York: Monacelli Press, 2005). Several of Krier's texts, while seen critically, were later published in *Oppositions*; see Leon Krier and Maurice Culot, "The Only Path for Architecture," *Oppositions* 14 (Fall 1978), 38–53; Leon Krier, "The Consumption of Culture," *Oppositions* 14 (Fall 1978), 54–59; "Vorwärts, Kamaraden, wir müssen zurück," *Oppositions* 24 (Spring 1981), 26–37. In the late 1970s, Krier again was contributing to the Institute, lecturing as part of the "Open Plan" series, and exhibiting there.

647 For *Oppositions*, thematic issues on "Post-Modernism" and "Modernism" respectively were discussed for a long time. *Oppositions* editors planned to write editorials under the title "Against Post-Modernism;" see Julia Bloomfield, communications to *Oppositions* editors, November 2, 1978, December 6, 1978 & September 15, 1980. Source: CCA Montréal, IAUS fonds: A.5-7. During this period, Bloomfield's salary was raised to US$ 3,000 per issue and quarterly payments were agreed, institutionalizing her position; further professionalization occurred when Bloomfield wrote the job description for the managing editor.

648 Vidler's editorial was republished in 1978; see Robert L. Delevoy, ed., *Rational Architecture* (Bruxelles: Archives d'Architecture Moderne), 28–31. Eisenman's, Gandelsonas's, and Vidler's editorials were translated into Spanish below and published in the architecture journal *Arquitecturas Bis 22* (May 1978). Eisenman's and Vidler's editorials were translated into German in 1980; see Gerald Blomeyer and Barbara Tietze, eds., *In Opposition zur Moderne. Aktuelle Positionen in der Architektur*. Bauwelt Fundamente 52 (Braunschweig and Wiesbaden: Friedrich Vieweg & Sohn, 1980), 108ff.

of The Great Society no longer applied and a critique of Corporate America had long since faded, some at least used *Oppositions* to voice criticism of reactionary developments in architecture. Nevertheless, real-world processes that conditioned and constrained the architectural profession were rarely addressed. For example, there was no discussion on the phasing out of the Urban Development Corporation or the competition for Roosevelt Island. Apart from selected examples of contemporary, if not avant-garde or radical practice, the major issues of the decade—the dismantling of the welfare state, the various economic and ecological crises, and the alignment of the state capitalist system with a neoliberal program, were clearly not of particular concern to the editors, or they possibly did not feel that *Oppositions* was the right medium for these topics. Instead, while the "Exhibition Program" expanded with a focus on monographic exhibitions, *Oppositions* supported a traditional and closed concept of the work, accompanied by a debate about the possibility of authorship and interpretation.[649] There was no denying that *Oppositions* was the responsibility of white men, while the women at the Institute did most of the work: historical and theoretical debates about the relevance of intersecting categories such as race, gender, and class (sexual orientation did not enter the pages of *October* until later) were surprisingly absent and silent in times of postmodernism and poststructuralism.

After Modern Architecture

In early February 1977, *Oppositions* invited editors of friendly architecture journals and magazines from Europe to a multi-day conference at the Institute under the title "After Modern Architecture." This conference, as part of the repositioning of *Oppositions*, represented another attempt to claim leadership in the international market in terms of the circulation of ideas, criticism, and authority in the coverage of postwar modernism and emerging postmodernism.[650] Next to Eisenman, Frampton, Gandelsonas, and Vidler, the so-called *Little Magazine Conference* was attended by editors from *Arquitecturas Bis* from Spain, *A.M.C.* from France, and *Controspazio* and *Lotus International* from Italy. "After Modern Architecture" was actually the follow-up to an earlier conference hosted by the editors of *Arquitecturas Bis* in Cadaques, near Barcelona, in September 1975.[651] After the first conference had established solidarity among

649 Barthes, [1967] 1977.

650 Ockman, 1988, 197ff. Apart from shorter reviews in the architecture press and *The New York Times*, there is no further coverage or historiography of *After Modern Architecture*. There are no major references to the conference, neither in CCA's IAUS fonds, nor in CCA's Peter Eisenman fonds. Some documents can be found only in private archives of participants, in the Robert A.M. Stern Archive at Yale University, and in the Robert Gutman Collection at Columbia University.

651 The first conference in Cadaques was attended by editors of *Arquitecturas Bis*, *Lotus*, and *Oppositions*, see Tomàs Llorens, "Arquitecturas Bis, Lotus, Oppositions: Convencion en Cadaqués. Septiembre de 1975," *Arquitecturas Bis* 10 (November 1975), 30–31. It is not possible to recon-

the editors, this second event was intended to consolidate and expand the common network and to compare the different editorial structures and policies.[652] To this end, the respective sociopolitical context in which the individual journals and magazines were produced was also to be reflected upon. *Oppositions* editors were working from the assumption that journals played an important role in the development of architectural ideas, both historically and culturally. Moreover, the conference was intended to offer participants the opportunity to network and explore the "possibility of future collaborations, exchange and republication." With this in mind, they had planned to record the results of the conference in a multilingual publication at MIT Press. To prepare for this, the participating editors were asked to compile a chronology of the architectural projects that had been featured in their journals over the past 30 years. In addition, all conference participants were invited to formulate their positions on contemporary architecture so that these could be sent out in advance as a basis for discussion.[653] In the run-up, a *Controspazio* editor had already voiced the criticism that the Institute was attempting to write a unifying history of European post-war architecture with "After Modern Architecture."[654]

The conference ultimately took place behind closed doors. Apart from an opening at MoMA and a reception at the Cooper Union, only Fellows, friends of the Institute, and donors, i.e., members of the Architects' Circle, as well as representatives of the architecture press and *The New York Times* were invited to attend the actual conference.[655] *Oppositions* editors had set out to achieve a great deal: a total of four sessions with presentations by the respective editors (*Oppositions* was introduced by Joan Ockman), followed by three thematic workshops in which conference participants discussed overarching methods and concepts of historiography and theory production such as "Historical continuity and discontinuity," "Progressist and non-progressist society," "The problem of cultural accessibility," on the one hand, and "Rationalism, realism

struct exactly who attended the second conference in New York. Listed as participants were: *Arquitecturas Bis* (Spain): Rafael Moneo, Oriol Bohigas, Frederico Correa, Helio Pinon; *A.M.C.* (France): Jacques Lucan; *Controspazio* (Italy): Alessandro Anselmi, Claudio D'Amato, Franco Purini; *Lotus* (Italy): Joseph Rykwert, Kenneth Frampton; *Oppositions*: Peter Eisenman, Kenneth Frampton, Mario Gandelsonas, Anthony Vidler. Of the editorial staff of the Swiss *Archithese*, Bruno Reichlin and Stanislaus von Moos were invited, but ultimately did not participate.

652 *Oppositions*, invitation to *Little Magazine Conference*, February 3–5, 1977. Source: Columbia University, Robert Gutman Collection.

653 Summaries of individual presentations were collected and mailed in advance. Source: Columbia University, Robert Gutman Collection.

654 *Controspazio*: Chronology. Source: Columbia University, Robert Gutman Collection.

655 Gutman was invited as a friend, Stern as a sponsor. Ada Louise Huxtable reported for *The New York Times*. *Controspazio* published two photographs of the conference with names of participants. Ockman highlighted Colin Rowe and Richard Meier as participants in the conference; see Ockman, 1988, 198.

and pragmatism," "Anthropocentric built forms vs. a non-anthropocentric conception of architecture," and "The Possibilities for a new typological structure as a strategy for a future architecture" on the other. The last workshop explicitly problematized the question of "Autonomy or the non-autonomy of architecture" for the first time at the Institute, paired with a discussion on "The role of criticism in architecture and the possibility for architecture to be criticism in itself."[656] In this context, *Oppositions* editors presented their editorials—sometimes simply reusing them, sometimes as a test lecture.[657] Eisenman, for example, used the occasion to further develop his theoretical reflections on a "post-functionalism," as he was primarily concerned with rethinking the role of the architect as author. For him "man is no longer the originating agent but rather he has a discursive and explanatory role vis-à-vis the making of the world," and he thus introduced the figure of the reader into architecture, recalling Roland Barthes' essay "The Death of the Author," an instant classic that had just been published in English in an anthology, and its critique of the authority of the author.[658] Eisenman, however, did not follow this poststructuralist approach to its logical conclusion, focusing more on the design process than on its use. The actual realization of the idea, as evidenced by *House VI*, which had just been widely reviewed in *The New York Times* and which he was to publish in the American architecture press that same year with further contributions curated by him, no longer played a role for him. And the peregrinations of the exhibition on *House X* after the client dropped out highlighted that he now favored the exhibition and art value of the model over the use value of the building. Frampton, on the other hand, in "Loss of Utopia," was the only *Oppositions* editor to speak about a new subject, namely the function and success of modernist utopias. He called for a contemporary avant-garde that was less hedonistic and more political.[659] If they had anything in common, then all *Oppositions* editors were ultimately interested, as was once again made clear here, in finding new criteria for the perception and appreciation of architecture, both modern and postmodern.

656 IAUS, program of the *Little Magazine Conference*. Source: Columbia University, Robert Gutman Collection.

657 *Oppositions* 8 with Vidler's editorial "The Third Typology" had not yet been published at that time.

658 Roland Barthes's essay "The Death of the Author" was first published in 1967 in the journal *Aspen*, 5/6 (1967), and subsequently in the 1977 essay collection *Image-Music-Text*. Barthes' critique of biographical interpretations of texts, applied to architecture, would have meant the architect playing a much smaller role in architecture history than the focus on heroes and the formation of the cult of stardom entails; the meaning of buildings would accordingly have been derived much more from their use.

659 Kenneth Frampton, "Abstract: 'The Loss of Utopia'," Source: Columbia University, Robert Gutman Collection.

Ada Louise Huxtable, an architecture critic for *The New York Times* who also participated in "After Modern Architecture," criticized the intellectual style of individual contributions in her review—curiously enough, one of the few she wrote at all about the activities at the Institute—and fundamentally questioned the meaning and purpose of this conference.[660] At the same time, however, Huxtable acknowledged the relevance of *Oppositions* as the Institute's mouthpiece; in her opinion, the projects presented there were stylistically definitive and would be copied on the architecture market over the next two decades. *Oppositions* was therefore of public interest. Other journals whose editors had attended the conference also reported on "After Modern Architecture:" Alessandra Latour wrote a short conference review for *Controspazio*, and *Arquitecturas Bis* reported in more detail, additionally publishing Spanish translations of three of the editorials from *Oppositions*.[661] The planned conference publication on which Eisenman and Gandelsonas had worked in 1977, along with Livio Dimitriu as an intern, was however never published. Nevertheless, by organizing the conference, the Institute contributed substantially to the historicization and theorization of postmodernism in journals on an international scale, which was henceforth subsumed under the headings of "autonomy" and "criticality," respectively, along with the associated cultural hegemony; the Institute's leadership even cited the conference in various grant applications. By either showing interest in contemporary architecture, less so in North America, but more in Europe and also increasingly in Asia, especially Japan,[662] or referring to its pioneering role in the debate on populism and historicism—and this was the other side of the coin, which it however cleverly exploited for its own benefit—the Institute demonstrated openness and topicality, and once again advertised on its own behalf. Different, even contradictory tendencies were discernible in the Institute's programming: while offering one of its most public events on urban culture free of charge, only to portray itself in the next moment as a venue that generated gratitude and distributed gratuities, alongside a professionalization and economization of cultural practice, the two theory (and history) laden journals *Oppositions* and *October* and other publications that were planned at the time came to be of central interest to Eisenman. In his 1977 "Director's Memo" addressed to the Board of Trustees, he expressed hopes the publications would have networking as well as discursive effects.[663] Ultimately, it was a matter of building and

660 Ada Louise Huxtable, "Architecture View: A Sense of Crisis About the Art of Architecture, Architecture in Crisis," *The New York Times* (February 20, 1977), 99.

661 Alessandra Latour, "Little Magazine Conference: 'After Modern Architecture'," *Controspazio 9* (June 1977), 62; *Arquitecturas Bis 22* (May 1978).

662 Arata Isozaki was a guest at the Institute at that time and they prepared "A New Wave of Japanese Architecture," an exhibition of young Japanese architects; later Japanese architects were featured as authors in *Oppositions*.

663 Peter Eisenman, "Director's Memo," January 11, 1977.

expanding his own reputation and that of the Institute within a complex economy of attention, which was of relevance for publication lists that functioned as a strong currency in the academic system.

Art I Theory I Criticism I Politics

For the Institute, and especially for Eisenman, whose support and sponsorship made the second journal based on the *Oppositions* model possible in the first place, the publication of *October* represented both a friendly service and a prestigious project.[664] The first issue of *October* appeared in the spring of 1976, after Rosalind Krauss and Annette Michelson, as editors, had previously been accommodated at the Institute on Eisenman's initiative following their expulsion from *Art Forum* the previous year. *October*, like *Oppositions* before it, was initially to be self-published, with the Institute acting as publisher. The title, according to an advertisement addressed to new subscribers, was a reference to Sergei Eisenstein's 1929 Russian avant-garde film. By their own admission, the goal of Krauss and Michelson was not "to perpetuate the mythology or hagiography of Revolution."[665] "It is rather to reopen an inquiry into the relationships between the several arts which flourish in our culture at this time, and in so doing, to open discussion of their role at this highly problematic juncture." With the main priorities indicated in the subheading "Art I Theory I Criticism I Politics," *October* investigated the structural and social relationships between artistic practice and political discourse. As much as this highly politicized stance set the *October* editors apart from their former employer *Art Forum*, which was by far more commercially oriented toward the galleries and art dealerships, Krauss and Michelson's interests were very different: contemporary visual art forms and Russian avant-garde film.

October was characterized above all by the broad concept of art that was just emerging at the time, and which included not only video, film, and photography, but also performance, music, and literature. The editorial work was also united by the reception of what was known as French Theory and psychoanalytic and feminist theory, whether as a toolkit of methods and concepts, as can be seen in citations and references, or by printing the relevant texts. In this regard, *October* and *Oppositions* editors shared an interest in examining modernism and contemporary practice, specifically formalism. But unlike *Oppositions*, *October* shaped a more critical, deliberative discourse of postmodernism, and subscribed to institutional critique, as both were practiced in the arts and art criticism. The first issue was in many ways groundbreaking. For

664 In CCA's IAUS fonds, there is a folder on *October*. Source: CCA Montréal, IAUS fonds: A.5-9. See Yves-Alain Bois, Hal Foster, and Rosalind Krauss, "New York–Paris," in *Clip Stamp Fold. The Radical Architecture of Little Magazines. 196X to 197X*, eds. Colomina, Beatriz and Craig Buckley (Barcelona: Actar, 2010), 36–45.

665 MIT Press, *October* subscriptions. Source: CCA Montréal, IAUS fonds: A.5-9.

example, it contained "Ceci n'est pas une pipe," by Michel Foucault, as well as essays by the editors: Krauss' "Video: The Aesthetics of Narcissism" exemplified by the video art of Vito Acconci, Nancy Holt, Bruce Nauman, Joan Jonas, Peter Campus, and "Gravity's Rainbow and the Spiral Jetty" by Jeremy Gilbert-Rolfe, who was actually the third editor next to Krauss and Michelson, a British painter based in New York, as well as an art critic and an educator, i.e., the first of a three-part essay (in collaboration with John Johnston), which involved a reading of the two works of art named in the title: the novel by Thomas Pynchon and Robert Smithson's land art. From a graphic point of view, *October*, which had a print run of 3,000 and was sold at a price of US$3, clearly classified itself as an art journal with its format, its single-column layout in justified type, and with Baskerville as its typeface, and its large illustrations, some of which were full-page and bleeding, printed right up to the edge; the cover, on the other hand, recalled Vignelli's design for the Institute with its title in capital letters, a large number for each issue, and otherwise only text, the names of the authors and titles of their essays, and also because of its specific color choice, red and black, reminiscent of Russian Constructivism. It was designed by Charles Read, a student of Gilbert-Rolfe at Princeton University, who was subsequently hired as the journal's graphic designer and remained on board for the first nine issues.

In the beginning, the editorial work on *October* was irregular, with two issues released in 1976 and in 1977. The journal was distributed exclusively through Jaap Rietman, an art bookstore in SoHo, which bought up a certain number; otherwise, only a few other bookstores, mainly in Manhattan, carried the journal, so it was initially read almost exclusively in New York. The first issue sold out completely, but after that advertising and sales were slow. Only *The New York Review of Books* advertised the journal. By the end of 1977, *October* had just 350 subscribers, although a subscription was offered quite cheaply at US$10 per year and US$18 for two years. The list of contributors to the first issue was impressive and included, in addition to the editors, Michel Foucault, Peter Handke, Noel Burch, Robert Morris, Hollis Frampton, Sergei Eisenstein, Jean Epstein, Hans Magnus Enzensberger, Yvonne Rainer, and Richard Forman. Krauss published her seminal essay, "Notes on Index," here—a two-part essay on developments in art, the first part introducing the concept of index using Marcel Duchamp as an example, and the second discussing the inaugural "Rooms" exhibition at P.S.1 in 1976. Overall, *October* was marked by a wide variety of formats: next to theory and criticism, it also featured philosophy and poetry, and texts by artists, especially filmmakers. But after only one year, *October* was at a crossroads, like *Oppositions* before it, as both production and funding had to be organized differently. The Institute continued to provide the framework for further institutionalization, which enabled the editorial team to apply for public funding and find an academic publisher as a partner. Eisenman personally championed *October*, bringing Krauss and Michelson together with Armand and

Celeste Bartos, as sponsors of the Institute, which was immediately followed by a US$10,000 gift from the private Gottesman Foundation to produce issue four.[666] While Frederieke Taylor was responsible for fundraising, grants, and donations in general at the Institute, Eisenman himself personally handled the accounting for *October*; the budget was now balanced with over US$17,500 in income from contributions and sales. After Gilbert-Rolfe left the editorial team after just three issues, Douglas Crimp, one of Krauss's students at CUNY's graduate school, joined the editorial team. Crimp initially worked as an editorial assistant for two issues and then served as managing editor responsible for the journal's editing and production.[667] In addition, as he had done previously for *Oppositions*, Eisenman worked to negotiate a contract with MIT Press in 1977 to improve the distribution and marketing of *October* in North America and to make the leap to Europe. On the basis of sympathetic advice from outside consultants. Frank Urbanowski, the head of the university publishing house, vigorously advocated for the conclusion of a contract: he expected nothing less from *October* than a substantial contribution to the cultural sphere.

Eisenman played a major role in the contract negotiations with MIT Press; when Frank Urbanowski sent a draft contract to him and Rosalind Krauss in late 1977, he himself revised it in writing.[668] However, Eisenman eventually dropped the Institute as a contractual partner, thus transferring editorial and financial responsibility directly to the two editors. Nevertheless, the Institute assumed a limited role in funding the journal, committing to acquiring grant money, again from the Gottesman Foundation, while waiving the obligatory 40% overhead to IAUS Central. *October* thus assumed a special position at the Institute, one that was even more extreme than *Oppositions*. The editorial staff worked completely on its own regarding the salaries for the editors and the fees for graphic designers, authors, and translators, the expenses for administration, telephone costs, reproductions, photographs, and the acquisition of publication rights. The Institute even agreed to pay the salary of a managing editor and to provide office space; Douglas Crimp, however, preferred to work from home in the long run, as the habitus at the Institute, which at that time was becoming an elite circle as a result of its 10th anniversary and with the expansion of the "Evening Program," was alien to him. That same year, Eisenman asked Krauss if she would write a text about his house designs for an issue of the Japanese magazine *Architecture + Urbanism* dedicated to him. The art critic prefaced

666 Bois, Foster, and Krauss, 2010, 40.

667 Mathias Danbolt, "Front Room—Back Room. An Interview with Douglas Crimp," *Trikster – Nordic Queer Journal*, no. 2, (2008), http://trikster.net/2/crimp/1.html (last accessed: May 31, 2023)

668 Frank Urbanowski, letter to Rosalind Krauss and Peter Eisenman, November 29, 1977, including a draft contract for *October*, 1977. Source: CCA Montréal, IAUS fonds: A.5-9.

her text, her only published commentary on Eisenman's architecture, with a personal note that the two of them were friends and had been going through a parallel development up to that point.[669] As an architecture review, the essay, which displayed a certain distance towards its subject, reads as an apt classification of Eisenman's oeuvre, at the interplay of textual and architectural production, as a representative of postmodernism. Krauss criticized the fact that *House I* and *House II* were still formalist, while *House VI* turned out to be "post-formalist"—alluding conceptually to Eisenman's theory of "post-functionalism." Nevertheless, only the long-standing friendship between the two explains why *October* and with it a certain discursive formation of art theory and criticism was professionalized through MIT Press, which was ultimately to outlive the Institute.

Critical Historiography

Published in 1977, *Oppositions* 8, a thematic issue on "Paris under the Academy," for which Anthony Vidler was responsible, marked a paradigm shift at the Institute in the critique of architectural modernism and postmodernism. Vidler, who had already contributed to the Institute's research projects as a Visiting Fellow in the early 1970s, and had contributed essays to *Oppositions* 1 and 5, but never previously played a decisive role, was now finally included in the circle of Fellows as editor. This *Oppositions* issue was his response to the controversial exhibition "The Architecture of the École des Beaux Arts," curated by Arthur Drexler at MoMA in the winter of 1975–76, which caused a stir in the architecture world with its large-format drawings. As MoMA curator, Drexler, who having made the founding of the Institute possible in the first place, had long acted as a trustee, had actually intended the long-planned exhibition to call for a differentiated approach to the architectural and urban legacy of the École des Beaux-Arts, but ultimately played into the hands of advocates and protagonists of a postmodern architectural language.[670] At the Institute, individual Fellows and representatives of the profession had already expressed their views on the MoMA exhibition at the "Forum" for the publication of *Oppositions* 4 at the end of January 1976, and for the most part, distanced themselves from Drexler's work.[671]

669 Krauss, 1987; The text, written in 1977, was first published in 1980, and then in Eisenman's 1977 monograph *Houses of Cards* in a slightly altered form, see Epp, 2007.

670 MoMA's exhibition "The Architecture of the Beaux Art" (October 29, 1975, to January 4, 1976) had apparently been in the planning stages since 1967. In an oral history interview, Drexler referred to the long planning period and his original intentions and characterized the opposition that formed as schizophrenic behavior: "This is why we did the show on the Ecole de Beaux Arts." See Arthur Drexler (interview). Source: The Museum of Modern Art, New York: Oral History Files; see also Felicity Scott, "When Systems Fail," in *Architecture or Techno-utopia. Politics after Modernism* (Cambridge: MIT Press, 2007), 59ff.

671 Ellis, 1976.

In his editorial for *Oppositions* 8, Vidler positioned himself clearly and meaningfully in relation to the postmodern ambivalences of architecture and the city, the operationalization of history and theory, and the role of architects as producers and critics.[672] Here, he outlined the object and approach of a genealogical historiography of architectural modernism: inspired by Michel Foucault's writings, he called for an analysis of the origins of modernist architecture and a critique of the conditions of its production, using the example of the urbanization of Paris under the influence of the École des Beaux-Arts. According to his reading, the heroic modernism of the twentieth century, through its opposition to academicism, would have completely transfigured the architecture of the nineteenth century. Vidler appreciated that the MoMA exhibition sought to dispel a dogmatic view of the École des Beaux-Arts: "'Post-modernism,' it is claimed, allows for an appreciation, if not enthusiastic espousal, of ornament, pattern, colors other than primaries, symmetry, monumental fantasy, even of the pure technique of rendering for its own sake; with the critique of functionalism, pure abstraction, and the machine utopia, realms of experience up to now forbidden by the stern purism of modernism are opened up."[673] Vidler saw the possibility for an unbiased history of nineteenth-century, as well as a critical history of twentieth-century architecture. "The exhibition emerged in fact as the Museum of Modern Arts's auto-critical act, exorcising in 1977 the Modern Movement principles it had so heartily embraced in 1932."[674] In the end, however, even he had to admit that his high expectations had not been met. *Oppositions* 8 thus called for a new historiography of modernity beyond a mere reversal of the previous reading.

In accordance with his humanistic, largely affirmative, and at least in parts critical approach to architecture history, Vidler was ultimately concerned with a better understanding of the modern and thus also the post-modern mindset in architecture. "If we are indeed entering a period of post-modern sensibility, then a clear understanding of modernism should be thought, one that begins to establish the ontological bases of its project rather than one that repeats the ideological polemics of intentions." Linking theory, history, and practice, Vidler viewed *Oppositions* 8 as a critique of a purely aesthetic and ideologically inflected attempt at explaining the MoMA exhibition and advocated a differentiated view without simple attributions. "This issue of *Oppositions* has been developed as a counter to those kinds of historical interpretations of nineteenth century architecture that rest solely on stylistic or ideological models of explanation." His ultimate aim, nevertheless, was to examine the experience of modernity and the development of a metropolis like Paris on two levels: in terms of a new architecture of bourgeois

672 Anthony Vidler, "Introduction: Academicism: Modernism," *Oppositions* 8 (Spring 1977), 2–5.

673 Ibid., 2.

674 Ibid., 2.

society and the new discipline of urban planning, and of a literary and visual representation of the city that emerged along with it, thus opening up to the humanities. Curiously, in addition to his own essay on "The Idea of Type," other contributions included literary scholar Peter Brooks' "The Text of the City," an essay on the invention of the nineteenth century by Honoré de Balzac, as well as urbanist Antoine Grumbach's "The Promenades of Paris," an essay on urban planning under Georges-Eugène Baron Haussmann. The essays in this issue were all about spaces, buildings, and processes that alluded to a nexus of bourgeoisie, architecture, and urban planning in nineteenth-century Paris, when liberalism and the emerging industrialization were increasingly gaining political prominence, and social change was upending established ways of behaving and thinking.

While Vidler was not able to accomplish everything he set out to do, he did succeed in conveying his main intention: that *Oppositions* 8 would provide not only a critique of the 1920s architectural avant-garde but also a better understanding of contemporary architectural practice.[675] He concluded his introduction by saying that his main concern was not "to find a new orthodoxy, nor to chronicle the events of the past as accomplished, knowable facts."[676] For him, historiography had another, in the words of Michel Foucault, genealogical task: "Rather, we hope to encourage the investigation of the recent past as an instrument for the analysis and criticism of the present, not once more as a fulfillment of the 'spirit of the age,' but now as an aid to understanding the impossible contradictions of our own practice." With such a critical understanding of historiography, Vidler complemented the approaches of *Oppositions* editors in terms of both methods and methodology, while Frampton and Gandelsonas were also interested in socio-political conditions. But although he cited Foucault's post-structuralist, ultimately post-Marxist philosophy of history, be it indirectly or directly, Vidler did not necessarily share its analytics of power. Rather, with his own research on the work of Claude-Nicolas Ledoux, Vidler went back to the beginnings of architectural modernism and developed an urbanist perspective on the architectural problem of typology that would be further negotiated in later issues of *Oppositions*.[677]

675 Stern, 1999, 69. Ralph Stern took a more general view of Vidler's approach to critical historiography.

676 Vidler, 1977, 5.

677 Anthony Vidler, "The Idea of Type," *Oppositions* 8 (Spring 1977), 94–115; Quatremere de Quincy, "Type," *Oppositions* 8 (Spring 1977), 146–150; Rafael Moneo, "On Typology," *Oppositions* 13 (Summer 1978), 22–45; see also Anthony Vidler, "On Type," *Skyline* (January 1979), 2; Anthony Vidler, "The 'Art' of History: Monumental Aesthetics from Winckelmann to Quatremere de Quincy," *Oppositions* 25 (Fall 1982), 52–67.

Labor, Work and Publishing

At the end of April 1977, a year after signing the contract with MIT Press and shortly before the publication of *Oppositions* 8, the editors began planning the third volume.[678] Issues 9 through 12 show that the Institute was continuing to work with a small circle of like-minded authors, making use of existing institutional, social, professional, and discursive networks. The slight delays to the publication of *Oppositions* 8 made it clear that the greatest difficulties were encountered in professionalizing the editorial work and financing the production. Income that had been firmly anticipated failed to materialize, not least because of the irregular publication schedule. After only three issues, MIT Press demanded that *Oppositions* be published quarterly as planned. The academic publisher was also concerned about economic efficiency; the circulation was therefore to be increased from 4,200 to 5,200 copies, starting with the third volume.[679] At the same time, *Oppositions* editors announced an increase in sponsorship dues to US$150 for individuals.[680] Despite the freedom of content, there were some initial disagreements between the Institute and the publisher, for example when MIT Press was planning a book-bound collected edition of *Oppositions* 5 to 8 to generate additional revenue; the project ultimately failed because the Institute demanded complete control over the graphic design and selection of paper. After Frampton returned to the Institute in 1977, he and Eisenman managed the editorial work on *Oppositions* alone at times. It was mainly Eisenman who championed the journal's programmatic and organizational concerns and continuously promoted it;[681] characteristically, for example, he responded to MIT Press' rebuke to be more disciplined in the future by formulating a letter to the head of the university publishing house, Frank Urbanowski, complaining about a lack of cooperation and that he was not receiving enough complimentary copies.[682] He even threatened that the Institute would not cede any more donations in the future. Frampton, on the other hand, was responsible for the time-intensive editorial work and the labor that went into the production of each individual issue, working closely with Julia Bloomfield on a day-to-day basis; they also shared an office during this time. He proofread incoming manuscripts and researched images for accepted essays. While Frampton was responsible for the editorial of *Oppositions* 9 and regularly contributed texts

678 IAUS, minutes of editorial meeting, April 26, 1977. Source: CCA Montréal, IAUS fonds: C.1-1/2.

679 MIT Press, budget for *Oppositions*, n.d. Source: CCA Montréal, IAUS fonds: A.5-7.

680 IAUS, draft letter to sponsors, n.d. Source: CCA Montréal, IAUS fonds: A.5-7.

681 Originally, Eisenman had planned for *Oppositions* 10 to appear, for the sake of the congruence of numbers, on the occasion of the Institute's tenth anniversary in the fall of 1977. For a time, he also planned an ominous "Black Issue," which, according to a concept paper, was to be devoted to the two main themes of "structure" and "metaphor;" see Ockman, 1988, 193. Like so many of his ideas, this one was not realized either.

682 The Institute had received only forty copies of each of the issues of *Oppositions* 5, 6, 7 and 8.

of his own, Eisenman kept to the background, interpreting his role as editor as encouraging others to write introductions, comments, and postscripts to the contributions to keep the debate alive. Eisenman thus exerted a strong influence, both directly and indirectly, on what was to be published in *Oppositions*, especially through his Italophilia and personal affinity with Tafuri, Rossi, and others, such as the not uncontroversial Italian rationalist Giuseppe Terragni, by publishing documents and texts. Not only did Eisenman share his interest in Terragni with Tafuri, but he also published a provocative feature in *Oppositions* on the fascist architect as another representative of modernism.

In addition, Eisenman had been in constant dialogue with Roger Conover at MIT Press since 1976 about publishing books through the university's publishing house. Eisenman's long-planned Terragni monograph was to be the first publication; he had even signed a contract—not as an author, but in his capacity as Institute director.[683] In addition, the Institute's application for an NEH Cultural Institution Grant in April 1977, which was primarily intended to raise funds for the continuation of the "Evening Program" and the transformation of the successful "Architecture" series into "Open Plan," already cited the production of *Open Plan Books* and *Documents* as a new publication series; despite the success of the application, however, this was ultimately not realized. At around the same time, Eisenman was also in conversation with Conover about another, longer-term publication project titled Oppositions Books. This was to be a book series of its own, meeting scholarly demands, in which the substantive emphases of *Oppositions*, the focus on the history of modernism and contemporary theory, as well as the editorial strategies of juxtaposing different positions were to be continued. Specifically, there were discussions about an English edition of Aldo Rossi's *L'architettura della città*, which at that time had already been translated into several languages, but not into English. As early as 1974, the Institute had offered MIT Press an English translation of *The Architecture of the City*, prepared by two former Research Associates, Victor Caliandro and Thomas Schumacher, who had previously worked on the "Streets Project" at the Institute, and their translation had even already been approved by the editorial board of the university's publishing house. However, due to personnel changes in management at MIT Press, the publication was postponed for an indefinite period. At the Institute, this long overdue title was now revisited, and other titles were later added to the list, with Rossi being built up, especially by Eisenman, as the Institute's central author, with the commission for a new manuscript for his *Scientific Autobiography*. In addition to international *Oppositions* authors, in particular Manfredo Tafuri, the editors, especially Kenneth Frampton and Anthony Vidler, were to be given an opportunity to publish here.

683 Apparently, MIT Press had already received 8,000 advance orders for Eisenman's *Terragni* monograph from a publisher in Europe in 1976.

Diverging Interests

By the third year of *Oppositions* at the latest, following *Oppositions* 7, the different interests of the individual editors were becoming increasingly apparent. This was particularly evident in the transatlantic dialogue with Europe as, on the one hand, texts by European theorists and historians were increasingly being published and, on the other, the origins of contemporary American architecture practice were repeatedly traced back to European modernism. In *Oppositions* 9 to 12, most of the texts in the "Theory" and "History" sections were written by architects and academics who taught at universities in North America and Europe, and in particular by historians at the IUAV; other authors were recruited from among the editors who had participated in the *Little Magazine Conference* in February 1977, especially from *Arquitecturas Bis.*[684] In addition, the "Documents" section featured extensive reprints of materials on the architectural avant-garde of Europe in the 1920s and 1930s, especially from the USSR and Italy: by Sergei Eisenstein, for example, as well as Nikolai Punin, Aleksej Aleksejevic, Sidov, Giuseppe Terragni, Gruppo Sette, and the American architect William S. Huff, who studied and taught at the HfG Ulm. An archive of this size had previously been difficult to access in North America or had not been translated. Back in New York, Frampton subsequently proved to be a tireless asset to the continued publication of *Oppositions*. Between 1977 and 1980 alone, he published five book reviews of historiographical interest, e.g., by and about Alison and Peter Smithson, Reyner Banham, Nikolai Miliutin, L'Architecture Vivante, and Alvar Aalto. Contrary to the editors' original intention in establishing the "Oppositions" section, reviews of contemporary American architecture practice were now underrepresented in the journal. Apart from more reviews of architecture by Robert Venturi, Richard Meier, and Michael Graves, the editors were unable to agree on any other current projects that they considered to be worthy of criticism.

Clearly, *Oppositions* also made institutional policy by establishing certain conditions and limits, building relationships, and making omissions. One example is *Oppositions* 10, whose publication date was given as fall 1977, but which did not appear until the following year. This issue was largely devoted to a single architect, Philip Johnson. While Johnson had been a regular benefactor of the journal since its inception and had helped establish the Architects' Circle as the Institute's philanthropic network, he did not play a major role in the Institute until 1978, on Eisenman's initiative. *Oppositions* 10, therefore, was neither financially disinterested nor editorially neutral. In the "Oppositions" section, a text by Eisenman was published in advance, which was to become the introduction to a publication on Johnson's texts and which was adorned with numerous quotations

684 The essays in the "Theory" section were written by Jorge Silvetti, Jacques Guillerme, Diana Agrest, and Alan Colquhoun; in the "History" section by Kurt Forster, Eric Dluhosch, Stanford Anderson, Manfredo Tafuri, Francesco Dal Co, and Sergio Polano.

from Johnson himself in bold print;[685] it also featured an extensive interview that Eisenman and Vidler had conducted with Johnson, in which they above all gave the latter the opportunity to distance himself from the International Style of the earlier days and to make a plea for decoration and eclecticism.[686] In the "Documents" section, numerous original drawings for the design of Johnson's *Glass House* (1948) were published, with an introduction by Robert Stern.[687] There was not even the most rudimentary attempt at a critical examination of the architect and his work, once postulated by the editors as the purpose of *Oppositions*. Instead, the journal served solely to celebrate the architect in several respects: first, because the issue represented an attempt to draw attention to arguably the most enigmatic, but also the most controversial exponent of postmodernism in the United States and to secure him one of the front seats in the history of American architecture by constant reporting on him across all their media formats and fostering a public and intellectual debate, and second, because the editors could thus gain Johnson's favor, possibly also an increase in their own standing, and ultimately win him over for a further, larger commitment to the Institute. For it was clear to everyone that in the New York architecture world of the late 1970s, if you wanted to build big, there was no getting around Philip Johnson: he held all the strings. Not surprisingly, *Oppositions* 10 had the highest circulation in the history of the journal, with over 5,000 copies printed.

Another example of the strategies by which the Institute's interests were furthered through the editorial design and policies of *Oppositions* was provided by the following, eleventh issue, whose publication date was given as winter 1977, but which did not appear until the end of 1978. The "Oppositions" section of this issue featured another seminal text by Tafuri, "Giuseppe Terragni: Subject and Mask," originally commissioned by Eisenman as an introduction to his own book on Terragni.[688] But with the publication of Eisenman's monograph drastically delayed, Tafuri had initially published the text in Italian and English in an issue of the bilingual journal *Lotus International* under the title "From the archives of modern architecture."[689] The republication of the text in *Oppositions* increased

685 Peter Eisenman, "Behind the Mirror: On the Writings of Philip Johnson," *Oppositions* 10 (Fall 1977), 1–13.

686 Philip Johnson, "Reflections. On Style and the International Style; On Postmodernism; On Architecture," *Oppositions* 10 (Fall 1977), 15–19.

687 Robert Stern, "The Evolution of Philip Johnson's Glass House, 1947–1948," *Oppositions* 10 (Fall 1977), 56–67.

688 Manfredo Tafuri, "Giuseppe Terragni: Subject and 'Mask'," trans. Diane Ghirardo, *Oppositions* 11 (Winter 1977), 1–25.

689 Manfredo Tafuri, "Il Soggetto a la Maschera. Una introduzione a Terragni / Giuseppe Terragni: Subject and Mask," *Lotus International*, no. 20 (September 1978), 5–31. After MIT Press had tried for some time to publish Eisenman's *Terragni* monograph, the forthcoming publication was still announced in the academic publisher's catalogue in the fall of 1979 but was eventually

its importance and that of its subject matter. Tafuri in turn used the text not only to showcase the development of individual projects in Terragni's oeuvre but to place his rationalist architecture in its historical context. In the contest between Eisenman and Tafuri for interpretative sovereignty, the Italian historian was critiquing a different Terragni from the one outlined by the American architect—or the one he had his students outline—to trace transformations in the design process, although Tafuri drew on the same vocabulary as Eisenman, albeit in a different semantic context. The version of the text published in *Oppositions* has been abridged and differs from the original in that the final paragraphs have been omitted. This omission may not have been due to spatial constraints, since it was possible to publish the text in full in *Lotus International*, but rather to the fact that it directly references Eisenman's book project. In the passage in question, Tafuri directly referred to Eisenman's way of reading Terragni—he called this "redesign"—and characterized him as the prototype of an American intellectual. Moreover, he criticized Eisenman and his questionable, clumsy practice as a theorist for approaching historical figures in his own, idiosyncratic way, entering into a dialogue, "and so to carry on transforming it, sectioning it, breaking it down and putting it together again."[690] Tafuri criticized Eisenman's formalism for being anti-historical and power-obsessed and described him as a master of simulation, who assembled remnants of modernist utopias in his own projects; he explicitly denied the accusation of being appropriated. Strikingly, when it was published in *Oppositions*, the essay appeared with an entirely new series of illustrations, largely from Eisenman's private archive: original drawings and photographs of Terragni's projects that Eisenman had found in the attic above the architect's studio in Como in the early 1960s during a Grand Tour of Italy he had undertaken with his former mentor Colin Rowe.[691]

The publication date of *Oppositions* 11 suggests that it appeared before the *Lotus International* issue—a crucial point in architecture historiography, which is concerned with originality and creativity even more than ambiguity and contextuality. The publication dates indicated on the cover, which were intended to preserve the illusion of regularity, took on a quality all of their own, both in institutional and discursive terms. For an architecture history of journal-making, it is

withdrawn from the program. Eisenman moved the book project, along with all the other Institute publications, to Rizzoli International; Conover did not hear of this until after the fact; see Peter Eisenman, *Giuseppe Terragni. Transformations, Decompositions, Critiques* (New York: Monacelli Press, 2003).

690 Tafuri, 1978, 29.

691 In an interview, Eisenman once stated that he had personally taken the documents published in *Oppositions*, which increase the significance of the essay, out of Italy in his Volkswagen; see Peter Eisenman (interview with Louis Martin), August 15, 2000, 19. Source: CCA Montréal, Oral History Project. Eisenman's story can be understood as another assertion of authorship, but could also be examined in terms of ownership, giving *Oppositions* 11 its own significance in terms of a debate about "evidence" and "narrative."

significant that from 1977 to 1980, despite Frampton and Bloomfield's best efforts, *Oppositions* ultimately only fulfilled its contractual obligations to the publisher to a limited extent. The editorial team was producing two to three issues a year and was by now more than a year behind schedule, which was increasingly becoming both a financial and a legal problem as commitments to subscribers and sponsors were made and issues and revenue failed to materialize, causing growing irritation, more so at MIT Press than at the Institute. To advance and diversify the editorial work, Frampton invited the Swiss architecture historian Kurt Forster, who after Yale University was now teaching at Stanford, to join the journal as its fifth editor, starting with *Oppositions* 12. Forster was no stranger to the Institute, as he had previously contributed an essay to *Oppositions* 9. Moreover, in 1978 he had commissioned Eisenman with *House 11a*, another paper architecture project, which was submitted to a competition announced by *Progressive Architecture* but not seriously pursued as a building project thereafter. At *Oppositions*, Foster was primarily responsible for German-language manuscripts, but he was far from enough of a regular at the Institute to make an impact.[692]

In general, by importing theory, history, and criticism as well as documents, and by circulating, valorizing, attributing, and appropriating knowledge, *Oppositions* certainly introduced a new corpus of texts into the American architecture debate and beyond, and on top of that, influenced the formation of a canon in architecture education. *Oppositions*' readers—primarily architects, students, and teachers—were introduced to approaches such as social theory and the critical philosophy of the Frankfurt School as well as French theory, poststructuralism, and deconstruction. Ultimately, only the English translation of Theodor W. Adorno's lecture "Functionalism Today" was published in *Oppositions* 17. Yet, even if the journal did not publish a single text by French authors like Michel Foucault, Jacques Derrida, or Roland Barthes (unlike *October*, but also *Skyline*), who were very much in vogue in North America at the time, especially in the arts and the humanities, their ideas were nevertheless quoted and cited there extensively, and appropriated by its authors, including Eisenman, Gandelsonas, Agrest, and Vidler, etc.[693] The fact that the readability of the philosophically and ideologically deliberative texts ranged from difficult

692 Potential textual contributions were discussed by the editorial staff, but in most cases rejected: Wolfgang Pehnt and Tillmann Buddensieg, and more specifically Stanislaus von Moos ("Synthesis and Utopia"), Werner Oechslin ("The Age of Philip Johnson," "New York's Projected Monument of Postmodernism," "Piranesi"), and Vittorio Lampugnani ("Die eigenwillige Muse," "Karlsruhe").

693 *Oppositions* editors also considered publishing texts by Roland Barthes, as well as Walter Benjamin and Martin Heidegger. The impact of the reception of European authors on academic and architecture debates can only be imagined and, at best, be measured by the publication of texts and footnotes. Meredith TenHoor once studied quotes and citations from authors associated with "French Theory" in *Oppositions* as part of the *Clip Stamp Fold* project.

to almost incomprehensible was not only due to poor translations, but can also only be explained, if only to a limited extent, by the fact that the introduction and establishment of new thinking, methods, and concepts is often paralleled by incomprehension. One thing is certain: over the years, the overall focus of *Oppositions* shifted more and more in the direction of historiography, not least due to the individual commitment and availability of its editors.[694] There was a strong focus on architecture from the Western world, from North America and Europe, and to some degree from Latin America and Japan, due to the editors' personal interests and biographical ties. Despite the geographies covered worldwide, and the critique of orientalism, i.e., of Western historical, cultural, and political perceptions of the East, propounded by Edward Said at the time, *Oppositions* and later Oppositions Books attest to the fact that the Institute did not attempt to write global architecture history or did so only to a limited extent.

4.2 Expanding the Portfolio

From 1978, at a time when the Institute's educational, cultural, and publishing work was increasingly shaped by an entrepreneurial spirit, the public programs endowed with a large budget based on funding from the National Endowment for the Humanities (NEH) were expanded, the "Exhibitions Program" was professionalized with funds from public and private foundations—especially the National Endowment for the Arts (NEA)—and the Fellow's text and editorial work was also stepped up. After *Oppositions* and *October*, further publication formats were planned, edited, produced, and published at the Institute by independent editorial teams. Taking advantage of synergies and networking actors, these new formats fit perfectly into the institutional identity of the Institute, both in terms of their aesthetics and their rhetoric and poetics. April 1978 thus saw the launch of the tabloid-like monthly architecture newspaper *Skyline*, edited for the first two years by Andrew MacNair who previously had been in charge of organizing the Institute's lecture series and exhibitions. *Skyline* was a much more popular format than *Oppositions*, with reviews and interviews that had not existed before in this form, and most importantly offered a calendar of events for New York's burgeoning architecture and design scene that also promoted the Institute's public events. In December 1978, the IAUS Exhibition Catalogues were launched with Kenneth Frampton as editor and Silvia Kolbowski as managing editor. Most of the exhibitions at the Institute were documented, archived, and catalogued in this series, with extensive material and accompanying essays. Soon, the exhibition catalogues became a product in their own right and additionally served to cross-finance the exhibition

694 Ockman, 1988.

operations as well as the Institute. Also in 1978, the first concrete plans began for the Oppositions Books series, with Eisenman and Frampton as editors-in-chief and Lindsay Stamm Shapiro as managing editor, but this had to wait and was not actually published until several years later. With Oppositions Books, the Institute aimed to publish translations of classics of architectural modernism, as well as key contemporary European works on theory and historiography, collections of essays by eminent contemporary American theorists and historians, and monographs by American architects, and make them available to a broad readership in an elaborately, even luxuriously designed large format. If *Oppositions* continued to be conceived and perceived as the Fellows' main outlet, its primacy was nevertheless eroded by the fact that it no longer was the Institute's only publication. As a result of this reorientation, modification, differentiation, and diversification, the Institute increasingly entered the American publication market, which until then had been clearly structured in the architecture segment by commercial publishers. As part of a larger discursive, institutional, cultural, and political strategy, the Institute's new publications were nevertheless independent productions that, depending on their format, assumed specific functions in education and debate, and in culture in general. In this way, whether directly or at least indirectly, the Institute contributed to expanding the market for architecture publications in the late 1970s and early 1980s with books on architecture, some of which were of quite high quality. With MIT Press and the New York office of Oxford University Press, the university presses also participated in this expansion on the East Coast, as did Rizzoli International as the Institute's commercial publisher and the recently founded, privately owned Princeton Architectural Press.

Initially, MIT Press remained Eisenman's first point of contact as Institute director, so that for a time the academic publisher marketed, advertised, and sold almost all of the Institute's print products.[695] In 1978, after years of waiting, MIT Press released another of the Institute's publications, the long-announced *On Streets*, for which Stanford Anderson, still a Fellow at the Institute since 1970, was editorially responsible.[696] *On Streets* became the Institute's first major book project, a comprehensive collection of essays on the subject of the urban street, dating back to the Institute's "Streets Project" (1970–1972), which had been commissioned by the U.S. Department of Housing and Urban Development. Originally conceived (and paid for) as a catalogue for a planned exhibition at MoMA that never materialized, the publication testified to the Institute's long-forgotten aspirations to produce new knowledge through its own research projects. The book included both historical and theoretical contributions by Fellows and

695 MIT Press had contracted *October* (from 1978) and the IAUS Exhibition Catalogues (1979) following *Oppositions* (1976). In addition, the university publisher eventually published Oppositions Books (1982).

696 Stanford Anderson, ed., *On Streets* (Cambridge: MIT Press, 1978).

Visiting Fellows, in addition to Anderson's own study, essays by Joseph Rykwert, Anthony Vidler, William Ellis, Peter Wolf, Diana Agrest, Robert Gutman, and Kenneth Frampton, as well as Research Associates Thomas Schumacher, Victor Caliandro, and Thomas Czarnowski. In addition, there were two specially commissioned guest essays by Gloria Levitas and Gary Winkel, one from an anthropological and the other from a sociological perspective, underscoring the interdisciplinary nature of the project. Here, the Institute finally published its almost historic, rather than applied, research on the revitalization of downtown Binghamton, NY, as well as Eisenman's design of two prototypes of a townhouse. By the time it was published, Anderson as editor for the Institute did not want *On Streets* to be understood in any way as a handbook, but as a genuinely scholarly publication.[697] Nevertheless, the texts represented for the most part the state-of-the-art American research on the subject of streets, street design, street culture, etc. in all their complexity; although in the end, Anderson himself had to admit that some of the contributions had already become outdated due to the long lead time. After the 1972 publication of *Learning from Las Vegas* by Robert Venturi, Denise Scott Brown and Steven Izenour, the American city was a hot topic in the architecture debate and architecture education on the East Coast.[698] With *On Streets*, which was less semiotic and cultural and much more anthropological and sociological, but ultimately formal, the Institute found itself in good company. In 1979, MIT Press also published *The American City. From Civil War to the New Deal*, a research edition, which had been compiled in the early 1970s by IUAV historians Giorgio Ciucci, Francesco Dal Co, Mario Manieri-Elia, and Manfredo Tafuri and was now available for the first time in an English translation.[699] However, compared with Rem Koolhaas' 1978 monograph *Delirious New York*, which he had researched at the Institute and produced with Eisenman's support, the Institute's publication seemed to have appeared at the wrong time, getting neither the same attention nor, most importantly, any more follow-up commissions for urban renewal projects.[700]

From 1978, with the expansion of its publication apparatus, the top floor office studios in the Institute's penthouse were transformed into proper writing and editorial offices within a short space of time. With *Oppositions* and

697 Stanford Anderson, "Preface," in *On Streets*, ed. Standford Anderson (Cambridge: MIT Press, 1978), VII–VIII.

698 Robert Venturi, Denise Scott Brown, and Steven Izenour, eds., *Learning from Las Vegas* (Cambridge: MIT Press, 1972); Martino Stierli, *Las Vegas im Rückspiegel. Die Stadt in Theorie, Fotografie und Film* (Zurich: gta Verlag, 2010).

699 Giorgio Ciucci, Francesco Dal Co, Mario Manieri-Elia, Manfredo Tafuri, eds., *The American City. From the Civil War to the New Deal*, trans. Barbara Luigia La Penta (Cambridge: MIT Press, [1973] 1979).

700 Koolhaas, 1978. Legend has it that the publication was made possible by a generous financial injection of US$ 10,000 from Philip Johnson, arranged by Eisenman.

October under contract, *Skyline* and the IAUS Exhibition Catalogues were initially self-published. As before, the editorial work was governed by the principle of self-exploitation. Because of the different text formats and editorial processes, from both an institutional and discursive perspective, the individual productions were defined by the complex networks: the productive but not always conflict-free collaborations and relationships between the Institute, the editors, editorial staff, the authors, translators, graphic designers, and potentially publishers, not to mention the readers. In addition to the creativity and intellectuality of its editors and authors, writing, editing, and ultimately publishing were always also about pursuing interests and realizing power strategies. It was more important for Eisenman, who exercised and enjoyed the rights, privileges, and benefits of Institute director, than for others to find suitable solutions for the Institute's publishing project with academic and later with commercial publishers. Eisenman maintained that after the Institute's tenth anniversary "major emphasis will be placed on the generation of critical and theoretical work."[701] For him, *Oppositions* and *October* remained the top priority as the Institute's "original" publications, and both journals remained exempt from overheads for IAUS Central. Yet the differentiation and diversification of the print products introduced new text formats and new visual and linguistic styles into architecture culture as particular forms of knowledge: on the one hand, the zeitgeisty star interview, previously celebrated in the art scene by Andy Warhol and *Interview Magazine*, and the literary book review, otherwise perfected by the *New York Review of Books*, mixed with current hype and gossip and garnished with sensationalist portrait photographs in *Skyline*, and on the other hand, the monographic essay on current projects and positions, like a work documentation of postmodernism rather than modernism, extensively illustrated with drawings and critically annotated in the IAUS Exhibition Catalogues. In light of the transformations in both the journal and book market and the art market, the Institute's entire publication portfolio vacillated not only between theory production and historiography, quality, and tabloid journalism, but also between acquisition, public relations, and marketing. The fact that network and mediation effects now played an increasingly important role, in addition to discourse production, was reflected in the Institute's new image brochure, produced at the end of 1978, where all the publications were subsumed under "Public Programs."[702] The buzzword Eisenman used was the "public environment," since the urban public in America was increasingly changing in the 1970s, as was the Institute's readership.[703] By combining quite different reaches and target audiences with its various publication formats, the Institute

701 Peter Eisenman, "Director's Memo" January 11, 1977.

702 IAUS brochure, 1978. Source: CCA Montréal, IAUS fonds: C.3-2.

703 Peter Eisenman, "Director's Memo," January 11, 1977.

expanded its sphere of influence and scope. With its expanded text and editorial work, the Institute as an educational and cultural institution propelled itself into a veritable monopoly position as gatekeeper or taste-maker in terms of the dissemination of not just information, but certain postmodern thinking styles and aesthetic *dispositifs* by celebrating cutting-edge and pioneering architects and building practices in its journals and newspaper, its exhibition catalogues and book series, while simultaneously promoting the circulation of ideas and criticism. A comparative reading of its publications—*Oppositions*, *October*, *Skyline*, IAUS Exhibition Catalogues, and Oppositions Books—shows that parallel developments of the Institute as an educational and cultural institution were characteristic of its success and responsible for its long-term legacy: the interplay of knowledge and cultural production and their dissemination, the openness towards other disciplines, art, and theory, such as the humanities, cultural studies, and social sciences, and the transatlantic, even global, dialogue with architects, theorists, historians, and critics. It was these three qualities that, according to Eisenman in his 1977 position paper, made the Institute stand out as a "cultural resource."[704]

October

The fifth issue of *October*, the first issue published by MIT Press in the summer of 1978, was a special issue on photography, with articles by Rosalind Krauss, Douglas Crimp, and Craig Owens, among others, as well as Humbert Damisch and Hollis Frampton. The contract between the two editors-in-chief and the publisher, which was signed on June 19 and 22, 1978, crossed the desk of Eisenman, who contributed significantly to the wording and content with many handwritten corrections.[705] The agreement secured professional production and distribution and, in return, committed the editors to a quarterly production schedule, similar to that of *Oppositions*. According to the contract, the two editors-in-chief waived their salaries until a circulation of 6,000 copies was reached. There were also changes at MIT Press, where Ann Reinke, as head of the journal department, was now responsible for both *Oppositions* and *October*; Institute's catalogue series also fell under her purview shortly thereafter, making her the point of contact at the academic publisher for nearly all of the Institute's publications. One of the reasons that *October* was able to operate as a financially independent production was that the Institute received a US $10,000 grant from the NEA for *October* in 1978–79, which was explicitly to be used for authors' fees and translation costs. Although no longer officially published by MIT Press, *October* remained part of the Institute's publication portfolio

704 Ibid.

705 Frank Urbanowski, letter to Peter Eisenman (including the contract between the Institute, MIT Press, and *October* editors), July 21, 1978. Source: CCA Montréal, IAUS fonds: A.5-9 / ARCH401775.

and was also listed in the IAUS brochure produced for promotional and fundraising purposes. When Crimp was promoted to managing editor with issue 7 in the winter of 1978, he was given a permanent position with an annual salary of US$8,000 and took over much of the editorial work, to which he made some substantial contributions. Soon the journal was no longer made at the Institute, but largely out of Crimp's apartment in downtown Manhattan, for private and professional reasons. For the Institute, its policies and practices, despite their reliance on theories of deconstruction and poststructuralism, were anything but versed in identity politics or discourses of sexuality and gender, and while the women among the Fellows and editorial staff were beginning to organize, analyzing and critiquing gender hierarchies and associated power structures, the identity models, gender constructs, and sexual orientations at the Institute were still largely based on traditional norms.

Compared to *Oppositions* and the Institute's other publications that were being developed at the time, *October* took a different editorial line, not only through its thematic focus but also in terms of the associated socio-political agenda. *October* 7 (winter 1978) was another special issue on "A Soviet Revolutionary Culture," edited by Annette Michelson, with one of her few written contributions. Ultimately, *October* remained primarily Krauss's project, as Michelson was mostly abroad at the time, which affected their working relationship and was a topic of conversation at the Institute when Krauss wrote to Eisenman in the spring of 1978 to complain that all the editorial work was falling to her. Eventually, Craig Owens joined *October* as associate editor. Owens, another of Krauss's students, had previously published contributions on performance and photography in the journal, had been an editor of *Skyline* in 1978, and had contributed exhibition reviews and other texts.[706] From 1979 to 1981, Owens, who was personally interested in a theory of signs, oversaw the production of several issues and during this time published a two-part essay, "The Allegorical Impulse," in *October* 12 and 13. Based on a review of artworks by Robert Smithson, the essay lays the foundations for a theory of postmodernism in art.[707] Ultimately, however, these structural and organizational changes in the *October* editorial team, which also affected its history and program, did not bring the hoped-for success, and the journal remained a loss-making business. Institutional documents show that in the fiscal years 1978–79 and 1979–80, *October* made heavy losses, with total liabilities exceeding US$70,000. Despite the professional production through MIT Press and the institutional footing at the Institute, the journal continued to be a low-budget project, given

706 Craig Owens, "Einstein on the Beach. The Primacy of Metaphor," *October* 4 (Fall 1977), 21–32; "Photography en abyme," *October* 5, (Summer, 1978) 73–88

707 Craig Owens, "The Allegorical Impulse," *October* 12 (Spring, 1980), 7–86; "The Allegorical Impulse. Part 2," *October* 13 (Summer, 1980), 58–80, see also Anders Stephanson, "Interview with Craig Owens," *Social Text*, no. 27 (1990), 55–71.

a circulation of just 1,300 copies as well as the handling of editorial salaries and authors' fees. The editors pointed the finger at MIT Press since the university's publishing house had apparently neglected to advertise or market the journal ever since the contract was signed. By the end of 1979, no contract had been signed for distribution and sales at newsstands or bookstores, and *October* was not distributed in Europe at all. But even at the Institute, print products were not treated equally. Krauss now officially complained to Eisenman because, unlike *Oppositions*, *October* was fulfilling its contract with MIT Press and producing four issues a year. To emphasize her point, she explained how *October* received a grant from the New York State Council on the Arts because it was considered the best small magazine in the United States, but at the same time she called it the "best-kept publishing secret." For four consecutive years, the editors were awarded a grant from the NEA, even receiving the maximum amount of US$15,000 in 1980; however, this could only reduce, not offset, the losses incurred in producing the journal.

Skyline

With the publication of *Skyline* in April 1978, the Institute entered new publishing territory. The monthly architecture newspaper, run by Andrew MacNair as editor from 1978 to 1980 and initially self-published with a circulation of five hundred copies, provided information about current cultural events, new buildings, and interesting people. Somewhat directly related to the Institute's expansion into a cultural institution, now competing not just with the programming at The Architecture League, but also exhibitions at MoMA, the new Architecture Room at P.S.1, and the commercial galleries specializing in architecture, *Skyline* was conceived as a more popular format to complement the two academic journals, *Oppositions* and *October*. *Skyline*, intended to be institutional rather than discursive, was initially designed as a letter-sized pamphlet that could contain double pages with event notices, further information, and a few illustrations based on a three-column grid; the League's postal newsletter may have served as a model. An initial mock-up model suggested that *Skyline* could have originally been typewritten, which promised time-efficient and cost-effective production in line with MacNair's DIY approach. Roles had yet to be assigned, and were approached in a playful rather than competitive manner: MacNair was initially listed as *Skyline*'s director, his assistant Mimi Shanley as managing editor, and Kenneth Frampton as editor in charge to ensure respectability and credibility.[708] When the Institute was awarded a one-time US$10,000 production grant from the New York Council on the Arts (NYSCA) for a calendar of events in the spring of 1978, the concept was quickly expanded, and *Skyline* was made into a tabloid format, again with Vignelli's help. In the newspaper,

708 In the first issue of *Skyline*, MacNair first gave himself the title "director," which was subsequently changed to "editor."

the calendar was now designed as a center fold with a double-page monthly overview of dates, initially exclusively in New York, which could be removed and hung up as a poster. Vignelli's approach and the established graphic identity, applied here to his preferred format, made the newspaper clearly identifiable to readers as printed matter from the Institute.[709] *Skyline*'s straightforward modernist layout, again based on a three-column grid, also meant that the newspaper could be set by hand by the editorial team themselves. In addition to the black title lettering, which was designed as a logotype in tightly set, boldly printed sans-serif black capital letters to recall the real Manhattan skyline, the black bar became the newspaper's most recognizable trademark. The horizontal bar, which originated from the paper's institutional identity, was designed as an eye-catcher, structuring not only all the information on the front page but also the calendar of events as an actual grid. On the single pages, too, the bar as a graphic element functioned both aesthetically and formally to organize the content: as a tab for all the texts and illustrations, large-scale photographs and architectural drawings, while also allowing for white spaces.

In view of the existing difficulties in producing *Oppositions* as well as *October* even four times a year, *Skyline* as a monthly tabloid was an extremely ambitious project at the Institute. The newspaper necessitated the development and testing of new publishing practices and organizational structures that allowed for much faster production, printing, distribution, and sales than had previously been the norm. Graphically, *Skyline* was laid out like a broadsheet tabloid with large, attention-grabbing headlines on the front page. The first issue had just eight pages and was built around a double-page calendar that listed cultural events in New York that were of interest to architects and designers, notably including Institute events. *Skyline* was a print publication produced to cross-promote events at the Institute—lecture series, "Open Plan" events, and exhibitions—in order to attract an ever-larger audience for the growing "Public Programs." The editorial of the very first issue, which was not signed, set the agenda and provided information about the functions of the newspaper and the ambitions of its editors: "*Skyline* is both a central information file for upcoming exhibitions, lectures, symposia, and publications and a platform for critical opinion about the events of the recent past. Thus, it should become an index to the condition, spirit, and direction of architecture."[710] While initially limited to New York, the newspaper was soon to expand geographically to cover the entire USA.

709 In our oral history interview, Vignelli highlighted that *Skyline* marked his return to the starting point of his career, newspaper design, in his view the supreme discipline in graphic design. For him, *Skyline*, was the most rewarding graphic job, compared to the other formats: the journal and the book.

710 Andrew MacNair, "Editorial," *Skyline* 1, no. 1 (April 1978), 2.

As a tabloid newspaper available on newsstands and in selected stores, as well as by subscription at a price of US$1, *Skyline* differed fundamentally from *Oppositions* in terms of aspirations and quality; the newspaper was in fact the diametric opposite of the journal in terms of form and content. For the *Skyline* editors, it was not about an international debate characterized by the juxtaposition of different positions. Instead, *Skyline* provided the Institute with a pluralism of listings, features, gossip, and hype, i.e., stories focusing on the human aspect. It was primarily a PR tool to report on people and events, with architecture coming only third place. "Today, there are more exhibitions with architectural themes than ever, and the teaching and study of architecture have been infused with new energy. The proliferation of the written word about architecture testifies to its popularity; new articles, magazines, books, and encyclopedias appear daily."[711] So while *Oppositions* stood for complex, intellectual topics and text-heaviness, *Skyline*, with its loud yet undogmatic approach, was the first architecture newspaper of its kind in the United States to advocate for a quickly written architecture journalism that was less serious in tone. "*Skyline* of course enters into this discursive mainstream. But it does so responsively and respondingly. Its hope is that, by channeling a mass of uncatalogued material through a central file, the significance of that material will become more apparent."[712] The *Skyline* editorial team, not least because of MacNair's playful approach, flirting with a certain kind of punk attitude, worked with a mostly refreshing but not always reliable mixture of actionism and dilettantism. This suited the zeitgeist in New York, where the alternative art and architecture scene, which had been given a new location in 1976 with P.S.1 in Queens, was just experiencing a peak, paralleling the subversive youth and music culture, and especially punk. In its discursive, cultural, social, and institutional function, *Skyline*, as *The New York Architecture and Design Calendar* (the newspaper's subheading), can—from an architecture history perspective—be read as a chronicle of the architecture and design culture of those years, at least as it was perceived from the perspective of the Institute (and also a chronicle of the Institute at that time), on the one hand, and on the other hand, as a specific mechanism for the constitution of a particular architecture and design scene.

As editor, financed by the NYSCA grant, with a small staff, and initially without many constraints and pressures, MacNair produced four pilot issues from April to August 1978, experimenting with the format and trying out various forms of editorial work. The editorial team included Craig Owens as second editor alongside MacNair for the first year, who contributed his expertise in the arts before joining *October* full-time in 1979, and Pilar Viladas, who served as *Skyline*'s managing editor for the first three issues and later went on to pursue

711 Ibid.

712 Ibid.

a career in journalism. In addition, two graphic designers from Vignelli's office, Lorraine Wild as an assistant and Jessica Helfant as an intern, helped design the issues; Vignelli then merely gave the go-ahead for printing. To build readership, the pilot issues were initially distributed free of charge throughout the New York metropolitan area. *Skyline* was printed just around the corner, within walking distance of the Institute, at Jae Kim Printing Company on 39th Street, which facilitated quick production. At first, William Eitner oversaw production; Brian Kay handled shipping and advertising. The format, distribution, and production met with success so in September 1978 *Skyline* began with a more or less regular production with up to ten issues per year, which had to compete on the market. In the first year, *Skyline* was largely financed by subscriptions. While MacNair and Owens contributed reviews and interviews, other articles—more news stories than academic writing—were written by Fellows and friends. MacNair produced primarily with staff from his circle of friends and acquaintances; as he repeatedly brought in new people during his time as editor-in-chief, personal continuity was thus only achieved for a few issues at a time. Professional distribution now made it possible for the architecture newspaper to be available for purchase in bookstores nationwide, drawing attention to the Institute and the local architecture and design scene. One of the merits of *Skyline* was that it featured emerging architects, thereby shaping what contemporary positions of architectural postmodernism were deemed relevant. However, the editors more than once had problems getting *Skyline* published on time at the beginning of the month, which had a negative impact on its function as a calendar of events and the Institute as host.

Since as a newspaper it displayed more creativity than intellectuality, with a focus on entertainment value rather than educational value, in addition to the novelty value of the calendar of events, *Skyline* mainly published smaller articles on topics relevant to architecture and design; in addition to exhibition and book reviews, these included, for example, articles on film sets and restaurant architecture, local cultural events in the arts, such as the New York Film and Theater Festival, or the latest postmodern trends in architecture. It was not until the editorial for the one-year anniversary appeared in April 1979 that the editors officially rejected the assumption that the title "Skyline" actually referred to Lewis Mumford's column in *The New Yorker* of the same name, published until 1963, in which the architecture critic discussed individual buildings or larger developments, and thus also to his criticism of architectural modernism; apparently they had been frequently asked about this and now felt compelled to issue a denial.[713] Nevertheless, from the outset, *Skyline* advocated for a broad notion of architecture

713 Donald Miller, *Lewis Mumford. A Life* (New York: Weidenfeld & Nicholson, 1989).

and encouraged a pluralistic view of contemporary architecture culture.[714] The "Skylights" section on the last page, for example, was initially a central column featuring short, sometimes polemical texts about events listed in the "Calendar" section or special cultural events and activities in Manhattan. While *Oppositions* increasingly involved professors of history and theory, rather than practicing architects, *Skyline*'s editorial policy was nowhere near as elitist or competitive in terms of the selection of authors and topics. However, just like the journal, the newspaper served a new economy of attention in architecture, popular and comprehensive, without regard to the already established positions. *Skyline* developed and distinguished itself mainly through two text formats: first, rather light interviews in the style of Warhol's booming lifestyle magazine *Interview Magazine*, which was launched in 1969 and which aimed to offer insights into the scene through its frequently unedited interviews with glamorous figures of the New York art world, and second, comparatively serious book and exhibition reviews, for which, like the "Reviews, Letters, Forum" section in *Oppositions*, the prestigious literary magazine *The New York Review of Books* once again served as a model. *Skyline* was clearly designed for light reading, even though the newspaper's readership was primarily a rather select circle, especially of local architects and designers, with a strong interest in the cultural life of the city.

Both Fellows and Visiting Fellows of the Institute, as well as experienced architecture critics, contributed to the first issues of *Skyline* with sometimes polemical, sometimes challenging texts. These also included the *Oppositions* editors, as well as the newspaper's two editors MacNair and Owens. While MacNair wrote about architecture exhibitions, Owens was responsible for art exhibitions. Exhibition reviews were also published, some of them quite inflammatory, even of the Institute's own events. For example, the exhibition "Projects, Sets, Arcadias," curated by Archigram members Peter Cook and Ron Herron at the Institute in 1978, was reviewed in the August issue with two texts by Reyner Banham and Livio Dimitriu.[715] Frampton penned several exhibition reviews, while at the same time almost single-handedly managing the *Oppositions* editorial office, launching the catalogue series, and, on top of that, heading the editorial office of Oppositions Books together with Eisenman.[716] Eisenman contributed two short texts to the pilot issues of *Skyline* in 1978. Under the pseudonym Ernesto di Casarotta—an allusion to Ernesto Rogers, the former editor

714 Patrick Pinnell, "Editorial," *Skyline* (April 1979), 2. On Lewis Mumford's column in *The New Yorker*, see Robert Wojtowicz, ed., *Sidewalk Critic. Lewis Mumford's Writings on New York* (New York: Princeton Architectural Press, 2000); see also Herbert Muschamp, "Sidewalk Critic. Lewis Mumford's Writings on New York," *Art Forum* (April 1999), 19–20.

715 Reyner Banham and Livio Dimitriu, "Peter Cook/Ron Herron: Arcadias/Insertions," *Skyline* (August 1978), 3.

716 Kenneth Frampton, "Drawings by Le Corbusier at the MoMA," *Skyline* (April 1978), 2; "Iceberg," *Skyline* (September 1979), 5; "Stellar Material: Eileen Gray at the Modern," *Skyline* (March 1980), 3.

(1953–1965) of the Italian *Casabella*—and mimicking the jargon of a sports reporter, he wrote about the relationship between the Institute and the IUAV on the one hand, and on the other hand, about the team of New York architects around Philip Johnson, to which he himself belonged.[717] As sociograms of the architecture field, described from the point of view of one of the protagonists (albeit under the protection of a pseudonym), these emotionalized texts demonstrate not only Eisenman's interest in gossip but also his strategic approach to acquiring and maintaining power, reordering the field without much ideological commitment, and inscribing himself without much affective involvement. The articles published in *Skyline* in general, and Eisenman's texts in particular, thus provide a good example and resource for a relational and network analysis of the dynamics of groups, organizations, and institutions. Even though *Skyline* clearly set itself apart from the Institute in its masthead, the Institute's interests and strategies were repeatedly reflected in its pages. For example, the May 1978 issue included a major interview with Philip Johnson, who was being systematically courted by Eisenman as an architect and as Institute director. With the publication of this interview, Johnson was given the opportunity to present and explain his design of the AT&T Building in detail at a time when the plans had just been made public. Then, when an exhibition of models and drawings of this postmodern skyscraper, the first to be built in New York after the financial and fiscal crisis, was presented at the Institute in the fall of 1978, *Skyline* issued several articles in advance and ensured that Johnson again became a topic of conversation. Finally, in the October 1978 issue, photographs of Johnson at the Institute were published. Using these tactics of familiarization, personalization, and scandalization, *Skyline* mixed information and entertainment, provided talking points, and advertised specific architecture firms.

Apart from that, *Skyline* also made a name for itself as a publishing platform for young authors; in the first year alone, almost sixty different authors contributed texts to the newspaper.[718] In addition to the editors, Livio Dimitriu, Lars Lerup, Herbert Muschamp, and Michael Sorkin published regularly in *Skyline*.[719] Moreover, *Skyline* provided young savages such as Rem Koolhaas or Bernard Tschumi, who had both spent a year at the Institute as Visiting Fellows in the mid-1970s, with another publishing opportunity after sporadic contributions to

717 Peter Eisenman [Ernesto di Casarotta, pseud.], "The Sound of Leather," *Skyline* (May 1978), 7; "Quarta Roma: Report from Rome," *Skyline* (August 1978), 6.

718 *Skyline*, (April 1979), 2.

719 Livio Dimitriu, "Report from Syracuse," *Skyline* (May 1978), 3; "Peter Cook/Ron Herron: Arcadias/Insertions," *Skyline* (August 1978), 3; "Swiss Transmissions and Exaggerations: An Interview with Mario Botta," *Skyline* (March 1980), 12–13; see Lars Lerup, "Gunnar Asplund," *Skyline* (September 1978), 4; "Report from San Francisco," *Skyline* (November 1978), 9; "Apropos Type: Patrick Henry Bruce and Aldo Rossi," *Skyline* (October 1979), 6; see Herbert Muschamp, "The Universal Style," *Skyline* (February 1980), 14–15; see Michael Sorkin, "Hollywood Matter," *Skyline* (September 1978), 9; "Cloning People," *Skyline* (November 1978), 11.

Oppositions. Koolhaas, for example, not only lobbied for the preservation of the landmarked hall in the Rockefeller Center in the "Skylights" section of the pilot issue in April 1978 under the title "The Birth of Radio City Hall," but in doing so advertised his book *Delirious New York*.[720] Tschumi, on the other hand, published the "Architectural Manifestos," his most current projects, in *Skyline*, as well as the "Manhattan Transcripts," and some of his Follies were also reported on there.[721] The two Italians Massimo Scolari and Giorgio Ciucci from the IUAV, who were guests at the Institute in the fall semester of 1978, also contributed their own drawings and texts to *Skyline*. Thus, *Skyline* could be read at any time as a reflection of the Institute's network at that particular moment, profiting from the resulting social and cultural capital. The Institute's transformation into a powerful cultural institution, its shift towards the establishment, and more than that, its transformation into a fashionable postmodern salon, expanding its sphere of influence with the "National Architecture Exchange" and variations on the "New Wave" series: all of these were accompanied by and accomplished through *Skyline*.

When the performance of *Skyline* was evaluated internally in early 1979, it was criticized for trying to come across as too intellectual and at the same time for not yet having found its own voice. Nevertheless, the publication of an architecture newspaper with its own calendar of events made the Institute less dependent on event announcements in *The New York Times* or the weekly neighborhood newspapers such as *The Village Voice* or *SoHo Weekly*. By publicizing its public events, lecture series, and exhibitions, the Institute succeeded in gaining a foothold in metropolitan urban culture through its media output and possibly reaching a larger audience. Soon the newspaper was available at one hundred and twenty-five outlets throughout the city, at newsstands and in bookstores, as well as in art galleries and selected shops, such as the flagship store of the trendy Milanese fashion label Fiorucci in Manhattan. In 1979, *Skyline* had a total circulation of 2,000 and nearly 1,200 subscribers. The newspaper, which was by now the central medium for topics related to architecture and design culture in New York, helped shape the Institute's hip and trendy image throughout the country.[722] And while *Skyline*, unlike other "little magazines," did not take up radical positions, as the arts and art criticism did, its main effect was to keep the Institute a topic of conversation in New York architecture circles and to attract public attention. Like the "Forum" section of *Oppositions*, *Skyline*

720 Rem Koolhaas, "The Birth of Radio City Music Hall," *Skyline* (April 1978), 7.

721 Bernard Tschumi, "Bernard Tschumi's Architectural Manifestos," *Skyline* (May 1979), 8–9; "Architectural Manifestos," *Skyline* (May/June 1980), 12.

722 Steven Holl, Alison Sky, Suzanne Stephens, "East Coast West Coast," in Colomina and Buckley, 2010, 70–81. For some reason, Andrew MacNair, who created Skyline, was neither invited to the roundtable nor interviewed as part of the Clip, Stamp, Fold project. Suzanne Stephens, however, did mention him.

engaged in politics with the photographs it printed, for example when the party celebrating *Skyline*'s first anniversary in April 1979 at the Rizzoli Gallery of the friendly commercial publisher and the coverage of it in the following issue were extensively documented with a photo spread of the party guests—including the funders and sponsors of the Institute, Fellows and friends, such as Gianfranco Monacelli, the publishing house's director—made it clear that the Institute was increasingly taking on the role of gatekeeper or taste-maker in the local architecture scene as well. In New York, the who's who of seeing and being seen ultimately defined who was part of the scene and who was not.

After one year, *Skyline*'s concept was modified slightly for the first time with the anniversary issue of April 1979 and adapted to the needs of the Institute. The newspaper, whose editorial staff was expanded to include Patrick Pinnell, was given the new subheading *The Architecture and Design Review*, significantly softening its local connection to New York and emphasizing its national stature.[723] In addition to an even greater focus on general interest interviews and reviews, the Calendar now included events across the East Coast of the USA—especially at the prestigious schools of architecture—which, in turn, allowed the Institute to manifest its close ties with them. Purely a city newspaper in its first year, in its second year *Skyline* became a review of the cultural life in architecture and design emanating from New York. The newspaper regularly announced or reviewed exhibitions in the major museums (The Museum of Modern Art, Solomon R. Guggenheim Museum), the commercial architecture galleries (Leo Castelli Gallery, Max Protetch Gallery), and the alternative self-managed art spaces (P.S.1 and Architecture Room). As part of its new alignment and new aspiration to become the leading medium for North American architecture culture, *Skyline*'s editorial team included high-profile topics that appealed to the general public, primarily through a strategic selection of authors. The April 1979 issue, for example, printed excerpts from talks by Charles Jencks, architecture historian and theorist, and Paul Goldberger, architecture critic at *The New York Times*, on the role and responsibilities of architecture journalism at a symposium in San Francisco; both were well-known beyond the field of architecture and had made a name for themselves primarily as apologists of a postmodern architecture.[724] Interestingly, in his editorial for the same issue, which consisted of introductory remarks to the feature on Jencks and Goldberger, Owens criticized their populist positions, since in his view they were presenting aesthetic arguments in their promotional and defensive pieces. Thus, *Skyline* was participating in the academic debate on postmodernism, albeit in a way that differed from that of *Oppositions* and *October*, by publishing popular and well-known authors while taking the liberty

723 Pinnell, who had previously taught in the Institute's "Undergraduate Program" as a tutor in the design studio during the fall 1978 semester and had a BA in literature, joined the editorial staff to ensure journalistic quality.

724 *Skyline* (April 1979).

of distancing itself from them, i.e., positioning itself as critical of the mainstream.

In its second year, *Skyline* repeatedly published reviews of books and exhibitions, some of them harsh polemics that were intended to shape public opinion. Once again, Frampton exemplified what he considered to be good journalism. For the April 1979 issue, he wrote no less than two texts on the recent MoMA exhibition "Transformations in Modern Architecture" (February 23 to April 24, 1979), curated by Arthur Drexler and the subject of much controversy at the Institute. "Transformations" proposed a particular interpretation of the heritage of architectural modernism, the global proliferation and corporatization of the International Style in the postwar period, and did not necessarily align with the architectural attitude held at the Institute—by historians and theorists as well as practitioners.[725] Here, Drexler exclusively presented realized projects in the form of photographs; this, above all, disqualified some representatives of what was known as paper architecture, as propagated by the New York Five around Eisenman.[726] While Frampton's first text, "Blow Up," was still a fairly objective review, his second piece, "Skylights: The Ins and Outs" was a revealing commentary in which he harshly criticized the exhibition's emphasis on images and hence the criteria for exclusion embedded in the curatorial concept; moreover, he attacked Drexler personally, accusing him of being motivated solely by sensationalism and of having betrayed his ideals.[727] *Skyline* then gave Drexler the opportunity to defend his exhibition against Frampton's criticism in an interview with MacNair.[728] In general, *Skyline* managed to publish regular reviews of current publications for a period of time. Pinnell wrote a review of Koolhaas' *Delirious New York*, for which *Skyline* also ran extra ads, Alan Plattus introduced Paul Goldberger's new architecture guide to Manhattan, and Peter Kaufman wrote a review of *The American City* coming out of the IUAV. These reviews were printed in *Skyline* rather than *Oppositions*, and it is particularly noticeable that many of the titles reviewed there were again penned by friends and authors associated with the Institute.[729] Next to the "Reviews" section, "Interviews" in the second

725 Arthur Drexler, *Transformations in Modern Architecture* [Exhib. Cat.] (New York: The Museum of Modern Art, 1979).

726 Apparently, none of John Hejduk's projects was shown in the exhibition, on the grounds that he did not fit into any category; see MoMA, "Transformations in Modern Architecture," Master Checklist and Press Release no. 7, February 21, 1979, https://www.moma.org/calendar/exhibitions/1773 (last accessed: May 31, 2023).

727 Kenneth Frampton, "Blow Up," *Skyline* (April 1979), 6; "Skylights: The Ins and Outs," *Skyline* (April 1979), 12. Frampton criticized Drexler for causing unnecessary turmoil with his curation and accused him of hysteria. This separation of information and opinion, by a single author, occurred only once.

728 Arthur Drexler (interview with Andrew MacNair), "Response. Arthur Drexler on Transformations," *Skyline* (Summer 1979), 6.

729 Patrick Pinnell, "Remifications," *Skyline* (March 1979), 5; Alan Plattus, "Manhattan Guides," *Skyline* (October 1979), 8; Peter Kaufman, "Italian Views of the American City," *Skyline* (May/June 1980), 17.

volume became the dominant format in *Skyline*.[730] Much like Warhol's *Interview Magazine*, which for a decade had published interviews with celebrities, artists, and musicians, the Institute's tabloid newspaper now also regularly interviewed well-known figures, mostly established architects who were often members of the Institute's Architects' Circle, and thus rewarded them for the financial support by putting them in the spotlight; in addition to Philip Johnson, for example, Cesar Pelli, Ulrich Franzen, Arata Isozaki, Aldo Rossi, John Hejduk, and Charles Gwathmey also found their way onto the pages of *Skyline*. Most of MacNair's conversations coincided with current building projects by the respective architects; his own interests also allowed him to interview contemporary designers and artists, such as the German stage, costume, and lighting designer Manuel Lütgenhorst, who first came to New York in 1978 and immediately earned a certain reputation on the scene by renovating the hip Studio 54,[731] or with the furniture designers and interior decorators Dino Gavina and Joseph d'Urso, both of whom worked for Knoll International, one of the main sponsors of *Skyline*—another example of the commercialization of architecture culture. In addition, MacNair also asked Robert Venturi for an interview to discuss his design for Knoll International's Manhattan showroom, with the tone of the conversation fluctuating cheerfully between attack and approval.[732] Oddly enough, the interview was for the first time accompanied by a caricature drawn by architect and artist Michael Mostoller, which made the point that the brand names of design classics now dominated a thoroughly commercialized architecture world, while actual design had long since receded into the background.

All of the Institute's publications, not just *Skyline*, bore witness to postmodernism with all its ambiguities and paradoxes, even though their editorial policies differed: while *Oppositions* in the late 1970s stood less for a theorizing and increasingly for a historicizing approach, and yet still struggled to discuss contemporary architecture or current publications, *Skyline* was able to establish itself as a popular format for popular content conveyed through popular forms of presentation. Frank Gehry, for example, who had been a successful architect in Los

730 The following interviews were published in *Skyline*, unless otherwise noted by Andrew MacNair: Philip Johnson, interview with Martha Carroll and Craig Owens (May 1978); Rouben Ter-Arutunian (September 1978); Ceasar Pelli (March 1979); Ulrich Franzen, interview with MacNair and Owens (April 1979); Arata Isozaki (May 1979); Arthur Drexler (Summer 1979); Aldo Rossi, interview with Diana Agrest (September 1979); Dino Gavina (October 1979); Joseph d'Urso, interview with Pilar Viladas (October 1979); John Hejduk, interview with Donald Wall and Nancy Ferrara (December 1979); Manuel Lütgenhorst (December 1979); Charles Gwathmey/Robert Siegel (February 1980); Robert Venturi (March 1980); Mario Botta, interview with Livio Dimitriu (March 1980); Coy Howard (March 1980).

731 Manuel Lütgenhorst (interview with Andrew MacNair), "Behind Studio 54," *Skyline* (November 1979), 17.

732 Robert Venturi (interview with Andrew MacNair), "Venturi and the Classic Modern Tradition," *Skyline* (March 1980), 4–5.

Angeles since the 1960s and was active in the vibrant art scene there, was discussed in an exhibition review in *Skyline*, but his architecture was not reviewed in *Oppositions*.[733] At the time, Gehry had just completed his private house, which differed from the approaches advocated by the Fellows in that, in addition to the strategy of the ready-made, it emphasized the idea of the frame, placing fragments of a timber-frame building, pergolas, and scaffolding in front of an existing residential building, while incorporating historical quotations. In another example of the policies surrounding architecture culture, *Oppositions* editors Vidler and Forster placed reviews of the exhibition "Lauretta Vinciarelli: Projects 1973–1978" (1979, at the Institute) and "Diana Agrest and Mario Gandelsonas: Architecture between Memory and Amnesia" (1978–79, Architecture Room of P.S.1) in *Skyline*, but their architecture received no attention in *Oppositions*.[734] In keeping with the tabloid format, *Skyline* could juxtapose different, at times opposing, positions. For example, in a two-part article, Pinnell discussed the architectural drawings and urban planning projects of the office Venturi and Rauch, and in doing so, featured Robert Venturi (but not Denise Scott Brown) in *Skyline*, while neither was discussed or published again in *Oppositions*.[735] That *Skyline*'s tabloidization of architecture discourse also offered both quality and controversy was once again evident when Rosemary Bletter reviewed a symposium on "Architectural Form and the Problems of Historicity," which engaged with the architecture of Michael Graves, along with critical commentary by Anthony Vidler and Alan Colquhoun.[736]

The issues of *Skyline* published in the fall of 1979, which announced and accompanied the coordinated exhibitions of Aldo Rossi's drawings at the Institute and the Max Protetch Gallery, were indicative of a new cult of personality that the Institute was embracing with its educational offerings, cultural productions, and publication formats. The self-created media hype ranged from the cover of the September 1979 issue, which featured Rossi's drawings for the San Cataldo Cemetery in Modena and its legendary Teatro del Mondo, to a pointed, if poorly edited, interview in which Rossi commented on architecture, politics, and film, and verbally applied his analogous approach to the American city; in the October 1979 issue, after the exhibitions closed, photographs of the vernissage party were published in the "Skylights" section, and the translation

733 A review of the inaugural exhibition at the Architecture Room of P.S.1 on Gehry, organized by Lindsay Stamm Shapiro, was published in *Skyline*; see Steven Harris, "202 Frank Gehry," *Skyline* (November 1978), 2. At the Institute, Gehry played only a minor role.

734 Vidler, 1979; Kurt Forster, "Between Memory and Amnesia," *Skyline* (January 1979), 4.

735 Patrick Pinnell, "On Venturi I: Drawing as Polemic," *Skyline* (December 1978), 5; "On Venturi II: Allegory and Kitsch," *Skyline*, (January 1979), 5. Scott Brown had been a partner in the firm since 1969, where she was responsible for urban design projects. This was not reflected in the name until 1989, when John Rauch resigned, and the office was renamed Venturi, Scott Brown and Associates.

736 Rosemarie Bletter, "About Graves," *Skyline* (Summer 1979), 2–3.

of a text by Manfredo Tafuri, "Theater of Memory," was reprinted in place of an exhibition review.[737] *Skyline* thus continued the media strategy already pursued with *Oppositions* of generating not only social and cultural but also symbolic capital from the envy of those who were not present by portraying the select circle of invited guests. This superficial, largely quite subjective approach was punctured by one of Mostoller's caricatures, whose sharply drawn commentary in this case illuminated the emerging phenomenon of celebrity culture in architecture embraced by Rossi. Mostoller depicted Rossi in multiple versions, as a copy of himself on the stage of architecture in the United States. At the Institute, it was precisely the interplay of pedagogical, cultural, and publishing practices that laid one of the cornerstones for the coming star cult, the excessive idolization, even glorification of a few, world-famous (mostly male) architects. This media culture that celebrated the genius of individual, often male, figures was a distinctive feature of architectural postmodernism, which was propelled by several major events in the 1980s: the first Biennale Architettura di Venezia (1980) headed by Paolo Portoghesi, the Internationale Bauausstellung IBA Berlin 84 under the dual direction of Josef Kleihues and Hardt-Waltherr Hämer (from 1980, culminating in 1984), and the "Deconstructivism" exhibition at MoMA curated by Philip Johnson and Mark Wigley (1987).[738]

Skyline was now in vogue and had become an important format for communicating and negotiating criteria for the perception and evaluation of contemporary architecture. The newspaper was subscribed to by libraries at leading universities and museums in New York, such as Columbia University and MoMA. It was also gaining recognition abroad; for example, Phyllis Lambert was an early subscriber to the newspaper for the Canadian Centre for Architecture in Montréal, which she founded in 1979. Most importantly, *Skyline*, by spreading gossip and generating media buzz, produced, reproduced, and represented the discursive and institutional networks that centered on the Institute, thus providing a good insight into its self-conception and self-image. Most importantly, the Institute increasingly used its monthly tabloid to advertise on its own behalf: it ran specially designed ads for its "Evening Program" and other print products, not just *Oppositions* and *October*. When the IAUS Exhibition Catalogues were introduced in late 1978, paralleling the professionalization of the "Exhibition Program," *Skyline* provided the ideal complement. With interviews, reviews, and, above all, the calendar of

737 Manfredo Tafuri, "The Theater of Memory," *Skyline* (October 1979), 7.

738 Szacka, 2016; While the Venice Biennale was only discussed at a "Forum" after Frampton had withdrawn his text contribution, Eisenman was the main contributor to the IBA Berlin 84 and "Deconstructivism" show from the circle of Fellows. With Eisenman and Frampton went to Berlin in 1984 as former Fellows at the invitation of the American Academy after the Institute, as it had existed for years, finally collapsed. See Senator für Bau- und Wohnungswesen, ed., *Idee Prozess Ergebnis. Die Reparatur und Rekonstruktion der Stadt* (Berlin: Frölich und Kaufmann, 1984); see also Philip Johnson and Mark Wigley, eds., *Deconstructivist Architecture* (New York: Museum of Modern Art, 1988).

events, *Skyline* guided a readership interested in architecture and design through New York's cultural life; the Institute portrayed itself as the main hub of architecture culture. The Institute's claim to national standing, expressed in "Open Plan," the "National Architecture Exchange," and the "New Wave" series was also demonstrated by the network of journalists that the *Skyline* editorial team maintained with other, new architecture newspapers such as *Archetype* from San Francisco, a network that also manifested itself in the placement of exchange ads.[739] While the American publication landscape in architecture had previously consisted primarily of book series, architecture press, and university journals, by the late 1970s it had been augmented by many smaller productions.[740] The summer 1979 issue of *Skyline* advertised university architecture journals, some of them new, such as *VIA*, *Perspecta*, and *The Harvard Architecture Review*.[741] The friendship between MacNair and Steven Holl, who had not only been a regular visitor to the Institute since moving to New York from the West Coast but also supported the production of *Skyline* and occasionally published pieces in the tabloid newspaper itself, helped establish a collegial relationship with the *Pamphlet Architecture* series that Holl was editing: small booklets featuring the designs of young architects.[742] One editorial strategy to extend the Institute's influence and reach beyond the East Coast was the introduction of the "Cross-Country" section with the October 1979 issue, which drew on a network of correspondents to report on buildings and cultural events from various North American cities.

Despite the editors' best efforts, the editorial work on *Skyline* proved to be difficult—and this was not only due to the inexperience of the editors and the strict publication schedule. After two years, newspaper-making at the Institute, working conditions, and decision-making processes were still precarious and marked

739 *Skyline* and *Archetype* ran exchange ads several times, e.g., in *Archetype* no. 1 through 4 and in *Skyline* (Summer 1979). The *Archetype* editorial staff included, among others, Andrew Batey, Demetra Bowles, and Henry Bowles; also Kurt Forster, who had newly joined *Oppositions* as editor, and Diane Ghirardo, who did translations for *Oppositions* and the Oppositions Books series, as well as Mark Mack, a friend of MacNair. In *Skyline*, *Archetype* was described as "the only non-New York architectural tabloid."

740 Colomina and Buckley, 2010.

741 Advertisements for *VIA IV*, *Perspecta*, no. 16, *The Harvard Architecture Review*, no. 1, *Skyline* (Summer 1979), 14.

742 Steven Holl, "USSR in the USA," *Skyline* (May 1979), 10; "Ungers at Columbia," *Skyline* (October 1979), 15. Holl launched *Pamphlet Architecture*, his own publication series, on December 30, 1977. Many of the architects featured were friends of MacNair's and part of *Skyline*'s extended circle. Among the first ten publications, in addition to Holl's projects, were designs by Mark Mack, Lars Lerup, Livio Dimitriu, Lebbeus Wood, Zaha Hadid, and Albert Sartoris. In the first *Pamphlet Architecture*, Holl wrote appreciatively of the Institute's influence not only on the New York architecture scene but on the American architecture world as a whole: "In New York, theorists rethought architecture education and founded the IAUS, analogous to London's Architectural Association. They first published Oppositions, edited by Peter Eisenman, Kenneth Frampton, and Mario Gandelsonas, to promote fresh creative ideas that were being ignored by the periodicals. New York thus became one of the most exciting architectural environments in the United States." See Steven Holl, ed., *Pamphlet Architecture*, no. 1 (New York: 1978).

by economic insecurity due to insufficient income and structural dependencies. The power imbalance became visible when MacNair complained several times to Eisenman, also in writing, that double standards were being applied to the Institute's publications. This was because, unlike *Oppositions*, *Skyline* editors were repeatedly told that the newspaper would have to be financially self-supporting and that it would have to pay forty percent of its revenue to IAUS Central as overhead. At this time, the newspaper format was considered a failure, at least economically. The problems with financing, management, and distribution remained unresolved, although several foundations provided funding in 1979, including NYSCA, the CBS Foundation, the J.M. Kaplan Fund, and The Gilman Paper Foundation, the private foundation of the largest paper manufacturer in the United States at the time. In the meantime, *Skyline* editors were able to report successes as private sponsors were secured and nationwide distribution was professionalized. But by early 1980 *Skyline* had accumulated debts totaling US$10,000, and salaries could no longer be paid. In the spring of 1980, MacNair sought professional outside advice from people willing to invest in the paper, working with Henry Hecker and Horace Havemeyer III to come up with a new business plan and various scenarios, such as launching a fully funded, fixed circulation publication or redesigning *Skyline* as a glossy magazine with a higher circulation and thus more financially strong advertisers. He even offered to acquire the copyright himself. But these efforts remained unsuccessful. Eisenman rejected all of MacNair's proposals on the grounds that they were in line with neither the Institute's goals nor its resources. While the production of *Oppositions* and *October* was largely covered by MIT Press by the end of the decade, Institute director Eisenman ultimately did not lobby hard enough for *Skyline* to find a publisher to include the architecture newspaper in their program; talks with Monacelli at Rizzoli International also ultimately failed to produce results. All this was to change.

Despite all the background difficulties, the production of *Skyline* continued and gave rise to discussion. One incident that was representative of the conflict potential in architecture culture was a "Letter to the Editor" written by Peter Fend and printed in the February 1980 issue.[743] Here, Fend reported on the groundbreaking "Real Estate Show," a politically charged exhibition that had been organized out of the emerging art scene in a vacant building on the Lower East Side and through which an artists' collective criticized real estate policies in the East Village and the role of the artist in the gentrification process: a topic that was not a concern at the Institute.[744] Fend, who had

743 Peter Fend, "Letter to the Editor," *Skyline* (February 1980), 2.

744 Alan Moore and Marc Miller, eds., "The Real Estate Show," in *ABC No Rio Dinero: The Story of a Lower East Side Art Gallery* (New York: ABC No Rio, 1985), 52–71; Kim Förster, "ABC No Rio: Architecture of Opposition," in *CinematoGraphies: Fictional Strategies and Visual Discourses in 1990s New York City*, eds. Günter H. Lenz, Dorothea Löbbermann, and Karl-Heinz Magister (Heidelberg: Universitätsverlag C. Winter, 2006), 97–120.

previously worked as an assistant to the late Gordon Matta-Clark and was one of the exhibition's organizers, made a sweeping accusation against all architects that they had so far only presented East Village residents "from above," with utopian designs for large-scale structures, an allusion to Paul Rudolph's design. By printing the letter, rather than commissioning a review, *Skyline* nevertheless supported Fend's call for ideas to contribute to a socially engaged architecture. Meanwhile, MacNair had begun to develop his own events outside the Institute, such as the exhibition "The Edge of Architecture 1980: Between Buildings and Bodies," which he organized at the Max Protetch Gallery in the spring of 1980 and advertised in *Skyline*.[745] By this time, his relationship with the Institute had come under lasting strain. Another incident that triggered this falling out and highlighted the power imbalance was a review of a symposium at the New York Institute for the Humanities that focused on the architecture of mental health facilities in light of the transformation of the hospital system in the United States, which was announced but never published. In his review "The Architecture of Confinement," commissioned by MacNair, NYU historian Thomas Bender also discussed Richard Meier's Bronx Development Center, which he criticized as being outdated due to changes in psychiatric practice shortly after its completion.[746] However, the text was withdrawn at the last minute, apparently at Meier's intervention, and not printed, which Bender interpreted as an act of censorship.[747] In the subsequent correspondence with a trustee, he condemned the economically driven decisions at the Institute but explicitly accepted those of MacNair as editor. Instead, he identified Meier as the main culprit, blaming him for the non-publication of his review. This incident, which went down in the Institute's archive as "The Bender Affair" and might thus serve as evidence of another, less celebratory history of the Institute, was to occupy Institute director Eisenman for more than a year and finally culminated in a rift between the Institute and the *Skyline* editor. The tense situation did not improve when a comic strip, the first of its kind, was published in the April 1980 issue under the newly introduced "Funny Pages" section.[748] In a sequence of twelve cartoons, it parodied the appearance of an architect who bore a strong resemblance to Eisenman in both appearance and demeanor, so that MacNair's clash was now being aired publicly.

In the course of these disputes, MacNair had already threatened to resign several times and finally did so in a letter to Peter Eisenman at the end of April

745 Muschamp, 1980.

746 Thomas Bender, "The Architecture of Confinement" (announced in *Skyline*, February 1980, unpublished). Thomas Bender's article was already set, see folder "The Thomas Bender Affair." Source: CCA Montréal, IAUS fonds: A.5-6.

747 Thomas Bender, letter to Armand Bartos, March 25, 1980, Source: CCA Montréal, IAUS fonds: A.5-6.

748 "R K Tecto-Comix," *Skyline* (April 1980), 18.

1980.[749] He also resigned from all the other posts he held at the Institute after six years as a Fellow, a major break in his career and a bitter loss for the Institute. Eisenman had previously offered him the post of director of public programs to keep him at the Institute. But MacNair cited various reasons, including institutional, financial, structural, and personal ones, and said he felt exploited and inadequately supported. The May/June 1980 issue of *Skyline*, for which Havemeyer III served as business consultant, James Saslow and Peter Lemos as associate editors, Margot Jacqz as managing editor, and Katherine Norment as editorial assistant, would be the last for the time being. The publication was suspended, at least temporarily, since Eisenman not only immediately sent a letter to all subscribers informing them of the suspension to avoid complaints and dissatisfaction, but he also immediately set out to find a new editor-in-chief and planned a relaunch with a professional editorial team, a commercial publisher, and secure financing. The format was too important a publicity tool for increasing the Institute's visibility to be abandoned. After the event, MacNair was engaged in the launch of two follow-up publications. First, he was involved in the creation of *Metropolis*, where he was slated to be editor-in-chief.[750] And when he was forced out here as well, abandoning the project before its first publication, he finally self-published *Express*, another architecture newspaper, in December 1980.

IAUS Exhibition Catalogues

With the expansion and professionalization of the Institute's "Exhibition Program" that began in 1978, the IAUS Exhibition Catalogues were introduced as another publication format. At Eisenman's initiative, Frampton was summarily appointed editor of the newly conceived publication series in his absence. Frampton was joined by Silvia Kolbowski as managing editor, who had initially worked at the Institute's reception desk and later assisted with the "Exhibitions Program" and the "New Wave" series. The publication not only promised to draw more attention to the Institute as a gallery space but also opened up the possibility of acquiring additional grants or donations through the catalogues to cross-fund operations.[751] Thus, beginning in the summer of 1978, Frampton and Kolbowski started collaborating on a new catalogue series with a supposedly simple concept. The publication even took on a historiographical function, as

749 Andrew MacNair, letter to Peter Eisenman, April 29, 1980. Source: CCA Montréal, IAUS fonds: A.5-6. Although MacNair had signed the letter, in our oral history interview MacNair did not confirm whether he drafted it himself or not.

750 *Skyline* can thus be seen as an indirect precursor to *Metropolis*, which first appeared in newspaper format in July 1981 and still exists.

751 By his own account, MacNair, having set up the exhibitions at the Institute since 1975 and having initially directed the program, had only produced self-made catalogues for some of the exhibitions by quickly photocopying materials from the shows and simply stapling them together. A first catalogue was already being planned for the 1976 "Idea as Model" exhibition, yet was not published until 1981.

the format was tasked with documenting the exhibitions running at the Institute for both professionals and posterity: a tangible product in contrast to the ephemeral nature of the "Exhibition Program." The catalogues were first advertised in 1978 on the poster for the "National Architecture Exchange," one of the offerings under the newly created outreach and publicity platform. They were thus another purchasable teaching and learning product produced at the Institute and distributed nationwide, along with the lecture tours, traveling exhibitions, and slide series. Advertised alongside the names of the architects exhibited were the names of the authors slated to provide introductions, in many cases Fellows or Visiting Fellows, who were listed as a mark of quality. Eight catalogues in total were offered for exhibitions that had been held in the previous three years: *The Architecture of O.M. Ungers* (with an introduction by Rem Koolhaas), *Idea as Model* (Richard Pommer), *Gwathmey/Siegel: Ten Years and Twenty-Four Houses* (Kenneth Frampton and Ulrich Franzen), *Robert Krier: Projects about Space* (Andrew MacNair), *Aldo Rossi in America, 1976, 1977, 1978* (Mario Gandelsonas), *Ivan Leonidov: Russian Constructivist, 1902–1959* (Gerrit Oorthuys), *The Princeton Beaux Arts. From Labatut to the Program of Geddes* (Anthony Vidler), and *Massimo Scolari: Architecture Between Memory and Hope* (Mario Gandelsonas).

It is noteworthy that this offer was made at a time when the catalogues had neither been issued nor published and was therefore a first step to drum up publicity and test demand, and buyers would thus have paid for them in advance. Despite their documentary nature, the IAUS Exhibition Catalogues, as a series, were clearly intended to establish a format that was distinct from *Oppositions*, with its own author base and budget plan; for the first time, the acronym "IAUS," which already graced the promotional and fundraising brochure, was now also used as a brand for one of the publications. As catalogues for past and present exhibitions of contemporary and, to a lesser degree, modernist architecture they promised to advance positions and projects of postmodernism. They also had an institutional function, as the exhibitions and the catalogue series not only depended on each other in terms of content but also built on each other for financial reasons. The concept was that each catalogue would print extensive visual materials (drawings, plans, and photographs) previously seen in the exhibition, and an introduction and further essays were planned for each: "critical and theoretical pieces that set the context for viewing architecture and express the didactic aims of the display."[752] Despite the educational goal, some catalogues ultimately took years to realize; and of the authors initially planned and already advertised, only a fraction ultimately wrote one of the planned introductions.

The catalogue series was launched before the end of 1978 with the exhibition and lecture series "A New Wave of Japanese Architecture," which was on

752 Frederieke Taylor, grant application to the NEH for a Challenge Grant, November 30, 1979 (CD-1444-81). Source: CCA Montréal, IAUS fonds: A.2-10.

view at the Institute before touring North America.[753] In formal terms, Catalogue 10, as it was officially numbered, was closely modeled on *Oppositions* in terms of format, graphics, and page layout. The catalogue series, also designed by Vignelli, once more corresponded to the Institute's graphic identity and yet, due to its cover design, could have been identified as an independent print product. Vignelli's cover design for the catalogue series, based on a three-column grid, was less obtrusive and eye-catching than that of the journal, with its fine black line drawings and a red serif font for the text against a creamy white background. In terms of content, the catalogue of the "Japanese New Wave" was a comprehensive document on contemporary architecture in Japan, consisting mainly of short programmatic texts and selected projects by eleven architects: Takefumi Aida, Tadao Ando, Hiromi Fujii, Hiroshi Hara, Osazmu Ishiyama, Arata Isozaki, Toyo Ito, Fumihiko Maki, Monta Mozuna, Minoru Takeyama, and Atelier Zo (in alphabetical order). Frampton introduced and classified the architects and their projects, and further formulated a definition of how the "Japanese New Wave" should be understood.[754] As editor, he not only positioned himself in terms of the architecture presented there in comparison to those contemporary attitudes familiar to American readers; he also wrote for the first time as an expert on Japanese architecture.[755] Frampton was clearly seeking to present the group exhibition promoted by the Institute across the country with the traveling exhibition and lecture series and shown at the Institute itself as the only true contemporary architectural avant-garde. He consistently wrote of "the New Wave" as if there was no other. For him, the young generation of Japanese architects differed fundamentally from the American generation that had dominated the early 1970s and had made architectural postmodernism acceptable in the United States—despite their dichotomous juxtaposition and media politics. The last page of Catalogue 10 once again publicized the Institute's new venture. Readers found a list announcing the publication of ten exhibition catalogues, documenting the Institute's exhibitions of 1976, 1977, and 1978.[756] While it took years to accomplish, this first exhibition catalogue—self-published and distributed by the Institute—was nonetheless a complete success. Within two months, over four hundred copies had already been sold in the USA.

753 IAUS, ed., *A New Wave of Japanese Architecture*, Catalogue 10 (New York: The Institute for Architecture and Urban Studies, 1978).

754 Kenneth Frampton, "The Japanese New Wave," in IAUS, 1978, 1–13.

755 Frampton later continued to push Japanese architecture in *Skyline* as well as in *Oppositions*. See Kenneth Frampton, "Modernism's Diffusion. Japan Diary, Summer 81, Part 1," *Skyline* (April 1982), 26–29; "Part 2," *Skyline* (May 1982), 26–29; "Part 3" *Skyline* (June 1982), 22–25. Frampton shared an interest with Eisenman in Arata Isozaki, who was a Visiting Fellow at the Institute.

756 Curiously, the numbering of the advertised IAUS Exhibition Catalogues corresponded to the chronology of the exhibition dates, but in the end did not coincide with the actual order of publication.

The Institute also negotiated with MIT Press to publish the IAUS Exhibition Catalogues in 1979. It was Taylor who, as director of development responsible for the acquisition of funds and public relations, now communicated with Reinke at the university's publishing house after the promising start of the catalogue series, which was no longer within the scope of Institute director Eisenman, and who in correspondence with Reinke praised the new series as a logical development of the already established "Exhibition Program."[757] Taylor indicated that, by January 1979, two other catalogues of past exhibitions were already being planned or in production: *Philip Johnson: Processes* and *Gwathmey/Siegel Architects: Twenty-four Residences*. Apart from the chronology of the exhibitions, these two publications were a top priority for the Institute, since Charles Gwathmey had been a trustee and President of the Institute since 1978, and Philip Johnson was a patron of the Institute. Additionally, Johnson, while having made few if any official appearances until the previous year, was about to pull the strings. More importantly, both architecture firms had already contributed to the production costs of their respective catalogues. Taylor highlighted the sales figures and pre-orders of the only title available to date to underscore the interest in the Institute's catalogues in the architecture books market. In addition, she sent a review of "A New Wave of Japanese Architecture" by Huxtable, one of the still rare interactions between the renowned architecture critic with the Institute, in which she specifically praised the care with which the editors had prepared the catalogue.[758]

Catalogue 9, *Philip Johnson: Processes*, published in the spring of 1979, then became the Institute's second catalogue, documenting the exhibition on Johnson's AT&T Building at the Institute in the fall of 1978.[759] It perpetuated the hype, the very mechanism of an attention economy that Johnson had cultivated throughout his life and that the Institute was now embracing as well. Even more than the exhibition, the catalogue recalled the controversial design for New York's first postmodern skyscraper shortly after it was commissioned, by paying tribute to the documentary role of architectural drawings and their capacity to provide insights into the design process. Much of the production budget of US$17,000 (of which Kolbowski as managing editor received US$3,000 and Frampton as editor received US$1,300) had been contributed by Johnson, signaling some degree of dependence on the part of the Institute. The catalogue, with a critical "preface" by Craig Owens, an "introduction" by Massimo Scolari, and several texts by Frampton, became an in-house test of character, especially since Frampton, in his text on the *Glass House*, did not pass up on the opportunity to criticize Johnson and his past fascist leanings, obviously not unknown at the Institute at the time, if only between the lines, whereas prior to that, *Oppositions* 10 had promoted

757 Frederieke Taylor, letter to Ann Reinke (MIT Press), January 30, 1979. Source: CCA Montréal, IAUS fonds: A.5-3 / ARCH401644.

758 Huxtable, 1979, D27.

759 IAUS, ed., *Philip Johnson: Processes. The Glass House, 1949 and The AT&T Headquarters, 1978*, Catalogue 9 (New York: The Institute for Architecture and Urban Studies, 1979).

Johnson almost uncritically and unquestioningly.[760] For Johnson, as was commonly known back then, apparently not only accompanied the invasion of Poland in the suite of German armed forces in September 1939 at the invitation of Joseph Goebbels' Ministry of Propaganda but also provided for the dissemination of Nazi propaganda in his articles and speeches. Frampton's article explicitly criticized Johnson's interior design for the utility core of his country house, which to him was reminiscent not only of ruins but of the Polish villages destroyed at the beginning of World War II.[761] Eisenman, on the other hand, took no responsibility as Institute director and did not cover for his editors, and when Johnson complained, pointed out that each of the Institute's programs was independent, as was the work of the program directors themselves.[762] He himself confined himself to vague allusions to Johnson's past in his preface to *Philip Johnson: Writings.*[763]

In 1979, the Institute managed to secure MIT Press for the IAUS Exhibition Catalogues. The decisive criterion for the further development of the series was now to select historically relevant or well-known contemporary architects with a broad appeal in order to attract public funding such as from NYSCA, the NEA, etc. to finance individual exhibitions and catalogues, as well as the operation of the Institute itself. In the spring of 1979, work was underway on the fourth title in the series, a catalogue on Ivan Leonidov, with Gerrit Oorthuys and Rem Koolhaas as potential authors. But initially, the editorial team faced financial difficulties and had to borrow money internally from other programs at the Institute, the "Exhibition Program" and the "National Architecture Exchange." Over the course of the year, Eisenman himself planned which new publications were conceivable and feasible for the Institute, including future catalogue titles (and thus, to some extent, new exhibitions). The crucial question in this regard was what financing might even be considered for possible productions; the handwritten list again testified to the patterns of thought that infused Eisenman's curation and his directing practice.[764] At the top of the list, in

760 Craig Owens, "Philip Johnson: History, Genealogy, Historicism," in IAUS, 1979, 1–11.

761 Kenneth Frampton, "The Glass House Revisited," in IAUS, 1979, 39–59, here 51.

762 Johnson's fascist past again became an issue at the Institute when, in September 1979, a staff member of MIT Press offered Johnson's early writings for publication. Source: CCA Montréal, IAUS fonds: A.5-7. However, *Oppositions* editors were obviously not interested in exposing this in their journalism.

763 Peter Eisenman, "Introduction," in *Philip Johnson: Writings,* eds. Peter Eisenman and Robert Stern (New York, Oxford University Press, 1979), 10–25. Instead, Eisenman used this sensitive information as leverage against Johnson, for instance when he conducted interviews with him in the early 1980s in which he, among other things, addressed this blind spot in Johnson's biography, threatening to make it public. Apparently, Eisenman was bought out, see Schulze, 1994, 372–376. Despite the overwhelming evidence, it is remarkable that it is only the subsequent generation of architecture historians who has studied Johnson's dissemination of fascist ideas and strategies of dealing with this past; see Varnelis, 1995.

764 Peter Eisenman, notes on IAUS Publications, 1979, Source: CCA Montréal, IAUS fonds: A.5-8 / ARCH401754.

alphabetical order, he placed a catalogue on the New York-based Austrian architect Raimund Abraham—who like himself taught at Cooper Union but had no other connection to the Institute—followed by catalogues on New York architects, notably Charles Gwathmey and John Hejduk, both longtime companions and erstwhile members of the New York Five, for which funding had already been secured. In a somewhat smaller type, Eisenman included catalogues on Rossi and Scolari, both of whom were given a second solo exhibition after 1976. The great promise associated with the series, in addition to symbolic gains for everyone: the architects, the Institute, and MIT Press, was further income from sales. In addition, Eisenman projected an exhibition and catalogue on Terragni; this was apparently one of the exhibitions he naturally considered particularly worthwhile. He also hoped to finance a retrospective including catalogue production on the glorious Texas Rangers, i.e., the group of educators around Bernard Hoesli, Colin Rowe, Robert Slutzky, and John Hejduk, who once taught at the University of Texas in the 1950s and, due to their influence, were now widely idolized, not only in New York architecture circles but internationally.[765]

The second part of Eisenman's list included exhibitions and catalogues as part of the "New Wave" series, first and foremost a "Swiss New Wave," for which he expected income from Pro-Helvetia, Swiss Air, and Swiss banks. Furthermore, young architects from Austria, France, Spain, and Argentina were to be featured by the Institute. The contract with MIT Press, similar to *Oppositions*, called for the production of four exhibition catalogues per year. When the Institute applied to the NEH in 1979 for funding to continue its public lecture series, Taylor also advertised these plans for further IAUS Exhibition Catalogues. Subsequently, the Institute organized, toured, and staged an "Austrian New Wave" in the spring of 1980, which was the first to be awarded grant money, including for a catalogue. The NEA, the Austrian Ministry for Education and the Arts and Ministry for Foreign Affairs, and the Zentralsparkasse and the Kommerzialbank Wien all contributed. Over the next few years, there was an abundance of ideas for exhibitions and catalogues, but funding remained a problem. After all, the Institute was counting on revenues of over US$100,000 for the 1979–80 fiscal year from the catalogue series alone. In 1979, the editors then invested in the production of another promising catalogue, this time on Rossi, which included the unique drawings on the analogous city he made during his stay in New York in 1976, along with an introduction by Eisenman and an exclusive text by Rossi himself.[766] The catalogue, which launched the collaboration on a coordinated exhibition at the Max Protetch Gallery in the fall of 1979, was to become a bestseller.[767] Another catalogue was to be produced for the Wallace Harrison retrospective at the Institute, planned for winter 1979–80, for which Rem Koolhaas was originally to be responsible

765 Caragonne, 1995.

766 IAUS, ed., *Aldo Rossi in America, 1976–1979*. Catalogue 2 (New York: MIT Press, 1979).

767 Kauffman, 2018, 236, 264. Not only the exhibitions but also the publications were instrumental in the expansion of the art market.

as external curator and editor. By the end of 1979, however, the Institute had only produced a total of three catalogues: in addition to *A New Wave of Japanese Architecture* and *Philip Johnson: Processes*, also Catalogue 2, *Rossi in America, 1976–1979*, the next big hype, now published for the first time by MIT Press, but not until after the exhibitions themselves. In general, the latest print products, actually conceived as catalogues for the exhibitions, were only rarely ready for the opening; in most cases, there were still texts missing. But when they were realized, they showcased the powerful interplay between culture and politics, between architecture, knowledge, and power that characterized all of the Institute's publications.

Oppositions Books

From 1978, the Institute's publication portfolio was set to be complemented by Oppositions Books as its own book series, with the English translation of Aldo Rossi's *The Architecture of the City* for the North American market as its first publication. In doing so, the translation of foreign-language books—particularly by authors from Europe—into English was intended to raise the culture of translation, which had previously been established and practiced with *Oppositions*, to a new level and turn it into the basis for an even more globalized debate on architecture. For the launch of Oppositions Books in 1978, the young architect Diane Ghirardo, who had been working intermittently as a translator of Italian texts for *Oppositions*, was commissioned to produce a new translation of Rossi's bestseller. The Italian architect, whose first monograph had already been translated into several languages, embarked on an updated introduction for a North American readership, as he had already done for other translations. A volume of essays by British architect, theorist, and historian Alan Colquhoun was also planned as the second contribution to the book series. Colquhoun, who in addition to his firm also taught at the Polytechnic of Central London and had repeatedly held visiting professorships in the United States since 1969, had two essays published in *Oppositions* 12 in 1978, both of which were indebted to his historical materialism: an architecture critique of the projects of Michael Graves and a theoretical essay on the modernist style, the relationship between form and function, and the legacy of the figurative tradition, thus revisiting the ideological discussion on "neo-realism" and "neo-rationalism" that had previously been conducted in the journal as contemporary positions.[768] By referencing the opposition between modern means of production and postmodern forms of expression, Colquhoun emphasized that "modern architecture was polemically committed to the transformation of the 'real' world."[769] Architecture, he diagnosed, had detached itself

768 Alan Colquhoun, "From Bricolage to Myth: or How to Put the Humpty Dumpty Together Again," *Oppositions* 12 (Spring 1978), 1–19. For a discussion of Colquhoun's politics, see also *Oase* 87 (2012): "Alan Colquhoun: Architect, Historian, Critic," especially in Owen Hatherley, "Two Notes on Alan Colquhoun," *Oase* 87 (2012), 87–98.

769 Alan Colquhoun, "Form and Figure," *Oppositions* 12 (Spring 1978), 28–37, here 37.

from its social role, becoming one of the arts—a role in which "'possible' and 'virtual' worlds are created." As far as the publication of his book of essays was concerned, however, Colquhoun was also negotiating with Conover at the same time, which made the matter more urgent for both parties, the Institute and MIT Press.

However, there was no budget for the Oppositions Books and only limited capacity on the part of the Fellows. When the Institute published its promotional and fundraising brochure at the end of 1978, with chapters on all the Fellows' fields of work, the Oppositions Books obviously had to be included, even though they did not de facto exist; in the brochure, they were nevertheless presented as a *fait accompli.*[770] The new book series was already announced in the introductory text about the history of the Institute, without a fixed publication date or a contract in place, let alone a definitive solution to financing the editorial work or the book production, including typesetting, printing, and distribution. In the search for funding and donations, the Institute's leadership simply declared the publication of the book series as a foregone conclusion that was certain to materialize in the future, without any planning certainty. It was not until 1979 that a more decisive approach was taken to the conception and planning of Oppositions Books. After Taylor attempted to obtain NEH funding for the translation of essays and books in the spring of 1979, without any notable success, Eisenman first presented a comprehensive concept for the further planning of *Oppositions Books* in July 1979.[771] He used the capital generated by *Oppositions* as his main argument and outlined the book series as a logical continuation, engaging with the same topics and implementing the same strategies and aims, only in a different format: "*Oppositions* began to have an effect not only in America but in Europe, beyond our most hopeful expectations." To him, the reasons were obvious: "Students and architects around the country began to talk about ideas. Other journals began carrying criticism. Theory courses began to appear in schools where previously there had been none. The *Oppositions* 'Forum,' public discussions of the journal began to be copied in other institutions along similar lines." He also emphasized the openness of the Institute: "The *Oppositions* 'Little Magazine Conference' After Modern Architecture spawned an entire historicizing tendency, post-modernism, which has become a keyword for journalists and architects alike." Finally, he derived the new aspiration for a book series at the Institute from the success of the journal: "In short, *Oppositions* has become a catalyst for a set of ideas and for discussion of architecture previously unseen in this country. It has also become an introduction to a present-day form of contemporary theory of architecture.

770 IAUS brochure, 1978. Source: CCA Montréal, IAUS fonds: C.3-2 & C.3-3. In CCA's IAUS fonds, there is no indication as to whether a contract between the Institute and MIT Press for the publication of the Oppositions Books already existed at that time.

771 Peter Eisenman, draft and concept of Oppositions Books, July 9, 1979. Source: CCA Montréal, IAUS fonds: A.5-8 / ARCH401742 & ARCH401744.

[...] But as such *Oppositions* has pointed the way for such a next step and has at the same time created its own audience for such a continuation: Oppositions Books." Following this line of reasoning, the book series was inevitable.

What was striking about Eisenman's conceptual planning was that he saw the book series as being quite similar to *Oppositions'* original purpose, although the journal had long since taken a different path. Accordingly, Oppositions Books would comprise four categories of texts: first, translations of "seminal texts unpublished in English," second, collections of "seminal texts never collected in one volume," third, previously "new texts or essays which have never been published before," and fourth, "commissioned texts which begin to explore the potentials of architecture theory and criticism." The concept paper subsequently illustrated that Eisenman's interest and motivation were the same as when *Oppositions* was founded in 1974, namely to pursue a linguistic or semiotic approach to architecture, which he viewed as a "humanistic discipline." A historiography of architectural modernism, as pursued with *Oppositions* by Frampton as well as Vidler, was not mentioned here, or at least not explicitly. In the concept, Eisenman also stated for the first time that a "critical" introduction was to be written for each title, the task of which was "to place the work into a critical matrix by locating the particular work or works in a context both of the author's other writing, to the time and place when it was written—a relationship to other significant texts. But also in relationship to the developing American context." This framing made it clear that Oppositions Books, according to the rhetoric of the concept paper, addressed a specific readership in the English-language book market, while Eisenman tried to convince potential partners, be they publishers or foundations, arguing that the Institute had already built up its target group with *Oppositions*.

Although *Oppositions* and Oppositions Books were two independent publications with separate budgets and different goals, there was some overlap in textual and editorial practice, as well as publishing. Eisenman and Frampton were responsible for the conceptualization of the book series, collaboration with MIT Press, and acquisition of funding, with Eisenman taking a more strategic approach and Frampton a more academic one. When Eisenman sat down to plan the IAUS Exhibition Catalogues in 1979, he was simultaneously outlining possible Oppositions Books titles.[772] The focus was not so much on the financial argument as on the names, i.e., the intellectual capital, of authors who had already been solicited or considered. Aldo Rossi was at the top of the list with two monographs, followed by Alan Colquhoun, Moisei Ginzburg, Soviet architect of the *Narkomfin* house and author, Colin Rowe, and Manfredo Tafuri. After Eisenman had secured the publication rights for Rossi's second publication, *A Scientific Autobiography*, for the Institute in early 1979, Rossi was fully absorbed that year and elevated to the status of key author at

772 Peter Eisenman, notes on IAUS Publications, 1979, Source: CCA Montréal, IAUS fonds: A.5-8 / ARCH401754.

Oppositions Books.[773] Eisenman's rather hurriedly scribbled list indicated that the conception of the book series focused less on the contributions to the architecture debate than on the names of the authors. He noted that he would share the work of writing the introductions and thus the responsibility for editing each title with Frampton. While Frampton subsequently—even before his own monograph *Modern Architecture. A Critical History* came out in 1980—oversaw the translations of modernist classics, Eisenman—neither of whose book-length publications *House X* nor *Giuseppe Terragni* had been published—was again, both in the book series and concurrently with the journals, primarily concerned with publishing contemporary Italian authors.[774] As instruments of branding and power politics, Oppositions Books did not differ significantly from the other publications issued at the Institute. However, the large format made it possible to focus attention on certain authors; there were to be monographs and essay collections by Fellows and friends that sought to inform the architecture debate and by extension architecture education. Frampton and Vidler, among others, were to receive their own publications.

MIT Press was ultimately won as a publishing partner for Oppositions Books, like *Oppositions*, *October*, and the IAUS Exhibition Catalogues before it. In 1979, they signed an initial agreement, providing for the publication of four titles per year. For Conover, who had been promoted to executive director at the university publishing house in 1978, this was a very attractive deal, since he still had to contract a certain number of titles each year, and the Institute provided a well-rehearsed editorial team with Eisenman and Frampton; in return, the Institute retained control over large parts of the production process. Unlike other monographs, MIT Press left the graphics, layout, and typesetting in the hands of the Institute. Vignelli was once again commissioned with the design of the book series to ensure that Oppositions Books was clearly recognizable as an Institute publication, even at first glance. This way, MIT Press could expand its focus on architecture books and significantly raise its profile in the New York architecture scene as well as the Institute's European network. In return, the university publisher agreed to pay at least part of the editorial staff's salary and production costs. Eisenman and Frampton, as editors-in-chief, each received a one-time fee of US$2,000 per title, as a bonus. In addition, MIT Press paid half the annual salary of a managing editor.

773 As early as 1973, Rossi had taken notes on both the drawings of the analogous city and the preparation of an autobiography of his projects. See Aldo Rossi, *I Quaderni azzurri* (Los Angeles: Getty Research Institute, 2000). Lobsinger, in her reading of *A Scientific Autobiography*, has pointed out that Rossi had already produced a first manuscript in 1975. See Lobsinger, 2002. Accordingly, Eisenman did not commission Rossi, but rather revisited an idea or drew on a manuscript.

774 Eisenman's first monograph, *Giuseppe Terragni. Transformations, Decompositions, Critiques*, with a foreword by Manfredo Tafuri, was already announced in the MIT Press catalogue in the fall of 1979.

At the Institute, on the other hand, the question of how it would pay for its share of the salary and production costs remained unresolved. The lack of funds was ultimately one of the reasons why the publication of the first of the Oppositions Books series was delayed for two more years. In 1979, Eisenman hired architect Lindsay Stamm Shapiro, who had previously curated exhibitions, as managing editor, without discussing this with Frampton beforehand. The plan was for Stamm Shapiro to be the only full-time employee, coordinating the translation, copy editing, and production of each title and communicating with the authors, publisher, and translators. When Ghirado's first translation of *The Architecture of the City* was available in the summer of 1979, Stamm Shapiro worked with Rossi on selecting the illustrations. At the time, she was working with Frampton on Colquhoun's collection of essays, the second publication in the series, creating footnotes and a bibliography, and researching illustrations at Columbia University's Avery Library. She also requested quotes for typesetting and printing Oppositions Books, even though neither the conceptual design of the book series nor the planning of additional titles had progressed. The editorial team was now working with a list of fourteen titles in all, the bulk of which, in addition to two translations—e.g., two commissioned works or first publications—was made up of ten essay collections, including one by Manfredo Tafuri.[775] Each title was already assigned an author who would write the introductions, with Eisenman himself taking on this responsibility for the two Rossi books. According to this list, Frampton was also slated to contribute a monograph to the series entitled *Architecture and Industrialized City*, which never materialized. But Eisenman's Terragni book no longer appeared in this context. The biggest problem was that the chronic underfunding of Oppositions Books almost forced the editors to adopt an amateurish approach. Stamm Shapiro was assigned only a small budget to acquire publication rights for targeted titles, let alone commission professional translations; for the most part, she worked with academics and especially students, who were cheaper but could not necessarily meet deadlines due to other commitments. She also earned comparatively less than her colleagues at *Oppositions* or the IAUS Exhibition Catalogues, which did not really speak in favor of the Institute as an employer.

With the conception of the book series, Eisenman and Frampton, who not only selected the authors to be celebrated but also lent legitimacy to the individual titles by either writing the introductions themselves or commissioning capable and favorably disposed authors to do so, ultimately assumed interpretative authority over approaches that were, methodologically and conceptually, more postmodern than poststructuralist (especially with the two Rossis), which they thus placed on a pedestal. Although they were responsible for the editorial and textual work on Oppositions Books, the contract with MIT Press asserted that

775 Oppositions Books, list of titles. Source: CCA Montréal, IAUS fonds: A.5-8.

final editorial control of the content and linguistic quality of all titles, including and especially the translations, ultimately fell to the publisher. Thus, by entering into the master agreement, MIT Press had not yet given the Institute general approval to print; the individual titles first had to be accepted by its editorial board. In the course of 1979, the editorial team worked hard to prepare concepts for the first batch of Oppositions Books: to request the first test translations of Ginzburg's monograph *Stil' i epokha* (1924) by Anatole Senkevitch Jr. and of the two collections of essays by Adolf Loos, *Ins Leere Gesprochen (*1897–1900) and *Trotzdem* (1900–1930) by Ernst Brandel, and then to obtain opinions from external experts.[776] Kurt Forster served as the Institute's external reviewer. Conover supervised this editorial procedure intensively, and Stamm Shapiro spoke with him on the phone almost every day. When the first four titles were presented to the editorial board in November 1979—the review of Rossi's *A Scientific Autobiography* was submitted later—all but the translations of the two Loos books were approved.[777] The university publishing house gave the Institute the go-ahead and awarded US$24,000 each to fund the production of the titles by Rossi, Colquhoun, and Ginzburg.[778] Conover had made a strong case for the Institute, even though Oppositions Books was a book series that did not necessarily promise commercial success. After all, it was he who, as head of the architecture division, had to guarantee that the book series would recoup MIT Press' expenses. In the end, only five titles were published; in addition to the two Rossi books, which were Eisenman's project, Frampton was in charge of the anthology of Colquhoun's texts, the translation of Ginzburg and eventually an anthology as a "best of" Loos were published, and while some books had already been previously accepted, or at least earmarked for publication, but were ultimately left to the Institute, the collaboration with MIT Press represented a new departure.

776 Individual texts by Moisei Ginzburg in a translation by Anatole Senkevitch Jr. and Adolf Loos in a translation by Ernst Brandel had already been planned for publication in *Oppositions* in 1975. The collaboration with Senkevitch was a guest editorial. He prepared a comprehensive concept for the translation of *Stil' i epokha*. He was also contractually assured that he would write the introduction to Oppositions Books.

777 Following the preliminary rejection of the two collections of essays by Adolf Loos by the editorial board of the MIT Press, the Institute, following Forster's recommendation, commissioned two Columbia University students, Jane Newman and John Smith, to translate it.

778 Alan Colquhoun, *Essays in Architectural Criticism* (Cambridge: MIT Press, 1981); Aldo Rossi, *A Scientific Autobiography*, trans. Lawrence Venuti (Cambridge: MIT Press, 1981); Aldo Rossi, *The Architecture of the City*, trans. Diane Ghirardo and Joan Ockman (Cambridge: MIT Press, 1982); Moisei Ginzburg, *Style and Epoch*, trans. Anatole Senkevitch. (Cambridge: MIT Press, 1983). Oppositions Books editors set the budget to produce a single title at US$ 30,000. This meant that they had to rely on an additional US$ 6,000 of the Institute's own capital per book.

4.3 Facing Increasing Bureaucratization

At the beginning of the new decade, work in the individual editorial offices of *Oppositions*, Oppositions Books, and the IAUS Exhibition Catalogues intensified. Additionally, the possible continuation of *Skyline*, which had been discontinued after Andrew MacNair's resignation, and the greater institutional involvement of *October*, which continued to be produced out of Douglas Crimp's home, required a complex publishing apparatus, subject to institutional, discursive, structural, and financial constraints. Publications were revalorized with the 1980–81 fiscal year, as the Institute faced yet another reinvention, which had been on the horizon for some time, when the NEH discontinued public funding for "Open Plan." For the first time in years, the Institute no longer had an "Evening Program," and the only series of public events was the "Exhibition Program." They financed operations primarily through revenue from architecture education, the commercially successful "Undergraduate Program," "Internship Program," and "Advanced Design Workshop." A Challenge Grant from the NEH was another source of revenue that influenced all activities, including the "Publication Program." This was a three-year grant, but it was conditional on the grant amount being matched on a one-to-three basis by private donations and other public funding. To this end, Institute director Eisenman devoted more time than in previous years to expanding its Architects' Circle and also made a compelling case for reaching out to the more financially powerful architecture establishment as well as the construction and real estate industries. Philip Johnson and Gerald Hines, who were appointed as new trustees on February 1, 1980, played an important role as intermediaries. Moreover, decisions about the future of the Institute were increasingly being made at the Century Association, a long-established society club not far away on 43rd Street that had over the years become a meeting place and power center for the New York architecture and construction world, a fact that was already being criticized in the architecture press at the time.[779]

Specifically, this reinvention led firstly to a structural transformation in the composition of the Board of Trustees and the hiring of new staff; in addition to Johnson and Hines, Douglas Banker, Eli Jacobs, Gerald McCue, Robert Meltzer, and John White were appointed trustees, as was Frederieke Taylor, who had resigned from her post as director of development in late 1979 to serve as executive director at the Lower Manhattan Cultural Council, and also retired from the Fellowship in the summer of 1980. She was replaced by Lynn Holstein. In addition, an associate director, Hamid Nouri, was hired in the fall of 1980 to handle financial operations and to serve as secretary and treasurer of the board. Secondly, the restructuring brought about a change in the Institute's collective

779 Plunz and Kaplan, 1984.

work, in terms of content, program, organization, and thus cultural policy. This meant not only a further professionalization of the Fellows' academic, journalistic, and publishing practice and an increasing bureaucratization of the editorial work but also an economization of each of the publications.[780] The focus of the Institute in 1980–81 was clearly on publications, as the main field of activity of both old and new Fellows, who were brought in for intellectually demanding and technically skilled textual and editorial work. The publishing houses involved played an important role in the production and distribution of the publications, and thus the redefinition of the culture and debate, the discipline and pedagogy of architecture. In addition to the Institute's longstanding collaboration with MIT Press, Rizzoli International now became the publisher of *Skyline*; a commercial enterprise that had previously attracted more readers with architecture monographs than with academic publications. After all, with the formalization of relations between editors, authors, translators, and publishers at the Institute, the instituted once and for all gained the upper hand over the instituting, and the formerly small productions now became professional commodities. Large parts of the Institute's activities were devoted to the acquisition of third-party funding and to major grant applications for selected exhibition and catalogue productions. With the relaunch of *Skyline* in 1981 it became clear that, next to its news, hype, and gossip function as a tabloid, the newspaper now had to assume an institutional function as a tool for acquisition. Institute director Eisenman's main focus, however, remained on *Oppositions* and Oppositions Books, because the impact and prestige of the two formats were particularly high, and because MIT Press now insisted on fulfillment of all contracts.

Fulfilling the Contracts

Since signing the contract with MIT Press in 1976, *Oppositions* editors had never really stuck to the agreed four issues per year. The practice of misdating issues could no longer hide this fact. The Institute, the editors, and the authors had benefitted from the fact that the journal was produced at the expense of the university publisher. For MIT Press, on the other hand, *Oppositions* continued to be a prestigious but increasingly costly investment. In order to restore some regularity to *Oppositions* and to reduce the debt accumulated by the delays at MIT Press, six issues were planned simultaneously in the spring of 1980. This immense undertaking was not helped by the fact that Eisenman also founded his own office with Jaquelin Robertson as a partner that year, as did Mario Gandelsonas and Diana Agrest. Nevertheless, *Oppositions* 15 to 20 were to appear within a year. Frampton

780 Ockman blamed the Institute's post-1980 development, which she summarized as "its bureaucratization, its cultivation as a fashionable salon and power base in New York, and its solicitation of mainstream patronage," for the decline of *Oppositions*, see Ockman, 1988, 199. There were indeed "internal and external transformations in the cultural climate," but the first indication became apparent, if not obvious, as early as the mid-1970s.

in particular, who had finally published *Modern Architecture. A Critical History* in 1980 after ten years of work, again assumed a central role with the conceptual design of two double issues, *Oppositions* 15/16 and 19/20 on Le Corbusier.[781] In these two issues, each of which was the size of a book, Frampton published the results of his many years of research, in order to, on the one hand, rehabilitate the French-Swiss master architect as a historical figure and one of the protagonists of European modernism and, on the other hand, formulate a critique of historicism and postmodernism by historicizing modernist construction. However, Frampton's own two essays "Le Corbusier and L'Esprit Nouveau" and "The Rise and Fall of the Radiant City" could be read not necessarily as a continuation of the polemic that characterized the architecture culture in the 1970s, but as a critical examination of the legacy of white architecture by addressing Le Corbusier's urban designs, intentions, influences, and conditions, and placing them in the larger contexts of the time.[782] Frampton's essays were printed along with contributions by other Fellows, including Eisenman on the *Maison Domino* and Forster on *Maison La Roche* and *Maison Jeanneret*.[783] The two double issues also provided opportunities for young scholars to publish recent research on Le Corbusier's life and work; e.g., Mary McLeod, who received her PhD on Le Corbusier from Princeton.[784] Featuring a list of authors, *Oppositions* heralded a new phase of Le Corbusier reception in the English-speaking architecture world, one that was topical and comprehensive and, simultaneously, responded to several monographic publications that had appeared since his death in 1965.[785] In addition, *Oppositions* 15/16 and *Oppositions* 19/20 published a range of documents, a text by Le Corbusier on the Weissenhofsiedlung in Stuttgart, a bibliography, and a review (the translations were by Joan Ockman and Christian Hubert).

Meanwhile, *Oppositions* 17, originally conceived by Gandelsonas as a special issue on "Architecture and Language," was more concerned with the Frankfurt School's critique of functionalism, and the social and economic conditions of modern architecture, art, and music. In his editorial "After Historicism," Vidler reflected on contemporary historiographical approaches. Gandelsonas, in "From Structure to Subject," one of his rare textual contributions to the journal,

781 Ockman referred to Frampton as the *Oppositions* "work horse;" see Ockman, 1988, 185.

782 Kenneth Frampton, "Le Corbusier and L'Esprit Nouveau," *Oppositions* 15/16 (Winter/Spring 1979), 13–58; "The Rise and Fall of the Radiant City," *Oppositions* 19/20 (Winter/Spring 1980), 2–25.

783 Peter Eisenman, "Maison Dom-ino," *Oppositions* 15/16 (Winter/Spring 1979), 119–128; see Kurt Forster, "Antiquity and Modernity in the La Roche-Jeanneret House of 1923," *Oppositions* 15/16 Winter/Spring 1979), 131–153.

784 Mary McLeod, "Le Corbusier and Algiers," *Oppositions* 19/20 (Winter/Spring 1980), 53–85.

785 Maurice Besset, *Qui était Le Corbusier?* (Geneva: Skira, 1968); Stanislaus von Moos, *Le Corbusier – Elements of a Synthesis* (Rotterdam: nai010 publisher, [1968] 2009); Charles Jencks, *Le Corbusier and the Tragic View of Architecture* (Cambridge: Harvard University Press, 1974).

addressed Eisenman's designs for *House VI* and *House X*, and expressed a critique of formalism derived from his analysis of precisely these structuralist approaches.[786] The "Theory" section was comprehensive, with an English translation of Theodor W. Adorno's seminal lecture "Functionalism Today" delivered at the Berlin Academy of the Arts on the occasion of the Werkbund Day of 1965, a postscript by Roberto Masiero, and a reprint of a discussion with Adorno from *Werk und Zeit*, the Werkbund's publication, as well as a historical text by Ernst Bloch on "Formative Education, Engineering Form and Ornament;" the "History" section included Tafuri's essay "The Historical Project" and a text by Oriol Bohigas on "Satoris. The First Classical of the Avant-Garde," which underscored historiographical tendencies. *Oppositions* 18 was a regular issue with no real thematic focus, with texts by William Ellis on "Type and Context in Urbanism. Colin Rowe's Contextualism," Christian Norberg-Schulz on "Kahn, Heidegger and the Language of Architecture," Elaine Hochman on "Confrontation: 1933; Mies van der Rohe and the Third Reich," and documents by Rudolph Schindler (introduced by Stephans Polyzoides) and Le Corbusier (Ivan Žaknić). Eisenman himself was responsible for the conceptual design of *Oppositions* 21, which again had a strong Italian focus with contributions by Giorgio Grassi and Massimo Cacciari of IUAV, and a text by Daniel Libeskind on Aldo Rossi's *Teatro del Mondo*.[787] Working on several issues at the same time posed logistical problems for the entire editorial team and for Bloomfield in particular; she was assigned two assistant managing editors, Jill Silverman and Kate Norment, to help her handle this extra workload. The race to catch up with the publishing backlog became tense when the production of the two double issues on Le Corbusier progressed further than that of *Oppositions* 17 and 18. In the end, this difficult undertaking, the schedule that had originally been agreed with MIT Press for all publications, was not entirely successful. The historiographical turn consequently resulted from the availability and commitment of *Oppositions* editors.

In 1980, Frampton increasingly began working with Silvia Kolbowski on the IAUS Exhibition Catalogues, which were to prove the Institute's most commercially successful print product. Again, the main concern was to meet the agreed four catalogues per year. Catalogue 7 on Gwathmey/Siegel Architects, honoring Charles Gwathmey's new role on the Board of Trustees, was the first to appear in 1980 and was partly financed by the architecture firm itself.[788]

786 Gandelsonas's essay was actually first published in *Architecture + Urbanism*; see Gandelsonas, 1979.

787 Frampton edited the essays by Thomas Hines on Richard Neutra and Stanford Anderson on Peter Behrens for *Oppositions* 21 in the spring of 1980. Later, they also planned a contribution by Vittorio Magnago Lampugnani, "Die eigenwillige Muse," but this had to be considerably revised and was subsequently not included at all.

788 IAUS, ed., *Five Houses. Gwathmey/Siegel Architects.* Catalogue 7 (New York: MIT Press, 1980).

Catalogue 1 on Massimo Scolari, also published in collaboration with the Max Protetch Gallery, for the first time explicitly indicated on the cover that the series was distributed by MIT Press.[789] In May 1980, Catalogue 13 on *A New Wave of Austrian Architecture* was published, comparatively close in time to the exhibition that was shown at the Institute in the spring of 1980 and was still touring the United States at the time of publication.[790] The exhibition and catalogue presented six Vienna-based architects and artists (Missing Link, Hermann Czech, Heinz Frank, Appelt-Kneissel-Prochazka, Heinz Tesar, and Rob Krier) each with several projects, texts, and biographies, introduced with essays by Friedrich Achleitner and Rudolf Kohoutek. Unlike other productions, with the exception of design and printing, all editing and text production had been carried out in Austria, with the Institute acting as a publishing house. Strikingly, it was the first catalogue in the series to feature a new cover design: the fine serif typeface of the titles had been replaced by large, bold letters in partly loud, partly pastel tones, giving the series a distinctly postmodern aesthetic. In addition, the fine line drawings inside which had previously characterized the series had now been replaced by color illustrations, drawings, and photographs. The new graphics also revealed a new strategy on the part of the Institute, and Vignelli was now more intent on putting his own recognizable stamp on each of the productions. A letter-size flyer was being produced to advertise a total of eight IAUS Exhibition Catalogues, but there were repeated delays. For example, Catalogue 11 on Wallace Harrison was postponed until further notice due to a lack of funds. After more than a year, Frampton and Kolbowski were still struggling with structural problems, but also with individual capacity, as very different priorities were set. With Catalogue 12 on John Hejduk, another catalogue had not been ready for the exhibition because Eisenman had not managed to deliver his introduction on time. Again, Kolbowski complained to Eisenman as author and Institute director about the resulting delay. The Institute's "Publication Program," with five different formats that required coordination of different schedules and diverse contributions from Fellows and external authors, was in danger of failing because of the discrepancy between aspirations and reality.

In 1980, after three years of planning and preparation, Eisenman and Frampton ultimately took the final steps towards launching Oppositions Books with Rossi's *The Architecture of the City*, the translation of which was largely completed, as a top priority; by now, MIT Press had invested a great deal of time and money in this book project. At the same time, the contract for *A Scientific Autobiography* had been signed in March 1980 and Lawrence Venuti of the Translation Center at Columbia University, who had done a test translation, had

789 IAUS, ed., *Massimo Scolari. Architecture. Between Memory and Hope.* Catalogue 1 (New York: MIT Press, 1980).

790 IAUS, ed., *Austrian New Wave.* Catalogue 13 (New York: The Institute for Architecture and Urban Studies, 1980).

already been commissioned to translate it. According to the contract, Rossi left the English-language publication rights to the two partners and also undertook to write an introduction to the planned Loos book. In a concept paper, Institute director Eisenman projected that during the 1980–81 fiscal year, work would be carried out on five Oppositions Books titles simultaneously.[791] At the same time, the editors pressed ahead with the conceptual work for other titles: for example, they commissioned the translation of the *Selected Writings of Arata Isozaki* and acquired the publication rights for texts by Theo van Doesburg. Another key book of the Oppositions Books series was Tafuri's *The Sphere and the Labyrinth*, a test translation for which was being prepared by Robert Connolly and Pellegrino D'Arcierno. Stamm Shapiro also sought and acquired the translation rights to this book, which contained texts on the modern avant-garde and its contemporary epigones, published in Italian by Einaudi, for the lump sum of US$900.

Then, in the fall of 1980, Tafuri's essay "The Historical Project" appeared in *Oppositions* 17.[792] In this text, which was actually the English translation of his introduction to *The Sphere and the Labyrinth*, presented here in the "History" section, Tafuri introduced a metatheoretical analysis of architecture discourse and institutions and, starting from a semiotic-linguistic interpretation of "architecture, language, techniques, institutions, historical space" committed to deconstruction, reflected at length on the question of "labor" in relation to "architectural writing." [793] "The historian is a worker 'in the plural,' like the subjects on which he labors."[794] He contrasted "operative criticism" that places itself in the service of the profession, as practiced at the Institute, with his own approach to critical historiography, with which, following poststructuralist theories, he focused primarily on an analysis of power.[795] Written as an introductory text, Tafuri called for the expansion of the "critical field," with architecture criticism now being called upon to start at another level of scale "from the analysis of the architectural object to the criticism of the global contexts that condition its configuration." The essay thus concluded Tafuri's own historiographical project of the 1970s, which had consisted of writing a history of "intellectual labor" of the professions of architecture and urban studies.[796] On the one hand, he was concerned with processes underlying the concrete material object and on the other, with the reception

791 IAUS, projected titles of Oppositions Books for 1980/81, n.d. Source: CCA Montréal, IAUS fonds: A.1-4.

792 Manfredo Tafuri, "The Historical Project," trans. Diane Ghirardo and Stephen Sartarelli, *Oppositions* 17, (Summer 1979), 55–75.

793 Ibid., 56.

794 Ibid., 66.

795 Ibid., 69.

796 Ibid., 71.

of buildings as final products. Reflecting on the contemporary practice of historians and theorists, this essay also represented the endpoint of Tafuri's engagement with Eisenman and the Institute that by now claimed a huge part of his publications in English. In addition to Tafuri, it was also thanks to Eisenman's still-intense Italian connections, especially to the IUAV, that two more monographs, by Massimo Cacciari and Giorgio Grassi, were being discussed for Oppositions Books in 1980, alongside their contributions in *Oppositions*, which had already been firmly scheduled.[797] Eisenman had contacted Francesco Dal Co to assist him in selecting the texts and as a potential author of the introductions. The idea was for Italian authors, more than from any other intellectual and cultural context, to become a figurehead of this book series; for them, the offer of a fully funded, high-quality English translation through the Institute and publication by a prestigious American university publisher must have been extremely attractive. However, Stamm Shapiro was often left almost entirely to her own devices. It was timely, therefore, that Joan Ockman graduated from Cooper Union in the summer of 1980 and not only rejoined the *Oppositions* editorial team but also took on a lot of work on Oppositions Books, initially handling the difficult translation of the Rossi book before becoming even more involved in the entire production in 1981.

Relaunch

While driving the Institute's publishing offensive forward with wholehearted personal enthusiasm in 1980, Institute director Eisenman also made *Skyline*'s relaunch a top priority. Only a short time after publication had been temporarily suspended following MacNair's resignation, he sought to continue the newspaper with a new, professional editorial staff, with Rizzoli as commercial publisher, and, above all, with secured financing. In June, in a memorandum to the trustees, he called *Skyline* "potentially" the Institute's "most important publication."[798] Obviously, Eisenman needed their support for a relaunch. He even linked the question of whether the publication should be continued to the very purpose of the Institute since "to give it up without exploring all options would seem to me to defeat the reason for the existence of the Institute." At a time when the Institute was facing profound institutional, financial, personnel, and programmatic transformations and needed not only to raise private funds but also to moderate a generational shift, Eisenman personally championed a

797 Cacciari had proposed his two books, *Metropolis* (1973) and *Oikos* (1975), for an English translation. In late 1980, Oppositions Books editors discussed using essays from *Oikos* and *Dallo Steinhof* and an article on Wittgenstein. A translation of *Metropolis* was published under the title *Architecture and Nihilism. On the Philosophy of Modern Architecture* by Yale University Press in 1996. A long version of the essay "Avantgarde and Continuity" by Grassi was initially under discussion. Later it was proposed to acquire the publication rights to *The Logical Construction of Architecture* (1967), but MIT Press was more interested in Grassi's more recent books.

798 Peter Eisenman, memo to the trustees regarding *Skyline*, June 9, 1980. Source: CCA Montréal, IAUS fonds: A.5-6.

relaunch of *Skyline*. The publication of *Skyline* was to be financed by donations from larger architecture firms, some of which already had philanthropic ties to the Institute through the Architects' Circle, as well as successful contractors, who in return were to be featured in interviews in addition to being named in the masthead. Eisenman was aware that the Institute would not initially make a profit on *Skyline*. He was, however, determined not to let the 1980–81 NEA grant lapse. Once again, Eisenman received support from Philip Johnson, who provided many contacts to sponsors, as well as his own name and financial resources, for a relaunch.

In his search for a new editor-in-chief, Eisenman turned to Suzanne Stephens on the recommendation of Robert Stern. Compared to other candidates, Stephens stood out because she had worked as an editor in the 1970s, first for *Architectural Forum* and then for *Progressive Architecture*. Importantly, she also had a keen insight into national and local building, cultural, and media politics, while maintaining a critical distance from New York architecture circles, including the Institute. When she agreed to take on the role of editor-in-chief in August 1980 after brief but very focused negotiations, she tied this to a number of conditions, including the hiring of an executive director.[799] In addition, Stephens secured the suspension of payment of the 40% overhead to IAUS Central for *Skyline* until further notice, as this was the only way for her to achieve financial independence for the newspaper. In making these demands, she demonstrated not only a strong sense of the Institute's politics but also negotiating skills, demanding, for example, a fixed annual salary for herself for a two-year period, as well as salaries for a managing editor, an assistant editor, and a copy editor, ultimately earning more than she had in her previous job. For the architecture newspaper to achieve a professional standard, she believed that this should be expressed in content and form, as well as in circulation, reach, and scope. Margot Jacqz, who remained managing editor, then took care of communications with subscribers and advertisers; after all, there were still almost 1,500 subscriptions to *Skyline*, including more than one hundred abroad. Even before she signed her contract, Stephens was already working on a budget plan and a new concept for the newspaper. At the time, Eisenman repeatedly emphasized *Skyline*'s function and its importance for the Institute in his concept papers.[800] For example, the Institute's publications had aroused interest in a new kind of architecture journalism, as *Skyline* had successfully communicated to a broader public. Eisenman displayed even more of an economic mindset as a publicist than before, in terms of attention and monetarization. *Skyline* was an ideal means for him to market architectural knowledge

799 Suzanne Stephens, letter to Peter Eisenman, August 26, 1980. Source: CCA Montréal, IAUS fonds: A.5-2.

800 Peter Eisenman, "Why the Institute?" & "Why Skyline?," n.d. Source: CCA Montréal, IAUS fonds: A.5-6.

and power, to present "a specifically New York view" of architecture culture. He also imagined how distressed New York architects would be if they were not included in *Skyline* and thus not the center of public interest. To Stephens, he made it clear that envy was a central mechanism that was to govern the editorial policy of the architecture newspaper in the future.[801] For fiscal year 1980–81, the Institute projected a budget of US$125,000 for *Skyline* alone, which included sales revenues and grants from private and public foundations.[802]

Renegotiations

At the same time, beginning in the fall of 1980, the Institute's new associate director Hamid Nouri entered into negotiations with MIT Press for Rizzoli to publish *Oppositions*; the university publisher responded calmly and agreed to let its commercial competitor handle distribution. However, working relations between MIT Press and the Institute were subsequently so badly damaged that the publisher made its first financial demands on the Institute in October 1980. Over the course of the winter, a long correspondence ensued. In a letter personally addressed to Eisenman, Frank Urbanowski, the head of MIT Press, revealed that they had invested more in the Institute in the past than in any other publication project.[803] "In total," he recounted, "there is approximately US$125,000 of MIT Press money currently advanced to support a combination of IAUS projects, including *Oppositions*, the catalogues, and the book series." The university publisher insisted on repayment of at least half of the debt. In addition, MIT Press pressed for fulfillment of the existing contract for another volume of *Oppositions* 21 through 24, especially since continued subscriptions provided a lucrative source of funds to cover the expenses already incurred.[804] Ultimately, MIT Press was pulling the plug on the project with these demands and declared *Oppositions* over after only six volumes.

Meanwhile, production of Oppositions Books continued to be slow; still, no books had been published in collaboration with MIT Press. At an editorial meeting in the fall of 1980, it became obvious that the editorial staff was simply overworked. After all, no one at the Institute except Frampton had experience in book production. The first translation of *The Architecture of the City* proved problematic and continued to be extremely time-consuming; there was also no money available for a new translation. An attempt was therefore made to publish the Colquhoun book first in the series. As the Institute continued its efforts to raise funds for Oppositions Books, Eisenman noted that Stamm Shapiro's salary

801 In our oral history interview, Stephens spoke about being tasked by Eisenman to produce envy.

802 IAUS, financial requirements of individual programs for fiscal year 1980/81, n.d. Source: CCA Montréal, IAUS fonds: A.2-11.

803 Frank Urbanowski, letter to Peter Eisenman, November 25, 1980. Source: CCA Montréal, IAUS fonds: A.5-7.

804 Ann Reinke, letter to Gianfranco Monacelli, December 12, 1980. Source: CCA Montréal, IAUS fonds: A.5-7.

was lower than that of the other managing editors, and so her annual salary was raised to US$18,000. To make matters worse, in the late fall of 1980 MIT Press refused to pay any more advances before the first title in the book series was published. Rossi's *A Scientific Autobiography* was still awaiting approval from the MIT Press' editorial board. Finally, a path opened up for Oppositions Books when the Institute submitted the book series to the Graham Foundation in Chicago on December 15, 1980.[805] Stamm Shapiro had compiled a comprehensive document for a grant application with texts by Eisenman and Frampton, which included a concept, a new list of now eleven titles, the current status of each title's editing, a timetable, a budget, summaries of four publications (Colquhoun, Rossi, Ginzburg, Loos), and detailed CVs of the two editors as qualifications. Here, Rossi's *The Architecture of the City* continued to be listed first, with the publication of an English translation fifteen years after the original described as long overdue. Attached to the book series proposal was a first draft cover for the Rossi book, prepared by Vignelli, which showed the Graham Foundation that they would be named exclusively in the imprimatur as a sponsor. MIT Press, on the other hand, was not mentioned as a publishing partner at this point; there had been negotiations with other publishers despite the existing contract.

With rhetoric that was both ambitious and lofty, the Institute sought to underscore the book series' eligibility for funding. The proposal stated that "Oppositions Books will function as a Great Books course, library, syllabus, and bibliography for the professional and the student."[806] Oppositions Books editors confidently claimed that they would produce future classics that would form a new compulsory canon of theory and history books, a task that in the United States had previously been undertaken primarily by the George Braziller publishing company, albeit with a different approach and focus; for example, with volumes on art history by Meyer Shapiro, on the architecture of modern masters such as Alvar Aalto, on contemporary positions such as Richard Buckminster Fuller or Oscar Niemeyer, on urban planning in specific periods or regions, or on the national architectural production of each decade. The editors specifically highlighted two titles, "the meditative autobiography and canonic study on the relationship of architecture and the city of Aldo Rossi, the rigorous and severe historical etudes of Manfredo Tafuri."[807] At the same time, they again argued that with this series, as with *Oppositions*, the Institute would provide teaching content for the newly created master's degree and doctoral programs at North American universities and that the central question of the target group was thus as good as resolved. "At the same time when burgeoning architecture history and theory

805 Peter Eisenman and Kenneth Frampton, "A Proposal for the Support of Oppositions Books," submitted to the Graham Foundation, December 15, 1980. Source: CCA Montréal, IAUS fonds: A.5-8.

806 Ibid., 4.

807 Ibid., 3.

courses throughout the nation require essential primary texts concerned with the theory and cultural history of architecture, this series has a ready-made and developing audience."[808] With its grant application to the Graham Foundation in the early 1980s, the Institute grandiosely portrayed itself as a discourse leader in the English-speaking, if not Western world, and also justified this move by citing university teaching aimed at the classics, especially at the University of Chicago and Columbia University. In his personally signed letter to Carter Manny, the director of the Graham Foundation, Eisenman used the example of Rossi's *A Scientific Autobiography* to emphasize the eligibility of commissioned publications and first publications and held out the future prospect of commissioning not only positions from Europe but increasingly exclusive publications by American architects. As examples, he listed Philip Johnson, Robert Venturi, Robert Stern, John Hejduk, and Michael Graves; all of them well-known and popular postmodernists whose publications should also be promising.[809] In his letter, Eisenman literally ingratiated himself to Manny by expounding on the great historical significance the book series would one day have: "Fifty years from now all historians will, I think, certainly appreciate your generosity and recognize this very natural partnership." When the Institute received a full grant just a month later, in mid-January 1981, Manny's response was equally rhetorical in emphasizing that *Oppositions Books* would become an extremely important project for architecture education.[810]

Publishing at the Institute took on a different status, despite the contract negotiations. With his lists of authors and titles, Eisenman knew how to harness and orchestrate the diverse capital of the networks attached to the Institute. The opportunities for editors, Fellows, and Visiting Fellows, as well as outside authors to publish books or journal articles resulted in a complex system of merits and awards for maintaining and creating commitments and connections. Thanks to his charisma, Eisenman also maintained the Institute's network of publishers and foundations. When Frampton was appointed director of publications in late 1980—a post created especially for him—Eisenman was rewarding him for his faith in the Institute and his loyalty.[811] Frampton was recompensed for his immense contribution to almost all of the Institute's publications, *Oppositions*, the IAUS Exhibition Catalogues, and the Oppositions Books; the

808 Ibid., 3.

809 Peter Eisenman, letter to Carter Manny, December 15, 1980. Source: CCA Montréal, IAUS fonds: A.5-8.

810 Carter Manny, letter to Peter Eisenman, January 16, 1981. Source: CCA Montréal, IAUS fonds: A.3-19. The Graham Foundation disbursed the seed capital in two installments, at the beginning of 1981 and of 1982.

811 Peter Eisenman, letter to Kenneth Frampton, December 1, 1980. Source: CCA Montréal, IAUS fonds: A.2-10.

expansion of the publication apparatus in the preceding years would have been impossible without him. Hence, it was more than justified that this position brought him an additional salary of US$12,500 annually from January 1981 as compensation. At the same time, however, the new position obliged Frampton to live up to the trust placed in him in the future. One advantage was certainly that his authority as a person and as an editor was recognized by all the managing editors active at the Institute.

Yet, despite the engagement and support of female staff, work at the Institute was still dominated by a hierarchical if not patriarchal structure that not only reflected North American society in the 1970s and 1980s but also represented gender roles in the field of architecture and in particular the male-dominated building world.[812] As the Institute's publication apparatus expanded, many women were now working on the editorial teams as managing editors and thus permanent staff, including Bloomfield at *Oppositions*, Jacqz at *Skyline*, Kolbowski at IAUS Exhibition Catalogues, and Stamm Shapiro and Ockman at Oppositions Books.[813] As an act of recognition, female editorial staff were gradually appointed to the rank of Fellows beginning in the 1980–81 fiscal year. However, in addition to Stephens as now editor-in-chief of *Skyline*, only two other women, Krauss and Michelson, had served on *October*'s editorial board from the beginning. As a result, few women at the Institute were given the chance to take on a truly senior position, and it was very difficult for the junior staff to rise in the Institute's power hierarchy. The women did provide quality editing and writing, but the ideas, concepts, and contacts came mostly from the male editors. Without the commitment of the female managing editors, however, the Institute would not have lived up to its former reputation as a think tank and current role in architecture discourse, and the tangible legacy in the form of publications would have been far less.

From the perspective of a cultural critique of the Institute's publishing activities and a sociology of architecture culture, its textual and editorial practices, which attest to the transformation of the culture industry in the field of architecture, especially the publishing market in North America, all formats produced at the Institute were transformed into commodities under Institute director Eisenman. As with the Institute's work on research and architectural projects, teaching, and cultural production in the broader sense, the long-practiced pragmatism was also apparent in the publications of the early 1980s. For the success of a publication was measured by the Institute's leadership not so much in terms of its reception and impact on architecture debate and education, but ultimately in terms of

812 Several women at the Institute initially worked for Richard Meier's firm before moving to the Institute, Julia Bloomfield at the front desk and Joan Ockman as an intern after graduation.

813 In an oral history interview, Eisenman boasts that there were numerous women working at the Institute.

the financial viability of its production and whether it yielded economic profits, whether each format could sustain itself in the expanding and yet competitive North American book and journal market. The situation worsened when Michael Leonard, MIT Press' deputy general manager, communicated to Eisenman in early 1981 that the university publisher was no longer willing to act as the Institute's bank for the publication of *Oppositions*. A repayment of the losses, as contractually agreed, was finally arranged through the Institute's lawyers. The debt, which now totaled over US$80,000, was to be repaid in two installments in 1982 and 1983. The somewhat one-sided collaboration, which had given the Institute enormous freedom, was thus terminated by MIT Press for economic reasons. The Institute now had to atone for years of mismanagement and budget deficits.

4.4 Embracing Commercial Benefits

The Institute had been trying for some time to move all its publications to Rizzoli International, which escalated matters. Eisenman expressed dissatisfaction towards the academic publisher about the low circulation of *Oppositions*—MIT Press had printed only 2,900 copies, half of which were for subscribers—and the unsolved distribution problems, as the journal was only irregularly marketed in Europe via de Boer, although there were separate agreements for England, France, Italy, and Japan. He had been friends with Rizzoli's director Gianfranco Monacelli for some time. The art book publisher, an American offshoot of the renowned traditional Italian publishing house from Milan, founded in 1927, which had maintained a second office in New York since 1964, had until then built its share with architecture monographs rather than academic journals. Nevertheless, the international distribution of *Oppositions* was of interest to the commercial publisher, so in September 1980 Monacelli made the Institute an offer to buy a print run of 6,000 copies of the journal per issue at a fixed price of US$25,000. This would have meant regular income. Eisenman also negotiated with Rizzoli about the Institute's other publications. Over the summer of 1981, he also negotiated with Monacelli about *Skyline* and finally managed to get Rizzoli to make an offer for the architecture newspaper too, according to which they would finance half of the production and, in addition, advertising and distribution.[814] The extent to which Eisenman once again combined his personal interests with those of the Institute was evident from the fact that, parallel to the institutional negotiations, he also succeeded in placing his own books with Rizzoli: first a publication on *House X* in 1982, and then his book on *Giuseppe Terragni*, albeit not until 2003.[815]

814 Gianfranco Monacelli, letter to Peter Eisenman, September 30, 1980. Source: CCA Montréal, IAUS fonds: A.5-6.

815 Peter Eisenman, *House X* (New York: Rizzoli International, 1982); *Giuseppe Terragni. Transformations, Decompositions, Critiques* (New York: Monacelli Press, 2003).

Although Eisenman threatened the existence of *Oppositions*, the Institute was now in a better position to renegotiate with MIT Press, since it had in Rizzoli a potential, financially strong partner as back-up. In the process, the conflicted and dispersed capital that took both embodied, objectified, and institutionalized forms at the Institute was reproduced in this dispute with MIT Press, and legitimized precarious cultural production for years. As a result, the Institute had to stay with MIT Press for *Oppositions* and Oppositions Books for the time being, at least until the contracts were fulfilled and its debts were paid.

In 1981, however, Rizzoli International initially took over the distribution of the IAUS Exhibition Catalogues and was eventually selected as the new publisher for the relaunch of *Skyline* that same year. The contract for the catalogues, which came into effect on January 1, 1981, covered the production of six catalogues per year, three in the spring and three in the fall, with a production cost of US$5,000 and an editorial salary of US$9,000 for each catalogue. Although this made the production of the catalogues a regular source of income, the Institute took a large risk by tying the catalogue series to the "Exhibition Program." The concept remained the same, namely that the catalogues would document historical, contemporary, and—new to the program—projected exhibitions in an increased print run of 3,000 to 4,000 copies. The first publication to be published by Rizzoli was Catalogue 14 on "Le Corbusier's Firminy Church," which accompanied the double exhibition curated by José Oubrerie at the Institute and Cooper Union in April 1981. The exhibition and the catalogue supported a kind of architectural fundraising at the Institute, as French architect Oubrerie was responsible for the completion of the extraordinary building in Firminy, France. *Oppositions* 19/20, Frampton's lavish double issue on Le Corbusier, was also published to coincide with the opening of the exhibition. With a total of seven issues in the 1980–81 fiscal year, *Oppositions* largely succeeded in fulfilling their contractual obligations to MIT Press and, for the first time, even made a small profit on the journal. The editors were optimistic and planned more individual editorials and contributions of their own for the next issues. In addition, the editorial team was rejuvenated when, after the departure of William Ellis, Alan Plattus took over the "Reviews, Letters, Forum" section, with book reviews again playing a larger role in the future.[816] With regular reviews of magazines and journals from Europe (*A.M.C.*, *Lotus*, *Rassegna*), *Oppositions* was to take on a somewhat different focus, position itself proactively, and assert itself as a central print medium. In addition, there were also more and more reviews of magazines and journals that had been produced at American universities in the meantime, such as *Harvard Architecture Review*, *Modulus*, *Perspecta*, and *VIA*. The production of Oppositions Books also took off in the spring of 1981. After the bold new layout

816 IAUS, book reviews in production, March 1981. Source: CCA Montréal, IAUS fonds: A.5-7.

for the catalogue series, Vignelli was now working on the graphic design for the book series, which, with its square format and gridded layout, borrowed heavily from *Oppositions* and the catalogues. The graphics had to be submitted to the publisher, although the contract actually stipulated that MIT Press would hand over the design completely to the Institute and only assume the costs and responsibility for printing. Significantly, Eisenman ended up fighting with Conover over the imprimatur, and the Institute once again overrode its partner by defining the collaboration as a hierarchical relationship and demoting the university publisher to a "publishing service."[817] But the new contract situation with Rizzoli soon proved as problematic as the one with MIT Press before it. As early as the spring of 1981, the new publisher refused to pay the agreed advances for the catalogues until Eisenman had delivered the artwork for his monograph.

One important publication for both the Institute and Eisenman was, however, Catalogue 3, "Idea as Model. 22 Architects 1976/80," which was published by Rizzoli in the summer of 1981, almost five years after the exhibition.[818] Eisenman had personally championed the publication. Unlike many others in the series, Catalogue 3 was ultimately more than an exhibition catalogue and featured photographs of not only the models and sculptures shown at the Institute in the winter of 1976–77 but also of newer ones, some of which had been made especially for the publication in 1980 by the architects involved at the time. However, the development of architectural thinking and design exhibited here is not the only reason why Catalogue 3 was less of a documentation and more of a document of both a history of ideas and the Institute. With an introduction by Christian Hubert, a young Fellow who was involved with the editorial team of the catalogue series, about the tasks performed by the architectural model in general and two texts by art critic Richard Pommer, an early review, written immediately after the exhibition, discussing the actual contributions, even rivaling conceptions of art and architecture vis-à-vis social responsibility, and an essay debating the profound changes in the art market with regard to the commercial uses of architectural models rather than their design purposes, the catalogue also described the reception history of "Idea as Model" over the five-year period and thus its cultural significance.[819] One striking aspect is that Eisenman's conceptually reasoned approach toward an autonomous architecture now took on a central role in the publication, in contrast to the exhibition, in which all models had been displayed side by side on an equal footing. Eisenman claimed authorship for the exhibition idea in the preface, which was ostensibly about the competition of ideas through models. In addition, the catalogue now also featured an interview

817 Peter Eisenman, letter to Roger Conover, April 21, 1981. Source: CCA Montréal, IAUS fonds: A.5-8.

818 IAUS, ed., *Idea as Model: 22 Architects 1976/1980*, Catalogue 3 (New York: Rizzoli International, 1981).

819 Pommer, 1981; see also Christian Hubert, "The Ruins of Representation," in IAUS, 1981, 17–27.

with him about his ideas on the function of architectural models, questions of representation, scale, and the relationship between models and reality.[820]

In 1981, in response to the new publishing regime, which called for the expansion of both the book series and the catalogue series, the editorial staff at the Institute was also restructured. Stamm Shapiro, who until then had worked exclusively as managing editor of Oppositions Books, was now also responsible for catalogue production, having taken over the management of the "Exhibition Program" that summer from Laurie Hawkinson, who had left the Institute. At the same time, Ockman was rising through the ranks to became executive editor of Oppositions Books as well as serving as an editorial consultant to the IAUS Exhibition Catalogues. Stamm Shapiro, meanwhile, had commissioned the typesetting for Alan Colquhoun's collection of essays and had already worked with Eisenman on the layout of Rossi's *A Scientific Autobiography*. Ockman revised the translation of *The Architecture of the City* and took over the editing of both the Loos and Ginzburg books. Moreover, on the initiative of Hamid Nouri, the editors were joined by two new assistant editors, Christopher Sweet and Thomas Mellins, whose salaries were again paid by Philip Johnson. However, production of the books and catalogues continued to progress slowly.

Skyline, Reissued

Skyline assumed more institutional importance than the other formats and its relaunch absorbed capacities at the Institute. To help counterbalance this, Eisenman wrote to seventeen architecture firms and contractors, in his words "leading members of the professional community"—mostly in Philip Johnson's name, sometimes in John Burgee's—in April 1981.[821] In the letter, which Eisenman rewrote several times and eventually cut and pasted together, he asked for donations of US$10,000 per year for a total of three years to build up a stock of capital for the relaunch. According to him, the hallmark of the new *Skyline* was to be that its editorial staff would be accountable to both the Institute and its sponsors but would ultimately operate independently. The campaign got off to a rather slow start, however, and by June, only four commitments had been made for this form of cultural sponsorship. Edward Saxe who, on Johnson's recommendation, had been advising the Institute's leadership on financial matters since early 1981, proposed that well-known American architects such as John Burgee, I.M. Pei, Cesar Pelli, and Kevin Roche, as well as a number of prominent and financially strong developers be added to the Board of Trustees in order to forge even stronger and more enduring links between

820 Eisenman had conducted the conversation with Lindsay Stamm Shapiro and her husband, the poet David Shapiro; see Peter Eisenman, "A Poetics of the Model: Eisenman's Doubt," (March 8, 1981) In IAUS, 1981, 121–125.

821 Peter Eisenman, draft letter to sponsors, n.d. Source: CCA Montréal, IAUS fonds: A.5-2 / ARCH401625.

the Institute and the establishment of the architecture and building world.[822] Johnson himself stepped forward as a trustee and patron when the Institute hosted a dinner in his name at the Century Association on July 20, 1981, to celebrate *Skyline*. Other dinners, such as those in the name of Gerald Hines, or in honor of *The New York Times* architecture critic Ada Louise Huxtable, were to follow. By promising patrons admission to this illustrious circle, the Institute ultimately built up a new philanthropic network, next to the Architects' Circle, that would become strategically important for *Skyline* and for the Institute itself. Contemporaries criticized this development and characterized the architecture newspaper as "the single most important media resource for the Inner Club."[823] Indeed, *Skyline* became a PR instrument for New York architects who rallied around Johnson in the Century Association.[824] The fact that the Institute was not averse to building bridges between architecture and the building industry—on the contrary—and its attempts to attract further sponsoring were also reflected in a conference on the subject of "Architecture, Development and the New Investment Pattern: Can They Co-exist?" which had been planned for some time and was finally scheduled for September 1981, that is at the same time as the relaunch of *Skyline*, under the direction of Gerald Hines and based on a concept by Jonathan Barnett.[825] Even if this conference ultimately did not take place, the planning alone highlights the economic promise of a construction and real estate industry that had gradually recovered by the early 1980s and its hoped-for impact on the architecture world.

Like Oppositions Books, the rebirth of *Skyline* was ultimately made possible by Philip Johnson, who once again paid the editor's salary, in this case, that of Suzanne Stephens, thus financing a key position in the Institute's publishing operations. During the preliminary negotiations, Stephens had her attorney draw up a contract that guaranteed her an income of US$30,000 per year—making her the Institute's top earner—and wide-reaching powers. Not only was she paid more than any other editor, but even more than Eisenman as Institute

822 Edward Saxe, memo to Bruce Brackenridge, April 15, 1981. Source: CCA Montréal, IAUS fonds: A.1-6. Saxe, who was previously deputy director and general manager at MoMA, initially advised the Institute without an official mandate.

823 Plunz and Kaplan, 1984.

824 *Skyline* was ultimately funded by a list of architecture firms: Daniel, Mann, Johnson & Mendenhall; Ulrich Franzen & Associates; Philip Johnson and John Burgee Architects; Paul Kennon / Caudill Rowlett Scott, Inc; Murphy/Jahn Architects/Engineers; I.M. Pei and Partners; Cesar Pelli Associates; Kevin Roche, John Dinkeloo & Associates; Paul Rudolph, Architect; The Skidmore, Owings & Merrill Foundation; Swanke, Hayden & Conell; see *Skyline* (October 1981); see also IAUS, list of sponsors. Source: CCA Montréal, IAUS fonds: A.5-6.

825 As designer for the CPC and head of the UDC, Barnett used to collaborate with the Institute; see Jonathan Barnett, proposal and revision of concept for conference on investment patterns, October 20, 1980, November 10, 1980, November 11, 1980, April 6, 1981 & April 10, 1981. Source: CCA Montréal, IAUS fonds: A. 5-11.

director; she was also given full editorial control over *Skyline*'s content while bearing no financial responsibility. Her two-year employment contract also covered her if the status of the architecture newspaper changed, such as in the event of discontinuation, merger, change of publisher, or change of name. She was also contractually assured that four full-time editorial positions would be available. This meant that *Skyline* had the only professional editorial staff at the Institute, and Stephens had full decision-making authority over all personnel: in addition to Margot Jacqz as managing editor, regular staff included Margot Norton, who worked freelance as a copy editor, and Heather Cogswell, who was initially editorial assistant; the fact that the editorial team was composed exclusively of women was a first in the Institute's history. As part of the professionalization, Stephens, who was provided with a fully equipped office at the Institute, was also able to ensure that—for the first time—authors were paid a fee of ten cents per written word. To ensure that the interests of all partners were safeguarded, an editorial board was set up at the Institute. This was composed of one representative each from *Skyline* (Suzanne Stephens), the Institute (Anthony Vidler), the sponsors (Henry Cobb), and the publisher (Gianfranco Monacelli), and served as a controlling body. Eisenman appointed Vidler, head of the editorial board, to act as an intermediary to ensure that the professional work and journalistic quality justified commercial distribution and financial support. Vignelli eventually also became a member of the editorial board. Over the summer, he also created the new graphic design of *Skyline* along with Michael Bierut, who worked as a junior designer at Vignelli Associates, further developing the old graphic design. As design director, Vignelli was responsible for the redesign of *Skyline*, ensuring that it was visually consistent with the Institute's identity—at least its brand identity if not its corporate identity—even as this evolved from a modernist approach to a postmodern look that subsequently built on the new *Skyline*. The relaunch was carried out professionally, not least because the conclusion of a contract with Rizzoli had ensured financial stability. On the publisher's side, David Morton was now responsible for the Institute's publications. Morton, as a former editor of *Progressive Architecture*, was well acquainted with Stephens and accordingly took personal responsibility for the newspaper, which now had a circulation of 5,000 copies and was distributed at an increased, but still affordable price of US$2.50. On September 29, 1981, the relaunch of *Skyline* was marked by a big release party at the Institute.

Running an Editorial Floor

Publishing at the Institute felt different then. In the summer of 1981, the *October* editorial staff, where Craig Owens had by then been replaced by Joan Copjec, moved back into the Institute's penthouse. At the beginning of the 1981–82 academic year, the Institute's upper floor was thus virtually a single editorial floor. Almost all publications now had their offices there and were connected by the bridge that ideally would have facilitated exchange between editorial offices. *Oppositions* alone was still edited on the 20th floor, although the journal had long since lost its key status. Finally, the restructuring of the Institute's publishing operations was accompanied by further personnel changes, as Silvia Kolbowski (September 1980), Joan Ockman (May 1981) and Rosalind Krauss (July 1981) were appointed Fellows. This meant that the editors, executive editors, and managing editors of IAUS Exhibition Catalogues, Oppositions Books, and *October* now had direct representation in the Fellowship, which underscored or reinforced their status and their voice in the Institute.[826] In early October 1981, the Institute's executive and operational structure was revised—it was divided into separate subdivisions for the first time—and the Fellows held four programmatic Institute meetings, each dealing with one of the four major areas of work, "Publication Programs," "Education Programs," "Public Programs," and "Development Programs," to set new strategies and goals for the next five years. In this context, Frampton, as director of publications, provided a status report on each of the Institute's publications, himself being responsible for *Oppositions*, the IAUS Exhibition Catalogues, and the Oppositions Books.[827] One point that was raised in this context was that *Oppositions* editors continued to be dissatisfied with the work of MIT Press, especially with its distribution abroad. Of particular concern was that the Institute was currently investing almost exclusively in the production of the journal but continued to see no significant revenue from academic publishing. In an effort to reduce the Institute's dependence on publishers, there was renewed talk of establishing its own publishing house as part of the search for new spaces. At the meeting, Frampton made a specific proposal to combine all steps of production under a single roof in the future to save costs through synergy effects. "Specifically the question was raised," the meeting minutes later state, "whether the Institute might do better, in the long term, to handle all aspects of publication, including in-house typesetting, graphics, and distribution."[828] For the current fiscal year, however, all five publications were contractually bound to the publishers, and Nouri pointed out that MIT Press and Rizzoli still had a better "selling name" than the Institute. The economic logic displayed by Nouri as controller became the measure of all things at the Institute, including publications in all respects. While contracts were supposed to guarantee maximization of profits, Nouri, in a departure from the previous model of self-exploitation practiced by the editorial offices, enforced that all managing editors would in the future receive an adequate salary to motivate them to continue doing good

826 Krauss as editor of *October* became more involved in meetings of the Fellows in the early 1980s, followed by other editorial staff in the following year: Joan Copjec, Douglas Crimp, and Annette Michelson were made Fellows in May 1982.

827 Kenneth Frampton, "Provisional Report on Publications," October 18, 1981; unofficial minutes, October 1, 1981 [sic!]; Marguerite McGoldrick, "Minutes of Fellows Meeting," October 8, 1981. Source: CCA Montréal, IAUS fonds: A.2-9.

828 Ibid.

work, and that they would also be provided with enough assistants to make their workloads manageable.

At these meetings, it was noted that the IAUS Exhibition Catalogues were quite commercially successful. In October 1981, Catalogue 8 on the Russian Constructivist Ivan Leonidov was published.[829] The Institute had thus published a total of ten catalogues, fulfilling the first half of the contract with Rizzoli. On the other hand, it was clear by now that the series would not be completed, as Catalogue 4, "The Princeton Beaux-Arts," had been canceled and Catalogue 11, "Wallace Harrison: Fifty Years of Architecture" had been postponed until further notice. At this point, four more productions on O.M. Ungers, Raymond Hood, Robert Krier, and William Lescaze were in the planning stages, and catalogues on the Office for Metropolitan Architecture and on Raimund Abraham were to follow. Oddly enough, more catalogues were produced than exhibitions. The Fellows consequently discussed a new relationship between exhibitions and catalogues to ensure the quantity and quality of the series, and the internal coordination of the "Public Program." Some criticized the series' historic focus and suggested it should document more contemporary projects. For the first time, there was a discussion about whether it was the Institute's goal to produce monographic exhibitions and catalogues. Group exhibitions such as "Idea as Model" or the "New Wave" series and the accompanying catalogues continued to be the Institute's flagships, but even they could not hide the fact that the "Exhibitions Program" focused more on the figure of the architect as artist than on pressing contemporary issues.

Even before these meetings, the *Oppositions* editorial team had again drawn up a very ambitious schedule and work plan in February 1981. According to this schedule, a total of nine issues were to be published in the coming months through May 1982, in order for the Institute to fulfill its contractual obligations and to make up for the journal's self-induced backlog. They outlined the dates and content of *Oppositions* 22 through *Oppositions* 30, the former quite detailed, the latter less so.[830] Essays by Vittorio Lampugnani and Werner Oechslin, by Massimo Cacciari, Giorgio Ciucci, Francesco Dal Co, and Giorgio Grassi, by Maurice Culot and Leon Krier, by Rafael Moneo and Ignasi de Solà-Morales, and by Alberto Perez Gomez and George Teyssot were discussed. The special issues planned by the editors at that time were: Forster working on an issue on "Monument," Frampton and Vidler on an issue on "Institutions, Power, and Architecture," and Eisenman on an issue on "Postmodernism." Vidler was also slated for special issues on "History and Practice" and "Nietzsche and Architecture;" Gandelsonas was no longer involved

829 IAUS, ed., *Ivan Leonidov. Russian Constructivist, 1902–1959*. Catalogue 8 (New York: Rizzoli International, 1981).

830 IAUS, editorial meeting agenda, content of *Oppositions* 22 to 30 including an "Updated List of Articles and Actions" February 24, 1981, Source: CCA Montréal, IAUS fonds: A.5-8 / ARCH 401765.

at all at this point. *Oppositions* 26/27, a double issue on housing, was scheduled for publication in March 1982. This issue grew out of the conference "Social Housing in Europe Between the Two World Wars" and was to be guest edited by Teyssot, a former *Oppositions* author and young academic who taught at the IUAV and espoused a type of historiography inspired by Foucault. The goal was for the journal to appear regularly for the first time in its history, starting with the August 1982 issue of *Oppositions* 29. Only then would the contractual requirements be fulfilled, and the Institute would finally have been able to part with its former publisher, the out-of-favor MIT Press, a plan that ultimately could not be sustained for a variety of reasons. For the sixth volume, *Oppositions* 21, 22, 23, and 24, which then appeared in 1981–82, reviews and essays by outside historians and theorists, some of whom had explicitly been earmarked as authors for Oppositions Books, were included, and with them the Institute's connections, to generate content.[831] Stanford Anderson repeatedly contributed texts: after an excerpt from his dissertation on the German architect Peter Behrens had been published in *Oppositions* 11 in 1978, further passages from the chapter on "Modern Architecture and Industry" appeared in *Oppositions* 21 and 23, edited by Frampton.[832] Frampton was also the only editor to contribute his own writings to *Oppositions* during this period, namely the essay "Louis Kahn and the French Connection."[833] Eisenman's Italian network led to the publication of a number of historiographical, theoretical, and critical texts in the "Theory," "History," "Documents," and "Reviews" sections, texts by Massimo Cacciari, Giorgio Ciucci, Francesco Dal Co, and Giorgio Grassi, authors associated with the IUAV, all translated by Stephen Sartarelli, some of which had to be heavily edited.[834] After the 1976 "Italian issue," the journal again took on a strong Italian focus, after architectural postmodernism gained a foothold in the global cultural world with the 1980 Venice Architecture Biennale. Finally, *Oppositions* 24 entered architecture history primarily through the juxtaposition of two articles, on the one hand, Leon Krier's essay on Albert Speer's architecture,

831 IAUS, minutes of editorial meeting for *Oppositions*, March 6, 1981. Source: CCA Montréal, IAUS fonds: A.1-4 / ARCH 401041.

832 Stanford Anderson, "Modern Architecture and Industry: Peter Behrens and the Cultural Policy of Historical Determinism," *Oppositions* 11 (Winter 1977), 52–71; "Modern Architecture and Industry: Peter Behrens, the AEG, and Industrial Design," *Oppositions* 21 (Summer 1980), 70–97; "Modern Architecture and Industry: Peter Behrens, and the AEG Factories," *Oppositions* 23 (Winter 1981), 53–83.

833 Kenneth Frampton, "Louis Kahn and the French Connection," *Oppositions* 22 (Fall 1980), 21–53; Frampton also published a review of *Hermann Muthesius. The English House*, "The Castellated Home," *Oppositions* 22 (Fall 1980), 106–113.

834 Giorgio Grassi, "Avant-Garde and Continuity," *Oppositions* 21 (Summer 1980), 25–33; Massimo Cacciari, "Eupalinos or Architecture," *Oppositions* 21 (Summer 1980) 106–115; Francesco Dal Co, "The Remoteness of die Moderne," *Oppositions* 22 (Fall 1980), 75–95 and "Notes Concerning the Phenomenology of the Limit in Architecture," *Oppositions* 23; (Winter 1981), 37–51; and Giorgio Ciucci, "The Invention of the Modern Movement," *Oppositions* 24 (Spring 1981), 69–91 (all translations: Stephen Sartarelli).

"Vorwärts, Kameraden, Wir Müssen Zurück," and on the other, Joan Ockman's response, "The Most Interesting Form of a Lie," a harsh critique of an appearance made by Krier at the Institute and of the architect's theory of a classical urbanism. With this contribution, Ockman, who had long worked only as an editor, finally made her debut as a promising young author.[835] But the plans were too ambitious, and *Oppositions* once again fell far short of its self-imposed goals due to other commitments, missing the opportunity to leave a lasting stamp on architecture debate, especially with a special issue on institutional critique in architecture.

To announce the forthcoming Oppositions Books—the first three had already been announced in the fall 1981 MIT Press catalogue—a poster was printed with six titles, in addition to those in production, including a collection of essays by Frampton titled *Labor, Work and Architecture*, in place of Tafuri.[836] After revising the schedule several times, Frampton was finally able to announce at the October 1981 Institute meeting that Colquhoun's collection of essays and Rossi's *The Architecture of the City* would appear in 1981. A major book launch and celebration with the two authors and invited guests was already scheduled for December 30, 1981. In the meantime, the Institute was deliberating who would be the best choice to review the books. At this point the editors were working on books planned for 1982; in addition to the translations of Loos and Ginzburg and the collections of essays by Frampton and Tafuri, these were currently the books by Massimo Cacciari and Giorgio Grassi, a translation of Theo van Doesburg, and a collection of essays by Arata Isozaki for 1983.[837] Eisenman and Frampton also sought to publish another book by Tafuri, a monograph with the title *Discordant Harmony*, as well as a new monograph by Colin Rowe titled *The Architecture of Good Intentions*. In addition, proposals had already been formulated for two more books by Francesco Dal Co and Vidler, but these were not yet in manuscript form and therefore not scheduled for a specific publication date, although Vidler stipulated that his book be published next year.[838] Authors from Switzerland (Werner Oechslin), France (Jean-Louis Cohen), and Japan (Koji Taki, Hiromi Fuji) were also discussed. In a

835 Krier, 1981; see also Joan Ockman, "The Most Interesting Form of Lie," *Oppositions* 24 (Spring 1981), 38–47.

836 In the fall of 1980, Tafuri had initially intended to publish a book titled *Toward an Ideology of Architecture*. Frampton suggested that his collection of essays *Labor, Work and Architecture* should include essays from *Oppositions* 1, 3, from *Architect's Yearbook* 12, from *Architectural Design* 38, 39, and from *L'Architecture d'Aujourd'hui*.

837 Joan Ockman, communication with Hamid Nouri, October 5, 1981. Source: CCA Montréal, IAUS fonds: A.5-8.

838 Oppositions Books editors had requested a publication by Vidler on the form of institutions and central themes of the late Enlightenment, titled *The Architecture of the Lodges* or *Institution & Utopia: Lodge*. Shortly thereafter, *Oppositions* 27 was planned as a special issue on "Institutions, Power, and Architecture."

guideline announced at an editorial meeting, Eisenman specified that of the four Oppositions Books per year, only one should be historiographical and another a translation of a classic; two books in the series, however, were to be monographs commissioned from contemporary authors, and he explicitly called for architects from the United States to be published as well.

But before Oppositions Books was actually launched on the market, Conover again called for urgent improvements in communications between the Institute and MIT Press, and for the academic publisher to be more involved in all decisions. Not surprisingly, MIT Press was still suspicious: despite years of delays in printing Rossi's *The Architecture of the City*, Eisenman still had not written his preface and introduction when the monograph was finally assigned to the typesetter in November 1981. Publication was eventually postponed yet again. Not only could the two Rossi books no longer appear simultaneously, as the author had requested, but they were eventually published in chronologically reversed order. In the meantime, Nouri had commissioned a feasibility study for the publication of Oppositions Books, which indicated that *The Architecture of the City* should now be printed third, after Colquhoun's collection of essays and Rossi's *A Scientific Autobiography*.

New Priorities in Publishing

The first issue of *Skyline* under Suzanne Stephens's direction in October 1981 was thirty-six pages long and displayed characteristics of a new, postmodern aesthetic. But there were changes in content as well. The bold graphic design of the cover, simply stating the names of the featured authors, interviewers, or interviewees, using oversized letters in 72pt font size, all set in the same size, headlines only, transformed the architecture newspaper into an advertising billboard on the newsstands. As the new editor-in-chief, Stephens reinterpreted the tabloid format created by MacNair, but retained the two central sections, reviews, and interviews, which continued to set the tone: on the one hand, through full-page or even longer book reviews, some of them fairly caustic, e.g., Vidler on Jencks' *Post-modern Classicism* or Stern on Frampton's *Modern Architecture. A Critical History*;[839] and, on the other hand, through exclusive, controversial interviews such as Eisenman speaking with the American writer Tom Wolfe, then known as the author of *From Bauhaus to Our House*, which had just been published in 1981.[840] In her programmatic editorial, Stephens wrote that she wanted *Skyline* to treat architecture as a cultural phenomenon, which was consistent with the Institute's policy as formulated in 1977.[841] With

839 Anthony Vidler, "Cooking up the Classics," *Skyline* (October 1981), 18–21; Robert Stern, "Giedion's Ghost," *Skyline* (October 1981), 22–25.

840 Peter Eisenman, "Interview. Tom Wolfe and Peter Eisenman. Part I," *Skyline* (October 1981), 12–14; "Part II," *Skyline* (November 1981), 3–4

841 Suzanne Stephens, "Skyline Rises Again," *Skyline* (October 1981), 2.

her experience and professionalism, Stephens succeeded in reviving *Skyline*, financed by the architecture establishment, as an established brand while still providing it with an independent voice within the Institute. She immediately set new standards by staging individual contributions as debates, juxtaposing pros and cons, and revisiting old conflict lines between Whites and Grays, true to the polemics cultivated at the Institute and especially with *Oppositions*. The intention was to keep readers engaged across issues by extending and continuing both reviews and interviews as a follow-up story, setting up cliffhangers, and printing a rebuttal or sequel in the following issue. Aside from these strategies, perhaps the biggest conceptual change was that Stephens replaced the double-page calendar with a luxuriously illustrated centerfold article. Longer essays of at least two pages, written either by Stephens herself, by Vidler, or by other notable authors on a topic of general interest, could be placed prominently here under an oversized headline. On the other hand, the monthly announcements of architecture, art, and cultural events that had made the *Architecture and Design Review* indispensable to receptive, culture-savvy architects and architecture lovers alike were now listed as events under the new *Dateline* column on the penultimate page. In general, the new graphic design of *Skyline* was now no longer elegant and restrained, but excessive and pompous. Building on Bierut's first design, Vignelli reinterpreted the newspaper and added characteristic features of a post-modern tabloid, especially with regard to the relation between text and image. The new layout was still based on a clear grid, but Vignelli now took a much more playful approach to the typographic elements, the choice of fonts and type sizes, and the function and arrangement of illustrations. While the old *Skyline* had featured the distinctive black bar, this was now also used as background for inverted headlines, for example, so that more black space was printed. Century Schoolbook was now replaced by Bodoni, a classicist typeface favored in postmodernist graphics and characterized by a greater contrast between base and hairlines. In addition, Vignelli now liked to use oversized type, especially for the headline of the centerfold. Content and form needed to work for the new *Skyline* to prevail against the new competition, as it now had to assert itself more strongly on the growing market and generate revenue through advertisement sales; three architecture newspapers were now published in New York alone, the other two being *Express* and *Metropolis*.[842]

From the outset, and despite its dependence on the Institute for contracts and funding, the far more professionalized editorial team managed to find its own voice. Stephens was able to draw on her own network of experienced architecture critics, such as Eleni Constantine and Martin Filler, both of whom were writing for *Progressive Architecture* at the time; ambitious young writers such as the young art historians Barry Bergdoll and Hal Foster also found a platform

842 Jane Kay Holtz, "Tabloid Trio. New Voices Speak up on Built Environment," *The Christian Science Monitor* (March 19, 1982), 15.

here.[843] The declared goal of reaching the broadest possible readership with an interest in architecture and the city was pursued with the introduction of additional columns. In the "City Reports," which aspired to a certain degree of investigative journalism, Stephens alternated with Jacqz in reporting on local building activity and current issues such as building restrictions and historic preservation, but also on architecture and urban policy debates of general interest; thanks to quality journalism with topical relevance and a clear stance, *Skyline* gained significance and, in contrast to *Oppositions*, was the only publication of the Institute to comment directly on issues of urban development and urban renewal in New York.[844] Another new feature was the "Obituaries" column, with articles on recently deceased architects; because of the generational shift in architecture, this read like a farewell to classic architectural modernism and its protagonists.[845] In addition, so-called "Insider's Guides" to architecture firms or schools of architecture on the East Coast were published: often anonymously written texts with apparently well-informed glimpses behind the scenes that nonchalantly revealed the networks of architecture education and practice, thus catering to *Skyline*'s target groups of young architects and architecture students as potential readers.[846] Stephens ultimately interpreted her new assignment creatively. *Skyline* now offered even more entertainment and human-interest stories: for example, when a list of the year's new publications—including, of course, those of the Institute—was presented as tips for Christmas gifts, or when fashion tips were offered with pointers to Johnson's optician or Eisenman's shoemaker. Eisenman himself encouraged Stephens to report on the lives of architects to provide readers with human interest. Gossip, i.e., informal, indirect communication about third parties, their character and social qualities, achievements, failures, and interpersonal relationships, played a key role in New York's architecture scene, which was governed by the laws of celebrity culture. Gossip could create or sustain celebrity; consistent disregard, on the

843 *Skyline* at that time published texts by Michael Kimmelman, later architecture critic for *The New York Times*, and Sylvia Lavin, architecture historian, who both wrote reviews as freelancers in March and April 1983.

844 Suzanne Stephens, "City Report: New York," published in *Skyline* from October to December 1981, see also Margot Jacqz, "City Report: New York," published in *Skyline* in October, and December 1981.

845 *Skyline* published obituaries of Robert Moses (October 1981), John Dinkeloo (October 1981), Alfred Barr (October 1981), Peter Collins (October 1981), Marcel Breuer (October 1981), Albert Speer (December 1981), Albert Mayer (December 1981), Wallace K. Harrison (January 1982), Richard Llewelyn Davis (January 1982), John Barrington Bayley (February 1982), Fazlur Kahn (May 1982), Bruce Goff (October 1982), O'Neil Ford (October 1982), Giovanni Muzio (November 1982).

846 *Skyline* published "Insider's Guides" to offices and schools: "Insider's Guide to Architectural Offices: Gwathmey and Siegel Architects," *Skyline* (October 1981); "Insider's Guide to Architectural Offices: Hardy Holzman Pfeiffer," *Skyline*, (November 1981); "Insider's Guide to Architecture Schools: Cornell," *Skyline* (December 1981), 27; "Insider's Guide to Architecture Schools: Columbia," *Skyline* (January 1982), 24; "Insider's Guide to Architecture Schools: Yale," *Skyline* (May 1982), 25; "Insider's Guide to Architecture Schools: Harvard," *Skyline* (June 1982), 15.

other hand, could destroy it. And although Stephens and her editorial team were criticized for this strategy, borrowed from the yellow press, the architecture newspaper's representation and reproduction of social relationships, pushed by Eisenman and endorsed by *Skyline*'s editorial board, was a central mechanism through which the Institute, as the "fame maker," influenced architecture education and practice, the cultural sector, and the art and architecture market.

Nevertheless, the new *Skyline* offered its readers a compelling read due to the enormous variety of topics, varied text formats, and writing styles appropriate to an entire range of target groups. At the beginning of 1982, it became clear why the new *Skyline* was an independent publication that not only offered shallow entertainment but also aspired to be a scholarly publication. The occasion was two milestone anniversaries: first, Columbia University was celebrating its 100th anniversary with the exhibition "The Making of an Architect, 1881–1981" and second, MoMA was celebrating the fiftieth anniversary of the groundbreaking 1932 exhibition "The International Style." In both cases, *Skyline* joined the celebrations with special issues.[847] In the January 1982 issue, Thomas Bender, as one of the two co-founders of the New York Institute for the Humanities, contributed again after "The Bender Affair" with an article entitled "Between Civic Culture and the Academy," in which he reviewed universities as sites for the production and consumption of discourse, arguing that throughout history, there has been a need for so-called "cultivators" to carry the results of research and teaching into the public sphere, without claiming that the New York Institute for the Humanities could be such a cultivator.[848] In the February 1982 issue, which also featured an Eisenman interview with Philip Johnson illustrated with a series of the latter's postmodern high-rise designs that underscored the architectural metamorphosis, Stephens recalled the scope of the exhibition once curated by Henry-Russell Hitchcock and Philip Johnson himself in her editorial "Looking back at 'Modern Architecture'," focusing in particular on curatorial and editorial positions, including differences in content and ideology. Stephens not only highlighted the role that American architecture played in the exhibition, compared with the publication accompanying it, but she also referenced the housing featured in the exhibition and the "social concern" that was evident here. In this way, *Skyline* congratulated not one but two established institutions in New York on their anniversaries, the university and the museum, with which the Institute had had a relationship throughout its existence, whether directly or indirectly, as an offshoot or provider, competitor or pioneer.

Eisenman's contribution to the new *Skyline* was regular interviews. He apparently chose his interviewees based on considerations of usefulness and current events: the two-part interview with Wolfe was followed by interviews

847 Suzanne Stephens, "Columbia Architecture at 100!" *Skyline* (January 1982), 16; "Looking Back at 'Modern Architecture.' The International Style Turns 50," *Skyline* (February 1982), 16–27.

848 Thomas Bender, "Culture of Cities: Between Civic Culture and the Academy: New York and Columbia in the 19th Century," *Skyline* (January 1982), 14–15.

with Robert Hughes, Paul Goldberger, Philip Johnson, Cesar Pelli, Henry Cobb, Robert Venturi, Gerald Hines, John Portman, Leon Krier, Robert Maxwell, and Richard Serra; i.e., he spoke mostly to people who were of general interest—not only to the public at the time but also the Institute.[849] In the introduction to the Johnson interview, which appeared in the MoMA issue of *Skyline*, Eisenman introduced an additional concept to the interview series by dividing his interviewees into four groups, according to which he sought to conduct a total of twelve interviews, with three "critics," "developers," "teachers," and "architects." In the end, he did not adhere to this concept either, which was not detrimental to the economy of attention; practicing architects were overrepresented, while university teachers were underrepresented. Moreover, it was noticeable that there was not a single woman among his interviewees. Instead, the interviews represented a text format that brought Eisenman as much attention as the people he interviewed. In this series of interviews, he repeatedly turned the spotlight on direct collaborators, both trustees of the Institute (in the case of Johnson and Hines), and sponsors of *Skyline* (Pelli, Cobb). At a time when Eisenman was working more and more as an architect with his own firm, for example on a housing project for the IBA Berlin 84, and at the same time withdrawing more and more from the Institute's operations, he used this platform to stage himself as a public figure before a larger audience, a kind of Andy Warhol of the architecture and construction world. Like the pop icon in the art and creative world before him, he carried a recording device with him everywhere he went. As part of the manifestation of his will and intellect, Dorothy Alexander, the Institute's in-house photographer since the mid-1970s, was commissioned to take not only portraits of the interviewees but also shots of significant, sometimes confrontational conversational situations, which were then used to illustrate the interviews. Eisenman, who obviously always had to be the center of attention, thus took his self-stylization as the mastermind of the Institute and the architecture scene to extreme heights. But it was above all Eisenman's selected interviews with the declared proponents and protagonists of postmodernism, e.g., Tom Wolfe, Robert Venturi, and Leon Krier, that were also received internationally and reprinted in other publications, sometimes in translation, which thus secured international attention for both *Skyline* and the Institute.

Vidler, on the other hand, not only contributed to *Skyline* with his writings on contemporary (James Stirling, Richard Meier, Rem Koolhaas/OMA), modernist (Adolf Loos), or classicist architects (John Soane), but also used his academic contacts to bring in scholars from other disciplines, such as the cultural

849 For *Skyline* Eisenman interviewed Tom Wolfe (October 1981) & (November 1981), Robert Hughes (December 1981), Paul Goldberger (January 1982), Philip Johnson (February 1982), Cesar Pelli (May 1982), Henry Cobb (June 1982), Robert Venturi (July 1982), Gerald D. Hines (October 1982), John Portman (January 1983), Leon Krier (February 1983), Robert Maxwell (March 1983), Richard Serra (April 1983).

historian Carl Schorske, or professor of comparative literature Peter Brooks, both of whom came from his immediate environment at Princeton University, as authors for the middle section. The new *Skyline* thus also bore his signature.[850] Vidler scored a particular coup when, on his initiative, the thematic focus of the March 1982 issue was devoted to Michel Foucault, thus introducing one of the most influential French philosophers of his time to the North American architecture debate in a high-profile and broad format.[851] This was indeed an impressive issue, first introducing Foucault's ideas and describing his engagement with certain building types and urban planning in the two-page article "Spatialization of Power" by Gwendolyn Wright and Paul Rabinow.[852] Wright and Rabinow outlined why Foucault-influenced poststructuralist philosophy was less interested in architecture than in urban space, and they themselves analyzed individual buildings and the built environment as a whole as technologies of power. The centerfold article of this issue, however, was an exclusive interview conducted by Rabinow with Foucault in Paris.[853] The interview, which was printed under the sweeping heading "Space, Knowledge, and Power," was one of the political philosopher's few concrete statements on the role of architecture in relation to issues of space and power. Here, contrary to the usual historiography of modernity, Foucault argued that while architects had understood and deployed their projects as a technique of government since the eighteenth century, with industrialization, they had increasingly lost control over built space to engineers, who played a greater role in the urbanization of territory. Foucault thus placed the role of architects in society into perspective but did not deny architecture its importance; on the contrary, he instead argued for the continued relevance of the profession in planning and housing.

In terms of a history of ideas (and possibly also the history of institutions), it is interesting that in the course of the conversation, Rabinow attempted to elicit from Foucault a direct statement on core issues of the architecture debate

850 Anthony Vidler, "Cooking up the Classics," *Skyline* (October 1981), 18–21; "Restructuring Modernism. The Architecture of James Stirling," *Skyline* (November 1981), 16–19; "Institutional Style. Deconstructing Modernism: Meier's Hartford Seminary," *Skyline* (March 1982), 21–23; "The Office for Metropolitan Architecture. The Irony of the Metropolis: Notes on the Work of OMA," *Skyline* (May 1982), 18–21; "The Big Greek Column Will be Built: Adolf Loos and the Sign of Classicism," *Skyline* (October 1982), 16–17; "Progress and Primitivism: The Roots of John Soane's Style," *Skyline* (November 1982), 32–33.

851 In New York, Foucault had previously been received in art and theory circles, but only rarely in architecture circles, e.g., in the context of the conference "Schizo-Culture" at Columbia University from November 13 to 16, 1975, organized by the journal *Semiotext(e)*; see Sylvère Lotringer and David Morris, eds., *Schizo-Culture. The Event/The Book* (Cambridge: MIT Press, 2014); see also Cusset, 2008. At the Institute, the French philosopher had previously been featured in *October* (but not in *Oppositions*) and had at most been cited or referenced.

852 Gwendolyn Wright and Paul Rabinow, "Spatialization of Power. A Discussion of the Work of Michel Foucault," *Skyline* (March 1982), 14–15.

853 Foucault, 1982.

of the time, such as the delineation of postmodernism in architecture and philosophy or the polarization of historicism and rationalism. Foucault, however, did not fall for the rhetorical trap of stylistic preferences that concealed bourgeois utopias and replied, entirely in line with deconstructivist, poststructuralist thinking, that he saw the task of philosophy as questioning any form of rationality, whereby he was fundamentally opposed to anything that claimed to be a return, be it historicism or playing with any kind of historical references. At the end of the highly readable conversation, translated by Christian Hubert, Foucault outlined his approach to historiography and epistemology, using the example of the fireplace as an architecture element to show how much the history of ideas, society, and technology are directly interrelated. Essentially, he stated that architecture interested him not in terms of its formal properties, but because it provided insights into social and political contexts. The reverse conclusion, that architecture represents a one-to-one reproduction of power hierarchies, was only valid in a few cases. Despite, or perhaps because of, the provocative heading—the word "power" was virtually shouted in large letters spanning the entire page—Foucault's statement is all the more remarkable; especially at a time when architecture was primarily about who would tell and publish the better story, and in an intellectual and institutional environment in which a battle was being waged over the correct historiography with *Oppositions*, and *Skyline* was suddenly producing theory.

Compared to Eisenman and Vidler, the other *Oppositions* editors contributed little to *Skyline*. Aside from the review of *Modern Architecture. A Critical History*, which was staged as a controversy between Frampton and Stern—the teaser and individual quotations from the articles in 36pt type testified to the fact that the two had been antagonistic, even outright shouting at each other—Frampton published only his three-part "Japan Diary" documenting his trip to the Far East in the summer of 1981 under the title "Modernist Diffusion" in the April, May, and June 1982 issues, again portraying himself as an admirer of and expert on Japanese architecture, a contemporary movement that he had previously presented as the only true alternative to American postmodernism.[854] Forster published a review of Frampton's *Modern Architecture, 1851–1919* in the July 1982 issue, which was strikingly more sympathetic than Stern's previous review of the monography had been.[855] Apart from this, *Skyline* provided a high-profile platform for Fellows, including support for their current architectural projects. When Gandelsonas and Agrest suffered a resounding defeat at the very beginning of their professional career in New York, as a postmodern

854 Frampton, 1981.

855 Kenneth Frampton, *Modern Architecture 1851–1919*, GA Document. Special Issue 2 (Tokyo: A.D.A. Edita, 1981); see also Kurt Forster, "Re modernism. Kurt W. Forster Reviews Kenneth Frampton's Latest," *Skyline* (July 1982), 27.

tower project for the historic district on the Upper East Side, a high-rise that was to be built over a landmarked house, failed to materialize due to historic preservation regulations, individual opposition, and a targeted media campaign, Stephens personally advocated for them; she presented the project in her "City Report" entitled "Tradition of the New" in the December 1981 issue, after it had been rejected by the Landmarks Preservation Commission. She described the planning process, the architecture of the three-tower project, and the controversy that it had sparked in detail.[856] In addition, *Skyline* took a stand in the controversy with another commentary by Aldo Rossi featured in the April 1982 issue, in which the Italian architect, as a friend and colleague, spoke out in favor of what he considered to be an excellent tower project, "a tower whose main virtue is that it interprets the history of the city."[857] *Skyline* thus used its media power, on the one hand, to explicitly take sides on behalf of Fellows and friends of the Institute, but on the other hand, also took a stand in a debate that was being conducted in the national daily and trade press.[858] The architecture reviews published in *Skyline* were another noteworthy feature—especially since Stephens herself taught a seminar on "American Architectural Criticism in Magazines and Newspapers, 1850 to The Present Day" at Barnard College starting in 1982—and reported with great regularity on new buildings that were planned or completed, for example, new museum buildings and repeatedly high-rise buildings.

International Circulation of Ideas

When the first two Oppositions Books, Alan Colquhoun's *Essays in Architectural Criticism. Modern Architecture and Historical Change* and Aldo Rossi's *A Scientific Autobiography* appeared in mid-January 1982, the Institute finally launched its own book series after five years in the making.[859] The discursive, educational, and institutional function of individual titles, and ultimately the historical significance of the overall book series are difficult to assess, and could be determined from reviews and references, reading lists and PhD dissertations, sales figures and reprints, etc.; it would be impossible to imagine the full extent, had it been implemented as originally planned. For the production of Oppositions Books was immediately interrupted by developments

856 Suzanne Stephens, "City Report: New York. Tradition of the New," *Skyline* (December 1981), 4–5.

857 Aldo Rossi, "On 22 East 71st Street," *Skyline* (April 1982), 2.

858 Paul Goldberger, "Debate Over Proposed 71st Street Tower," *The New York Times* (November 10, 1981); Pilar Viladas, "Right Building, Wrong Block," *Progressive Architecture* (January 1982), 33–34; George Lewis, "Chapter Active on Upper East Side District," *Occulus*, no. 8 (May 1982), 5; see also Diana Agrest and Mario Ganelsonas, "Manhattan Additions I," *Architectural Design*, no. 52 (May/June 1982), 44–48. This contained excerpts from letters of support sent to Landmarks Preservation Committee by Samuel Brody, John Hejduk, and Jaquelin Robertson, and Anthony Vidler.

859 Colquhoun, 1981; Rossi, 1981.

and events at the Institute in 1982, caused by Peter Eisenman's resignation as Institute director in June 1982, the subsequent break-up of the Fellowship and discontinuation of the editorial work, and further restructuring, that brought all publication activities to a halt. Finally, under Eisenman and Frampton's editorship, only five Oppositions Books that had already been started were ultimately published: Rossi's *The Architecture of the City*, finally, in April 1982, and in the fall of 1982, the English translation of Moisei Ginzburg's *Style and Epoch* and the collection of essays by Adolf Loos, compiled from two volumes, under the title *Spoken into the Void*. The editors' introductory comments, whether printed as "Forewords," "Prefaces," or "Introductions," influenced the reception and interpretation of each title, distinguished the authors, and played an important role in the international circulation of ideas—in all five cases these were books by European authors for the English-speaking world.[860] This sociology of introductions, however, did not refer solely to the intellectual transfer of knowledge or appropriation of cultural, symbolic, or even economic capital, as is quite common in book series, but also to discursive properties and interpersonal relations in the form of controversies, polemics, disputes, etc., and thus complex network effects between the Institute as publisher, the authors of the introductions, and the actual authors of the individual Oppositions Books.[861]

The battle for attention in publishing was evident in the prefaces to the English translation of *L'Architettura della Città*. The book was finally introduced with two texts by Eisenman, an "Editor's Preface" and an "Editor's Introduction," as well as one by Rossi himself, an "Introduction to the First American [sic!] Edition," which he had already written in 1978, a good three years before publication. Eisenman, who had been promoting the Italian architect for years, once again slipped into the role of expert on Rossi to achieve fame for himself. Rossi, on the other hand, when he wrote his introduction in 1978, as he did with every new edition, explicitly wanted the updated translation to be understood as another chapter on the American city, even if he only alluded to

860 In an essay on the sociology of culture, Bourdieu called for a sociology of prefaces and introductions; regarding the republication of texts in translation and the selection of the authors for introductions, see Pierre Bourdieu, "The Social Conditions of the International Circulation of Ideas," in *Bourdieu: A Critical Reader*, ed. Richard Shusterman (Oxford: Blackwell Publishers, 1999), 220–228.

861 For example, in the case of Ginzburg, Anatole Senkevitch, as guest editor (and translator), was enabled to write the introduction himself. In addition, in the case of Rossi's *A Scientific Autobiography*, Vincent Scully was requested as an external author, but Rossi expressly did not want an introduction, so that the already commissioned text was summarily turned into an afterword. Rossi perhaps profited most from the attention, after all; in the end two out of five Oppositions Books were authored by him, and he wrote the introduction to the Loos book. See Adolf Loos, *Spoken Into the Void: Collected Essays* 1897–1900, trans. Jane O. Newman and John H. Smith (Cambridge: MIT Press, 1982).

this in most of his remarks.[862] After repeatedly visiting New York, he was enthusiastic about primarily the architecture and atmosphere of Manhattan: "Perhaps no urban construct in the world equals that of a city like New York. New York is a city of monuments, such as I did not believe could exist."[863] He later added in *A Scientific Autobiography* that for him New York represented the confirmation of his theses from *L'Architettura della Città*, again without elaborating further.[864] Although he flattered his North American readers, especially in the second half of this unique text, he did not go into more detail about New York or any other American metropolis.[865]

Rossi, however, had evidently lost sight of his urban geographic perspective during the 1970s and had since abandoned a critique of the socio-economic conditions of urban landscapes. As late as 1966, when he first published *L'Architettura della Città*, he had diagnosed the "blighted zones" as a typical problem of the modern capitalist city.[866] When the two Oppositions Books were published a good fifteen years later, however, his attention was focused almost exclusively on the primary elements of architecture; as a practicing architect, he had apparently lost interest in analyzing the city from a Marxist perspective.[867] Beginning in the spring of 1976, Rossi shuttled back and forth between Milan and New York with increasing frequency; since then, his drawings exhibited American motifs and were now characterized not only by transhistorical but also by transcultural references.[868] His depictions of the analogous city were followed by architectures of Broadway and Wall Street, the Twin Towers of the World Trade Center; above all, his fictional skyline was now defined by the wooden water reservoirs above the rooftops of Manhattan. After excursions to Maine, typical New England lighthouses also appeared again and again as a central motif. In his writings as much as in his drawings, Rossi indicated how much the visual impressions of his stays in the United States served him as a source of inspiration. Even though he repeatedly referenced New York in *A Scientific Autobiography*, Rossi only explicitly referred to

862 Aldo Rossi, "Introduction to the First American Edition," in *The Architecture of the City* (Cambridge: MIT Press, 1982), 13–19.

863 Ibid., 15.

864 Rossi, 1981, 76.

865 Rossi admits that up to this point he had not written about the American city and architecture. In a few passages, he discusses the atmospheric qualities of the lighthouses in Massachusetts and Maine, Broadway in New York, the widows' walkways in New England, and the piers on Manhattan's Westside; see Rossi, 1981, 52, 58, 64, 65, 75f.

866 Rossi, 1982, 50.

867 In Rossi's biography, the publication dates of his two monographs, 1966 and 1981, are considered cornerstones of a work phase. After his time at the ETH Zurich (1972–1975), his stays in New York from 1976 onward represent an incisive experience; see Angelika Schnell, "Von Jörn Janssen zu Rossi—Eine hochschulpolitische Affäre an der ETH Zürich," *Arch+*, no. 215 (2014), 16–23.

868 IAUS, 1979.

the complex reality of cities experienced in everyday life in a single passage: "The line of the karst plateau corresponds to the skyline of New York, a city which is something like a mountain with stratifications where the built structures represent, more than anywhere else, the social, ethnic, and economic tangle of the city."[869] With this equation of natural conditions and urban geography, however, which was more of a poem than an analysis, Rossi naturalized any intersectional discrimination; his interest in the American city was clearly superficial. Instead, Rossi had discovered the American art market for his drawings, at least since the 1979 exhibitions at the Institute and the Max Protetch Gallery, and now the publishing market for his books, and in his autobiographical writing he even briefly addressed his biographical project at the time, the boosting of his career in the USA: "If I were to speak now of my American work or 'formation,' I would be digressing too far from the scientific autobiography of my projects and would be entering into a personal memoir, or a geography of my experience. I will say only that in this country, analogies, allusions, or call them observations, have produced in me a great creative desire and also, once again, a strong interest in architecture."[870]

Obviously, Rossi's work was marketed and sold by the Institute, and all the resulting hype, the cult of personality, contributed to the success of the cultural production and publishing there for several years, just as Rossi profited from his new fame in the USA. It was clearly in Rossi's interest to publish his two Oppositions Books as special editions; he was involved in the selection of images and even reissued some of his drawings. Nevertheless, Rossi failed to mention the Institute—and Eisenman in particular—in his autobiography. Vidler was the only Fellow mentioned by name.[871] In the end, Rossi tried to distance himself from the Institute's idiosyncratic reading of his drawings and writings and to resist outright appropriation, indicating that he felt he had been misinterpreted. Eisenman, on the other hand, in his creative misreading of *The Architecture of the City*, referred to the North American version as an "analogous artifact," borrowing directly from Rossi. The analogous character of *The Architecture of the City*, which incorporates diverse references and plays with different cross-references, is obvious: first, because it contains various prefaces and updates, and second because Eisenman added his own "Editor's Introduction," which he gave the witty title "The Houses of Memory: The Texts of Analogue."[872] Here, he purported to historicize Rossi's monograph: "The task of this preface then is to locate this book for an American audience not only in its own tradition, in the

869 Rossi, 1981, 64.

870 However, Rossi ultimately did not say what significance his numerous stays in New York, his teaching at Cooper Union, and the two exhibitions at the Institute and Max Protetch Gallery actually had for him, see ibid., 64.

871 Ibid., 68. Apparently, Vidler had given Rossi a book as a reference to his Teatro del Mondo.

872 Eisenman, 1982.

context of Italian theoretical writings by architects, but also in the more contemporary context of Italy of the 1960s and 1970s [sic!]."[873] His remarks, however, made it clear that Eisenman—and herein lies the misinterpretation—preferred his own reading and had no further interest in introducing readers to Rossi's theory.[874] In light of this, his introduction should not be understood as a classification or contextualization, despite the fact that in the grant application to the Graham Foundation he had claimed that this was the task of every introduction. Moreover, Eisenman did not really situate the book within current American debates either.[875] Instead, in his own introduction, he offered a glimpse into his own thought processes: "My own introduction [...] is in certain ways not only about this book but also about the Rossi that this book anticipates." Ultimately, he was less concerned with *The Architecture of the City* as theory, or with Rossi as architect and author, and more with his own creative, rather than critical writing: "My own introduction attempts to enter into this memory and in this sense serves as a kind of analogy of an analogy, a creation of yet another artifact with its own history and memory." This rhetoric served Eisenman for the acquisition of power rather than knowledge. He appropriated not only Rossi's notion of the 'analogous,' but also that of the 'collective,' and used it, emptied of its original meaning, in his introduction when he compared this publication with the previous Italian edition: "[T]his book is similarly, and even to a greater degree, a 'collective' artifact." Remarkably, Eisenman even put the word "collective" in quotation marks to emphasize the quote.[876] Accordingly, a socio-political or even critical concept of the collective apparently did not exist in his vocabulary, since Oppositions Book*s* could be regarded both as being the cultural product of a labor process and the result of a production of knowledge.[877] What mattered was that by americanizing the ideas formulated in *The Architecture of the City*, that is, by producing an "American Rossi," so to speak, Eisenman could adopt or even ignore them without presenting a theory of his own.

Eisenman's introduction to *The Architecture of the City* made Oppositions Books polemical rather than pedagogical. Unlike Rossi, he had never really been

873 Ibid., n.p.

874 Lobsinger's reading of *The Architecture of the City* contextualized the publication in architecture and urbanist debates; see Lobsinger, 2006; see also Pier Vittorio Aureli, "Rossi. The Concept of Locus as a Political Category of the City," in *The Project of Autonomy. Politics and Architecture Within and Against Capitalism* (New York: Princeton Architectural Press, 2008), 53–69.

875 In the late 1970s and early 1980s, Eisenman had used the Institute's various publications, especially *Oppositions* and the IAUS Exhibition Catalogues, to publish texts on notable contemporary architects (Philip Johnson, Michael Graves, Aldo Rossi, John Hejduk) in which he formulated text blocks of his own theory.

876 Peter Eisenman, "Editor's Preface," In Rossi, 1982, n.p.

877 The introduction failed to mention some of the people who worked on the book's production; see ibid. Importantly, the imprimatur stated that Eisenman and Rossi both revised *The Architecture of the City* for publication by MIT Press.

interested in the city, as his contributions to the Institute's early research and building projects show, nor in the cultural memories or the political concerns of its inhabitants, which for Rossi were configured as a collective and expressed in monuments.[878] More importantly, Eisenman's prefaces placed him on the grand stage of the international architecture world as a creative and intellectual. By adorning his introduction with quotes from Jacques Derrida and Sigmund Freud, he implied that he was well-read in two of the hottest theories of the day: deconstruction and psychoanalysis. Strictly speaking, however, he offered neither a deconstructionist nor a psychoanalytic interpretation of *The Architecture of the City*. Instead, he made biased judgments about how Rossi interpreted his role, accusing him of "disillusionment and anger" about the proper way to deal with the legacy of modernism.[879] "For Rossi's generation, it was no longer possible to be a hero, no longer possible to be an idealist; the potentials for such memories and fantasies had been taken forever." Eisenman discredited Rossi's writings, which were simultaneously historiographical and biographical, academic and poetic, and characterized him pejoratively as "unheroic and autonomous." In his view, Rossi was still in a process of self-discovery when he wrote *The Architecture of the City*: "Rossi's psychological subject—the autonomous researcher—still continues to seek his own home in the collective house of the city."[880] And further: "The shadow of the humanist poet hovers continuously behind the figure of the autonomous researcher."[881] Eisenman had solved this dilemma, which arose from "modern architecture's failure," for himself after he set himself apart from the corporate architecture of the International Style by placing the design process at the center of his work and negating the socio-political moments of utopia. Building on his 1977 theoretical construct "Post-Functionalism," Eisenman's introduction to *The Architecture of the City* also postulated an autonomy of architecture, both from the subject as agent and from the concrete object, yet his approach was equally devoid of history and place, and thus diametrically opposed to Rossi's.

Taken together, these two introductions to *The Architecture of the City* by Eisenman and Rossi testify to an increasing depoliticization that is representative of the globalized architecture culture and debate. As a paradigm of a postmodern discursive formation, they celebrate what Derrida may have meant by "*différance*," the arbitrary, even unconsidered juxtaposition of supposed pairs of opposites (in this case Rossi/Eisenman) on the one hand, and on the other

878 Martin points out that neither of the two concepts of the city developed by Rossi and Eisenman in their dialogue were in keeping with the times, Martin, 2010, 7–9. Earlier, Tafuri had criticized Eisenman's concern with the urban renewal project of the 1967 exhibition "The New City," as well as the Institute's housing project, as being merely about forms and not about urban problems or a political agenda; see Tafuri, 1976, 49.

879 Eisenman, 1982, 4.

880 Ibid., 10.

881 Ibid., 11.

a permanent displacement or emptying out of what might ultimately have been meant by architecture and the city, the architect and practice, at the time of writing.[882] Eisenman's argument took up Rossi's concept of typology, which he had originally adopted from Marxist literary theory in order to bring together form and content.[883] However, for his own purposes, he reinterpreted the concept as subjectless: "Rossi, however, discovers in typology the possibility of invention precisely because type is now both process and object." According to this conservative, reduced understanding of typology, the architectural object would have had to analyze and reinvent itself. Ultimately, Eisenman was not interested in fundamentally questioning the architect as an authorial subject. Thus, his understanding of architecture became entangled in contradictions, for example when he reversed the relationship between process and product and affirmed Rossi's view that an architectural drawing, "and not its built representation, becomes architecture."[884] As a building theorist, Eisenman thus provided arguments for a conceptual reassessment of the tools of design when these had long been established on the art market; as a conceptual architect, meanwhile, he began to serve the architecture market. The two Oppositions Books published by Rossi at MIT Press can thus be understood not only as the culmination and beacon of the Eisenman-driven hype surrounding Rossi in the United States but also as part of his systematic and deliberate self-promotion through the Institute's publications.[885]

Following its successful departmentalization, 1982 was a pivotal year for the Institute in publishing, with the individual publications taking up a great deal of resources. While the contract with MIT Press for *Oppositions* was terminated in the spring, effective October after issue 24, to the regret of the academic publisher's management, Eisenman proudly announced to the April annual meeting of the Board of Trustees that a contract had been successfully concluded with Rizzoli International, which was expected to bring the Institute US$105,000 per year for *Oppositions* alone.[886] *October*, which in its first five years had

882 Jacques Derrida, "Die différance," in *Postmoderne und Dekonstruktion. Texte französischer Philosophen der Gegenwart*, ed. Peter Engelmann (Ditzingen: Reclam, 2004), 76–113.

883 Lobsinger, in her reading of *A Scientific Autobiography*, points out that Rossi's notion of typology referred not to Quatremere de Quincy, but to Georg Lukácz; see Lobsinger, 2002, 47ff.

884 Eisenman, 1982, 10–11.

885 McLeod already referred to the specific context in which Rossi's urbanist theory emerged and its emphasis on the collective and public space in an early review of *The Architecture of the City* published in 1984; see Mary McLeod, "The Architecture of the City," *Design Book Review*, 3 (Winter 1984), 50. Later reviews did not make the connection between the Rossi hype at the Institute and the subsequent commodification of his architectural drawings and writings; see Botond Bognar, "Rossi's Ultimate Dilemma?" *Journal of Architectural Education* 41, no. 2 (Winter 1988), 56–59. For *A Scientific Autobiography*, MIT Press was the point of contact for international publishers for German, French, and Spanish translations.

886 Peter Eisenman, "Director's Report," April 12, 1982. Source: CCA Montréal, IAUS fonds: A.1-13.

been distinguished by contemporary theoretical approaches, in particular an institutional critique directed against art institutions (the artist studio, gallery, museum, patronage, etc.) and their role in the market economy, especially real estate development in SoHo, and a blend of deconstructivism, psychoanalysis, and feminism, was by now more fully integrated into the Institute, with Joan Copjec, Douglas Crimp, and Annette Michelson also being elected Fellows in May 1982.[887] *Skyline*, the publication with the Institute's largest budget in fiscal year 1981–82, had received a grant from the J.M. Kaplan Fund, on the basis of which Peter Freiberg had been hired as editor for the new "City Report" section and continued to report critically and in detail on current building activity in New York, especially around Times Square.[888] Gradually, however, individual sponsors began to cut off their support, and so by the summer of 1982 the architecture newspaper was facing major financial difficulties; with debts amounting to US$23,500, it could no longer afford to pay the salaries of its editorial staff. Stephens offered to suspend work on the July 1982 issue until a solution could be found, but Eisenman declined, and *Skyline* continued to be produced.

Throughout 1982, production of IAUS Exhibition Catalogues was moving ahead at full speed, although there was no budget for additional salaries and internal coordination was inadequate. In addition to the *October* editors, Stamm Shapiro and Hubert were also elected Fellows in May 1982. Furthermore, the editorial staff, which now included Deborah Berke, who had previously taught in the Institute's "High School Program," continued to work simultaneously on catalogues for long-completed exhibitions as well as for current ones. 1982, for example, saw the publication of Catalogue 6 for the exhibition on the German architect O.M. Ungers, which had been shown at the Institute five years earlier (May 1977), after he had completed his professorship at Cornell University, and of Catalogue 5 for the exhibition of the Vienna-based Luxembourg architect Robert Krier (April to May 1977) who, together with his brother Leon, was one of the best-known proponents of the European city and European postmodernism. Catalogue 15 on the American architect Raymond Hood (1981), on the other hand, which resulted from a larger historical exhibition and publication project directed by Robert Stern and was prepared in collaboration with Thomas Catalano, was a novelty, since it was published without an exhibition at the Institute.[889] Catalogue 16 was published to accompany another historical exhibition, this time on the

887 Annette Michelson, Rosalind Krauss, Douglas Crimp, Joan Copjec, "Introduction," in *October. The First Decade, 1976–1986* (Cambridge: MIT Press, 1986), IX–XII.

888 Peter Freiberg, "City Report: New York. 42nd Street Redevelopment," *Skyline* (May 1982), 3; "City Report: New York. Theaters," *Skyline* (November 1982), 28; see also Susana Torre, "Times Square. At the Crossroads," *Skyline* (December 1982), 18–22.

889 Stern had originally planned an exhibition on Raymond M. Hood at the Institute for 1981, but this was eventually displayed at a branch of the Whitney Museum in midtown Manhattan; see Carol Willis, "Review of Raymond Hood," *Skyline* (July 1982), 10–11.

Swiss-American architect William Lescaze (April to June 1982), who had once been involved in the *Modern Architecture* exhibition at MoMA and had realized modernist office and residential buildings in the USA in the 1930s. This was published to coincide with the exhibition.[890] Finally, 1982 also saw the publication of Catalogue 17 for the traveling exhibition on Japanese architect Kazuo Shinohara (December 1981 to January 1982) from UQAM in Montréal, and Catalogue 18 for the group exhibition *New West Coast Architecture. California Counterpoint* (1982), which marked the first time that the Institute was to display positions from the West Coast on a larger scale. The catalogues for the double exhibition of the Office for Metropolitan Architecture at the Institute and the Rizzoli Gallery (March to May 1982) and the exhibition on the Argentine architect Clorindo Testa (November to December 1981), on the other hand, were not produced.[891] No serious attempt was made to develop other, previously considered exhibition and catalogue projects on Hans Hollein, Raimund Abraham, Kevin Roche, John Dinkeloo and Associates, or Gregorio Grassi. Although the editorial team could not quite keep up with the new Rizzoli contract, and the publication could barely finance itself, the series made it to a total of sixteen volumes, including some outstanding titles. They are the most tangible document and lasting legacy of the "Exhibition Program" at the Institute, in effect, an institutional archive of some of the most widely recognized postmodern positions of the 1970s and 1980s and some of the most forgotten protagonists of architectural modernism from the United States.

Oppositions 25, a special issue on the theme "Monument/Memory," the first to be edited by Kurt Forster, appeared in the fall of 1982 as the first issue of the eighth volume at Rizzoli. It was slightly different in appearance, with a high gloss cover. Forster had built the issue around the English translation of a 1903 essay "The Modern Cult of Monuments. Its Character and Origin" by Austrian art historian and monument conservator Alois Riegl. This applied less to creative works, be they artistic or literary, but rather to monuments that had been officially designated landmark sites, once erected and now preserved.[892] In this mediation on works of art and historical value, *Oppositions* 25 represented a very different view of history than the one held by Frampton and Vidler as editors, namely that of a European style of modernism that discussed both models and precursors of a modernist movement or era, or even those positions that had previously been held by the "Venice School" around Tafuri. In

890 The exhibition on William Lescaze was taken over by Syracuse University and subsequently went on tour; the exhibition "Le Corbusier's Firminy Church" was shown in 1981–1982 at six other venues: Zolla-Lieberman Gallery, Chicago; Harvard GSD, Cambridge; Louisiana Tech, Ruston; Rice University, Houston; Carnegie-Mellon University, Pittsburgh; University of Minnesota, Minneapolis.

891 Vidler, 1982.

892 Alois Riegl, "The Modern Cult of Monuments: Its Character and Its Origin," trans. Kurt Forster and Diane Ghirardo, *Oppositions* 25 (Fall 1982), 21–51.

his editorial, however, Forster, drawing on Riegl and his concept of the will to art, rejected eclecticism as a central defining characteristic of a postmodern age and thus of the contemporary, playful, and at times ironic view of history.[893] The new *Oppositions* in its graphic variations—the cream-colored title, the simple journal cover—indicated a trend toward a different, more conservative form of postmodern discourse on history, but also testified to a thoroughly economized future of publishing at the Institute which reflected recent developments.[894] Despite all the changes and uncertainties, however, 1982 also brought cause for celebration: *Oppositions* was awarded a gold medal by the American Institute for Architects, the highest honor bestowed by the association every year on architects whose work had a lasting impact on architecture. Because of its merits, the journal continued to be considered the Institute's flagship publication, the primary medium of architecture debate (rather than practice) in North America, even as other publications competed for its market position.

893 Kurt Forster, "Monument/Memory and the Mortality of Architecture," *Oppositions* 25 (Fall 1982), 2–15.

894 Postmodern architecture was characterized precisely by the fact that architecture historians had set themselves the goal of reviving old forms of historiography; see Angelika Schnell, "What is Meant by History?" *Oase* 87 (2012): "Alan Colquhoun: Architect, Historian, Critic," 58–76, here 59.

THE PHILIP JOHNSON CENTER FOR ARCHITECTURE

Philip Johnson is undisputably America's leading architect. As critic, philosopher and practitioner, he has exerted enormous influence over our perceptions of design, always setting excellence as the single criterion by which a work must be judged. For many he is the great prophet of our time, for the Institute for Architecture he has been the guiding spirit. It is, therefore, most fitting that the future home of the Institute for Architecture and Urban Studies be within the newly created Philip Johnson Center for Architecture.

The Philip Johnson Center for Architecture will have its own charter, board of governors and may, at its discretion, play host to any number of architectural entities in the future. From its inception, it will act as the new home of the Institute for Architecture and Urban Studies.

The present intention is to acquire a landmark building situated in the Murray Hill section of Manhattan to be named the Philip Johnson Center for Architecture. The building, which was erected for the Lanier family in 1904, is in the Beaux Arts style of McKim, Mead and White. It is located one block away from the Morgan Library and within easy access of Grand Central Station, buses, subways and the Queens Midtown Tunnel. It is an ideal location for students, scholars and the general public alike.

Behind its impressive facade, the building contains some 17,000 square feet of usable area on seven stories, including a basement. This building would almost double the present floor space of the Institute for Architecture and Urban Studies, thus not only meeting the current needs of that institution, but also providing for its future growth as well as anticipating possible additional affliations. The building is situated in an area where there are a number of other cultural and academic facilities which could futher amplify the building's capacity.

Among the essential public facilities to be provided by the new building, the following resources are of particular importance to the development and maintenance of the Institute for Architecture and Urban Studies programs:

* A 150 seat auditorium
* A main floor exhibition space
* A library
* An archive
* A slide library
* An advanced study and research center
* Augmented administrative, editorial, research and educational facilities.

Fig. 125

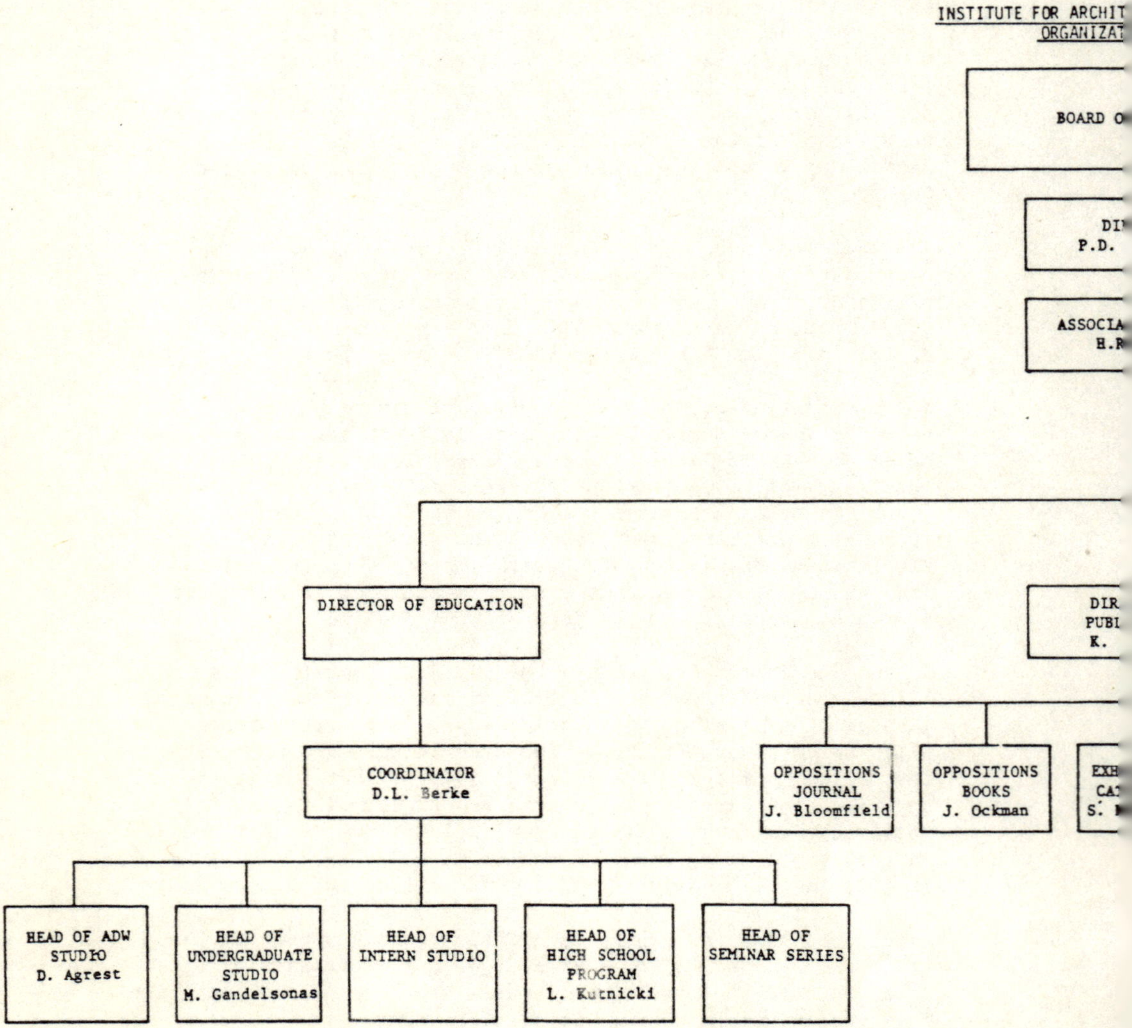
INSTITUTE FOR ARCHIT
ORGANIZAT
BOARD O
DI
P.D.
ASSOCIA
H.R
DIRECTOR OF EDUCATION
DIR
PUBL
K.
COORDINATOR
D.L. Berke
OPPOSITIONS
JOURNAL
J. Bloomfield
OPPOSITIONS
BOOKS
J. Ockman
EXH
CA
S.
HEAD OF ADW
STUDIO
D. Agrest
HEAD OF
UNDERGRADUATE
STUDIO
M. Gandelsonas
HEAD OF
INTERN STUDIO
HEAD OF
HIGH SCHOOL
PROGRAM
L. Kutnicki
HEAD OF
SEMINAR SERIES

Fig. 126

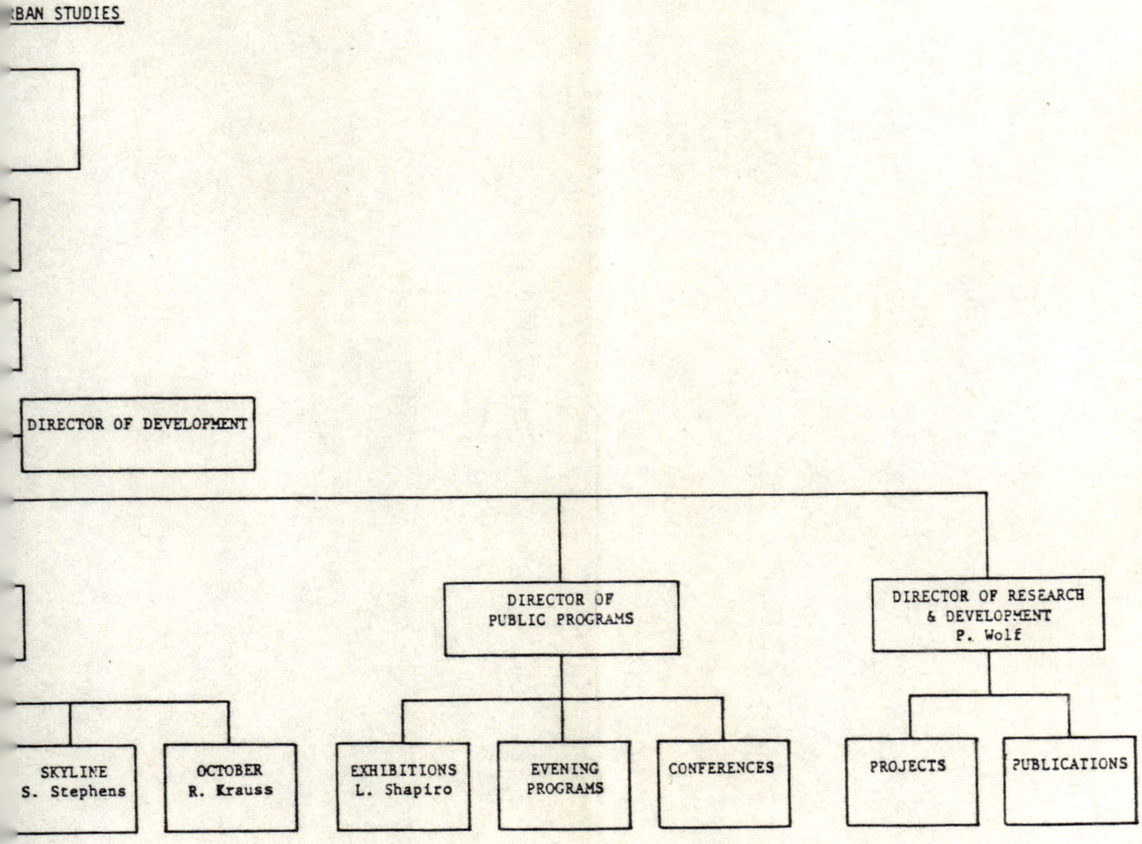

RBAN STUDIES
DIRECTOR OF DEVELOPMENT
DIRECTOR OF PUBLIC PROGRAMS
DIRECTOR OF RESEARCH & DEVELOPMENT P. Wolf
SKYLINE S. Stephens
OCTOBER R. Krauss
EXHIBITIONS L. Shapiro
EVENING PROGRAMS
CONFERENCES
PROJECTS
PUBLICATIONS

Fig. 127

The Institute for Architecture and Urban Studies
Eight West Fortieth Street, New York, N.Y. 10018
Telephone: 212 398 9474
1982
1967

Fig. 128

The IAUS at 15

Left to right: Kenneth Frampton, Peter D. Eisenman, Edward L. Saxe.

An exhibition celebrating the **15th Anniversary of the Institute for Architecture and Urban Studies** opened on November 15 and will run into 1983. The exhibition was organized by Margot Jacqz with the assistance of all those at the Institute.

Margot Jacqz and Kenneth Frampton

15th Anniversary party and exhibit

The Institute for Architecture and Urban Studies was founded in 1967 in conjunction with the Museum of Modern Art and Cornell University. Its founding followed an exhibition of urban design proposals held at the Museum of Modern Art and titled "The New City: Architecture and Urban Renewal." It featured the work of a group of young architects and planners seeking creative alternatives to traditional forms of education and practice. A number of groups emerged out of the exhibition, one of which became the Institute for Architecture and Urban Studies, under the direction of Peter Eisenman.

From the outset it was intended that the Institute remain entirely independent of any exisiting school or agency so that it could develop a unique form of educational structure in which students and teachers would work together in an effort to achieve a synthesis between the theoretical world of the university and the real problems confronting urban centers throughout the country. The Institute was thus initially structured as an "atelier," with teachers and students working on a number of projects that were successively commissioned by a number of government agencies, including HUD, UDC, and NIMH. In the first year many of the Institute students came from the Cornell University Urban Design Program headed by Colin Rowe. The Institute's first practical assignment—a design study of Kingsbridge Heights in the Jerome Avenue area of the Bronx—came from the New York City Planning Commission.

The intent to develop a theory of urban form—one that would be capable of influencing the actual planning process—culminated in 1973 with the design and exhibition of two "Low-Rise High-Density" housing prototypes and projects, one of which was for the construction of 650 dwelling units in the Brownsville section of Brooklyn. The realization of this scheme was nationally acknowledged as an important contribution to the investigation of such low-rise models.

The student-faculty dialogue established through these projects and the development of a new attitude towards architectural education was to be the basis for what has since become a number of different educational programs currently in progress at the Institute. These programs have been devised to serve a broad spectrum of students in an effort to fill a wide range of study between pre-architectural training and a consideration of architecture as a humanistic discipline. In 1974 the Undergraduate Program was formed for college students, and in 1976 the Design and Study Options program, now known as the Advanced Design Workshop, was formally established for students enrolled in six-year professional programs elsewhere. Around the same time, the Internship Program was also started as a course of study for students of varying backgrounds who wish to enter the realm of architecture. Last but not least, mention should be made of the High School Program, which has been held regularly in the Institute as an auxiliary weekend and summer course since 1975.

To further engage the public-at-large in the discussion of architecture, the Institute also established an Evening Program of lectures in 1974 as a natural extension of the "in-house" lectures that had been taking place since its foundation. At the same time, beginning with a Russian Constructivist exhibition in 1971, a series of exhibitions was initiated to illustrate various aspects of the current state of architectural design and research. Since that time the Institute has mounted some 32 exhibitions, including seminal shows such as "Idea as Model," the Japanese and Austrian "New Wave," and the work of individual architects—among them Scolari, Rossi, the Kriers, Leonidov, Ungers, and Hejduk. The evening lecture program became consolidated in 1977 under the title of Open Plan and, with strong support from the National Endowment for the Humanities, was developed into an integrated three-year, inter-disciplinary public education program.

To further disseminate its concerns among those outside the organization, the Institute has also undertaken a number of publications: *Oppositions*, an internationally respected journal of architectural theory, history, and criticism; *October*, a quarterly of theory and criticism on the arts within their social and political context; *Skyline*, a monthly review of news, books, events, and discussion in architecture; Exhibition Catalogues, which provide documentation and elaboration on work presented at the Institute; and the new series of Oppositions Books, which extend *Oppositions*' coverage of theory and criticism of modern architecture.

In June of 1982, Peter Eisenman resigned as director of the Institute and will now serve as Vice Chairman of the Board of Trustees. Under the new leadership of Edward L. Saxe as President and Kenneth Frampton as Director of Programs, the Institute is looking forward to strengthening both its civic role and its reputation as a center for advanced research and a forum for architectural debate through the reactivation of public programs embracing a variety of disciplines. A renewed emphasis will be placed on endeavors in the field of urban studies in both Institute publications and educational programs. It will remain an organization whose influence is felt not only throughout the world of architecture but also in the lay community, supporting a unique variety of activities.

Notes from the Sidelines

As a member of the "trade" architectural press for the first fourteen of the fifteen years the Institute has been in business, I have been able to observe in an interested but reasonably detached manner the various transformations and developments occurring inside the organization. But more importantly, during that time I have witnessed the impact of the Institute on the architectural community of professionals, journalists, and academics, as well as lay people. For a small independent educational-, research-, and publications-oriented organization, the Institute's influence has been enormous: It has coalesced and consolidated a critical discourse in architecture. It gave substance to intellectual explorations occurring at the time, and directly or indirectly influenced actual designs produced by the major architectural firms of the period.

When the IAUS was founded in 1967, its planning and urban design activities were most evident to outsiders. In the 1960s Jane Jacobs' pragmatic and empirically-based conclusions about street life, and the need for designing within existing urban patterns—for acknowledging the presence of the pedestrian and the importance of the public spaces between buildings—was being much touted. The Institute was soon submitting such values to intensive and systematic investigation, greatly influenced by the historically-based analyses of Colin Rowe and his contextually-oriented theories.

Planning projects undertaken by the Institute in its early years included research on various sections of the city for the City Planning Commission and an analysis of new towns alternatives (the "New Urban Settlements" project) in 1970 with Emilio Ambasz and Peter Eisenman, co-directors, and Kenneth Frampton and Susana Torre, director and coordinator of the analytic phases. Probably the best known example of the Institute's planning and architectural research work, however, was the "Low-Rise High-Density" housing prototypes Frampton and the IAUS developed with the Urban Development Corporation in 1973. The prototype, built in Brooklyn in 1979, recognized the value of low-rise living for low-income family dwellings, incorporating "house-like" features missing in large-scale projects, such as double exposures, private gardens, and individual entrances to the units. Like the "case-study" street scheme for Binghamton published in *On Streets* (edited by Stanford Anderson for the IAUS and published in 1978 by MIT Press), the "Low-Rise High-Density" housing addressed the problems of the pedestrian-oriented *place* where architectural elements of housing and urban design elements of the street would intersect to foster social interaction. While these and other planning studies, reflecting the concerns of the day, advanced certain innovations, or at least refinements of prevailing strategies, it is difficult to make absolute claims for the ultimate impact of the investigations at this point.

One can see however, the effect of the Institute's other more publicly-directed activities. Through its symposia, exhibits, lectures, and publications, the Institute managed to foster a climate of debate in the 1970s when there was virtually none. In its forums and "Open Plan" lectures, with its introduction of *Oppositions* and other publications, the IAUS was to spur a renewed interest in architectural theory, history, and criticism. Many observers have complained that the language of the debate was sometimes hard to decipher, but nevertheless theoretical ideas of significance were being heard or read by those of us in the "normal" architectural world. Soon ideas and arguments by Mario Gandelsonas and Diana Agrest on meaning in architecture, or by Anthony Vidler and Rafael Moneo on typology, by Kenneth Frampton on social and political determinants of form, were being bruited about. Journalists and academics outside the Institute began to listen and to look.

The intellectual discussions, exhibits and publications, criticized for being "hermetic" and "high-design," nevertheless spurred the "nationalization" of architectural discourse. Other architects in other cities began organizing their own lectures, exhibits, and conferences. The mid-1970s appearance of the "L.A. Silvers" or the "Chicago Seven" and the recent revival of the Chicago Architectural Club are only some of the well-known manifestations of architects debating each other, and eventually communicating with the public. Even architects designing large-scale work were influenced by that discussion, if not by some of the work being produced within the Institute's walls. The fact that the Agrest/Gandelsonas tower form of 1981, based on the investigation of the formal, symbolic and urbanistic analyses of 1920s skyscrapers, can now be spotted in designs coming out of the larger offices underscores the implications of that influence.

Besides generating debate on the domestic front, the Institute has religiously brought the news of other architectural developments in England, Italy, Spain, Japan, and elsewhere to the architectural community in New York. Its exhibits have antedated the publication of this work in the major professional press by several years: Architects such as Arata Isozaki and Aldo Rossi were showing their work at the Institute long before it began to appear in four-color spreads in the U.S.

And now with these successes behind it, the Institute faces a new stage of development. At fifteen it has reached maturity, but hardly its dotage. It is important that the Institute maintain its past diversity of activity and the strength of its intellectual leadership, while influencing an even larger sector of the public. The path widens. — **S S**

Fig. 129

Fig. 130

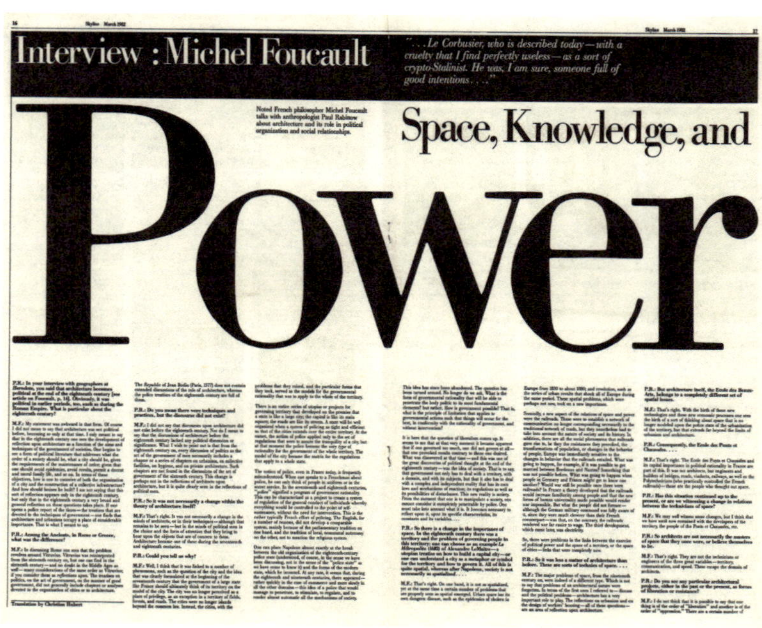

Interview : Michel Foucault

". . . Le Corbusier, who is described today—with a cruelty that I find perfectly useless—as a sort of crypto-Stalinist. He was, I am sure, someone full of good intentions. . . ."

Noted French philosopher Michel Foucault talks with anthropologist Paul Rabinow about architecture and its role in political organization and social relationships.

Space, Knowledge, and

Power

Fig. 131

The Institute for Architecture and Urban Studies
8 West 40 Street, NY, NY 10018
212-398-9474
ON STYLE
Two Evenings at the IAUS
December 6, 1982
6:30 pm
Presentation of The Portland Building
by Michael Graves
Exhibition
Working drawings, sketches and photographs
of The Portland Building
December 8, 1982
6:30 pm
Panel Discussion
"On Style: A Critical Assessment"
Vincent Scully
Kurt Forster
Paul Goldberger
Mario Gandelsonas
Rosalind Krauss
Douglas Crimp
Tony Vidler, Moderator

Fig. 132

The Institute for Architecture and Urban Studies
8 West 40 Street, New York, New York 10018
212-398-9474
THE HIGH SCHOOL ARCHITECTURE STUDIO
For the first time, the High School Program in Architecture offers a series of five week courses that are designed to be taken as a series but may also be taken individually. The curriculum is geared to students of all backgrounds with an interest in the composition of their physical environment. All three courses are run as studios with supplementary lectures, discussions, and field trips. The program is taught by Institute faculty with the aid of guest architects and critics.
The High School Program will meet at the Institute for Architecture and Urban Studies, 8 West 40 Street, NYC, on Saturdays, from 10:30 to 3:30. Each five week course costs $75. For further information and registration please contact Ms. Linda Dukess Bernstein at 212-398-9474.
The Institute for Architecture and Urban Studies is a non-profit organization offering a wide variety of programs and publications in the fields of architecture and design.
The Courses
Five In Five
November 13-December 18
A five week course devoted to the study of pivotal figures in the emergence of modern architecture. Each session will use the work of a famous practitioner to illustrate a major principle of design and theory of the twentieth century. The cast of heroes includes Frank Lloyd Wright, Mies van der Rohe, Le Corbusier, Louis Kahn, and Alvar Aalto.
Five Again
January 15-February 12
An examination of the figures who are presently shaping our physical environment. The course will investigate the contributions of contemporary architects.
The Muses + Two
March 12-April 16
A study of the relation between architecture and the arts of poetry, painting, drama, dance, and film.
Design: Massimo Vignelli

Fig. 133

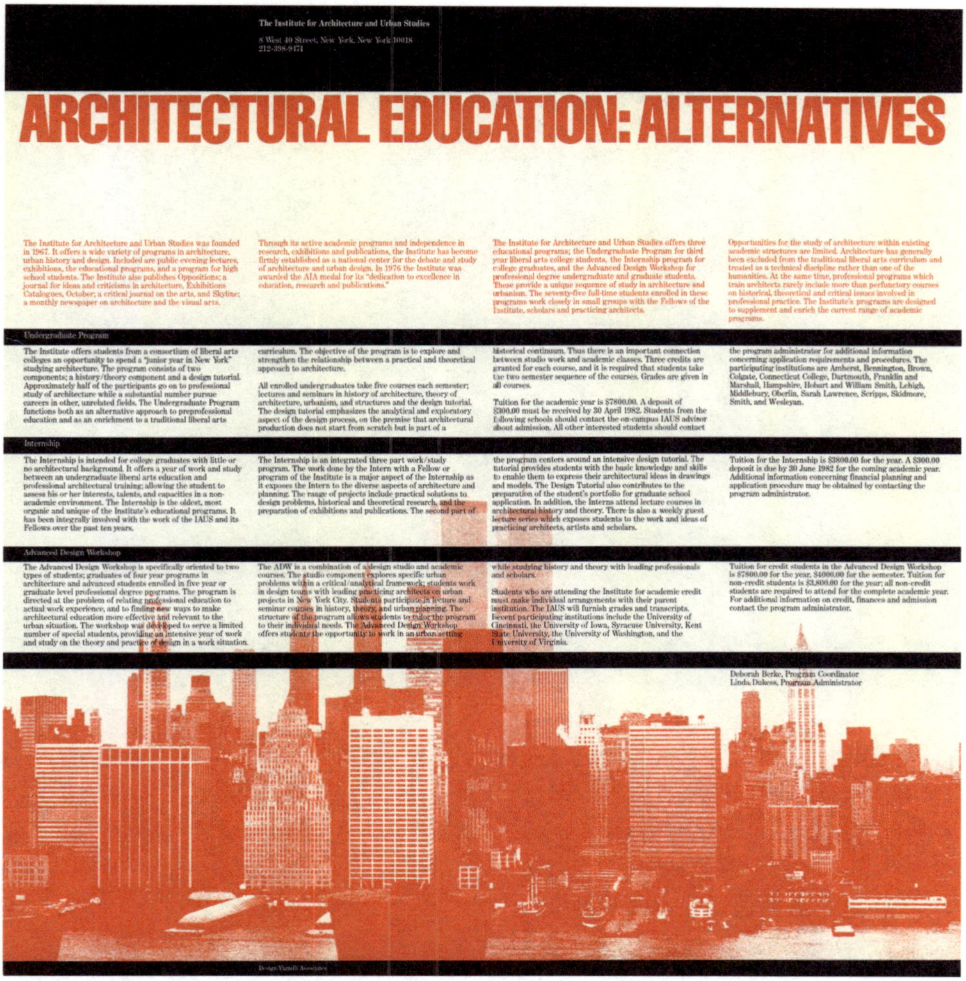
The Institute for Architecture and Urban Studies
8 West 40 Street, New York, New York 10018
212-398-9474
ARCHITECTURAL EDUCATION: ALTERNATIVES
The Institute for Architecture and Urban Studies was founded in 1967. It offers a wide variety of programs in architecture, urban history and design. Included are public evening lectures, exhibitions, the educational programs, and a program for high school students. The Institute also publishes Oppositions; a journal for ideas and criticisms in architecture, Exhibitions Catalogues, October; a critical journal on the arts, and Skyline; a monthly newspaper on architecture and the visual arts.
Through its active academic programs and independence in research, exhibitions and publications, the Institute has become firmly established as a national center for the debate and study of architecture and urban design. In 1976 the Institute was awarded the AIA medal for its "dedication to excellence in education, research and publications."
The Institute for Architecture and Urban Studies offers three educational programs; the Undergraduate Program for third year liberal arts college students, the Internship program for college graduates, and the Advanced Design Workshop for professional degree undergraduate and graduate students. These provide a unique sequence of study in architecture and urbanism. The seventy-five full-time students enrolled in these programs work closely in small groups with the Fellows of the Institute, scholars and practicing architects.
Opportunities for the study of architecture within existing academic structures are limited. Architecture has generally been excluded from the traditional liberal arts curriculum and treated as a technical discipline rather than one of the humanities. At the same time, professional programs which train architects rarely include more than perfunctory courses on historical, theoretical and critical issues involved in professional practice. The Institute's programs are designed to supplement and enrich the current range of academic programs.
Undergraduate Program
The Institute offers students from a consortium of liberal arts colleges an opportunity to spend a "junior year in New York" studying architecture. The program consists of two components; a history/theory component and a design tutorial. Approximately half of the participants go on to professional study of architecture while a substantial number pursue careers in other, unrelated fields. The Undergraduate Program functions both as an alternative approach to preprofessional education and as an enrichment to a traditional liberal arts curriculum. The objective of the program is to explore and strengthen the relationship between a practical and theoretical approach to architecture.
All enrolled undergraduates take five courses each semester; lectures and seminars in history of architecture, theory of architecture, urbanism, and structures and the design tutorial. The design tutorial emphasizes the analytical and exploratory aspect of the design process, on the premise that architectural production does not start from scratch but is part of a historical continuum. Thus there is an important connection between studio work and academic classes. Three credits are granted for each course, and it is required that students take the two semester sequence of the courses. Grades are given in all courses.
Tuition for the academic year is $7800.00. A deposit of $300.00 must be received by 30 April 1982. Students from the following schools should contact the on-campus IAUS advisor about admission. All other interested students should contact the program administrator for additional information concerning application requirements and procedures. The participating institutions are Amherst, Bennington, Brown, Colgate, Connecticut College, Dartmouth, Franklin and Marshall, Hampshire, Hobart and William Smith, Lehigh, Middlebury, Oberlin, Sarah Lawrence, Scripps, Skidmore, Smith, and Wesleyan.
Internship
The Internship is intended for college graduates with little or no architectural background. It offers a year of work and study between an undergraduate liberal arts education and professional architectural training; allowing the student to assess his or her interests, talents, and capacities in a non-academic environment. The Internship is the oldest, most organic and unique of the Institute's educational programs. It has been integrally involved with the work of the IAUS and its Fellows over the past ten years.
The Internship is an integrated three part work/study program. The work done by the Intern with a Fellow or program of the Institute is a major aspect of the Internship as it exposes the Intern to the diverse aspects of architecture and planning. The range of projects include practical solutions to design problems, historical and theoretical research, and the preparation of exhibitions and publications. The second part of the program centers around an intensive design tutorial. The tutorial provides students with the basic knowledge and skills to enable them to express their architectural ideas in drawings and models. The Design Tutorial also contributes to the preparation of the student's portfolio for graduate school application. In addition, the Interns attend lecture courses in architectural history and theory. There is also a weekly guest lecture series which exposes students to the work and ideas of practicing architects, artists and scholars.
Tuition for the Internship is $3800.00 for the year. A $300.00 deposit is due by 30 June 1982 for the coming academic year. Additional information concerning financial planning and application procedure may be obtained by contacting the program administrator.
Advanced Design Workshop
The Advanced Design Workshop is specifically oriented to two types of students; graduates of four year programs in architecture and advanced students enrolled in five year or graduate level professional degree programs. The program is directed at the problem of relating professional education to actual work experience, and to finding new ways to make architectural education more effective and relevant to the urban situation. The workshop was developed to serve a limited number of special students, providing an intensive year of work and study on the theory and practice of design in a work situation.
The ADW is a combination of a design studio and academic courses. The studio component explores specific urban problems within a critical/analytical framework; students work in design teams with leading practicing architects on urban projects in New York City. Students participate in lecture and seminar courses in history, theory, and urban planning. The structure of the program allows students to tailor the program to their individual needs. The Advanced Design Workshop offers students the opportunity to work in an urban setting while studying history and theory with leading professionals and scholars.
Students who are attending the Institute for academic credit must make individual arrangements with their parent institution. The IAUS will furnish grades and transcripts. Recent participating institutions include the University of Cincinnati, the University of Iowa, Syracuse University, Kent State University, the University of Washington, and the University of Virginia.
Tuition for credit students in the Advanced Design Workshop is $7800.00 for the year, $4000.00 for the semester. Tuition for non-credit students is $3,800.00 for the year; all non-credit students are required to attend for the complete academic year. For additional information on credit, finances and admission contact the program administrator.
Deborah Berke, Program Coordinator
Linda Dukess, Program Administrator
Design: Vignelli Associates

Fig. 134

The Institute for Architecture
and Urban Studies
Educational Programs
1984–1985

Fig. 135

The Institute for Architecture and Urban Studies

19 Union Square West, New York, NY 10003
Telephone: 212 206-1116

PRESENT IAUS FACULTY

1984 - 1985

Lynne Breslin	A.B. Harvard University; M. Arch., M.A. Princeton
Deborah Gans	A.B. Harvard University; M. Arch. Princeton
Paul Gates	B. Arch., U.S.C.; M. Arch. Princeton
Christian Hubert	B.A. Columbia University; M. Arch. Harvard
Michael Manfredi	B. Arch. University of Notra Dame; M. Arch. Cornell
David Mohney	A.B. Harvard University; M. Arch. Princeton
Michael Monsky	B.A. Northeastern University; B. Arch. Cooper Union; M. Arch. Yale University
Steven K. Peterson	B. Arch. Cornell University; M. Arch. Cornell
Stephen Potters	B. Arch. Cornell University; M. Arch. Cornell
Joel Sanders	B.A. Columbia University; M. Arch. Columbia Univ.
Pat Sapinsley	B.A. Hampshire College; M. Arch. Harvard
Jon Michael Schwarting	B. Arch. Cornell University; M. Arch. Cornell
Graham Shane	Diploma, Architectural Association; M. Arch. Cornell University; Ph.D. Cornell
Robert Silman	B.A. Cornell University; BCE New York University; MCE New York University
Andrea Simitch	B. Arch. Cornell University
Jerri Smith	B. Arch. Ohio State University; M. Urban Design Cornell University

Fig. 136

News

The Institute for Architecture and Urban Studies

1

February/March 1985
Volume I, Number 1

The IAUS newsletter is intended to keep the arts and architectural communities informed of Institute events and programs. Any letters or announcements will be welcomed and printed in future issues, space allowing. Please address all correspondence to: Julie Carlson, IAUS, 19 Union Square West, New York, NY 10003.

DESIGNER: Michael Bierut, Vignelli Associates
EDITOR: Julie Carlson

Program Notes

IAUS PUBLIC PROGRAMS
This spring, the Institute is offering the most extensive series of lectures, panel discussions, seminars, and forums since 1980. The series will continue in the spirit of early IAUS programs, praised by Paul Goldberger for bringing "the gospel of architecture to the average man on the street." Leading architects, urban designers and planners, and interested participants will gather to discuss and debate issues inherent in the relationship between architecture and the city. The 1985 Spring Public Programs will focus in particular on the treatment of public spaces in the public realm. Future series will draw thematic inspiration from the design issues of the moment, as well as from subjects of continuing significance.
The exchange of ideas and information is central to the mission of the IAUS Public Programs; time will be reserved at the end of each program for participants to air their reactions and for panelists and lecturers to respond to questions and engage in debate and discussion. Series subscriptions will be sold in advance. Seating is limited so please mail checks in advance to IAUS, 19 Union Square West, New York, NY 10003. Please write the title of the series on the face of the check. If available, seats for individual lectures will be sold through phone reservation. Please call 212-206-1116 for reservations or information. All programs take place at 6:30 pm at the IAUS, 19 Union Square West, unless otherwise indicated.

IAUS JOINT VENTURES: THE CITY AND THE GARDEN
Five films produced by Edmund Bacon on the history of the city will be shown at the Urban Center on March 18, April 1, and April 15, 1985. On alternate weeks the series will meet at IAUS for related lectures concerning the history of public space in both the garden and the city. This series is being co-sponsored by the American Society of Landscape Architects.
$40 for the series, $7 for each evening
March 11: Claudio Lazzaro at IAUS
March 18: Edmund Bacon at the Urban Center
March 25: Steven Peterson at IAUS
April 1: Film at the Urban Center
April 8: Joanna Dougherty at IAUS
April 15: Film at the Urban Center
April 22: Anne Griswold Tyng at IAUS

IAUS OPEN REVIEW: NEW YORK PUBLIC SPACE
The simultaneous development of 42nd Street, Bryant Park, Battery Park, and Union Square will change the character of some of New York's most familiar public spaces and will have a tremendous impact on future developments. The Institute is providing the opportunity for the principal architects and designers of some of these projects, along with critics from various publications and architecture schools, to discuss important design issues raised by these proposals. Sessions will be conducted in the spirit of an open critical review.
$30 for the series, $10 for each evening
March 20: Bryant Park: Daniel Beiderman, John Belle, Bronson Binger, Hugh Hardy, Mary McLeod, Carter Wiseman
March 27: 42nd Street: John Burgee, Philip Johnson, William Pederson, Jaquelin Robertson, Robert Slutzky, Robert Stern
April 10: Union Square: Jonathan Barnett, David Birney, Hui Mei Grove, Fred Koetter, James Polshek, Terrance Williams, Peter Wolf
April 17: Rector Park at Battery Park City: Amanda Burden, William Conklin, Stan Eckstut, Ulrich Franzen, John James, Steven Peterson, James Polshek, Colin Rowe, Peter Samton
April 24: Commentary on the Public Realm: Michael Dennis, Robert Maxwell, Colin Rowe, William Taylor

IAUS LECTURES: LIFE AND DEATH IN VENICE 1780–1980
By Anthony Vidler, Professor of Architectural History and Theory, Princeton University and IAUS.
$28 for the series, $7 per evening
February 7, 14, 21, 28
March 7

IAUS LECTURES: URBAN ARCHITECTURE
Eleven architects have been invited to discuss their recent projects and to examine the particular relationship between their work and the surrounding urban context.
$55 for the series, $7 for each evening
January 31: Frank Gehry
February 6: Stanley Tigerman
March 4: Paul Kennon
March 14: Tod Williams
March 21: Beyhan Karahan
April 4: Tom Beeby
April 11: Fred Koetter
April 18: Victor Caliandro
April 25: Michael Dennis
April 29: Graham Gund
May 2: John Burgee

IAUS PUBLIC FORUM: FIFTEEN FACADES
The facade is the primary surface which defines urban space. The fifteen panelists in this series will each select an important New York City or other historical facade to be discussed and critically evaluated. Three panelists each evening will explain, defend, and otherwise justify their choices.
$25 for the series, $7 for each evening
February 13: Steven Peterson, Thomas Schumacher, Robert Slutzky
February 20: Barbara Littenberg, Michael Schwarting, Bartholomew Voorsanger
February 27: Alan Chimacoff, Richard Haas, William Pederson
March 6: Edward Mills, Graham Shane, Jerry Wells
March 13: Peter Eisenman, Michael Graves, Robert Maxwell

15

NEW PROGRAMS
The Institute is introducing two new programs to provide postgraduates with the opportunity for further exploration of urban issues.
The Program in Advanced Study in Architecture and Urban Design is offered to practicing architects who wish to pursue advanced work on issues generated by urban architecture. "This program is unique," says Steven Peterson. "Historically architecture and urban design evolved as two distinct departments. This separation of purpose results in overspecialization. Architects are not taught to conceptualize the whole city, and yet they are the very specialists who need to integrate environmental considerations into their designs." Visiting critics participating in the program will include: Michael Graves, Princeton University, and Robert Krier, University of Vienna.
The Program in Advanced Theory and Criticism is intended for students interested in developing a critical facility in the analysis of urban issues. Peterson comments, "This new program will examine the assumptions, methods, and criterea employed in architectural judgment. Architecture is a fashionable topic, yet its current vogue is not synonomous with a widespread knowledge or sound judgment about architectural design. The program in theory and criticism is intended to advance the dialogue concerning architecture and urban design." Colin Rowe from Cornell University, Alan Colquhoun from Princeton University, and Arthur Drexler, Director of the Department of Architecture and Design at the Museum of Modern Art will be among the visiting critics taking part in this program.

Events

Philip Johnson and John Burgee, IAUS Trustees

The newly occupied AT&T building, the landmark corporate headquarters designed by IAUS trustees John Burgee and Philip Johnson, opened its doors on October 24 for a cocktail reception hosted by the Institute. One hundred and fifty Institute benefactors and affiliates, including architects, members of the arts community, and IAUS trustees admired the white marble expanses, toured the executive enclaves located under the Chippendale pediment, and mingled in the flower-filled reception hall.

Arthur Drexler, Director, Department of Architecture and Design, MOMA, and Armand Bartos, former IAUS Trustee

Steven Peterson, IAUS Director and Trustee, Edward Mills, Voorsanger and Mills Associates, Architects, and Cesar Pelli, IAUS Trustee

John Burgee, Chairman and President of the Board at the Institute, introduced Steven K. Peterson, who made some short remarks and thanked Mr. Burgee and Mr. Johnson for their continued largess and support over the years. Mr. Burgee and Mr. Johnson also spoke, commenting briefly on the design history of the building. It was a spirited occasion, bringing together IAUS contributors and trustees to celebrate the history and accomplishments of the Institute.

Massimo Vignelli, Richard Meier, and Philip Johnson, IAUS Trustees

Charles Gwathmey, IAUS Trustee, and William McMinn, Dean, College of Architecture, Art, and Planning, Cornell University

Photographs by Dorothy Alexander

IAUS

Fig. 137

Coda: Institutional Legacy and Critical History

Even during its lifetime, the Institute for Architecture and Urban Studies not only made history but also wrote its own: be it with its promotional material which listed its achievements for the purposes of public relations and media outreach, or with an exhibition for its fifteenth anniversary shown at the Institute's premises at 8 West 40th Street in November 1982 after the once so charismatic Peter Eisenman, the Institute's founder and longtime director, stepped down earlier that year.[895] Exhibits included the Institute's research and design projects, posters of its events—both lecture series and exhibitions—and various printed matter—early brochures and especially the covers and single pages of publications—as well as architectural projects stemming from the education program.[896] The fundamental differences between the possible narratives about the Institute were highlighted in the December 1982 issue of *Skyline* which, below a triptych of three portraits—in the middle Eisenman, now Vice President of the Board of Trustees, flanked by Kenneth Frampton, who took over the Institute's day-to-day management as director of programs in June 1982, and Edward Saxe, who was briefly the President and CEO of the Institute in 1982–83 and for the past year had been tasked to look after its economic well-being and financial survival,

895 On the naturalization of making history, see Tomàs Llorens, "On Making History," in Ockman, ed., 1985, 24ff. Eisenman launched a successor grouping to the Institute with a two-day symposium at the University of Pennsylvania in Charlottesville, mysteriously named "P3," of selected practicing architects—the very weekend that the Institute's anniversary was to be celebrated. See Robertson, 1985; Robertson and Tigerman 1991.

896 In the run-up to the anniversary exhibition, the editors of the IAUS Exhibition Catalogues had already spoken out against a catalogue that would have been nothing short of a requiem for them.

featured two articles celebrating its anniversary.[897] Under the heading "The IAUS at 15," Margot Jacqz and Frampton jointly wrote a matter-of-fact report on the Institute's various fields of activities, i.e., about what it had intended to accomplish as a group and what it actually had accomplished over the past years by focusing on architectural movements and their manifestos (as Frampton, the historian, had done in *Modern Architecture*), while Suzanne Stephens, who as the editor of *Skyline* was in charge of both quality and entertainment journalism, wrote about the Institute's influence on architectural discourse, its outstanding personalities, and their individual contributions—also with the goal of maximizing the journal's readership to make it commercially viable. At this point, the Institute's story was not yet over, even if its fate seemed sealed by the fundamental fifteen-year conflict between all the discontinuities of bureaucracy and charisma, institutionalization and consolidation, professionalized business, and generational change.[898] While categories of critical theory such as race, class, gender, and sexuality were kept out of these kinds of institutional accounts, the institutional legacy, continuation, and influence of the Institute's hard-won position was not just about "scattered elements of building knowledge and notions of design," but the authority that came with "the whole process of symbolization, mythical transposition, taste, style, and fashion."[899]

Building Institution thus expands conventional narratives in architecture history on knowledge production. A critical history of the Institute would relate this to the architectural community in New York, the USA, and across the world, as reflected in the Institute's educational and cultural offerings, and especially in its publications since the mid-1970s, all of which were put to the test in the early 1980s with the elimination of public cultural funding in the summer of 1980, the greater role given to patronage, the demand for publishing commitments, and the shift to commercial publishers. Eventually it was the Institute's publications (more than the profession) that laid the groundwork for widespread impact on the discipline and still resonate in architecture and architectural education today (and the debates about criticality and post-criticality, pragmatism and dogmatism of the 1980s, 1990s and 2000s were arguably a symptom rather than a cure): First, a juxtaposition of very different but ultimately

897 Margot Jacqz and Kenneth Frampton, "The IAUS at 15," *Skyline* (December 1982), 33; Suzanne Stephens, "Notes from the Sidelines," *Skyline* (December 1982), 33.

898 In her history of Oppositions, Joan Ockman characterizes the large-scale development of the Institute as a history of "bureaucratization," see Ockman, 1988. While at the Institute there was evidence of bureaucratization, be it political or financial, since the beginning, with the granting of non-profit status, the negotiation of the Fellowship, and the accountability to third parties, she thus addresses, whether consciously or unconsciously, sociologist Max Weber's threefold definition of types of rule, and the transition from traditional, or charismatic to legal-rational authority. See Weber, 2019.

899 Demetri Porphyrios, "On Critical History," in Ockman, ed., 1985, 16ff. For Porphyrios, critical history examines "the process of naturalization of architectural ideology into myth" and is structured "by relations invested in institutions."

self-centered and self-serving theories (and to a lesser extent histories) competing for intellectuality and debatability; second, a diffuse concept of research that is absorbed in the dominance of research and curation as practices; and third, the lack of specific content, as certain issues that architecture has had to address—domination along one or more axes of inequality, oppression, power, prejudice, stratification, and subordination; housing that has been privatized and urbanization that has been economized; or more global, bio- and geopolitical trends such as environmental degradation, resource scarcity, population growth and labour migration—were ultimately largely ignored.[900]

Archives, Discontinuities, and Institutional Endings

Building Institution concludes with a look at the Institute's gradual decline, the waning and eventually cessation of its institutional operations, following an analysis of its social constructedness in terms of its founding narratives and mythmaking, the specificity and ephemerality of all the projects, programs, and products—both realized and unrealized—that were undertaken under Eisenman's lead, its creation and repeated reinvention and restructuring, always building on a degree of administration, the composition and re-composition of the Board of Trustees, depending on its shifting institutional mission, and the establishment and expansion of the Fellowship, which was awarded on merit and characterized by a system of roles and assignments. In conclusion, the epistemic shifts, that emerged at the Institute, the historical ruptures and breaks, following Michel Foucault, and the series of contested institutional endings that ultimately led to its closure in the spring of 1985, allow us to better understand the transition from one era to another and to draw general conclusions about the conditions and constraints of the very institution of architecture, explicitly of architectural culture today, in terms of knowledge, power, and subjectivities.[901]

900 A special issue on "History/Theory" was published on *e-flux Architecture* in the fall of 2018, in collaboration with the Institute for the History and Theory of Architecture (gta), ETH Zurich, with some contributions by American architecture historians addressing the discursive legacy of the Institute implicitly, if not explicitly: see Reinhold Martin, "On the Uses and Disadvantages of Architecture for History" *e-flux Architecture* (November 2, 2018): "History/Theory," last accessed: May 31, 2023, www.e-flux.com/architecture/history-theory/225181/on-the-uses-and-disadvantages-of-architecture-for-history; Mark Jarzombek, "The School of Architectural Scandals" *e-flux Architecture* (October 29, 2018): "History/Theory," last accessed: May 31, 2023, www.e-flux.com/architecture/history-theory/225182/the-school-of-architectural-scandals; Joan Ockman, "Slashed" *e-flux Architecture* (October 27, 2018): "History/Theory," last accessed: May 31, 2023, www.e-flux.com/architecture/history-theory/159236/slashed.

901 Even the grand narratives of former Fellows, give different ending dates. See Frank, 2010. Also, in the title of Diana Agrest's film *The Making of an Avant-garde: The Institute for Architecture and Urban Studies 1967–1984 (2013), suggests that the Institute lasted until 1984*, the year of her own departure. Other historical accounts have used this date without further looking into the matter.

There is some evidence in established and private archives, however, that the Institute underwent a slow decline that stretched out over a longer period of time; the new decade saw a gradual waning of interest and commitment of the veteran Fellows who had shaped its agenda, venture, and output in the 1970s. Before this decline became palpable, however, the Institute faced another major reinvention in the early 1980s, when many commendable contributors were inducted into the circle of Fellows and tasked with managing the now complex publication apparatus, with *Oppositions*, *October*, *Skyline*, the IAUS Exhibition Catalogues, and the Oppositions Books series (finally launched in early 1982), which shaped the discursive and material conditions of the production, use, and dissemination of knowledge, power, and institutionality, and when, at the same time, the next generation of institutional talent was being encouraged and called upon to take responsibility.[902] The changes on the management level and concerning the organizational structure were no less significant, as evidenced by the minutes of the Institute's meetings as well as those of the Board of Trustees.[903] Not only was Saxe, who as deputy director and general manager had formerly advised MoMA but was otherwise not experienced in the field, appointed to executive management, but the existing leadership for development, public relations, and outreach was replaced and restaffed. The appointment of Philip Johnson as a member of the board in 1980—at the same time as his first official appearance—followed by his office partner John Burgees in 1982 was decisive in terms of structures of power, along with the political and financial rationality at play. However, the motives and processes, interests and responsibilities for the Institute's transformation as a cultural institution, and the changes in the institutional and cultural production contexts that ultimately led to the demise of the Institute are difficult to reconstruct from the documentation kept in the archives.[904]

902 Porsché, Scholz, and Singh, 2022.

903 Suzanne Stephens published a brief commentary titled “Skyline Rises II” in “The Byline” section of the October 1982 issue of *Skyline* on the occasion of the newspaper's two year anniversary, in which she informed the readers of the recent change in the Institute's direction, which she welcomed and considered beneficial in terms of the strength of its internal structure and effectiveness in informing, if not influencing, decision-making processes in architecture and urbanism; see Suzanne Stephens, “Skyline Rises II,” *Skyline* (October 1982), 34. Eisenman, Frampton, and Saxe were subsequently added to *Skyline*'s editorial board.

904 In the IAUS fond at the CCA in Montréal, which presents itself, by title alone, an official archive, there are some gaps on the events of 1981–82 and 1982–83 compared to the scope of the founding years, not to mention completeness. One reason for this is that the IAUS fonds was compiled and bequeathed by Eisenman personally and thus actually belong to the Eisenman fonds. That the IAUS holdings are nevertheless administered independently can be understood as an argument for the independence of the Institute vis-à-vis Eisenman, even if archival questions of provenance remain unanswered. While the route to the CCA is established, it remains unclear how the Institute's archive came into Eisenman's hands. According to oral history, Eisenman took a large number of documents with him, when he stepped back as Institute director. Whether this occurred while he was still at the Institute or only after the Institute moved from 8 West 40th Street to 19 Union Square West in 1983 is not clear.

There are several sets of documents that are particularly revealing in terms of institutional history: these include a folder on the so-called "Philip Johnson Center for Architecture," a last major project planned under Eisenman in fiscal year 1981–82 but ultimately not realized, which was intended to transform the Institute into a more enduring institution.[905] According to the concept papers, the Philip Johnson Center was to create an umbrella that would have housed "The Archive of American Architecture," "The Center for Advanced Studies in Architecture," and "The Library for Primary Sources of Modernism," while also being home to "The Institute for Architecture and Urban Studies" itself.[906] While the Institute had for years been discussing whether to become a registered school of architecture with accredited degrees and to establish an academic library, not least because the "Educational Programs" had formed the backbone of the Institute's operations since the academic year 1974–75 and covered the fixed costs for rent and staff, it was now looking nervously and somewhat enviously toward Montréal, where the architect and philanthropist Phyllis Lambert had in 1979 just founded the Canadian Centre for Architecture (which, however, would first be accommodated in existing office spaces—the actual building designed by Peter Rose was not ready until 1987). Other documents indicate that the Institute had already found suitable Manhattan premises and had a "lovely landmarked building" at 123 East 35 Street in mind.[907] Naming the center after Philip Johnson, who before officially serving as a trustee had remained largely in the background while possibly acting as the main donor (and making only a limited appearance in the books), was a strategic choice.[908] It helped to open the doors to the corporate world; after all, Johnson, considered to be "indisputably America's leading architect," was the key power broker in the New York architectural world, holding court at the Four Seasons Restaurant on Park Avenue in the Seagram Building.[909] The fact that Johnson's fascist past had become known at the time did not matter here—on the contrary.[910] As part of this capital campaign, Eisenman then approached Houston-based developer Gerald Hines, among others, upon Johnson's recommendation.[911] At the same time, the Institute's Board of Trustees, and with it, its connections to social, political, and economic affairs, was restructured and expanded to include

905 IAUS, project description for "The Philip Johnson Center for Architecture," 1981, Source: CCA Montréal, IAUS fonds: A.1-13 / ARCH263662.

906 Cynthia Warwick Kemper, letter to Mrs. Armand P. Bartos, April 9, 1982, Source: CCA Montréal, IAUS fonds: A.1-13.

907 Minutes of the Board of Trustees meeting, February 22, 1982, Source: CCA Montréal, IAUS fonds: A.1-13.

908 Johnson as the Institute's "prime benefactor and *éminence grise*" had already been honored, not without controversy at the time. See Sorkin, 1978 (1991).

909 Warwick Kemper, 1982.

910 Varnelis, 1995.

911 Thomas Weaver and Peter Eisenman, "Peter Eisenman in conversation with Thomas Weaver," *AA Files*, no. 74 (2017),: 150–172.

commercially successful American architectural firms and clients and an impressive roster of international architectural stars.[912] In the end, however, despite Hines' substantial donation of US$1.4 million and other fundraising efforts, the Institute was apparently unsuccessful in raising the US$10 million it had sought for the purchase price, US$5 million for the building and US$400.000 operating costs annually, and Eisenman's most ambitious institutional plan, except for perhaps the establishment of satellites of the Institute in various North American cities, failed. Ultimately, plans for the Philip Johnson Center and the Institute's hopes of managing the turnaround and realizing its long-term goals had to be buried—certainly a loss for New York, if not American and even global architecture culture. It is hard to imagine how the Institute's institutional legacy would have manifested itself and be perceived today compared to other institutions besides the CCA, e.g., the Getty Center for the History of Art and the Humanities in Los Angeles, 1983 (now the Getty Research Institute) or the Deutsches Architekturmuseum in Frankfurt, 1984 (DAM). However, it would be years before the Institute's overwhelming whiteness at the time was fully exposed.[913]

Other documents show that, after the appointment of Hamid-Reza Nouri as the Institute's associate director and Lynn Holstein as a new director of development (having been in office since 1976, Frederieke Taylor had resigned from her post as well as her role as Institute Fellow in 1981), fundraising became a branch of its own and now defined every other activity. The Institute's successful association between architecture culture and commercialization was exemplified by Eisenman's almost single-handedly pushing through a 1981 relaunch of *Skyline*, which he vaunted as the Institute's most important publication (even more than *Oppositions* or the Oppositions Books series), not only as an architectural newspaper but also as a fundraising tool. In addition, a series of official and unofficial minutes from meetings of the Institute's Fellows held in the fall of 1981 testify to the fact that the Institute was not yet defunct, but had grand ambitions, as this was when the institution, which had in practice already been in place for five years, was for the first time divided into four functional departments (or "silos"): "Publication," "Education," "Public," and "Development."[914] These meetings once again addressed the truly big issues, e.g., the transformation of the Institute's publishing activities into a full-fledged publishing house, the professionalization of the "Internship Program" (under Mario Gandelsonas), the continuation of the "Advanced Design Workshop" (under Diana Agrest), the role of the lecture series and exhibitions and their funding, the resumption of commissioned work, and, by extension, the establishment of a research library, etc.

912 As of October 1982, the board also included John Burgee, Henry Cobb, Arata Isozaki, Phyllis Lambert, Cesar Pelli, Kevin Roche, Aldo Rossi, and James Stirling. The acquisition of donations was neatly recorded in an index card system that resides in the IAUS fonds at the CCA.

913 Linder, 1996. In 1996, an entire issue of *Any* magazine, produced out of Eisenman's office, was devoted to "Whiteness."

914 Minutes of Fellows' meetings, fall 1981. Source: CCA Montréal, IAUS fonds: A.2-9.

The IAUS fonds at the CCA also holds several miscellaneous documents (applications, flyers, posters, and press releases) in a folder about what was known as the "Young Architects' Circle," a group that almost had parity and operated on an equal footing.[915] Part of the program they curated, generously sponsored by Walter Chatham, one of the Institute's trustees, in the spring of 1981 consisted of an event series, under the title "ReVisions," organized and administered by Joan Ockman and Christian Hubert, of twelve Monday evenings held at the Institute. Another part was the announcement of an architectural competition for individuals aged thirty-five or under for an intervention in Columbus Circle in midtown Manhattan (the winning entry was Elizabeth Diller's installation of 2,500 traffic cones, each spaced four meters apart). This was, according to participants, followed by the formation of the Young Architects' Circle as a reading group, which held its meetings outside of the Institute, in private SoHo lofts, where it focused on post-structuralist, post-Marxist theory.[916] For in the spring of 1982, at a time when individual Fellows were starting to voice internal complaints about Institute matters apparently being settled at the "Century Club" and wondering what direction they wanted the Institute to take,[917] the Young Architects' Circle organized a symposium at the Institute on the topic of "Architecture and Ideology: Notes on Material Criticism."[918] In order to avoid casting themselves in a polemical role of the postmodern era, they invited three speakers—Demetri Porphyrius, Tomàs Llorens, and Fredric Jameson—all of whom were working on an ideological criticism of architecture, i.e., the relation of critical history to practice, the limits of positivist and structuralist architectural theory, and the question of whether a new architecture culture could contribute to society and social change.[919] All these different, if not opposing developments of institutional continuation, or even institutional opening came to an abrupt end when Eisenman surprisingly stepped down from his post as the Institute's director in June 1982, ostensibly in response to outside pressure. As a

915 "Young Architects' Circle," 1981. Source: CCA Montréal, IAUS fonds: A.5-10.

916 Members of ReVisions were: Deborah Berke, Walter Chatham, Alan Colquhoun, Pe'era Goldman, Denis Hector, Christian Hubert, Michael Kagan, Beyhan Karahan, Mary McLeod, Joan Ockman, Alan Plattus, Michael Schwarting, Bernard Tschumi, Lauretta Vinciarelli. See Joan Ockman, ed., *Architecture Criticism Ideology* (New York: Princeton Architectural Press, 1985).

917 Minutes of a Fellows' meeting, November 5, 1981. Source: CCA Montréal, IAUS fonds: A.2-9. The most outspoken voices here were Kenneth Frampton, Silvia Kolbowski, and Rosalind Krauss. For a critique of the Century Association being turned into a powerhouse of New York's architecture community, see Plunz and Kaplan, 1984.

918 Margot Norton and Margot Jacqz, "Lecture Notes: The Other Day," *Skyline* (May 1982), 32.

919 Mary McLeod, "Introduction," in Ockman, ed., 1985, 7–11. In 1985, the three papers were published in the anthology *Architecture Criticism Ideology*, edited by Joan Ockman; see Porphyrios, 1985, 16–21; Llorens, 1985, 24–47; Fredric Jameson, "Architecture and the Critique of Ideology," in Ockman, ed., 1985, 51–87. Another relevant publication was to emerge from the Young Architect's Circle, published by the newly founded Princeton Architectural Press, which was based in New York long after the Institute had ceased to exist. Colomina, ed., 1988.

result, the Institute was beset by power struggles and disputes over Eisenman's unresolved succession. Just shortly after celebrating its fifteenth anniversary in November 1982, the Institute disintegrated within the space of only a few months. This was triggered by the mass resignation of Fellows, old and new, a development from which it would never fully recover, and the backstory of which can only be speculated upon in oral history.

The transformations and conflicts of this period, in which the establishment prevailed, can be inferred from the public events organized by the Institute, such as the "On Style" lecture, featuring Michael Graves on his iconic, postmodernist *Portland Building* in December 1982 or Gwathmey Siegel Architects on their Beach Houses in February 1983.[920] Issues of *Skyline* are another historical source, not only in terms of its coverage of the Institute's anniversary, but also reviews and interviews, columns that were formerly quite specific to the architecture scene, announcements of recent events at the Institute, and ultimately the declaration of the reconstitution of the Board of Trustees.[921] *Skyline*, which was published until May 1983, was a vehicle for institutional communication and eventually became an archive of the paradigm shift to which the Institute had contributed:[922] the differentiation, marketization, and commercialization of architecture culture, education, and practice, the triumph of "starchitecture" (a process in which the Institute was not uninvolved) and the increasing dominance of the archetype of the architect as developer in the world of construction, the transformation of New York, especially the sanitization of Times Square as an entertainment district, and the resurgence of conservatism in the United States, especially under the new Ronald Reagan administration after the January 1981 election. Further developments, especially those leading to the ultimate decline of the Institute, are however difficult to reconstruct from archival records. 1983 saw a new start for the Institute at a new address, 19 Union Square—the graduated rent of the old lease had become a huge, even fatal burden—with Diana Agrest, Mario Gandelsonas, Rosalind Krauss, and Anthony Vidler as the remaining Fellows. While the successful, income-generating Educational Programs continued under the lead of Gandelsonas as director of education, now with a strong preponderance of faculty members from Princeton, not least as a source of revenue, the Institute's publications, except for a final twenty-sixth issue of *Oppositions*

920 Suzanne Stephens, "At the Institute: The Portland Building Analyzed," *Skyline* (January 1983), 20–21; "Gwathmey/Siegel's Beach House. Discussed at IAUS," *Skyline* (March 1983), 8–9.

921 In the *Skyline* issue of November 1982, a news item was inserted in the "Dateline" section that read like an official report on reorganization and restructuring; see "Dateline: The Institute for Architecture an Urban Studies," *Skyline* (November 1982), 34.

922 Usually, the postmodern paradigm shift is illustrated by architectural projects presented in the context of the two major events of the time, the 1980 Venice Architecture Biennale and the IBA International Building Exhibition Berlin 84 (from 1979 onwards, and extended to 1987); by the time Eisenman and Frampton, like other former Institute Fellows, visited Berlin in 1983 at the invitation of the American Academy, the Institute, as it was known, had already ceased to exist.

and the successful continuation of *October*, were halted (Manfredo Tafuri's *The Sphere and the Labyrinth*, originally contracted for Oppositions Books, was finally published in 1987 by MIT Press).

In 1984–85, the Institute underwent a final, comprehensive redesign under Steven Peterson as new Institute director and with an ambitious program that included the resumption of the events series—now on urban topics—as well as exhibitions and new plans for publishing; but this attempt to rebuild the institution failed. Even though barely a stone was left unturned, the institutional graphic identity, which continued to be the responsibility of Massimo Vignelli and was partly designed by his employee Michael Bierut, remained its cornerstone. Ultimately, in the last two to three years before its final dissolution, the Institute was unable to regain the importance it had assumed under Eisenman. But historiography was astonishingly silent on the end of the Institute, and while only scattered traces can be found about the academic and fiscal years 1983–84 and 1984–85, details about institutional practices, discourses, and materializations can still be gleaned from oral history, i.e., from interviews with individuals who had been involved.[923] If one thing is certain, it is that the events came to a head in the spring of 1985 when, after public criticism of their Times Square Center project at an Institute event moderated by architect and critic Michael Sorkin, Johnson and Burgee—the latter serving as the Institute's president since 1983—terminated their financial support, which for the last years had been vital to the Institute's livelihood.[924] Finally, in May of 1985, the Institute declared bankruptcy and closed its doors forever.[925]

Evidence, Narrative, and Research Contribution

Postmodernism, as architectonic expression, discursive formation, and material culture from the 1960s to 1980s, is one of the fields extensively explored and written about in architecture history. Writings on its protagonists, their projects and positions, housing and planning, schools and pedagogy, books, periodicals, and exhibitions, drawings, and models highlight the thematic strands that reference the postmodern paradigm shift. In addition, there is a well-established

923 Minutes of the meetings of the Board of Trustees can be found in the Vignelli Center for Design Studies at the Rochester Institute of Technology. In general, oral history is a historiographic method that serves everyday rather than institutional or cultural history and contributes to the history of empowerment (as opposed to disempowerment). In the case of the Institute, the limitations and possibilities must also be reflected upon in terms of faded memory, identity, and experience.

924 Joshua Leon, "The Times Square Postmodern," *Urban Omnibus*, September 30, 2015, https://urbanomnibus.net/2014/09/times-square-postmodern/ (last accessed: May 31, 2023). An article written by Sorkin in the *Village Voice* described the events, while calling for a profound reform of the Institute. See Sorkin, 1985 (1991), 102.

925 There is much speculation and rumor about the whereabouts of the Institute's archive from that period, whatever it contained. Some say it was sunk in the East River; others that it was presented to the bankruptcy trustee and auctioned off to the highest bidder.

body of work on social and cultural change in the United States primarily in literary and cultural studies, sociology, and geography, which includes important work by Fredric Jameson, John McHale, and David Harvey.[926] While there have been biographical studies on the architecture historian and critic Manfredo Tafuri at the Istituto di Architettura di Venezia (IUAV) and the architect and educator Alvin Boyarsky at the Architectural Association (AA) in London, the last decade has seen several isolated and promising approaches, both historical and critical, through the study of the architect Peter Eisenman at the Institute.[927] What makes Jameson's critique of ideology so relevant within the Institute's history is that he concluded the paper he presented at the "ReVisions" symposium at the Institute in 1982 by stating that he did not engage in moralizing judgments that stem from the opposition between dialectical thought and aesthetics, but instead demanded that any position on postmodernism, including that of the historian and the critic (and he explicitly mentioned Tafuri here), be seen as a product of the times and that it must therefore begin with self-criticism.[928] And yet the Institute's history was rarely viewed in the context of the changes of the 1970s, the breakdown of the promise of modern architecture, or the new revisionism of neoliberal politics, along with the processes of de-bureaucratization, the withdrawal of the state, and the outsourcing of state services from the public sphere in the 1980s.[929]

Building Institution, conceived as a collective biography, has undertaken the historiographic challenge of examining the Institute in retrospect as a complex entity: how it was created when the opportunity arose, and how it was characterized, transformed, and resisted over the seventeen years of its existence, in terms of the discourses and materializations related to the four major institutional roles of "project office," "architecture school," "cultural space," and "publishing imprint"—an almost impossible undertaking. The highly detailed historical analysis, while quite difficult to untangle, does allow us to focus not only on one aspect and/or to highlight a single person, e.g., the autonomous practices of theory production or historiography expressed in publications, or the strategic orientation of the pedagogical experiment (if not how it adjusted to the changing conditions).[930] Rather, this book, as an institutional and cultural history that employs both socio-analysis and discourse analysis, explores the multifaceted institutional project of Eisenman and his followers that is paradigmatic of the larger changes of the mid-1970s, especially after its initial intent

926 Harvey, 1989; McHale, 1976, Jameson, 1984.

927 When Eisenman resigned, there was a search for his successor, and next to Daniel Libeskind one of the candidates who was contacted was Alvin Boyarsky.

928 Jameson, 1985, 87.

929 Leach, 2005; 2007, 2014; Sunwoo, 2009; 2012,

930 On the synergies of teaching and publishing, see Martin, 2010, 66; on the Institute's production of theory, see Allais, 2012; on the Institute's pedagogical experiment, see Esther Choi, "Life, in Theory," in Colomina, et al., 2022, 146–149.

to make an impact—as a group, as an organization, even as an institution—on the architecture community in New York through research and design, and eventually by making, exhibiting, and realizing a prototype for low-rise housing. The archival research is what makes it possible to question not only how the Institute portrayed itself, i.e., as working as an interface between theory and practice, as a think tank, or as an educational alternative in architecture. A key research contribution of *Building Institution* has been to explore the formation of the Institute itself, how it was made and unmade through everyday practice and the circulation of all sorts of texts, beginning with the name "Institute for Architecture and Urban Studies," be it in terms of orientation, committees, organization, attribution, resources, reputation, etc. In addition to the Institute's own agency, how it was socially embedded and contextually dependent, its relationships with other social institutions—next to planning authorities and ministries, universities, and museums, these also increasingly included the art and culture scene and the publishing industry with its publication and distribution channels—are central to the study of institutions, power, and architecture as exemplified by the Institute.

As a contribution to the institutional history of architecture, *Building Institution* was written on the basis of diligent research and due care in the complex documentation and multi-layered narrative regarding the institutional agenda, goals, and responsibilities of the Institute. Unlike the long narratives from the circles of former Institute Fellows that previously dominated the subject, testifying to the fact that romantic transfiguration always plays a role alongside the need for biographical work and coming to terms with the past, this book draws on exhaustive archival research at the various institutions involved (and myriad oral history interviews). It is based on the study of original documents that provide insight into the mix of multiple interests and stakeholders, both institutional and personal, the everyday practices of the Institute's leadership, its Fellows, Visiting Fellows, staff, students, and interns at various points in time. Chapter one thus explored how the Institute was initially legitimized through its collaborators and networks, which encompassed not only the Museum of Modern Art and the Department of Architecture at Cornell University, but also the State University of New York and the Graham Foundation for Advanced Studies in the Fine Arts. Eventually, the biographical and the institutional were strangely combined in Eisenman's initial claims of radicalism, as asserted in a *New York Times* article (and subsequently unquestioningly promulgated).

Teaching and learning at the Institute, on the other hand, which was shaped by Eisenman's persona, fluctuated between formalist and contextual, sociological and art historical approaches, in accordance with the preferences of the faculty's most dedicated members. While it was grounded in reality by performing commissioned work for public authorities, the New York City Planning Commission, the Urban Development Corporation of New York State, and the U.S. Department of Housing and Urban Development, and by contributing to downtown revitalization as well as to a solution to the housing crisis, at least on paper, this book has shown how fragile the construction was in the early years, and that the Institute was repeatedly doomed to failure, not least because of the clash of strong personalities. And unlike the short narratives that shape the historiography of the Institute today, which, beyond friendships and intrigues, misunderstandings and conflicts, testify to the fact that even scholarly work runs the risk of devoting itself to personal attacks rather than institutional critique and of taking too simplistic a view of the institutional order under consideration, this study differentiates between the preconditions, ideas, and interests of the Fellows involved, who at some point demanded rights and assumed duties. The development of the Institute's organization and structure, depending on monetary as well as non-monetary resources, and mediated by the Board of Trustees, especially the establishment of IAUS Central as an accounting office in the fiscal year 1972–73 under Peter Wolf as the new, second partner of the Institute's dual leadership, provided transparency and obligated the Institute's administration to accountability.

Building Institution has shown how, through its relationship with other institutions, the Institute took advantage of all the capital that came with the positions held by its Fellows at universities and colleges on the East Coast, whether at Columbia University, Cooper Union, MIT, or Princeton University and how, along with the redisciplining of architecture at the established schools of architecture there, it sought to gain—i.e., support, rather than subvert—hegemonical power over the institutional order.[931] Chapter two then discussed the intellectualization, i.e., the academization, if not scientification of architecture, as Fellows, with the launch of the journal *Oppositions*, semanticized, historicized, and aestheticized developments in contemporary and modern architecture, which informed the development of curricula and new doctoral programs. At the Institute, history and theory (along with planning, construction, and design, with a focus on semiotics and typology; urbanism was added later) were, according to the syllabi that can be accessed as historical documents, taught from the 1974–75 academic year, with educational offerings related to the development of a new kind of network of liberal arts colleges, led by Sarah Lawrence College, where one of the relevant archives can be found, as part of the internship offerings with which the Institute positioned itself as an entry point to graduate schools, or the continuing education offerings in the spirit of "life-long learning" in cooperation with the New School—all of which contributed to the redefinition of architecture as one of the humanities. The Institute's history shows that even though it never offered accredited degrees, it received institutionalized recognition in 1976 when it was awarded the AIA Medal, the highest honor bestowed by the American Institute for Architecture.

931 Regarding this relationship of architecture culture to hegemonic power, see Porphyrios, 1985, 16.

From an interdisciplinary perspective, drawing on concepts and methods of institutional and cultural sociology, and literary and cultural studies, one important contribution of this study to the history of institutions in architecture, if not the institution of architecture itself, is to have analyzed the impact and relevance of the Institute from its reinvention in 1974–75 as a "cultural space," with a variety of cultural offerings, in the interplay of propaedeutic and adult education, that oscillate between high and popular culture, instituting and instituted practices. At the time, Tafuri, as participant and observer, already highlighted the emergence of a new type of institution, designed more for entertainment than for anything else, and above all, new mechanisms of production, use, and circulation.[932] At a time when the construction sector was strongly affected by the fiscal and financial crisis in New York (before major commissions were awarded for a new generation of skyscrapers), the Institute was exemplary for this, but it was never, in the interplay of material and immaterial culture, really made the subject of historical research. Notwithstanding, education and culture were at the time viewed together in sociology as core areas of the post-industrial knowledge and information society, which was characterized by the transition from the production of goods, in this case architectural production, to a service economy, or architectural reproduction.[933] Pierre Bourdieu's concept of cultural production, developed in relation to developments in art and literature in nineteenth-century bourgeois Paris at the transition to modernity and applied to architecture culture in 1970s New York, offers a useful approach for addressing not only the discursive but also the material conditions of the broader paradigm shift to postmodernity.[934]

Chapter three, in this sense, focused on the Evening Program curated at the Institute, which included lecture series of both an academic and a more popular nature, and its "Exhibition Program", which was successively professionalized—the production and reception of both of which can be reconstructed through concepts, minutes, reports, posters, and flyers, and, in the best case, through publications of the works shown, drawings or models, and reviews in the daily and professional press.[935] While Robert Stern, then president of The Architectural League, became Eisenman's main collaborator at the Institute, offering his own lecture series and attracting a specific audience, the introduction and expansion of Vignelli's graphic design, now encompassing all of

932 Tafuri, 1976, 1987.

933 McHale, 1976; see also Daniel Bell, *The Coming of Post-Industrial Society* (New York: Basic Books, 1973), Alain Touraine, *The Post-Industrial Society. Tomorrow's Social History: Classes, Conflicts and Culture in the Programmed Society* (New York: Random House, 1971).

934 Bourdieu, 1983 (1994).

935 Like educational programs, the events themselves, the lecture series, and exhibitions, can only be partially reconstructed for posterity in terms of what was ultimately conceived, presented, and exhibited.

the Institute's work, and the use of event photographs by Dorothy Alexander, which made up a large part of the institutional visual language, offer further approaches to the intertwining of institutional identity, politics, self-image, and self-representation. With funding, particularly from the National Endowments for the Arts and the Humanities, the "Architecture" and "Open Plan" lecture series were for a long time organized by Andrew MacNair (in dialogue with senior Fellows), who also curated the first solo and group exhibitions. This study has made it possible to see the exhibitions at the Institute in relation to those at other New York institutions, MoMA and commercial galleries, such as Max Protetch and Leo Castelli, and new spaces that specialized in architecture, such as the Architecture Room at P.S.1, which had a greater proximity to alternative art spaces. Specifically, in terms of institutional administration, *Building Institution* shows how the Institute, through the work of MacNair and especially Taylor as director of development, financed itself and cross-funded programs through revenue from tuition, private and public grants, and increasingly patronage in the form of individual, institutional, and corporate sponsorship. Not only did the Institute (and individual Fellows) celebrate itself with events and publications—as evidenced by articles, reviews and interviews, as well as the society photographs taken at the release parties and published in *Oppositions*, or at exhibition openings in *Skyline*, both of which created publicity, i.e., the changing social relationship of marketing, and the politics of envy under capitalism.[936] Ultimately, these cultural productions demonstrated that the Institute was already operating a symbolic economy that produced attention and stars, both architects and people engaged in architecture history, theory, and criticism. This is one main research contribution of this book, as the development has previously only been discussed in relation to trends in deconstructivist architecture in the 1980s, drawing on Pierre Bourdieu's (and indirectly on Karl Marx's) notions of capital.[937] In this respect, the tenth anniversary of the Institute in 1977 represented a turning point in its history.

In addition to being the most comprehensive study undertaken on the Institute's publications to date, another contribution of *Building Institution* is chapter four's investigation into the production, use, and circulation of its

936 The last issue of *Skyline*, which appeared in April 1983, is notable here since it featured two articles, a preprint of one book and a book review of another, both of which exemplified central modes of discourse of New York architecture culture that had been successfully implemented at the Institute with its publications since 1973: first, opposition, i.e. ideas and criticism based on speech and counter-speech as a basic discursive configuration, and second, hype, an exaggerated communication of certain positions as a defining rhetorical stylistic device. One was an excerpt from *Hype* by Steven Aronson, published in conjunction with the release of the book, which encompassed an architecture chapter originally titled "Philip Loves Them, Philip Loves Them Not," based on an interview that Aronson had conducted with Philip Johnson; see Steven Aronson, "Philip's List," *Skyline* (April 1983), 18–19; see also Steven Aronson, *Hype* (New York: William Morrow and Company, 1983).

937 Franck, 1998; 2000.

entire portfolio and, above all, the collaboration with academic and commercial publishers (after Wittenborn Art Books, its collaborators were MIT Press with Roger Conover from 1976, later Rizzoli International with Gianfranco Monacelli from 1980, and to a certain extent Princeton Architectural Press with Kevin Lippert). This immense corpus of texts, next to the impact on biographies, both individual and collective, is certainly one of its lasting institutional legacies, and with it the Fellows' profound influence on the architectural practice, thought, and aesthetics of at least an English-speaking readership. In this sense, the emphasis on (or challenge to) the importance of theory and history in architectural debate should be understood as a symptom rather than a reflection, for the Institute's publication apparatus was becoming increasingly multi-layered in the second half of the 1970s, encompassing documents, criticism, interviews, reviews, gossip, hype, etc., with implications for the understanding and practice of culture and institution. However, this study has shown that at no point did the Institute take the step of operating as a publishing house itself, although *Oppositions*, *October*, *Skyline*, and the IAUS Exhibition Catalogues (but not Oppositions Books) were initially self-published, self-produced, and self-distributed (while all being anything but micro-productions). And yet the Institute's publishing, as well as its other institutional practices, are neither standardized mass productions of a culture industry, as discussed by Theodor W. Adorno and Max Horkheimer, two representatives of the Frankfurt School of social theory and critical philosophy, in relation to developments in film, radio and print in the USA in the first half of the 20th century, nor to developments in television and advertising in the post-war period[938]—on the contrary, limited editions of books and other printed matter such as posters were sometimes produced as collector's items.[939]

Thus, *Building Institution* has ultimately also shown the extent to which the Institute not only related to the museum and university landscape of the New York metropolitan region, but also, through its various publications,

938 Adorno and Horkheimer, [1944] 1972, 120–176.

939 The April 1983 issue of *Skyline* also featured a review by Brendan Gill of Eisenman's first monograph, *House X*, which was eventually published in 1982; see Brendan Gill, "On Reading. Peter Eisenman's House X," *Skyline* (April 1983), 33; see also Eisenman, 1982. Initially full of praise for the book as an aesthetic object of material culture, particularly for Vignelli's graphic design, Gill aimed for nuanced criticism. However, unimpressed with the design of *House X*, he lambasted Eisenman's writing as "highfalutin nonsense," particularly for his conception of the city. This was a devastating judgment, underscored by Gill's declaration that Eisenman's statements about the suburbanization and automobilization of the United States, namely that the American city was based on tabula rasa planning and that the automobile had emerged from urban space, were a distortion of history that completely ignored white settler colonialism and the significance of the automobile industry for rural spaces. The publication of *House X*, however, reproduced Eisenman's unresolved contradiction between theory and practice, the seductive projects on the one hand, whether as drawings or models, and the disconcerting texts on the other.

documented and reflected, if not contemporary building activity as intended, then at least local art and cultural activity, and the extent to which the Institute influenced changes in the publication landscape in North America. Previously afforded less consideration in architecture history, *October*, *Skyline*, and the IAUS Exhibition Catalogues can be understood as chronicles, while the editorials of *Oppositions* issues and the prefaces of Oppositions Books in particular can be read as important sources of insight into the social construction of what was considered architecturally valuable, culturally acceptable, and institutionally powerful.[940] *Building Institution*, however, in its structure and scope, ultimately suggests that it would be too short-sighted, despite all the correct and justified criticism, to identify the Institute's institutional legacy merely in terms of a particular institutional figure or a single publication or event, journal or exhibition, at best as a case study in institutional practice in architecture. By the early 1980s at the latest, this development, the turn to the architecture establishment, the ultimately failed transition to a veritable institution, and moreover, a postmodernization between simulation and spectacle that requires institutional critique from the perspective of a sociology of institutions and culture in terms of the commodification of education and culture was truly celebrated at the Institute as the new functional elite in North American architecture. As a contribution to the broadening of architecture history, indeed the architecture humanities, with a critical, interdisciplinary outlook on the role of institutions, organizations, and groups in architecture, and the basis for not only the processes of urbanization that determine social life but, as we know today, more sustainable social-environmental relations, this book offers insights into the ideas about architecture that have been powerful in New York as well as a globalized architecture culture, shaping research and design, education, culture, and publishing for the last half a century.

940 The "distinctions" in cultural consumption and artistic taste that Bourdieu discussed at length in his 1979 monograph to be published the following year in an English translation, were once again evident here, as the new connection between architecture culture and celebrity culture first emerged, along the difference of elite culture and popular culture, prominence and populism, etc.; see Pierre Bourdieu, *Distinction. A Social Critique of the Judgement of Taste* (Cambridge: Harvard University Press, [1979] 1984).

Acknowledgements

Building Institution presents research that has been underway for over fifteen years and began with my dissertation at the Institute for the History and Theory of Architecture (gta) at ETH Zurich (2007–11). During this time, the work took me to the Canadian Centre for Architecture in Montréal several times as a research student, to Columbia University in New York as a visiting researcher for a year, and to various archives and libraries (in addition to the Avery Library, also at the Massachusetts Institute of Technology, these include Princeton University, Yale University, and Sarah Lawrence College in New York State), and still bears traces of the network of all the intellectual and interpersonal relationships and interactions I enjoyed there. Since then, more than a decade has passed before the release of this publication, and in addition to the subject and method, i.e., institutional history, or indeed institutional analysis based on diligent archival research and a comprehensive oral history, this circumstance alone makes it almost impossible to fully acknowledge everyone for their contributions.

The book came about in a number of different institutional and thus also political contexts, and the project was shaped by the corresponding interests and values, opportunities and offerings: the Institute gta, where I was employed for three more years as a postdoctoral researcher (2012–15) following my dissertation, and where I was able to produce a first draft; the CCA, where I subsequently worked for three years as Associate Director of Research (2016–18) and contributed, among others, to the exhibition "Educating Architects. Four Courses by Kenneth Frampton" (2017), as well as revising my manuscript once again after engaging in further archive research, especially in the recently acquired and not yet catalogued Kenneth Frampton Fonds; and the University of Manchester (since 2019), where, as a member of the Manchester Architecture Research Group, I was given the opportunity to finally translate my manuscript from German into English and transform it into a final version ready for publication.

As is so often the case in life and work in institutions, the production of this book was characterized by a process of converging and fading interests, of received and withdrawn verbal promises, of different capacities and needs, of opening up new paths and bearing the consequences. At the end of this long journey, I would therefore like to thank all those people who contributed critically and constructively and whose names I can still recall at this stage of writing (some of whom are listed chronologically, others alphabetically) and who have made this book publication possible in very different roles, functions, and relationships.

Above all, I would like to express my gratitude to:

Supervisors and reviewers of the original research project and those who were involved in various capacities and acted as advisors and mentors: Laurent Stalder (ETH, first supervisor), Günter Lenz (Humboldt University of Berlin, second supervisor) and Mark Jarzombek (MIT, third supervisor), as well as Axel Sowa (RWTH

Aachen University, external review); also Ákos Moravánsky (ETH), to whom I am indebted for his excellent book recommendations (e.g. "How New York Stole the Idea of Modern Art" and "Please Kill Me. The Uncensored Oral History of Punk"); Giovanna Borasi (CCA) for her support and encouragement during my time there; and finally, Albena Yaneva (UoM/MARG) for reviewing the book and ensuring that it performs well in the UK Research Excellence Framework.

Protagonists of the Institute and other persons involved who feature and have a voice in this collective biography: above all Peter Eisenman for his support and help from the very beginning in 2006, for introducing me to certain circles and making telephone calls that facilitated the first round of conversations and interviews, and especially for introducing me to Robert Gutman; Kenneth Frampton for repeatedly engaging with me and inviting me to Columbia University in 2009–10, which made my research in New York possible in the first place; also Suzanne Frank and Andrew MacNair, with whom I had a long relationship and became friends; and all the members of the inner and outer circle, Fellows, Trustees, Visiting Fellows, staff, students and interns and contemporaries, with whom I was able to talk about their experiences, memories, and perceptions of the North American and East Coast architectural milieu in the course of my oral history.

All those in Zurich and Berlin who have supported this project at various stages and contributed in many ways with questions and suggestions, proofreading, comments, etc.: early fellow sufferers and longstanding collaborators, Achim Pietzcker, Elke Beyer, Anke Hagemann; everyone who shared a workspace at the Ideenshop/Haus des Reisens in Berlin in 2010–11, and Stefanie Schlüter and Nadja Geer, for reading and commenting on chapters of my dissertation; my later colleagues at the Institute gta, Cornelia Escher, Moritz Gleich, Gregor Harbusch, Anne Kockelkorn, Caspar Schärer, Sabine von Fischer, for daily exchanges; Diogo Seixas Lopez, for mutual interest, the numerous visits to the HIL building, and interchange of ideas; as well as the participants in the Doctoral Program in History and Theory, especially Charlotte Malterre-Barthes, Laila Seewang, and Dubravka Sekulić, for numerous conversations.

Colleagues and confidants in New York who took the time to contribute to this project by discussing it with me, and who also served as inspiration, reference, and role models through their work: Cesare Birignani, Craig Buckley, Ralph Ghoche, John Harwood, Olympia Kazi, Reinhold Martin, Mary McLeod, Joaquim Moreno, Felicity Scott, Meredith TenHoor, Mark Wigley, and especially "common room" (Lars Fischer as one of the founders and members, alongside Maria Ibañez de Sendadiano, Todd Rouhe, and Geoff Han) for sharing their desks, networks, and knowledge.

Institutions, schools, museums and archives where I conducted my research and the archivists who processed and made available the countless boxes and folders of documents that made my work possible in the first place: the CCA, in particular Collection and the Study Room under the direction of Martien de Vletter (first Anne-Marie Sigouin and Renata Guttman, and later Mathieu Pomerleau, Colin MacWirther, and Mar Golzáles Palacios), Publication under the direction of Albert Ferré, who showed interest in my book and even supported it for a certain time by providing an editorial review and a test translation, as well as the great research team working on the "Educating Architects" exhibition, Jacqueline Meyer as researcher and Tania Tovar Torres as curatorial intern; also the archivists of the American Institute of Architects (AIA), the New York State Council on the Arts (NYSCA), and the National Endowments for the Arts and the Humanities (NEA and NEH), whose names are regrettably not known to me.

Organizers, moderators, respondents, and peers at academic conferences that provided opportunities to present research and receive valuable feedback, as well as for networking and conversation: the 2009 Graduate Student Workshop "What Site?," hosted by the History, Theory and Criticism of Architecture and Art Program at the School for Architecture and Planning, MIT (Mark Jarzombek, Felicity Scott); the 2009 Dissertation Colloquium of the Temple Hoyne Buell Center for the Study of American Architecture, Columbia University (Reinhold Martin, with Lucia Allais, John Harwood, and Timothy Hyde); the 2012 international workshop "The Making of Architects / Architecture in the Making" of the Architectural Sociology Working Group, TU Darmstadt (Monika Grubbauer and Silke Steets); the 2012 symposium "The Göhner Case. On the Crisis of Large-Scale Housing Production in the 1970s," ETH Zurich (Anne Kockelkorn, Susanne Schindler, Fabian Furter, Patrick Schoeck-Ritschard, and Philip Ursprung); the 2013 conference "Architecture Education Goes Outside Itself. Crossing Borders, Breaking Barriers," School of Design, University of Pennsylvania (Joan Ockman, David Leatherbarrow, Daniel Barber); the 2013 conference "ISSUES? Concerning the Project of Peter Eisenman," Faculty of Architecture, University of Belgrade (Vladan Djokić and Petar Bojanić); the 2014 conference of the European Architectural History Network (EAHN), Politecnico di Torino; the 2016 J. Irwin Miller Symposium "Learning/Doing/Thinking: Educating Architects in the 21st Century," Yale

School of Architecture (Eeva-Liisa Pelkonen, Anya Bokov); the fifth thematic 2017 EAHN conference "Tools of the Architect," TU Delft; the 2019 symposium "Wer bestimmt die Architektur? Netzwerke, Wettbewerbe und der öffentliche Diskurs," TU Darmstadt (Oliver Elser, Carsten Ruhl, and Daniela Ortiz dos Santos).

Teachers, mediators, and facilitators, as well as students and the public, whose lectures, seminars and workshops, and events helped me to sharpen my arguments, clarify terms, and explain facts: Todd Rouhe and Lars Fischer ("Provisional Practices," Barnard College); Craig Buckley ("Images after Images: Case Studies in Architecture and Print," Columbia University); Carsten Ruhl ("Modern Architecture in Italy: From Futurism to Neorationalism," Bauhaus-Universität Weimar); Nikolaus Kuhnert, Anh-Linh Ngo (Arch+ features 19: "Die Netzwerke des Peter Eisenman," Haus Ungarn, Berlin); André Bideau ("Die 1970er. Umbrüche in Architektur und Städtebau," University of Basel); Liane Lefaivre (University of Applied Arts, Vienna); Patrizia Bonifazio ("Urban and Planning History," Politecnico di Milano); Andreas Rumpfhuber (Architektursalon, Depot, Vienna); Gaia Caramellino and Ivo Covic ("History, Theory+Practice" lectures, Politecnico di Milano); Christophe Van Gerrewey (film screening of "The Making of an Avant-garde. The Institute for Architecture and Urban Studies," EPFL); Léa-Catherine Szacka and Salomon Frausto ("Berlage Afternoon Symposium," TU Delft).

Editors and publishers of architectural journals, online magazines, exhibition catalogues, and anthologies who have supported, promoted, and disseminated this research: Reto Geiser (*Explorations in Architecture: Teaching, Design, Research*, 2008); Nancy Levinson (*Places*, 2010); Beatriz Colomina and Craig Buckley (*Clip, Stamp, Fold. The Radical Architecture of Little Magazines* 196X–197X, 2010); Susanne Schindler and Axel Sowa (*Candide 5*, 2012); Hannes Meyer (*Archithese 6*, 2012); common room (*Arts for Living*, 2013); Nikolaus Kuhnert and Anh-Linh Ngo (*Arch+ features 19*, 2013); Giovanna Borasi (*The Other Architect*, 2015); Véronique Patteeuw and Léa-Catherine Szacka (*Mediated Messages: Periodicals, Exhibitions and the Shaping of Postmodern Architecture*, 2018); Sylvia Lavin (*Architecture Itself and Other Postmodern Myths*, Leipzig: Spector Books, 2020); Peter Laurence (*Histories of Architecture Education in the United States*, 2023).

Various sponsors and the publisher who ultimately made this book in its present form possible and provided for the necessary professionalism through their generous financial contributions: after a joint research grant from CCA and ETH and a personal grant from the Swiss National Science Foundation (SNSF), initially a publication grant from the Graham Foundation for Advanced Studies in Fine Arts, Chicago (2013) for copyrights and reproductions; another seed grant from SEED, UoM from the post-pandemic recovery fund, which enabled a final research trip to the USA in 2022; an open access publication grant, again from the SNSF (2023), for the copyediting and proofreading by Nicola Heine, the graphic design by Till Sperrle, the image editing by Achim Hatzius, the typesetting by Jan Gerbach, the production management by Die Produktion, and the widest possible dissemination and accessibility; Transcript Verlag, who in the end made the best publication possible, Jörg Gleiter for making contact; Johanna Tönsing for the project support; Gero Wierichs for the production management.

Research institutions, museums, libraries, collections, and archives that provided images and various documents (plans, photography, and graphics) for publication, some of them free of charge, and the archivists who handled my multiple and often complex requests professionally and promptly, sometimes with great enthusiasm, dedication, and goodwill: Natalia Sciarini, Matthew Rowe, and Moses Goren as research assistant at Yale University's Beinecke Library (where Dorothy Alexander's slide archive is located); Jennifer Whitlock at the Vignelli Center for Design Studies, Rochester Institute for Technology; Céline Pereira at the CCA; and all the individuals who also provided documents from their private collections without hesitation: Cameron Armstrong and Deborah Berke for the student photos, Suzanne Frank for the institutional photos, Theodore Liebman for the construction photos, and Suzanne Stephens for the editorial photos.

Above all, my family and friends, longstanding allies and companions who have provided practical and emotional support over the years (despite several changes of publisher): my parents, Ursula and Joachim Förster, the latter of whom passed away far too soon and was thus unable to witness the completion of my dissertation or the publication of this book; my brothers Kai Förster and Klas Förster and their families—Klas also being my portrait photographer; Sascha Roesler, who has followed the dissertation and then the book project from the beginning and repeatedly commented on it over the years with consummate Swiss precision; Lars Fischer, who has always actively supported me, who was the only person to have read and comment on the revised manuscript, and who finally provided me with the title; Anuschka Scheibel, who patiently listened and kept me focused over the last decade.

My most sincere thanks are therefore owed to all those who have accompanied, backed, and collaborated with me on this project, the dissertation and the book, or parts of it. Those who have supported me and encouraged me to remain critical of institutions rather than simply conforming to and assimilating into hierarchies. Those who view themselves as belonging to the camp of the instituting rather than the instituted. Those who work to promote a critical understanding of architecture and its relationship to the world, and who leave nothing unquestioned. Those who are more interested in knowledge than power. Those who continue to form groups in the name of autonomy and self-determination and to pursue alternative architectural practices instead of chasing the latest hype and celebrating stardom. Those who, despite all the temptations of neoliberal capitalism, do not allow themselves to be co-opted so easily. Those authors, historians, and critics whose work made the existence of *Building Institution* possible the first place.

My interest in institution building through architectural and cultural production began in the mid-2000s, when it became fashionable to found or re-found our own groups, organizations, and institutions, to address the actors and networks (and more specifically periodicals, exhibitions, artefacts, and concepts) of postmodernism, and has now culminated in this book at a time when the institution of architecture itself, the office, the curriculum are being called into question and perceived as in need of repair in the face of the polycrises that humanity is facing in very different ways. Given the task of combating the climate emergency and loss of biodiversity, however, it is hard to tell where to start, because we now know that transformation must be both rapid and comprehensive. Once again, the multifaceted, problematic, and ambivalent relationships with institutions must be explored anew, and it is up to everyone to decide for themselves whether to embark on the journey through existing institutions or to find other paths.

Abbreviations

AA Architectural Association
ACSA Association of Collegiate Schools of Architecture
AIA American Institute of Architects
CASE Conference of Architects for the Study of the Environment
CIAM Congrès Internationaux d'Architecture Moderne
CCA Canadian Centre for Architecture
CPC New York City Planning Commission
FHA Federal Housing Association
GLCA Great Lakes Colleges Association
HEW U.S. Department of Health, Education and Welfare
HFA the State Housing Finance Agency
HUD U.S. Department of Housing and Urban Development
IAUS Institute for Architecture and Urban Studies
IUAV Istituto Universitario di Architettura di Venezia
MIT Massachusetts Institute of Technology
MGPV Marcus Garvey Park Village
MoMA Museum of Modern Art
NEA National Endowment for the Arts
NEH National Endowment for the Humanities
NIMH National Institute of Mental Health
NYCHA New York City Housing Authority
NYSCA New York State Council on the Arts
SUNY State University of New York
UDC New York State Urban Development Corporation
UDG Urban Design Group

Bibliography

Publications of the Institute

Anderson, Stanford, ed. *On Streets*. Cambridge: MIT Press, 1978.

Casabella, no. 359/360 (November/December 1971): "The City as an Artifact."

Colquhoun, Alan. *Essays in Architectural Criticism*. Cambridge: MIT Press, 1981.

Ginzburg, Moisei. *Style and Epoch*. Translated by Anatole Senkevitch. Cambridge: MIT Press, 1983.

Hejduk, John. *The Silent Witness and Other Poems*. New York: The Institute for Architecture and Urban Studies, 1980.

IAUS, ed. *Aldo Rossi in America, 1976–1979*. Catalogue 2. New York: MIT Press, 1979.

IAUS, ed. *Architecture Education U.S.A.: Issues, Ideas, and People. A Conference to Explore Current Alternatives*. New York: The Institute for Architetcure and Urban Studies, 1971.

IAUS, ed. *Art and Architecture. USSR. 1917–32*. New York: Wittenborn Art Books, 1971.

IAUS, ed. *Austrian New Wave*. Catalogue 13. New York: The Institute for Architecture and Urban Studies, 1980.

IAUS, ed. *California Counterpoint. New West Coast Architecture 1982*. Catalogue 18. New York: Rizzoli International, 1982.

IAUS, ed. *Five Houses: Gwathmey/Siegel Architects*. Catalogue 7. Cambridge: MIT Press, 1980.

IAUS, ed. *Idea as Model. 22 Architects 1976/80*, Catalogue 3. New York: Rizzoli International, 1981.

IAUS, ed. *Ivan Leonidov. Russian Constructivist, 1902–1959*. Catalogue 8. New York: Rizzoli International, 1981.

IAUS, ed. *John Hejduk. 7 Houses*. Catalogue 12. Cambridge: MIT Press, 1980.

IAUS, ed. *Kazou Shinohara*. Catalogue 17. New York: Rizzoli International, 1982.

IAUS, ed. *Kingsbridge Heights*. New York: The Institute for Architecture and Urban Studies, 1968.

IAUS, ed. *Le Corbusier's Firminy Church*. Catalogue 14. New York: Rizzoli International, 1981.

IAUS, ed. *Massimo Scolari. Architecture. Between Memory and Hope*. Catalogue 1. New York: MIT Press, 1980.

IAUS, ed. *New Urban Settlements. Analytical Phase*. New York: The Institute for Architecture and Urban Studies, December 1970.

IAUS, ed. *A New Wave of Japanese Architecture*, Catalogue 10. New York: The Institute for Architecture and Urban Studies, 1978.

IAUS, ed. *O.M. Ungers. Works in Progress 1976–1980*. Catalogue 6. New York: Rizzoli International, 1981.

IAUS, ed. *Philip Johnson: Processes. The Glass House, 1949 and The AT&T Headquarters, 1978*, Catalogue 9. New York: The Institute for Architecture and Urban Studies, 1979.

IAUS, ed. *Raymond M. Hood*. Catalogue 15. New York: Rizzoli International, 1982.

IAUS, ed. *Robert Krier. Urban Models*. Catalogue 5. New York: Rizzoli International, 1982.

IAUS, ed. *UDC Report: The Generation of Low Rise High Density Housing Criteria*. New York: The Institute for Architecture and Urban Studies, 1972.
IAUS, ed. *Union Square Park. Project Development: Phase 3. Report to the Manhattan Community Board* 5. New York, June 30, 1974.
IAUS, ed. *William Lescaze*. Catalogue 16. New York: Rizzoli International, 1982.
Loos, Adolf. *Spoken Into the Void: Collected Essays 1897–1900*. Translated by Jane O. Newman and John H. Smith. Cambridge: MIT Press, 1982.
October (1976–1985).
Oppositions 1–26 (1973–1984).
Rossi, Aldo. *A Scientific Autobiography*. Translated by Lawrence Venuti. Cambridge: MIT Press, 1981.
Rossi, Aldo. *The Architecture of the City*. Translated by Diane Ghirardo and Joan Ockman. Cambridge: MIT Press, 1982.
Skyline (1978–1980, 1981–1983).

Primary Sources

Agrest, Diana. "The Architecture of the City: An Interview with Aldo Rossi." *Skyline* (September 1979), 4–5.
Agrest, Diana, and Mario Ganelsonas. "Manhattan Additions I." *Architectural Design*, no. 52 (May/June 1982), 44–48.
"Aldo Rossi: Two Exhibits." *Progressive Architecture* (October 1979), 21, 23.
Ambasz, Emilio. "I The University of Design and Development," "II Manhattan: Capital of the Twentieth Century," "III The Designs of Freedom," *Casabella*, no. 359/360 (November/December 1971): "The City as an Artifact," 87–99.
Ambasz, Emilio. "A Selection of Working Fables." *Oppositions* 4 (October 1974), 65–74.
Anderson, Stanford. "Modern Architecture and Industry: Peter Behrens and the Cultural Policy of Historical Determinism." *Oppositions* 11 (Winter 1977), 52–71.
Anderson, Stanford. "Modern Architecture and Industry: Peter Behrens, the AEG, and Industrial Design." *Oppositions* 21 (Summer 1980), 70–97.
Anderson, Stanford. "Modern Architecture and Industry: Peter Behrens, and the AEG Factories." *Oppositions* 23 (Winter 1981), 53–83.
Anderson, Stanford. "Preface." In *On Streets*, ed. Stanford Anderson. Cambridge: MIT Press, 1978, VII–VIII.
Anderson, Stanford. "Studies toward an Ecological Model of the Urban Environment." In *On Streets*, ed. Stanford Anderson. Cambridge: MIT Press, 1978, 267–307.
Arch+, no. 37 (April 1978): "Der Tod der Architektur"
Architectural Design 47, no. 5 (1977): "OMA."
L'Architecture d'Aujourd'hui, no. 186 (August/September 1976): "New York in White and Gray."
Architecture + Urbanism, no. 52 (April 1975): "White and Gray: Eleven Modern Architects."
Architecture + Urbanism, no. 112, (January 1980): "Peter Eisenman."
Architecture + Urbanism, no. 114, (March 1980): "Diana Agrest, Mario Gandelsonas."
Archithese, no. 17 (1976): "Metropolis 1. New York: ein europäischer Mythos."
Archithese, no. 18 (1976): "Metropolis 2. New York, oder die architektonische Vermittlung einer Explosion."
Archithese, no. 20 (1976): "Metropolis 3. Amerikanismus, Skyscraper und Ikonografie."
Architectural Forum (May 1973): "Five on Five."
Aronson, Steven. "Philip's List." *Skyline* (April 1983), 18–19.
Arquitecturas Bis, no. 22 (May 1978).
Art Forum (October 1976).
Banham, Reyner. "Vitruvius over Manhattan." *New Society* (December 7, 1967), 827–828.
Banham, Reyner, and Livio Dimitriu. "Peter Cook/Ron Herron: Arcadias/Insertions." *Skyline* (August 1978), 3.
Barbanel, Josh. "The Low-Rise Solution for the Poor." *The New York Times* (May 7, 1978), R1.
Barnett, Jonathan. *Urban Design as Public Policy. Practical Methods for Improving Cities*. New York: Architectural Record Books, 1974.
Bender, Thomas. "The Architecture of Confinement." (announced in *Skyline*, February 1980, unpublished).
Bender, Thomas. "Culture of Cities: Between Civic Culture and the Academy: New York and Columbia in the 19th Century." *Skyline* (January 1982), 14–15.
Besset, Maurice. *Qui était Le Corbusier?* Geneva: Skira, 1968.
Bletter, Rosemarie. "About Graves." *Skyline* (Summer 1979), 2–3.
Cacciari, Massimo. "Eupalinos or Architecture." Translated by Stephen Sartarelli. *Oppositions* 21 (Summer 1980) 106–115.

Cacciari, Massimo. *Architecture and Nihilism. On the Philosophy of Modern Architecture.* New Haven: Yale University Press, 1996.
Cacciari, Massimo, Franco Rella, Manfredo Tafuri, George Teyssot. *Il Dispositivo Foucault.* Venice: Cluva, 1977.
"Citation: Architectural Design (House 11a)." *Progressive Architecture* (January 1979), 84–85.
Ciucci, Giorgio. "The Invention of the Modern Movement." Translated by Stephen Sartarelli. *Oppositions* 24 (Spring 1981), 69–91.
Ciucci, Giorgio, Francesco Dal Co, Mario Manieri-Elia, Manfredo Tafuri, eds. *The American City. From the Civil War to the New Deal.* Translated by Barbara Luigia La Penta. Cambridge: MIT Press, [1973] 1979.
Colquhoun, Alan. *Essays in Architectural Criticism.* Cambridge: MIT Press, 1981.
Colquhoun, Alan. "Form and Figure." *Oppositions* 12 (Spring 1978), 28–37.
Colquhoun, Alan. "From Bricolage to Myth: or How to Put the Humpty Dumpty Together Again." *Oppositions* 12 (Spring 1978), 1–19.
CPC. *Plan for New York City. A Proposal: 3 Brooklyn* (New York: City Planning Commission, 1969).
Dal Co, Francesco. "Notes Concerning the Phenomenology of the Limit in Architecture." Translated by Stephen Sartarelli. *Oppositions 23* (Winter 1981), 37–51.
Dal Co, Francesco. "The Remoteness of die Moderne." Translated by Stephen Sartarelli. *Oppositions 22* (Fall 1980), 75–95.
"Dateline: The Institute for Architecture an Urban Studies." *Skyline* (November 1982), 34.
de Quincy, Quatremere. "Type." *Oppositions* 8 (Spring 1977), 146–150.
Design Quarterly, no. 78/79, 1970: "Conceptual Architecture."
Design Quarterly, no. 101/102 (October 1976): "The River: Images of the Mississippi." [Exhib. Cat.].
Diller, Elizabeth, Diane Lewis, and Kim Shkapich, eds. *Education of an Architect.* New York: Rizzoli International, 1988.
Dimitriu, Livio. "Report from Syracuse." *Skyline* (May 1978), 3.
Dimitriu, Livio. "Swiss Transmissions and Exaggerations: An Interview with Mario Botta." *Skyline* (March 1980), 12/13.
Drexler, Arthur. *Transformations in Modern Architecture* [Exhib. Cat.] New York: The Museum of Modern Art, 1979.
Drexler, Arthur (interview with Andrew MacNair). "Response. Arthur Drexler on Transformations." *Skyline* (Summer 1979), 6.
Eisenman, Peter. "Behind the Mirror: On the Writings of Philip Johnson." *Oppositions* 10 (Fall 1977), 1–13.
Eisenman, Peter. "Cardboard Architecture." *Architecture + Urbanism* 3, no. 35 (November 1973), 185–189.
Eisenman, Peter. "Cardboard Architecture: castelli di carte." *Casabella*, no. 386 (February 1974), 17–31.
Eisenman, Peter. "Castelli di Carte: Due Opere di Peter Eisenman." *Casabella*, no. 374 (February 1973), 17–31.
Eisenman, Peter. "Commentary." *Oppositions* 9 (Summer 1977), 19–20.
Eisenman, Peter. "A Critical Practice: American Architecture in the Last Decade of the Twentieth Century." In *Education of an Architect*, eds. Elizabeth Diller, Diane Lewis, and Kim Shkapich. New York: Rizzoli International, 1988, 190–193.
Eisenman, Peter. "Dall' oggetto alla relazionalità: la casa del Fascio di Terragni." *Casabella*, no. 344, (1970), 38–41.
Eisenman, Peter. "Editor's Preface." In *The Architecture of the City*, Aldo Rossi. Cambridge: MIT Press, 1982, n.p.
Eisenman, Peter. "Five Easy Pieces: Dialectical Fragments Toward the Decomposition and Reintegration of Suburbia." In *Environment, Participation, Cultural Structures: General Catalogue.* Second Volume, ed. La Biennale di Venezia. Venice: Alfieri Edizioni d'arte, 1976, 256.
Eisenman, Peter. *The Formal Basis of Modern Architecture.* Baden: Lars Müller Publishers, 2006.
Eisenman, Peter. *Die formale Grundlegung der modernen Architektur.* Zurich: gta Verlag, 2005.
Eisenman, Peter. "From Golden Lane to Robin Hood Gardens; Or If You Follow the Yellow Brick Road, It May Not Lead to Golder's Green." *Architectural Design* 42, no. 9 (September 1972), 557–573, 588–592.
Eisenman, Peter. "From Golden Lane to Robin Hood Gardens; Or If You Follow the Yellow Brick Road, It May Not Lead to Golder's Green." *Oppositions* 1 (September 1973), 27–56.
Eisenman, Peter. "From Object to Relationship II: Casa Giuliani Frigerio. Giuseppe Terragni." *Perspecta*, no. 13–14 (1971), 36–75.
Eisenman, Peter. *Giuseppe Terragni. Transformations, Decompositions, Critiques.* New York: Monacelli Press, 2003.
Eisenman, Peter. "House I." *Architecture + Urbanism* 3, no. 11 (November 1973), 177–184.
Eisenman, Peter. "House II Transformations." In *Idea as Model. 22 Architects 1976/80*, Catalogue 3, ed. IAUS. New York: Rizzoli International, 1981, 34/35.

Eisenman, Peter. "House III." *Architecture + Urbanism* 3, no. 11 (November 1973), 193.
Eisenman, Peter. *House X*. New York: Rizzoli International, 1982.
Eisenman, Peter. "The House of the Dead as the City of Survival." In *Aldo Rossi in America: 1976–1979*, Catalogue 2, ed. IAUS. Cambridge: MIT Press, 1979, 4–15.
Eisenman, Peter. *Houses of Cards*. New York: Oxford University Press, 1987.
Eisenman, Peter. "The Houses of Memory: The Texts of Analogue." In *The Architecture of the City*, Aldo Rossi. Cambridge: MIT Press, 1982, 2–11.
Eisenman, Peter (in conversation with Alvin Boyarsky). "The Institute in Theory and Practice" (January 20, 1976) In *Supercritical: Peter Eisenman & Rem Koolhaas*, Brett Steele. London: AA Publications, 2007, 83–87.
Eisenman, Peter. "Interview. Tom Wolfe and Peter Eisenman. Part I." *Skyline* (October 1981), 12–14.
Eisenman, Peter. "Interview. Tom Wolfe and Peter Eisenman. Part II." *Skyline* (November 1981), 3–4.
Eisenman, Peter. "Interview. Robert Hughes and Peter Eisenman." *Skyline* (December 1981), 10–13.
Eisenman, Peter. "Interview. Paul Goldberger and Peter Eisenman." *Skyline* (January 1982), 6–9.
Eisenman, Peter. "Interview. Philip Johnson and Peter Eisenman." *Skyline* (February 1982), 14–17.
Eisenman, Peter. "Interview. Cesar Pelli and Peter Eisenman." *Skyline* (May 1982), 23–25.
Eisenman, Peter. "Interview. Henry Cobb and Peter Eisenman." *Skyline* (June 1982), 12–14.
Eisenman, Peter. "Interview. Robert Venturi and Peter Eisenman." *Skyline* (July 1982), 12–15.
Eisenman, Peter. "Interview. Gerald D. Hines and Peter Eisenman." *Skyline* (October 1982), 18–21.
Eisenman, Peter. "Interview. John Portman and Peter Eisenman." *Skyline* (January 1983), 12–15.
Eisenman, Peter. "Interview. Leon Krier and Peter Eisenman." *Skyline* (February 1983), 12–15.
Eisenman, Peter. "Interview. Robert Maxwell and Peter Eisenman." *Skyline* (March 1983), 12–15.
Eisenman, Peter. "Interview. Richard Serra and Peter Eisenman." *Skyline* (April 1983), 14–17.
Eisenman, Peter. "Introduction." In *Philip Johnson: Writings*, eds. Peter Eisenman and Robert Stern. New York: Oxford University Press, 1979, 10–25.
Eisenman, Peter. "Maison Dom-ino." *Oppositions* 15/16 (Winter/Spring 1979), 119–128.
Eisenman, Peter. "Notes on Conceptual Architecture: Towards a Definition." *Design Quarterly*, no. 78/79 (1970): "Conceptual Architecture," 1–5.
Eisenman, Peter. "Notes on Conceptual Architecture: Towards a Definition." *Casabella*, no. 359/360, (November/December 1971): "The City as an Artifact," 48–58.
Eisenman, Peter (in conversation with Lindsay Stamm Shapiro and David Shapiro). "A Poetics of the Model: Eisenman's Doubt." (March 8, 1981) In *Idea as Model. 22 Architects 1976/80*, ed. IAUS. New York: Rizzoli International, 1981, 121–125.
Eisenman, Peter. "Post-Functionalism." *Oppositions* 6 (Fall 1976), n.p.
Eisenman, Peter. "Preface" to "The Education of Reality." In *Architectural Education USA: Issues, Ideas, and People. A Conference to Explore Current Alternatives*, IAUS. New York: The Institute for Architecture and Urban Studies, 1971.
Eisenman, Peter [Ernesto di Casarotta, pseud.]. "Quarta Roma: Report from Rome." *Skyline* (August 1978), 6.
Eisenman, Peter. "Real and English: Destruction of the Box I." *Oppositions* 4 (October 1974), 5–34.
Eisenman, Peter [Ernesto di Casarotta, pseud.]. "The Sound of Leather." *Skyline* (May 1978), 7.
Eisenman, Peter. "Transformations, Decompositions and Critiques: House X." In *House X*. New York: Rizzoli International, 1982, 33ff.
Eisenman, Peter et al. "Eisenman (and Company) Respond." *Progressive Architecture* (February 1995), 88–91.
Eisenman, Peter et al. *Five Architects*. New York: Wittenborn Art Books, 1972.
Eisenman, Peter, and Michael Graves. "Jersey Corridor Project." *Life Magazine* (December 24, 1965): "The U.S. City: Its Greatness Is at Stake."
Eisenman, Peter, and Robert Stern, eds. *Philip Johnson: Writings*. New York: Oxford University Press, 1979.
Eisenman, Peter, Kenneth Frampton, Mario Gandelsonas. "Editorial." *Oppositions* 1 (September 1973), n.p.
Eisenman, Peter, Kenneth Frampton, Mario Gandelsonas. "Editorial." *Oppositions* 2 (January 1974), n.p.
Eisenman, Peter, Kenneth Frampton, Mario Gandelsonas. "Editorial." *Oppositions* 3 (May 1974), n.p.
Eisenman, Peter, William Gass, Robert Gutman. "House VI. Residence. Critique of

Weekend House by Philosopher, Sociologist and Architect Himself." *Progressive Architecture* (June 1977), 57–67.
Ellis, William. "Beaux." *Oppositions* 5 (Summer 1976), 131–134.
Ellis, William. "Type and Context in Urbanism: Colin Rowe's Contextualism." *Oppositions* 18 (Fall 1979), 2–27.
"Europe-America. Historical Centre – Suburbia." In *Environment, Participation, Cultural Structures: General Catalogue.* Second Volume, ed. La Biennale di Venezia. Venice: Alfieri Edizioni d'arte, 1976, 235–264.
IAUS. "Application of the Prototype to Community Board 2. Fox Hills, Staten Island, New York." In *Another Chance for Housing, Low-Rise Alternatives*, ed. MoMA [Exhib. Cat.] New York: The Museum of Modern Art, 1973, 28–37.
IAUS. "Application of the Prototype to the Marcus Garvey Park Village Urban Renewal Plan, Brownsville, New York." In *Another Chance for Housing, Low-Rise Alternatives*, ed. MoMA [Exhib. Cat.] New York: The Museum of Modern Art, 1973, 20–27.
IAUS. "Demonstration Project: Streets in the Central Area of a Small American City." In *On Streets*, ed. Stanford Anderson. Cambridge: MIT Press, 1978, 339–375.
Fend, Peter. "Letter to the Editor." *Skyline* (February 1980), 2.
Forster, Kurt. "Antiquity and Modernity in the La Roche-Jeanneret House of 1923." *Oppositions* 15/16 Winter/Spring 1979), 131–153.
Forster, Kurt "Between Memory and Amnesia." *Skyline* (January 1979), 4.
Forster, Kurt. "Monument/Memory and the Mortality of Architecture." *Oppositions* 25 (Fall 1982), 2–15.
Forster, Kurt. "Re modernism. Kurt W. Forster Reviews Kenneth Frampton's Latest." *Skyline* (July 1982), 27.
Foucault, Michel. "Space, Knowledge, and Power." *Skyline* (March 1982), 16–20.
Frampton, Kenneth. "America 1960–1970. Notes on Urban Images and Theory." *Casabella*, no. 359/360 (November/December 1971): "The City as an Artifact," 25–40.
Frampton, Kenneth. "Apropos Ulm: Curriculum and Critical Theory." *Oppositions* 3 (May 1974), 17–36.
Frampton, Kenneth. "Blow Up." *Skyline* (April 1979), 6.
Frampton, Kenneth. "Casabella and the Reading of History." *Oppositions* 24 (Spring 1981), 92–101.
Frampton, Kenneth. "The Castellated Home: Review of Herman Muthesius, The English House." *Oppositions* 22 (Fall 1980), 106–113.
Frampton, Kenneth. "Criticism: Eisenman's House I." *Architecture + Urbanism* 3, no. 11 (November 1973), 190–192.
Frampton, Kenneth. "Drawings by Le Corbusier at the MoMA." *Skyline* (April 1978), 2.
Frampton, Kenneth. "The Evolution of Housing Concepts: 1870–1970." In *Another Chance for Housing, Low-Rise Alternatives*, ed. MoMA. [Exhib. Cat.] New York: The Museum of Modern Art, 1973, 6–11.
Frampton, Kenneth. "Five Architects." *Lotus International*, no. 9 (1976), 136–151 (English original 231–233).
Frampton, Kenneth. "George Wittenborn. 1905–1974." *Oppositions* 4 (October 1974), 1–4.
Frampton, Kenneth. "The Glass House Revisited." In *Philip Johnson: Processes. The Glass House, 1949 and The AT&T Headquarters*, 1978, Catalogue 9, ed. IAUS. New York: The Institute for Architecture and Urban Studies, 1979, 39–59.
Frampton, Kenneth. "Iceberg." *Skyline* (September 1979), 5.
Frampton, Kenneth. "Industrialization and the Crisis of Architecture." *Oppositions* 1 (September 1973), 57–82.
Frampton, Kenneth. "The Japanese New Wave." In *A New Wave of Japanese Architecture*, Catalogue 10, ed. IAUS. New York: The Institute for Architecture and Urban Studies, 1978, 1–13.
Frampton, Kenneth. "Labour, Work and Architecture." In *Meaning in Architecture*, eds. George Baird and Charles Jencks. New York: Braziller, 1969, 150–168.
Frampton, Kenneth. *Labour, Work and Architecture.* London: Phaidon, 2003.
Frampton, Kenneth. "Le Corbusier and L'Esprit Nouveau." *Oppositions* 15/16 (Winter/Spring 1979), 13–58.
Frampton, Kenneth. "Louis Kahn and the French Connection." *Oppositions* 22 (Fall 1980), 21–53.
Frampton, Kenneth. "Maison VI." *L'Architecture d'Aujourd'hui*, no. 186 (August/September 1976), 63–66.
Frampton, Kenneth. *Modern Architecture 1851–1919*, GA Document. Special Issue 2. Tokyo: A.D.A. Edita, 1981.
Frampton, Kenneth. *Modern Architecture. A Critical History.* London: Thames and Hudson, 1980.
Frampton, Kenneth. "Modernism's Diffusion. Japan Diary, Summer 81, Part 1." *Skyline* (April 1982), 26–29.
Frampton, Kenneth. "Modernism's Diffusion. Japan Diary, Sommer 81, Part 2." *Skyline* (May 1982), 26–29.

Frampton, Kenneth. “Modernism’s Diffusion. Japan Diary, Sommer 81, Part 3.” *Skyline* (June 1982), 22–25.

Frampton, Kenneth. “On Charles Jencks and Nathan Silver’s Adhocism: The Case for Improvisation.” *Oppositions* 3 (May 1974), 104–105.

Frampton, Kenneth. “On Reading Heidegger.” *Oppositions* 4 (October 1974), n.p.

Frampton, Kenneth. “Prospects for a Critical Regionalism.” *Perspecta*, no. 20, (1983), 147–162.

Frampton, Kenneth. “The Rise and Fall of the Radiant City.” *Oppositions* 19/20 (Winter/Spring 1980), 2–25.

Frampton, Kenneth. “Skylights: The Ins and Outs.” *Skyline* (April 1979), 12.

Frampton, Kenneth. “Stellar Material: Eileen Gray at The Modern.” *Skyline* (March 1980), 3.

Frampton, Kenneth. “Six Points Towards and Architecture of Resistance. Towards a Critical Regionalism.” In *The Anti-Aesthetic. Essays on Postmodern Culture*, ed. Hal Foster. Seattle: Bay Press, 1983, 16–31.

Frampton, Kenneth. “Two or Three Things I Know About Them: A Note on Manhattanism.” *Architectural Design* 47, no. 5 (1977): “OMA,” 315–318.

Frampton, Kenneth. “U.D.C. Low Rise High Density Housing Prototype.” *L’Architecture d’Aujourd’hui*, no. 186 (August/September 1976), 15–21. (English translation on pages XXXVII–XX).

Frampton, Kenneth, and Alessandra Latour. “Notes on American Architectural Education from the End of the Nineteenth Century until the 1970’s.” *Lotus International*, no. 27 (1980): “Architecture in the American University,” 5–39.

Freiberg, Peter. “City Report: New York. 42nd Street Redevelopment.” *Skyline* (May 1982), 3.

Freiberg, Peter. “City Report: New York. Theaters.” *Skyline* (November 1982), 28.

Fried, Joseph S. “Low-Rise Development Project Begun in Brownsville by U.D.C.” *The New York Times* (June 12, 1973), 49.

Gandelsonas, Mario. “From Structure to Subject. The Formation of an Architectural Language.” *Oppositions* 17 (Summer 1979), 6–29.

Gandelsonas, Mario. “Linguistics in Architecture” & “Due opere di Peter Eisenman: Castelli di carte.” *Casabella*, no. 374 (February 1973), 17–31.

Gandelsonas, Mario. “Neo-Functionalism.” *Oppositions* 5 (Summer 1976), n.p.

Gandelsonas, Mario. “On Reading Architecture.” *Progressive Architecture* (March 1972), 68–88.

Gandelsonas, Mario. “On Reading Architecture.” *Architecture + Urbanism* 2, no. 9 (September 1972), 51–69.

Gandelsonas, Mario. “On Reading Architecture II [House IV].” *Architecture + Urbanism* 4, no. 39 (March 1974), 89–100.

Gandelsonas, Mario, and Diana Agrest. “Semiotics and Architecture: Ideological Consumption of Theoretical Work.” *Oppositions* 1 (September 1973), 93–100.

Goldberger, Paul. “Architects Will End Ban on Advertising.” *The New York Times* (May 25, 1978), A20.

Goldberger, Paul. “Architectural Drawings at the Modern.” *The New York Times* (March 14, 1975), 24.

Goldberger, Paul. “Architectural Drawings Raised to an Art.” *The New York Times* (December 12, 1977), 50.

Goldberger, Paul. “Architecture: Harrison Retrospective.” *The New York Times* (January 2, 1980), C 15.

Goldberger, Paul. “Architecture: Houses Designed by John Hejduk.” *The New York Times* (January 32, 1980), C15.

Goldberger, Paul. “Architecture’s ‘5’ Make Their Ideas Felt.” *The New York Times* (November 26, 1973), 33.

Goldberger, Paul. “Debate Over Proposed 71st Street Tower.” *The New York Times* (November 10, 1981).

Goldberger, Paul. “How Architects Develop Ideas.” *The New York Times* (December 27, 1976), 58.

Goldberger, Paul. “How the Cityscape fared in 1977.” *The New York Times* (January 5, 1977), *The Home Section*, C1 & C11.

Goldberger, Paul. “Institute of Architects Keeps Bans on Advertising and Contracting.” *The New York Times* (June 9, 1977), 45.

Goldberger, Paul. “Midtown Architecture Institute Flowering as a Student Mecca.” *The New York Times* (October 30, 1975), 41 & 77.

Goldberger, Paul. “An Overview of Japanese Architecture.” *The New York Times* (December 22, 1978), C26.

Goldberger, Paul. “Young Architects Display Recent Visionary Designs.” *The New York Times* (September 19, 1975), 25.

Goldberger, Paul. “Young Summer Class Students Learn Architecture by ‘Building’.” *The New York Times* (July 27, 1976), 33.

Gordon Matta-Clark: One for All – All for One. Düsseldorf: Städtische Kunsthalle Düsseldorf, 1979.

Grassi, Giorgio. “Avant-Garde and Continuity.” Translated by Stephen Sartarelli. *Oppositions* 21 (Summer 1980), 25–33.

Harris, Steven. “202 Frank Gehry.” *Skyline* (November 1978), 2.

Hatch, C. Richard. “The Museum of Modern Art Discovers Harlem.” *The Architectural Forum* (March 1967), 38–47.
Hatch, C. Richard. “Planning for Change. Towards Neighborhood Design and Urban Politics in the Public. Schools.” *Perspecta*, no. 11 (1967), 43.
Hatch, C. Richard. “Renewal in Harlem.” *Zodiac*, no. 17 (1967), 196–198.
Holl, Steven. “Ungers at Columbia.” *Skyline* (October 1979), 15.
Holl, Steven. “USSR in the USA,” *Skyline* (May 1979), 10.
Holl, Steven, ed. *Pamphlet Architecture*, no. 1. New York: 1978.
Holl, Steven, Alison Sky, Suzanne Stephens. “East Coast West Coast.” In *Clip Stamp Fold. The Radical Architecture of Little Magazines*. 196X to 197X, eds. Beatriz Colomina and Craig Buckley. Barcelona: Actar, 2010, 70–81.
“Housing: How Edward Logue Does It.” *Newsweek* (November 6, 1972).
Hubert, Christian. “The Ruins of Representation.” In *Idea as Model: 22 Architects 1976/1980*, Catalogue 3, ed. IAUS. New York: Rizzoli International, 1981, 17–27.
Huxtable, Ada Louise. “Another Chance for Housing.” *The New York Times* (June 24, 1973), 125.
Huxtable, Ada Louise. “Architecture View: A Sense of Crisis About the Art of Architecture, Architecture in Crisis.” *The New York Times* (February 20, 1977), 99.
Huxtable, Ada Louise. “The Austere World of Rossi.” *The New York Times* (October 1979), D31.
Huxtable, Ada Louise. “The Japanese New Wave.” *The New York Times* (January 14, 1979), D27.
Huxtable, Ada Louise. “John Hejduk: A Mystic and Poet.” *The New York Times* (February, 3 1980), D25.
Huxtable, Ada Louise. “Poetic Visions of Design for the Future.” *The New York Times* (April 27, 1975), 142.
Huxtable, Ada Louise. “Planning the New City. Modern Museum Exhibits Projects that Link Esthetics and Sociology.” *The New York Times* (January 24, 1967), 39 & 45.
Huxtable, Ada Louise. “Reexamining Wallace Harrison.” *The New York Times* (January 6, 1980), D23.
Jacob, Mary Jane. *Gordon Matta-Clark: A Retrospective*. Chicago: Museum of Contemporary Art 1985.
Jacqz, Margot. “City Report: Biltmore is now Less.” *Skyline* (October 1981), 4.
Jacqz, Margot. “City Report: New York. Goings On on 42nd St.” *Skyline* (December 1981), 3.
Jacqz, Margot, and Kenneth Frampton. “The IAUS at 15.” *Skyline* (December 1982), 33.
Jameson, Fredric. “Architecture and the Critique of Ideology.” In *Architecture Criticism Ideology*, ed. Joan Ockman. New York: Princeton Architectural Press, 1985, 51–87.
Jencks, Charles. *The Language of Post-Modern Architecture*. New York: Rizzoli International, 1977.
Jencks, Charles. *Le Corbusier and the Tragic View of Architecture*. Cambridge: Harvard University Press, 1974.
Johnson, Philip. “Reflections. On Style and the International Style; On Postmodernism; On Architecture.” *Oppositions* 10 (Fall 1977), 15–19.
Johnson, Philip. “Rejected Architects: The Berlin Building Exposition of 1931, Architecture of the Third Reich.” *Oppositions* 2 (January 1974), 81–94.
Johnson, Philip (interview w/ Martha Carroll and Craig Owens). “Skylights: Philip Johnson on Philip Johnson.” *Skyline* (May 1978), 7–8.
Johnson, Philip, and Mark Wigley, eds. *Deconstructivist Architecture*. New York: Museum of Modern Art, 1988.
Kaplan, Samuel. “Bridging the Gap from Rhetoric to Reality. The New York State Urban Development Corporation.” *Architectural Forum* (November 1969), 70–73.
Kaplan, Samuel. “Defensible Space.” *The New York Times* (April 29, 1973), 489.
Kaufman, Peter. “Italian Views of the American City.” *Skyline* (May/June 1980), 17.
Killian, Tom. “MoMA on Housing: Nothing New.” *Architecture Plus* (August 1973), 68.
Koolhaas, Rem. “The Architects’ Ball – A Vignette, 1931.” *Oppositions* 3 (May 1974), 91–96.
Koolhaas, Rem. “The Birth of Radio City Music Hall.” *Skyline* (April 1978), 7.
Koolhaas, Rem. “Delirious New York. A Manifesto of Manhattanism.” *Progressive Architecture* (December 1978), 70–78.
Koolhaas, Rem. *Delirious New York. A Retroactive Manifesto for Manhattan*. New York: Oxford University Press, 1978.
Koolhaas, Rem. “‘Life in the Metropolis’ or ‘The Culture of Congestion’.” *Architectural Design* 47, no. 5 (1977): “OMA,” 319–325.
Koolhaas, Rem, and Gerrit Oorthuys. “Ivan Leonidov’s Dom Narkomtjazjprom, Moscow.” *Oppositions* 2, (January 1974), 95–103.

Krauss, Rosalind. "Death of a Hermeneutic Phantom. Materialization of the Sign in the Work of Peter Eisenman." In *Houses of Cards*, ed. Peter Eisenman. New York: Oxford University Press, 1987, 166–184.
Krauss, Rosalind. "The Fountainhead." *Oppositions* 2 (January 1974), 61–70.
Krier, Leon. "The Consumption of Culture." *Oppositions* 14 (Fall 1978), 54–59.
Krier, Leon. "Vorwärts, Kameraden, wir müssen zurück." *Oppositions* 24 (Spring 1981), 26–37.
Krier, Leon, and Maurice Culot. "The Only Path for Architecture." *Oppositions* 14 (Fall 1978), 38–53.
Latour, Alessandra. "Little Magazine Conference: 'After Modern Architecture'." *Controspazio* 9 (June 1977), 62.
Le Corbusier. *When the Cathedrals Were White*. New York: Reynal and Hitchcock, 1947.
Lerup, Lars. "Apropos Type: Patrick Henry Bruce and Aldo Rossi." *Skyline* (October 1979), 6.
Lerup, Lars. "Gunnar Asplund." *Skyline* (September 1978), 4.
Lerup, Lars. "Report from San Francisco." *Skyline* (November 1978), 9.
Lewis, George. "Chapter Active on Upper East Side District." *Occulus* 8 (May 1982), 5.
Liebman, Theodore. "Learning from Experience. The Evolution of Housing Criteria." *Progressive Architecture* (November 1974), 70–77.
Liebman, Theodore. "The UDC and the Evolution of a Housing Policy." In *Another Chance for Housing, Low-Rise Alternatives*, ed. MoMA [Exhib. Cat.] New York: The Museum of Modern Art, 1973, 12–13.
Llorens, Tomàs. "Arquitecturas Bis, Lotus, Oppositions: Convencion en Cadaqués. Septiembre de 1975." *Arquitecturas Bis* 10 (November 1975), 30–31.
Llorens, Tomàs. "Making History." In *Architecture Criticism Ideology*, ed. Joan Ockman. New York: Princeton Architectural Press, 1985, 24–47.
Logue, Edward. "Introduction." In *Another Chance for Housing, Low-Rise Alternatives*, ed. MoMA [Exhib. Cat.] New York: The Museum of Modern Art, 1973, 4–5.
Lotus International, no. 27, 1980: "Architecture in the American University. Cornell, Cooper Union, Columbia."
Lütgenhorst, Manuel (interview with Andrew MacNair). "Behind Studio 54." *Skyline* (November 1979), 17.
McLeod, Mary. "The Architecture of the City." *Design Book Review*, 3 (Winter 1984), 50.
McLeod, Mary. "Introduction." In *Architecture Criticism Ideology*, ed. John Ockman. New York: Princeton Architectural Press, 1985, 7–11.
McLeod, Mary. "Le Corbusier and Algiers." *Oppositions* 19/20 (Winter/Spring 1980), 53–85.
Michelson, Annette, Rosalind Krauss, Douglas Crimp, Joan Copjec. "Introduction." In *October. The First Decade, 1976–1986*. Cambridge: MIT Press, 1986, IX–XII.
Moneo, Rafael. "Aldo Rossi: The Idea of Architecture and the Modena Cemetery." Translated by Angela Giral. *Oppositions* 5 (Summer 1976), 2–21.
Moneo, Rafael. "On Typology." *Oppositions* 13 (Summer 1978), 22–45.
Moore, Alan, and Marc Miller, eds. "The Real Estate Show." In *ABC No Rio Dinero: The Story of a Lower East Side Art Gallery*. New York: ABC No Rio, 1985, 52–71.
Morgen, James. "MoMA on Housing: Nothing New." *Architecture Plus* (August 1973), 68.
Morton, David. "Low-rise, High-density. UDC/IAUS Publicly Assisted Housing." *Progressive Architecture* (December 1973), 56–63.
Muschamp, Herbert. "The Universal Style." *Skyline* (February 1980), 14–15.
MoMA, ed. *Another Chance for Housing, Low-Rise Alternatives* [Exhib. Cat.] New York: The Museum of Modern Art, 1973.
MoMA, ed. *The New City: Architecture and Urban Renewal* [Exhib. Cat.] New York: The Museum of Modern Art, 1967.
Neue Grafik / New Graphic Design / Graphisme actuel (1959–1965).
Nevins, Deborah, eds. *The Roosevelt Island Housing Competition*. New York: Wittenborn Art Book, 1975.
Norton, Margot, and Margot Jacqz. "Lecture Notes: The Other Day." *Skyline* (May 1982), 32.
Ockman, Joan. "The Most Interesting Form of Lie." *Oppositions* 24 (Spring 1981), 38–47.
Ockman, Joan, ed. *Architecture Criticism Ideology*. New York: Princeton Architectural Press, 1985.
Owens, Craig. "The Allegorical Impulse." *October* 12 (Spring, 1980), 67–86.
Owens, Craig. "The Allegorical Impulse. Part 2." *October* 13 (Summer, 1980), 58–80.
Owens, Craig. "Einstein on the Beach. The Primacy of Metaphor." *October* 4 (Fall 1977), 21–32.
Owens, Craig. "Philip Johnson: History, Genealogy, Historicism." In *Philip Johnson: Processes. The Glass House, 1949 and The AT&T Headquarters*, 1978, Catalogue 9, ed. IAUS. New York: The Institute for Architecture and Urban Studies, 1979, 1–11.
Owens, Craig. "Photography en abyme." *October* 5, (Summer, 1978) 73–88
Pangaro, Anthony, and Kenneth Frampton. "Low Rise High Density: Issues and Criteria."

In *Another Chance for Housing, Low-Rise Alternatives*, ed. MoMA. [Exhib. Cat.] New York: The Museum of Modern Art, 1973, 16–17.
Papademetriou, Peter. “Good-bye Five: Work by Young Architects.” *Progressive Architecture* (November 1975), 20–21.
Pawley, Martin. “Universitas. Martin Pawley Reports on the University that Never Was.” *Architectural Design* (April 1972), 214–215.
Pinnell, Patrick. “Editorial.” *Skyline* (April 1979), 2.
Pinnell, Patrick. “On Venturi I: Drawing as Polemic.” *Skyline* (December 1978), 5.
Pinnell, Patrick. “On Venturi II: Allegory and Kitsch.” *Skyline* (January 1979), 5.
Pinnell, Patrick. “Remifications.” Skyline (March 1979), 5.
Plattus, Alan. “Manhattan Guides.” *Skyline* (October 1979), 8.
Plunz, Richard. “The Four Typologies. The ‘Master of Architecture’ Program at Columbia University.” *Lotus International*, no. 27 (1980), 106–126.
Plunz, Richard, and Kenneth Kaplan. “On ‘Style’.” *Precis* (Fall 1984), 33–43.
“Political Progress.” *Architectural Forum*, (May 1968), 37–38.
Pommer, Richard. “The Idea of ‘Idea as Model’.” In *Idea as Model: 22 Architects 1976/1980*, Catalogue 3, ed. IAUS. New York: Rizzoli International, 1981, 3–9.
Pommer, Richard. “Post-script to a Post-mortem.” In *Idea as Model: 22 Architects 1976/1980*, Catalogue 3, ed. IAUS. New York: Rizzoli International, 1981, 10–15.
Ponzini, Davide, and Michele Nastasi. *Starchitecture. Scenes, Actors, and Spectacles in Contemporary Cities*. New York: The Monacelli Press, 2016.
Porphyrios, Demetrios. “On Critical History.” In *Architecture Criticism Ideology*, ed. Joan Ockman. New York: Princeton Architectural Press, 1985, 16–21.
Porphyrios, Demetrios. “Pandora’s Box. An Essay on Metropolitan Portraits.” *Architectural Design* 47, no. 5 (1977): “OMA,” 357–362.
Progressive Architecture (March 1976): “Housing: High-rise vs. Low-rise.”
Progressive Architecture (October 1979): “Low-rise Housing.”
“R K Tecto-Comix.” *Skyline* (April 1980), 18.
Riegl, Alois. “The Modern Cult of Monuments: Its Character and Its Origin.” Translated by Kurt Forster and Diane Ghirardo. *Oppositions* 25 (Fall 1982), 21–51.
Roberts, Steven L. “School Is Formed for Urban Design.” *The New York Times* (October 15, 1967), 52.
Robertson, Jaquelin. *The Charlottesville Tapes. Transcripts of the conference held at the University of Virginia School of Architecture, November 12–13*, 1982. New York: Rizzoli International, 1985.
Robertson, Jaquelin, and Stanley Tigerman, eds. *Der Postmoderne Salon: Architekten über Architekten*. Basel: Birkhäuser Verlag, 1991.
Rossi, Aldo. “The Blue of the Sky.” Translated by Marlène Barsoum, Livio Dimitriu. *Oppositions* 5 (Summer 1976), 31–34.
Rossi, Aldo. *A Scientific Autobiography*. Cambridge: MIT Press, 1981.
Rossi, Aldo. *The Architecture of the City*. Cambridge: MIT Press, 1982.
Rossi, Aldo. “Introduction to the First American Edition.” In *The Architecture of the City*. Cambridge: MIT Press, 1982, 13–19.
Rossi, Aldo. “On 22 East 71st Street.” *Skyline* (April 1982), 2.
Rowe, Colin. “Architectural Education in the USA: Issues, Ideas, and People. A Conference to Explore Current Alternatives.” *Lotus International*, no. 27 (1980), 42–46.
Rowe, Colin. “Character and Composition, or: Some Vicissitudes of Architectural Vocabulary in the Nineteenth Century.” *Oppositions* 2 (January 1974), 41–60.
Rowe, Colin. “Nicollet Island, Minneapolis.” In *As I was Saying, Volume 3: Urbanistics*. Cambridge: MIT Press, 1996, 121–126.
Rudofsky, Bernard. *Streets for People. A Primer for Americans*. Garden City: Doubleday,1969.
Rykwert, Joseph. *Adam’s House. Papers on Architecture*. New York: The Museum of Modern Art, 1972.
Rykwert, Joseph. “The Institute for Architecture and Urban Studies.” *Casabella*, no. 359/360 (November/December 1971): “The City as an Artifact,” 100–103.
Rykwert, Joseph. “The Street. The Use of Its History.” In *On Streets*, ed. Stanford Anderson. Cambridge: MIT Press, 1978, 14–26.
Scott Brown, Denise. “Learning from Pop.” *Casabella*, no. 359/360 (November/December 1971): “The City as an Artifact,” 15–24.
Scott Brown, Denise. “Reply to Kenneth Frampton.” *Casabella*, no. 359/360 (November/December 1971): “The City as an Artifact,” 41–47.
Senator für Bau- und Wohnungswesen, ed. *Idee Prozess Ergebnis. Die Reparatur und Rekonstruktion der Stadt*. Berlin: Frölich und Kaufmann, 1984.
“Skylights: Rossi Opening Crowds.” *Skyline* (October 1979), 15.

Sorkin, Michael. "Cloning People." *Skyline* (November 1978), 11.
Sorkin, Michael. "Hollywood Matter." *Skyline* (September 1978), 9.
Sorkin, Michael. "Philip Johnson: The Master Builder as a Self-Made Man." *Village Voice* (October 30, 1978), 61–62; republished in *Exquisite Corpse: Writing on Buildings*. New York: Verso, 1991, 7–14.
Sorkin, Michael. "Reforming the Institute." *The Village Voice* (April 30, 1985), 102; republished in *Exquisite Corpse: Writing on Buildings*. New York: Verso, 1991, 110–113.
Stephens, Suzanne. "Architecture for Undergraduates." *Progressive Architecture* (March 1975), 23.
Stephens, Suzanne. "At the Institute: The Portland Building Analyzed." *Skyline* (January 1983), 20–21.
Stephens, Suzanne. "Building Commentary. Franzen's Champion H.Q." *Skyline* (October 1981), 16.
Stephens, Suzanne. "City Report: New York. Portman Hotel: Vive la Résistance!" *Skyline* (November 1981), 5.
Stephens, Suzanne. "City Report: New York. Air Thickens over St. Barts." *Skyline* (November 1981), 6.
Stephens, Suzanne. "City Report: New York. Tradition of the New." *Skyline* (December 1981), 4–5.
Stephens, Suzanne. "City Report: New York. The Sky is not the Limits. Midtown Zoning." *Skyline* (October 1981), 5–7.
Stephens, Suzanne. "City Report: New York. Whitewash on White Way." *Skyline* (October 1981), 4.
Stephens, Suzanne. "City Report: New York. Zoning for Context: Upper Madison." *Skyline* (December 1981), 6.
Stephens, Suzanne. "Columbia Architecture at 100!" *Skyline* (January 1982), 16.
Stephens, Suzanne. "Compromised Ideal: Marcus Garvey Park Village, Brooklyn, NY." *Progressive Architecture* (October 1979): "Low-rise Housing," 50–53.
Stephens, Suzanne. "From piano to forté. Interior Architecture." *Progressive Architecture* (February 1975), 60–67.
Stephens, Suzanne. "Gwathmey/Siegel's Beach House. Discussed at IAUS." *Skyline* (March 1983), 8–9.
Stephens, Suzanne. "It's all in the Family." *Architectural Forum* (July/August 1973), 25 & 27.
Stephens, Suzanne. "Looking Back at 'Modern Architecture.' The International Style Turns 50." *Skyline* (February 1982), 16–27.
Stephens, Suzanne. "Notes from the Sidelines." *Skyline* (December 1982), 33.
Stephens, Suzanne. "Polemicist-theorist." *Progressive Architecture* (May 1977), 68.
Stephens, Suzanne. "Skyline Rises Again." *Skyline* (October 1981), 2.
Stephens, Suzanne. "Skyline Rises II." *Skyline* (October 1982), 34.
Stephens, Suzanne. "This Side of Habitat." *Progressive Architecture* (July 1975), 58–63.
Stephens, Suzanne et al. "Role-Models. Multiple Protagonists." *Progressive Architecture* (May 1977), 59–71.
Stern, Robert. "The Evolution of Philip Johnson's Glass House, 1947–1948." *Oppositions* 10 (Fall 1977), 56–67.
Stern, Robert. "Giedion's Ghost." *Skyline* (October 1981), 22–25.
Stern, Robert. *New Directions in American Architecture*. New York: Braziller, 1969.
Stern, Robert. "Stompin at the Savoye." *Architectural Forum* (May 1973): "Five on Five," 46–48.
Stern, Robert. "Yale 1950–1965." *Oppositions* 4 (October 1974), 35–62.
Stetson, Damon. "A Most Unusual Strike. Bread-and-Butter Issues Transcended by Educational and Racial Concerns." *The New York Times* (September 14, 1968), 19.
Tafuri, Manfredo. "American Graffiti. Five x Five = Twenty-five." Translated by Victor Caliandro. *Oppositions* 5 (Summer 1976), 35–72.
Tafuri, Manfredo. *Architecture and Utopia*. Cambridge: MIT Press, 1976.
Tafuri, Manfredo. "L'Architecture dans le Boudoir. The Language of Criticism and the Criticism of Language." Translated by Victor Caliandro. *Oppositions* 3 (May 1974), 37–62.
Tafuri, Manfredo. "The Ashes of Jefferson." In *The Sphere and the Labyrinth: Avant-Gardes and Architecture from Piranesi to the 1970s*. Translated by Pellegrino d'Arcierno and Robert Connolly. Cambridge: MIT Press, 1987, 291–303.
Tafuri, Manfredo. "Les cendres de Jefferson." *L'Architecture d'Aujourd'hui*, no. 186 (August/September 1976): "New York in White and Gray," 53–72.
Tafuri, Manfredo. "Design and Technological Utopia." In *Italy: The New Domestic Landscape*, ed. Emilio Ambasz. New York: The Museum of Modern Art, 1972, 388–404.
Tafuri, Manfredo. "The Dialectic of the Avant-Garde: Piranesi and Eisenstein." Translated by Marlène Barsoum and Liviu Dimitriu. *Oppositions* 11 (Winter 1977), 72–80.
Tafuri, Manfredo. "Giuseppe Terragni: Subject and 'Mask'." Translated by Diane Ghirardo. *Oppositions* 11 (Winter 1977), 1–25.
Tafuri, Manfredo. "Giuseppe Terragni: Subject and 'Mask'." In *Giuseppe Terragni. Transformations, Decompositions,*

Critiques, ed. Peter Eisenman. New York: Monacelli Press, 2003, 273–293.

Tafuri, Manfredo. "The Historical Project." Translated by Diane Ghirardo and Stephen Sartarelli. *Oppositions* 17 (Summer 1979), 55–75.

Tafuri, Manfredo. "Die Kritik der Architektursprache und die Sprache der Architekturkritik." Translated by Michael Haase, Marc Fester, Nicolaus Kuhnert. *Arch+*, no. 37 (April 1978): "Der Tod der Architektur," 4–16.

Tafuri, Manfredo. "Peter Eisenman: The Meditations of an Icarus." In *Houses of Cards*, ed. Peter Eisenman. New York: Oxford University Press, 1987, 167–187.

Tafuri, Manfredo. "Il Soggetto a la maschera. Una introduzione a Terragni / Giuseppe Terragni: Subject and Mask." *Lotus International*, no. 20 (September 1978), 5–31.

Tafuri, Manfredo. *The Sphere and the Labyrinth: Avant-Gardes and Architecture from Piranesi to the 1970s*. Translated by Pellegrino d'Arcierno and Robert Connolly. Cambridge: MIT Press, 1987.

Tafuri, Manfredo. "The Theater of Memory." *Skyline* (October 1979), 7.

Tafuri, Manfredo. *Theories and History of Architecture*. London: Granada, 1980.

Telstch, Kathleen. "Brownsville to get 50-Block Renewal." *The New York Times* (June 20, 1969), 1 & 75.

"Three Institutes Are Formed to Study Urban Problems." *Architectural Record* (December 1967), 54

Time Magazine (January 8, 1979): "U.S. Architects. Doing their Own Thing."

Tschumi, Bernard. "Architectural Manifestos." *Skyline* (May/June 1980), 12.

Tschumi, Bernard. "Bernard Tschumi's Architectural Manifestos." *Skyline* (May 1979), 8–9.

Torre, Susana. "Times Square. At the Crossroads." *Skyline* (December 1982), 18–22.

Torre, Susana. *Women in American Architecture: A Historic and Contemporary Perspective*. New York: Whitney Library of Design, 1977.

UDC, *Planned Unit Development*. New York: City Planning Department, 1968.

Venturi, Robert. *Complexity and Contradiction in Architecture*. New York: MoMA, 1966.

Venturi, Robert (interview with Andrew MacNair). "Venturi and the Classic Modern Tradition." *Skyline* (March 1980), 4–5.

Venturi, Robert, Denise Scott Brown, Steven Izenour, eds. *Learning from Las Vegas*. Cambridge: MIT Press, 1972.

Vidler. Anthony. "The 'Art' of History: Monumental Aesthetics from Winckelmann to Quatremere de Quincy." *Oppositions* 25 (Fall 1982), 52–67.

Vidler, Anthony. "The Big Greek Column Will be Built: Adolf Loos and the Sign of Classicism." *Skyline* (October 1982), 16–17.

Vidler, Anthony. "Cooking up the Classics." *Skyline* (October 1981), 18–21.

Vidler, Anthony. "The Idea of Type." *Oppositions* 8 (Spring 1977), 94–115.

Vidler, Anthony. "Introduction: Academicism: Modernism." *Oppositions* 8 (Spring 1977), 2–5.

Vidler, Anthony. "Institutional Style. Deconstructing Modernism: Meier's Hartford Seminary." *Skyline* (March 1982), 21–23.

Vidler, Anthony. "The Office for Metropolitan Architecture. The Irony of the Metropolis: Notes on the Work of OMA." *Skyline* (May 1982), 18–21.

Vidler, Anthony. "On Type." *Skyline* (January 1979), 2.

Vidler, Anthony. "Progress and Primitivism: The Roots of John Soane's Style." *Skyline* (November 1982), 32–33.

Vidler, Anthony. "Restructuring Modernism. The Architecture of James Stirling." *Skyline* (November 1981), 16–19.

Vidler, Anthony. "The Third Typology." *Oppositions* 7 (Winter 1976), 1–4; republished in *Rational Architecture*, ed. Robert L. Delevoy. Bruxelles: Archives d'Architecture Moderne, 1978, 28–31.

Vignelli, Massimo. *Grids. Their Meaning and Use for Federal Designers*. U.S. Government Printing Office: Federal Design Library, December 1978.

Viladas, Pilar. "Right Building, Wrong Block." *Progressive Architecture* (January 1982), 31–32.

von Eckardt, Wolf. "Toward New Horizons." *The Washington Post* (June 24, 1973).

von Moos, Stanislaus. *Le Corbusier – Elements of a Synthesis*. Rotterdam: nai010 publisher, [1968] 2009.

White, Norval and Elliot Willensky, eds. *The AIA Guide to New York City*. 2nd ed. New York: Macmillan Publishers, 1978.

White, Norval and Elliot Willensky, eds., *The AIA Guide to New York City*. 3rd ed. New York: Macmillan Publishers, 1988.

Willis, Carol. "Review of Raymond Hood." *Skyline* (July 1982), 10–11.

Wittkower, Rudolf. *Architectural Principles in the Age of Humanism*. London: Warburg Institute / University of London, 1949.

Wolf, Peter. *The Future of the City: New Directions in Urban Planning*. New York: Whitney Library of Design, 1974.

Secondary Literature

Adorno, Theodor W. *Aesthetic Theory.* Minneapolis: University of Minnesota Press, [1970] 1997.
Adorno, Theodor W., and Max Horkheimer. "The Culture Industry: Enlightenment or Mass Deception." In *Dialectic of Enlightenment.* New York: Herder and Herder, [1944] 1972, 120–176.
Agrest, Diana, dir. *The Making of an Avant-garde. The Institute for Architecture and Urban Studies 1967–1984* (2013).
Akcan, Esra. "Manfredo Tafuri's Theory of the Architectural Avant-garde." *The Journal of Architecture* 7 (Summer 2002), 135–167.
Albertsen, Niels, and Bülent Diken. "Artworks' Networks. Field, System or Mediators?" *Theory, Culture & Society* 21, no. 3 (2004), 35–58.
Ambasz, Emilio. *Design Vignelli.* New York: Rizzoli International, 1981.
Ambasz, Emilio, ed. *The Universitas Project. Solutions for a Post-technological Society.* New York: The Museum of Modern Art, 2006.
An Architektur, no. 19–21 (2006): "Community Design: Involvement and Architecture in the US since 1963."
Anderson, Stanford. "Architectural History in Schools of Architecture." *Journal of the Society of Architectural Historians* 58, no. 3 (September 1999), 282–290.
Anderson, Stanford. "CASE and MIT. Engagement." In *A Second Modernism. MIT, Architecture and the 'Techno-Social' Moment*, ed. Arindam Dutta. Cambridge: MIT Press, 2013, 578–651.
Anderson, Stanford. "HTC at MIT: Architectural History in Schools of Architecture." In *Architektur weiterdenken. Werner Oechslin zum 60. Geburtstag*, eds. Sylvia Claus, Michael Gnehm, Bruno Maurer, Laurent Stalder. Zurich: gta Verlag, 2004, 330–338.
Any, no. 16 (1996): "Whiteness."
Any, no. 25/26 (2000): "Being Manfredo Tafuri: Wickedness, Anxiety, Disenchantment."
Allais, Lucia. "The Real and the Theoretical, 1968." *Perspecta* no. 42 (2010): "The Real," 27–41.
Allen, Stan, and Hal Foster. "A Conversation with Kenneth Frampton." *October*, no. 106 (Fall 2003), 35–58.
Ammon, Francesca. *Bulldozer. Demolition and Clearance of the Postwar Landscape.* New Haven: Yale University Press, 2016.
"Architecture I" In *Exhibit A: Exhibitions That Transformed Architecture 1948–2000*, ed. Eeva-Liisa Pelkonen, 196, London: Phaidon, 2018.
Aronson, Steven M. L. *Hype.* New York: William Morrow and Company, 1983.
Ault, Julie. *Alternative Art, New York, 1965–1985: A Cultural Politics Book for the Social Text Collective.* Minneapolis: University of Minnesota Press, 2002.
Aureli, Pier Vittorio. *The Project of Autonomy. Politics and Architecture Within and Against Capitalism.* FORuM Project Publications. New York: Princeton Architectural Press, 2008.
Baird, George. "Les Extremes. Qui se Touchant." *Architectural Design* 47, no. 5 (1977): "OMA," 326–327.
Baird, George. "A Reflection on the End of Assemblage." *Assemblage*, no. 41 (April 2001), 11.
Baird, George, and Charles Jencks, eds. *Meaning in Architecture.* New York: Braziller, 1969.
Banham, Reyner, ed. *The Aspen Papers: Twenty Years of Design Theory from the International Design Conference in Aspen.* New York: Praeger, 1974.
Barthes, Roland. "Death of the Autor." *Aspen Magazine*, no. 5–6 (1967); republished in *Image-Music-Text.* New York: Hill and Wang, 1977, 142–148.
Barthes, Roland. *Writing Degree Zero.* New York: Hill and Wang [1953] 1968.
Bell, Daniel. *The Coming of Post-Industrial Society.* New York, Basic Books, 1973.
Benjamin, Walter. "The Author as Producer." *New Left Review* 62, no. 1 (July-August 1970), 83–96.
Best, Steven, and Douglas Kellner, *The Postmodern Turn.* New York: Gilford Press, 1997.
Bédard, Jean François. *Cities of Artificial Excavation. The Work of Peter Eisenman, 1978–1988.* New York, Rizzoli International, 1994.
Binkiewicz, Donna. *Federalizing the Muse: United States Policy and the National Endowment for the Arts, 1965–1980.* Chapel Hill: University of North Carolina, 2004.
Birignani, Cesare. "Feature: Talking Heads. Team Vitruvius." *The Architects' Newspaper* (April 6, 2011), https://www.archpaper.com/2011/04/talking-heads/ (last accessed: May 31, 2023).
Bleecker, Samuel. *The Politics of Architecture. A Perspective on Nelson A. Rockefeller.* New York: The Rutledge Press, 1981.
Blomeyer, Gerald, and Barbara Tietze, eds. In *Opposition zur Moderne. Aktuelle Positionen in der Architektur.* Bauwelt Fundamente 52. Braunschweig and Wiesbaden: Friedrich Vieweg & Sohn, 1980.

Bloom, Harold. *A Map of Misreadings*. New York: Oxford University Press, 1975.
Bloom, Nicholas Dagen, and Matthew Gordon Lasner, eds. *Affordable Housing in New York. The People, Places, and Policies That Transformed a City*. Princeton: Princeton University Press, 2015.
Bloomfield, Julia. "A Tale of Two Institutes: Thoughts on Publication Worlds." In *Art History on the Move: Festschrift für Kurt W. Forster*, eds. Nanni Baltzer, Jacqueline Burckhardt, Marie Stauffer, Philip Ursprung. Zurich: Diaphanes, 2010, 66–83.
Bognar, Botond. "Rossi's Ultimate Dilemma?" *Journal of Architectural Education* 41, no. 2 (Winter 1988), 56–59.
Bois, Yves-Alain, Hal Foster, and Rosalind Krauss. "New York—Paris." In *Clip Stamp Fold. The Radical Architecture of Little Magazines. 196X to 197X*, eds. Beatriz Colomina and Craig Buckley. Barcelona: Actar, 2010, 36–45.
Bourdieu, Pierre. *Distinction, A Social Critique of the Judgement of Taste*. Cambridge: Harvard University Press, [1979] 1984.
Bourdieu, Pierre. *The Field of Cultural Production. Essays on Art and Literature*. New York: Columbia University Press, 1994.
Bourdieu, Pierre. "The Field of Cultural Production, or: the Economic World Reversed." *Poetics* 12, no. 4–5 (1983), 311–356.
Bourdieu, Pierre. "The Field of Cultural Production, or: the Economic World Reversed." In *The Field of Cultural Production. Essays on Art and Literature*. New York: Columbia University Press, 1994, 29–73.
Bourdieu, Pierre. "The Forms of Capital." In *Handbook of Theory and Research for the Sociology of Education*, ed. John G. Richardson. New York: Greenwood Press, 1986, 241–258.
Bourdieu, Pierre. "The Market of Symbolic Goods." *Poetics* 14, no. 1–2, ([1971] 1983), 13–44.
Bourdieu, Pierre. "The Social Conditions of the International Circulation of Ideas." In *Bourdieu: A Critical Reader*, ed. Richard Shusterman. Oxford: Blackwell Publishers, 1999, 220–228.
Bourdieu, Pierre. "Social Space and Symbolic Power." *Sociological Theory* 7, no. 1 (Spring 1989), 14–25.
Brilliant, Eleanor. *The Urban Development Corporation. Private Interests and Public Authority*. Lexington: Lexington Books, 1975.
Bröckling, Ulrich. *Das unternehmerische Selbst: Soziologie einer Subjektivierungsform*. Frankfurt am Main: Suhrkamp, 2007.
Caragonne, Alexander. *The Texas Rangers: Notes from the Architectural Underground*. Cambridge: MIT Press, 1995.
Castells, Manuel. "The Service Economy and Postindustrial Society: A Sociological Critique." *International Journal of Health Services* 6, no. 4 (1976), 595–607.
Chemla, Karine, François Jullien, Jacqueline Pigeot. *Die Kunst, Listen zu erstellen*. Berlin: Merve, 2004.
Choi, Esther. "Life, in Theory." In *Radical Pedagogies*, eds. Beatriz Colomina, Ignacio Gonzalez Galán, Evangelos Kotsioris, and Anna-Maria Meister. Cambridge: MIT Press, 2022, 146–149.
Clutter, McLain. *Imaginary Apparatus. New York City and Its Mediated Representation*. Zurich: Park Books, 2015.
Collier Hegener, Karen, and David Clarke, eds. *Architecture Schools in North America*. Princeton: Peterson's Guides, 1976.
Colomina, Beatriz, ed. *ArchitectuReproduction*. New York: Princeton Architectural Press, 1988.
Colomina, Beatriz, and Craig Buckley, eds. *Clip Stamp Fold. The Radical Architecture of Little Magazines. 196X to 197X*. Barcelona: Actar, 2010.
Colomina, Beatriz, and Evangolos Kotsioris, with Ignacio Gonzalez Galán, und Anna-Maria Meister. "The Radical Pedagogies Project." *Volume* 45 (2015): "Learning, Insert," 2–5.
Colomina, Beatriz, with Esther Choi, Ignacio Gonzalez Galán, and Anna-Maria Meister. "Radical Pedagogies in Architectural Education." *Architectural Review* (September 28, 2012), http://www.architectural-review.com/essays/radical-pedagogies-in-architectural-education/8636066.article (last accessed: May 31, 2023).
Colomina, Beatriz, Ignacio Gonzalez Galán, Evangelos Kotsioris, and Anna-Maria Meister. *Radical Pedagogies*. Cambridge: MIT Press, 2022.
Conradi, Jan. Unimark International: *The Design of Business and the Business of Design*. Baden: Lars Müller Publishers, 2009.
The Cooper Union. *Education of an Architect: A Point of View*. New York: The Cooper Union, 1971.
Crimmel, Henry. *The Liberal Arts College and the Ideal of Liberal Arts Education*. Lanham: University Press of America, 1993.
Crow, Thomas. *Modern Art in Common Culture*. New Haven: Yale University Press, 1996.

Cuff, Dana. *In Architecture. The Story of Practice*. Cambridge: MIT Press, 1991.

Cuff, Dana, and John Wriedt, eds. *Architecture from the Outside in. Selected Essays by Robert Gutman*. New York: Princeton Architectural Press, 2010.

Cusset, Francois. *French Theory: How Foucault, Derrida, Deleuze, & Co. Transformed the Intellectual Life of the United States*. Minneapolis: University of Minnesota Press, 2008.

Danbolt, Mathias. "Front Room—Back Room. An Interview with Douglas Crimp." *Trikster – Nordic Queer Journal*, no. 2 (2008), http://trikster.net/2/crimp/1.html (last accessed: May 31, 2023)

Davidson, Cynthia. *Eisenman/Krier: Two Ideologies*. New York: Monacelli Press, 2005.

Davis II, Charles L. "An Appeal to Protest." *Harvard Design Magazine* 44 (2018): "Seventeen," 182–188.

Derrida, Jacques. "Die différance." In *Postmoderne und Dekonstruktion. Texte französischer Philosophen der Gegenwart*, ed. Peter Engelmann. Ditzingen: Reclam, 2004, 76–113.

Deutsche, Rosalyn. "The Threshold of Democracy." In *Urban Mythologies. The Bronx Represented since the 1960s*, ed. John Alan Farmer. New York: Bronx Museum of the Arts, 1999, 94–101.

Dutta, Arindam, ed. *A Second Modernism. MIT, Architecture and the 'Techno-Social' Moment*. Cambridge: MIT Press, 2013.

The Edward John Noble Foundation *1940–2000. A Report at Sixty Years*. New York: The Edward John Noble Foundation, 2000.

Eisenman, Peter (interview with Beatriz Colomina and Urtzi Grau). In *Clip Stamp Fold. The Radical Architecture of Little Magazines*. 196X to 197X, eds. Beatriz Colomina and Craig Buckley. Barcelona: Actar, 2010, 261–264.

Eisenman, Peter. "The Agency Interview: The Institute for Architecture and Urban Studies." *Perspecta*, no. 45, (2012): "Agency," 59–66.

Eisenman, Peter, Kenneth Frampton, Mario Gandelsonas, Anthony Vidler. "New York—Barcelona—Milan." In *Clip Stamp Fold. The Radical Architecture of Little Magazines. 196X to 197X*, eds. Beatriz Colomina and Craig Buckley. Barcelona: Actar, 2010, 58–69.

Elser, Oliver, and Peter Cachola Schmal, eds. *Das Architekturmodell. Werkzeug, Fetisch, kleine Utopie*, Ausstellungskatalog DAM. Zurich: Scheidegger und Spiess, 2012.

Epp, Colin Brent. "The Education of Rosalind Krauss, Peter Eisenman and Other Americans: Why the Fantasy of Postmodernism Still Remains." PhD diss., University of British Columbia, 2007.

Fezer, Jesko. "Soft Cops und Anwaltsplanung: Planungsbeteiligung oder die Politik der Methode (1962–1973)." In *Wer gestaltet die Gestaltung? Praxis, Theorie und Geschichte des partizipatorischen Designs*, eds. Claudia Mareis, Matthias Held, Gesche Joost. Bielefeld: transcript, 2015, 43–64.

Fleck, Ludwik. *The Genesis and Development of a Scientific Fact*. Chicago: University of Chicago Press [1935] 1979.

Förster, Kim. "ABC No Rio: Architecture of Opposition." In *CinematoGraphies: Fictional Strategies and Visual Discourses in 1990s New York City*, eds. Günter H. Lenz, Dorothea Löbbermann, and Karl-Heinz Magister. (Heidelberg: Universitätsverlag C. Winter, 2006), 97–120.

Förster, Kim. "Alternative Educational Programs in Architecture: The Institute for Architecture and Urban Studies." In *Explorations in Architecture. Teaching, Design*, Research, ed. Reto Geiser. Basel: Birkhäuser, 2008, 26–27.

Förster, Kim. "The Housing Prototype of The Institute for Architecture and Urban Studies. Negotiating Housing and the Social Responsibility of Architects Within Cultural Production." *Candide*, no. 5 (March 2012), 57–92.

Förster, Kim. "The Institute for Architecture and Urban Studies: Wallace K. Harrison. New York Architect (Press Release)." In *Architecture Itself*, ed. Sylvia Lavin. Leipzig: Spector Books, 2019, 121–124.

Förster, Kim. "Institutionalizing Postmodernism: Reconceiving the Journal and the Exhibition at the Institute for Architecture and Urban Studies in 1976." In *Mediated Messages: Periodicals, Exhibitions*, and the Shaping Postmodern Architecture, eds. Véronique Patteeuw and Léa-Catherine Szacka. London: Bloomsbury, 2018, 213–229.

Förster, Kim. "Die Netzwerke des Peter Eisenman." *Arch+* features 19: supplement to *Arch+*, no. 210 (2013).

Förster, Kim. "A Postmodern School of Architecture. Education at the Institute for Architecture and Urban Studies." In *Histories of Architecture Education in the United States*, ed. Peter L. Laurence. New York and London: Routledge, 2023, 98–117.

Förster, Kim. "From Remment to Rem. A Quite Literary Story of Someone Who Made It in New York." *Clog* (June 2014), 32–33.

Förster, Kim. "Teaching Architecture, or, 'How to Create Spaces for Teenagers?'" In *Arts for Living*, eds. common room and Kim Förster. New York: common books, 2013, 63–108.

Foster, Hal. "1976." In *Art Since 1900. Modernism, Antimodernism, Postmodernism*, eds. Hal Foster, Rosalind Krauss, Yve-Alain Bois, Benjamin H. D. Buchloh, David Joselit. London: Thames & Hudson, 2006, 620–623.

Foucault, Michel. *The Archaeology of Knowledge*. New York: Pantheon Books, [1969] 1972.

Foucault, Michel. *The Order of Things*. New York: Vintage Books, [1966] 1970.

Foucault, Michel. "What is an Author?" In *Language, Counter-Memory, Practice. Selected Essays and Interviews*, ed. Donald F. Bouchard. Ithaca: Cornell University Press, 1977, 113–138.

Foucault, Michel, and Gilles Deleuze. "Intellectuals and Power." In Language, Counter-Memory, Practice. *Selected Essays and Interviews*, ed. Donald F. Bouchard. Ithaca: Cornell University Press, 1977, 205–217.

Franck, Georg. *Mentaler Kapitalismus. Eine politische Ökonomie des Geistes*. Munich and Vienna: Carl Hanser Verlag, 2005.

Franck, Georg. *Ökonomie der Aufmerksamkeit*. Munich and Vienna: Carl Hanser Verlag, 1998.

Franck, Georg. "Ökonomie der Aufmerksamkeit." In *Perspektiven metropolitaner Kultur*, ed. Ursula Keller. Frankfurt: Suhrkamp, 2000, 101–118.

Frank, Suzanne. *IAUS. An Insider's Memoir (with 27 Other Insider's Accounts)*. New York: self-published, 2010.

Frank, Suzanne. "Institute for Architecture and Urban Studies. New York, New York." In *Encyclopedia of 20th Century Architecture*, ed. R. Stephen Sennott. New York: Fitzroy Dearborn, 2006, 677–678.

Freemark, Yonah. "Roosevelt Island: Exception to a City in Crisis." *Journal of Urban History* 37, no. 3 (May 2011), 355–383.

Friedman, Martin. "The River: Images of the Mississippi, Walker Art Center, Minneapolis, United States of America." *Museum International* 32, no. 1/2 (1980), 15–21.

Freeman, Belmont. "The Moment for Something to Happen." *Places* (January 13, 2014), https://placesjournal.org/article/the-moment-for-something-to-happen/ (last accessed: May 31, 2023).

Gass, William. "House VI." *Progressive Architecture* (June 1977), 60–64.

Geddes, Robert, and Bernard Spring. *A Study for Environmental Design*. Princeton: Princeton University, 1967.

Ghirardo, Diane. "Eisenman's Bogus Avantgarde." *Progressive Architecture* (November 1994), 70–73.

Ghirardo, Diane. "Manfredo Tafuri and the Architectural Theory in the U.S., 1970–2000." *Perspecta* no. 33 (2002): "Mining Autonomy," 38–47.

Gill, Brendan. "On Reading. Peter Eisenman's House X." *Skyline* (April 1983), 33.

Gill, Brendan. "The Talk of the Town. Partygoing." *The New Yorker* (November 28, 1977), 45.

Gottmann, Jean. *Megalopolis. The Urbanized Northeastern Seaboard of the United States*. New York: Twentieth Century Fund, 1961.

Grafton, Anthony. "The Death of the Footnote (Report on an Exaggeration)." *The Wilson Quarterly* 21, no. 1, (Winter 1997), 72–77.

Greenberg, Miriam. "The Battle to Brand New York: 1975–1985." In *Branding New York*. New York: Routledge, 2008, 131–260.

Grima, Joseph, José Esparza, Chong Cuy, Charles Sneath, Suzannah Bohlke, Cesar Cotta, Pernilla Ohrstedt, Danny Wills, eds. *Storefront Newsprints 1982–2009*. New York: Storefront for Art and Architecture, 2009.

Guilbaut, Serge. *How New York Stole the Idea of Modern Art. Abstract Expressionism, Freedom, and the Cold War*. Chicago: University of Chicago Press, 1983.

Gutman, Robert. "Architecture: The Entrepeneurial Profession." *Progressive Architecture* (May 1977), 55–58.

Gutman, Robert. "House VI." Progressive Architecture (June 1977), 64–67.

Gutman, Robert. "Urban Studies as a Field of Research." *The American Behavioral Scientist 6*, no. 6 (1963): "Urban Studies," 11–16.

Gutman, Robert, and David Popenoe, eds. *The American Behavioral Scientist 6*, no. 6 (1963): "Urban Studies".

Hartung, Martin. "Architectural criticism on the art market: a US- American debate." *trans*, no. 31 (2017), 119–129.

Hartung, Martin. "Idea as Model." In *Exhibit A: Exhibitions That Transformed Architecture 1948–2000*, ed. Eeva-Liisa Pelkonen. London: Phaidon, 2018, 196–201.

Harvey, David. *The Condition of Postmodernity: An Enquiry into the Origins of Cultural Change*. Cambridge: Blackwell, 1989.

Harwood, John. "How Useful? The Stakes of Architectural History, Theory and Criticism at MIT, 1945–1976." In *A Second Modernism. MIT, Architecture and the 'Techno-Social' Moment*, ed. Arindam Dutta. Cambridge: MIT Press, 2013, 106–143.

Hatherley, Owen. "Two Notes on Alan Colquhoun." *Oase* 87 (2012): "Alan Colquhoun: Architect, Historian, Critic," 87–98.

Hays, K. Michael. "The Oppositions of Autonomy and History." In *Oppositions Reader*. New York: Princeton Architectural Press, 1998, IX–XV.

Hays, K. Michael, ed. *Oppositions Reader*. New York: Princeton Architectural Press, 1998.

Heynen, Hilde. *Architecture and Modernity: A Critique*. Cambridge: MIT Press, 1999, 128–147.

Holl, Steven, ed. *Pamphlet Architecture: Nos. 1–10*. New York: Princeton Architectural Press, 1999.

Holtz, Jane Kay. "Tabloid Trio. New Voices Speak up on Built Environment." *The Christian Science Monitor*, (March 19, 1982), 15.

Horlitz, Sabine. "The Case of Pruitt-Igoe: On the Demolition of the US Public Housing Complex in St. Louis, 1972." *Candide*, no. 10 (2016), 61–84.

Horlitz, Sabine. "The Construction of a Blast. The 1970s Urban Crisis and the Demolition of the Pruitt-Igoe Public Housing Complex." In *Crisis, Rupture and Anxiety*, eds. Will Jackson. Newcastle upon Tyne: Cambridge Scholars Publishing, 2012, 17–38.

Horlitz, Sabine. "Pruitt-Igoe: Ikone des Scheiterns? Planungsparadigmen, Lenkungsmodelle und Rezeption des US-amerikanischen Sozialwohnungsprojektes." PhD diss., FU Berlin, 2014.

Hustwit, Gary, dir. *Helvetia* (2007).

Jacobs, Jane. *The Death and Life of Great American Cites*. New York: Random House, 1961.

Jameson, Fredric. "Postmodernism, or, The Cultural Logic of Late Capitalism." *New Left Review*, no. 146 (July/August 1984), 53–92.

Jameson, Fredric. *Postmodernism, or, The Cultural Logic of Late Capitalism*. Durham: Duke University Press, 1991.

Jameson, Fredric. "Modernity versus Postmodernity in Peter Eisenman." In *Cities of Artificial Excavation. The Work of Peter Eisenman, 1978–1988*, ed. Jean François Bédard. New York, Rizzoli International, 1994, 27–37.

Jarzombek, Mark. "The School of Architectural Scandals." *e-flux Architecture* (October 29, 2018): "History/Theory," www.e-flux.com/architecture/history-theory/225182/the-school-of-architectural-scandals/ (last accessed: May 31, 2023).

Jasper, Michael. "Working It Out: On John Hejduk's Diamond Configurations." *Architectural Histories* 2, no. 1, (2014), https://journal.eahn.org/articles/10.5334/ah.cb/ (last accessed: May 31, 2023).

Jay, Martin. *The Dialectical Imagination. A History of the Frankfurt School and the Institute of Social Research, 1923–1950*. Berkeley: University of California Press, 1973.

Kaji-O'Grady, Sandra. "The London Conceptualists: Architecture and Performance in the 1970s." *Journal of Architectural Education* 61, no. 4 (2008), 43–51.

Kalpakci, Andreas. "Making CIAM: The Organizational Techniques of the Moderns, 1928–1959" PhD diss., ETH Zurich, 2017.

Kauffman, Jordan. *Drawing on Architecture. The Object of Lines, 1970–1990*. Cambridge: MIT Press, 2018.

Knoblauch, Joy. "Defensible Space and the Open Society." *Aggregate*, Volume 2, (March 2015), http://we-aggregate.org/piece/defensible-space-and-the-open-society (last accessed: May 31, 2023).

Knoblauch, Joy. "Going Soft: Architecture and the Human Sciences in Search of New Institutional Forms (1963–1974)." PhD diss., Princeton University, 2012.

Koolhaas, Rem. "Why I wrote Delirious New York and Other Textual Strategies." *Any*, no. 0 (May/June 1993), 42–43.

Kuhn, Thomas S. *The Structure of Scientific Revolutions*. Chicago: University of Chicago Press: 1962.

Kuhn, Thomas S. "Introduction: A Role for History" & "The Nature and Necessity of Scientific Revolutions." In *The Structure of Scientific Revolutions*. Chicago: University of Chicago Press: 1962, 1–9 & 92–110.

Lamb, Charles. *Housing Segregation in Suburban America Since 1960: Presidential and Judicial Politics*. Cambridge: Cambridge University Press, 2005.

Lapassade, George. *Gruppen, Organisationen, Institutionen*. Stuttgart: Klett Verlag, [1967] 1972.

Larrabee, Eric, and Massimo Vignelli. *Knoll Design*. New York: Harry N. Abrams, Inc. Publishers, 1981.

Lavin, Sylvia. "IAUS. Institute for Architecture and Urban Studies." *Log*, no. 13/14 (Fall 2008), 53–66.

Lazzarato, Maurizio. "Immaterial Labor." In *Radical Thought in Italy. A Potential Politics*, eds. Paolo Virno and Michael Hardt. Minneapolis: University of Minnesota Press, 1996, 133–147.

Leach, Andrew. "Choosing History. Manfredo Tafuri, Criticality and the Limits of Architecture." *The Journal of Architecture* 10, no. 3 (2005), 235–244.

Leach, Andrew. "Imagining Critique, or the Problematic Legacy of the Venice School." In *The Missed Encounter of Radical Philosophy with Architecture*, ed. Nadir Lahiji. London: Bloomsbury, 2014, 95–112.

Leach, Andrew. *Manfredo Tafuri. Choosing History*. Gent: A&S/books, 2007.
Leon, Joshua K., "The Times Square Postmodern," *Urban Omnibus*, September 30, 2015, https://urbanomnibus.net/2014/09/times-square-postmodern/ (last accessed: May 31, 2023).
Liebman, Kimberly, Laren Tenney, and Susan Saegert, "Good Design Alone Can't Change Society: Marcus Garvey Village (Brownsville, Brooklyn) after Thirty Years." *Planners Network* (Summer 2005), https://www.plannersnetwork.org/2005/07/good-design-alone-cant-change-society-marcus-garvey-village-brownsville-brooklyn-after-thirty-years/ (last accessed: May 31, 2023).
Linder, Mark. "'Entropy Colorized:' The Gray Decades, 1966–96." *Any*, no. 16 (1996): "Whiteness," 45–49.
Lobsinger, Mary Louise. "That Obscure Object of Desire: Autobiography and Repetition in the Work of Aldo Rossi." *Grey Room*, no. 8 (Summer 2002), 38–61.
Lobsinger, Mary Louise. "The New Urban Scale in Italy: On Aldo Rossi's L'architettura della città." *Journal of Architectural Education* 59, no. 3 (February 2006), 28–38.
Lotringer, Sylvère and David Morris, eds. *Schizo-Culture. The Event/The Book*. Cambridge: MIT Press, 2014.
Mallgrave, Harry Francis. *An Introduction to Architectural Theory. 1968 to the Present*. Oxford: Wiley-Blackwell, 2011.
Mallgrave, Harry Francis. *Modern Architectural Theory. A Historical Survey, 1673–1968*. Cambridge: Cambridge University Press, 2005.
Martin, Louis. "Notes on the Origins of Oppositions." In *Architectural Periodicals in the 1960s and 1970s. Towards a Factual, Intellectual and Material History*, eds. Alexis Sornin, Hélène Jannière, and France Vanlaethem. Montréal: IRHA Institut de recherche en histoire de l'architecture, 2008, 147–169.
Martin, Louis. "The Search for a Theory in Architecture. Anglo-American Debates, 1957–1976." PhD diss., Princeton University, 2002.
Martin, Reinhold. "On the Uses and Disadvantages of Architecture for History." *e-flux Architecture* (November 2, 2018): "History/Theory," www.e-flux.com/architecture/history-theory/225181/on-the-uses-and-disadvantages-of-architecture-for-history/ (last accessed: May 31, 2023).
Martin, Reinhold. *Utopia's Ghost. Architecture and Postmodernism, Again*. Minneapolis: University of Minnesota Press, 2010.
Marx, Karl. *Das Kapital. Kritik der politischen Ökonomie*. London: Faber & Faber, [1867] 2007.
McHale, John. *The Changing Information Environment*. Boulder: Westview Press, 1976.
McLeod, Mary. "The End of Innocence. From Political Activism to Postmodernism." In *Architecture School: Three Centuries of Educating Architects in North America*, ed. Joan Ockman. Cambridge: MIT Press, 2012, 160–201.
Miller, Donald. *Lewis Mumford. A Life*. New York: Weidenfeld & Nicholson, 1989.
Mogilevich, Mariana. "Designing the Urban: Space and Politics in Lindsay's New York." PhD diss., Harvard University, 2012.
Moody, Kim. *From Welfare State to Real Estate: Regime Change in New York City from 1974 to the Present*. New York: The New Press, 2007.
Müller-Brockmann, Josef. *Gestaltungsprobleme des Grafikers*. Heiden: Arthur Niggli, 1961.
Müller-Brockmann, Josef. *Raster Systeme für die visuelle Gestaltung*. Heiden: Arthur Niggli, 1981.
Müller, Lars. *Josef Müller-Brockmann. Pioneer of Swiss Graphic Design*. Baden: Lars Müller, 2001.
Mumford, Lewis. *The Culture of Cities*. New York: Harcourt, Brace & Co., 1938.
Mumford, Lewis. "What is a City?" *Architectural Record* 82 (November 1937), 92–96.
Muschamp, Herbert. "Sidewalk Critic. Lewis Mumford's Writings on New York." *Art Forum* (April 1999), 19–20.
Newman, Oscar. "Defensible Space." *Progressive Architecture* (October 1972), 92–105.
Newman, Oscar. *Defensible Space. Crime Prevention Through Urban Design*. New York: Macmillan, 1972.
Nowotny, Stefan, and Gerald Raunig. *Instituierende Praxen. Bruchlinien der Institutionskritik*. Vienna: transversal texts, 2016.
Oase 87 (2012): "Alan Colquhoun: Architect, Historian, Critic."
Ockman, Joan. "'Boudoir-Architektur' als Anschauungsmaterial: Manfredo Tafuri und New York." *werk, bauen + wohnen* (September 1995), 41–48.
Ockman, Joan. "Resurrecting the Avant-Garde: The History and Programme of Oppositions." In *ArchitectuReproduction*, ed. Beatriz Colomina. New York: Princeton Architectural Press, 1988, 180–199.
Ockman, Joan. "Slashed." *e-flux Architecture* (October 27, 2018): "History/Theory," www.e-flux.com/architecture/history-theory/159236/slashed/ (last accessed: May 31, 2023).

Ockman, Joan. “Venice and New York.” *Casabella* 59, no. 619/20, (1995), 56–73.

Ockman, Joan, ed. *Architecture School: Three Centuries of Educating Architects in North America*. Cambridge: MIT Press, 2012.

Oechslin, Werner. “‘Out of History’? ‘Formal Basis of Modern Architecture.’” In *Die formale Grundlegung der modernen Architektur*, Peter Eisenman. Zurich: gta Verlag, 2005, 12–61.

O’Neill, Paul. “The Curatorial Turn: From Practice to Discourse.” In *The Biennial Reader*, eds. Jelena Filipovic, et al. Osfildern: Hatje Cantz, 2010, 240–259.

O’Neill, Paul, Lucy Steeds, and Mick Wilson. *How Institutions Think: Between Contemporary Art and Curatorial Discourse*. Cambridge: MIT Press, 2017.

Pecora, Vincent. “Eisenman and Friends.” *Progressive Architecture* (May 1995), 13–15, 26.

Pecora, Vincent. “Towers of Babel.” In *Out of Site. A Social Criticism of Architecture*, ed. Diane Ghirardo. Seattle: Bay Press, 1991, 46–76.

Perspecta no. 33 (2002): “Mining Autonomy.”

Petit, Emmanuel, ed. *Philip Johnson: The Constancy of Change*. New Haven: Yale University Press, 2009.

Plunz, Richard. *A History of Housing in New York City*. New York: Columbia University Press, 1990.

Plunz, Richard, ed. *Housing Form and Public Policy in the United States*. New York: Greenwood Publishing Group, 1980.

Porsché, Yannik, Ronny Scholz, and Jaspal Naveel Singh, eds. *Institutionality. Postdisciplinary Studies in Discourse*. Cham: Palgrave Macmillan, 2022.

Pritchett, Wendell. “A Modern Ghetto? Brownsville since 1970.” In *Brownsville, Brooklyn. Blacks, Jews, and the Changing Face of the Ghetto*. Chicago: University of Chicago Press, 2003, 239–270.

Pritchett, Wendell. “Which Urban Crisis? Regionalism, Race, and Urban Policy, 1960–1974.” *Journal of Urban History* (January 2008), 266–286.

Privens, Francis F., and Richard A. Cloward. “Die Krise der Stadt am Beispiel New Yorks.” *Arch+*, no. 38 (May 1978): “‘Amerikanisierung’ der Städte?,” 24–27.

Purcell, Kerry William. *Josef Müller-Brockmann*. London: Phaidon, 2006.

Raunig, Gerald. *Fabriken des Wissens: Streifen und Glätten 1*. Zurich: diaphanes, 2012.

Raunig, Gerald. *Industrien der Kreativität: Streifen und Glätten 2*. Zurich: diaphanes, 2012.

Reckwitz, Andreas. *Die Erfindung der Kreativität. Zum Prozess gesellschaftlicher Ästhetisierung*. Frankfurt: Suhrkamp, 2012.

Richards, William. *Revolt and Reform in Archi-tecture’s Academy. Urban Renewal, Race and the Rise of Design in the Public Interest*. New York and London: Routlegde, 2017.

Riley, Terence, ed. *The Changing of the Avant-Garde: Visionary Architectural Drawings from the Howard Gilman Collection*. New York: MoMA, 2002.

Rispoli, Ernesto Ramon. *Ponti sull’Atlantico. L’Institute for architecture and urban studies e le relazioni Italia-America (1967–1985)*. Quodlibet: Macerata, 2012.

Roberts, Sam, ed. *America’s Mayor: John V. Lindsay and the Reinvention of New York*. New York: Columbia University Press, 2010.

Rosati, Lauren, and Mary Anne Staniszewski, eds. *Alternative Histories. New York Art Spaces 1960 to 2010*. Cambridge: MIT Press, 2012.

Rossi, Aldo. *I Quaderni azzurri*. Los Angeles: Getty Research Institute, 2000.

Sassen, Saskia. *The Global City. New York, London, Tokyo*. New York: Princeton University Press, 1992.

Saunders, Francis Stoner. *The Cultural Cold War: The CIA and the World of Arts and Letters*. New York: The New Press, 2000.

Schindler, Susanne. “The Housing that Model Cities Built. Context, Community, and Capital in New York City, 1966–76.” PhD diss., ETH Zurich, 2018.

Schindler, Susanne, and Juliette Spertus. “A Few Days in the Bronx: From Co-op City to Twin Parks.” *Urban Omnibus* (July 25, 2012), https://urbanomnibus.net/2012/07/a-few-days-in-the-bronx-from-co-op-city-to-twin-parks/ (last accessed: May 31, 2023).

Schnell, Angelika. “Von Jörn Janssen zu Rossi—Eine hochschulpolitische Affäre an der ETH Zürich.” *Arch+*, no. 215 (2014): “Hardcore-Architektur 2 - Politik versus Form,” 16–23.

Schnell, Angelika. “What is Meant by History?” *Oase*, no. 87 (2012): “Alan Colquhoun: Architect, Historian, Critic,” 58–76.

Schröder, Wilhelm Heinz. “Kollektive Biographien in der historischen Sozialforschung: Eine Einführung.” In *Lebenslauf und Gesellschaft. Zum Einsatz von kollektiven Biographien in der historischen Sozialforschung*, ed. Schröder. Stuttgart: Klett-Cotta, 1985, 7–17.

Schulze, Franz. *Philip Johnson. Life and Work*. New York: Alfred A. Knopf, 1994.

Schuman, Anthony. “Community Engagement. *Architecture’s Evolving Social Vocation.*”

In *Architecture School: Three Centuries of Educating Architects in North America*, ed. Joan Ockman. Cambridge: MIT Press, 2012, 252–259.
Schuman, Anthony. "The Pedagogy of Engagement: Some Historical Notes." *An Architektur*, no. 19–21, 2006: "Community Design: Involvement and Architecture in the US since 1963."
Schwarting, Michael. "The Institute of [sic!] Architecture and Urban Design [sic!], New York City—1967: The Museum of Modern Art exhibition: The New City: Architecture and Urban Renewal—1967." *Arc 2 citta* (July 10, 2012), http://www.arcduecitta.it/2012/07/archduecitta-magazine/ (last accessed: May 31, 2023).
Schwarz, Ullrich. "Another look-anOther gaze. Zur Architekturtheorie von Peter Eisenman" In *Aura und Exzess. Zur Überwindung der Metaphysik der Architektur*, ed. Schwarz. Vienna: Passagen Verlag, 1995, 11–34.
Schwartzer, Mitchell. "History and Theory in Architectural Periodicals. Assembling Oppositions." *Journal of the Society of Architectural Historians* 3 (September 1999), 342–348.
Scott, Felicity. *Architecture or Techno-utopia. Politics after Modernism*. Cambridge: MIT Press, 2007.
Scott, Felicity. "On the 'Counter-Design' of Institutions: Emilio Ambasz's Universitas Symposium at MoMA." *Grey Room*, no. 14 (Winter 2004), 46–77.
Searle, John. "What is an Institution?" *Journal of Institutional Economics* 1, no. 1 (2005), 1–22.
Sennett, Richard. *The Fall of Public Man. On the Social Psychology of Capitalism*. New York: Alfred A. Knopf, 1976.
Sherer, Daniel. "Architecture in the Labyrinth. Theory and Criticism in the Unites States: 'Oppositions,' 'Assemblage,' 'Any' (1973–1999)." *Zodiac*, no. 20 (1999), 36–63.
Somol, Robert. *Autonomy and Ideology: Positioning an Avant-garde in America*. New York: Monacelli Press, 1997.
Sorkin, Michael. *Exquisite Corpse: Writing on Buildings*. New York: Verso, 1991.
Spatial Information Design Lab. *Architecture and Justice*. New York: The Architectural League, 2006, www.spatialinformationdesignlab.org/MEDIA/PDF_04.pdf (last accessed: May 31, 2023).
Stanford Encyclopedia of Philosophy, "Social Institutions," last modified April 9, 2019, http://plato.stanford.edu/entries/social-institutions/ (last accessed: May 31, 2023).
Steele, Brett. Supercritical: Peter Eisenman & Rem Koolhaas. London: AA Publications, 2007.
Stein, Franziska. "Peter Eisenman: House II (Falk House)." In *Das Architekturmodell. Werkzeug, Fetisch, kleine Utopie*, eds. Oliver Elser and Peter Cachola Schmal. Zurich: Scheidegger und Spiess, 2012, 250–254.Stephanson, Anders. "Interview with Craig Owens." *Social Text*, no. 27 (1990), 55–71.
Stern, Ralph. "Oppositions Revisited—The Oppositions Reader." *Kritische Berichte*, no. 3 (1999), 65–72.
Stern, Robert, Thomas Mellins, David Fishman, eds. New York 1960. *Architecture and Urbanism Between the Second World War and the Bicentennial*. New York: The Monacelli Press, 1995.
Stierli, Martino. *Las Vegas im Rückspiegel. Die Stadt in Theorie, Fotografie und Film*. Zurich: gta Verlag, 2010.
Sunwoo, Irene. "From the 'Well-Laid Table' to the 'Market Place:' The Architectural Association Unit System." *Journal of Architectural Education* 65, no. 2 (March 2012), 24–41.
Sunwoo, Irene. "Pedagogy's Progress: Alvin Boyarsky's International Institute of Design." *Grey Room*, no. 34 (Winter 2009), 28–57.
Szacka, Léa-Catherine. "Debates on Display at the 1976 Venice Biennale." In *Place and Displacement: Exhibiting Architecture*, eds. Thordis Arrhenius, Mari Lending, Wallis Miller, Jérémie Michael McGowan. Zurich: Lars Müller, 2014, 97–112.
Szacka, Léa-Catherine. *Exhibiting the Postmodern. The 1980s Venice Biennale*. Venice: Marsilio, 2016.
Tafuri, Manfredo. "L'Architecture dans le Boudoir. The Language of Criticism and the Criticism of Language." Translated by Victor Caliandro. *Oppositions* 3 (May 1974), 37–62.
Tafuri, Manfredo. "The Ashes of Jefferson." In *The Sphere and the Labyrinth: Avant-Gardes and Architecture from Piranesi to the 1970s*. Translated by Pellegrino d'Arcierno and Robert Connolly. Cambridge: MIT Press, 1987, 291–303.
Tafuri, Manfredo. "Les cendres de Jefferson." *L'Architecture d'Aujourd'hui*, no. 186 (August/September 1976): "New York in White and Gray," 53–72.
Tafuri, Manfredo. "European Graffiti. Five x Five = Twenty-five." Translated by Victor Caliandro. *Oppositions* 5 (Summer 1976), 35–72.
Tafuri, Manfredo. "The Historical Project." Translated by Diane Ghirardo and Stephen Sartarelli. *Oppositions* 17 (Summer 1979), 55–75.
Tafuri, Manfredo. "Die Kritik der Architektursprache und die Sprache der

Architekturkritik." Translated by Michael Haase, Marc Fester, Nicolaus Kuhnert. *Arch+*, no. 37 (April 1978): "Der Tod der Architektur," 4–16.

Tafuri, Manfredo. *The Sphere and the Labyrinth: Avant-Gardes and Architecture from Piranesi to the 1970s*. Translated by Pellegrino d'Arcierno and Robert Connolly. Cambridge: MIT Press, 1987.

Tafuri, Manfredo. *Theories and History of Architecture*. London: Granada, 1980.

Taylor, Brian Brace. "Self Service Skyline." *L'Architecture d'Aujourd'hui*, no. 186 (August/September 1976), 42–46.

Taylor, Frederieke. "Appendix C: Frederieke Taylor on Exhibitions." In *IAUS. An Insider's Memoir (with 27 Other Insider's Accounts)*, Suzanne Frank. New York: self-published, 2010, 315–322.

Thelin, John. *A History of American Higher Education*. Baltimore: Johns Hopkins University Press, 2004.

Touraine, Alain. *The Academic System in American Society*. New York: McGraw-Hill, 1974.

Touraine, Alain. *The Post-Industrial Society. Tomorrow's Social History: Classes, Conflicts and Culture in the Programmed Society*. New York: Random House, 1971.

Tzonis, Alex, and Liane Lefaivre. "The Narcissist Phase in Architecture." *The Harvard Architecture Review*, no. 1 (Spring 1978), 53–63.

Tzonis, Alex, and Liane Lefaivre. "Die narzisstische Phase der Architektur." *Arch+*, no. 42 (December 1978), 51–57.

Ursprung, Philip. "Blinde Flecken der 1970er Jahre: Gordon Matta-Clarks 'Window Blow-out'." In *Reibungspunkte, Ordnung und Umbruch in Architektur und Kunst*, Festschrift für Hubertus Günther, eds. Hanns Hubach, Barbara von Orelli-Messerli, Tadej Tassini. Petersberg: Michael Imhof Verlag, 2008, 293–300.

Varnelis, Kazys. "The Education of the Innocent Eye." *Journal of Architectural History* (May 1998), 212–223.

Varnelis, Kazys. "Philip Johnson's Empire: Network Power and the AT&T Building." In *Philip Johnson: The Constancy of Change*, ed. Emmanuel Petit. New Haven: Yale University Press, 2009, 120–135.

Varnelis, Kazys. "The Spectacle of the Innocent Eye: Vision, Cynical Reason, and the Discipline of Architecture in Postwar America." PhD diss., Cornell University, 1994.

Varnelis, Kazys. "'We Cannot Not Know History:' Philip Johnson's Politics and Cynical Survival." *Journal of Architectural Education* 49, no. 2 (November 1995), 92–104.

Vervoort, Stefaan. "Scale Models and Postmodernism: Revisiting Idea as Model (1976–81)." *Architectural Theory Review* 24, no. 3 (2020), 224–240.

Vidler, Anthony (interview with Beatriz Colomina and Daniel Lopez-Perez). In *Clip Stamp Fold. The Radical Architecture of Little Magazines. 196X to 197X*, eds. Beatriz Colomina and Craig Buckley. Barcelona: Actar, 2010, 537–541.

Vidler, Anthony. *Histories of the Immediate Present*. Cambridge: MIT Press, 2008.

Vignelli, Massimo. "On Rational Design." (interview with Steven Heller and Elinor Pettit) In *Design Dialogues*, eds. Steven Heller and Elinor Pettit. New York: Allworth Press, 1998, 3–7.

Walker, Stephen. *Gordon Matta-Clark: Art, Architecture and the Attack on Modernism*. London: I.B. Tauris, 2009.

Watson, Nadia. "The Whites vs. the Grays: Re-Examining the 1970s Avant-Garde." *Fabrications* (July 2005), 55–69.

Weaver, Thomas, and Peter Eisenman. "Peter Eisenman in conversation with Thomas Weaver." *AA Files*, no. 74 (2017), 150–172,

Weber, Max. *Economy and Society*. Cambridge and London: Harvard University Press, [1922] 2019.

Ween, Camilla. "Fiddling While New York Burns." *Architectural Design* 46, no. 10 (October 1976), 630.

White, Hayden. *Metahistory. The Historical Imagination in Nineteenth-Century Europe*. Baltimore: Johns Hopkins University Press, 1973.

Wigley, Mark. *Cutting Matta-Clark. The Anarchitecture Investigation*. Zurich: Lars Müller, 2018.

Williams, Raymond. "Institution" In *Keywords. A Vocabulary of Culture and Society*. New York: Oxford University Press, 1976, 139–140.

Wojtowicz, Robert, ed. *Sidewalk Critic. Lewis Mumford's Writings on New York*. New York: Princeton Architectural Press, 2000.

Wright, Gwendolyn, and Paul Rabinow. "Spatialization of Power. A Discussion of the Work of Michel Foucault." *Skyline* (March 1982), 14–15.

Oral History Interviews

Diana Agrest
Stan Allen
Emilio Ambasz
Andrew Anker
Samuel Anderson
Stanford Anderson

George Baird
Jonathan Barnett
Andrew Bartle
Michael Bierut
Thomas Bender
Vincent Benedetto
Barry Bergdoll
Deborah Berke
Rosemarie Bletter
David Buege
Colin Campbell
Walter Chatham
Kees Christiaanse
Nathaniel Coleman
Roger Conover
Joan Copjec
Douglas Crimp
Peggy Deamer
Livio Dimitriu
Linda Dukes
Peter Eisenman
William Ellis
Kurt Forster
Kenneth Frampton
Suzanne Frank
Mario Gandelsonas
Deborah Gans
Paul Goldberger
Peter Greenberg
Robert Gutman
Sarah Halliday
Laurie Hawkinson
Henry Hecker
Jessica Helfand
Christian Hubert
Thomas Hut
Margot Jacqz
Louise Joseph
Brian Kaye
Kevin Kennon
Jonathan Kirschenfeld
Silvia Kolbowski
Randall Korman
Rosalind Krauss
Lawrence Kutnicki
Robert Lane
John Leeper
Theodore Liebman
Kevin Lippert
Peter Lynch
Andrew MacNair
Mary McLeod
Jay Measley
Richard Meier
Thomas Mellins
Robert Meltzer
David Mohney
Elizabeth Moule
Hamid Nouri
Joan Ockman
Miguel Oks
Kyong Park
Patrick Pinnell
Alan Plattus
Stephen Potters
Tim Prentice
George Ranalli
Mark Robbins
Joseph Rykwert
Pat Sapinsley
Michael Schwarting
Richard Sennett
Lindsay Stamm Shapiro
Caroline Sidnam
Robert Silman
Carla Skodinski
Michael Sorkin
Suzanne Stephens
Robert Stern
Jon Stouman
Mimi Shanley Taft
Frederieke Taylor
Bernard Tschumi
Anthony Vidler
Massimo Vignelli
Lauretta Vinciarelli
Laura Waltz
Peggy Weil
Richard Wengenroth
Lorraine Wild
Terrance Williams
Peter Wolf
Stuart Wrede

Index

Illustrations

Fig. 1 Provisional charter of the Institute for Architecture and Urban Studies, September 29, 1967, ARCH401124, IAUS fonds © CCA.
Fig. 2 Prospectus/poster for the Institute with insignia (Cesariano's Vitruvian man), 1967–68, ARCH153783, IAUS fonds © CCA.
Fig. 3 Cover of the Institute's *New Urban Settlements*-report, self-published, December 1970 (design: Robert Slutzky), Avery Library, GSAPP, Columbia University.
Fig. 4 IAUS, HUD Urban Renewal Demonstration Project Budget for the Institute for Architecture and Urban Studies, October 14, 1969, PI II.B.579, MoMA Archives © MoMA, New York/Scala, Florence.
Fig. 5 Minutes of meeting with annotations by Peter Eisenman, October 5, 1971, ARCH401120, IAUS fonds © CCA.
Fig. 6 Cover of *Casabella* no. 359/360, ed. by the Institute (November/December 1971).
Fig. 7 Collage/group photograph of Fellows, Staff, and Research Associates, *Casabella* no. 359/360, (November/December 1971), 101.
Fig. 8 Cover of *Art and Architecture. USSR. 1917–32*-exhibition catalogue, ed. IAUS (New York: Wittenborn Art Books, 1971).
Fig. 9 Poster for IAUS Spring Lecture Series, 1973, ARCH252116, IAUS fonds © CCA.
Fig. 10 Poster for the IAUS Student Internships, ca. 1972, ARCH252117, IAUS fonds © CCA.
Fig. 11 Photograph of Peter Eisenman (with Silvia Kolbowski) in the Institute's entrance area (photographer: unknown), private archive of Deborah Berke.
Fig. 12 Photograph of the Institute's main hall, view from the balcony (photographer: Suzanne Frank), private archive of Suzanne Frank.
Fig. 13 Kenneth Frampton, edited draft poster for the Institute's evening lectures "Habitat & Urban Form," Spring 1972, ARCH279005, Kenneth Frampton fonds © CCA.
Fig. 14 IAUS, "Streets-Project," diagram of organizational structure, phase III, ca. 1972, ARCH272397, IAUS fonds © CCA.
Fig. 15 IAUS, proposed financial scheme, ca. 1972, ARCH263663, IAUS fonds © CCA.
Fig. 16 IAUS, Specifications for the construction of Low-Rise High-Density housing based on the Institute's model study, 1972, ARCH274452, IAUS fonds © CCA.
Fig. 17 Photograph of a model of the low-rise, high-density housing prototype for the MoMA exhibition, 1973, ARCH288337, Kenneth Frampton fonds © CCA.
Fig. 18 Photograph of a model of the prototype application in an urban setting for the MoMA exhibition, 1973, ARCH288338, Kenneth Frampton fonds © CCA.
Fig. 19 Kenneth Frampton, Street unit and mews entry for the Low-Rise High-Density MoMA exhibition, 1973, ARCH288311, Kenneth Frampton fonds © CCA.
Fig. 20 Kenneth Frampton, Typical street unit for the Low-Rise High-Density MoMA

exhibition 1973, ARCH288310, Kenneth Frampton fonds © CCA.
Fig. 21 Kenneth Frampton, Typical street units and plan of mews entry for the Low-Rise High-Density MoMA exhibition, 1973, ARCH288312, Kenneth Frampton fonds © CCA.
Fig. 22 Cover of *Another Chance for Housing*-exhibition catalogue, ed. by MoMA (New York: The Museum of Modern Art, 1973).
Fig. 23 Installation view of the exhibition "Another Chance for Housing," MoMA, New York, June 12 to August 19, 1973 (photographer: George Cserna), IN 1037.7, MoMA Archives © MoMA, New York/Scala, Florence.
Fig. 24 Rendering of Fox Hills, Staten Island (design: Craig Hodgetts), *Another Chance for Housing*-exhibition catalogue, ed. by MoMA (New York: The Museum of Modern Art, 1973), 26.
Fig. 25 Rendering of Marcus Garvey Park Village (design: Craig Hodgetts), *Another Chance for Housing*-exhibition catalogue, ed. by MoMA (New York: The Museum of Modern Art, 1973), 20.
Fig. 26 Photograph of construction site of Marcus Garvey Park Village, ca. 1975 (photographer: Theodore Liebman), private archive of Theodore Liebman.
Fig. 27 Photograph of mews, Marcus Garvey Park Village, ca. 1976 (photographer: Theodore Liebman), private archive of Theodore Liebman.
Fig. 28 Peter Eisenman, draft for commited and projected income and dept repayment schedule from July 1, 1975 to June 30, 1976, ARCH401157, IAUS fonds © CCA.
Fig. 29 Front side of the Institute's prospectus for Educational Programs, 1975, ARCH401149_R, IAUS fonds © CCA.
Fig. 30 Back side of the Institute's prospectus for Educational Programs, 1975, ARCH401149_V, IAUS fonds © CCA.
Fig. 31 Minutes of Fellows' meeting, April 25, 1975, ARCH401158, IAUS fonds © CCA.
Fig. 32 Model photograph of students' work, advised by Colin Rowe and Judith diMaio, for the exhibition "The River Images of the Mississippi," Walker Art Center, Minneapolis, MN, October 3, 1976 to January 9, 1977 © Walker Art Center.
Fig. 33 Installation view of the exhibition "The River. Images of the Mississippi," Walker Art Center, Minneapolis, MN, October 3, 1976 to January 9, 1977 (photographer: unknown) © Walker Art Center.
Fig. 34 Photograph of the Institute's main hall during crit of the Undergraduate Program in Architecture, ca. 1976 (photographer: Cameron Armstrong), private archive of Cameron Armstrong.
Fig. 35 Photograph of students from the Undergraduate Program in Architecture, ca. 1976 (photographer: Cameron Armstrong), private archive of Cameron Armstrong.
Fig. 36 Photograph of crit of the Undergraduate Program in Architecture, ca. 1976 (photographer: Cameron Armstrong), private archive of Cameron Armstrong.
Fig. 37 Flyer for the High School Program "Manhattan: Capitol of the Seventies," 1976 (design: Andrew MacNair) © Walker Art Center.
Fig. 38 Photograph of a students' trip to the Metropolitan Museum (Lawrence Kutnicki sitting in between with sunglasses) in the High School Program, October 1977 (photographer: Deborah Berke), private archive of Lawrence Kutnicki.
Fig. 39 Photograph of a students' trip to the Rockefeller Center (Deborah Berke on the left in the brown coat) in the High School Program (photographer: Lawrence Kutnicki), private archive of Deborah Berke.
Fig. 40 Photograph of a students' trip within Manhattan (with Deborah Berke and Lawrence Kutnicki) in the High School Program, May 1978 (photographer: unknown), private archive of Lawrence Kutnicki.
Fig. 41 Photograph of students in the High School Program "Architecture and the Arts," August 1978 (photographer: unknown), private archive of Deborah Berke.
Fig. 42 Photograph of students in the High School Program Architecture and the Arts," August 1978 (photographer: unknown), private archive of Deborah Berke.
Fig. 43 Application for a NEH Learning Institute Program, 1977 © NEH.
Fig. 44 Diagram of the NEH Learning Institute Program, 1977 © NEH.
Fig. 45 Photograph of opening of the exhibition "Princeton's Beaux Arts and Its New Academicism," January 18 to February 27, 1977 (photographer: Dorothy Alexander), Alexander Photographs, Beinecke, Yale © Dorothy Alexander.
Fig. 46 Poster for the Advanced Design Workshop, 1979–80 (design: Massimo Vignelli), VCDS0001 IAUS 007, Vignelli papers, Vignelli Center for Design Studies, RIT.

Fig. 47 Photograph of architecture crits (Charles Gwathmey pointing, with Diana Agrest, Lars Lerup and Robert Stern sitting) in the Advanced Design Workshop, 1980–81 (photographer: Deborah Berke), private archive of Deborah Berke.
Fig. 48 Photograph of architecture crits in the Advanced Design Workshop, 1980–81 (photographer: Deborah Berke), private archive of Deborah Berke.
Fig. 49 Statement of Trustees' authority, January 13, 1977, ARCH401096, IAUS fonds © CCA.
Fig. 50 Poster for the lecture series "Architecture," Fall 1974 (design: Massimo Vignelli), VCDS0001 IAUS 022, Vignelli papers, Vignelli Center for Design Studies, RIT.
Fig. 51 Poster for the lecture series "Architecture 2," Spring 1975 (design: Massimo Vignelli), VCDS0001 IAUS 023, Vignelli papers, Vignelli Center for Design Studies, RIT.
Fig. 52 Poster for the lecture series "Architecture 3," Fall 1975 (design: Massimo Vignelli), VCDS0001 IAUS 027, Vignelli papers, Vignelli Center for Design Studies, RIT.
Fig. 53 Poster for the lecture series "Architecture 4," Spring 1976 (design: Massimo Vignelli), VCDS0001 IAUS 024, Vignelli papers, Vignelli Center for Design Studies, RIT.
Fig. 54 Poster (red) for the lecture series "Architecture 5," Fall 1976 (design: Massimo Vignelli), VCDS0001 IAUS 025, Vignelli papers, Vignelli Center for Design Studies, RIT.
Fig. 55 Poster (black/red) for the lecture series "Architecture 5," Fall 1976 (design: Massimo Vignelli), VCDS0001 IAUS 026, Vignelli papers, Vignelli Center for Design Studies, RIT.
Fig. 56 Ad for IAUS Exhibitions Program, 1975 (design: Andrew MacNair) © AIA.
Fig. 57 Ad for the exhibition "Kinetic Systems. Sculpture by Tim Prentice," May 22 to June 15, 1975 (design: Tim Prentice) © AIA.
Fig. 58 Flyer for "Good-bye Five," September 16 to October 1, 1975 (design: Andrew MacNair) © AIA.
Fig. 59 Poster for the exhibition "Idea as Model," December 16, 1976 to January 14, 1977 (design: Michael Graves), Museum für Gestaltung Zürich, Plakatsammlung, ZHdK, 3DK-0003
Fig. 60 Poster for lecture series "City as Theater," Spring 1977 (design: Massimo Vignelli), VCDS0001 IAUS 017, Vignelli papers, Vignelli Center for Design Studies, RIT.
Fig. 61 Poster for "Saturday Seminars," Spring 1978 (design: Massimo Vignelli), VCDS0001 b110 IAUS 013, Vignelli papers, Vignelli Center for Design Studies, RIT.
Fig. 62 Peter Wolf, memorandum about staffing and allocation of tasks, June 9, 1977, ARCH401152, IAUS fonds © CCA.
Fig. 63 Poster for "National Tour," ca. 1978 (design: Massimo Vignelli), VCDS0001 IAUS 020, Vignelli papers, Vignelli Center for Design Studies, RIT.
Fig. 64 Photograph of Peter Eisenman in conversation with Philip Johnson, October 26, 1976 (photographer: Dorothy Alexander), Alexander Photographs, Beinecke, Yale © Dorothy Alexander.
Fig. 65 Photograph of Trustees' meeting the day before the celebrations for the tenth anniversary, November 10, 1977 (photographer: Dorothy Alexander), Alexander Photographs, Beinecke, Yale © Dorothy Alexander.
Fig. 66 Poster for the tenth anniversary, November 11, 1977 (design: Massimo Vignelli), VCDS0001 IAUS 001, Vignelli papers, Vignelli Center for Design Studies, RIT.
Fig. 67 Cover of IAUS Brochure, ca.1978 (design: Massimo Vignelli), VCDS0001 b110 IAUS 006, Vignelli papers, Vignelli Center for Design Studies, RIT.
Fig. 68 Fellows' portraits for IAUS Brochure, ca. 1978 (design: Massimo Vignelli), VCDS0001 b110 IAUS 006b, Vignelli papers, Vignelli Center for Design Studies, RIT.
Fig. 69 Organizational diagram for IAUS Brochure, ca. 1978 (design: Massimo Vignelli), VCDS0001 b110 IAUS 006a, Vignelli papers, Vignelli Center for Design Studies, RIT.
Fig. 70 Poster for lecture series "Open Plan 77," Fall 1977 (design: Massimo Vignelli), VCDS0001 IAUS 008a, Vignelli papers, Vignelli Center for Design Studies, RIT.
Fig. 71 Poster for lecture series "Open Plan 78," Spring 1978 (design: Massimo Vignelli), VCDS0001 IAUS 029b, Vignelli papers, Vignelli Center for Design Studies, RIT.
Fig. 72 Poster for lecture series "Open Plan 78," Fall 1978 (design: Massimo Vignelli), VCDS0001 IAUS 009, Vignelli papers, Vignelli Center for Design Studies, RIT.
Fig. 73 Poster for lecture series "Open Plan 79," Spring 1979 (design: Massimo Vignelli), VCDS0001 IAUS 030, Vignelli papers, Vignelli Center for Design Studies, RIT.
Fig. 74 Poster for lecture series "Open Plan 79," Fall 1979 (design: Massimo Vignelli),

VCDS0001 IAUS 010, Vignelli papers, Vignelli Center for Design Studies, RIT.

Fig. 75 Poster for lecture series "Open Plan 80," Spring 1980, (design: Massimo Vignelli), VCDS0001 IAUS 011, Vignelli papers, Vignelli Center for Design Studies, RIT.

Fig. 76 Poster for the travelling exhibition "A New Wave of Japanese Architecture," September 25 to November 14, 1978 (design: Massimo Vignelli), ARCH253031, IAUS fonds © CCA.

Fig. 77 Cover of the IAUS Exhibition Catalogue 10, *A New Wave of Japanese Architecture*, 1978 (design: Massimo Vignelli), VCDS0001 b110 IAUS 017, Vignelli papers, Vignelli Center for Design Studies, RIT.

Fig. 78 Leaflet for the "Japanese New Wave"-travelling exhibition, ca. 1978 (design: unknown), VCDS0001 IAUS 033, Vignelli papers, Vignelli Center for Design Studies, RIT.

Fig. 79 Poster for the exhibition "Philip Johnson," September 12 to October 31, 1978 (design: Massimo Vignelli), VCDS0001 IAUS 016, Vignelli papers, Vignelli Center for Design Studies, RIT.

Fig. 80 Cover of the IAUS Exhibition Catalogue 9, *Philip Johnson*, 1978 (design: Massimo Vignelli), VCDS0001 b110 IAUS 003, Vignelli papers, Vignelli Center for Design Studies © RIT.

Fig. 81 Back side of flyer for the exhibition opening of "Philip Johnson," September 12, 1978 (design: Massimo Vignelli), VCDS0001 b110 IAUS 110b, Vignelli papers, Vignelli Center for Design Studies © RIT.

Fig. 82 Front side of flyer for the exhibition opening of "Philip Johnson," September 12, 1978 (design: Massimo Vignelli), VCDS0001 b110 IAUS 110a, Vignelli papers, Vignelli Center for Design Studies © RIT.

Fig. 83 Photograph of opening of the "Wallace Harrison" exhibition, December 9, 1979 (photographer: Dorothy Alexander), Alexander Photographs, Beinecke, Yale © Dorothy Alexander.

Fig. 84 Poster for exhibition "Ivan Leonidov. Russian Visionary Architect," February 1 to February 21, 1978 (design: Massimo Vignelli), VCDS0001 IAUS 002, Vignelli papers, Vignelli Center for Design Studies, RIT.

Fig. 85 Poster for exhibition "Le Corbusier's Saint-Pierre de Firminy," April 29 to June 13, 1981 (design: Massimo Vignelli), VCDS0001 IAUS 013, Vignelli papers, Vignelli Center for Design Studies, RIT.

Fig. 86 Poster for exhibition "Clorindo Testa. Architecture and Personal Mythology," November 9 to 25, 1981 (design: Massimo Vignelli), VCDS0001 IAUS 014, Vignelli papers, Vignelli Center for Design Studies, RIT.

Fig. 87 Poster for the exhibition "Kazou Shinohara. 11 Houses (1971–1976)," December 8, 1981 to January 8, 1982 (design: Massimo Vignelli), VCDS0001 IAUS 015, Vignelli papers, Vignelli Center for Design Studies, RIT.

Fig. 88 Poster for the exhibition "Office for Metropolitan Architecture," March 12 to May 28, 1983 (design: Massimo Vignelli), ARCH288305, IAUS fonds © CCA.

Fig. 89 Peter Eisenman, notes titled "RE:FORM A Magazine toward rational architecture," ARCH401716, IAUS fonds © CCA.

Fig. 90 Peter Eisenman, notes for content of *Oppositions* 5, 6, 7 and 8, ca. 1976, ARCH401325, IAUS fonds © CCA.

Fig. 91 Poster for *Oppositions* 5–8, ca. 1976 (design: Massimo Vignelli), VCDS0001 IAUS 003, Vignelli papers, Vignelli Center for Design Studies, RIT.

Fig. 92 Poster for *Oppositions* 1–4, ca. 1976 (design: Massimo Vignelli), VCDS0001 IAUS 028, Vignelli papers, Vignelli Center for Design Studies, RIT.

Fig. 93 Cover for *Oppositions* 5, 1976 (design: Massimo Vignelli), VCDS0001 b069 IAUS 001, Vignelli papers, Vignelli Center for Design Studies, RIT.

Fig. 94 Collage for *Oppositions* 5, 1976, 36 (design: Massimo Vignelli).

Fig. 95 IAUS, Invitation to subscribe to *Oppositions* sent to Peter Eisenman, October 15, 1976, ARCH288309, IAUS fonds © CCA.

Fig. 96 Photograph of *Oppositions* launch event "Forum 6," October 26, 1976 (photographer: Dorothy Alexander), Alexander Photographs, Beinecke, Yale © Dorothy Alexander.

Fig. 97 Cover for *Skyline*, April 1978 (design: Massimo Vignelli), VCDS0001 Skyline 100, Vignelli papers, Vignelli Center for Design Studies, RIT.

Fig. 98 Centerfold calendar in *Skyline*, April 1978, 4–5 (design: Massimo Vignelli), VCDS0001 Skyline 300, Vignelli papers, Vignelli Center for Design Studies, RIT.

Fig. 99 Peter Eisenman, notes on IAUS publications, ca. 1979, ARCH401754, IAUS fonds © CCA.

Fig. 100 Flyer for IAUS Exhibition Catalogues, ca. 1980, ARCH401633, IAUS fonds © CCA.

Fig. 101 Design process for IAUS Exhibition Catalogue 7, *Gwathmey/Siegel Architects* (design: Massimo Vignelli), VCDS0001 IAUS 223, Vignelli papers, Vignelli Center for Design Studies, RIT.
Fig. 102 Cover for IAUS Exhibition Catalogue 7, *Gwathmey/Siegel Architects* (design: Massimo Vignelli), VCDS0001 b110 IAUS 002, Vignelli papers, Vignelli Center for Design Studies, RIT.
Fig. 103 Cover for IAUS Exhibition Catalogue 2, *Rossi in America* (design: Massimo Vignelli), VCDS0001 b110 IAUS 004, Vignelli papers, Vignelli Center for Design Studies, RIT.
Fig. 104 Cover for IAUS Exhibition Catalogue 1, *Massimo Scolari* (design: Massimo Vignelli), VCDS0001 b110 IAUS 014, Vignelli papers, Vignelli Center for Design Studies, RIT.
Fig. 105 Cover for IAUS Exhibition Catalogue 3, *Idea as Model* (design: Massimo Vignelli), VCDS0001 b110 IAUS 015, Vignelli papers, Vignelli Center for Design Studies, RIT.
Fig. 106 Cover for IAUS Exhibition Catalogue 5, *Robert Krier* (design: Massimo Vignelli), VCDS0001 b110 IAUS 016, Vignelli papers, Vignelli Center for Design Studies, RIT.
Fig. 107 Cover for IAUS Exhibition Catalogue 12, *John Hejduk* (design: Massimo Vignelli), VCDS0001 b110 IAUS 018, Vignelli papers, Vignelli Center for Design Studies, RIT.
Fig. 108 Cover for IAUS Exhibition Catalogue 13, *Austrian New Wave* (design: Massimo Vignelli), VCDS0001 b110 IAUS 019, Vignelli papers, Vignelli Center for Design Studies, RIT.
Fig. 109 Cover for IAUS Exhibition Catalogue 16, *William Lescaze* (design: Massimo Vignelli), VCDS0001 b110 IAUS 020, Vignelli papers, Vignelli Center for Design Studies, RIT.
Fig. 110 Cover for IAUS Exhibition Catalogue 17, *Kazuo Shinohara* (design: Massimo Vignelli), VCDS0001 b110 IAUS 021, Vignelli papers, Vignelli Center for Design Studies, RIT.
Fig. 111 Peter Eisenman, draft description of Oppositions Books, July 9, 1979, ARCH401742, IAUS fonds © CCA.
Fig. 112 List of prospective series of Oppositions Books w/ 14 titles, ARCH288307, IAUS fonds © CCA.
Fig. 113 Publication Schedule for Oppositions Books, October 1981 to October 1983, ARCH288308, IAUS fonds © CCA.
Fig. 114 Kenneth Frampton, proposal for an Oppositions Book with annotations by Peter Eisenman, December 3, 1981, ARCH401766, IAUS fonds © CCA.
Fig. 115 Anthony Vidler, proposal for an Oppositions Book, ca. 1981, ARCH401769, IAUS fonds © CCA.
Fig. 116 Table setting for dinner for *Skyline* sponsors given by Philip Johnson with annotations by Peter Eisenman, July 20, 1981, ARCH264853, IAUS fonds © CCA.
Fig. 117 Photograph of dinner for *Skyline* sponsors, Century Association, New York, 20 July 1981 (photographer: Bettina Cirone), private archive of Suzanne Stephens.
Fig. 118 Design process for *Skyline*, October 1981 (design: Massimo Vignelli), VCDS0001 poS5 IAUS 002, Vignelli papers, Vignelli Center for Design Studies, RIT.
Fig. 119 Cover for *Skyline*, October 1981 (design: Massimo Vignelli), VCDS0001 Skyline 200, Vignelli papers, Vignelli Center for Design Studies, RIT.
Fig. 120 Flyer for release party "Skyline Returns!," September 29, 1981 (design: Massimo Vignelli), VCDS0001 b110 IAUS 012, Vignelli papers, Vignelli Center for Design Studies, RIT.
Fig. 121 Poster for six Oppositions Books, ca. 1981 (design: Massimo Vignelli), VCDS0001 IAUS 021, Vignelli papers, Vignelli Center for Design Studies, RIT.
Fig. 122 Photograph of Peter Eisenman in conversation with Aldo Rossi, Oppositions Books release, March 31,1982 (photographer: Dorothy Alexander), Alexander Photographs, Beinecke, Yale © Dorothy Alexander.
Fig. 123 Book cover of Aldo Rossi, *A Scientific Autobiography* (design: Massimo Vignelli), VCDS0001 b031 IAUS 001, Vignelli papers, Vignelli Center for Design Studies, RIT.
Fig. 124 Agenda for *Oppositions* editors meeting with attachments, February 24, 1981, ARCH401765_004, IAUS fonds © CCA.
Fig. 125 Project description for the Philip Johnson Center for Architecture, ca. 1981, ARCH401091, IAUS fonds © CCA.
Fig. 126 Organizational chart of the Institute, ca. 1981–82, VCDS0001 b544 IAUS 002, Vignelli papers, Vignelli Center for Design Studies, RIT.
Fig. 127 Photograph of panel discussion of the symposium "Architecture and Ideology," March 16, 1982, featuring Demetri Porphyrius, Fredric Jameson, and Tomàs Llorens, moderated by Mary McLeod (photographer: Dorothy Alexander), Alexander Photographs, Beinecke, Yale © Dorothy Alexander.

Fig. 128 Poster for the fifteenth anniversary, November 15, 1982 (design: Massimo Vignelli), VCDS0001 IAUS 012, Vignelli papers, Vignelli Center for Design Studies, RIT.

Fig. 129 Coverage on fifteenth anniversary in *Skyline*, December 1982, 33, VCDS0001 Skyline 500, Vignelli papers, Vignelli Center for Design Studies, RIT.

Fig. 130 Design process for "Space, Knowledge, and Power," *Skyline*, March 1982 (design: Massimo Vignelli), VCDS0001 poS5 IAUS 001, Vignelli papers, Vignelli Center for Design Studies, RIT.

Fig. 131 Spread of "Space, Knowledge, and Power," *Skyline*, March 1982, 16–17 (design: Massimo Vignelli), VCDS0001 Skyline 400, Vignelli papers, Vignelli Center for Design Studies, RIT.

Fig. 132 Poster for "On Style," December 6 & 8, 1982 (design: Massimo Vignelli), VCDS0001 IAUS 032, Vignelli papers, Vignelli Center for Design Studies, RIT.

Fig. 133 Poster for "The High School Architecture Studio," 1982–83 (design: Massimo Vignelli), VCDS0001 IAUS 031, Vignelli papers, Vignelli Center for Design Studies, RIT.

Fig. 134 Poster for "Architectural Education: Alternatives," 1982–83 (design: Massimo Vignelli), VCDS0001 IAUS 018, Vignelli papers, Vignelli Center for Design Studies, RIT.

Fig. 135 Cover page of brochure *Educational Programs 1984–1985* (design: Massimo Vignelli), VCDS0001 b110 IAUS 001, Vignelli papers, Vignelli Center for Design Studies, RIT.

Fig. 136 List of IAUS Faculty, 1984–85, VCDS0001 b544 IAUS 005c, Vignelli papers, Vignelli Center for Design Studies, RIT.

Fig. 137 Front page of *IAUS News*, February/March 1985 (design: unknown), VCDS0001 b544 IAUS 003c, Vignelli papers, Vignelli Center for Design Studies, RIT.

Figures 56, 57, 58
The American Institute of Architects Archives, Washington, D.C.

Figures 45, 64, 65, 83, 96, 122, 127
Dorothy Alexander Photographs, General Collection, Beinecke Rare Book and Manuscript Library, Yale University, New Haven, CT.

Figures 1, 2, 4, 5, 9, 10, 14, 15, 16, 28, 29, 30, 31, 49, 62, 76, 88, 89, 90, 95, 99, 100, 111, 112, 113, 114, 115, 116, 124, 125
IAUS Fonds (Gift of Peter Eisenman), Canadian Centre for Architecture, Montréal.

Figures 13, 17, 18, 19, 20, 21
Kenneth Frampton Fonds (Gift of Kenneth Frampton), Canadian Centre for Architecture, Montréal.

Figures 4, 23
The Museum of Modern Art, New York/Scala, Florence.

Figures 43, 44
National Endowment for the Humanities, Washington, DC.

Figure 59
Plakatsammlung, Museum für Gestaltung, Zurich University of the Arts (ZHdK), Zurich.

Figures 46, 50, 51, 52, 53, 54, 55, 60, 61, 63, 66, 67, 68, 69, 70, 71, 72, 73, 74, 75, 77, 78, 79, 80, 81, 82, 84, 85, 86, 87, 91, 92, 97, 98, 101, 102, 103, 104, 105, 106, 107, 108, 109, 110, 118, 119, 120, 121, 123, 126, 128, 129, 130, 131, 132, 133, 134, 135, 136, 137
Massimo and Lella Vignelli papers, Vignelli Center for Design Studies, Rochester Institute of Technology, Rochester, NY.

Figures 32, 33, 37
Walker Art Center, Minneapolis, MN.